Mineralogy and Geochemistry of Gems

Mineralogy and Geochemistry of Gems

Special Issue Editors

Panagiotis Voudouris
Stefanos Karampelas
Vasilios Melfos
Ian Graham

MDPI • Basel • Beijing • Wuhan • Barcelona • Belgrade

Special Issue Editors
Panagiotis Voudouris
National and Kapodistrian
University of Athens
Greece

Stefanos Karampelas
Bahrain Institute for Pearls &
Gemstones (DANAT)
Bahrain

Vasilios Melfos
Aristotle University of
Thessaloniki
Greece

Ian Graham
The University of New South
Wales
Australia

Editorial Office
MDPI
St. Alban-Anlage 66
4052 Basel, Switzerland

This is a reprint of articles from the Special Issue published online in the open access journal *Minerals* (ISSN 2075-163X) (available at: https://www.mdpi.com/journal/minerals/special_issues/Gems).

For citation purposes, cite each article independently as indicated on the article page online and as indicated below:

LastName, A.A.; LastName, B.B.; LastName, C.C. Article Title. *Journal Name* **Year**, *Article Number*, Page Range.

ISBN 978-3-03928-076-6 (Pbk)
ISBN 978-3-03928-077-3 (PDF)

Cover image courtesy of Bahrain Institute for Pearls & Gemstones (DANAT).

Contents

About the Special Issue Editors

Panagiotis Voudouris is Professor of Mineralogy at the Department of Geology and Geoenvironment, National and Kapodistrian University of Athens (Greece). He graduated in 1986 at this University and received his Doctorate (1993) in Mineralogy and Economic Geology from the University of Hamburg (Germany), where he also served as a Visiting Professor. His main research interests are in the fields of ore mineralogy, genesis of magmatic–hydrothermal ore deposits, mineralogy and genesis of gemstones, and oxidation zones of ore deposits. He is member of the Editorial Board of the journal *Minerals* and member of the Scientific Board of the Mineralogy and Petrology Museum, National and Kapodistrian University of Athens. He has published 72 scientific papers on international refereed journals, is the co-author of 4 IMA-approved new mineral species, and has co-authored three books. The International Mineralogical Association has adopted a new mineral name in his honor (Voudourisite).

Stefanos Karampelas is Research Director for DANAT (Bahrain Institute for Pearls and Gemstones). He is also Lecturer for the Advanced Gemmology Diploma and his research is focused on gem formation and treatment as well as their characterization using nondestructive methods. He is frequently visiting gem mines and pearl producing areas around the globe and often delivering lectures to international scientific conferences and gemmological meetings. Stefanos is also a member of the Commission of Gem Materials, delegate for the International Gemmological Conference as well as of the Editorial Board of *Gems and Gemology* and Associate Editor of *The Journal of Gemmology*.

Vasilios Melfos is Associate Professor in Economic Geology-Geochemistry, at the Department of Mineralogy, Petrology, Economic Geology, School of Geology, Faculty of Sciences, Aristotle University of Thessaloniki. His research activity mainly refers to issues of economic geology, geochemistry, mineralogy and petrography, and involves mineralogical, geochemical, and fluid inclusion studies of magmatic–hydrothermal mineralizations, provenance determination of the raw materials of archaeological artifacts, and environmental geochemistry and mineralogy in sediments and water of streams and rivers. He has been responsible for the Microthermometric Laboratory at the University of Thessaloniki for over 30 years. He is involved in several collaborative research groups worldwide.

Ian Graham is a Senior Lecturer in Mineralogy and Petrology within the PANGEA Research Centre, School of Biological, Earth and Environmental Sciences, University of New South Wales Sydney, Sydney, NSW, Australia, and a Distinguished Visiting Professor at the China University of Mining and Technology. His research activity covers a broad spectrum though is currently mainly focused on nontraditional critical element resources of China, intraplate basaltic volcanism, genesis of sapphires and rubies, mineralogical and geochemical vectoring towards ore deposits, and epithermal mineral systems. He is involved with a number of collaborative research projects worldwide. Ian Graham is also an Associate Editor of *Mineralogical Magazine* and the *International Journal of Coal Geology*.

Preface to "Mineralogy and Geochemistry of Gems"

Gems have been used in the manufacture of jewellery and as ornaments since antiquity. Nowadays, gemmology, i.e., the study of gem materials, is one of the most rapidly expanding fields in the earth sciences. Recent statistics have shown that about 15 billion Euros are annually at stake. This Special Issue emphasises the recent advances in both fundamental and applied studies of gems based on different aspects of research, which can be used to constrain the conditions of their formation. Given that gems are rarely available for scientific research, the present compilation of 20 publications offers very good examples of the application of various methods for their study.

Panagiotis Voudouris, Stefanos Karampelas, Vasilios Melfos, Ian Graham
Special Issue Editors

Editorial

Editorial for Special Issue "Mineralogy and Geochemistry of Gems"

Panagiotis Voudouris [1,*], Stefanos Karampelas [2], Vasilios Melfos [3] and Ian Graham [4]

[1] Faculty of Geology and Geoenvironment, National and Kapodistrian University of Athens, GR-15784 Athens, Greece

[2] Bahrain Institute for Pearls & Gemstones (DANAT), WTC East Tower, P.O. Box 17236 Manama, Bahrain; Stefanos.Karampelas@danat.bh

[3] Faculty of Geology, Aristotle University of Thessaloniki, 54124 Thessaloniki, Greece; melfosv@geo.auth.gr

[4] PANGEA Research Centre, School of Biological, Earth and Environmental Sciences, University of New South Wales, Sydney, NSW 2052, Australia; i.graham@unsw.edu.au

* Correspondence: voudouris@geol.uoa.gr

Received: 4 December 2019; Accepted: 11 December 2019; Published: 12 December 2019

Gems are materials used for adornment or decoration that must satisfy several criteria where they must be aesthetic and visually appealing; relatively rare; hard and tough enough to resist "normal" wear; and able to withstand corrosion by skin contact and cosmetics [1]. Gems have been used since antiquity thus, gemology, the science dealing with gems, is positioned between academia and industry. As an applied science, in gemology, the instruments used should be non- or micro-destructive and their cost should be reasonable (both in terms of equipment and time consumption [2,3]). Gemology can also contribute to the development of pure science and in some cases, destructive techniques may have to be used [1–3]. This special issue presents recent advances on the study of various types of gems based on a variety of research (e.g., geology, trace element geochemistry, inclusion studies, geochronology, spectroscopy). It includes 20 articles by around 100 researchers from over 30 different institutions situated in 20 countries from around the globe. These articles will hopefully contribute to our better understanding of the formation of gems.

Pegmatites are known to be a source of several (>50) gem-quality minerals [4,5]; and in the present Special Issue, three works related to gems and pegmatites are published [6–8]. The first work by Strmić Palinkaš et al. [6] is concerned with a detailed analysis of the economically important gem-bearing Boqueirão granitic pegmatite situated in the Borborema Pegmatitic Province (BPP) from Rio Grande do Norte in Northeast Brazil. The Boqueirao granitic pegmatite is classified as a member of the Lithium-Caesium-Tantalum (LCT) pegmatite family and was emplaced during a late stage of magmatic activity in the late Cambrian. The second work by Huong et al. [7] presents a comparison of the trace element geochemistry of danburite, a gem mineral commonly formed within transition zones of metacarbonates and pegmatites as a late magmatic accessory phase from Mexico, Tanzania, and Vietnam. Differences in the rare earth elements (REE) concentrations of danburite from the different localities were observed and these show that trace element variations reflect different degrees of involvement of metacarbonates and pegmatites among different locations. The third work by Diella et al. [8] is a study of the first gem-quality multicoloured tourmalines found in the Alps hosted in LCT pegmatites of the Adamello massif, Italy. Tourmaline is considered as an important recorder of its geological formation [9,10] and the results of the study may contribute to understanding the evolution of the pegmatites in this massif.

Giuliani et al. [11] and Karampelas et al. [12] present studies on emeralds; the bluish-green to green to yellowish-green variety of beryl coloured by chromium and sometimes vanadium. The first group of authors in their review proposed an enhanced classification for emerald deposits based on the geological environment (magmatic or metamorphic), host-rock types (mafic-ultramafic rocks, sedimentary rocks, and granitoids), degree of metamorphism, styles of mineralization (veins, pods, metasomatites, shear zone) as well as the type of gem-forming fluids and their temperature, pressure, and composition [11]. Karampelas et al. [12] present an applied study that provides a chemical and spectroscopic analysis of gem-quality emeralds from the most important sources (i.e., Afghanistan (Panjsher Valley), Brazil (Itabira), Colombia (Coscuez), Ethiopia (Shakisso), Madagascar (Mananjary), Russia (Ural mountains), Zambia (Kafubu) and Zimbabwe (Sandawana)). Their study demonstrates how these different analyses can collectively be used to distinguish them from one another (i.e., geographic gem determination).

Rubies and sapphires are coloured gem varieties of corundum and can be found in various places around the globe. Ruby is the red variety and sapphire is the blue variety, while all other coloured corundums are called fancy sapphires and need a colour prefix (e.g., pink sapphire, yellow sapphire). A comparative study of ruby chemistry and inclusions between Myanmar and eastern Australia is presented by Sutherland et al. [13] and clearly shows that although having formed in different parts of the world, at different times and under different tectonic settings, unusual Ga-rich rubies occur in both regions, indicating primary generation involving magmatic processes. Sorokina et al. [14] focus on the genesis of gem-quality sapphires from the Ilmen Mountains (South Urals, Russia) found in situ within ultramafites. These sapphires were formed together with a spinel–chlorite–muscovite rock during the metasomatic alteration of orthopyroxenites at a temperature around 700–750 °C and pressure of about 1.8–3.5 kbar. Despite their metasomatic genesis, Laser Ablation-Inductively Coupled Plasma-Mass Spectrometry (LA-ICP-MS) analyses of these blue sapphires showed that they clearly fall into the range of "metamorphic" sapphires (e.g., Ga/Mg < 2.7). Filina et al. [15] presented data on another occurrence of sapphires in the same region of the South Urals occurring in anorthosites (kyshtymites). Syngenetic zircon inclusions of magmatic origin were found in these sapphires; however, a few of the measured elements using LA-ICP-MS show that some of these sapphires also fall into the range for "metamorphic" sapphires (e.g., Ga/Mg < 0.8). Formation of anorthosites (kyshtymites) is still debatable and two possible scenarios (magmatic and metamorphic-metasomatic) are proposed. Voudouris et al. [16] give an overview of gem corundum deposits from Greece where they occur within clearly diverse geological settings. For instance, pink sapphires to rubies from Paranesti (Drama area; found in boudinaged lenses of Al-rich metapyroxenites alternating with amphibolites and gneisses), pink to purple to blue sapphires from Gorgona (Xanthi area; occurring within marble layers alternating with amphibolites), and sapphires from the central part of Naxos Island (associated with desilicated granite pegmatites intruding ultamafic lithologies, a.k.a. plumasites) can be classified as metamorphic in origin. On the other hand, blue sapphires from the southern part of the Naxos and Ikaria Islands (both occurring in fissures within metabauxites hosted in marbles) display atypical magmatic signatures, indicating a likely hydrothermal origin. In these four papers [13–16], it is evident that some ratios used to separate gem corundum from different geological environments (e.g., Ga/Mg ratio [17]) should be applied with caution. They possibly need revision using solely data obtained from gem-quality samples, in addition, the data should be grouped carefully according to colour. Corundum oxygen isotope values have been used to determine the likely geological origin of gem-quality corundum as well as to constrain the likely geological environment of samples collected from secondary alluvial deposits [18–22]. There are two contributions on the use of in situ oxygen isotope analysis on gem corundum from a primary and secondary occurrence to help better understand their origin [23,24]. The first by Wang et al. [23] is on fingerprinting rubies from Paranesti in northern Greece and importantly suggests that this method can be used to distinguish between two similar occurrences only 500 metres apart. The second paper by Graham et al. [24] is on the use of in situ oxygen isotopes to help in determining the genesis and evolution of alluvial sapphires from the Orosmayo region (Jujuy Province, NW Argentina) and importantly shows that there is a wide, although systematic, range in oxygen isotope values, which can be explained by differing degrees of interaction between mantle-derived magmas, lower crustal felsic magmas, and, most likely, both mantle- and crustal-derived metasomatic fluids.

Pearls are biogenic gems and are of historic and present-day importance [1,25]. A study of a large number of natural and cultured pearls found in various bivalves from saltwater and freshwater environments using LA-ICP-MS and X-ray luminescence is given by Karampelas et al. [26]. LA-ICP-MS can be used to accurately separate freshwater from saltwater samples using manganese, barium, sodium, magnesium, and strontium ratios, and in some cases, even to identify their host bivalve species. Additionally, X-ray luminescence reactions of the studied samples have confirmed a correlation between yellow-green intensity and manganese content in aragonite. It is suggested that orange luminescence, observed in a few freshwater samples under X-rays, is due to a different coordination of Mn^{2+} in vaterite as compared to aragonite.

Zircon is another important mineral geologically [27,28], which can be of gem-quality. Studies on gem-quality zircons from two occurrences are also included in this Special Issue [29,30]. Data by Piilonen et al. [29] on zircon xenocrysts from alkali basalts in Ratanakiri Province (Cambodia) suggest that their genesis involved zirconium-saturated, aluminium-undersaturated, carbonatitic-influenced, low-degree partial melting (<1%) of peridotitic mantle at ca. 60 km beneath the Indochina terrane. Data by Bui Thi Sinh et al. [30] on zircon crystals (up to 3 cm long) from placer deposits in the Central Highlands of Vietnam suggest a genesis from carbonatite-dominant melts as a result of partial melting of a metasomatized lithospheric mantle source as well as resorption and re-growth processes.

The paper by Dill [31] is an excellent review of gem placer deposits, outlining their processes of formation, controls on deposition, and concludes with a new classification system. Štubňa et al. [32] describe some relatively small (< 0.2 ct) and rare gem-quality demantoid (i.e., yellowish-green to green coloured andradite garnet) from serpentinized harzburgites situated in Dobšiná, Slovakia. Curtis et al. [33] in their work reviewed the current opal classification including additional data on samples from new localities. Classification of opal-A, opal-CT, and opal-C as well as transitional types can use XRD and infrared spectroscopy with the aid of Raman spectroscopy and nuclear magnetic resonance (NMR).

Klemme et al. [34] focus on the genesis of prase (green-coloured quartz) and amethyst from Serifos Island (Cyclades, Greece). The stable oxygen and hydrogen isotopic composition of both quartz varieties suggest a mixing of magmatic and meteoric (and/or marine) fluids. Large (up to several centimetres) and vividly coloured Mn-rich minerals (kyanite, green andalusite, garnet -grossular, and spessartine- and red-epidote) of "near" gem-quality from Thassos Island (Rhodope, Greece) are described by Tarantola et al. [35]. They also show that the orange colour of kyanite from Thassos is due to Mn^{3+}; this is the second reported occurrence (after Loliondo, Tanzania) of such kyanite. Voudouris et al. [36] offer an overview of collector and gem-quality mineral occurrences from Greece, relating them to various geological environments such as regional metamorphic-metasomatic, alpine-type fissures, plutonic-subvolcanic intrusions and pegmatites, zones of contact metamorphism, and peripheral volcanic rocks.

This Special Issue is a good example of the growing number of scientists working and collaborating on various gem-related topics around the world. We hope that this issue will shed light on various aspects of gemology, enhance scientific debate, and attract more scientists from various disciplines to become involved in this field of research.

Acknowledgments: The Guest Editors would like to sincerely thank all authors, reviewers, the editor Theodore J. Bornhorst and the editorial staff of *Minerals* for their timely efforts to successfully complete this Special Issue. Danat (Bahrain Institute for Pearls & Gemstones) is also thanked for providing one of the images (inclusions hosted in Russian emerald) for the cover page of this Special Issue book.

Conflicts of Interest: The authors declare no conflicts of interest.

References

1. Fritsch, E.; Rondeau, B. Gemology: The developing science of gems. *Elements* **2009**, *5*, 147–152. [CrossRef]
2. Gramaccioli, C.M. Application of mineralogical techniques to gemmology. *Eur. J. Mineral.* **1991**, *3*, 703–706. [CrossRef]

3. Devouard, B.; Notari, F. Gemstones: From the naked eye to laboratory techniques. *Elements* **2009**, *5*, 163–168. [CrossRef]
4. Simmons, W.B.; Pezzotta, F.; Shigley, J.E.; Beurlen, H. Granitic pegmatites as sources of colored gemstones. *Elements* **2009**, *8*, 281–288. [CrossRef]
5. London, D. Ore-forming processes within granitic pegmatites. *Ore Geol. Rev.* **2018**, *101*, 349–383. [CrossRef]
6. Strmić Palinkaš, S.; Palinkaš, L.; Neubauer, F.; Scholz, R.; Borojević Šoštarić, S.; Bermanec, V. Formation Conditions and $^{40}Ar/^{39}Ar$ Age of the Gem-Bearing Boqueirão Granitic Pegmatite, Parelhas, Rio Grande do Norte, Brazil. *Minerals* **2019**, *9*, 233. [CrossRef]
7. Huong, L.T.-T.; Otter, L.M.; Förster, M.W.; Hauzenberger, C.A.; Krenn, K.; Alard, O.; Macholdt, D.S.; Weis, U.; Stoll, B.; Jochum, K.P. Femtosecond Laser Ablation-ICP-Mass Spectrometry and CHNS Elemental Analyzer Reveal Trace Element Characteristics of Danburite from Mexico, Tanzania, and Vietnam. *Minerals* **2018**, *8*, 234. [CrossRef]
8. Diella, V.; Pezzotta, F.; Bocchio, R.; Marinoni, N.; Cámara, F.; Langone, A.; Adamo, I.; Lanzafame, G. Gem-Quality Tourmaline from LCT Pegmatite in Adamello Massif, Central Southern Alps, Italy: An Investigation of Its Mineralogy, Crystallography and 3D Inclusions. *Minerals* **2018**, *8*, 593. [CrossRef]
9. Dutrow, B.L.; Henry, D.J. Tourmaline: A geologic DVD. *Elements* **2011**, *7*, 301–306. [CrossRef]
10. van Hinsberg, V.; Henry, D.J. Tourmaline as a petrologic forensic mineral: A unique recorder of its geologic past. *Elements* **2011**, *7*, 327–332. [CrossRef]
11. Giuliani, G.; Groat, L.A.; Marshall, D.; Fallick, A.E.; Branquet, Y. Emerald Deposits: A Review and Enhanced Classification. *Minerals* **2019**, *9*, 105. [CrossRef]
12. Karampelas, S.; Al-Shaybani, B.; Mohamed, F.; Sangsawong, S.; Al-Alawi, A. Emeralds from the Most Important Occurrences: Chemical and Spectroscopic Data. *Minerals* **2019**, *9*, 561. [CrossRef]
13. Sutherland, F.L.; Zaw, K.; Meffre, S.; Thompson, J.; Goemann, K.; Thu, K.; Nu, T.T.; Zin, M.M.; Harris, S.J. Diversity in Ruby Geochemistry and Its Inclusions: Intra- and Inter- Continental Comparisons from Myanmar and Eastern Australia. *Minerals* **2019**, *9*, 28. [CrossRef]
14. Sorokina, E.S.; Rassomakhin, M.A.; Nikandrov, S.N.; Karampelas, S.; Kononkova, N.N.; Nikolaev, A.G.; Anosova, M.O.; Somsikova, A.V.; Kostitsyn, Y.A.; Kotlyarov, V.A. Origin of Blue Sapphire in Newly Discovered Spinel–Chlorite–Muscovite Rocks within Meta-Ultramafites of Ilmen Mountains, South Urals of Russia: Evidence from Mineralogy, Geochemistry, Rb-Sr and Sm-Nd Isotopic Data. *Minerals* **2019**, *9*, 36. [CrossRef]
15. Filina, M.I.; Sorokina, E.S.; Botcharnikov, R.; Karampelas, S.; Rassomakhin, M.A.; Kononkova, N.N.; Nikolaev, A.G.; Berndt, J.; Hofmeister, W. Corundum Anorthosites-Kyshtymites from the South Urals, Russia: A Combined Mineralogical, Geochemical, and U-Pb Zircon Geochronological Study. *Minerals* **2019**, *9*, 234. [CrossRef]
16. Voudouris, P.; Mavrogonatos, C.; Graham, I.; Giuliani, G.; Melfos, V.; Karampelas, S.; Karantoni, V.; Wang, K.; Tarantola, A.; Zaw, K.; et al. Gem Corundum Deposits of Greece: Geology, Mineralogy and Genesis. *Minerals* **2019**, *9*, 49. [CrossRef]
17. Peucat, J.J.; Ruffault, P.; Fritsch, E.; Bouhnik-La Coz, M.; Simonet, C.; Lasnier, B. Ga/Mg ratios as a new geochemical tool to differentiate magmatic from metamorphic blue sapphire. *Lithos* **2007**, *98*, 261–274. [CrossRef]
18. Giuliani, G.; Fallick, A.E.; Garnier, V.; France-Lanord, C.; Ohnenstetter, D.; Schwarz, D. Oxygen isotope composition as a tracer for the origins of rubies and sapphires. *Geology* **2005**, *33*, 249–252. [CrossRef]
19. Rossman, G.R. The geochemistry of gems and its relevance to gemology: Different traces, different prices. *Elements* **2009**, *5*, 159–162. [CrossRef]
20. Giuliani, G.; Ohnenstetter, D.; Fallick, A.E.; Groat, L.; Fagan, J. The geology and genesis of gem corundum deposits. In *Geology of Gem Deposits*, 2nd ed.; Groat, L.A., Ed.; Mineralogical Association of Canada: Quebec City, QC, Canada, 2014; pp. 29–112.
21. Vysotskiy, S.V.; Nechaev, V.P.; Kissin, A.Y.; Yakovenko, V.V.; Ignat'ev, A.V.; Velivetskaya, T.A.; Sutherland, F.L.; Agoshkov, A.I. Oxygen isotopic composition as an indicator of ruby and sapphire origin: A review of Russian occurrences. *Ore Geol. Rev.* **2015**, *68*, 164–170. [CrossRef]
22. Wong, J.; Verdel, C. Tectonic environments of sapphire and ruby revealed by a global oxygen isotope compilation. *Int. Geol. Rev.* **2018**, *60*, 188–195. [CrossRef]
23. Wang, K.K.; Graham, I.T.; Martin, L.; Voudouris, P.; Giuliani, G.; Lay, A.; Harris, S.J.; Fallick, A. Fingerprinting Paranesti Rubies through Oxygen Isotopes. *Minerals* **2019**, *9*, 91. [CrossRef]

24. Graham, I.T.; Harris, S.J.; Martin, L.; Lay, A.; Zappettini, E. Enigmatic Alluvial Sapphires from the Orosmayo Region, Jujuy Province, Northwest Argentina: Insights into Their Origin from in situ Oxygen Isotopes. *Minerals* **2019**, *9*, 390. [CrossRef]

25. Gauthier, J.P.; Karampelas, S. Pearls and corals: "Trendy biomineralizations". *Elements* **2009**, *5*, 179–180. [CrossRef]

26. Karampelas, S.; Mohamed, F.; Abdulla, H.; Almahmood, F.; Flamarzi, L.; Sangsawong, S.; Alalawi, A. Chemical Characteristics of Freshwater and Saltwater Natural and Cultured Pearls from Different Bivalves. *Minerals* **2019**, *9*, 357. [CrossRef]

27. Harley, S.L.; Kelly, N.M. Zircon: Tiny by timely. *Elements* **2007**, *3*, 13–18. [CrossRef]

28. Hanchar, J.M.; Hoskin, P.W.O. Zircon. *Rev. Mineral. Geochem.* **2003**, *53*, 500.

29. Piilonen, P.C.; Sutherland, F.L.; Danišík, M.; Poirier, G.; Valley, J.W.; Rowe, R. Zircon Xenocrysts from Cenozoic Alkaline Basalts of the Ratanakiri Volcanic Province (Cambodia), Southeast Asia—Trace Element Geochemistry, O-Hf Isotopic Composition, U-Pb and (U-Th)/He Geochronology—Revelations into the Underlying Lithospheric Mantle. *Minerals* **2018**, *8*, 556. [CrossRef]

30. Bui Thi Sinh, V.; Osanai, Y.; Lenz, C.; Nakano, N.; Adachi, T.; Belousova, E.; Kitano, I. Gem-Quality Zircon Megacrysts from Placer Deposits in the Central Highlands, Vietnam—Potential Source and Links to Cenozoic Alkali Basalts. *Minerals* **2019**, *9*, 89. [CrossRef]

31. Dill, H.G. Gems and Placers—A Genetic Relationship Par Excellence. *Minerals* **2018**, *8*, 470.

32. Štubňa, J.; Bačík, P.; Fridrichová, J.; Hanus, R.; Illášová, Ľ.; Milovská, S.; Škoda, R.; Vaculovič, T.; Čerňanský, S. Gem-Quality Green Cr-Bearing Andradite (var. Demantoid) from Dobšiná, Slovakia. *Minerals* **2019**, *9*, 164. [CrossRef]

33. Curtis, N.J.; Gascooke, J.R.; Johnston, M.R.; Pring, A. A Review of the Classification of Opal with Reference to Recent New Localities. *Minerals* **2019**, *9*, 299. [CrossRef]

34. Klemme, S.; Berndt, J.; Mavrogonatos, C.; Flemetakis, S.; Baziotis, I.; Voudouris, P.; Xydous, S. On the Color and Genesis of Prase (Green Quartz) and Amethyst from the Island of Serifos, Cyclades, Greece. *Minerals* **2018**, *8*, 487. [CrossRef]

35. Tarantola, A.; Voudouris, P.; Eglinger, A.; Scheffer, C.; Trebus, K.; Bitte, M.; Rondeau, B.; Mavrogonatos, C.; Graham, I.; Etienne, M.; et al. Metamorphic and Metasomatic Kyanite-Bearing Mineral Assemblages of Thassos Island (Rhodope, Greece). *Minerals* **2019**, *9*, 252. [CrossRef]

36. Voudouris, P.; Mavrogonatos, C.; Graham, I.; Giuliani, G.; Tarantola, A.; Melfos, V.; Karampelas, S.; Katerinopoulos, A.; Magganas, A. Gemstones of Greece: Geology and Crystallizing Environments. *Minerals* **2019**, *9*, 461. [CrossRef]

Article

Formation Conditions and ^{40}Ar/^{39}Ar Age of the Gem-Bearing Boqueirão Granitic Pegmatite, Parelhas, Rio Grande do Norte, Brazil

Sabina Strmić Palinkaš [1,*], Ladislav Palinkaš [2], Franz Neubauer [3], Ricardo Scholz [4], Sibila Borojević Šoštarić [5] and Vladimir Bermanec [2]

[1] Department of Geosciences, Faculty of Sciences and Technology, UiT-The Arctic University of Norway in Tromsø, N-9037 Tromsø, Norway

[2] Department of Geology, Faculty of Science, University of Zagreb, Horvatovac 95, HR-10000 Zagreb, Croatia; lpalinkas@geol.pmf.hr (L.P.); vladimir.bermanec@public.carnet.hr (V.B.)

[3] Department of Geography and Geology, Paris-Lodron-University of Salzburg, Hellbrunner Str. 34, A-5020 Salzburg, Austria; franz.neubauer@sbg.ac.at

[4] Departamento de Geologia, Escola de Minas, Universidade Federal de Ouro Preto, Ouro Preto MG-31400-000, Brazil; r_scholz_br@yahoo.com

[5] Faculty of Mining, Geology and Petroleum Engineering, University of Zagreb, Pierottijeva 6, HR-10000 Zagreb, Croatia; sibila.borojevic-sostaric@rgn.hr

* Correspondence: Sabina.s.palinkas@uit.no; Tel.: +47-77-625-177

Received: 10 February 2019; Accepted: 4 April 2019; Published: 15 April 2019

Abstract: The Boqueirão granitic pegmatite, alias Alto da Cabeça pegmatite, is situated in Borborema Pegmatitic Province (BPP) in Northeast Brazil. This pegmatitic province hosts globally important reserves of tantalum and beryllium, as well as significant quantities of gemstones, including aquamarine, morganite, and the high-quality turquoise-blue "Paraíba Elbaite". The studied lithium-cesium-tantalum Boqueirão granitic pegmatite intruded meta-conglomerates of the Equador Formation during the late Cambrian (502.1 ± 5.8 Ma; ^{40}Ar/^{39}Ar plateau age of muscovite). The pegmatite exhibits a typical zonal mineral pattern with four defined zones (Zone I: muscovite, tourmaline, albite, and quartz; Zone II: K-feldspar (microcline), quartz, and albite; Zone III: perthite crystals (blocky feldspar zone); Zone IV: massive quartz). Huge individual beryl, spodumene, tantalite, and cassiterite crystals are common as well. Microscopic examinations revealed that melt inclusions were entrapped simultaneously with fluid inclusions, suggesting the magmatic–hydrothermal transition. The magmatic–hydrothermal transition affected the evolution of the pegmatite, segregating volatile compounds (H_2O, CO_2, N_2) and elements that preferentially partition into a fluid phase from the viscous silicate melt. Fluid inclusion studies on microcline and associated quartz combined with microthermometry and Raman spectroscopy gave an insight into the P-T-X characteristics of entrapped fluids. The presence of spodumene without other $LiAl(SiO_3)_2$ polymorphs and constructed fluid inclusion isochores limited the magmatic–hydrothermal transition at the gem-bearing Boqueirão granitic pegmatite to the temperature range between 300 and 415 °C at a pressure from 1.8 to 3 kbar.

Keywords: gem-bearing pegmatite; fluid inclusions; P-T-X equilibria; spodumene; Ar/Ar dating

---

1. Introduction

Pegmatites, plutonic igneous rocks characterized by extremely coarse crystals with a systematically variable size, represent an important source of industrial minerals (feldspars, quartz, spodumene, petalite), hi-tech mineral commodities (e.g., Li, Cs, Be, Nb, Ta, Sn), and gemstones. The most common pegmatite-hosted gem minerals are colored varieties of beryl (aquamarine, heliodor, and morganite),

Li-rich tourmaline (elbaite-rossmanite and liddicoatite), blue and sherry topaz, transparent varieties of spodumene (kunzite), low-iron spessartine, and optical-grade quartz [1].

The Boqueirão granitic pegmatite, also known as the Alto da Cabeça pegmatite, is situated in the northernmost part of the Serra das Queimadas Mountains, the State of Rio Grande do Norte, Northeast Brazil (Figure 1). The pegmatite is hosted by Borborema Pegmatitic Province (BPP), which represents one of the world's most important sources of tantalum and beryllium, as well as of gemstones, including aquamarine, morganite, and the high-quality turquoise-blue "Paraíba Elbaite" [2–5].

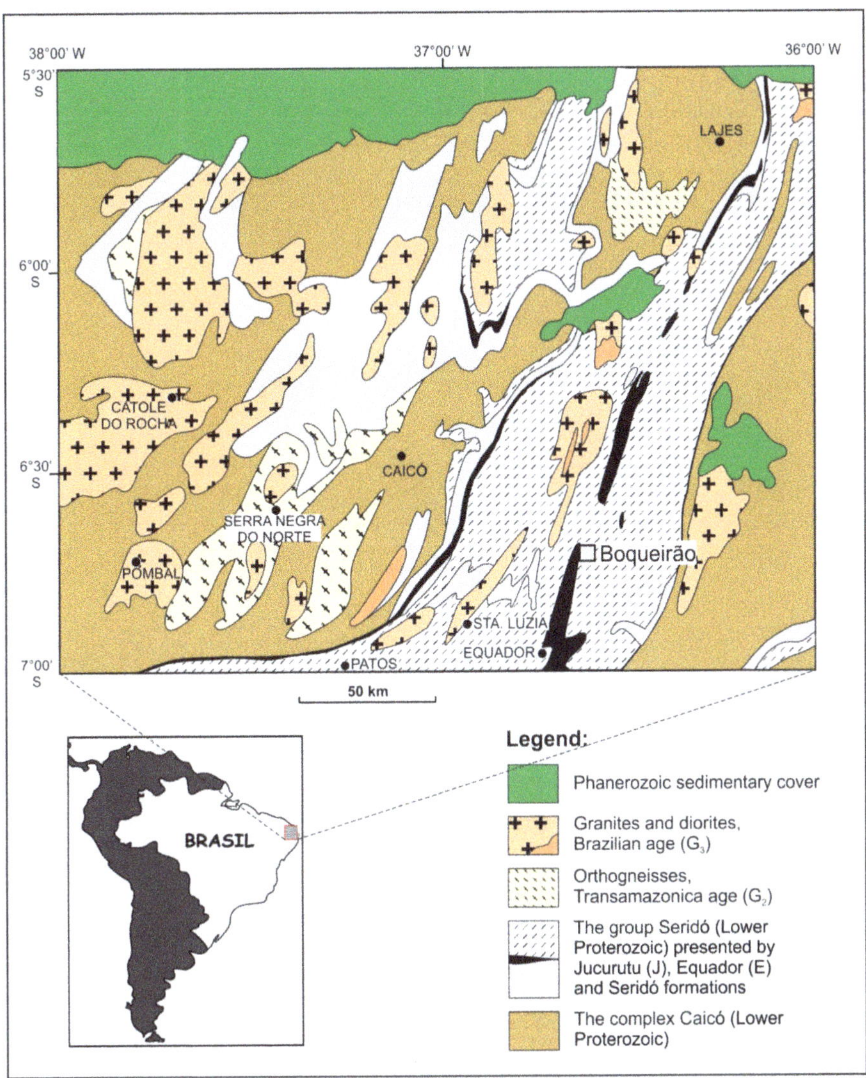

Figure 1. A simplified geological map of the Borborema Pegmatitic Province with the location of the Boqueirão granitic pegmatite, Parelhas, Rio Grande do Norte, Brazil. Adapted with permission from [6].

Mineralogical and geochemical features, characterized by a strong enrichment in numerous incompatible elements, including Li, Rb, Cs, Be, Sn, Ta, Nb (with Ta > Nb), B, P, and F, classify the

Boqueirão granitic pegmatite into the lithium-cesium-tantalum (LCT) family [5–9]. According to previously published data, the BPP pegmatites were formed in a relatively narrow range of pressure (2.1 to 4 kbar) but within a wide range of temperatures (390–900 °C) [10–12].

This study brings new fluid inclusion and ^{40}Ar/^{39}Ar data from the gem-bearing Boqueirão granitic pegmatite. The fluid inclusion data combined with the calculated thermodynamic equilibria of the established mineral paragenesis shed light on the formation conditions of the Boqueirão granitic pegmatite, whereas the ^{40}Ar/^{39}Ar dating confirmed the late Cambrian age of pegmatite emplacement. A particular focus has been given to the recorded magmatic–hydrothermal transition and its potential role in the evolution of the studied pegmatite.

2. Geological Setting

The Borborema Pegmatitic Province (6°–7°S, 36°15′–36°45′W) covers the southern part of the Meso- to Neo-Proterozoic Seridó Foldbelt (Figure 1; [13]). The Seridó Foldbelt comprises a basal volcano-sedimentary sequence (Jucurutu Formation), a quartzite–metaconglomerate complex (Equador Formation), and a turbidite–flysch sequence (Seridó Formation). The area is metamorphosed up to the upper amphibolite facies (Abukuma type) and retro-metamorphosed into the upper greenschist-facies grade [14]. Four generations of granite intrusions in the area have been described [15]. The formation of the pegmatites is related to granites of the late- to post-orogenic phase [16], labelled as G4 granites [15,17].

The Boqueirão granitic pegmatite intruded into meta-conglomerates of the Equador Formation [17]. The central part of the pegmatite exhibits a typical zonal mineral distribution. From its margin to the center, the pegmatite consists of (Figure 2): Zone I with comb-textured muscovite and/or tourmaline intergrown with medium-grained albite and quartz (Figure 3a–c); Zone II hosting homogeneous medium-grained K-feldspar (microcline) accompanied by quartz and albite; Zone III composed almost exclusively of large perthite crystals (blocky feldspar zone); and Zone IV, i.e., a monomineralic nucleus of massive milky and/or rose quartz [6,18]. The contact between Zones III and IV was a preferred site for deposition of decimetric to metric, irregular pockets of medium- to fine-grained cleavelandite (albite), muscovite, and lepidolite selvages with some phosphates and disseminated ore minerals (Figure 3c). Individual beryl, spodumene, tantalite, and cassiterite crystals have also been found at the boundary, with the roots of the crystals in the blocky feldspar zone and the tips growing idiomorphically into the former open space, now filled by the massive quartz core (Figure 3d). Occurrences of beryl and tourmaline mineralization at the boundaries between Zones II and III (Figure 3e,f), and spodumene and/or cassiterite in Zones I and II are common ([6] and references therein). The U-Pb dating of manganocolumbite and ferrocolumbite constrained the time of pegmatite emplacement between 509 and 515 Ma [19].

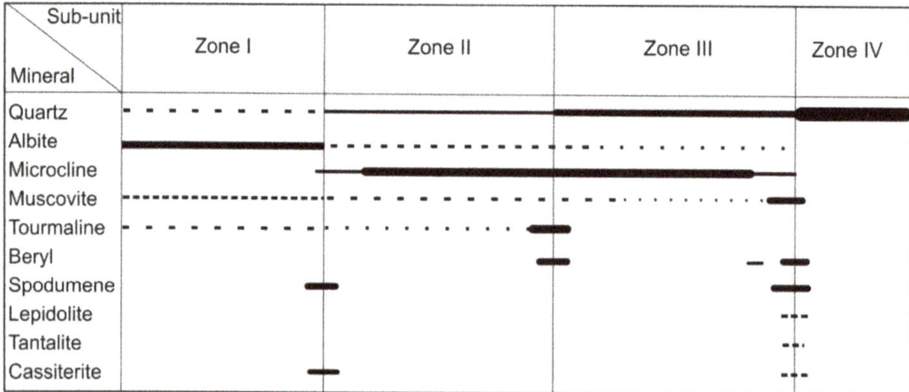

Figure 2. The paragenetic sequence of the gem-bearing Boqueirão granitic pegmatite.

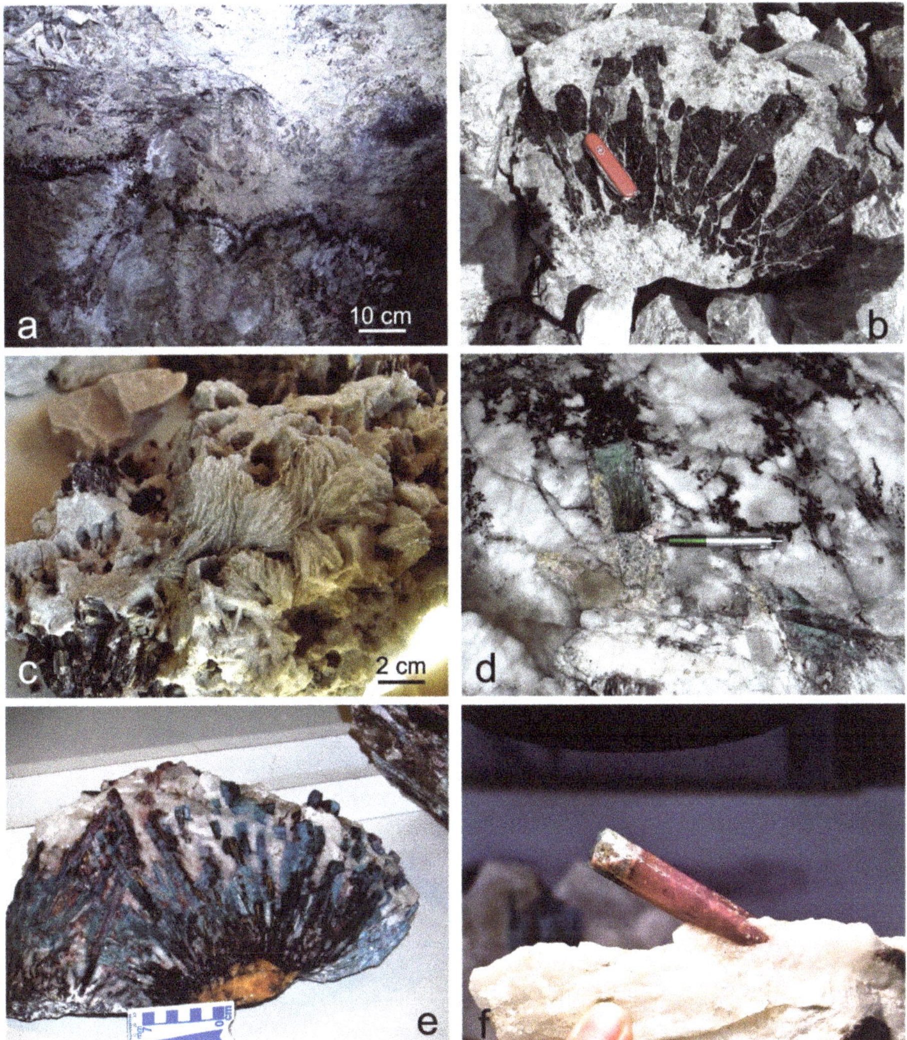

Figure 3. Macrophotographs showing: (**a**) A several-centimeters-thick tourmaline front, Zone I; (**b**) Comb-textured tourmaline intergrown within medium-grained albite and quartz, Zone I; (**c**) Mineral association composed of cleavelandite (albite), muscovite, and lepidolite; (**d**) Beryl crystals embedded within a lepidolite envelopment at the contact between Zones III and IV; (**e**) Transition from schorl to blue-colored tourmaline; (**f**) Red-colored tourmaline from the Boqueirão granitic pegmatite.

3. Materials and Methods

A fluid inclusion study was carried out on microcline and quartz crystals collected from Zone II. Muscovite grains gathered from Zone I were suitable for $^{39}Ar/^{40}Ar$ dating.

Petrographic and microthermometric measurements of fluid inclusions were performed at the University of Zagreb. Double polished, 0.1–0.3 mm thick, transparent mineral wafers were studied. Measurements were carried out on Linkam THMS 600 (Linkam Scientific Instruments Ltd., Tadworth, UK) stages mounted on an Olympus BX 51 (Olympus, Tokyo, Japan) using 10× and 50× Olympus long-working distance objectives. Two synthetic fluid inclusion standards (SYN FLINC; pure H_2O

and mixed H_2O–CO_2) were used to calibrate the equipment. The precision of the system was <2.0 °C for homogenization temperatures, and <0.2 °C in the temperature range between −60 and +10 °C. Microthermometric measurements were conducted on carefully defined fluid inclusion assemblages, representing groups of inclusions that were trapped simultaneously. The fluid inclusion assemblages were identified based on petrography prior to heating and freezing. If all of the fluid inclusions within the assemblage showed similar homogenization temperatures, the inclusions were assumed to have trapped the same fluid and to have not been modified by leakage or necking; these fluid inclusions thus record the original trapping conditions [20–22]. The salinity of aqueous inclusions was calculated from the final ice melting temperature using the BULK computer program [23]. Calculations are based on purely empirical best–fits, with no fundamental thermodynamic modeling involved [24]. The salinity of aqueous-carbonic inclusions was calculated according to [25]. Isochores were calculated by the ISOC computer program [23].

Raman spectroscopy of fluid inclusions, used for the semiquantitative analysis of entrapped volatiles, was performed on a Dilor LabRAM instrument (Horiba, Kyoto, Japan). Investigations were carried out at Department of Mineralogy and Petrology, Montanuniversität Leoben. A laser beam was focused through an Olympus BX 40 microscope (Olympus, Tokyo, Japan) onto the fluid inclusion of interest. The objective lenses of 50× and 100× magnification, combined with a confocal optical arrangement, enable a spatial resolution in the order of a cubic micrometer. A frequency-doubled Nd-YAG green laser (532 nm, 100 mW) was employed.

The $^{40}Ar/^{39}Ar$ analysis of muscovite was carried out at the ARGONAUT laboratory at the Department of Geography and Geology, Paris-Lodron-University of Salzburg. Mineral concentrates of muscovite were packed in aluminum-foil, sealed in quartz vials, and irradiated in the MTA KFKI reactor (Budapest, Hungary) for 16 h. The neutron fluence was monitored with DRA1 sanidine standard for which a $^{40}Ar/^{39}Ar$ plateau age of 25.03 ± 0.05 Ma has been reported [26]. Analyses were performed using a defocused (~1.5 mm diameter) 25 W CO_2-IR laser operating at the wavelengths between 10.57 and 10.63 μm. Gas cleanup was performed using two Zr–Al SAES getters (Milan, Italy). Ar-isotopes were measured on the VG ISOTECHTM NG3600 mass spectrometer (Isotopx, Middlewich, UK) on an axial electron multiplier in a static mode. Intensities of the peaks were back-extrapolated over 16 measured intensities to the time of gas admittance either with a straight line or a curved fit. Intensities were corrected for system blanks, background, post-irradiation decay of ^{37}Ar, and interfering isotopes. Isotopic ratios, ages, and uncertainties for individual steps were calculated following the suggestions by [27] and [28] using decay factors reported by [29]. The calculation of the plateau age was carried out using ISOPLOT/EX [28].

To avoid $^{40}Ar/^{39}Ar$ dating of hydrothermally altered muscovite, prior to the $^{40}Ar/^{39}Ar$ analysis an aliquot of the muscovite sample was analyzed by applying the X-ray powder diffraction (XRD) technique. The XRD analysis was conducted at the University of Zagreb on a Philips PW 3040/60 X'Pert PRO powder diffractometer (45 kV, 40 μA), with CuK-monochromatized radiation (λ = 1.54056 Å) and θ–θ geometry. The area between 4 and 63° 2θ, with 0.02 steps, was measured with a 0.5 primary beam divergence.

4. Results

4.1. Petrography of Fluid Inclusions

Microscopic examinations, performed at the room temperature on double-side-polished microcline and quartz wafers from Zone II, distinguished four types of inclusions:

Type I. Two phase-aqueous fluid inclusions (FIs) are mostly irregular, but some of them show progressive formation of negative crystal forms. The degree of fill (F), around 0.9, is fairly uniform (Figure 4a–c).

Type II. Aqueous-carbonic FIs show mostly irregular forms. This type of inclusion is characterized by the presence of two immiscible liquid phases (L_1 and L_2) and a vapor (V) phase. The F value varies slightly around 0.7 (Figure 4a,d). L_1 represents an aqueous solution whereas L_2 is composed of liquid CO_2.

Type III. Monophase elongated or slightly irregular gas inclusions (Figure 4e).

Type IV. Melt inclusions are mostly polyphase (Figure 4f), but they do not undergo any phase transition in the temperature range between −180 and +600 °C.

All four types of inclusions occur together in numerous inclusion assemblages, reflecting a melt-fluid immiscibility during crystallization of Zone II microcline and quartz.

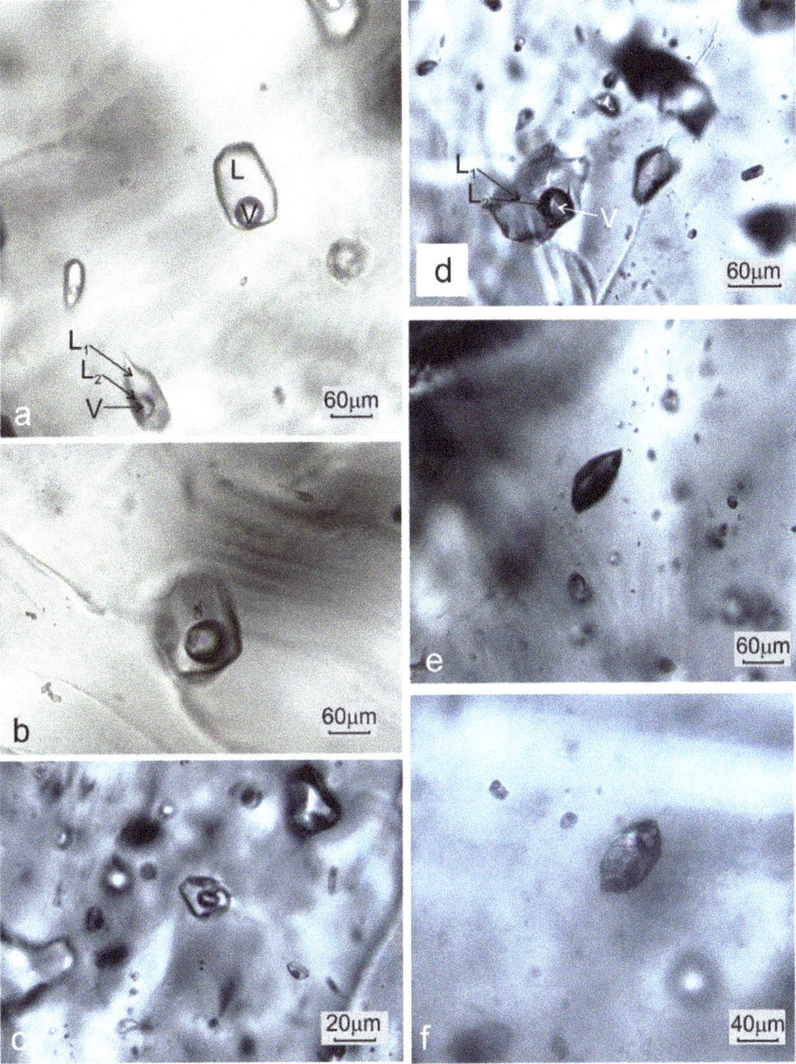

Figure 4. Macrophotographs of (**a**) coexisting two-phase, L + V, aqueous fluid inclusions and three-phase, $L_1 + L_2 + V$, aqueous-carbonic inclusions hosted by quartz; (**b**) two-phase, L + V, aqueous fluid inclusions seldom contain accidentally entrapped solid phases; (**c**) two-phase, L + V, aqueous fluid inclusions in microcline; (**d**) aqueous-carbonic inclusions in microcline; (**e**) a vapor-only inclusion hosted by microcline.

4.2. Microthermometry of Fluid Inclusions

Microthermometric data were collected from Type I and Type II inclusions hosted by microcline and quartz (Figure 5). Monophase fluid inclusions (Type III) as well as melt inclusions (Type IV) do not show any phase transition in the temperature range between −180 and +600 °C.

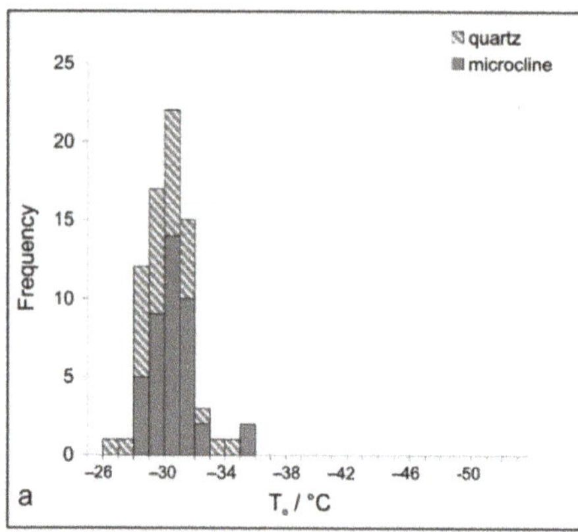

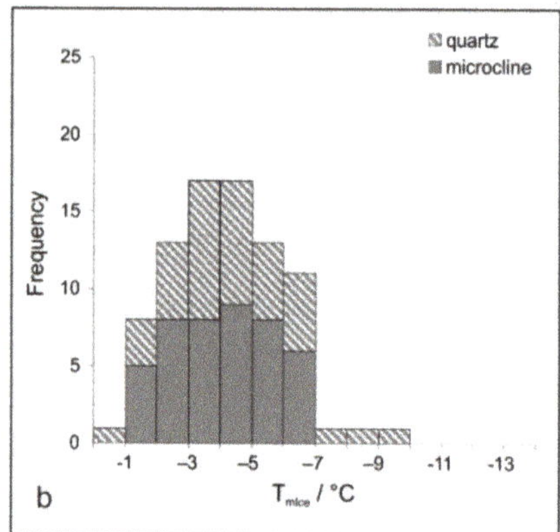

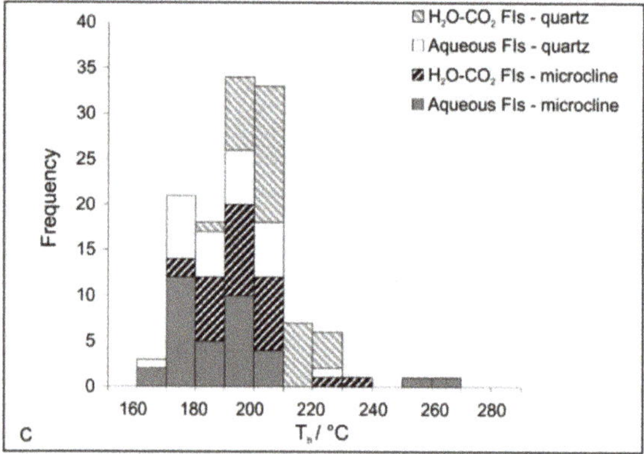

Figure 5. Histograms showing frequency distributions of fluid inclusion data: (**a**) Eutectic temperature (*Te*); (**b**) Final melting temperature of ice (T_{mIce}); (**c**) Homogenization temperature (T_h).

Microcline; Type I. A colorless frozen content of the aqueous FIs, formed at moderately low temperatures (generally around −40 °C), could be observed only by distortion and shrinkage of the vapor bubble. The initial melting temperature (eutectic temperature, T_e) is observed in an interval between −25.2 and −32.7 °C (Figure 5a) and the final ice melting (T_{mIce}) occurs between −1.8 and −6.2 °C (Figure 5b), reflecting the apparent salinity of 3.1–9.5 wt.% equ. NaCl. Homogenization (T_h) follows the disappearance of the vapor phase at 170–220 °C (Figure 5c). The bulk fluid density, calculated utilizing the equation of state proposed by [30] through the BULK software [23], spans from 0.867 to 0.965 g/cm^3. Type II carbonic-aqueous FIs are composed of a low-density gaseous phase and higher-density aqueous and carbonic liquid phases. The initial freezing of the aqueous phase was recorded around −40 °C, and complete freezing occurs at around −100 °C. The first melting of solid CO_2 occurs mostly in the temperature range between −60.5 and −67.8 °C, which is assigned to the presence of other volatiles (such as CH_4 and N_2; [31]). Melting of the aqueous part of inclusions was observed in a wide range of temperatures between −40 and −2 °C. The final melting of clathrate (T_{mClath}) spans between 8.0 and 9.0 °C and reflects salinities ranging from 2.5 to 4.0 wt.% equ. NaCl. Homogenization of the carbonic phase proceeds in two ways, L + V→V and L + V→L. Critical phenomena have not been observed. The data gather around 29 °C in either way of homogenization. Total homogenization (T_h) into the liquid phase occurred in the interval between 170 and 230 °C (Figure 5c). Bulk fluid densities, calculated according to the equation of state from [32] revised by [33], fall in the range between 0.808 and 0.893 g/cm^3.

Quartz, Type I: Measurements performed on aqueous FIs within quartz samples yielded microthermometric data that overlap with those gathered from microcline (Figure 5a–c). In contrast, Type II (carbonic-aqueous) FIs in quartz show some differences compared to the same type of FIs hosted by microcline. The initial melting of CO_2 occurs in the interval between −57.1 and −65.0 °C. The recorded T_{mClath} between 7.0 and 9.0 °C points to the salinity between 2.5 and 6.1 wt.% equ. NaCl. Homogenization of the carbonic phase proceeds mostly, at around 20 °C, into a liquid phase (Figure 5e). A critical phenomenon was observed only in one inclusion at +19.9 °C. The Th values have been recorded in an interval between 185 and 225 °C (Figure 5c).

4.3. Raman Spectroscopy of Fluid Inclusions

Raman spectroscopy measurements performed on aqueous FIs (Type I) within microcline and quartz samples recognized only the presence of water. The measurements of Type II FIs suggest CO_2 as the major non-H_2O volatile component (Figure 6a). In addition, variable amounts of N_2 were detected (Figure 6b). However, in several cases, the measured peak areas were too small for reliable quantification. A simple formula based on Placzek's polarizability theory was applied to derive quantitative molar fractions of species present in FIs [34–38]. The molar fraction of N_2 ranges up to 9 mole % (Table 1). No difference between quartz and microcline samples was observed.

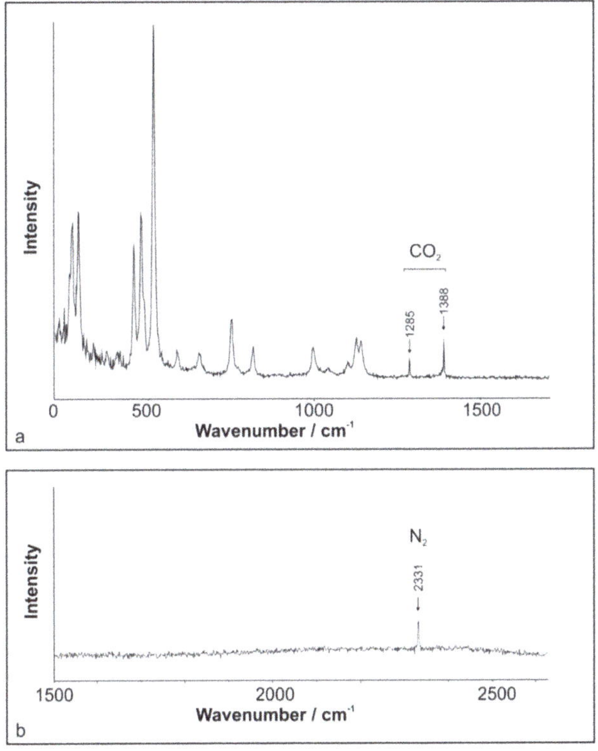

Figure 6. Raman spectra of the vapor bubble from an aqueous-carbonic inclusion reveal the presence of (**a**) CO_2 and (**b**) N_2.

Table 1. Microthermometric results, Raman data, and calculated bulk compositions of selected carbonic-aqueous fluid inclusions.

Sample	FI	Microthermometry (°C)				Raman (mol %)		Bulk Composition (mol %)			Density
		T_{mice}	T_{mclath}	T_{hCO2}	Mode	CO_2	N_2	H_2O	CO_2	N_2	(g·cm^{-3})
					Microcline						
m-1	1	−2.5	8.1	26.2	V	93.2	6.8	85.4	11.6	0.7	0.9366
m-1	2	−3.1	8.6	28.3	V	96.5	3.5	84.4	12.6	0.2	0.9576
m-1	3	−4.3	7.8	27.5	L	98.0	2.0	83.4	12.6	0.1	0.9691
m-1	4	−2.0	8.9	27.9	V	94.7	5.3	86.7	10.9	0.6	0.9253
m-2	1	−2.7	8.7	30.0	L	99.5	0.5	84.3	13.2	<0.05	0.9671
m-2	2	−3.3	9.2	28.0	L	97.8	2.2	84.2	12.7	0.2	0.9608
m-3	1	−4.0	8.1	28.6	V	92.1	7.9	83.6	12.1	0.6	0.9621
m-3	2	−2.9	7.6	29.1	V	99.9	0.1	84.1	13.2	<0.01	0.9694
m-3	3	−3.1	8.8	28.8	V	93.3	6.7	85.3	11.5	0.4	0.9359
					Quartz						
q-1	1	−4.0	8.2	20.1	V	93.4	6.6	83.7	12.0	0.8	0.9673
q-1	2	−3.1	8.9	18.5	V	99.7	0.3	82.8	14.4	<0.05	0.9950
q-1	3	−2.9	8.4	17.9	V	92.6	7.4	83.5	13.0	0.9	0.9749

4.4. The $^{40}Ar/^{39}Ar$ Age

A concentrate of a few relatively large flakes (<0.5 mm) of white mica (muscovite) free of any inclusions was selected for dating. The XRD pattern of the analysed muscovite is presented in Figure 7. The experimental results of $^{40}Ar/^{39}Ar$ dating are given in Table 2. The argon release pattern of the muscovite sample (Figure 8) shows a slightly disturbed U-shaped pattern with a plateau age of 502.5 ± 5.8 Ma, constituting together 88.9% of ^{39}Ar released (Steps 4 to 10). Low-energy Steps 1 to 3 yielded an addition of excess argon, whereas a significant increase in some ages (Steps 6 and 8) is attributed to an internally inhomogeneous distribution of argon in muscovite. The $^{37}Ar_{Ca}$ values are low (Table 2) and the variation of $^{37}Ar_{Ca}$ is small. In contrast, the chlorine-derived $^{38}Ar_{Cl}$ values are relatively high (Table 2), showing that muscovite did grow under some saline conditions. We consider the plateau age of 502.5 ± 5.8 Ma to be geologically significant and to date the cooling of pegmatite through the argon retention temperature, which is experimentally determined at 425 ± 25 °C in slowly cooling terranes [39].

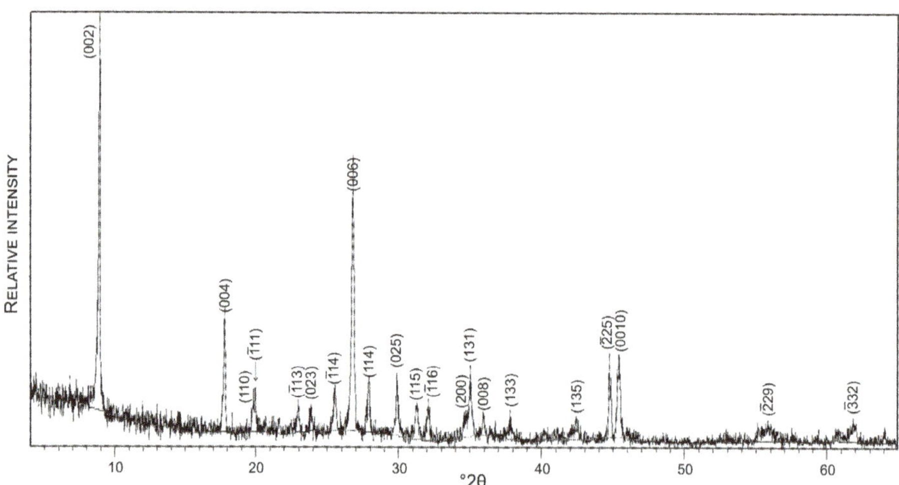

Figure 7. XRD pattern of muscovite dated by the of $^{40}Ar/^{39}Ar$ technique.

Table 2. The experimental results of ^{40}Ar/^{39}Ar analytical data.

J-Value:		0.01850	+/−	0.00019											
Step	^{36}Ar Meas.	$\pm\sigma_{36}$	^{37}Ar Decay Corr.	$\pm\sigma_{37}$	^{38}Ar Meas.	$\pm\sigma_{38}$	^{39}Ar Decay Corr.	$\pm\sigma_{39}$	^{40}Ar Meas.	$\pm\sigma_{40}$	^{40}Ar*/^{39}Ar$_K$	$\pm\sigma$	%^{40}Ar*	%^{39}Ar	Age (Ma)	$\pm$(Ma) 1-Sigma Abs.
1	62.9	9.3	56	25	120	7.9	7820	66.0	1.96×10^5	215	22.69	0.40	90.5	2.0	632.4	10.9
2	24.7	7.2	55	19	120	13	4070	26.2	9.59×10^4	183	21.74	0.54	92.4	1.1	610.0	13.9
3	56.1	12	52	29	360	14	33,100	68.6	6.27×10^5	253	18.44	0.11	97.3	8.6	529.6	5.5
4	17.7	8.7	46	22	940	15	90,500	119	1.59×10^6	916	17.49	0.04	99.6	23.5	505.7	4.6
5	99.8	1.3	37	25	1100	13	111,000	22.0	1.95×10^6	1340	17.24	0.04	98.4	28.9	499.5	4.6
6	27.8	18	63	32	130	11	9500	68.3	1.83×10^5	274	18.36	0.58	95.5	2.5	527.5	15.3
7	17.2	7.0	10	25	1100	18	108,000	142	1.88×10^6	1630	17.28	0.03	99.7	28.2	500.5	4.6
8	37.5	15	67	27	170	11	12,300	53.1	2.39×10^5	289	18.48	0.38	95.3	3.2	530.5	10.6
9	28.8	11	120	32	260	14	23,300	54.9	4.24×10^5	338	17.85	0.14	98.0	6.1	514.7	5.9
10	51.1	13	120	39	120	13	6060	43.0	1.26×10^5	187	18.25	0.67	88.0	1.6	524.7	17.3
Total	−	−	−	−	−	−	−	−	−	−	−	−	−	100.0	506.3	4.6

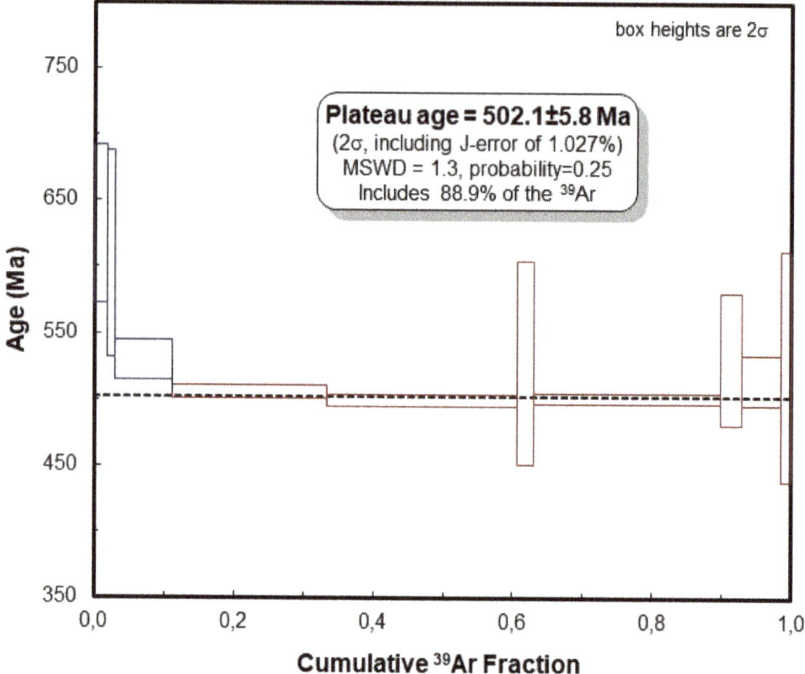

Figure 8. ^{40}Ar/^{39}Ar apparent age spectra of the coarse-grained muscovite sample. Laser energy increases from left to right. Vertical width of bars represents the 2σ error and includes the error of the J-value. Steps 4–10 are used for calculation of the plateau age.

5. Discussion

Fluid inclusions hosted by pegmatite crystals may provide a snapshot of P-T-X conditions at the time of their entrapment [40]. Numerous studies on the LCT-type of pegmatites worldwide have outlined the following types of inclusions: (1) melt inclusions; (2) saline aqueous fluid inclusions; and (3) CO_2-enriched fluid inclusions [41–45]. Classical theories of pegmatite genesis emphasize the importance of volatiles during crystallization of granitic pegmatites from coexisting aluminosilicate melt and hydrous fluids [46,47]. According to the model proposed by [48], aluminosilicate melts can produce highly evolved pegmatitic liquids via continuous crystallization under particular kinetic conditions. The ubiquitous crystallization commenced by formation of schorl (tourmaline), which buffered the boron content in the hydrous silicic melt, produces a high amount of exsolution of hydrous fluid [48]. The "boron quenching" in turn was enhanced by consequent crystallization of elbaite. The sink of boron, due to tourmaline crystallization and removal of the Li-alkali borate fluxing component, caused the separation of silicate and aqueous fluids producing supersaturation of alkali aluminosilicates (albite, microcline, quartz) and oxides, and their massive growth. Experimental studies support the coexistence of alumino-silicate melt, hydrous fluid, and hydrosaline fluid during the late stage of magma evolution [49].

Fluid inclusion assemblages, composed of melt inclusions and aqueous and CO_2-bearing fluid inclusions in minerals from Zone II of the gem-bearing Boqueirão granitic pegmatite, indicate the magmatic–hydrothermal transition. Similar phenomena have been recorded in other granitic pegmatites worldwide [50,51]. Melt inclusions represent entrapped remains of the silicate-rich melt, whereas fluid inclusions contain the fluid phase exsolved during the crystallization process. Additionally, coexistence of aqueous and aqueous-carbonic inclusions as well as their overlapping homogenization temperatures reflect an immiscibility between the low-salinity, low-density CO_2-rich fluid phase and the higher-salinity, higher-density fluid phase during the magmatic–hydrothermal transition. The recorded three-phase immiscibility (silicate melt–low salinity and low density carbonic-aqueous fluid–moderate salinity and moderate density fluid) affected the fate of metals in the evolving pegmatite. The majority of lithophile metals stay in the silicate melt, but some elements preferentially enter immiscible fluids. The partitioning is strongly controlled by the salinity of the exsolving fluids [52]. Pb, Zn, Ag, and Fe preferentially partition into fluids with a higher salinity, whereas Mo, B, As, Sb, and Bi prefer low-salinity fluids. In contrast, Li and Sn do not show systematic variations in their partition coefficients with the salinity of the fluids [53]. Partitioning of P between silicate melts and exsolving fluids strongly depends on pressure and temperature. Regardless, partition of flux elements (e.g., B, Li, and P) into exsolving fluids together with the loss of H_2O during the magmatic–hydrothermal transition may increase the viscosity of evolving silicate melts and affect textural features of pegmatites [54].

The lithium aluminosilicate phase diagram has been used as a petrogenetic grid for lithium-rich pegmatites [55]. Spodumene and petalite are stable Li-phases in a quartz-saturated system up to a temperature of 700 °C, which sets the upper limit on the crystallization conditions for Li-rich pegmatites [56]. However, the Boqueirão granitic pegmatite is characterized by the presence of spodumene as the only $LiAl(SiO_3)_2$ polymorph, which reflects the minimum formation pressure of 1.6 kbar (Figure 9).

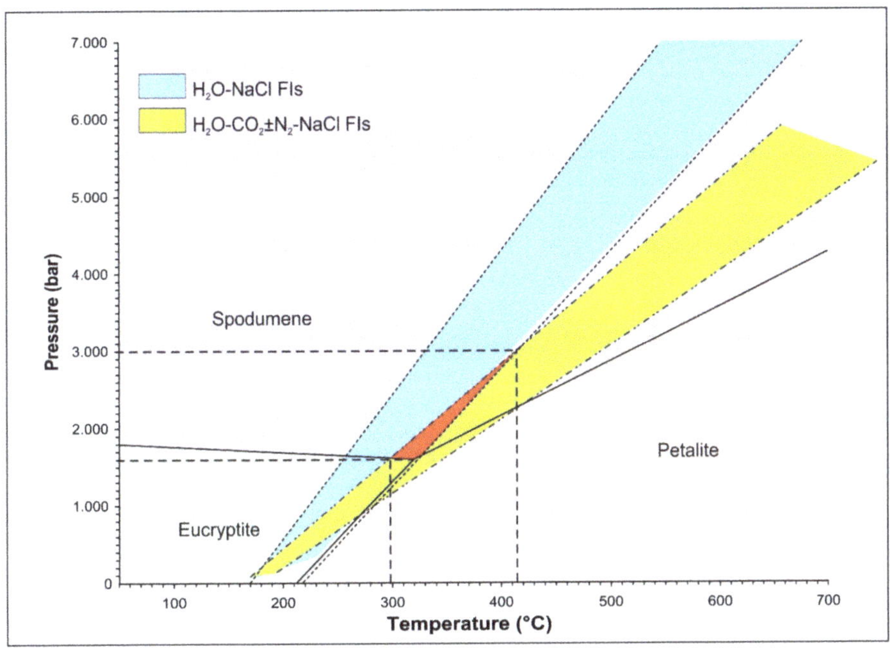

Figure 9. The isochores for aqueous (H_2O–NaCl FIs) and aqueous-carbonic (H_2O–CO_2 ± N_2-NaCl FIs) fluid inclusions extrapolated across the stability field of spodumene suggest a formation temperature in the range between 300 and 415 °C at a pressure from 1.8 to 3 kbar.

Isochores, constructed for the coexisting aqueous and aqueous-carbonic fluid inclusions (calculated using the equation of state proposed by [57] for the NaCl–H_2O system and [32] revised by [33] for the H_2O–CO_2–CH_4–N_2–NaCl system) extrapolated across the stability field of spodumene suggest that the magmatic–hydrothermal transition associated with the formation of the Boqueirão granitic pegmatite occurred at a temperature from 300 to 415 °C and a pressure between 1.8 and 3 kbar (Figure 9).

According to the $^{40}Ar/^{39}Ar$ plateau age of muscovite, the Boqueirão granitic pegmatite crystallized simultaneously with late stage of magmatic activity in the BPP (511–500 Ma; e.g., [58]).

6. Conclusions

According to its mineralogical and geochemical characteristics, the Boqueirão granitic pegmatite has been classified as a member of the LCT pegmatite family, broadly widespread over the BPP. The pegmatite shows a zonal structure with significant enrichment on incompatible elements from its outer rim toward the massive quartz core.

Coexistence of melt inclusions and aqueous and CO_2-enriched fluid inclusions suggests the magmatic–hydrothermal transition that resulted with segregation of two liquid phases (low salinity and low density versus moderate salinity and moderate density) from the silicate melt. The fluid inclusion data together with the well-defined stability of $LiAl(SiO_3)_2$ polymorphs over the P-T area can be used as an indicator of formation conditions for this type of granitic pegmatite. The fluid inclusion data obtained from the Boqueirão granitic pegmatite accompanied by the P-T stability of spodumen revealed that the magmatic–hydrothermal transition occurred in the temperature range between 300 and 415 °C at a pressure ranging from 1.8 to 3 kbar.

The $^{40}Ar/^{39}Ar$ plateau age of muscovite, at 502.5 ± 5.8 Ma, sets the Boqueirão granitic pegmatite to the late stage of magmatic activity in the BPP.

Author Contributions: Conceptualization, S.S.P., L.P. and V.B.; Methodology, F.N., L.P., R.S. and S.S.P.; Formal Analysis, S.S.P. and S.B.Š.; Writing-Original Draft Preparation, all coauthors; Writing-Review & Editing, S.S.P., F.N. and R.S.; Visualization, S.S.P.; Funding Acquisition, S.S.P., L.P. and V.B.

Funding: This research was supported by the Croatian Ministry of Sciences, Technology, and Sports (Projects No. 119-0000000-1158 and 119-0982709-1175).

Acknowledgments: Early versions of this paper greatly benefited from comments by David London and Rainer Thomas. Ronald J. Bakker is appreciated for help in obtaining the Raman spectroscopy data. Special thanks go to Panagiotis Voudouris and Vasilios Melfos for handling the manuscript as well as to the anonymous reviewers whose comments substantially improved the quality of this paper. The publication charges for this article have been funded by a grant from the publication fund of UiT The Arctic University of Norway.

Conflicts of Interest: The authors declare no conflict of interest.

References

1. London, D. Ore-forming processes within granitic pegmatites. *Ore Geol. Rev.* **2018**, *101*, 349–383. [CrossRef]
2. Pough, M.W. Simpsonite and the Northern Brazilian pegmatite region. *Bull. Geol. Soc. Am.* **1945**, *56*, 505–514. [CrossRef]
3. Beurlen, H. The mineral resources of the Borborema Province in Northeastern Brazil and its sedimentary cover: A review. *J. Am. Earth Sci.* **1995**, *8*, 365–376. [CrossRef]
4. Magyar, M.J. Tantalum. In *Mineral Commodity Summaries 2007*; U.S. Geological Survey: Reston, VA, USA, 2007; pp. 164–165.
5. Beurlen, H.; Barreto, S.; Martin, R.; Melgarejo, J.; Da Silva, M.R.R.; Souza Neto, J.A. The Borborema pegmatite province. NE Brazil revisited. *Estud. Geológicos* **2009**, *19*, 62–66.
6. Beurlen, H.; Da Silva, M.R.R.; Thomas, R.; Soares, D.R.; Olivier, P. Nb–Ta–(Ti–Sn) oxide mineral chemistry as tracer of rare-element granitic pegmatite fractionation in the Borborema Province, Northeastern Brazil. *Miner. Depos.* **2008**, *43*, 207–228. [CrossRef]
7. Černy, P. Rare element granitic pegmatites. Part 1: Anatomy and internal evolution of pegmatite deposits. *Geosci. Can.* **1991**, *18*, 49–67.
8. Černy, P.; Ercit, T.S. The Classification of Granitic Pegmatites Revisited. *Can. Miner.* **2005**, *43*, 2005–2026. [CrossRef]
9. Soares, D.R.; Beurlen, H.; Barreto, S.D.B.; Da Silva, M.R.R.; Ferreira, A.C.M. Compositional variation of tourmaline-group minerals in the borborema pegmatite province, northeastern brazil. *Can. Miner.* **2008**, *46*, 1097–1116. [CrossRef]

10. Beurlen, H.; da Silva, M.R.; de Castro, C. Fluid inclusion microthermometry in Be–Ta–(Li–Sn)-bearing pegmatites from the Borborema Province, northeast Brazil. *Chem. Geol.* **2001**, *173*, 107–123. [CrossRef]

11. Thomas, R.; Davidson, P.; Beurlen, H. Tantalite-(Mn) from the Borborema Pegmatite Province. northeastern Brazil: Conditions of formation and melt- and fluid-inclusion constraints on experimental studies. *Miner. Depos.* **2011**, *46*, 749–759. [CrossRef]

12. Strmić Palinkaš, S.; Wegner, R.; Čobić, A.; Palinkaš, L.A.; Barreto, S.D.B.; Váczi, T.; Bermanec, V. The role of magmatic and hydrothermal processes in the evolution of Be-bearing pegmatites: Evidence from beryl and its breakdown products. *Am. Mineral.* **2014**, *99*, 424–432. [CrossRef]

13. Scorza, E.P. Província Pegmatítica da Borborema. DNPM. *Div Geol Min Boletim. Rio de Janeiro* **1944**, *112*, 57.

14. Lima, E.S. Metamorphism and Tectonic Evolution in the Seridó Region. Northeastern Brazil. Ph.D. Thesis, Universty of California Los Angeles, Los Angeles, CA, USA, 1986; p. 215.

15. Jardim de Sá, E.F.; Legrand, J.M.; McReath, I. Estratigrafia de rochas granitóides na Região do Seridó (RN-PB) com base em critérios estruturais. *Rev. Bras. Geoc.* **1981**, *11*, 50–57. [CrossRef]

16. Araújo, M.N.C.; Silva, F.C.A.; Jardim, E.F. Pegmatite emplacement in the Seridó Belt. Northeastern Brazil: Late stage tectonics of the Brasiliano Orogen. *Gondwana Res.* **2001**, *4*, 75–85.

17. DA Silva, M.R.R.; Höll, R.; Beurlen, H. Borborema Pegmatitic Province: Geological and geochemical characteristics. *J. S. Am. Earth Sci.* **1995**, *8*, 355–364. [CrossRef]

18. Johnston, J.R. Beryl–tantalite pegmatites of Northeastern Brazil. *Geol. Soc. Am. Bull.* **1945**, *56*, 1015–1070. [CrossRef]

19. Baumgartner, R.; Romer, R.L.; Möritz, R.; Sallet, R.; Chiaradia, M. Columbite-tantalite-bearing granitic pegmatites from the serido belt, northeastern brazil: Genetic constraints from U-Pb dating and Pb isotopes. *Can. Miner.* **2006**, *44*, 69–86. [CrossRef]

20. Goldstein, R.H. Fluid inclusions in sedimentary and diagenetic systems. *Lithos* **2001**, *55*, 159–193. [CrossRef]

21. Goldstein, R.H.; Reynolds, T.J. Systematics of fluid inclusions in diagenetic minerals. *Soc. Sediment. Geol. Short Course* **1994**, *31*, 199.

22. Bodnar, R.J. Introduction to fluid inclusions. In *Short Course, Fluid Inclusions: Analysis and Interpretation*; Mineralogical Association of Canada: Ottawa, ON, Canada, 2003; Volume 32, pp. 1–8.

23. Bakker, R.J.; Package, F. Computer programs for analysis of fluid inclusion data and for modelling bulk fluid properties. *Chem. Geol.* **2003**, *194*, 3–23. [CrossRef]

24. Naden, J. *Calcic Brine: A Microsoft Excel 5.0 Add-in for Calculating Salinities from Microthermometric Data in the System NaCl–CaCl$_2$–H$_2$O*; Brown, P.E., Hagemann, S.G., Eds.; PACROFI VI: Madison, WI, USA, 1996; pp. 97–98.

25. Collins, P.L. Gas hydrates in CO$_2$-bearing fluid inclusions and the use of freezing data for estimation of salinity. *Econ. Geol.* **1979**, *74*, 1435–1444. [CrossRef]

26. Wijbrans, J.R.; Pringle, M.S.; Koopers, A.A.; Schveers, R. Argon geochronology of small samples using the Vulkaan argon laserprobe. *Proc. K. Ned. Akad. Van Wet.* **1995**, *98*, 185–218.

27. Mcdougall, I.; Harrison, M.T. *Geochronology and Thermochronology by the $^{40}Ar/^{39}Ar$ Method*, 2nd ed.; Oxford University Press: Oxford, UK, 1999; p. 26.

28. Ludwig, K.R. *Users Manual for ISOPLOT/Ex 3.75c: A Geochronological Toolkit for Microsoft Excel*; Special Publication; Berkeley Geochronology Center: Berkeley, CA, USA, 2012.

29. Steiger, R.H.; Jäger, E. Subcommission on geochronology: Convention on the use of decay constants in geo- and cosmochronology. *Earth Planet. Sci. Lett.* **1977**, *36*, 359–362. [CrossRef]

30. Zhang, Y.G.; Frantz, J.D. Determination of the homogenization temperatures and densities of supercritical fluids in the system NaCl-KCl-CaCl$_2$-H$_2$O using synthetic fluid inclusions. *Chem. Geol.* **1987**, *64*, 335–350. [CrossRef]

31. Burrus, R.C. Hydrocarbon fluid inclusions in studies of sedimentary diagenesis. In *Fluid Inclusions: Applications to Petrology*; Short Course Notes; Mineralogical Association of Canada: Toronto, ON, Canada, 1981; Volume 6, pp. 138–156.

32. Bowers, T.S.; Helgeson, H.C. Fortran programs for generating fluid inclusion isochores and fugacity coefficients for the system H$_2$O-CO$_2$-NaCl at high pressures and temperatures. *Comput. Geosci.* **1985**, *11*, 203–213. [CrossRef]

33. Bakker, R.J. Adaptation of the Bowers and Helgeson (1983) equation of state to the H$_2$O–CO$_2$–CH$_4$–N$_2$–NaCl system. *Chem. Geol.* **1999**, *154*, 225–236. [CrossRef]

34. Schrötter, H.W.; Klöckner, H.W. Raman scattering cross-sections in gases and liquids. In *Raman Spectroscopy of Gases and Liquids*; Weber, A., Ed.; Springer: Berlin, Germany, 1979; pp. 123–166.
35. Wopenka, B.; Pasteris, J.D. Limitation to quantitative analysis of fluid inclusions in geological samples by laser Raman microprobe spectroscopy. *Appl. Spectrosc.* **1986**, *40*, 144–151. [CrossRef]
36. Wopenka, B.; Pasteris, J.D. Raman intensities and detection limits of geochemically relevant gas mixtures for a laser Raman microprobe. *Anal. Chem.* **1987**, *59*, 2165–2170. [CrossRef]
37. Dubessy, J.; Poty, B.; Ramboz, C. Advances in C-O-H-N-S fluid geochemistry based on micro-Raman spectroscopic analysis of fluid inclusions. *Eur. J. Mineral.* **1989**, *1*, 517–534. [CrossRef]
38. Burke, E.A. Raman microspectrometry of fluid inclusions. *Lithos* **2001**, *55*, 139–158. [CrossRef]
39. Harrison, T.M.; Célérier, J.; Aikman, A.B.; Hermann, J.; Heizler, M.T. Diffusion of ^{40}Ar in muscovite. *Geochim. Cosmochim. Acta* **2009**, *73*, 1039–1051. [CrossRef]
40. Roedder, E. Fluid inclusions. *Rev. Mineral.* **1984**, *12*, 644.
41. Thomas, A.; Spooner, E. The volatile geochemistry of magmatic H_2O-CO_2 fluid inclusions from the Tanco zoned granitic pegmatite, southeastern Manitoba, Canada. *Geochim. Cosmochim. Acta* **1992**, *56*, 49–65. [CrossRef]
42. London, D. Geochemical features of peraluminous granites. pegmatites. and rhyolites as sources of lithophile metal deposits. Magmas, Fluids and Ore Deposits. In *Magmas, Fluids, and Ore Deposits*; Short Course Handbook; Mineralogical Association of Canada: Nepean, ON, Canada, 1995; Volume 23, pp. 175–202.
43. Webster, J.D.; Thomas, R.; Rhede, D.; Förster, H.-J.; Seltmann, R. Melt inclusions in quartz from an evolved peraluminous pegmatite: Geochemical evidence for strong tin enrichment in fluorine-rich and phosphorus-rich residual liquids. *Geochim. Cosmochim. Acta* **1997**, *61*, 2589–2604. [CrossRef]
44. Thomas, R.; Webster, J.D.; Heinrich, W. Melt inclusions in pegmatite quartz: Complete miscibility between silicate melts and hydrous fluids at low pressure. *Contrib. Mineral. Petrol.* **2000**, *139*, 394–401. [CrossRef]
45. Sirbescu, M.-L.C.; Nabelek, P.I. Crystallization conditions and evolution of magmatic fluids in the Harney Peak Granite and associated pegmatites, Black Hills, South Dakota—Evidence from fluid inclusions. *Geochim. Cosmochim. Acta* **2003**, *67*, 2443–2465. [CrossRef]
46. Jahns, R.H.; Burnham, C.W. Experimental studies of pegmatite genesis: I. A model for the derivation and crystallization of granitic pegmatites. *Econ. Geol.* **1969**, *64*, 843–864. [CrossRef]
47. Veksler, I.V. Liquid immiscibility and its role at the magmatic–hydrothermal transition: A summary of experimental studies. *Chem. Geol.* **2004**, *210*, 7–31. [CrossRef]
48. London, D. Magmatic-hydrothermal transition in Tanco Pegmatite. *Am. Mineral.* **1986**, *71*, 76–395.
49. Veksler, I.V.; Thomas, R.; Schmidt, C. Experimental evidence of three coexisting immiscible fluids in synthetic granitic pegmatite. *Am. Mineral.* **2002**, *87*, 775–779. [CrossRef]
50. Sirbescu, M.-L.C.; Nabelek, P.I. Crustal melts below 400 °C. *Geology* **2003**, *31*, 685–688. [CrossRef]
51. Kaeter, D.; Barros, R.; Menuge, J.F.; Chew, D.M. The magmatic–hydrothermal transition in rare-element pegmatites from southeast Ireland: LA-ICP-MS chemical mapping of muscovite and columbite–tantalite. *Geochim. Cosmochim. Acta* **2018**, *240*, 98–130. [CrossRef]
52. Webster, J.D. The exsolution of magmatic hydrosaline chloride liquids. *Chem. Geol.* **2004**, *210*, 33–48. [CrossRef]
53. Zajacz, Z.; Halter, W.E.; Pettke, T.; Guillong, M. Determination of fluid/melt partition coefficients by LA-ICPMS analysis of co-existing fluid and silicate melt inclusions: Controls on element partitioning. *Geochim. Cosmochim. Acta* **2008**, *72*, 2169–2197. [CrossRef]
54. Nabelek, P.I.; Whittington, A.G.; Sirbescu, M.L.C. The role of H_2O in rapid emplacement and crystallization of granite pegmatites: Resolving the paradox of large crystals in highly undercooled melts. *Contrib. Mineral. Petrol.* **2010**, *160*, 313–325. [CrossRef]
55. London, D.; Burt, D.M. Lithium aluminosilicate occurrences in pegmatites and the lithium aluminosilicate phase diagram. *Am. Mineral.* **1982**, *67*, 483–493.
56. London, D. Experimental phase equilibria in the system $LiAlSiO_4$-SiO_2-H_2O: A petrogenetic grid for lithium-rich pegmatites. *Am. Mineral.* **1984**, *69*, 995–1004.

57. Potter, R.W., II. Pressure corrections for fluid-inclusion homogenization temperatures based on the volumetric properties of the system NaCl-H_2O. *J. Res. U. S. Geol. Surv.* **1977**, *5*, 603–607.
58. Corsini, M.; De Figueiredo, L.L.; Caby, R.; Féraud, G.; Ruffet, G.; Vauchez, A. Thermal history of the Pan-African/Brasiliano Borborema Province of northeast Brazil deduced from $^{40}Ar/^{39}Ar$ analysis. *Tectonophysics* **1998**, *285*, 103–117. [CrossRef]

Article

Femtosecond Laser Ablation-ICP-Mass Spectrometry and CHNS Elemental Analyzer Reveal Trace Element Characteristics of Danburite from Mexico, Tanzania, and Vietnam

Le Thi-Thu Huong [1,*], Laura M. Otter [2,3,*], Michael W. Förster [3], Christoph A. Hauzenberger [1], Kurt Krenn [1], Olivier Alard [3,4], Dorothea S. Macholdt [2], Ulrike Weis [2], Brigitte Stoll [2] and Klaus Peter Jochum [2]

[1] NAWI Graz Geocentre, University of Graz, 8010 Graz, Austria;
christoph.hauzenberger@uni-graz.at (C.A.H.); kurt.krenn@uni-graz.at (K.K.)

[2] Climate Geochemistry Department, Max Planck Institute for Chemistry, 55128 Mainz, Germany;
d.macholdt@mpic.de (D.S.M.); ulrike.weis@mpic.de (U.W.); Brigitte.stoll@mpic.de (B.S.);
k.jochum@mpic.de (K.P.J.)

[3] Department of Earth and Planetary Sciences, Macquarie University, Sydney NSW 2109, Australia;
michael.forster@hdr.mq.edu.au (M.W.F.); olivier.alard@mq.edu.au (O.A.)

[4] Géosciences Montpellier, UMR 5243, CNRS & Université Montpellier, 34095 Montpellier, France

* Correspondence: thi.le@uni-graz.at (L.T.-T.H.); laura.otter@hdr.mq.edu.au (L.M.O.)

Received: 7 May 2018; Accepted: 28 May 2018; Published: 29 May 2018

Abstract: Danburite is a calcium borosilicate that forms within the transition zones of metacarbonates and pegmatites as a late magmatic accessory mineral. We present here trace element contents obtained by femtosecond laser ablation-inductively coupled plasma (ICP)-mass spectrometry for danburite from Mexico, Tanzania, and Vietnam. The Tanzanian and Vietnamese samples show high concentrations of rare earth elements ($\sum$REEs 1900 $\mu g \cdot g^{-1}$ and 1100 $\mu g \cdot g^{-1}$, respectively), whereas Mexican samples are depleted in REEs (<1.1 $\mu g \cdot g^{-1}$). Other traces include Al, Sr, and Be, with Al and Sr dominating in Mexican samples (325 and 1611 $\mu g \cdot g^{-1}$, respectively). Volatile elements, analyzed using a CHNS elemental analyzer, reach <3000 $\mu g \cdot g^{-1}$. Sr and Al are incorporated following $Ca^{2+} = Sr^{2+}$ and $2\ B^{3+} + 3\ O^{2-} = Al^{3+} + 3\ OH^{-} + \square$ (vacancy). REEs replace Ca^{2+} with a coupled substitution of B^{3+} by Be^{2+}. Cerium is assumed to be present as Ce^{4+} in Tanzanian samples based on the observed Be/REE molar ratio of 1.5:1 following $2\ Ca^{2+} + 3\ B^{3+} = Ce^{4+} + REE^{3+} + 3\ Be^{2+}$. In Vietnamese samples, Ce is present as Ce^{3+} seen in a Be/REE molar ratio of 1:1, indicating a substitution of $Ca^{2+} + B^{3+} = REE^{3+} + Be^{2+}$. Our results imply that the trace elements of danburite reflect different involvement of metacarbonates and pegmatites among the different locations.

Keywords: danburite; trace elements; REE; femtosecond LA-ICP-MS; CHNS elemental analyzer; pegmatites; skarn

---

1. Introduction

Danburite crystallizes in the orthorhombic system and has the formula $CaB_2Si_2O_8$. Its structure consists of a tetrahedral framework with boron and silicon orderly distributed in different tetrahedral sites. The framework of corner-sharing Si_2O_7 and B_2O_7 groups are interconnected by Ca atoms [1,2]. According to previous studies [3,4], the structural unit of danburite contains two tetrahedrally coordinated cations (T1: B and T2: Si), one calcium, and five oxygen atoms, among which O1, O2, and O3 are bonded to both B and Si, while O4 and O5 are bridging oxygens of the Si_2O_7 and B_2O_7 groups, respectively.

Danburite is one of the few boron minerals that are valued as gemstones. After its discovery in Danbury, Connecticut, USA, colorless gem-quality danburite has been subsequently found in Japan, Mexico, Russia, Sri Lanka, and Switzerland [5]. Exceptionally rare is yellow danburite, which so far has been reportedly found only in Madagascar, Tanzania, Myanmar, and Vietnam [6,7]. The important geological environments that are known to have produced gem-quality danburite specimens include pegmatites and metacarbonates associated with hydrothermal activity [8–10].

Previous studies presenting danburite compositions were generally limited to its major element geochemistry (e.g., [11,12]) due to the lack of microanalytical reference materials for boron minerals. While Huong et al. [7] overcame this issue by applying femtosecond laser ablation-inductively coupled plasma-mass spectrometry (LA-ICP-MS), which allows virtually matrix-independent calibration, their study was confined to a regional scale. Here, we present state-of-the-art femtosecond LA-ICP-MS determination of major and trace element concentrations in danburite from three distinct worldwide distributed occurrences (Tanzania, Vietnam, and Mexico). The aims of this study are (1) to investigate the geochemical differences of danburites from different locations (Tanzania, Vietnam, and Mexico) and rock types (pegmatites and skarn), (2) to elucidate their potential for further provenance discrimination, and (3) to understand the incorporation of trace elements into the danburite structure.

2. Materials and Methods

2.1. Sample Material

For this study, we selected 6 danburite samples from 3 deposits in Mexico, Tanzania, and Vietnam (2 samples from each deposit), representing different geological environments. The Mexican danburites were collected from the polymetallic skarn deposit (sulfides of Ag, Pb, Cu, and Zn) in the Charcas mining district, San Luis Potosi. The area is characterized by marine siliciclastic and volcaniclastic rocks, with 2 domains (east and west) separated by a regional fault [9]. In the region of San Luis Potosí, numerous volcanic systems and igneous rocks are associated with different mineral deposits. The Charcas deposit has a large Ca–B metasomatic envelope composed of early datolite and later danburite. Other minerals associated with danburite include calcite, apophyllite, stilbite, chalcopyrite, sphalerite, and citrine. The samples appear as colorless, transparent, prismatic euhedral crystals and are up to 6 cm in length.

The Tanzanian danburite originates from the central zone of a pegmatite mostly as yellowish, fine-grained, massive, opaque aggregates, but occasionally also as larger single crystals with color and transparency. The mine is referred to by the locals as "Munaraima" and is situated in Eastern Tanzania, at the edge of the Uluguru Mountains [10] near the village of Kivuma. The region around Kivuma is dominated by a metasedimentary sequence including metapelites, gneisses, and spinel and ruby-bearing marbles, which underwent granulite facies metamorphism during the East African Orogen at ~640 Ma [13,14]. Tonalitic dikes and pegmatites, commonly found in this area, intruded the basement rocks during slow cooling of the whole area. The contact zone of marble and pegmatite is dominated by a mineral assemblage consisting of microcline (variety amazonite), blue quartz, kyanite, and dravite, while the core complex is mainly composed of massive quartz and schörl [10]. For this study, small, anhedral, transparent yellow danburite crystals ranging from 0.5 to 1 cm in size were selected.

The Vietnamese danburite samples (1–1.5 cm) appear as yellow, transparent, broken, and slightly rounded crystals and have been found in a placer deposit (Bai Cat) in the Luc Yen mining area, Yen Bai province, Northern Vietnam [7,15]. The geology of Luc Yen is dominated by metamorphic rocks, mainly granulitic gneisses, mica schists, and marbles, which are associated with the large-scale Ailao Shan–Red River shear zone. Locally, aplitic and pegmatitic dykes occur [16]. Danburite crystals are associated with ruby, sapphire, spinel, topaz, and tourmaline in the Bai Cat placer deposit, which is surrounded by marble units. While the primary formations of ruby, sapphire, and spinel in Luc Yen are associated with metamorphosed limestones, those of tourmaline and topaz originate from pegmatite bodies. Besides tourmaline and topaz, these pegmatites contain orthoclase, smoky quartz, lepidolite, and beryl. Danburite crystals have not yet been discovered in situ, hence their genetic relationship with the Luc Yen pegmatites is not verified. However, fluid inclusion studies of Luc Yen danburites indicate a pegmatitic origin [15].

2.2. Analytical Methods

Chemical data for major elements were obtained by electron microprobe at the Institute of Geosciences, Johannes Gutenberg University Mainz, by laser ablation-inductively coupled plasma-mass spectrometry (LA-ICP-MS) at the Max Planck Institute for Chemistry, Mainz, and by CHNS Elemental Analyzer at Macquarie University.

Electron probe micro-analysis (EPMA) was performed at the University of Mainz with a JEOL JXA 8200 Superprobe instrument equipped with 5 wavelength-dispersive spectrometers, using 15 kV acceleration voltage and 12 nA filament current. Calcium and silicon were analyzed with wollastonite as a standard material.

LA-ICP-MS data for a total of 55 elements were obtained using an NWRFemto femtosecond laser operating at a wavelength of 200 nm in combination with a ThermoFisher Element2 single-collector sector-field ICP mass spectrometer (see Table 1). Pre-ablation cleaning was performed using a spot size of 65 μm, 80 μm/s scan speed, and 50 Hz pulse repetition rate at 100% energy output to remove any superficial surface residue. Thereafter, samples were ablated using line scans of 300 μm length at a spot size of 55 μm and a scan speed of 5 μm/s. These parameters resulted in an energy density of ca. 0.51 J/cm^2 at the sample surface, and the pulse repetition rate was set to 50 Hz. Since there is no matrix-matched calibration material for Ca–B silicates available, we applied a laser device that produces pulses at 150 fs, enabling virtually matrix-independent calibration [17]. The glass microanalytical reference material NIST SRM 610 was used as calibration material in the evaluation process, where ^{43}Ca was used as internal standard. Reduction of data and elimination of obvious outliers were performed following a programmed routine in Microsoft Excel described in Jochum et al. [18].

All samples were additionally analyzed for their H, C, N, and S contents in a vario EL cube elemental analyzer (Elementar, Langenselbold, Germany). For analysis, 50 to 100 mg samples were packed in Sn-foils (no flux added) and were ignited in an oxygen–He gas atmosphere furnace at around 1150 °C. The produced gases were then trapped and released in a set of chromatographic columns for the sequential analysis of N (no trapping), then C, H, and S. Each sample was measured for 9 min, and released gases were sequentially analyzed with a thermal conductivity detector. Sample measurements were repeated 3 times for each sampling location, and all values were calibrated against the reference materials BAM-U110, JP-1, and CRPG BE-N (Table 2). Analytical uncertainties were evaluated from reference material values, which were found to lie within 16% and 25% for C and H of the data tabulated in the GeoReM database [19].

Table 1. Operating conditions of the femtosecond laser ablation-inductively coupled plasma-mass spectrometry (fs-LA-ICP-MS) system.

Operating Conditions of NWRFemto200 Laser System	
Wavelength λ (nm)	200
Fluence (J·cm^{-2})	0.51
Pulse length (fs)	150
Pulse repetition rate (Hz)	50
Laser energy output (%)	100
Spot size (μm)	55
Line length (μm)	300
Scan speed (μm·s^{-1})	5
Warm-up time (s)	28
Dwell time (s)	60
Washout time (s)	30

Table 1. *Cont.*

Operating Conditions of the Element2 Mass Spectrometer	
RF power (W)	1055
Cooling gas (Ar) flow rate ($L \cdot min^{-1}$)	16
Auxiliary gas (Ar) flow rate ($L \cdot min^{-1}$)	1.19
Additional gas (He) flow rate ($L \cdot min^{-1}$)	0.7
Sample gas (Ar) flow rate ($L \cdot min^{-1}$)	0.7
Sample time (s)	0.002
Samples per peak	100
Mass window (%)	10
Time per pass (s)	2
Scan mode (Escan/Bscan)	both
Mass resolution	300

Table 2. Reference materials for CHNS analyzer.

BE-N Altered Basalts (SARM)	H TCD	C TCD	N TCD	S TCD	S IR
n	14	20	17	21	8
Average ($\mu g \cdot g^{-1}$)	2771 ± 534	2301 ± 147	197 ± 42	301 ± 37	298 ± 23
RSD %	19	6	21	12	8
BAM-U110					
n	13	18	18	17	–
Average ($\mu g \cdot g^{-1}$)	$12{,}258 \pm 1758$	$72{,}340 \pm 2640$	4237 ± 165	9114 ± 1082	–
RSD %	14	4	4	12	–
JP-1 Peridotite massif (JGS)					
n	4	12	14	14	14
Average ($\mu g \cdot g^{-1}$)	3195 ± 170	763 ± 82	91 ± 23	27 ± 14	26 ± 7
RSD %	5	11	26	51	27

n denotes the number of measurement performed; average refers to arithmetic means of the n values measured; RSD % is relative standard deviation expressed in %. "TCD" refers to the thermal conductivity detector and "IR" to the infrared detector devices.

3. Results

The chemical composition of the danburite samples from Mexico, Tanzania, and Vietnam are presented in Table 3. The major element mass fractions of B, Ca, and Si are close to the stoichiometric composition, i.e., 28.32 wt % B_2O_3 (calculated from 87,890 ppm B obtained by ICP-MS), 22.81 wt % CaO, and 48.88 wt % SiO_2, respectively.

The trace elements Li, Sc, Ga, Se, Rb, Zr, Nb, Ag, Cd, Sn, Cs, Hf, Ta, W, Ir, Pt, Au, Tl, Bi, and U have concentrations below the detection limit in all samples (see detection limits in footnote of Table 3). A C1-chondrite–normalized plot of rare earth element (REE) mass fractions (normalizing data from [20]) displays a strong enrichment of light rare earth elements (LREEs: La, Ce, Pr, Nd, Sm, Eu) compared to heavy rare earth elements (HREEs: Gd, Tb, Dy, Ho, Er, Tm, Yb, Lu) in danburite from Tanzania and Vietnam, while the samples from Mexico are mostly below detectability (Figure 1). Europium shows a negative anomaly of the same order for samples from Tanzania and Vietnam, while no other strong anomaly is observed (e.g., Ce). Total lanthanide content of Tanzanian samples is up to 1900 $\mu g \cdot g^{-1}$, hence the mass fractions of LREE exceed those of HREE by a 500-fold enrichment. The Vietnamese samples are different, with a total REE content of around 1000 $\mu g \cdot g^{-1}$, hence the mass fractions of LREE exceed those of HREE by a 200-fold enrichment. The Mexican danburites appear to be REE-poor, with total REE contents below 1 $\mu g \cdot g^{-1}$ for La and Ce, while the remaining REEs from Nd to Lu are below the detection limits. Possible quadrivalent trace elements such as Ti, Hf, and Zr were also below the detection limits. Thorium and Pb show low mass fractions of 0.1 and 11 $\mu g \cdot g^{-1}$ in the Mexican danburites, and 0.6 and 8 $\mu g \cdot g^{-1}$ in the Vietnamese and Tanzanian danburites, respectively.

The trivalent element Al shows highly varying concentrations among the different deposits and is anti-correlated with $\sum$REE, thus also with Be (see Figure 2A,B), while no correlation was observed between Al and Sr (Figure 2C). The highest mass fractions of Al are found in the Mexican samples (325 μg·g^{-1}) and the lowest in the Tanzanian samples (84 μg·g^{-1}) (Table 3). Strontium is highest (1611 μg·g^{-1}) in the Mexican samples and lowest (66 μg·g^{-1}) in the Vietnamese samples (Figure 2D). Manganese yields up to 18 μg·g^{-1} in the Vietnamese danburites, while the Mexican and Tanzanian samples have Mn contents below the detection limits. In general, transition metals (e.g., Fe and Ni) have extremely low concentrations in all danburite samples. The three danburite origins can also be separated from each other using the mass fractions of Y (Figure 2E). Beryllium is found to be highest in the Tanzanian danburite (up to 178 μg·g^{-1}) and lowest in the Mexican samples (3 μg·g^{-1}). In addition, low concentrations of less than 1, 9, and 2 μg·g^{-1} of the elements Ba, Mg, and Cu, respectively, are identified in all samples regardless of their origin. The positive correlation between Be and $\sum$REE is exceptionally strong for all three deposits (Figure 3A), and due to varying mass fractions of all shown elements, this plot enables excellent discrimination among the three deposits. Univalent elements such as Li, Na, and, K fall below the detection limits.

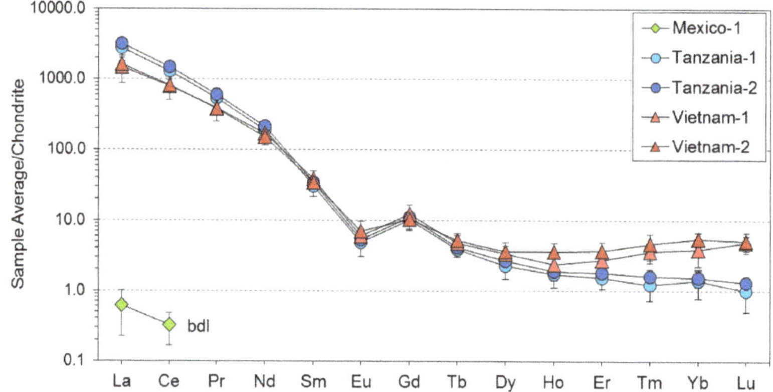

Figure 1. Rare earth element (REE) mass fractions in danburite from Mexico, Tanzania, and Vietnam. All values are presented as averages and normalized to C1-chondrite (data from [20]). Samples from Tanzania and Vietnam have a high abundance of REEs, especially LREEs, while samples from Mexico are largely devoid of these elements (bdl, below detection limit, from Pr to Lu).

The contents of the light volatile elements H, C, N, and S are generally low for both unpowdered and powdered samples (see Table 4). Samples from both Tanzania and Mexico exhibit mass fractions of H, C, N, and S of <10, <200, <100, and <15 μg·g^{-1}, respectively. The samples from Vietnam show significantly higher mass fractions for H, C, N, and S in unpowdered specimens, reaching values of up to 300, 5600, 1000, and 15 μg·g^{-1}, respectively. However, most elements are present in lower concentrations in the powdered sample set (H, C, and N of 40, 3000, and 520 μg·g^{-1}, respectively). Higher values for powdered samples from Mexico and Tanzania are likely attributed to the significantly increased surface-to-volume ratio facilitating higher adhesion of atmospheric gases. Significantly higher values for H and C in all Vietnamese samples agree well with a previous study by Huong et al. [15], who characterized a high abundance of primary CO_2-bearing fluid inclusions in these samples, which likely accounts for the elevated concentrations of both elements, while fluid inclusion is not present in the Mexican or Tanzanian specimens. High N mass fractions are in the expected range of metamorphosed sediments, which are involved in danburite formation and are known to contain ~200–3000 μg·g^{-1} N for a typical metamorphic gradient of 500–700 °C (e.g., [21]). Overall lower values of H, C, and N in the powdered Vietnamese sample set support the observation that these elements are derived from fluid inclusions and were lost during crushing in the agate mortar. Nevertheless, powdered samples were still found to contain up to 40 $\mu g \cdot g^{-1}$ structurally bound H (equivalent to 0.036 wt % H_2O+), which is close to the value of 0.04 wt % H_2O+ determined by IR spectroscopy as published in [4].

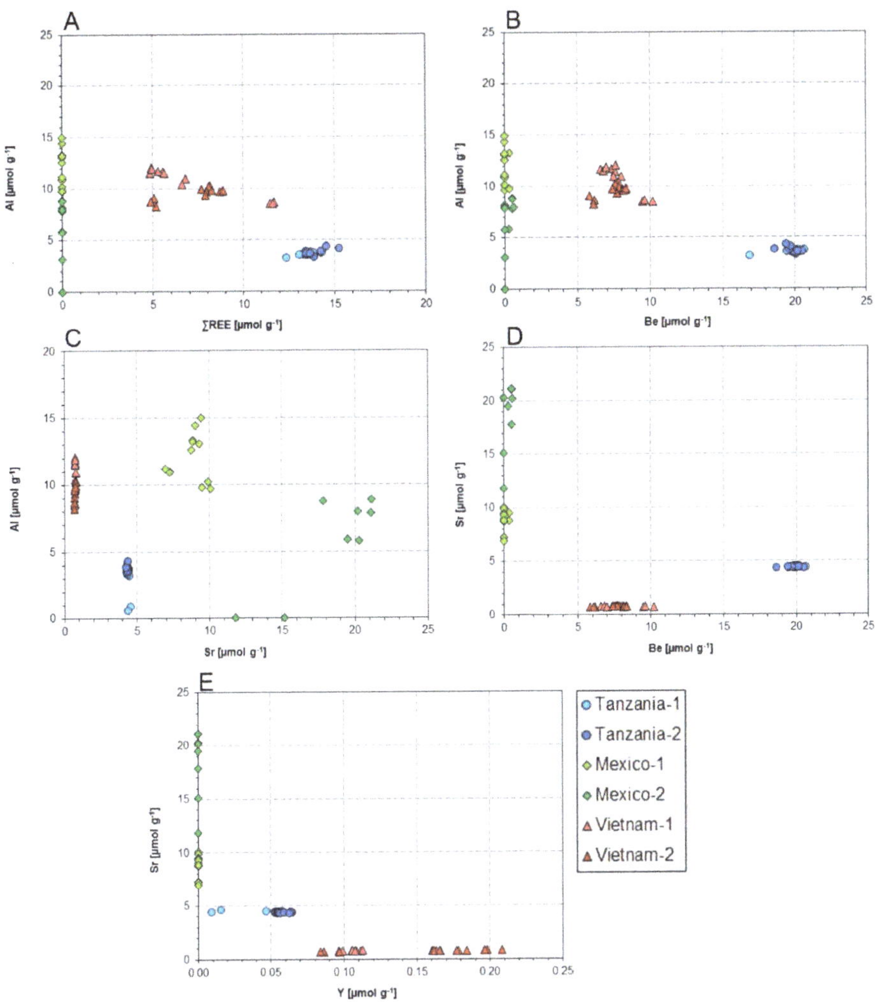

Figure 2. Molar abundance of Al versus (**A**) $\sum$REE, (**B**) Be, and (**C**) Sr, and Sr versus (**D**) Be and (**E**) Y. Aluminum is negatively correlated with (**A**) REEs, (**B**) Be, and (**D**) partly Sr, while Sr correlates negatively with (**D**) Be and (**E**) Y.

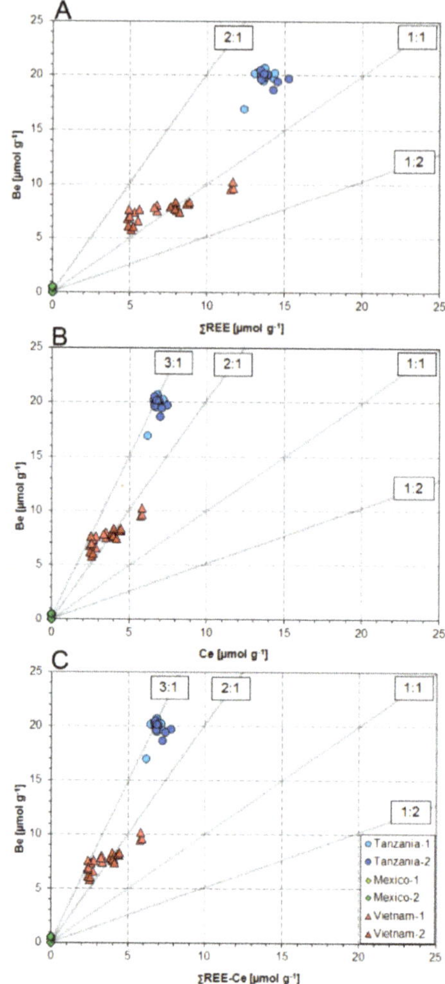

Figure 3. (**A**) Molar abundance of Be shows a linear correlation with $\sum$REE and illustrates that incorporation of REEs in the danburite structure is accompanied by Be. The Be/$\sum$REE ratio in Tanzanian samples is approximately 1.5:1, while it varies in Vietnamese samples from ca. 1:1 to ca. 1.5:1. Correlations between Be and $\sum$REE are ideally suited to distinguish danburite sampling locations. However, the simple substitution equation $Ca^{2+} + B^{3+} = REE^{3+} + Be^{2+}$, where the Be/$\sum$REE ratio is 1:1, is not sufficient to explain the varying ratios of Be and $\sum$REE. (**B**) Molar abundance of Be and Ce shows a linear correlation, implying that incorporation of Ce into the danburite structure is accompanied by Be. The Be/Ce ratio in Tanzanian samples is approx. 3:1, while in Vietnamese samples it varies from ca. 2:1 to ca. 3:1. The equation $2\,Ca^{2+} + 3\,B^{3+} = Ce^{4+} + REE^{3+} + 3\,Be^{2+}$, where the Be/Ce ratio is 3:1, explains the Tanzanian and most of the Vietnamese cases very well. Therefore, we argue here that Ce occurs not only as Ce^{3+}, but also as Ce^{4+} in Tanzanian and Vietnamese danburite and the substitution mechanism $2\,Ca^{2+} + 3\,B^{3+} = Ce^{4+} + REE^{3+} + 3\,Be^{2+}$ takes places in these samples. (**C**) Be/($\sum$REE − Ce) ratios show similar behavior to Be/Ce ratios. As the Be/($\sum$REE − Ce) ratios vary from 2:1 to 3:1, both substitution mechanisms should take place: $Ca^{2+} + B^{3+} = REE^{3+} + Be^{2+}$, $2\,Ca^{2+} + 3\,B^{3+} = Ce^{4+} + REE^{3+} + 3\,Be^{2+}$.

Table 3. Average chemical composition (in $\mu g \cdot g^{-1}$ and $\mu mol \cdot g^{-1}$) and relative standard deviation (RSD, %) obtained from danburite of three different occurrences. All concentrations were obtained by fs LA-ICP-MS (n = 12 line scans), except CaO and SiO$_2$, which were evaluated with Electron Probe Microanalyzer (EPMA) (averaged from n = 20 spot analyses per sample).

Element	Isotope Used	L.O.D.	Charcas, San Luis Potosi, Mexico						Morogoro, Tanzania						Luc Yen, Vietnam					
			Mex-1			Mex-2			Tanz-1			Tanz-2			Viet-1			Viet-2		
			Ø ($\mu g \cdot g^{-1}$)	Ø ($\mu mol \cdot g^{-1}$)	RSD (%)	Ø ($\mu g \cdot g^{-1}$)	Ø ($\mu mol \cdot g^{-1}$)	RSD (%)	Ø ($\mu g \cdot g^{-1}$)	Ø ($\mu mol \cdot g^{-1}$)	RSD (%)	Ø ($\mu g \cdot g^{-1}$)	Ø ($\mu mol \cdot g^{-1}$)	RSD (%)	Ø ($\mu g \cdot g^{-1}$)	Ø ($\mu mol \cdot g^{-1}$)	RSD (%)	Ø ($\mu g \cdot g^{-1}$)	Ø ($\mu mol \cdot g^{-1}$)	RSD (%)
CaO*	–	–	**22.47**	**0.04**	**0.49**	**22.43**	–	**0.71**	**22.63**	**0.04**	**0.60**	**22.48**	**0.04**	**0.57**	**22.32**	**0.03**	**0.51**	**22.40**	**0.03**	**0.37**
B	11	10	86218	7975	6.67	90471	8368	4.05	86280	7981	1.56	87270	8072	2.69	85698	7927	3.13	89138	8245	3.19
SiO$_2$ *	–	–	**48.31**	**0.8**	**0.40**	**48.81**	**0.8**	**0.33**	**48.80**	**0.8**	**0.36**	**48.88**	**0.8**	**0.38**	**48.81**	**0.8**	**0.43**	**48.43**	**0.8**	**0.37**
La	139	0.01	0.15	0.0011	63.5	0.01	0.0001	124.0	675	4.9	35.3	807	5.8	4.7	368	2.7	42.3	400	2.9	22.0
Ce	140	0.01	0.20	0.0014	48.1	0.05	0.0004	10.00	827	5.9	34.7	964	6.9	3.4	508	3.6	36.9	525	3.8	19.6
Pr	141	0.03	<0.03	–	–	<0.03	–	–	51.4	0.37	35.1	59.5	0.42	4.0	37.4	0.27	35.3	36.5	0.26	17.5
Nd	143	0.08	<0.08	–	–	<0.08	–	–	91.1	0.63	34.5	106	0.74	4.9	80.9	0.56	32.2	72.8	0.51	16.2
Sm	147	0.05	<0.05	–	–	<0.05	–	–	4.65	0.03	28.7	5.68	0.04	9.6	5.94	0.04	27.1	5.23	0.03	18.7
Eu	151	0.05	<0.05	–	–	<0.05	–	–	0.28	0.002	37.7	0.31	0.002	14.1	0.34	0.002	28.5	0.41	0.003	38.2
Gd	157	0.04	<0.04	–	–	<0.04	–	–	2.05	0.013	28.1	2.38	0.015	10.9	2.46	0.016	37.1	2.13	0.014	28.3
Tb	159	0.01	<0.01	–	–	<0.01	–	–	0.15	0.001	22.4	0.16	0.001	15.5	0.18	0.001	29.5	0.20	0.001	26.1
Dy	163	0.05	<0.05	–	–	<0.05	–	–	0.58	0.004	35.7	0.68	0.004	7.80	0.85	0.005	30.5	0.92	0.006	33.7
Ho	165	0.01	<0.01	–	–	<0.01	–	–	0.10	0.001	35.9	0.11	0.001	11.3	0.13	0.001	40.3	0.20	0.001	31.1
Er	167	0.04	<0.04	–	–	<0.04	–	–	0.25	0.001	30.0	0.28	0.002	38.6	0.46	0.003	41.0	0.61	0.004	33.3
Tm	169	0.02	<0.02	–	–	<0.02	–	–	0.03	0.0002	40.5	0.04	0.0002	49.3	0.09	0.0005	31.4	0.12	0.0007	40.1
Yb	173	0.04	<0.04	–	–	<0.04	–	–	0.23	0.0013	44.4	0.26	0.0015	32.5	0.64	0.0037	42.1	0.90	0.0052	22.4
Lu	175	0.02	<0.02	–	–	<0.02	–	–	0.02	0.0001	50.3	0.03	0.0002	43.5	0.12	0.0007	23.7	0.13	0.0007	32.2
Al	27	10	325	12.0	14.9	171	6.4	40.5	84.3	3.1	36.2	101	3.7	7.58	288	10.7	12.8	257	9.5	6.41
As	69	1	46.6	0.6	8.74	7.05	0.1	19.0	<1	0.01	56.5	<1	0.01	67.8	1.13	0.02	103	1.10	0.01	46.0
Ba	135, 137	0.1	0.50	0.004	54.1	0.19	0.001	36.60	0.29	0.002	25.06	0.22	0.002	14.77	0.48	0.003	89.66	0.63	0.005	27.03
Be	9	3	3.06	0.3	12.5	4.94	0.5	26.0	156	17.4	32.8	178	19.8	2.42	71.7	8.0	14.09	67.0	7.4	12.2
Cr	53	5	<5	–	–	<5	–	–	<5	–	–	<5	–	–	6.44	0.12	43.8	<5	–	–
Cu	65	1	1.56	0.02	65.3	<1	–	–	<1	–	–	<1	–	–	1.10	0.02	78.2	2.00	0.03	56.4
Fe	57	20	<20	–	–	<20	–	–	<20	–	–	<20	–	–	<20	–	–	<20	–	–
K	39	7	<7	–	–	<7	–	–	<7	–	–	<7	–	–	<7	–	–	<7	–	–
Mg	25	9	<9	–	–	<9	–	–	<9	–	–	<9	–	–	<9	–	–	<9	–	–
Mn	55	1	<1	–	–	<1	–	–	<1	–	–	<1	–	–	18.4	0.34	8.93	14.57	0.27	14.3
Na	23	50	<50	–	–	<50	–	–	<50	–	–	<50	–	–	<50	–	–	<50	–	–
Ni	62	18	<18	–	–	<18	–	–	<18	–	–	<18	–	–	<18	–	–	<18	–	–
Pb	207, 208, 209	0.1	0.27	0.0013	49.2	0.23	0.0011	40.0	7.46	0.0360	33.8	8.05	0.0389	8.94	11.1	0.0534	5.05	10.6	0.0514	5.60

Table 3. *Cont.*

Element	Isotope Used	L.O.D.	Charcas, San Luis Potosi, Mexico						Morogoro, Tanzania						Luc Yen, Vietnam					
			Mex-1			Mex-2			Tanz-1			Tanz-2			Viet-1			Viet-2		
			∅ (µg·g⁻¹)	∅ (µmol·g⁻¹)	RSD (%)	∅ (µg·g⁻¹)	∅ (µmol·g⁻¹)	RSD (%)	∅ (µg·g⁻¹) ∅ (µmol·g⁻¹)		RSD (%)	∅ (µg·g⁻¹)	∅ (µmol·g⁻¹)	RSD (%)	∅ (µg·g⁻¹)	∅ (µmol·g⁻¹)	RSD (%)	∅ (µg·g⁻¹)	∅ (µmol·g⁻¹)	RSD (%)
Sb	121, 123	1	13.0	0.1	81.9	35.8	0.3	75.6	<1 –	–	<1	–	–	<1	–	–	<1	–	–
Sr	88	0.1	767	8.8	12.09	1611	14.8	40.6	387 4.4	1.44	381	4.4	1.02	66.5	0.8	2.81	66.1	0.8	5.08
Th	232	0.01	<0.01	–	–	<0.01	–	–	0.47 0.0020	41.3	0.66	0.0028	8.45	0.11	0.0005	51.7	0.11	0.0005	44.6
Ti	49	3	<3	–	–	<3	–	–	<3 –	–	<3	–	–	<3	–	–	<3	–	–
V	51	0.5	0.71	0.01	36.54	0.98	0.02	19.1	0.84 0.02	10.8	0.83	0.02	11.3	0.73	0.01	33.8	0.95	0.02	22.5
Y	89	0.1	<0.1	–	–	<0.1	–	–	4.21 0.05	35.1	5.23	0.1	6.19	10.6	0.1	21.2	14.1	0.2	27.9
Zn	67	10	59.6	0.9	34.6	27.5	0.4	53.1	15.8 0.2	32.5	15.5	0.2	41.7	63.2	1.0	39.3	89.9	1.4	49.0

* (wt %), excluded due to concentrations below limits of detection (L.O.D.) in µg·g⁻¹ in all samples: Li (<5), Sc (<1), Ga (<0.5), Se (<10), Rb (<0.5), Zr (<0.1), Nb (<0.01), Ag (<0.1), Cd (<1), Sn (<1), Cs (<0.1), Hf (<0.01), Ta (<0.01), W (<0.01), Ir (<0.01), Pt (<0.01), Au (<0.1), Tl (<0.1), Bi (<0.01), and U (<0.01).

Table 4. Light volatile elements in unpowdered and powdered danburite samples provided as $\mu g \cdot g^{-1}$; calculated H$_2$O+ values are given in wt %.

Sample Location	Unpowdered Samples					Powdered Samples				
	H	eq. H$_2$O+ (wt %)	C	N	S	H	eq. H$_2$O+ (wt %)	C	N	S
Mexico	<10	<0.01	200 ± 5	50 ± 10	13 ± 1	31 ± 10	0.028	130 ± 50	230 ± 130	25 ± 7
Tanzania	<10	<0.01	120 ± 40	100 ± 50	8 ± 1	14 ± 1	0.012	70 ± 40	100 ± 50	22 ± 3
Vietnam	300 ± 50	0.244	5600 ± 20	1000 ± 10	15 ± 3	40 ± 10	0.036	3000 ± 40	520 ± 20	19 ± 11

4. Discussion

4.1. Substitution Mechanisms of REEs, Be, and Sr in Danburite Structure

Referring to the similarity in ionic size and charge, eightfold-coordinated Ca^{2+} (1.12 Å) can be replaced to a certain extent by Sr^{2+} (1.26 Å). This substitution commonly takes place in danburite from all deposits and is mostly observed in the Mexican samples, where the concentration of Sr reaches 1611 $\mu g \cdot g^{-1}$, followed by the Tanzanian and Vietnamese samples, with Sr concentrations up to 387 $\mu g \cdot g^{-1}$ and 66 $\mu g \cdot g^{-1}$, respectively (Table 3).

$$Ca^{2+} = Sr^{2+} \tag{1}$$

Another substitution in danburite is the replacement of Ca^{2+} by REE^{3+} (here we presume that all REEs are trivalent, with the exception of Ce, which can be quadrivalent under strongly oxidizing conditions). It is obvious that danburite from all deposits prefer to incorporate LREE over HREE by a 200- to 500-fold enrichment. The REE^{3+} have decreasing radii with respect to increasing atomic number, i.e., from La (1.16 Å) to Ce (1.15 Å) to Lu (0.98 Å). Moreover, LREE radii are more compatible with the eightfold-coordinated Ca^{2+} lattice site. This explains why LREEs, especially La and Ce, are preferentially incorporated in the danburite lattice. The negative Eu anomaly observed in the Vietnamese and Tanzanian danburite is in accordance with a general depletion of Eu in highly oxidized magma, such as granites and pegmatites [22].

The substitution of Ca^{2+} by a REE^{3+} requires charge compensation and is therefore coupled with the substitution of B^{3+} (0.11 Å) by Be^{2+} (0.27 Å) and/or Si^{4+} (0.26 Å) by Al^{3+} (0.39 Å). These coupled substitutions theoretically allow all sites to be filled and charges to be balanced accordingly:

$$Ca^{2+} + B^{3+} = REE^{3+} + Be^{2+} \tag{2}$$

$$Ca^{2+} + Si^{4+} = REE^{3+} + Al^{3+} \tag{3}$$

An omission-style substitution of Ca^{2+} by a trivalent REE^{3+} is also suitable to gain charge balance:

$$3Ca^{2+} = 2REE^{3+} + \square \ (vacancy) \tag{4}$$

However, the positive correlation of molar abundance of REE and Be suggests that Equation (2) is the dominating process of REE incorporation into the danburite lattice (Figure 3A). The remaining substitution Equations (3) and (4) are theoretically possible, but are not supported by the datasets, which show, e.g., a negatively correlated relationship of Al with $\sum$REE (Figure 2A). In the Vietnamese samples, the Be/REE ratio is at an approximate 1:1 trend. In the Tanzanian samples, all the values are approximately equal to 1.5:1 (Figure 3A). A Be/REE ratio that is equal to or higher than 1:1 indicates that REEs are fully coupled with Be; subsequently, the two other forms of substitution, Ca^{2+} + Si^{4+} = REE^{3+} + Al^{3+} (3) and 3Ca^{2+} = 2REE^{3+} + $\square$ (4), are subordinate mechanisms. However, Equation (2) implies a Be/REE ratio equal to 1:1, rather than the 1.5:1 ratio measured in the Tanzanian samples. Hence, the excessive molar abundance of Be over REE in the Tanzanian samples needs to be explained by another substitution process with different ratios for Be and REEs or by an REE-independent substitution mechanism. This first hypothesis leads to the suggestion that

Ce occurs not only as Ce^{3+}, but also Ce^{4+} in the samples (with ionic sizes of 1.15 Å and 0.97 Å, respectively). The existence of Ce^{4+} in geological materials has been observed in various studies [23–26]. Hence, we extend substitution mechanism (2) to account for the probable presence of quadrivalent Ce:

$$2Ca^{2+} + 3B^{3+} = Ce^{4+} + REE^{3+} + 3Be^{2+} \tag{5}$$

Equation (5) is an example of a substitution mechanism where the ratio of $\sum$REEs (all REE^{3+} and Ce^{4+}) to Be is equal to 3:2 (or 1.5:1). According to Equation (5), the ratios Be/Ce and Be/($\sum$REEs − Ce) are both 3:1. Our chemical data (Figure 3B,C) show that the Be/Ce and Be/($\sum$REEs − Ce) ratios in the Tanzanian samples are approximately 3:1 and 1.5:1. Therefore, we assume that mechanisms (2) and (5) take place predominantly in the Vietnamese and Tanzanian samples, respectively, with Ce likely present as Ce^{3+} and Ce^{4+}, respectively. This might suggest that REE uptake into Tanzanian danburite occurs at elevated oxygen fugacity compared to Mexican and Vietnamese danburite.

4.2. Substitution Mechanisms Involving OH and Al in the Danburite Lattice

The presence of Be may also be the result of a REE-independent substitution of B^{3+} by Be^{2+} coupled with the substitution of O^{2-} by OH^{-}:

$$B^{3+} + O^{2-} = Be^{2+} + OH^{-} \tag{6}$$

The presence of OH^{-} species in the danburite lattice was indicated in [4,7] by means of FTIR spectroscopy. The bridging oxygen O5 in the B_2O_7 group is an ideal candidate for partial OH^{-} replacement, which allows the presence of low amounts of OH^{-} in danburite. However, a coupled incorporation of Be^{2+} and OH^{-} was not observed in our study (Figure 4A). Beran [4] proposed a coupled 1:1 substitution of Si^{4+} and O^{2-} by Al^{3+} and OH^{-} to charge balance OH incorporation. However, a direct substitution (1:1) was not confirmed by either dataset in the present study. Instead, we observed a positive correlation of Al^{3+} with OH^{-} (Figure 4B) in a 1:3 ratio, which suggests a coupled incorporation of Al^{3+} and OH^{-}, substituting for B^{3+} and O^{2-}, respectively:

$$2B^{3+} + 3O^{2-} = Al^{3+} + 3OH^{-} + \square \text{ (vacancy)} \tag{7}$$

Regarding the possibility of Al incorporation into danburite, it should be noted that the geochemical behavior of B^{3+} and Al^{3+} is very similar; however, they differ in radius size, with 0.11 Å for B^{3+} and 0.39 Å for Al^{3+} when in a tetrahedral coordination environment. Hence, a simple substitution mechanism such as $B^{3+} = Al^{3+}$ would not be possible. Although a substitution of Al^{3+} with B^{3+} has been observed in the system albite $NaAlSi_3O_8$ - $NaBSi_3O_8$ reedmergnerite [27], as well as in a synthetic phlogopite $KMg_3(BSi_3)O_{10}(OH)_2$ [28], this process seems not to be valid for the danburite datasets (Figure 4A,B). Substitution mechanism (7) explains the Mexican samples as well, where the Al concentration is high and both REE and Be concentrations are low.

In general, the four main substitution mechanisms discussed above take place with different priority in the three studied locations. The substitutions of Ca^{2+} by Sr^{2+} and $2B^{3+}$ by Al^{3+} are more common in the Mexican samples, while the substitutions of Ca^{2+} by REEs in the forms $Ca^{2+} + B^{3+} = REE^{3+} + Be^{2+}$ and $2Ca^{2+} + 3B^{3+} = Ce^{4+} + REE^{3+} + 3Be^{2+}$ are more common in the Vietnamese and Tanzanian samples, respectively.

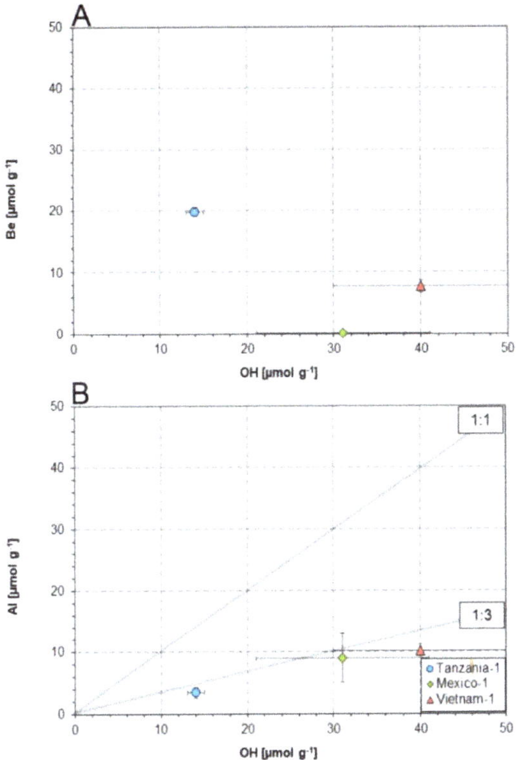

Figure 4. (**A**) Beryllium and OH correlate negatively, which suggests that incorporation into the danburite structure is not coupled. (**B**) However, Al shows a positive correlation with OH at a 1:3 ratio for all sampling locations, making a coupled incorporation of Al with OH likely.

4.3. Constraints on the Geochemical Formation Environment of Danburite

Since danburite samples from Tanzania and Vietnam show a similar strong enrichment in LREE, the reported low values in the Mexican samples must therefore mean either a deficit of REE in the source material or REEs were already sequestered in datolite, which co-occurs with Mexican danburite. However, the depletion of REEs in the Mexican samples is accompanied by exceedingly high amounts of Sr (ca. 1000 $\mu g \cdot g^{-1}$ on average), which is an independent indicator of a different source composition with a higher component of biogenic calcareous sediments in the source material of this location. Biogenic limestone is known to contain high Sr values by being virtually free of REE [29,30]. In comparison, samples from Vietnam and Tanzania exhibit high REE, Y, and Be coupled with low Sr, thereby representing a composition that results from the involvement of highly differentiated late-stage silicic magmas (i.e., pegmatites). This is in agreement with the observed negative Eu anomalies in these samples, which are characteristic for late magmas, where the depletion in Eu is driven by fractional crystallization of plagioclase, which is commonly found to incorporate high amounts of Eu^{2+} [22]. This is in agreement with the nature of the outcrop and mineral assemblage in which the danburite was found. The high compatibility of LREE in the danburite lattice must therefore only be limited by the availability from the source material (i.e., highest in Tanzanian and lowest in Mexican samples). High contents of REE are in accordance with the significantly low amounts of N in the Tanzanian samples, due to the incompatibility of N in highly fractionated magmatic rocks, indicating a strong pegmatite component in the Tanzanian samples. Charge compensation and a

Be/$\sum$REEs ratio of 1.5:1 indicate the presence of Ce^{4+} (Equation (5)), additionally supporting a highly oxidized pegmatitic source for the Tanzanian samples. Nitrogen mass fractions generally increase with decreasing temperature and increasing involvement of metamorphic rocks (e.g., [21]), which suggests that the Vietnamese and Mexican samples were either formed at greater distance from the pegmatite or sourced from a higher proportion of recycled metamorphic rocks. Hence, we conclude that even though danburite always forms in a transition zone of metacarbonates and pegmatites, there are significant geochemical differences, i.e., $\sum$REE, Be, Sr, Al, and OH in danburite, that directly reflect the different proportions and compositions of these source materials. Thus, our results suggest that trace element concentrations are suitable for determining the origins and locations of danburite crystals (i.e., for gem-testing laboratories).

5. Conclusions

In this study, we investigated trace element variations of danburite from three different locations in Mexico, Tanzania, and Vietnam. The most important trace elements in danburite that reflect their provenance include REEs, Sr, Al, Be, and, to a lesser extent, Mn, Zn, and Y. Mexican samples are fairly devoid of REEs, while Tanzanian samples contain up to 1900 $\mu g \cdot g^{-1}$ and Vietnamese samples have intermediate total values of around 1100 $\mu g \cdot g^{-1}$. LREEs are more abundant than HREEs in all danburite samples, showing a 200- to 500-fold relative enrichment. Strontium and Al are more enriched in Mexican danburite than in Tanzanian and Vietnamese danburite, with mass fractions up to 1611 and 325 $\mu g \cdot g^{-1}$, respectively.

Based on fs-LA-ICP-MS and CHNS analysis, we identified four mechanisms of trace element substitution in danburite: Two replacements of Ca^{2+} by Sr^{2+} ($Ca^{2+} = Sr^{2+}$) and of B^{3+} by Al^{3+}, which are coupled with an incorporation of OH^- for O^{2-}: $2B^{3+} + 3O^{2-} = Al^{3+} + 3\,OH^- + \square$, are dominant in the Mexican samples. The incorporation of REE^{3+} for Ca^{2+} coupled with a simultaneous replacement of B^{3+} by Be^{2+} is present in the Vietnamese and Tanzanian samples. Different valance states of Ce are present in the Vietnamese and Tanzanian samples, leading to two different substitutions: $Ca^{2+} + B^{3+} = REE^{3+} + Be^{2+}$ and $2Ca^{2+} + 3B^{3+} = Ce^{4+} + REE^{3+} + 3Be^{2+}$, respectively.

The observed significant differences in trace element abundance not only suggest a high potential for provenance discrimination, but also provide information on the contrasting source compositions of the three deposits. The formation of danburite generally involves both metacarbonates and pegmatites as source materials. Different proportions of these two source components were involved in the formation of danburite at the three locations, and likely explain the observed trace element variations. Low REE and Be coupled with high Sr, Al, N, and OH in the Mexican samples indicate a dominant biogenic metacarbonate component, while the Vietnamese and Tanzanian samples show high REE and Be coupled with low Sr, Al, N, and OH, characteristic of a predominantly pegmatitic source. The 200- to 500-fold enrichment of LREE over HREE in the Tanzanian and Vietnamese samples results from the preferential replacement of Ca ions by similarly sized LREE ions. The negative Eu anomaly, which is characteristic of highly fractionated igneous rocks, is characteristic of Vietnamese and Tanzanian danburite and supports the predominance of the pegmatitic source at these locations.

Author Contributions: L.T.-T.H., L.M.O., K.P.J., C.A.H., and K.K. designed and coordinated the study. K.P.J. and C.A.H. supervised the project. L.T.-T.H., L.M.O., and K.K. collected and prepared the samples. M.W.F. prepared epoxy mounts and performed electron probe microanalyses. L.M.O., D.S.M., B.S., and U.W. collected and evaluated trace element concentrations. M.W.F. and O.A. collected and evaluated light volatile element concentrations and calibrated the CHNS analyzer. L.T.-T.H., L.M.O., and M.W.F. wrote the first drafts of the manuscript. K.P.J., C.A.H., K.K., and O.A. carefully edited the final version. All authors contributed to the final version and gave their approval for submission.

Funding: Vietnamese danburite samples were collected during a field trip funded by a NAFOSTED Project, grant number 1055.99-2013.13. L.T.-T.H is grateful to the ASEAN-European Academic University Network, the Austrian Federal Ministry of Science, Research and Economy, and the Austrian Agency for International Cooperation in Education and Research for financial support.

Acknowledgments: Lauren Gorojovsky is acknowledged for preparation and analyses of reference materials for the CHNS elemental analyzer, and we kindly acknowledge the insightful comments by two anonymous reviewers.

Conflicts of Interest: The authors declare no conflict of interest.

References

1.   Lindbloom, J.T.; Gibbs, G.V.; Ribbe, P.H. Crystal Structure of Hurlbutite—Comparison with Danburite and Anorthite. *Am. Mineral.* **1974**, *59*, 1267–1271.
2.   Best, S.P.; Clark, R.J.H.; Hayward, C.L.; Withnall, R. Polarized single-crystal Raman spectroscopy of danburite, $CaB_2Si_2O_8$. *J. Raman Spectrosc.* **1994**, *25*, 557–563. [CrossRef]
3.   Sugiyama, K.; Takéuchi, Y. Unusual thermal expansion of a B–O bond in the structure of danburite $CaB_2Si_2O_8$. *Z. Kristallogr. -Cryst. Mater.* **1985**, *173*, 293–304. [CrossRef]
4.   Beran, A. OH groups in nominally anhydrous framework structures: An infrared spectroscopic investigation of danburite and labradorite. *Phys. Chem. Miner.* **1987**, *14*, 441–445. [CrossRef]
5.   Hurwit, K.N. Gem Trade Lab Notes: Golden yellow danburite from Sri Lanka. *Gems Gemol.* **1986**, *22*, 47.
6.   Chadwick, K.M.; Laurs, B.M. Gem News International: Yellow danburite from 346 Tanzania. *Gems Gemol.* **2008**, *44*, 169–171.
7.   Huong, L.T.-T.; Otter, L.M.; Häger, T.; Ullmann, T.; Hofmeister, W.; Weis, U.; Jochum, K.P. A New Find of Danburite in the Luc Yen Mining Area, Vietnam. *Gems Gemol.* **2016**, *52*. [CrossRef]
8.   De Vito, C.; Pezzotta, F.; Ferrini, V.; Aurisicchio, C. Nb–Ti–Ta oxides in the gem-mineralized and "hybrid" Anjanabonoina granitic pegmatite, central Madagascar: A record of magmatic and postmagmatic events. *Can. Mineral.* **2006**, *44*, 87–103. [CrossRef]
9.   Cook, R.B. Connoisseur's: Danburite, Charcas, San Luis Potosí, Mexico. *Rocks Miner.* **2003**, *78*, 400–403. [CrossRef]
10. Hintze, J. Safari njema—AFRIKANISCHES TAGEBUCH (I): Reise zu den gelben Danburiten von Morogoro, Tansania. *Lapis Die Aktuelle Monatsschrift Fuer Liebhaber Und Sammler Von Mineralien Und* **2010**, *35*, 25.
11. Dyar, M.D.; Wiedenbeck, M.; Robertson, D.; Cross, L.R.; Delaney, J.S.; Ferguson, K.; Francis, C.A.; Grew, E.S.; Guidotti, C.V.; Hervig, R.L. Reference minerals for the microanalysis of light elements. *Geostand. Geoanal. Res.* **2001**, *25*, 441–463. [CrossRef]
12. Ottolini, L.; Cámara, F.; Hawthorne, F.C.; Stirling, J. SIMS matrix effects in the analysis of light elements in silicate minerals: Comparison with SREF and EMPA data. *Am. Mineral.* **2002**, *87*, 1477–1485. [CrossRef]
13. Balmer, W.A.; Hauzenberger, C.A.; Fritz, H.; Sutthirat, C. Marble-hosted ruby deposits of the Morogoro Region, Tanzania. *J. Afr. Earth Sci.* **2017**, *134*, 626–643. [CrossRef]
14. Möller, A.; Mezger, K.; Schenk, V. U–Pb dating of metamorphic minerals: Pan-African metamorphism and prolonged slow cooling of high pressure granulites in Tanzania, East Africa. *Precambrian Res.* **2000**, *104*, 123–146. [CrossRef]
15. Huong, L.T.-T.; Krenn, K.; Hauzenberger, C. Sassolite- and CO_2-H_2O-bearing Fluid Inclusions in Yellow Danburite from Luc Yen, Vietnam. *J. Gemmol.* **2017**, *35*, 544–550. [CrossRef]
16. Garnier, V.; Ohnenstetter, D.; Giuliani, G.; Maluski, H.; Deloule, E.; Trong, T.P.; Van, L.P.; Quang, V.H. Age and significance of ruby-bearing marble from the Red River Shear Zone, northern Vietnam. *Can. Mineral.* **2005**, *43*, 1315–1329. [CrossRef]
17. Jochum, K.P.; Stoll, B.; Weis, U.; Jacob, D.E.; Mertz-Kraus, R.; Andreae, M.O. Non-Matrix-Matched Calibration for the Multi-Element Analysis of Geological and Environmental Samples Using 200 nm Femtosecond LA-ICP-MS: A Comparison with Nanosecond Lasers. *Geostand. Geoanal. Res.* **2014**, *38*, 265–292. [CrossRef]
18. Jochum, K.P.; Stoll, B.; Herwig, K.; Willbold, M. Validation of LA-ICP-MS trace element analysis of geological glasses using a new solid-state 193 nm Nd: YAG laser and matrix-matched calibration. *J. Anal. At. Spectrome.* **2007**, *22*, 112–121. [CrossRef]
19. Jochum, K.P.; Nohl, U.; Herwig, K.; Lammel, E.; Stoll, B.; Hofmann, A.W. GeoReM: A new geochemical database for reference materials and isotopic standards. *Geostand. Geoanal. Res.* **2005**, *29*, 333–338. [CrossRef]
20. Palme, H.; Jones, A. Solar system abundances of the elements. *Treat. Geochem.* **2003**, *1*, 711.
21. Plessen, B.; Harlov, D.E.; Henry, D.; Guidotti, C.V. Ammonium loss and nitrogen isotopic fractionation in biotite as a function of metamorphic grade in metapelites from western Maine, USA. *Geochim. Cosmochim. Acta* **2010**, *74*, 4759–4771. [CrossRef]
22. Fowler, A.D.; Doig, R. The significance of europium anomalies in the REE spectra of granites and pegmatites, Mont Laurier, Quebec. *Geochim. Cosmochim. Acta* **1983**, *47*, 1131–1137. [CrossRef]

23. Braun, J.-J.; Pagel, M.; Muller, J.-P.; Bilong, P.; Michard, A.; Guillet, B. Cerium anomalies in lateritic profiles. *Geochim. Cosmochim. Acta* **1990**, *54*, 781–795. [CrossRef]

24. Takahashi, Y.; Shimizu, H.; Kagi, H.; Yoshida, H.; Usui, A.; Nomura, M. A new method for the determination of CeIII/CeIV ratios in geological materials; application for weathering, sedimentary and diagenetic processes. *Earth Planet. Sci. Lett.* **2000**, *182*, 201–207. [CrossRef]

25. Taunton, A.E.; Welch, S.A.; Banfield, J.F. Microbial controls on phosphate and lanthanide distributions during granite weathering and soil formation. *Chem. Geol.* **2000**, *169*, 371–382. [CrossRef]

26. Bao, Z.; Zhao, Z. Geochemistry of mineralization with exchangeable REY in the weathering crusts of granitic rocks in South China. *Ore Geol. Rev.* **2008**, *33*, 519–535. [CrossRef]

27. Wunder, B.; Stefanski, J.; Wirth, R.; Gottschalk, M. Al-B substitution in the system albite ($NaAlSi_3O_8$)-reedmergnerite ($NaBSi_3O_8$). *Eur. J. Mineral.* **2013**, *25*, 499–508. [CrossRef]

28. Stubican, V.; Roy, R. Boron substitution in synthetic micas and clays. *Am. Mineral.* **1962**, *47*, 1166.

29. Chen, C.; Liu, Y.; Foley, S.F.; Ducea, M.N.; He, D.; Hu, Z.; Chen, W.; Zong, K. Paleo-Asian oceanic slab under the North China craton revealed by carbonatites derived from subducted limestones. *Geology* **2016**, *44*, 1039–1042. [CrossRef]

30. Gozzi, F.; Gaeta, M.; Freda, C.; Mollo, S.; Di Rocco, T.; Marra, F.; Dallai, L.; Pack, A. Primary magmatic calcite reveals origin from crustal carbonate. *Lithos* **2014**, *190*, 191–203. [CrossRef]

Article

Gem-Quality Tourmaline from LCT Pegmatite in Adamello Massif, Central Southern Alps, Italy: An Investigation of Its Mineralogy, Crystallography and 3D Inclusions

Valeria Diella [1,*], Federico Pezzotta [2], Rosangela Bocchio [3], Nicoletta Marinoni [1,3], Fernando Cámara [3], Antonio Langone [4], Ilaria Adamo [5] and Gabriele Lanzafame [6]

[1]   National Research Council, Institute for Dynamics of Environmental Processes (IDPA), Section of Milan, 20133 Milan, Italy; nicoletta.marinoni@unimi.it
[2]   Natural History Museum, 20121 Milan, Italy; fpezzotta@yahoo.com
[3]   Department of Earth Sciences "Ardito Desio", University of Milan, 20133 Milan, Italy; rosangela.bocchio@unimi.it (R.B.); fernando.camara@unimi.it (F.C.)
[4]   National Research Council, Institute of Geosciences and Earth Resources (IGG), Section of Pavia, 27100 Pavia, Italy; langone@igg.cnr.it
[5]   Italian Gemmological Institute (IGI), 20123 Milan, Italy; ilaria.adamo@guest.unimi.it
[6]   Elettra-Sincrotrone Trieste S.C.p.A., Basovizza, 34149 Trieste, Italy; gabriele.lanzafame@elettra.eu
*   Correspondence: valeria.diella@cnr.it; Tel.: +39-02-50315621

Received: 12 November 2018; Accepted: 7 December 2018; Published: 13 December 2018

Abstract: In the early 2000s, an exceptional discovery of gem-quality multi-coloured tourmalines, hosted in Litium-Cesium-Tantalum (LCT) pegmatites, was made in the Adamello Massif, Italy. Gem-quality tourmalines had never been found before in the Alps, and this new pegmatitic deposit is of particular interest and worthy of a detailed characterization. We studied a suite of faceted samples by classical gemmological methods, and fragments were studied with Synchrotron X-ray computed micro-tomography, which evidenced the occurrence of inclusions, cracks and voids. Electron Microprobe combined with Laser Ablation analyses were performed to determine major, minor and trace element contents. Selected samples were analysed by single crystal X-ray diffraction method. The specimens range in colour from colourless to yellow, pink, orange, light blue, green, amber, brownish-pink, purple and black. Chemically, the tourmalines range from fluor-elbaite to fluor-liddicoatite and rossmanite: these chemical changes occur in the same sample and affect the colour. Rare Earth Elements (REE) vary from 30 to 130 ppm with steep Light Rare Earth Elemts (LREE)-enriched patterns and a negative Eu-anomaly. Structural data confirmed the elbaitic composition and showed that high manganese content may induce the local static disorder at the O(1) anion site, coordinating the Y cation sites occupied, on average, by Li, Al and Mn^{2+} in equal proportions, confirming previous findings. In addition to the gemmological value, the crystal-chemical studies of tourmalines are unanimously considered to be a sensitive recorder of the geological processes leading to their formation, and therefore, this study may contribute to understanding the evolution of the pegmatites related to the intrusion of the Adamello pluton.

Keywords: granitic pegmatite; gem-quality tourmaline; Adamello Massif; Central Alps; Italy

---

1. Introduction

"Tourmaline" is considered one of the most beautiful gemstones, because it occurs in a large spectrum of colours, as well as in multi-coloured crystals [1]. In Italy, gem-quality tourmaline is known to be from the historic locality, at present rather exhausted, of Elba Island (Tyrrhenian sea), from which it derived the root-name elbaite, given to Li-bearing and sodium-rich tourmalines of the alkali group [2]. For a very long time, Elba Island has remained the only known locality providing gem-quality tourmaline crystals in Italy. Nevertheless, since 2001, an additional significant locality has been discovered and described in the Italian Alps, which is the Adamé valley, along the western border of the late Alpine Adamello tonalite massif, in the central Southern Alps [3].

In the Adamé valley, gem-quality tourmaline occurs in multi-coloured elongated crystals up to 7 cm in length in the miarolitic cavities of LCT pegmatite (i.e., a pegmatite with prevailing minerals of lithium, caesium and tantalum) hosted in contact with metamorphic sandstones, belonging to a Permian-Mesozoic sedimentary sequence. The locality is included in the Adamello Park natural reservation. The miarolitic pegmatite has been found on the steep slope of Forcel Rosso pass, disrupted in large blocks in an ancient landslide, at an altitude of about 2600 m. Tourmaline group minerals occur together with albite, smoky quartz, K-feldspar, mica (lepidolite–muscovite) and other accessory minerals such as fluorapatite, fluorite, and several Nb–Ta–Sb oxides, all very interesting from a scientific and collector point of view. In consideration of the significant potential of the pegmatite for mineral production, the locality underwent, under strict protection, an official detailed field investigation for scientific purposes, with the supervision of one of the authors (F.P.). The first results [3] show that the tourmaline crystals have a zoned composition, mostly characterized by fluor-elbaite, with fluor-liddicoatite developing at the antilogous pole of the crystals. Colours range from colourless to yellow, pink, orange, light blue, green, amber, brownish-pink, purple and black (Figure 1). Gem-quality tourmalines are present as sectors of larger crystals, and rough fragments can reach a maximum weight of 2–3 g.

Figure 1. Crystals of pink (3.5 cm long) and green (5 cm long) tourmalines from Valle Adamé, Adamello Massif. Natural History Museum of Milan collection. Photos by R. Appiani.

The collected materials are preserved in the mineralogical collections of the Museum of Natural History of Milan and at the Museum of the Adamello Natural Park. These gem tourmalines have not entered the market, because the deposit is located within the protected areas of the Adamello Natural Park and collecting is allowed only for scientific purposes. At present, only a few gems have been faceted (Figure 2).

Figure 2. Cut tourmalines analysed in this study. (**a**) 3.97 × 8.31 × 2.61 mm; (**b**) 4.73 × 6.58 × 3.35 mm; (**c**) 4.61 × 6.36 × 3.26 mm, (**d**) 5.72 × 7.16 × 4.03 mm, corresponding to samples 13, 14, 15 and 16 described below in Table 1. Photos by M. Chinellato and F. Picciani.

As mentioned above, tourmaline represents one of the most beautiful gemstones and may be characterized through multi-methodological methods including non-invasive and non-destructive techniques, such as Fourier Transform Infrared Spectroscopy and Raman Spectroscopy, when a conservative treatment of sample is required ([4] and references therein). Many studies recognize a significant petrological interest of this mineral, as it can contribute to the reconstruction of the evolution of the history of a crystalline basement, e.g., [5].

In view of this, and considering that the Adamello Massif is one of the most geologically studied portions of the Alpine chain, our study aims to provide a complete mineralogical characterization of these recently discovered gem tourmalines, as well as an interesting contribution to the knowledge of local pegmatite mineralization. We investigated a suite of cut and rough samples selected from the mineralogical collections of the Natural History Museum of Milan, through gemmological analyses, electron microprobe chemical analyses (EPMA) and laser ablation-inductively coupled plasma–mass spectrometry (LA-ICP-MS). The unit cell parameters and refinement of crystal structure of three selected tourmalines were carried out by the single-crystal X-ray diffraction method. Selected rough samples have also been examined by means of synchrotron X-ray computed micro-tomography (X-μCT), which made it possible to obtain a 3D visualisation of the inner objects (pore, inclusions, etc.) within a volume [6,7]. This technique does not require any sample preparation and makes it possible to overcome the problem of deriving the 3D results from the traditional 2D data obtained by traditional imaging techniques such as optical and scanning electron microscopy. In the present study, a fully characterisation of inclusions in terms of size, shape and orientation was performed.

2. Background Information

Tourmaline is a ring-silicate crystallizing in the acentric $3m$ point group (ditrigonal pyramidal) with the $R3m$ space group and a general formula of $XY_3Z_6T_6O_{18}(BO_3)_3V_3W$ [8]. The unit cell consists of a six-fold ring of tetrahedra (T sites, occupied primarily by Si) on top of a concentric arrangement of three Y-site and six Z-site octahedra. The X site is nine-coordinated and situated centred over of the six-fold ring and can be occupied by Na^+, Ca^{2+}, K^+ cations or be vacant. The Y and Z sites may contain a large variety of cations such as Mg^{2+}, Fe^{2+}, Fe^{3+}, Li^+, Al^{3+}, Mn^{2+}, Cu^{2+}, Ti^{4+}, Cr^{3+}, V^{3+} and Al^{3+}, Mg^{2+}, Fe^{2+}, Fe^{3+}, Cr^{3+}, V^{3+}, respectively. Three planar triangular boron-centred polyhedra (BO_3) are further present, which are roughly coplanar with the groups of octahedra in the unit cell, and roughly perpendicular to the c axis. The V and W sites are anion sites that are occupied by OH^- or O^{2-} (or both) at the V site, and OH^-, F^-, or O^{2-} at the W site. The Cl contents are generally negligible. Given the range of elements and heterovalent substitutions that can be accommodated by the tourmaline structure, the high number of different end-member compositions constitute a super group consisting of quite a large number of species, as approved by the International Mineralogical Association's Commission on New Minerals, Nomenclature and Classification [2,9–11].

3. Geological Setting

The Tertiary Adamello batholith outcrops over an area of ca. 670 km^2 in the Central Southern Alps (Northern Italy) (CNR, Italian Geological Map Adamello-Presanella, 1:50,000). It consists of four composite plutons decreasing in age from south (ca. 42 Ma) to north (ca. 31 Ma) and is composed mainly of granitoid rocks (granodiorite, tonalite, quartz diorite) with minor amounts of diorite and gabbro [12–14]. The plutons intrude into the South-Alpine crystalline basement and its Permo-Mesozoic sedimentary cover (Figure 3). The studied tourmaline crystals were collected in pegmatite dykes occurring in the thermo-metamorphic contact aureole between the igneous rocks and the surrounding sediments, at the northern slope of the Foppa mountain (2752 m) in a gully of Forcel Rosso (or "Vallone del Forcel Rosso").

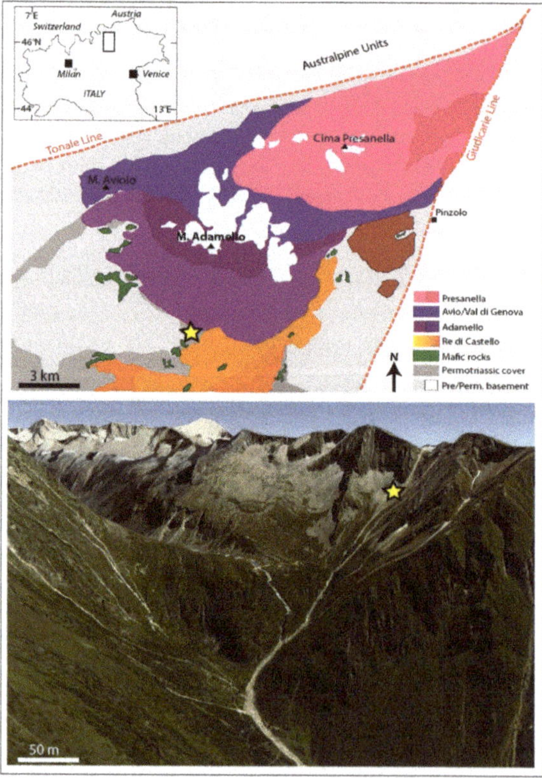

Figure 3. Geological sketch map of the Adamello pluton and a view of the area of Forcel Rosso. Photo by F. Pezzotta. The stars indicate the location of the studied pegmatite dikes.

The pegmatites are sub-horizontal and hosted in meta-sandstones enriched in accessory tourmaline. The larger veins are up to approximately 1.5 m in thickness, and some tens of meters long. Despite their relatively small size, these pegmatites display a quite strong asymmetric zoning with a layered lower unit composed by fine to medium grain size assemblage of feldspars and quartz, with accessory muscovitic mica, schorl and spessartine, and an upper coarse-grained unit composed of an assemblage of quartz and feldspars, rich in schorl and muscovitic mica, occasionally displaying cores with milky white quartz masses, lepidolite overgrowth at the rim of muscovite blades, granular lepidolite masses and multi-coloured tourmalines. Miarolitic cavities are not common, but they can be locally abundant, with a diameter achieving several decimetres in length, at the coarse-grained cores of the veins.

Accessory minerals in cavities include pink to green fluor-apatite, purple and pale green fluorite, late-stage calcite, several pyrochlore group minerals and various Ta–Nb oxides [3].

4. Materials and Methods

Sixteen samples of tourmaline from different miarolitic cavities were selected for this study, and they are described in Table 1.

Table 1. Description of the studied tourmalines.

Sample	Cavity Name/Description *	Colour	Analytical Techniques	
1	Cavity 1	Pink	Electron microprobe (WDS) (°)	Single crystal XRD (°)
2	Cavity boulder 2	Light blue	Electron microprobe (WDS) (°)	Single crystal XRD (°)
3	Pizio cavity	Colourless to brown	Electron microprobe (WDS) (°)	
4: sample cut perpendicular to the *c*-axis	Pizio cavity	Pink green	Electron microprobe (WDS) (°)	
5	Cavity 3	Light green yellow	Electron microprobe (WDS) (°)	
6	Cavity 4	Yellow to colourless	Electron microprobe (WDS) (°)	
7	Cavity 5		Synchrotron X-ray computed μ-tomography	
8	Cavity 6		Synchrotron X-ray computed μ-tomography	
9	"Black quartz" cavity	Pink green	Electron microprobe (WDS) (°)	
10: double terminated sample	Inv.# M36742 cavity	Browm green	Electron microprobe (WDS) (°)	
11: homologous terminated sample	Inv.# M36742 cavity	Green to colourless	Electron microprobe (WDS) (°)	
12: sample cut perpendicular to the *c*-axis	Inv.# M36742 cavity	Blue	Electron microprobe (WDS) (°)	Single crystal XRD (°)
13	Cut stone (0.731 ct) from pocket	Green brown	Specific gravity	Refractive index
14	Cut stone (0.848 ct) from pocket	Green brown	Specific gravity	Refractive index
15	Cut stone (0.838 ct) from pocket	Green brown	Specific gravity	Refractive index
16	Cut stone (1.225 ct) from pocket	Pale green	Specific gravity	Refractive index

* Cavity numbers refer to the unpublished field note-book made during the collecting campaigns performed by the Natural History Museum of Milan in 2001 and 2002. The reported inventory numbers are the ones given to specimens catalogued in the Mineralogical Collections of the Natural History Museum of Milan. ° Wavelength Dispersive Spectroscopy (WDS); X-ray Diffractometry (XRD).

Four faceted samples (13–16) were examined by standard gemmological methods at the Italian Gemmological Institute in Milan in order to determine their refractive index, hydrostatic specific gravity and microscopic features. The refractive indices were measured with a Krüss refractometer (A. Krüss Optronic, Hamburg, Germany) using ordinary light source with a sodium filter (589 nm) and a methylene iodide as a contact liquid (n = 1.80). A Mettler hydrostatic balance was used to determine the specific gravity in bi-distilled water.

Two grains (7 and 8) were used for synchrotron X-ray computed micro-tomography. The samples, with a size of about 1.5 mm, were imaged at the SYRMEP beamline of ELETTRA synchrotron facility (Trieste, Italy). The computer micro-tomography experiments were performed using a polychromatic X-ray beam (white beam mode. With this configuration, the outcome beam from the storage ring was intercepted before the monochromator and pre-filtered with 1.5 mm of Si and 1.0 mm of Al. A water-cooled, 16 bit, 2028 × 2048 pixel microscope Charge-Coupled Device (CCD) camera, coupled with a 100 micron thick LuAG scintillator screen was used as detector. The mean energy was set to 28 keV. Pixel size was set at 2 µm/pixel, yielding a field of view of 5 × 5 mm.

The software suite Syrmep Tomo Project (STP) [15,16] was used to reconstruct two-dimensional axial slices from the sample projections, applying the filtered back projection algorithm [17,18]. Before image reconstruction, a single-distance phase retrieval algorithm was applied to the projection images [19] using the STP software and setting the δ/β ratio to 15 (see [15,16] for details about the phase retrieval process).

Ten of the selected fragments, 1–6 and 9–12, were embedded in epoxy resin, polished and prepared for EMPA and LA-ICP-MS analysis. Optical images of the samples were performed in advance using a stereo microscope Leica M205 C in reflected light. The backscattered electron images and quantitative chemical analyses of major and minor elements were obtained with the JEOL JXA-8200 electron microprobe in wavelength dispersion mode (EMPA-WDS) at the laboratory of the Department of Earth Sciences of the University of Milan under the following conditions: 15 kV accelerating voltage, 5 nA beam current, and a count time of 60 s on peak and 30 s on the background, with a 1 µm diameter beam. The Kα wavelengths and natural standards of pure metals were used for calibration: F (hornblende), Ti (ilmenite), Mn and Zn (rhodonite), K (K-feldspar), Na (omphacite), Fe (fayalite), Ca, Si and Al (grossular), Mg (olivine), Cr and V (pure elements). The raw data were corrected for matrix effects using a conventional $\Phi\rho Z$ routine in the JEOL software package. Both Fe and Mn were calculated as Fe^{2+} and Mn^{2+}. The structural formula was calculated on the basis of 31 anions (O, OH and F), and Li_2O, B_2O_3 and H_2O were calculated based on the assumed elbaite stoichiometry.

Rare earth and selected trace elements were determined by laser ablation–inductively coupled plasma–mass spectroscopy (LA-ICP-MS) at the IGG-CNR Laboratory of Pavia. The instrument consisted of a Quantel Brilliant 213 nm Nd:YAG laser (New Wave Research) coupled to a Perkin Elmer DRCe quadrupole ICPMS. The spot size was 55 µm, using NIST SRM 610 glass as an external standard and Si as an internal standard, as analysed by microprobe. Precision and accuracy estimated on the basaltic glass standard BCR2 and NIST612 were better than 10%.

Three fragments (~500 µm) from samples 1, 2, and 12 were cut for single crystal X-ray diffraction measurements. Data were collected with an Xcalibur-Oxford Diffraction diffractometer (Oxford Diffraction Ltd., Abingdon, UK) equipped with a CCD, using graphite-monochromatized MoKα radiation and operated at 50 kV and 30 mA. To maximize the reciprocal space coverage, a combination of ω and φ scans was used, with a step size of 1° and an exposure time per frame of 3–5 s. Intensity data were then integrated and corrected for Lorentz polarization effects, using the computer program CrysAlis [20]. An empirical absorption correction was applied using CrysAlis [20]. The structure was refined using SHELX-97 [21]. Scattering curves for neutral chemical species were used at all sites. No peaks larger than ±0.9 e$^-$/Å^3 were present in the final difference—the Fourier maps of the electron density and residual maxima and minima were equally balanced. The final agreement index (R_1) was 0.016–0.018/0.017–0.019 (obs/all) for 97 refined parameters.

5. Results

5.1. Gemmological Properties and 3D Visualisation of the Tourmaline Inclusions

From a gemmological point of view, the four faceted tourmalines are all transparent, with a colour ranging from green to brown (Figure 2). The specific gravity results from 3.09 to 3.13 g/cm^3, which is typical of elbaitic tourmaline [22]. The optic character is uniaxial negative, and the refractive indices are n_ω = 1.640 and n_ε = 1.620 with a birefringence of 0.020, which is in agreement with the values of elbaite [22]. The samples show a strong dichroism, in the green and brown hues, and they are all inert to ultraviolet radiation. Microscopic observations revealed that the samples contain a few inclusions, typically fluid inclusions, often with a fringed aspect, also called "thrichites" in gemology [23].

More details were provided by the 3D visualisation, which made it possible to display internal features and to quantify the porosity, cracks and voids of fragments from the prismatic portion of samples 7 and 8.

The volume rendering of the tourmaline sample is reported in Figure 4a, whereas the cross sectional slices, displaying the general appearance of the microstructural features in the grains, are highlighted in Figure 4b,c. In particular, the reconstructed 3D images show that tourmalines are characterised by pores and cracks, the detail of whose volume rendering is displayed in Figure 4d.

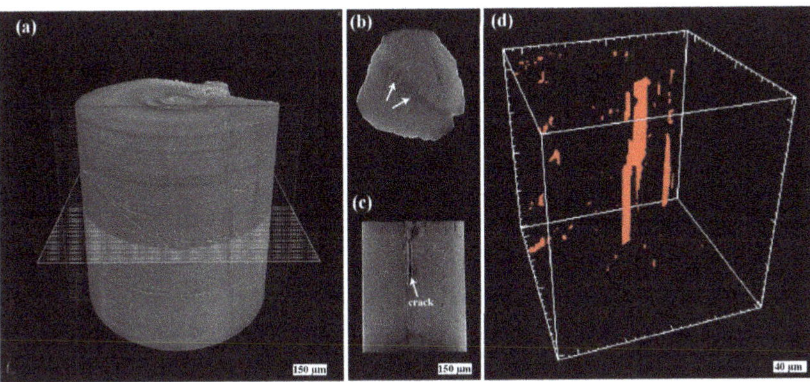

Figure 4. Synchrotron X-ray computed micro-tomography of tourmaline 7: (**a**) volume rendering of the sample; (**b**) reconstructed axial slice; (**c**) the cross section of the stack of reconstructed slices; (**d**) volume rendering of pores and cracks, a detail.

The total measured porosity accounts for about 0.9 vol.% and is given as the sum of the voids/pores (0.1 vol.%) and cracks (0.8 vol.%), respectively. Please note that the distinction between voids and cracks, which are characterised by the same grey scale values, was performed by considering some morphological features as described in [24]. In particular, the voids appear as spherical bubbles with a small surface/volume ratio (normalized to a sphere) and a low aspect ratio as well. On the contrary, cracks show a medium to medium-high surface/volume ratio, together with high values of aspect ratio.

The rendering pointed out that the voids appear spherical in shape, with an equivalent diameter ranging from 10 to 15 µm. Morphometric data showed that they are homogeneously distributed within the entire volume and are mostly isolated without any preferred orientation.

Tourmaline crystals show a predominance of crack spacing from the centre to the outer part of the crystals, with a non-homogenous spatial distribution and a heterogeneous size ranging from 20 to 200 µm. The fractures are mostly flat with a regular surface and appear oriented preferentially parallel to the z-axis. Finally, the results of the skeleton analysis pointed out that the crack connectivity is null, because they lack a connected network.

5.2. Chemical Composition

Electron microprobe analyses (EMPA) and backscattered electron (BKSE) images were performed to chemically characterize 10 of the selected samples. Analytical profiles from core to rim or from rim to rim were carried out with step width depending on the crystal size. LA-ICP-MS analyses were performed on the same samples in the areas previously analysed by EMPA-WDS.

The values of Li_2O measured by LA-ICP-MS indicated that tourmalines contain an average of 1.8 wt % of Li_2O (ranging from 0.7 to 2.7 wt %), and the result is comparable to that calculated by stoichiometry from the results of the electron microprobe analyses. The studied tourmaline samples exhibit extremely convoluted chemical zoning and are very similar in composition, all classifiable as Li-bearing tourmalines: elbaite ($Na(Li_{1.5},Al_{1.5})Al_6(BO_3)_3Si_6O_{18}(OH)_4$), rossmanite ($\gamma(LiAl_2)Al_6(BO_3)_3Si_6O_{18}(OH)_4$), fluor-elbaite ($Na(Li_{1.5},Al_{1.5})Al_6(BO_3)_3Si_6O_{18}(OH)_3F$) and fluor-liddicoatite ($Ca(LiAl_2)Al_6(BO_3)_3Si_6O_{18}(OH)_3F$). Minor foitite ($\gamma Fe^{2+}_2Al)Al_6(BO_3)_3Si_6O_{18}(OH)_4$), a rare tourmaline iron rich in Y-site and with more than 50% vacant X-sites [25], was also found.

The complete set of chemical analyses is graphically plotted in the ternary diagram of X-site composition (Figure 5).

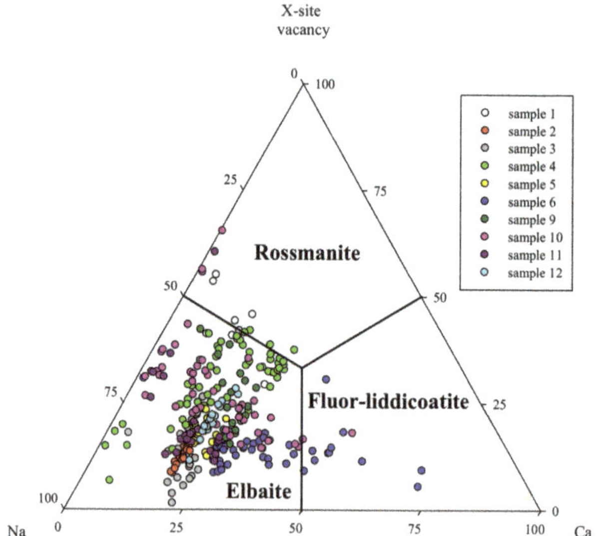

Figure 5. Ternary diagram of X-site composition of all analysed tourmalines.

Most analyses fall in the elbaite/fluor-elbaite field, although a few correspond to rossmanite (samples 1, 4, 10, 11) and fluor-liddicoatite (samples 6, 10). Foitite was found at the analogous pole of the crystals of tourmalines 10, 11 and as fibrous overgrowth on sample 9b. The foitite analyses are not reported in the diagram where they would fall in the rossmanite field due to their >50% vacant in X-site [25].

Selected electron microprobe analyses are reported in Table 2.

Table 2. Representative electron microprobe analyses of the studied tourmalines.

Sample	1				2	3a	3b	4							5
	Pink Elbaite	Pink Rossmanite	Pink Elbaite	Pink Rossmanite	Light Blue Elbaite	Colourless Elbaite	Brown Elbaite	Rim-Green Elbaite	Core-Pink Elbaite	Core-Pink Elbaite	Core-Pink Elbaite	Inter-Pink Elbaite	Rim-Green Elbaite	Elbaite	Light Green Elbaite
					Average (20 pts)	Average (10 pts)	Average (15 pts)								Average (20 pts)
Point Number °	4	6	8	9				16	10	24	23	56	54	33	
SiO$_2$ (wt %)	36.34	36.63	36.28	37.06	36.2	36.17	36.54	36.99	36.79	36.67	36.59	37.04	37.28	36.71	36.29
TiO$_2$	bdl	bdl	0.06	bdl	0.03	0.03	0.07	0.04	bdl	bdl	0.04	bdl	0.02	0.24	0.03
B$_2$O$_3$ *	11.06	11.18	11.02	11.27	10.78	10.87	10.77	10.98	10.94	10.93	10.91	11.04	11.05	10.64	10.9
Al$_2$O$_3$	43.3	44.73	42.95	44.91	38.83	40.28	37.86	40.32	41.41	41.36	41.34	42.14	40.72	36.01	40.16
V$_2$O$_3$	0.02	bdl	bdl	bdl	0.01	0.02	0.02	0.02	0.02	bdl	bdl	0.01	bdl	0.04	0.01
Cr$_2$O$_3$	bdl	bdl	bdl	bdl	0.02	0.01	bdl	bdl	bdl	0.03	0.01	0.03	bdl	0.01	bdl
FeO	0.02	bdl	bdl	0.03	1.00	0.73	1.43	0.83	bdl	bdl	0.01	0.01	0.52	4.75	0.77
MgO	0.03	bdl	bdl	0.01	0.02	0.02	0.01	0.01	bdl	0.01	0.01	0.01	bdl	bdl	0.01
MnO	0.52	0.21	0.55	0.27	3.79	2.79	3.99	2.82	0.16	0.16	0.13	0.12	2.02	2.45	2.93
ZnO	0.09	bdl	bdl	bdl	0.05	0.03	0.08	bdl	0.14	0.08	0.05	0.03	0.10	0.50	0.05
CaO	0.95	0.28	0.98	0.27	1.08	0.98	1.23	0.78	1.52	1.65	1.57	1.23	1.27	0.36	1.18
Li$_2$O *	1.87	1.74	1.91	1.75	1.73	1.77	1.76	1.76	2.12	2.16	2.15	2.09	1.95	1.5	1.80
Na$_2$O	1.38	1.39	1.42	1.35	2.19	2.05	2.25	1.98	1.26	1.29	1.31	1.39	1.72	2.73	1.98
K$_2$O	0.01	bdl	0.01	0.01	0.02	0.02	0.01	0.01	0.01	0.01	0.02	bdl	bdl	0.03	0.01
H$_2$O*	3.53	3.83	3.53	3.78	3.12	3.27	3.13	3.24	3.22	3.24	3.17	3.34	3.26	3.00	3.24
F	0.61	0.07	0.57	0.24	1.27	1.00	1.25	1.16	1.17	1.11	1.25	0.98	1.16	1.41	1.10
Total	99.73	100.07	99.27	100.94	100.13	100.02	100.48	100.95	98.77	98.7	98.57	99.45	101.08	100.39	100.47
O=F	0.26	0.03	0.24	0.1	0.53	0.42	0.53	0.49	0.49	0.47	0.53	0.41	0.49	0.59	0.46
Total *	99.47	100.04	99.03	100.84	99.6	99.6	99.96	100.46	98.27	98.23	98.04	99.04	100.59	99.79	100
Structural formula based on 31 anions (O, OH, F)															
Si	5.7	5.692	5.723	5.714	5.835	5.783	5.905	5.856	5.846	5.831	5.829	5.831	5.864	5.996	5.788
Al	0.28	0.308	0.277	0.286	0.165	0.217	0.097	0.144	0.154	0.169	0.171	0.169	0.136	0.004	0.212
T sum	6.000	6.000	6.000	6.000	6.000	6.000	6.000	6.000	6.000	6.000	6.000	6.000	6.000	6.000	6.000
B	3.000	3.000	3.000	3.000	3.000	3.000	3.000	3.000	3.000	3.000	3.000	3.000	3.000	3.000	3.000
Al (Z)	6.000	6.000	6.000	6.000	6.000	6.000	6.000	6.000	6.000	6.000	6.000	6.000	6.000	6.000	6.000
Cr	-	-	-	-	-	-	-	-	-	-	-	-	-	-	-
Al	1.727	1.884	1.709	1.875	1.211	1.373	1.081	1.379	1.601	1.583	1.59	1.65	1.413	0.928	1.336
Ti	-	-	0.007	0.001	0.004	0.004	0.012	0.005	-	-	0.004	-	0.002	0.03	0.004
V	0.003	-	-	-	0.001	0.001	0.002	0.002	0.003	-	-	0.001	-	0.006	0.001
Cr	-	-	-	-	0.002	0.001	0.001	-	bdl	0.004	0.002	0.003	-	0.001	-
Mg	0.007	0.001	-	0.002	0.004	0.004	0.003	0.003	bdl	0.003	0.003	0.002	-	-	0.003
Mn	0.069	0.028	0.073	0.035	0.518	0.377	0.527	0.378	0.022	0.021	0.017	0.016	0.269	0.339	0.396
Fe^{2+}	0.003	-	-	0.003	0.134	0.097	0.231	0.11	0.001	-	0.001	0.001	0.069	0.649	0.102
Zn	0.01	-	-	-	0.006	0.004	0.013	-	0.016	0.009	0.006	0.003	0.012	0.061	0.006
Li *	1.181	1.087	1.212	1.084	1.12	1.139	1.129	1.122	1.358	1.38	1.375	1.325	1.235	0.987	1.151
Y sum	3.000	3.000	3.000	3.000	3.000	3.000	3.000	3.000	3.000	3.000	3.000	3.000	3.000	3.000	3.000
Ca	0.159	0.046	0.165	0.044	0.187	0.168	0.201	0.132	0.259	0.281	0.268	0.207	0.214	0.064	0.202
Na	0.42	0.419	0.434	0.404	0.684	0.635	0.707	0.608	0.388	0.396	0.406	0.424	0.525	0.865	0.613
K	0.002	-	0.002	0.003	0.004	0.003	0.003	0.003	0.002	0.002	0.004	-	-	0.005	0.002
Vacancy	0.418	0.535	0.399	0.55	0.125	0.194	0.089	0.257	0.351	0.32	0.322	0.368	0.261	0.066	0.183
X sum	1.000	1.000	1.000	1.000	1.000	1.000	1.000	1.000	1.000	1.000	1.000	1.000	1.000	1.000	1.000
OH	3.695	3.967	3.715	3.884	3.353	3.492	3.363	3.419	3.412	3.442	3.37	3.512	3.423	3.272	3.445
F	0.305	0.033	0.285	0.116	0.647	0.508	0.637	0.581	0.588	0.558	0.63	0.488	0.577	0.728	0.555

Table 2. *Cont.*

Sample		6			9a	9b				10					11	12
							Analogous Green			Antilogous Dark Green						
	Light Yellow Elbaite	Light Yellow Elbaite	Light Yellow Fluor-Liddicoatite	Colourless Flour-Liddicoatite	Pink Elbaite	Green Elbaite	Foitite	Elbaite	Elbaite	Elbaite	Fluor-Liddicoatite	Elbaite	Elbaite	Rossmanite	Green Elbaite	Blue Elbaite
					Average (9 pts)	Average (10 pts)									Average (21 pts)	Average (20 pts)
Point Number °								4	10	12	36	42	43	45		
SiO$_2$ (wt %)	35.70	36.00	36.71	37.50	37.12	37.32	35.82	37.38	38.13	37.51	37.60	37.82	37.13	37.06	36.95	37.10
TiO$_2$	0.14	0.17	0.09	0.09	0.01	0.17	bdl	bdl	bdl	0.18	0.03	0.05	0.20	0.07	0.18	0.17
B$_2$O$_3$ *	10.84	10.83	10.94	10.99	11.07	11.00	10.39	10.92	11.09	11.02	10.99	11.04	10.98	10.76	10.96	10.93
Al$_2$O$_3$	40.53	39.68	39.90	39.43	41.59	39.78	33.55	39.33	39.60	39.74	39.03	40.09	39.60	37.57	39.86	39.30
V$_2$O$_3$	0.01	0.04	0.01	bdl	0.01	0.02	bdl	0.03	bdl	0.10	0.02	bdl	0.04	bdl	0.01	bdl
Cr$_2$O$_3$	bdl	bdl	bdl	0.01	0.01	0.01	0.01	0.01	bdl	bdl	0.02	bdl	0.05	bdl	0.01	0.01
FeO	0.34	0.54	0.22	0.16	0.08	0.71	11.52	1.68	0.44	0.85	0.34	0.88	1.25	3.40	1.18	0.66
MgO	0.01	0.01	0.01	bdl	0.01	0.01	2.57	0.03	0.01	0.01	0.02	0.05	0.04	0.02	0.03	0.03
MnO	2.38	2.51	1.29	0.99	1.60	2.05	0.17	3.68	1.82	1.82	1.45	2.89	1.87	5.67	2.31	3.10
ZnO	bdl	bdl	0.08	bdl	0.03	0.06	0.10	0.12	0.12	0.02	0.08	0.09	0.19	0.18	0.02	0.03
CaO	1.49	1.58	2.67	2.94	1.16	1.53	0.02	0.13	1.96	1.62	3.05	0.27	1.47	0.07	1.00	1.21
Li$_2$O *	1.90	1.97	2.31	2.44	1.96	2.05	0.14	1.45	2.20	2.05	2.38	1.71	1.98	0.91	1.87	1.88
Na$_2$O	1.81	1.97	1.32	1.05	1.59	1.77	1.48	1.97	1.50	1.54	0.99	2.10	1.84	1.37	2.08	1.88
K$_2$O	bdl	0.02	0.01	bdl	0.01	0.01	0.01	0.01	bdl	0.01	0.01	bdl	0.01	0.02	0.02	0.02
H$_2$O *	3.29	3.17	3.23	3.21	3.39	3.25	3.58	3.77	3.45	3.22	3.18	3.81	3.28	3.71	3.33	3.22
F	0.95	1.19	1.16	1.23	0.91	1.14	bdl	bdl	0.79	1.23	1.29	bdl	1.08	bdl	0.95	1.17
Total	99.39	99.67	99.94	100.03	100.54	100.88	99.37	100.52	101.11	100.90	100.49	100.81	101.00	100.82	100.77	100.70
O=F	0.40	0.50	0.49	0.52	0.38	0.48	bdl	bdl	0.33	0.52	0.54	bdl	0.45	bdl	0.40	0.49
Total *	98.99	99.17	99.46	99.52	100.16	100.40	99.37	100.52	100.78	100.38	99.95	100.81	100.55	100.82	100.37	100.21
Structural formula based on 31 anions (O, OH, F)																
Si	5.726	5.779	5.831	5.931	5.830	5.894	5.994	5.948	5.978	5.918	5.946	5.952	5.875	5.985	5.858	5.898
Al	0.274	0.221	0.169	0.069	0.170	0.106	0.006	0.052	0.022	0.082	0.054	0.048	0.125	0.015	0.142	0.102
T sum	6.000	6.000	6.000	6.000	6.000	6.000	6.000	6.000	6.000	6.000	6.000	6.000	6.000	6.000	6.000	6.000
B	3.000	3.000	3.000	3.000	3.000	3.000	3.000	3.000	3.000	3.000	3.000	3.000	3.000	3.000	3.000	3.000
Al (Z)	6.000	6.000	6.000	6.000	6.000	6.000	6.000	6.000	6.000	6.000	6.000	6.000	6.000	6.000	6.000	6.000
Cr	-	-	-	-	-	-	-	-	-	-	-	-	-	-	-	-
Al	1.387	1.287	1.301	1.282	1.528	1.299	0.612	1.323	1.296	1.307	1.220	1.387	1.260	1.136	1.305	1.263
Ti	0.017	0.020	0.011	0.010	0.001	0.020	-	-	-	0.021	0.004	0.006	0.024	0.008	0.022	0.020
V	0.002	0.005	0.001	-	0.001	0.002	-	0.004	-	0.012	0.003	-	0.005	-	0.001	0.001
Cr	-	-	-	0.001	0.001	0.001	0.002	0.001	-	-	0.003	-	0.006	-	0.001	0.001
Mg	0.004	0.003	0.003	0.001	0.001	0.003	0.641	0.008	0.002	0.002	0.005	0.012	0.009	0.006	0.008	0.007
Mn	0.323	0.341	0.173	0.133	0.213	0.275	0.024	0.496	0.242	0.243	0.194	0.385	0.251	0.776	0.310	0.417
Fe^{2+}	0.046	0.072	0.029	0.021	0.010	0.093	1.612	0.224	0.058	0.112	0.045	0.116	0.165	0.459	0.157	0.087
Zn	-	-	0.010	-	0.004	0.007	0.013	0.014	0.014	0.002	0.010	0.010	0.022	0.022	0.003	0.003
Li *	1.223	1.271	1.473	1.552	1.240	1.299	0.096	0.930	1.389	1.300	1.517	1.083	1.259	0.594	1.193	1.200
Y sum	3.000	3.000	3.000	3.000	3.000	3.000	3.000	3.000	3.000	3.000	3.000	3.000	3.000	3.000	3.000	3.000
Ca	0.256	0.272	0.454	0.498	0.195	0.259	0.003	0.022	0.329	0.274	0.517	0.046	0.249	0.012	0.171	0.207
Na	0.563	0.613	0.407	0.321	0.485	0.542	0.480	0.608	0.456	0.471	0.303	0.641	0.564	0.429	0.640	0.581
K	0.001	0.003	0.002	0.000	0.001	0.002	0.002	0.002	-	0.001	0.002	-	0.003	0.003	0.003	0.004
Vacancy	0.181	0.112	0.137	0.180	0.317	0.197	0.515	0.368	0.214	0.254	0.179	0.313	0.184	0.555	0.187	0.209
X sum	1.000	1.000	1.000	1.000	1.000	1.000	1.000	1.000	1.000	1.000	1.000	1.000	1.000	1.000	1.000	1.000
OH	3.519	3.396	3.417	3.385	3.547	3.429	4.000	4.000	3.608	3.386	3.355	4.000	3.460	4.000	3.522	3.412
F	0.481	0.604	0.583	0.615	0.453	0.571	-	-	0.392	0.614	0.645	-	0.540	-	0.478	0.588

* calculated on the assumed elbaite stoichiometry. bdl = below the detection limit (0.01 wt % for all oxides). ° point number corresponding to the number in the sample images, pts = points.

Colour changes are generally controlled by a combination of concentration and oxidation states of chromophore elements such as Ti, V, Fe and Mn [26]: in samples 3, 10 and 11, a higher content of iron is detected in the portion typically brown-black. Sample 10 (Figure 6a,b) is the only one including the analogous and antilogous pole and shows compositional variations with changes in Mn, Fe (Figure 6c) and Ca. This last displays the highest values in the central part of the crystal (Figure 6d). Titanium ranges from 0 to 0.04 a.p.f.u., whereas V and Cr are very low or below the detection limit. Rossmanite is present both at the analogous and antilogous poles (Figure 6b).

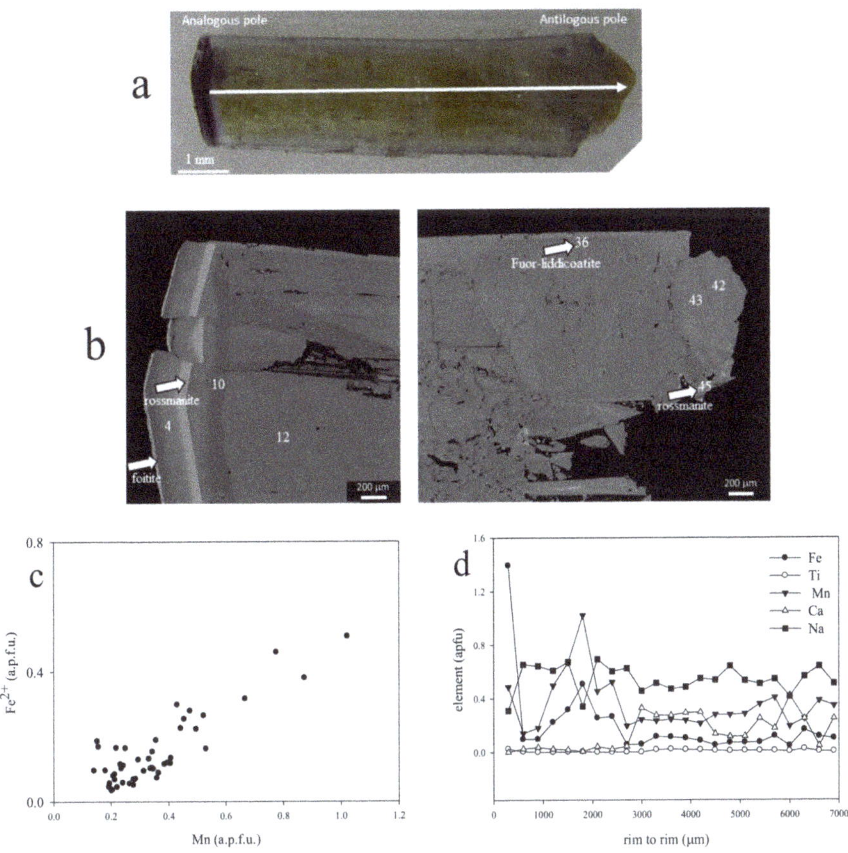

Figure 6. Optical microscope (**a**), BKSE (**b**) images, (**c**) variation of Fe vs. Mn and (**d**) an example of zoning patterns of different elements of sample 10. The points in BKSE images correspond to the points reported in Table 2.

Sample 3 (Figure 7a), along the *c*-axis, shows a colourless (a) and a brown (b) zone with an iron content of up to 0.106 and up to 0.762 a.p.f.u., respectively. Sample 6 shows an elbaite composition with a decrease of Mn and Na together with an increase of Ca from light yellow to near colourless zones where a fluor-liddicoatite composition is identified (Figure 7b,c). Ti and Fe are very low, and therefore manganese content seems to be mainly responsible of the colour changes.

Samples 2 and 12 have a homogeneous composition with the higher content of Mn that may induce the deep blue colour (up to 0.548 and 0.443 a.p.f.u., respectively).

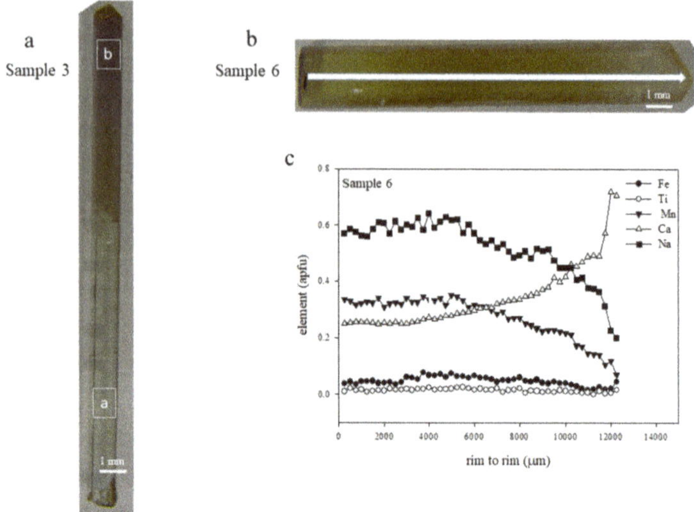

Figure 7. Optical microscope image of sample 3 (**a**) and sample 6 (**b**); zoning patterns of different elements of sample 6 (**c**).

The pink tourmaline (sample 1, Figure 8a) is nearly devoid of iron and presents the lowest Mn content with two different compositional zones: one with higher Ca + Mn (on average, 0.27 a.p.f.u., i.e., pts 4 and 8), the second one with lower Ca + Mn (on average, 0.12 a.p.f.u.), Mn < 0.04 a.p.f.u. and higher mole fraction of rossmanite (i.e., pts 6 and 9). Lighter-coloured areas included in rossmanite resulted in being muscovite (Figure 8b).

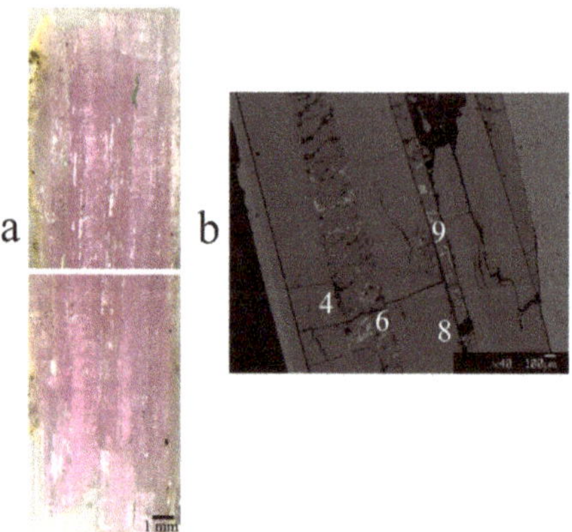

Figure 8. Optical microscope (**a**) and BKSE (**b**) images of sample 1. The points in BKSE image correspond to the points reported in Table 2. Points 6 and 9 correspond to rossmannite.

Sample 4 (Figure 9a) occurs in the same geode of sample 3 and has been cut perpendicular to the *c*-axis. The slice reveals a complex growth history characterized by a central part, pink in colour, depleted in Mn and Fe, but enriched in Ca with respect to the green rim, with a composition ranging from fluor-liddicoatite-rich elbaite to elbaite. The lighter-coloured zone at the bottom of the sample (Figure 9b, e.g., point 33, analysis 33 in Table 2) presents an enrichment in Fe^{2+} with respect to Mn^{2+} (Figure 9c). In Figure 9d, an example of the variation from rim to rim of these elements, including Ti, Ca and Na, is shown.

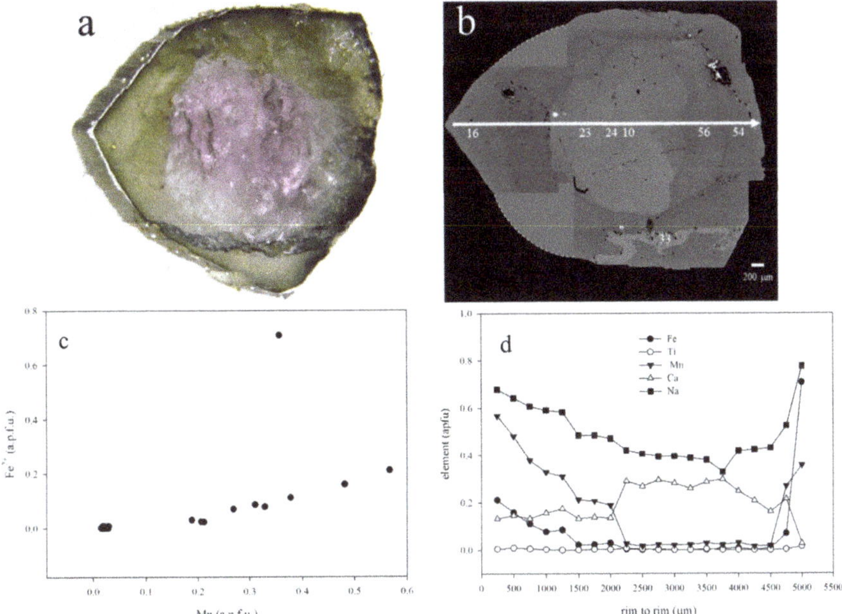

Figure 9. Optical microscope (**a**) and BKSE (**b**) images, (**c**) variation of Fe vs. Mn and (**d**) an example of zoning patterns of different elements of sample 4. The points in BKSE image correspond to the points reported in Table 2.

Sample 9 (Figure 10a,b) comes from a geode containing black quartz, zircon and uraninite and displays a change from pink (sample 9a) to green colour (sample 9b) that seems to be due to manganese, iron and titanium enrichment. In sample 9b, a thin overgrowth with foitite composition (see Figure 10c and Table 2) is present.

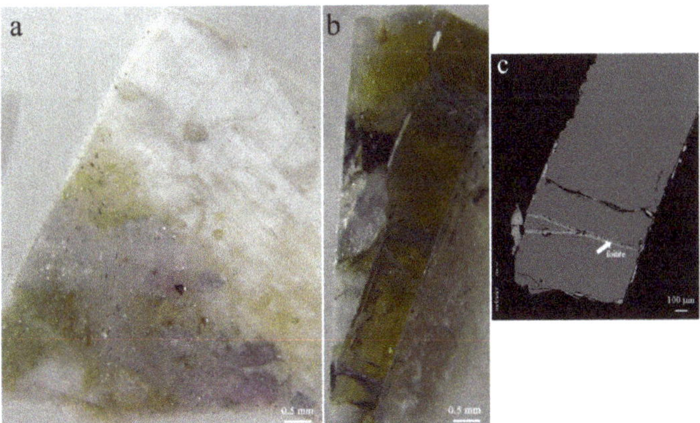

Figure 10. Optical microscope images of sample 9a (**a**), 9b (**b**) and BSKE image of sample 9b (**c**) where the arrow indicates the foitite.

The averaged LA-ICP-MS analyses of selected trace elements: Be, Sc, V, Cu, Ga, Ge, Sr, Y, Ta, Pb, Th, U and REE of the tourmalines are reported in Table 3.

Table 3. LA–ICP–MS analyses of trace elements in studied tourmalines (ppm).

Sample	1		2		3a		3b		4		5	
	Average (4 pts)	st dev	Average (4 pts)	st dev	Average (4 pts)	st dev	Average (5 pts)	st dev	Average (8 pts)	st dev	Average (5 pts)	st dev
Li	8546.20	463.86	7985.55	182.95	7859.06	75.06	7919.98	324.03	8899.59	1020.41	8073.58	322.46
Be	52.19	14.18	37.62	2.83	28.92	2.96	21.29	8.82	99.38	122.07	31.05	9.50
Sc	3.03	0.55	1.83	0.28	1.82	0.35	2.79	0.45	1.63	0.52	2.12	0.16
V	0.38	0.32	1.08	0.20	0.82	0.12	13.95	7.92	8.56	5.26	3.17	4.77
Cu	196.20	27.01	19.23	1.05	16.10	1.41	16.04	12.69	71.89	62.62	22.16	3.19
Ga	181.73	28.06	40.83	2.51	45.27	1.04	59.90	13.21	78.28	26.48	44.19	6.70
Ge	9.11	2.22	20.47	1.68	19.85	1.52	15.65	2.90	15.91	2.78	21.57	2.77
Sr	0.73	0.23	31.66	1.23	19.88	1.72	231.22	390.52	93.98	234.91	29.42	3.66
Y	0.25	0.17	0.46	0.05	0.35	0.03	0.24	0.15	0.22	0.08	0.39	0.20
Ta	4.63	3.57	1.07	0.14	0.28	0.01	0.56	0.31	3.18	5.43	0.44	0.40
Pb	11.94	4.76	43.78	2.22	35.05	1.19	2882.66	3775.66	184.51	453.38	163.22	71.23
Th	3.58	2.95	0.79	0.06	0.55	0.06	0.16	0.14	6.98	15.66	0.52	0.32
U	0.36	0.29	0.02	0.00	bdl		0.01	0.01	0.79	1.15	0.03	
La	44.13	18.50	24.29	1.46	13.58	0.35	22.86	14.36	9.50	4.54	17.07	4.19
Ce	46.68	19.98	49.72	3.99	30.16	0.30	48.87	26.13	15.89	8.20	37.41	11.14
Pr	2.35	1.08	5.13	0.38	2.87	0.10	4.94	2.44	1.44	0.85	3.60	1.08
Nd	3.05	1.40	14.54	1.49	7.46	0.24	13.43	6.78	3.42	2.55	9.15	2.88
Sm	0.53	0.36	3.14	0.69	1.96	0.17	2.74	1.23	1.17	0.41	2.29	0.95
Eu	bdl		0.12	0.03	0.12	0.04	0.13	0.05	0.05	0.02	0.10	0.04
Gd	0.27	0.02	0.93	0.19	0.64	0.13	0.68	0.40	0.42	0.16	0.72	0.38
Tb	0.05	0.04	0.11	0.02	0.09	0.02	0.08	0.05	0.05	0.02	0.09	0.04
Dy	0.15	0.17	0.36	0.11	0.24	0.05	0.24	0.17	0.14	0.04	0.26	0.17
Ho	bdl		bdl		0.03	0.02	0.02	0.01	0.02	0.01	0.02	0.01
Er	0.20	0.26	bdl		0.05	0.001	bdl		0.06	0.03	bdl	
Tm	bdl		0.02	0.005	0.01	0.005	0.01	0.00	bdl		bdl	
Yb	0.10	0.10	bdl				bdl		0.17	0.15	bdl	
Lu	bdl		bdl		bdl		bdl		bdl		0.01	
Σ REE	97.50		98.36		57.21		94.00		32.34		70.72	

Table 3. *Cont.*

Sample	6		9a		9b		10		11		12	
	Average (7 pts)	st dev	Average (3 pts)	st dev	Average (3 pts)	st dev	Average (8 pts)	st dev	Average (5 pts)	st dev	Average (4 pts)	st dev
Li	9993.23	1095.08	8494.71	249.68	8767.86	145.96	8444.96	1786.74	7175.63	2373.44	8271.11	399.03
Be	20.20	5.78	23.17	2.81	27.37	7.68	22.63	13.36	14.62	11.00	19.95	2.56
Sc	1.96	0.40	2.96	0.25	3.09	0.24	2.26	0.46	3.47	0.84	1.65	0.47
V	0.37	0.09	0.48	0.18	8.88	3.03	0.84	0.63	1.21	0.54	0.54	0.09
Cu	21.30	9.49	148.85	58.16	31.74	3.36	24.96	16.28	17.92	12.74	30.46	6.96
Ga	68.20	25.78	121.62	49.09	35.29	1.08	104.43	26.62	91.33	12.88	45.58	5.77
Ge	12.01	2.63	7.05	0.84	16.59	1.50	6.97	2.13	7.10	1.54	11.74	3.80
Sr	104.28	95.41	4.37	1.46	168.96	25.82	30.01	25.73	175.48	356.03	16.88	4.55
Y	0.36	0.05	0.46	0.08	0.48	0.20	0.65	0.62	0.31	0.11	0.21	0.12
Ta	1.11	0.36	1.06	0.39	1.02	0.43	1.04	1.47	0.16	0.10	0.56	0.25
Pb	135.44	129.63	13.36	1.50	64.72	10.31	40.34	60.43	881.33	710.51	27.57	6.20
Th	0.23	0.09	1.07	0.40	0.49	0.16	1.55	1.67	0.52	0.25	0.32	0.12
U	0.03	0.02	0.06	0.01	bdl		0.06	0.02	0.06	0.06	0.02	0.02
La	22.46	4.14	23.09	7.25	33.99	10.92	18.37	19.28	9.75	8.73	14.74	5.05
Ce	46.61	14.32	49.63	12.21	66.11	21.77	35.44	44.29	20.35	18.36	31.63	11.11
Pr	4.65	1.67	4.84	0.58	6.41	2.04	3.98	5.18	2.03	1.80	3.15	1.06
Nd	12.45	5.70	11.78	0.79	18.64	5.80	13.82	16.80	7.01	4.65	8.26	2.74
Sm	2.83	1.40	3.86	0.57	3.76	1.18	4.74	5.45	1.73	1.19	1.91	0.70
Eu	0.13	0.05	0.07	0.03	0.32	0.05	0.13	0.16	0.11	0.03	0.09	0.02
Gd	0.84	0.44	1.31	0.35	1.49	0.59	1.66	1.44	0.50	0.38	0.49	0.16
Tb	0.08	0.02	0.15	0.02	0.17	0.02	0.21	0.18	0.07	0.05	0.07	0.03
Dy	0.24	0.12	0.48	0.17	0.41	0.14	0.61	0.61	0.21	0.10	0.16	0.05
Ho	bdl		0.02	0.01	0.04	0.02	0.06	0.04	0.03	0.03	bdl	
Er	0.05	0.03	0.05	0.04	0.07	0.03	bdl		0.05	0.02	0.08	0.04
Tm	0.01	0.01	0.02	0.01	bdl		0.03	0.01	0.02	bdl	bdl	
Yb	bdl		bdl		bdl		bdl		0.15	0.20	0.10	0.01
Lu	0.02	0.01	bdl		bdl		0.02	0.01	0.04	0.05	bdl	
Σ REE	90.37		95.29		131.42		79.06		42.06		60.67	

bdl = below detection limit, pts = points, st dev = standard deviation.

In general, the concentration of trace elements is variable and does not show any significant trend or correlation with the colour changes. The exceptions are represented by the higher values of Pb (up to 880 and 2880 ppm in sample 11 and 3b, respectively) corresponding to Mn (Fe)-enriched points and of Cu (up to 196 ppm) in samples 1, 4, 9a, determined in the pink part of the crystals. Variable concentration of Ga (41–182 ppm) and Sr (1–231 ppm) was observed. The Y content, generally correlated with HREE, is very low, less than 1 ppm in all samples.

The REE content in tourmalines from granitic pegmatites is generally low (<30 ppm) while in the examined samples, the total REE content ranges from 30 to 130 ppm with light REE-enrichment and Ce being the most abundant element. Chondrite normalized REE patterns (plotted as mean values for each sample in Figure 11) display a general depletion in the medium and heavy, with respect to the light, rare earths. The negative Eu anomaly is probably related to the local depletion of Eu^{2+} content in the melt due to its consumption during the growth of K-feldspar, an important carrier of Eu^{2+} in magmatic rocks.

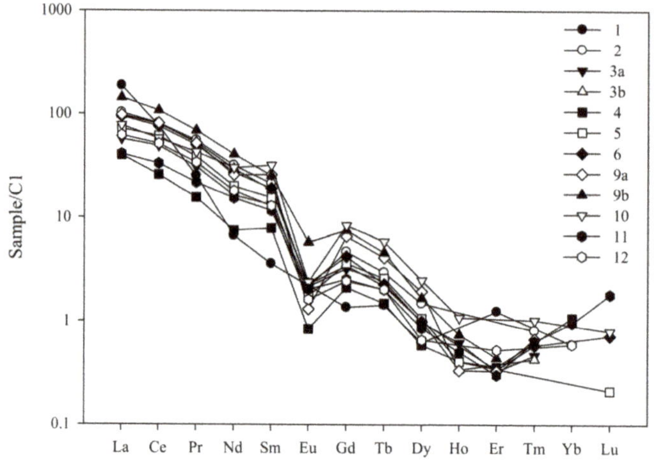

Figure 11. REE patterns normalized to the C1 values reported in [27] for all the analysed samples.

5.3. X-Ray Crystal Structure Refinement

The crystallographic data obtained by single crystal refinement confirm that the examined tourmalines belong to the *R3m* space group with the cell parameters in the range of elbaite species. Absolute structure parameter ranged from 0.01(11) to −0.10(10) and secondary extinction coefficient ranged from 0.0024(2) to 0.0049(3) (Table 4).

The selected interatomic distances, geometrical parameters and refined and observed (from chemical analyses) site-scattering values are given in Tables 5 and 6, respectively. The final atom coordinates and equivalent displacement parameters, as well as the complete set of crystallographic data (crystallographic information files and lists of observed and calculated structure factors), have been deposited in supplementary electronic material (Tables S1 and S2a–c).

The structural data confirm elbaitic compositions: <Y–O> ranges from 1.989 to 2.030 Å compared to a calculated value of 2.015 Å for an ideal elbaite (using ionic radii of [11]; <Z–O> ranges from 1.907 to 1.908 Å compared to a <^{Z}Al–O> grand mean value of 1.906 Å (Figure 3 of [28]); site scattering at X sites (8.73–12.43 electrons per formula unit or a.p.f.u., Table 6) is compatible with a dominant Na occupancy. The high quality of the reported structure refinements makes it possible to discuss the site assignment of cations among the different sites of the studied crystals. Following [11] and using his Equation (4) [^{Z}Al = −0.1155 + 1.1713·$^{[6]}$Al − 0.0522·$^{[6]}$Al^2; Al = Al − ^{T}Al (a.p.f.u.)], it is possible to estimate the ^{Z}Al occupancy just from chemical data. Applying this equation to the data reported in Table 2 produces a slight ^{Z}Al deficiency of 0.126–0.439 a.p.f.u.. However, this is in contrast with single crystal XRD data that provide refined scattering values of 12.85(5)–12.96(3) a.p.f.u. for the Z site, implying a maximum of 0.12 ^{Z}Li atoms per formula unit or just pure Al Z sites, considering 3σ. It is, therefore, highly improbable that any Mn^{2+} (or Fe^{2+}) could have been disordered into the Z-sites. The agreement between observed (SC-XRD) and calculated (EMPA-WDS) site scattering at X-sites is poorer (9–10%, Table 6), probably due to the high chemical variability of crystals, because the occupancy of X site is the main chemical vector observed in the studied tourmalines.

It is worthwhile to note the high values of U equivalent for the O(1) (from 0.0291(7) to 0.057(2) $Å^2$, compared to the mean value of 0.007–0.010 $Å^2$ for the other anion sites; see Tables S1 and S2a–c). This is very probably due to static disorder at the O(1) anion site. All three studied crystals show large and flat [parallel to (0001)] thermal ellipsoids, making the estimation of the electron density at the O(1) site inaccurate (see Figure 12). We tried a split model, but it was unsuccessful. A similar delocalization of electron density was reported by [29] (compare their Figure 1a with our Figure 12b) for manganese-bearing elbaitic compositions. Burns and co-workers [29] interpreted the large anisotropic displacements as positional disorder, rather than thermal vibration, due to the 6 possible local arrangements at the three Y sites around the O(1) site; the three principal Y cations (Al, Li and Mn^{2+}) have very different ionic radii (0.547(3), 0.751(9), 0.809(1) Å, respectively; values from [11]). Considering the composition of the Y sites (close to AlLiMn) of our tourmalines, the ideal average trimer would promote a distorted environment for the O(1), confirming in a new set of samples the behaviour already reported by [29].

Table 4. Crystal data and structure refinement for tourmalines 1, 2, 12.

Sample	1	2	12
Temperature	293(2) K	293(2) K	293(2) K
Wavelength	0.71073 Å	0.71073 Å	0.71073 Å
Crystal system	Trigonal	Trigonal	Trigonal
Space group	$R3m$	$R3m$	$R3m$
Unit cell dimensions	$a = 15.8283(3)$ Å	$a = 15.8951(8)$ Å	$a = 15.8909(2)$ Å
	$c = 7.09392(18)$ Å	$c = 7.1216(4)$ Å	$c = 7.1163(3)$ Å
Volume	1539.16(5) $Å^3$	1558.22(15) $Å^3$	1556.26(7) $Å^3$
Z	3	3	3
Absorption coefficient	1.004 mm^{-1}	1.056 mm^{-1}	1.020 mm^{-1}
$F(000)$	1395	1430	1422
Crystal size (mm^3)	0.29 × 0.61 × 0.47	0.35 × 0.51 × 0.65	0.40 × 0.53 × 0.70
θ range for data collection	3.23 to 36.11°.	3.22 to 29.04°.	3.22 to 35.95°.
Index ranges	$-26 \le h \le 26$, $-24 \le k \le 24$, $-9 \le l \le 9$	$-16 \le h \le 13$, $-20 \le k \le 20$, $-9 \le l \le 9$	$-25 \le h \le 26$, $-25 \le k \le 25$, $-9 \le l \le 9$
Reflections collected	14,418	3806	14,500
Independent reflections	1571	914	1261
$R(int)$	0.0277	0.0216	0.0334
Completeness to θ = 35.95°	89.30%	95.20%	72.70%
Refinement method	Full-matrix least-squares on F2	Full-matrix least-squares on F2	Full-matrix least-squares on F2
Data/restraints/parameters	1571/1/97	914/1/97	1261/1/97
Goodness-of-fit on F^2	1.161	1.076	1.09
Final R indices [$I > 2\sigma (I)$]	R1 = 0.0159 wR2 = 0.0400	R1 = 0.0175 wR2 = 0.0452	R1 = 0.0179 wR2 = 0.0448
R indices (all data)	R1 = 0.0168 wR2 = 0.0403	R1 = 0.0177 wR2 = 0.0453	R1 = 0.0186 wR2 = 0.0451
Absolute structure parameter	0.04(7)	0.01(11)	−0.10(10)
Extinction coefficient	0.0049(3)	0.0041(3)	0.0024(2)
Largest diff. peak and hole	0.461 and −0.439 e.$Å^{-3}$	0.530 and −0.497 e.$Å^{-3}$	0.889 and −0.709 e.$Å^{-3}$

Table 5. Bond lengths [Å] and angles [°] for tourmalines 1, 2, 12.

Sample	1	2	12
T–O(6)	1.6092(9)	1.6049(15)	1.6021(14)
T–O(7)	1.6103(7)	1.6134(13)	1.6110(10)
T–O(4)	1.6208(5)	1.6263(8)	1.6260(7)
T–O(5)	1.6363(5)	1.6412(9)	1.6389(8)
$<T\text{–}O>$	1.619	1.621	1.62
V (Å^3)	2.173	2.181	2.173
TQE	1.0017	1.0023	1.0022
TAV	6.812	9.502	9.0511
B–O(2)	1.3630(18)	1.352(4)	1.361(3)
B–O(8) ($\times$2)	1.3798(10)	1.386(2)	1.3850(15)
$<B\text{–}O>$	1.374	1.375	1.377
X–O(2) ($\times$3)	2.4570(18)	2.426(3)	2.425(2)
X–O(5) ($\times$3)	2.7422(13)	2.748(2)	2.7492(17)
X–O(4) ($\times$3)	2.8095(13)	2.810(2)	2.8078(18)
$<X\text{–}O>$	2.67	2.661	2.661
V (Å^3)	31.14	31.419	31.086
Y–O(2) ($\times$2)	1.9620(9)	1.9824(15)	1.9812(14)
Y–O(6) ($\times$2)	1.9591(9)	2.0150(15)	2.0129(13)
Y–O(1)	1.9577(14)	2.028(2)	2.024(2)
Y–O(3)	2.1346(14)	2.171(2)	2.1678(19)
$<Y\text{–}O>$	1.989	2.032	2.03
V (Å^3)	10.125	10.796	10.759
OQE	1.0252	1.0253	1.0253
OAV	79.39	79.54	79.7
Z–O(6)	1.8643(8)	1.8506(15)	1.8514(12)
Z–O(7)	1.8816(8)	1.8823(14)	1.8834(12)
Z–O(8)	1.8877(8)	1.8865(14)	1.8847(11)
Z–O(8)	1.9008(8)	1.9114(14)	1.9098(11)
Z–O(7)	1.9415(7)	1.9551(14)	1.9551(11)
Z–O(3)	1.9647(6)	1.9617(11)	1.9612(10)
$<Z\text{–}O>$	1.907	1.908	1.908
V (Å^3)	9.039	9.079	9.075
OQE	1.0154	1.0137	1.0137
OAV	52.08	45.82	45.75
O(3) –H(3)	0.80(3)	0.71(4)	0.72(4)

Table 6. Observed (Single Crystal XRD) and calculated (EMPA) site scattering.

Site	(Electrons Per Site, eps) for Tourmalines 1, 2, 12		
	1	2	12
X (obs)	8.73	12.43	11.81
X (calc)	7.95	11.34	10.61
Y (obs)	9.49	11.98	11.26
Y (calc)	9.25	11.97	11.12
Z (obs)	12.96	12.83	12.85
Z (calc)	13	13	13

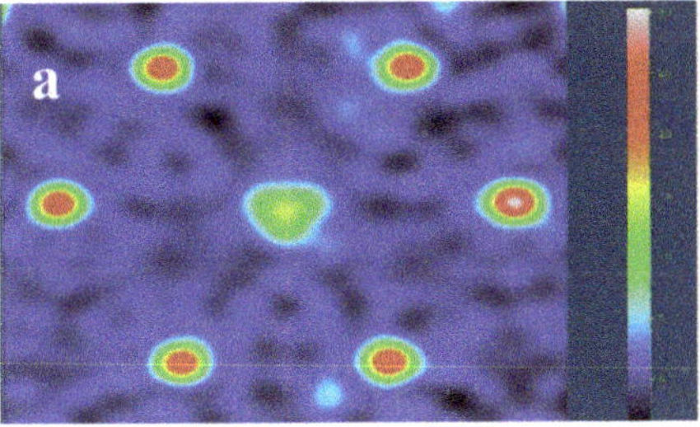

Figure 12. Sample 2: Electron density at O(1). The high displacement parameters and oblate behaviour of the O(1) anion site in the three samples are related to the trilobated shape of the maxima at that position. This is ascribable to local static disorder due to the presence of three different size cations as Al < Li < Mn^{2+} in the three Y coordinating sites for every O(1) anion site. (**a**) is a F_o Fourier synthesis; (**b**) is a F_o–F_c Fourier synthesis.

6. Discussion and Conclusions

The tourmalines discovered for the first time in the miarolitic LCT pegmatites at the western border of Adamello Massif may be considered gem materials due to their attractive pink to green-brown hue and transparency. Crystals providing the gem-quality requirements are available, despite the presence of cracks and voids well evidenced by synchrotron X-ray computed micro-tomography.

The studied tourmalines resulted principally in fluor-elbaite, along with minor fluor-liddicoatite, foitite and rossmanite that likely represent the final stages of tourmaline compositional evolution.

The chemical variations, reflecting the environment physicochemical changes during their growth, resulted both in a zoning from nearly black to green to blue to pink elbaite to rossmanite (sample 10) and in a reverse geochemical trend at the latest stages by a pink core, colourless to white midsection and green rim in sections cut perpendicularly to the *c*-axis (sample 4). In the non-homogeneous samples two different zones, possibly indicating the occurrence of a different and separated generations of tourmaline, can be distinguished: the first can be described as having high Ca + Mn or with a continuous fractionation trend of Mn vs. Fe, and the second one with poor Ca + Mn (sample 1) or with a marked enrichment in Fe (sample 4 and 10).

Mn and Fe are the main factors controlling the colour of the Li-rich tourmaline. Iron is always lower than manganese, increases in the darker zone and is virtually absent in the near colourless and pink tourmalines. One of the hypotheses on the intensifying pink hues in tourmaline is the presence of manganese in the 3+ oxidation state. Despite the uranium and thorium contents being very low (6.98 and 0.79 ppm), the presence, as in sample 9, of uraninite and its natural γ-radiation could be responsible for the oxidization change from Mn^{2+} to Mn^{3+}. From our results in tourmalines 1 and 4, it seems that a very low amount (<0.08 a.p.f.u.) of Mn may also be enough to give the bright pink colour. Cu^{2+} in combination with other cations may modify the resulting colour in tourmalines [1] and, interestingly, the pink parts of samples 1, 4, 9a, corresponding to Ti, Fe and Mn depleted points, contain a relatively high content of Cu.

The REE content is slightly higher than the literature data for tourmalines from relatively primitive NYF (Niobium, Yttrium, Fluorine) and mixed NYF-LCT pegmatites [30]. The chondrite normalized REE patterns are very similar and present a negative Eu^{2+} anomaly, probably due to the local depletion of Eu^{2+} in the melt.

The structural data confirm elbaitic compositions of the examined samples. The positional disorder found at the O(1) anion positions may be due to the high content of manganese entering into the Y sites together with cations of very different charge and radius (Li and Al).

The presence of gem-quality tourmalines in miarolitic cavities suggests the enrichment in volatiles and other exotic elements in pegmatite melt, as well as a shallow level formation in the thermo-metamorphic aureole of the Adamello pluton. The pegmatitic liquids could be exsolved from the granitoid magmas during the latest stages of crystallization of the pluton or, alternatively, generated by partial melting of the including metasedimentary sequence.

Further field exploration could turn up additional reserves of gem-quality tourmalines, and their study could contribute to a better understanding of the formation environment of the pegmatitic swarms at the border of Adamello Massif.

Supplementary Materials: The following are available online at http://www.mdpi.com/2075-163X/8/12/593/s1, Table S1: Atom coordinates of the three tourmaline crystals of this study, Table S2a: Anisotropic displacement parameters ($\text{Å}^2 \times 10^3$) for tourmaline 1. The anisotropic displacement factor exponent takes the form: $-2\pi^2[h^2a^{*2}U^{11} + \ldots + 2\,h\,k\,a^*\,b^*\,U^{12}]$, Table S2b: Anisotropic displacement parameters ($\text{Å}^2 \times 10^3$) for tourmaline 2. The anisotropic displacement factor exponent takes the form: $-2\pi^2[h^2a^{*2}U^{11} + \ldots + 2\,h\,k\,a^*\,b^*\,U^{12}]$, Table S2c: Anisotropic displacement parameters ($\text{Å}^2 \times 10^3$) for tourmaline 12. The anisotropic displacement factor exponent takes the form: $-2\pi^2[h^2a^{*2}U^{11} + \ldots + 2\,h\,k\,a^*\,b^*\,U^{12}]$.

Author Contributions: V.D. performed EPMA analyses; F.P. collected the research material; N.M. and G.L. performed the Synchrotron X-ray computed micro-tomography; F.C. performed the SC-XRD analysis; A.L. collected LA-ICP-MS data; I.A. characterized the gemmological materials; V.D., F.P., R.B., N.M., F.C. analysed the data and wrote the paper. All authors contributed to the final version and gave their approval for submission.

Funding: This research received no external funding.

Acknowledgments: We are thankful to Jan Cempirek for fruitful suggestions on electron microprobe data.

Conflicts of Interest: The authors declare no conflict of interest.

References

1. Pezzotta, F.; Laurs, B.M. Tourmaline: The kaleidoscopic gemstone. *Elements* **2011**, *7*, 333–338. [CrossRef]
2. Henry, D.J.; Novák, M.; Hawthorne, F.C.; Ertl, A.; Dutrow, B.L.; Uher, O.; Pezzotta, F. Nomenclature of the tourmaline-supergroup minerals. *Am. Mineral.* **2011**, *96*, 895–913. [CrossRef]
3. Pezzotta, F.; Guastoni, A. Adamello: La pegmatite LCT della valle Adamè. *Riv. Mineral. Ital.* **2002**, *3*, 127–142.
4. Mercurio, M.; Rossi, M.; Izzo, F.; Cappelletti, P.; Germinario, C.; Grifa, C.; Petrelli, M.; Vergara, A.; Langella, A. The characterization of natural gemstones using non-invasive FT-IR spectroscopy: New data on tourmalines. *Talanta* **2018**, *178*, 147–159. [CrossRef]
5. Van Hinsberg, V.J.; Henry, D.J.; Marschall, H.R. Tourmaline: An ideal indicator of its host environment. *Can. Mineral.* **2011**. [CrossRef]
6. Marinoni, N.; Voltolini, N.; Mancini, L.; Vignola, P.; Pagani, A.; Pavese, A. An investigation of mortars affected by alkali-silica reaction by X-ray synchrotron microtomography: A preliminary study. *J. Mater. Sci.* **2009**, *44*, 5815–5823. [CrossRef]
7. Marinoni, N.; Voltolini, N.; Broekmans, M.A.T.M.; Mancini, L.; Monteiro, P.J.M.; Rotiroti, N.; Ferrari, E.; Bernasconi, A. A combined synchrotron radiation micro computed tomography and micro X-ray diffraction study on deleterious alkali-silica reaction. *J. Mater. Sci.* **2015**, *50*, 7985–7997. [CrossRef]
8. Hawthorne, F.C.; Henry, D.J. Classification of the minerals of the tourmaline group. *Eur. J. Mineral.* **1999**, *11*, 201–215. [CrossRef]
9. Bosi, F.; Andreozzi, G.B.; Skogby, H.; Lussier, A.J.; Abdu, Y.; Hawthorne, F.C. Fluor-elbaite, $Na(Li_{1.5}Al_{1.5})Al_6(Si_6O_{18})(BO_3)_3(OH)_3F$, a new mineral species of the tourmaline supergroup. *Am. Mineral.* **2013**, *98*, 297–303. [CrossRef]
10. Bosi, F.; Skogby, H.; Ciriotti, M.E.; Gadas, P.; Novák, M.; Cempírek, J.; Všianský, D.; Filip, J. Lucchesiite, $CaFe_3^{2+}Al_6(Si_6O_{18})(BO_3)_3(OH)_3O$, a new mineral species of the tourmaline supergroup. *Miner. Mag.* **2017**. [CrossRef]
11. Bosi, F. Tourmaline crystal chemistry. *Am. Mineral.* **2018**, *103*, 298–306. [CrossRef]
12. Dupuy, C.; Dostal, J.; Fratta, M. Geochemistry of the Adamello Massif (Northern Italy). *Contrib. Mineral. Pet.* **1982**, *80*, 41–48. [CrossRef]
13. Zhang, C.; Gieré, R.; Stünitz, H.; Brack, P.; Ulmer, P. Garnet-quartz intergrowths in granitic pegmatites from Bergell and Adamello, Italy. *Schweiz. Mineral. Petrogr. Mitt.* **2001**, *81*, 89–113. [CrossRef]

14. Fiedrich, A.M.; Bachmann, O.; Ulmer, P.; Deering, C.D.; Kunze, K.; Leuthold, J. Mineralogical, geochemical, and textural indicators of crystal accumulation in the Adamello Batholith (Northern Italy). *Am. Mineral.* **2017**, *102*, 2467–2483. [CrossRef]

15. Brun, F.; Mancini, L.; Kasae, P.; Favretto, S.; Dreossi, D.; Tromba, G. Pore 3D: A software library for quantitative analysis of porous media. *Nucl. Inst. Methods Phys. Res. A* **2010**, *615*, 326–332. [CrossRef]

16. Brun, F.; Massimi, L.; Fratini, M.; Dreossi, D.; Billé, F.; Accardo, A.; Pugliese, R.; Cedola, A. SYRMEP Tomo Project: A graphical user interface for customizing CT reconstruction workflows. *Adv. Struct. Chem. Imag.* **2017**, *3*, 1–9. [CrossRef]

17. Herman, G.T. *Image Reconstruction from Projections*; Academic Press: New York, NY, USA, 1980.

18. Kak, A.C.; Slaney, M. *Principles of Computerized Tomographic Imaging*; IEEE, Institute for Electrical and Electronic Engineers Press: Piscataway, NY, USA, 1988.

19. Paganin, D.; Mayo, S.C.; Gureyev, T.E.; Miller, P.R.; Wilkins, S.W. Simultaneous phase and amplitude extraction from a single defocused image of a homogeneous object. *J. Microsc.* **2002**, *206*, 33–40. [CrossRef]

20. Agilent CrysAlis Computer Program, Agilent Technologies, XRD Products 2012. Available online: http://www.agilent.com/chem (accessed on 11 December 2018).

21. Sheldrick, G.M. Crystal structure refinement with SHELXL. *Acta Crystallogr.* **2015**, *C71*, 3–8.

22. O'Donoghue, M. *Gems*, 6th ed.; Butterworth-Heinemann: Oxford, UK, 2006.

23. Gübelin, E.J.; Koivula, J.I. *Photo Atlas of Inclusions in Gemstones*; Opinio Publishers: Basel, Switzerland, 2005; Volume 2.

24. Voltolini, M.; Marinoni, N.; Mancini, L. Synchrotron X-ray computed microtomography investigation of a mortar affected by alkali silica reaction: A quantitative characterization of its microstructural features. *J. Mater. Sci.* **2011**, *46*, 6633–6641. [CrossRef]

25. MacDonald, D.J.; Hawthorne, F.C.; Grice, J.D. Foitite, a new alkali deficient tourmaline: Description and crystal structure. *Am. Mineral.* **1993**, *78*, 1299–1303.

26. Brown, C.D.; Wise, M.A. Internal zonation and chemical evolution of the Black Mountain granitic pegmatite, Maine. *Can. Mineral.* **2001**, *39*, 45–55. [CrossRef]

27. Anders, E.; Grevesse, N. Abundances of the elements: Meteoritic and solar. *Geochim. Cosmochim. Acta* **1989**, *53*, 197–214. [CrossRef]

28. Bosi, F.; Andreozzi, G.B. A critical comment on Ertl et al. (2012): "Limitations of Fe^{2+} and Mn^{2+} site occupancy in tourmaline: Evidence from Fe^{2+}- and Mn^{2+}-rich tourmaline". *Am. Mineral.* **2013**, *98*, 2183–2192. [CrossRef]

29. Burns, P.C.; MacDonald, D.J.; Hawthorne, F.C. The crystal-chemistry of manganese-bearing elbaite. *Can. Mineral.* **1994**, *32*, 31–41.

30. Čopjaková, R.; Škoda, R.; Vašinová Galiová, M.; Novák, M. Distributions of Y + REE and Sc in tourmaline and their implications for the melt evolution; examples from NYF pegmatites of the Třebíč Pluton, Moldanubian Zone, Czech Republic. *J. Geosci.* **2013**, *58*, 113–131. [CrossRef]

Review

Emerald Deposits: A Review and Enhanced Classification

Gaston Giuliani [1,*], Lee A. Groat [2], Dan Marshall [3], Anthony E. Fallick [4] and Yannick Branquet [5]

[1] Université Paul Sabatier, GET/IRD et Université de Lorraine, CRPG/CNRS, 15 rue Notre-Dame des Pauvres, BP 20, 54501 Vandœuvre cedex, France

[2] Department of Earth, Ocean and Atmospheric Sciences, University of British Columbia, Vancouver, BC V6T 1Z4, Canada; groat@mail.ubc.ca

[3] Department of Earth Sciences, Simon Fraser University, Burnaby, BC V5A 1S6, Canada; marshall@sfu.ca

[4] Isotope Geosciences Unit, S.U.E.R.C., Rankine Avenue, East Kilbride, Glasgow G75 0QF, Scotland, UK; fallickt@gmail.com

[5] Institut des Sciences de la Terre d'Orléans (ISTO), UMR 7327-CNRS/Université d'Orléans/BRGM, 45071 Orléans, France; yannick.branquet@univ-orleans.fr

[*] Correspondence: giuliani@crpg.cnrs-nancy.fr; Tel.: +33-3-83594238

Received: 12 January 2019; Accepted: 9 February 2019; Published: 13 February 2019

Abstract: Although emerald deposits are relatively rare, they can be formed in several different, but specific geologic settings and the classification systems and models currently used to describe emerald precipitation and predict its occurrence are too restrictive, leading to confusion as to the exact mode of formation for some emerald deposits. Generally speaking, emerald is beryl with sufficient concentrations of the chromophores, chromium and vanadium, to result in green and sometimes bluish green or yellowish green crystals. The limiting factor in the formation of emerald is geological conditions resulting in an environment rich in both beryllium and chromium or vanadium. Historically, emerald deposits have been classified into three broad types. The first and most abundant deposit type, in terms of production, is the desilicated pegmatite related type that formed via the interaction of metasomatic fluids with beryllium-rich pegmatites, or similar granitic bodies, that intruded into chromium- or vanadium-rich rocks, such as ultramafic and volcanic rocks, or shales derived from those rocks. A second deposit type, accounting for most of the emerald of gem quality, is the sedimentary type, which generally involves the interaction, along faults and fractures, of upper level crustal brines rich in Be from evaporite interaction with shales and other Cr- and/or V-bearing sedimentary rocks. The third, and comparatively most rare, deposit type is the metamorphic-metasomatic deposit. In this deposit model, deeper crustal fluids circulate along faults or shear zones and interact with metamorphosed shales, carbonates, and ultramafic rocks, and Be and Cr ($\pm$V) may either be transported to the deposition site via the fluids or already be present in the host metamorphic rocks intersected by the faults or shear zones. All three emerald deposit models require some level of tectonic activity and often continued tectonic activity can result in the metamorphism of an existing sedimentary or magmatic type deposit. In the extreme, at deeper crustal levels, high-grade metamorphism can result in the partial melting of metamorphic rocks, blurring the distinction between metamorphic and magmatic deposit types. In the present paper, we propose an enhanced classification for emerald deposits based on the geological environment, i.e., magmatic or metamorphic; host-rocks type, i.e., mafic-ultramafic rocks, sedimentary rocks, and granitoids; degree of metamorphism; styles of minerlization, i.e., veins, pods, metasomatites, shear zone; type of fluids and their temperature, pressure, composition. The new classification accounts for multi-stage formation of the deposits and ages of formation, as well as probable remobilization of previous beryllium mineralization, such as pegmatite intrusions in mafic-ultramafic rocks. Such new considerations use the concept of genetic models based on studies employing chemical, geochemical, radiogenic, and stable isotope, and fluid and solid inclusion fingerprints. The emerald occurrences and deposits are classified into two main types: (Type I)

Tectonic magmatic-related with sub-types hosted in: (IA) Mafic-ultramafic rocks (Brazil, Zambia, Russia, and others); (IB) Sedimentary rocks (China, Canada, Norway, Kazakhstan, Australia); (IC) Granitic rocks (Nigeria). (Type II) Tectonic metamorphic-related with sub-types hosted in: (IIA) Mafic-ultramafic rocks (Brazil, Austria); (IIB) Sedimentary rocks-black shale (Colombia, Canada, USA); (IIC) Metamorphic rocks (China, Afghanistan, USA); (IID) Metamorphosed and remobilized either type I deposits or hidden granitic intrusion-related (Austria, Egypt, Australia, Pakistan), and some unclassified deposits.

Keywords: emerald deposits; classification; typology; metamorphism; magmatism; sedimentary; alkaline metasomatism; fluids; stable and radiogenic isotopes; genetic models; exploration

---

1. Introduction

Emerald is the green gem variety of the mineral beryl, which has the ideal formula of $Be_3Al_2SiO_{18}$. It is considered one of the so-called precious gems and in general the most valuable after diamond and ruby. The color of emerald is of greater importance than its clarity and brilliance for its valuation on the colored gem market. In the Munsell color chart, emerald exhibits a green color palette that is the consequence of peculiarities of its formation in contrasting environments (Figure 1). The pricing of emerald is unique in terms of the color and weight in carats. In 2000, an exceptional 10.11 ct Colombian cut gem was sold for US$1,149,850 [1]. In October 2017, Gemfield's auction of Zambian emeralds generated revenues of US$21.5 million and companies placed bids with an average value of $66.21 per carat [2]. This auction included the 6100 ct. *Insufu* rough emerald extracted from the Kagem mine in 2015. Lastly, on the 2th of October, the remarkable emerald called *Inkalamu* ("the lion elephant") was extracted from the same mine [3]. Other giant crystals have been discovered in Colombia, such as *el Monstro* (16,020 ct) and *Emilia* (7025 ct), both from the Gachalá mines. In 2017, a large piece of biotite schist with several emerald crystals was discovered in the Carnaíba mine, Brazil; the specimen weighed 341 kg with 1.7 million ct of emerald, of which 180,000 ct were of gem quality. The specimen has been valued at approximately US$309 million [4].

The present article assesses the state of our knowledge of emerald then and now, through several questions regarding their locations on the planet; their crystal chemistry; pressure-temperature conditions of crystallization; the source of the constituent elements, i.e., beryllium (Be), chromium (Cr,) and vanadium (V); and their age of formation; and also proposes a new classification scheme.

Exploration beyond the 21st century may require a comprehensive data base of the typology of emerald deposits to understand why some emerald occurrences are economic in terms of quantity and quality and most are not. These future efforts will improve exploration guidelines in the field, including plate tectonics and its consequences in terms of modeling our landscape through time and within the Wilson cycle of continents.

Figure 1. Emeralds' worldwide photographs: (**a**) Emerald crystals on quartz and adularia, Panjshir Valley, Afghanistan, 6.6 × 4.4 cm. Specimen: Fine Art Mineral. Photograph: Louis-Dominique Bayle, le Règne Minéral; (**b**) emerald on pyrite, Chivor, Colombia, 3.9 × 2.6 cm. Collection MulitAxes. Photograph: Louis-Dominique Bayle, le Règne Minéral; (**c**) Emerald in quartz, Kagem mine, Zambia, longest crystal: 7.9 × 1.2 × 1.2 cm. The Collector's Edge. Photograph: Louis-Dominique Bayle, le Règne Minéral; (**d**) emerald in quartz vein and minor potassic feldspar, Dyakou, China, longest crystal: 1.5 cm. Specimen DYKO6-zh. Photograph: Dan Marshall; (**e**) emerald in plagioclase, Carnaíba, Brazil, longest crystal: 6 cm. Photograph: Gaston Giuliani.

2. Worldwide Emerald Deposits

Emerald is rare, but it is found on all five continents (Figure 2). Colombia, Brazil, Zambia, Russia, Zimbabwe, Madagascar, Pakistan, and Afghanistan (Figure 3) are the largest producers of emerald [5]. Emerald deposits occur mainly in the Precambrian series in Eastern Brazil, Eastern Africa, South Africa, Madagascar, India, and Australia, and younger volcano-sedimentary series or ophiolites in Bulgaria, Canada, China, India, Pakistan, Russia, and Spain. Colombian emerald deposits, which produce most of the world's high-quality emeralds, are unique in that they are located in sedimentary rocks, i.e., the Lower Cretaceous black shales (BS) of the Eastern Cordillera basin. Other deposits are hosted in Alpine-type veins, also called Alpine-type clefts. Emerald is found in veins and cavities in the European Alps (Binntal) as well as in the United States (Hiddenite). Nigerian emeralds are unique and are located in pegmatitic pods.

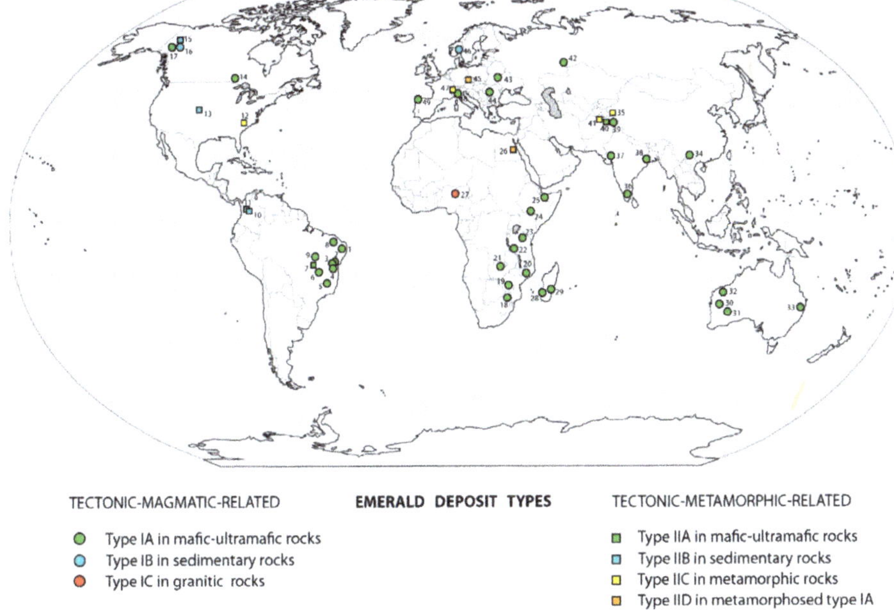

Figure 2. The location of emerald deposits and occurrences worldwide reported following their geological types, i.e., tectonic magmatic-related (type I) and tectonic metamorphic-related (type II): *Brazil*: 1. Fazenda Bonfim; 2. Socotó; 3. Carnaíba; 4. Anagé, Brumado; 5. Piteiras, Belmont mine, Capoierana, Santana dos Ferros; 6. Pirenópolis, Itaberai; 7. Santa Terezinha de Goiás; 8. Tauá, Coqui; 9. Monte Santo. *Colombia*: 10. Eastern emerald zone (Gachalá, Chivor, Macanal); 11. Western emerald zone (Yacopí, Muzo, Coscuez, Maripi, Cunas, La Pita, La Marina, Peñas Blancas). *United States*: 12. Hiddenite; 13. Uinta. *Canada*: 14. Dryden; 15. Mountain River; 16: Lened; 17. Tsa da Gliza. *South Africa*: 18. Gravelotte. *Zimbabwe*: 19. Sandawana, Masvingo, Filibusi. *Mozambique*: 20. Morrua. *Zambia*: 21. Kafubu, Musakashi. *Tanzania*: 22. Sumbawanga; 23. Manyara. *Ethiopia*: 24. Kenticha (Halo-Shakiso). *Somalia*: 25. Boorama. *Egypt*: 26. Gebels Sikaït, Zabara, Wadi Umm Kabu. Nigeria: 27. Kaduna. *Madagascar*: 28. Ianapera; 29. Mananjary. *Australia*: 30. Poona; 31. Menzies; 32. Wodgina; 33. Emmaville, Torrington. *China*: 34. Dyaku; 35. Davdar. *India*: 36. Sankari Taluka; 37. Rajasthan (Bubani, Rajgarh, Kaliguman); 38: Gubaranda (Orissa state). *Pakistan*: 39: Khaltaro; 40. Swat valley. *Afghanistan*: 41. Panjshir valley. Russia: 42. Urals (Malyshevo). *Ukraine*: 43. Wolodarsk. *Bulgaria*: 44. Rila. *Austria*: 45. Habachtal. *Norway*: 46. Eidswoll. *Switzerland*: 47. Binntal. *Italia*: 48. Val Vigezzo. *Spain*: 49. Franqueira.

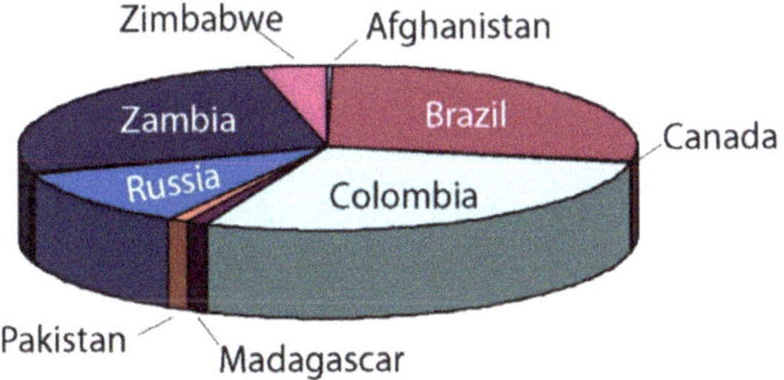

Figure 3. Emerald production worldwide in 2005.

3. Crystal Chemistry of Emerald

Beryl is hexagonal and crystallizes in point group $6/m\,2/m\,2/m$ and space group $P6/m2/c2/c$. It is a cyclosilicate mineral. The crystal structure, as shown in Figure 4, is characterized by six-membered rings of silica tetrahedra lying in planes parallel to (0001). The Al or Y site is surrounded by six O atoms in octahedral coordination, and both the Be and silica (Si) sites are surrounded by four O atoms in tetrahedral coordination. The SiO_4 tetrahedra polymerize to form six-membered rings parallel to (001); stacking of the rings results in large channels parallel to c. The channels are not uniform in diameter; in fact, they consist of cavities with a diameter of approximately 5.1 Å separated by bottlenecks with a diameter of approximately 2.8 Å. The channels can be occupied by alkali ions (such as Na^+, Li^+, K^+, Rb^+, Cs^+, etc.) whose presence is required to balance reductions in positive charges when cation substitutions occur in the structure. Neutral H_2O and CO_2 molecules [6] and noble gases, such as argon, helium [7], xenon, and neon [8], are generally also present in variable amounts in the channels.

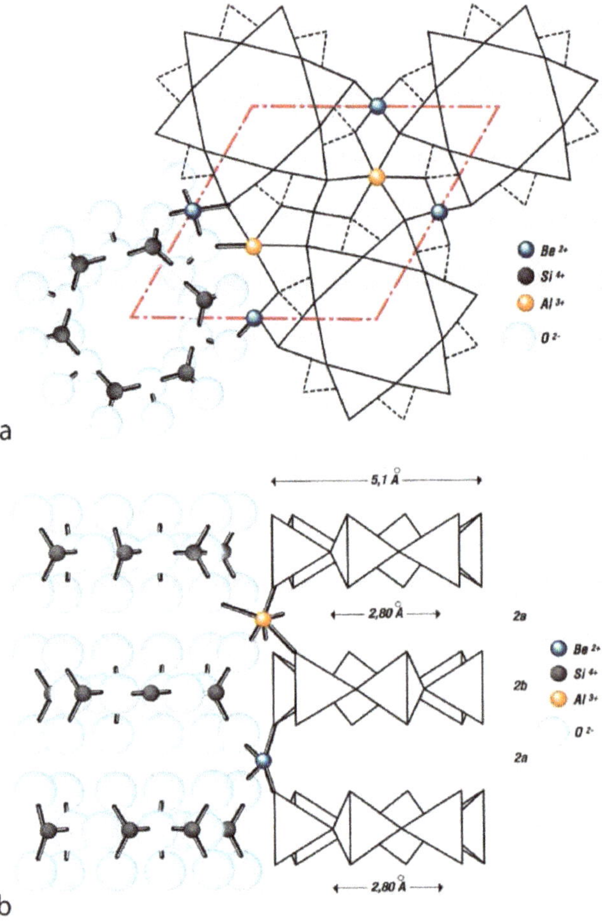

Figure 4. Structure of beryl [9] in: (**a**) Apical view: Hexagonal silicate rings stacked parallel to the *c* axis (normal to the drawing) are held together by Al^{3+} (octahedral site) and Be^{2+} (tetrahedral site). The radii of the ions are respected in the drawing; (**b**) lateral view perpendicular to the *c* axis showing the hexagonal silicate rings and the bottleneck (2*b* site) and the open cage (2*a* site) structure.

Emerald was defined by [10] as "the yellowish green, green or bluish green beryl which reveals distinct Cr and/or V absorption bands in the red and blue violet ranges of their absorption spectra". Quantitatively, Cr and V substitutions range between 25 [6] and 34,000 ppm [11] for Cr and from 34 ppm [12] to 10,000 ppm [13] for V. Emerald crystals typically exhibit a prismatic habit (Figure 5) characterized by eight faces and their corresponding growth sectors: Six {10$\bar{1}$0} first order prismatic faces and two pinacoidal {0001} faces. Small additional {10$\bar{1}$2} and {11$\bar{2}$2}faces can also be present.

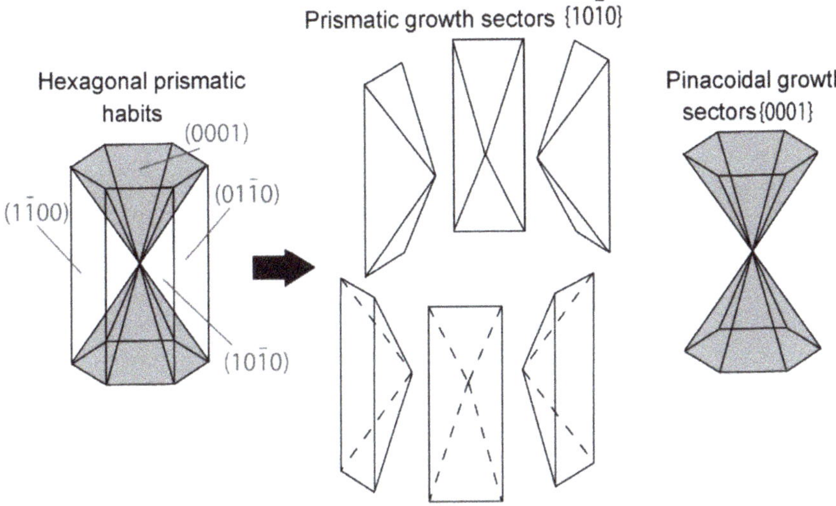

Figure 5. Habits of emerald crystals, characterized by eight main faces and their corresponding growth sectors: Six {10$\bar{1}$0} first order prismatic faces and two pinacoidal {0001} faces.

Representative emerald compositions from the literature are listed in Table 1. Most substitutions occur at the Y site. Figure 6 shows Al versus the sum of other Y-site cations for 499 emerald compositions from the literature; as expected, they show an inverse relationship. Figure 7 shows a slight deviation from a 1:1 correlation between Mg + Mn + Fe and the sum of monovalent cations. This graph suggests that, to achieve charge balance, the substitution of divalent cations for Al at the Y site is coupled with the substitution of a monovalent cation for a vacancy at a channel site. There are two sites in the channels; these are referred to as the 2*a* (at 0,0,0.25) and 2*b* (at 0,0,0) positions. Artioli et al. [14] suggested that in alkali- and water-rich beryls, H_2O molecules and the larger alkali atoms (Cs, Rb, K) occupy the 2*a* sites and Na atoms occupy the smaller 2*b* positions, but in alkali- and water-poor beryl, both Na atoms and H_2O molecules occur at the 2*a* site and the 2*b* site is empty. The amount of water in beryl can be difficult to determine, but Giuliani et al. [15] derived the following equation from existing experimental data for emerald:

$$H_2O \text{ (in wt.\%)} = [0.84958 \times Na_2O \text{ (in wt.\%)}] + 0.8373. \tag{1}$$

This equation has been improved [16] and is best defined by the relationship:

$$H_2O = 0.5401 \ln Na_2O + 2.1867a. \tag{2}$$

Table 1. Representative composition with the structural formulas of emerald samples from the recent literature (in wt.%) [1].

Elementss	1	2	3	4	5	6	7	8	9	10	11	12	13	14	15	16
SiO_2	64.50	64.94	63.58	65.98	67.13	66.74	66.30	66.04	64.71	64.75	65.57	65.57	66.49	66.28	67.13	64.39
TiO_2	0.00	0.00	n.d.	0.01	0.00	0.00	0.01	0.01	n.d.	n.d.	0.04	0.01	0.00	0.06	0.01	0.01
Al_2O_3	11.54	10.61	17.36	16.81	18.77	18.67	18.60	18.17	17.23	14.83	17.24	14.51	18.29	18.10	16.52	14.48
Sc_2O_3	n.d.	n.d.	n.d.	0.06	0.01	0.00	0.05	0.01	0.03	n.d.	n.d.	n.d.	n.d.	n.d.	n.d.	0.02
V_2O_3	0.15	0.10	0.02	0.17	0.05	0.01	0.10	0.01	0.23	0.03	0.07	0.00	0.17	0.05	0.00	0.03
Cr_2O_3	1.83	3.39	0.25	0.35	0.20	0.26	0.02	0.05	0.01	0.57	0.07	0.09	0.05	0.24	0.37	0.17
Fe_2O_3	1.34	1.27	n.d.	n.d.	n.d.	n.d.	n.d.	n.d.	n.d.	n.d.	n.d.	n.d.	n.d.	n.d.	n.d.	n.d.
La_2O_3	0.06	0.12	n.d.	n.d.	n.d.	n.d.	n.d.	n.d.	n.d.	n.d.	n.d.	n.d.	n.d.	n.d.	n.d.	n.d.
Ce_2O_3	0.02	0.00	n.d.	0.02	n.d.	n.d.	n.d.	n.d.	n.d.	n.d.	n.d.	n.d.	n.d.	n.d.	n.d.	n.d.
BeO	13.40	13.45	12.85	13.65	13.99	13.91	13.82	13.73	13.54	13.50	13.47	13.59	13.83	13.78	13.80	13.37
MgO	3.49	3.36	0.58	0.95	0.04	0.03	0.06	0.15	0.88	2.24	0.87	2.63	0.05	0.16	0.87	2.38
CaO	0.21	0.17	n.d.	0.00	0.00	0.00	0.01	0.01	0.02	0.02	0.00	0.06	0.00	0.00	0.00	0.04
MnO	0.10	0.00	n.d.	0.03	0.00	0.01	0.01	0.00	n.d.	0.01	0.00	0.01	0.00	0.00	0.01	0.01
FeO	n.d.	n.d.	0.37	0.17	0.21	0.19	0.12	0.15	0.26	0.65	0.33	0.96	0.71	0.40	0.14	0.63
Li_2O	n.d.	n.d.	0.28	0.03	n.d.	n.d.	n.d.	n.d.	n.d.	n.d.	0.11	0.02	0.01	0.01	0.01	n.d.
Na_2O	0.83	0.89	1.13	0.62	0.05	0.06	0.08	0.27	0.82	1.76	0.82	1.84	0.04	0.10	0.70	1.86
K_2O	2.04	1.88	0.05	0.03	0.01	0.02	0.00	0.01	0.02	0.03	0.00	0.24	0.02	0.05	0.00	0.04
Rb_2O	n.d.	n.d.	n.d.	n.d.	n.d.	n.d.	n.d.	0.00	n.d.	0.14	n.d.	n.d.	n.d.	n.d.	n.d.	n.d.
Cs_2O	0.02	0.02	0.13	n.d.	n.d.	n.d.	0.01	0.11	0.10	0.03	n.d.	n.d.	n.d.	n.d.	n.d.	n.d.
H_2O	2.09	2.12	2.25	1.93	0.57	0.67	0.82	1.48	2.08	2.49	2.08	2.52	0.45	0.94	1.99	2.52
Total	101.62	102.32	98.85	100.81	101.04	100.57	100.01	100.20	99.94	101.05	100.67	102.05	100.11	100.17	101.55	99.96
apfu																
Si^{4+}	6.012	6.030	5.963	6.015	5.992	5.991	5.991	6.005	5.970	5.991	5.998	6.012	5.998	6.001	6.066	6.013
Ti^{4+}	0.000	0.000	0.000	0.001	0.000	0.000	0.001	0.001	0.000	0.000	0.003	0.001	0.000	0.004	0.001	0.001
Al^{3+}	1.268	1.161	1.919	1.806	1.975	1.975	1.981	1.947	1.873	1.617	1.858	1.568	1.945	1.931	1.759	1.594
Sc^{3+}	0.000	0.000	0.000	0.005	0.001	0.000	0.004	0.001	0.002	0.000	0.000	0.000	0.000	0.000	0.000	0.002
V^{3+}	0.011	0.007	0.002	0.012	0.004	0.001	0.007	0.001	0.017	0.002	0.005	0.000	0.012	0.004	0.000	0.002
Cr^{3+}	0.135	0.249	0.019	0.025	0.014	0.018	0.001	0.004	0.001	0.042	0.005	0.007	0.004	0.017	0.026	0.013
Fe^{3+}	0.094	0.089	0.000	0.000	0.000	0.000	0.000	0.000	0.000	0.000	0.000	0.000	0.000	0.000	0.000	0.000
La^{3+}	0.002	0.004	0.000	0.000	0.000	0.000	0.000	0.000	0.000	0.000	0.000	0.000	0.000	0.000	0.000	0.000
Ce^{3+}	0.001	0.000	0.000	0.001	0.000	0.000	0.000	0.000	0.000	0.000	0.000	0.000	0.000	0.000	0.000	0.000

Table 1. *Cont.*

Elementss	1	2	3	4	5	6	7	8	9	10	11	12	13	14	15	16
Be^{2+}	3.000	3.000	2.894	2.989	3.000	3.000	3.000	3.000	3.000	3.000	2.960	2.993	2.996	2.996	2.996	3.000
Mg^{2+}	0.485	0.465	0.081	0.129	0.005	0.004	0.008	0.020	0.121	0.309	0.119	0.359	0.007	0.022	0.117	0.331
Ca^{2+}	0.021	0.017	0.000	0.000	0.000	0.000	0.001	0.001	0.002	0.002	0.000	0.006	0.000	0.000	0.000	0.004
Mn^{2+}	0.008	0.000	0.000	0.002	0.000	0.001	0.001	0.000	0.000	0.001	0.000	0.001	0.000	0.000	0.001	0.001
Fe^{2+}	0.000	0.000	0.029	0.013	0.016	0.014	0.009	0.011	0.020	0.050	0.025	0.074	0.054	0.030	0.011	0.049
Li^{+}	0.000	0.000	0.106	0.011	0.000	0.000	0.000	0.000	0.000	0.000	0.040	0.007	0.004	0.004	0.004	0.000
Na^{+}	0.150	0.160	0.205	0.110	0.009	0.010	0.014	0.048	0.147	0.316	0.145	0.327	0.007	0.018	0.123	0.337
K^{+}	0.243	0.223	0.006	0.003	0.001	0.002	0.000	0.001	0.002	0.004	0.000	0.028	0.002	0.006	0.000	0.005
Rb^{+}	0.000	0.000	0.000	0.000	0.000	0.000	0.000	0.000	0.000	0.008	0.000	0.000	0.000	0.000	0.000	0.000
Cs^{+}	0.001	0.001	0.005	0.000	0.000	0.000	0.000	0.004	0.004	0.001	0.000	0.000	0.000	0.000	0.000	0.000

[1] Compositions renormalized on the basis of 3 (Be + Li) and 18 O per formula unit. H_2O = 0.5401 ln Na_2O + 2.1867 [16]. **1**. Ianapera, Madagascar. Ultramafic host, Type-3 core [11]. **2**. Ianapera, Madagascar. Ultramafic host, Type-3 rim [11]. **3**. Taylor 2, Canada. Average of 51 analyses with Li = (Na + K + Cs) - (Mg + Fe), which assumes all Fe as Fe^{2+} [17]. **4**. Davdar, China. Average of 48 analyses; Ce_2O_3 by neutron activation, LiO 0.03 wt.% by fusion with Na_2O_2 [18]. **5**. Emmaville-Torrington, Australia. "Line 1"; average of 27 analyses [19]. **6**. Emmaville-Torrington, Australia. "Line 2"; average of 31 analyses [19]. **7**. Byrud, Norway. Average of 38 analyses [20]. **8**. Poona, Australia. Average of 37 analyses [16]. **9**. Lened, Canada. Average of 88 analyses [21]. **10**. Fazenda Bonfim, Brazil. Average of approximately 130 analyses [22]. **11**. Malyshevsk, Russia. Average of five analyses [23]. **12**. Alto Ligonha, Mozambique. Average of three analyses [23]. **13**. Jos, Nigeria. Average of five analyses [23]. **14**. Sumbawanga, Tanzania. Average of five analyses [23]. **15**. Muzo, Colombia. Average of three analyses [23]. **16**. Ethiopia [this study, average of 32 analyses].

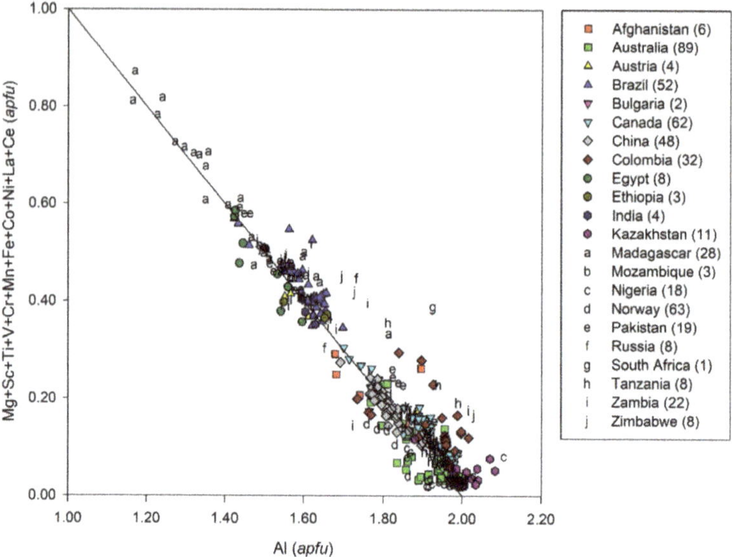

Figure 6. Al versus the sum of other *Y*-site cations, in atoms per formula unit, for 499 emerald analyses from the literature. The number of analyses per country is given in parentheses in the legend. Sources of data: [11–15], [16] (average of 37 analyses), [17–20], [21] (average of 88 analyses), [22] (average of approximately 130 analyses), [23–43], [44] (Kazakhstan values are averages of 11 analyses), [45] (average of 10 analyses), [46], [47] (two averages of five analyses each), [48–51], [52] (average of 55 analyses), and this study.

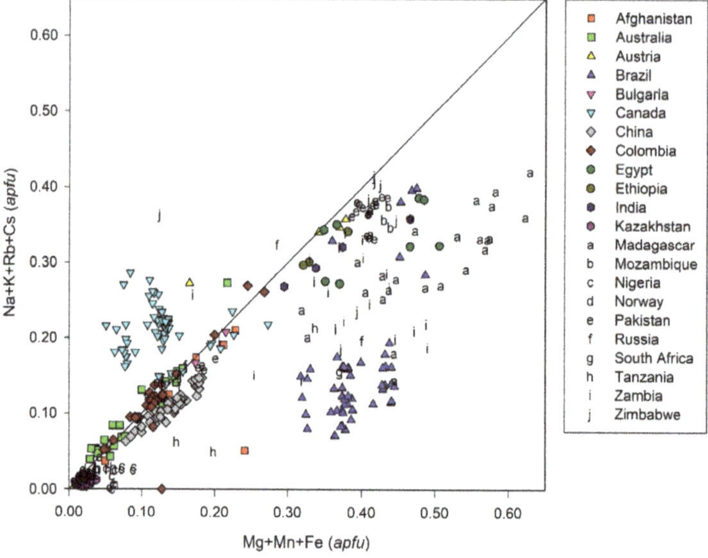

Figure 7. Mg + Mn + Fe vs. monovalent channel-site cations, in atoms per formula unit, for analyses from the literature. Sources of data are the same as in Figure 6.

Unfortunately, it is difficult to obtain accurate analyses of Be, Li, and ferric-ferrous ratios in beryl. Both Be and Li are too light to measure with the electron microprobe, which cannot distinguish between

Fe^{3+} and Fe^{2+}. Thus, most published analyses of beryl are renormalized on the basis of 18 O and 3 Be atoms per formula unit and Fe is generally reported as exclusively ferric or ferrous. Points that lie to the right of the 1:1 line in Figure 6 indicate that some of the Fe in a given emerald is present as Fe^{3+}. Likewise, points that lie above the line could suggest the presence of Li$^+$, which may substitute for Be^{2+} at the Be site. Charge balance is maintained by adding a monovalent cation to a channel site. Beryllium can be analyzed with LA-ICP-MS (Laser Ablation–Inductively Coupled Plasma–Mass Spectrometry), SIMS (Secondary Ionisation Mass Spectrometry, or "ion microprobe" [23]), or by using wet chemical techniques. However, there is so much Be in the structure that the accuracy of such analyses is suspect. Lithium may be analyzed by the same techniques as Be, but because the concentrations are much lower, the accuracy would presumably be better. The amount of Li can also be estimated in Fe-free beryl from the number of monovalent cations at the channel sites.

The main substituents for Al at the Y site are plotted as oxides in Figure 8a. Magnesium is the main substituent in emeralds from most localities. The elements responsible for most of the variation in color in emerald crystals are plotted as oxides in Figure 8b. In most cases, the Cr$_2$O$_3$ content is much greater than that of V$_2$O$_3$; the main exceptions are for samples from the Lened occurrence in Canada [21,47], the Davdar occurrence in China [18], the Muzo mine in Colombia [1], the Mohmand district in Pakistan [1], and Eidswoll in Norway [13]. The accuracy of data obtained by LA-ICP-MS is primarily dependent on the standards used for the analysis. Currently, the NIST 610 and 612 glasses are used for calibration standards, but it is likely that emerald standards are necessary to obtain accurate data for trace elements. Beryl has a wide stability field; the lower limit in the presence of water is between 200 °C and 350 °C, depending on the pressure and coexisting minerals [53]. However, the high-temperature stability and melting relationships remain unclear, partly because beryl may contain significant amounts of H$_2$O at the channel sites, and water has a significant effect on stability [52]. The effect of other channel constituents, such as Na, may be similar. Although thermodynamic data exist for beryl, the lack of experimental data for anything more complex than the BASH (BeO-Al$_2$O$_3$-SiO$_2$-H$_2$O) system can be a barrier to understanding the formation of natural occurrences, such as those where Be-bearing minerals occur in metamorphic rocks [52].

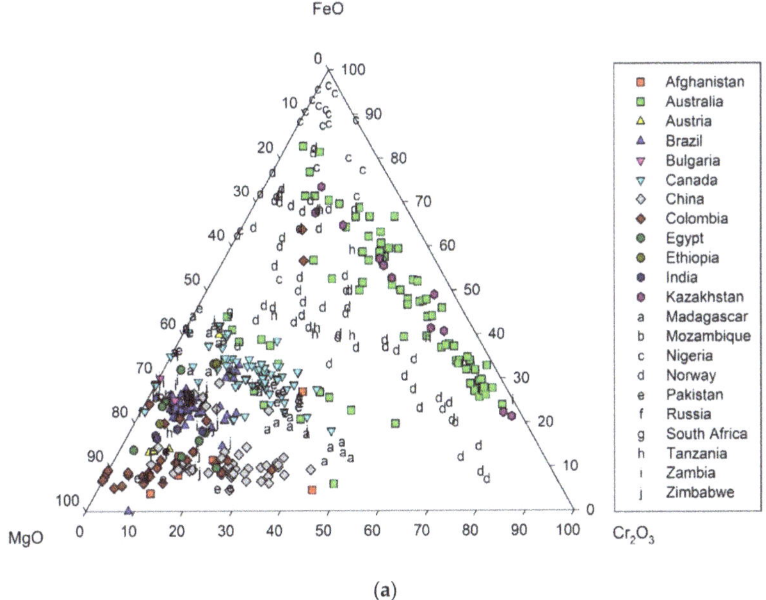

(**a**)

Figure 8. *Cont.*

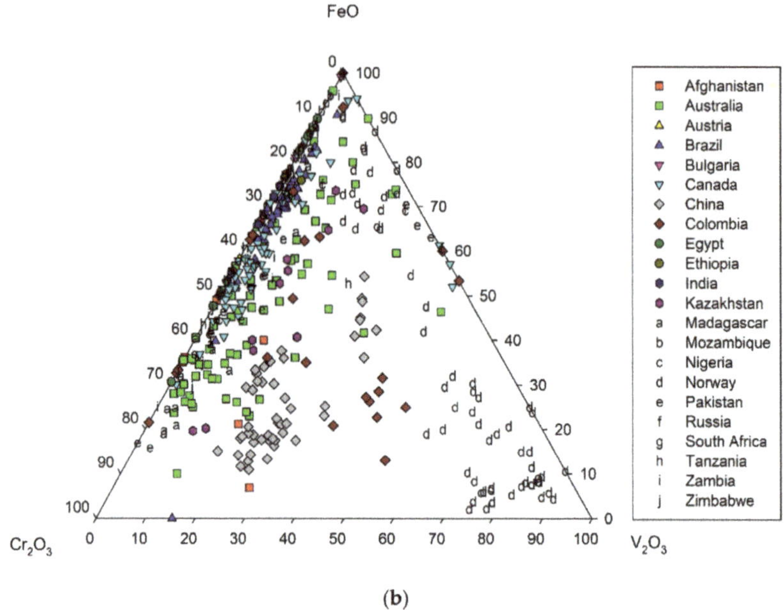

(**b**)

Figure 8. (**a**) Plot of emerald compositions in terms of FeO-MgO-Cr$_2$O$_3$ (wt.%). Data from the literature (with all Fe as FeO). Sources of data are the same as in Figure 6. The diagram is after [39]. (**b**) Plot of emerald compositions in terms of FeO-Cr$_2$O$_3$-V$_2$O$_3$ (wt.%). Data from the literature (with all Fe as FeO). Sources of data are the same as in Figure 6. The diagram is after [39].

4. Sources of Be, Cr, and V: The Formation of Emerald

The sources of Be, Cr, and V of emerald. The Cr-V-bearing beryl is rare because (i) its constituent metals have opposite affinities and behavior in the continental and oceanic lithospheres, and (ii) there is only 2 ppm of Be in the upper continental crust. Beryllium is a lithophile element, which has a strong affinity for oxygen producing beryl and chrysoberyl (BeAlO$_4$) at higher temperatures and at medium to low temperatures joins with silica to form silicate minerals, such as beryl, bertrandite [Be$_4$Si$_2$O$_7$(OH)$_2$], and/or Be-bearing micas. These minerals have a relatively low-density (ρ_{beryl} = 2.76) and are concentrated in the crust (ρ_{crust} = 2.7). Chromium and V are high-density transition metals that generally concentrate in the core and the mantle of the earth. Although Cr and V show both lithophile and siderophile affinities, they tend to form solid solutions with iron (Fe) to form, e.g., chromite (FeCr$_2$O$_4$) at high temperatures. Progressive changes in tectonic processes since the Archean preserved metamorphosed mafic-ultramafic rocks (M-UMR) and ancient mantle-related terranes in the upper continental crust. Consequently, Cr and V are more common than Be in the upper continental crust, with concentrations of ~100 ppm each.

Beryllium is present in crustal granites and associated dyke swarms of pegmatites, aplites, and quartz veins. The felsic rocks (FELSR), with more than 70% silica (SiO$_2$), are intrusive into the continental lithosphere. The highest Be concentration occurs in granites and rhyolites (~5 ppm). The two-mica granites have concentrations between 5 and 10 ppm, while specialized granites have more than 200 ppm. Beryllium is carried by Be-minerals, but also by feldspar and micas. Chromium and V are more highly concentrated in M-UMR of the oceanic lithosphere that contain less than 50% SiO$_2$ (Figure 9a). These rocks, which are generally termed peridotites, are often metamorphosed into serpentinites containing up to 14 wt.% H$_2$O (Figure 9a,b), or into talc-chlorite-carbonate schists. Chromium is dominantly sourced from chromite and V is sourced from V-bearing magnetite or coulsonite (Fe^{2+}V^{3+}$_2$O$_4$). Nevertheless, Cr-V and Be can be present in sedimentary rocks and are highly concentrated in shale (Figure 9c). Indeed, the weathering and erosion of M-UMR and FELSR on the continents delivers to the sea fine-grained clastic sediments composed of mud with Cr-V-Be-bearing grains and organic matter, as well as tiny fragments of quartz and calcite.

Figure 9. Potential geological sources of chromium (Cr) and vanadium (V) necessary for emerald formation in mafic-ultramafic rocks and black shales: (**a**) A lherzolite (Lrz) that is transformed into serpentinite (Srp), region of Montemaggiore on the Corsican cape, France; (**b**) Photomicrograph (plane polarized light) of an antigorite vein (V2) cross cutting lizardite veins (V1) and olivine (Ol). Fine-grained magnetite (Mag) is present in the centre of V1; V2 boundaries contain minute secondary olivine grains. Photo: B. Debret; (**c**) Colombian black shale crosscut by pyrite (Py) -bearing calcite (Ca) veinlets from the Coscuez mine. Photograph: G. Giuliani; (**d**) Pie diagram representation of the concentration (in %) and relative distribution of beryllium (Be) between the different phases extracted from a black shale containing 4 ppm of ^{9}Be, Coscuez mine [54]. OM: Organic matter; [Fe,Mn(O,OH)] = oxy-hydroxides of iron and manganese.

In anoxic and reducing environments, the un-oxidized organic matter imparts a dark colour to the black shale (BS). The Be contents of BS for Colombian deposits vary between 2 and 6 ppm, while Cr-V concentrations range between 100 and 1000 ppm. At the Coscuez emerald deposit (Figure 9d), the Be content of the BS is around 5 ppm. Beryllium mobility is principally associated with the breakdown of iron-manganese oxy-hydroxide phases. The amount of Be that can be mobilized (~0.7 ppm) may represent up to 18 wt.% of the total Be contained in the BS [54].

Emerald formation requires Be and Cr-V inter-reservoir circulation of fluids to mobilize these elements. Metasomatism and fluid/rock interaction is the main mechanism for element mobilization in sedimentary (black shales) or granitic rocks for Be and metamorphosed-mafic-ultramafic rocks for Cr and V. These mineralizing fluids are trapped by emerald within the large channels parallel to the *c* axis (molecular components) or in fluid inclusions (several constituents in different systems, i.e., H_2O-$NaCl$-CO_2-($\pm H_2S$)-($\pm N_2$)-($\pm CH_4$)-(K, Mg, Be, F, B, Li, SO_4, P, Cs)) within primary growth planes and secondary trails of fractures. Fluid inclusions are important fingerprints for emerald, and microthermometry, Raman spectrometry, and mass spectrometry for O-H isotope signatures make it possible to determine the nature and origin of the fluids [16,55].

5. Classification of Emerald Deposits

5.1. Genetic Classifications

The genetic classification schemes developed for emerald deposits in the 21st century were reviewed by [1]. The classification schemes are ambiguous and not particularly useful when it comes to understanding the mechanisms and conditions that led to the formation of an emerald deposit [52]. There is no ideal combination of mineralogical, chemical, geochemical, and physical parameters or combinations of these data with the age of the formation of the deposits and O-isotope composition of emerald [56].

Dereppe et al. [57] used artificial neural networks (ANN) to classify emerald deposits based on 450 electron microprobe analyses of emeralds from around the world. They defined five categories of deposits with "bad scores". These important misclassifications affected essentially the shear zone and thrust-related deposits in mafic–ultramafic rocks and oceanic suture rocks, such as in Santa Terezinha de Goiás, Brazil, and either Panjsher valley (Afghanistan) or Swat valley (Pakistan).

Schwarz and Giuliani [58,59] recognized two main types of emerald deposit: Those related to granitic intrusions (type I) and those where mineralization is mainly controlled by tectonic structures, such as a fault or a shear zone (type II). Most emerald deposits fall into the first category and are subdivided based on the presence or absence of biotite schist at the contact. Type II deposits are subdivided into schist without pegmatite and black shale with veins and breccia. However, a number of emerald deposits of type I have been influenced by syntectonic events (e.g., Carnaíba, Brazil; Poona, Australia; Sandawana, Zimbabwe) or remain unclassified, such as the Gravelotte (Leydsdorp) deposit in South Africa [58].

Schwarz et al. [60] classified emerald deposits based upon their appearance in the field following several sketched geological profiles drawn by G. Grundmann: (i) Pegmatites without phlogopite schist; (ii) pegmatite and greisen with phlogopite schist; (iii) schist without pegmatite with (iiia) phlogopite schists, (iiib) carbonate-talc schist and quartz lens, (iiic) phlogopite schist and carbonate-talc schist; and (iv) black shale with breccia and veins.

Barton and Young [53] divided emerald deposits into those with a direct igneous connection and those where such a connection was indirect or absent. Further subdivisions were done based on the chemistry of the magma and/or the nature of the emerald-bearing metasomatic rocks (greisen and vein-like, skarn type, biotite schist, and vein-related). However, a number of emerald deposits cannot be unambiguously classified using this scheme (Kaduna in Nigeria, Swat valley in Pakistan and Eastern desert deposits in Egypt).

Schwarz and Klemm [61] used LA-ICP-MS to obtain approximately 2600 spot analyses of 40 major and trace elements from ca. 650 emerald samples from 21 different occurrences worldwide. The classification of the deposits was "non-genetic descriptive", but was used instead as a geographic fingerprint. Nowadays, the gemological laboratories routinely use LA-ICP-MS for deciphering the geographic origin of individual stones [12,62–65], which is important for provenance and international trade certification. New standards and protocols for emerald analysis are being applied in gem testing laboratories for geographic and geological applications [66].

Aurisicchio et al. [23] obtained major and trace element data for emerald from several world deposits and reported important modifications regarding the origin of Be in some type II deposits (in the classification proposed by Schwarz and Giuliani [58], i.e., type I granitic intrusions- and type II tectonic-related emerald deposits).

5.2. A Revised Classification for Emerald Deposits

We start by asking why a reclassification is desireable. Emerald deposits are relatively rare and form in a limited number of geological settings. Existing classification systems or models used to describe emerald formation are too restrictive and attempts at reclassification have proved inadequate to date.

Emerald is a medium to high temperature by-product of the Earth system and is hosted by diverse rock types of different ages. The first task of a field geologist is to identify the emerald-bearing rocks and to define the geological environment as magmatic, sedimentary, or metamorphic. Then, each emerald occurrence worldwide is linked to a major geological and tectonic regime related to the movement of lithospheric plates. Regional stresses in the lithosphere generate deformation subsystems with characteristic geometry, chemistry, and geology. The systems are either (i) closed (diabatic) systems with an exchange of heat with the surrounding rock and with very limited or absent fluid-rock interaction, or (ii) open systems with the migration of material and fluids along shear-zones and faults, resulting in the mobilization and re-precipitation of elements and tectonic melange zones. The tectonic systems are then characterized as compressive or extensional sub-systems with variable amounts of vertical and strike-slip movement encompassing large areas within orogens. Consequently, a more inclusive classification system for emerald deposits should consider tectonic, magmatic, and metamorphic-related types, geological environment, magmatic/metamorphic/sedimentary host rocks, and metasomatic conditions.

So, what would an enhanced classification look like? Giuliani et al. [55] classified emerald deposits into three broad types based on worldwide production in 2005 (see Figure 3): (i) The magmatic-metasomatic type (Ma) accounting for about 65% of the production; and (ii) the sedimentary-metasomatic (Se) and metamorphic metasomatic (Me) types, for 28% and 7%, respectively.

Marshall et al. [16] proposed an enhanced classification for emerald deposits based on the Me, Ma, and Se models, but also including the temperature of formation. They examined the possibility, at deeper crustal levels and high grade metamorphism, of the possible remobilization of previous beryl or emerald occurrences and partial melting of metamorphic rocks blurring the distinction between Me and Ma types.

In this work, we propose classifying emerald occurrences into two main types (Table 2):

(Type I) Tectonic magmatic-related with sub-types hosted in:

    (IA)   Mafic-ultramafic rocks (Brazil, Zambia, Russia, and others);
    (IB)   Sedimentary rocks (China, Canada, Norway, Kazakhstan Australia);
    (IC)   Granitic rocks (Nigeria).

(Type II) Tectonic metamorphic-related with sub-types hosted in:

    (IIA)   M-UMR (Brazil, Austria);
    (IIB)   Sedimentary rocks-black shale (Colombia, Canada, USA);
    (IIC)   Metamorphic rocks (China, Afghanistan, USA);
    (IID)   Metamorphosed type I deposits or hidden-granitic intrusion-related (Austria, Egypt, Australia, Pakistan), and some unclassified deposits.

6. Different Types of Emerald Deposits

We will now examine the main emerald deposits worldwide using the above classification.

Table 2. New typological classification of the emerald occurrences and deposits worldwide.

Type of Deposit	Tectonic-Metamorphic-Related					Tectonic-Magmatic-Related		
Geological Environment	SEDIMENTARY	Metamorphic				Granitic		
Metamorphic Conditions	Anchizone to Greenschist facies	Greenschist to granulite facies				Greenschist to granulite facies		
Host-rocks	*Sedimentary Rocks*	*Metamorphic rocks*				*Mafic-UltraMafic Rocks (M-UMR)*	*Sedimentary Rocks (SR)*	*Granitoids*
Type	**TYPE IIB** carbonate platform sediments	**TYPE IIC** Metamorphism of SR	Migmatites	**TYPE IIA** Metamorphism of M-UMR	**TYPE IID** Metamorphosed Type IA, Mixed IA and IIA in M-UMR, and unknown	**TYPE IA** pegmatite- aplite- quartz- greisen veins, pods, metasomatites	**TYPE IB**	**TYPE IC**
Mineralization Style	veins and/or metasomatites	veins	veins	shear zone	shear zone, metasomatites, veins, boudins, fault	veins and/or metasomatites, skarns		pods
Origin of the Fluid	Metasomatic-hydrothermal	Metasomatic-hydrothermal	Hydrothermal	Metamorphic-metasomatic	(Magmatic- Metasomatic) with a metamorphic remobilization	Metasomatic-hydrothermal		
Deposits	Colombia (Eastern and western emerald zones) Canada (Mountain River) USA (Uinta (?): question about the presence of emerald)	China (Davdar) Afghanistan (Panjsher)	USA (Heddenite)	Austria (Habachtal) Brazil (Itaberai, Santa Terezinha de Goiás) Pakistan (Swat-Mingora–Gujar-Kili, Barang)	Austria (Habachtal) ?: Probably metamorphic remobilization of type IA deposit Brazil (Santa Terezinha de Goias) ?: Probably related to hidden granitic intrusive cut by thrust and emerald-bearing shear-zone Pakistan (Swat-Mingora)? Probably related to undeformed hidden y intrusives Australia (Poona): Probably metamorphic remobilization of Tye IA deposit Egypt (Djebel Sikait, Zabara, Umm Kabu): Probably metamorphic remobilization of Type IA deposit Zambia (Musakashi): Unknown genesis, vein style, fluid inclusion indicates affinities with Types IIB and IIC	Brazil (Carnaíba, Socotó, Itabira, Fazenda Bonfim, Pirenópolis, etc.) Canada (Tsa da Gliza, Taylor 2) Bulgaria (Rila) Urals (Malysheva, etc.) Pakistan (Khaltaro) Afghanistan (Tawak) India (Rajhastan, Gubaranda) South Africa (Gravelotte) Zambia (Kafubu, etc.) Tanzania (Manyara, Sumbawanga) Mozambique (Rio Maria, etc.) Australia (Menzies, Wodgina, etc.) Ethiopia (Kenticha) Madagascar (Ianapera, Mananjary) Zimbabwe (Sandawana, Masvingo, Filibusi) Somalia (Boorama) Ukraine (Wolodarsk)	Norway (Eidsvoll) China (Dyakou) Canada (Lened) Australia (Emmavile, Torrington) Kazakstan (Delgebetey)	Nigeria (Kaduna)

6.1. Tectonic Magmatic-Related (Type I)

Type I deposits are found in all five continents (Figure 2). The main geological environment is characterized by the presence of aluminous to peraluminous granitoids formed in continental collision domains. The collision increases the continental lithosphere thickness, resulting in increased pressure and temperature, and producing a zone of higher grade of metamorphism and partial melting of rocks, forming felsic magmas. These continental collisions have occurred at different geological times throughout the Wilson cycles of the supercontinents, with the formation of emerald deposits (Figure 10; [67]) during the following orogenies: Eburnean or Transamazonian (2.0 Ga (giga-annum)), Pan-African/Brasiliano (490–520 Ma (mega-annum)), Hercynian (300 Ma), Uralian (299–251 Ma), Yenshan (125–110 Ma), and Himalayan (25–9 Ma). In contrast, the emerald deposit at Kaduna in Nigeria is associated with Mesozoic (213–141 Ma) ring complexes with peralkaline granites formed in a volcanic to subvolcanic continental environment [68]. In the case of the Byrud mine in Norway, the emerald is related to Permo-Triassic intrusions associated with the evolution of the Oslo Paleorift. This rift was characterized by a succession of volcanic rocks and the emplacement of batholiths and the intrusion of syenitic dykes and sills [69].

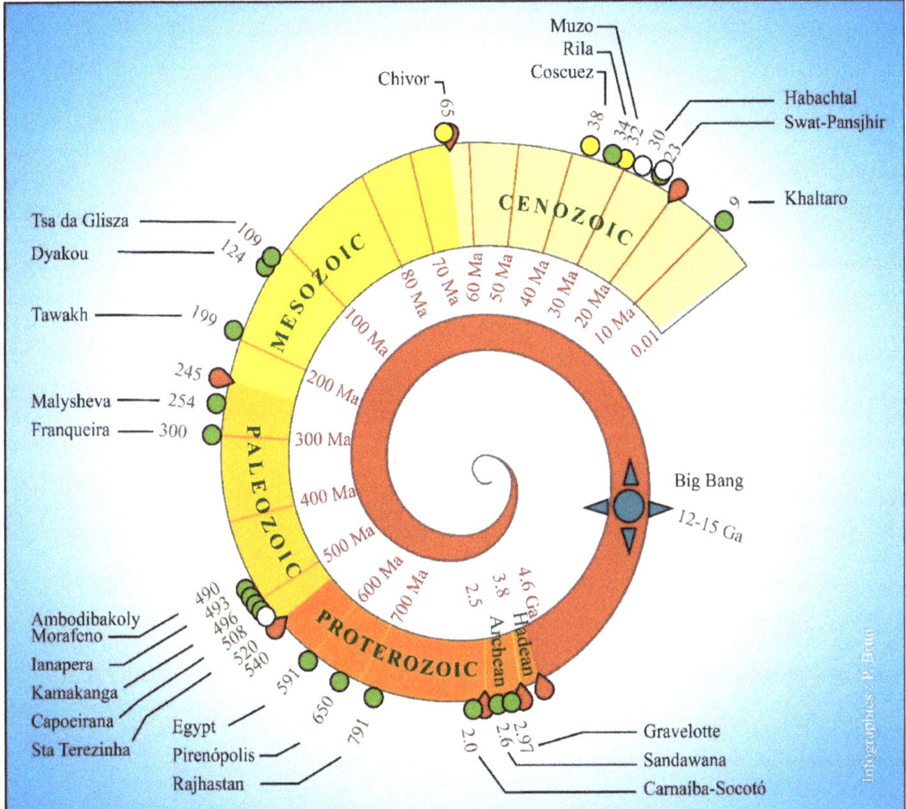

Figure 10. The spiral time of emerald. The oldest emerald formed during Archean times (2.97 Ga) in South Africa (Gravelotte deposit) and the youngest during Cenozoic times (9 Ma) in Pakistan (Kaltharo deposit). Nomenclature of the circles: Red = limit of geological era, green circle = emerald deposits related to the Tectonic magmatic-related types (types IA, IB, and IC); yellow and white circles = Tectonic metamorphic-related types (types IIA, IIB, IIC, and IID) with deposits hosted either in mafic-ultramafic rocks (white circle) or in sedimentary rocks (yellow circle).

Type I emerald deposits, in terms of quantity and gem quality, remain the world's most important emerald deposits. Three sub-types are distinguished according to the emerald host-rocks.

6.1.1. Sub-Type IA: Tectonic Magmatic-Related Emerald Deposits Hosted in M-UMR (Brazil, Zambia, Russia, and Others; See Table 2 and Figure 2)

These deposits produce high volumes of emerald. Why?

There are numerous Be- and/or Be-Ta-Nb-Sn-W-bearing granites of different ages accompanied by subordinate amounts of M-UMR in the continental crust.

There are large volumes of metamorphosed M-UMR in the Precambrian series of the continental crust, such as the recently discovered southern Ethiopian emerald deposit (Halo-Shakiso) hosted in the Archean and Proterozoic volcano-sedimentary series. It is associated with pegmatites and quartz veins similar to the proximal Ta-Nb-Be-bearing pegmatites from the nearby Kenticha mine.

At the regional scale and in several emerald mining districts, there are extended and continuous zones of mineralization related to granites, pegmatites, and quartz veins, as in the Kafubu mining area in Zambia where the mining licences extend for approximately 15 km of the strike length [50,70]. The development of modern mining on a large scale, as done by the company, Gemfields Group Ltd., through underground and huge open-cast mining, allows for the production of large quantities of high-quality commercial-grade gems.

Another example is the mining district of Itabira-Nova Era in Minas Gerais, located in Quadrilátero Ferrifero, Brazil (Figure 11). In 1978, the future deposit of Itabira was discovered on a private property called the "Itabira farm" [71]. After three years of development, the mine was mechanized and named the Belmont mine [72]. In 1988, emeralds were discovered at Capoierana near Nova Era, 10 km southeast of the Belmont mine [73]. In 1998, the deposit of Piteiras was officially discovered. The emerald deposit is located 15 km southeast of Itabira city, between the Belmont and Capoierana mines. The mineralized zone extends for over 5 km and the mines are located on two thrust zones (Brasiliano orogenesis) juxtaposing ultramafic rocks and highly deformed granites called "Borrachudos". At the regional scale, an extended emerald mining district was defined based on the geological continuity of the thrusts [74]; this district accounts for the majority of the Brazilian production.

In this type of deposit, granitic magmas have intruded M-UMR within volcano-sedimentary sequences or greenstone belts (Figure 12) and fluids expelled from the granite circulated within the thermal aureole, sometimes mixing with metamorphic fluids, and reacted with the pegmatite or quartz veins and M-UMR, forming emerald. The deposits are associated with vast amounts of fluid-rock interactions producing intense K-Na-Mg metasomatism of M-UMR and granite. This genetic pathway is the most common ([1,16,22,75,76]) and is characterized by emeralds contained in magnesium-rich micaceous rocks known as phlogopite schists or phlogopitites, or "glimmerite" in the Russian literature, "black wall zones" in [77], or "sludite" in Brazil. Generally, the pegmatite and the M-UMR experienced intense fluid infiltration and metasomatism where: (i) The M-UMR is converted into the emerald-bearing micaceous-host rock; (ii) the granitic rock itself is transformed into a feldspar-rich rock called plagioclasite comprised of albite-andesine feldspar. Quartz is dissolved and the pegmatite becomes a desilicated pegmatite (Figure 13). The metasomatic processes are highly variable and dependent upon P-T conditions, the extent of fluid circulation, and the timing between granitic intrusions and regional deformation.

At Carnaíba, the vein-like metasomatic rocks, in which phlogopite is by far the most abundant mineral, have a longitudinal extension that may extend several hundreds of meters, whereas their thickness does not exceed a few metres. They display a clear zoning and are organized, in many places in the M-UMR, symetrically around intrusive aplo-pegmatite dikelets, which channel the fluids [78].

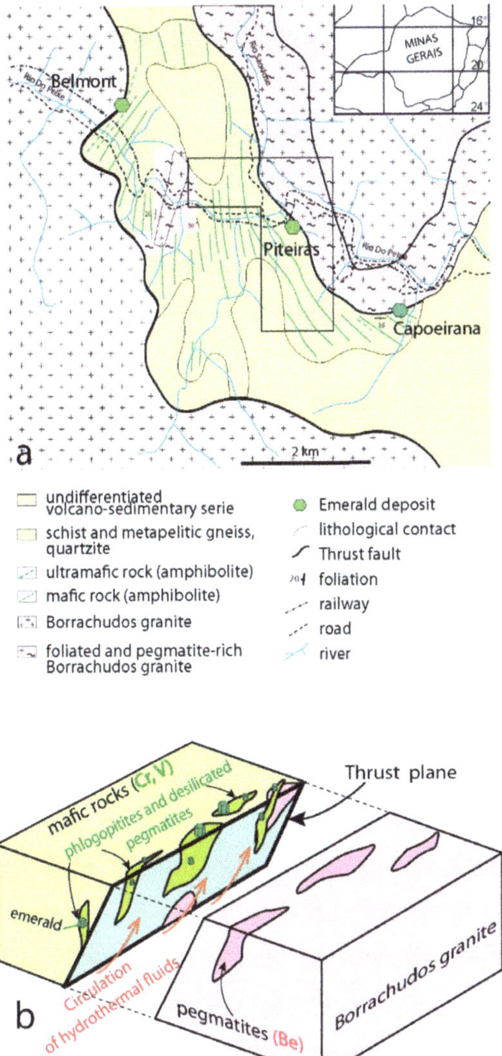

Figure 11. The emerald mining district of Itabira-Nova Era in Minas Gerais, Brazil, with the mines of Belmont, Piteiras, and Capoierana: (**a**) Geological map of the extended regional mining district. The mining concession of Piteiras is bounded by the polygon area plotted in the figure. Three deposits are currently mined. All occur in the same geological context. Ultramafic bodies (chromium—Cr, vanadium—V, and magnesium-rich-bearing rocks) are in contact with the deformed granites of Borrachudos (beryllium—Be, aluminium-silica-rich pegmatites), modified after [73]; (**b**) schematic diagram representing the probable formation of the Piteiras-Capoeirana-Belmont emerald deposits. During the Brasiliano orogencsis (508 Ma), the hydrothermal fluid (red arrows) circulated along the thrust planes, altering granites and associated Be-bearing pegmatites and Cr-V-bearing mafic-ultramafics. The hydrothermal fluids interacted with both rocks and were enriched in all the elements necessary for emerald crystallization. Emerald crystallized in plagioclasite (desilicated pegmatite) and phlogopitite.

Figure 12. Idealized model for the Tectonic magmatic-related emerald type. The model is based on the emplacement in the crust of a granitic massif, with its pegmatite and aplite dikes and their tourmaline- (Tr) or beryl- (Brl) bearing quartz veins, intruding mafic (metabasalt) and/or ultramafic rocks (metaperidotite, serpentinite). The fluid circulations from the granite into the surrounding rocks and granitic dykes (arrows), preferentially along the contacts between the pegmatite or aplite or quartz veins and the regional rocks, transform the mafic rocks into a magnesium-rich biotite schist and the pegmatite into an albite-rich plagioclasite. Emerald (Em) and apatite (Ap) precipitates in the rocks affected by the fluid/rock interaction. It can precipitate in the pegmatite, aplite, plagioclasite, and quartz veins and their adjacent phlogopite schist zones.

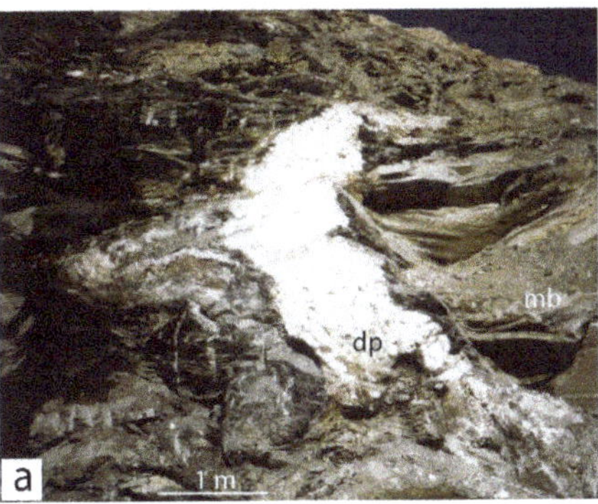

Figure 13. Desilicated emerald-bearing pegmatites: (**a**) Desilicated pegmatites crosscutting metabasites (mb) from the Kafubu mine (Zambia). Photograph: Dietmar Schwarz; (**b**) desilicated pegmatite from the Carnaíba mine, Brazil. At the contact of a pegmatite (pg) and a serpentinite, the metasomatic fluid dissolved quartz, mobilized Cr and V from the chromite-bearing serpentinite, and Be from the beryl-bearing pegmatite to form emerald (Em). The fluid transformed the serpentinite into a phlogopitite (ph) and the pegmatite in an albite-rich rock (ab). The three arrows indicate the limit of the dissolution of quartz (q) from the pegmatite. Photograph: Gaston Giuliani.

These metasomatic rocks, called exo-F-phlogopitites (1.3 to 4 wt.% F), formed a metasomatic column that consists of seven zones, from the central desilicated granitic vein (zone 7) to the enclosing serpentinite (Figure 14):

Zone 6: Coarse-grained F-phlogopite + apatite + emerald + quartz;
Zone 5: Fine-grained F-phlogopite + apatite + emerald;
Zone 4: F-phlogopite + spinel (chromite + magnetite);
Zone 3: F-phlogopite + spinel + amphibole (actinolite + tremolite);
Zone 2: F-phlogopite + spinel + ampbibole + talc; and
Zone 1: spinel + amphibole (or dolomite) + talc + serpentine + chlorite.

From the inner to the outer zones, (i) phlogopite composition evolves with continuous or discontinuous decrease in the Al and Fe contents with an increase in the Si and Mg contents and K/Al ratio; (ii) amphibole evolves from actinolite to tremolite; and (iii) spinel from Al-chromite to Cr-magnetite.

The aplopegmatite dikelets were transformed into plagioclasites (with disseminated phlogopite and little hornblende) with irregular commonly centimeter-thick phlogopite rims, embayments, and veinlets called endo-phlogopitites by analogy with the endo-skarns [78].

The strong chemical gradients in the different zones of the metasomatic column are characterized by a change in the composition of phlogopite (Figure 15; [78,79]). From zone 6 to zone 2, the Al and Fe contents decrease while Si and Mg increase. This evolution is discontinuous and two abrupt changes are observed, one at the front of zones 6/5, where the evolution corresponds to an increase in Si and Mg and a decrease in the Al, Fe, and Na contents and the other within zone 4. Such pattern of evolution are coherent with infiltration metasomatism. The potassium content remains constant in zones 6 and 5, suddenly decreases within zone 4, and remains constant up to zone 2.

Over the whole column, the F content of phlogopite is in the range of 1.3 to 4 wt.% and the highest values are observed in the phlogopite of the outer zone that has the highest Mg/(Fe + Mg) ratios, in agreement with the so-called "Fe-F avoidance effect" [80].

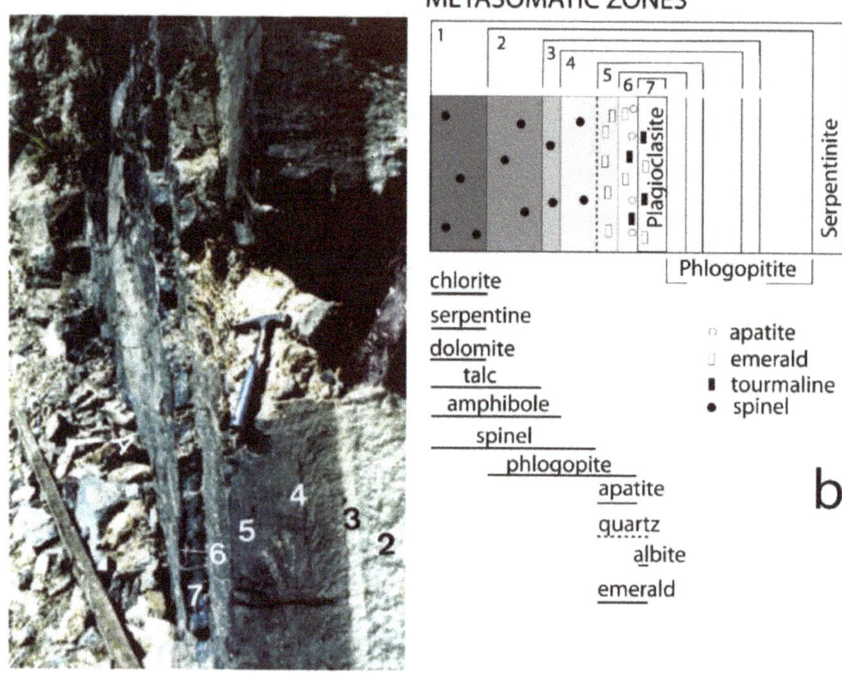

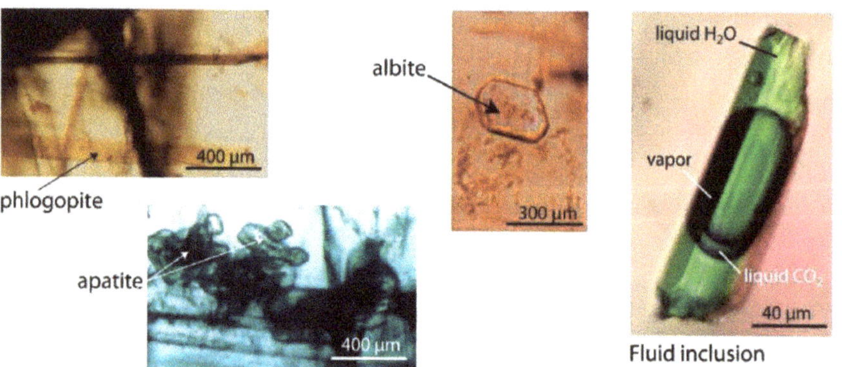

Figure 14. The metasomatism related to fluid circulation between aplopegmatite dikes and mafic-ultramafic rocks: (**a**) Metasomatic column formed by different mineralogical zones (zones 7 to 2) developed around a central pegmatite vein crosscutting serpentinites, Braúlia prospecting pit, Carnaíba; (**b**) vein-like metasomatic rocks and their mineralogical composition. The pegmatite is transformed into plagioclasite (zone 7) and the metasomatic rocks consist of six zones, from the desilicated pegmatite to the enclosing serpentinite (zones 6 to 2). Emerald is found in the plagioclasite and in zones 6 (coarse-grained phlogopitite) and 5 (fine-grained phlogopitite). Zone 1 is the protolith formed by the serpentinite (not seen on the photograph); (**c**) solids and fluid inclusion in emeralds from zones 7 and 6 (Photographs: Dietmar Schwarz and Gaston Giuliani).

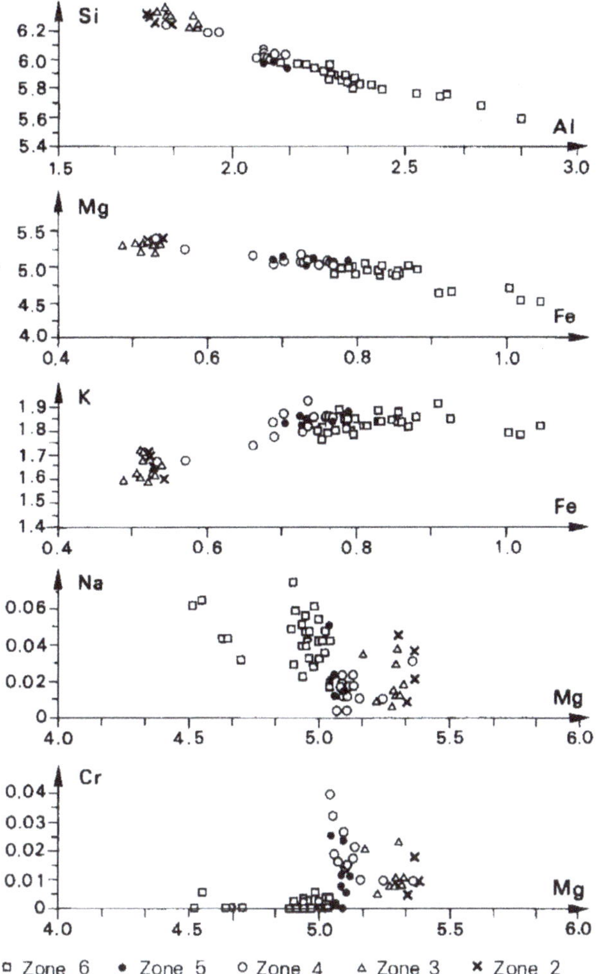

Figure 15. Evolution of the phlogopite composition in the phlogopitites of the different metasomatic zones (zones 6 to 2, see Figure 14 for the repartition of the different zones from the plagioclasite to the serpentinite) at Braúlia prospecting pit, Carnaíba mine, Brazil. Composition calculated in atom per formula (electron microprobe data from [79]). From the inner to the outer zones, phlogopite composition evolves with a continuous or discontinuous decrease in the Al and Fe contents and an increase of the Si and Mg contents and K/Al ratio, corresponding to changes in the petrographic habitus of the phlogopite.

Chromium contents are usually very low (<0.05 wt.%) in zones 6 and 5 and present a peak (up to 0.7 wt.%) at the front of zones 5/4, at the place where the first crystals of chromite are observed. In the outer zones 4 to 2, the Cr contents of phlogopite are nearly constant (0.25 wt.%). Chromium dispersed in the original rock can be concentrated at the front advancing outwards, like a "sweep effect" [81]. The almost uniform Cr content of phlogopite in the outer zones, 4 to 2, may result from the establishment of a exchange equilibrium between mica and chromite, Cr behaving as an inert component.

Mass transfer calculations show that the infiltration metasomatism in serpentinites is accompanied by important transfers of material, concerning mainly Al, K, F, Si, Mg, Ca, and H_2O (Figure 16). The strong enrichment in K, Rb, Li, Be, Nb, P, and F suggests that the metasomatic fluids have a magmatic origin, since some of these elements (including Rb, Nb, and Be) behave incompatibly during magmatic differentiation.

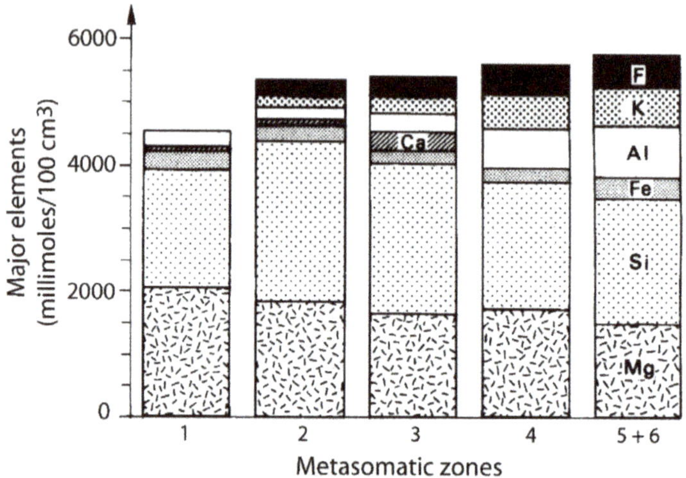

Figure 16. Mass balance calculations realized on the metasomatic column at Bode prospecting pit, Carnaíba mine, Bahia, Brazil (modified from [79]). Chemical mass balance was realized on major elements' molar composition (in millimoles/100 cm^3) of the different metasomatic zones (1 to 5 + 6). It indicates the supply of many elements, such as K, Si, Al, and F, over the whole column, and Fe, in the inner most zones of 6 and 5, the leaching of Mg and Ca from the outer to the inner zones.

The association of phlogopitite–plagioclasite corresponds to a "bimetasomatic system" in the sense given by [82]. The hydrothermal solutions that developed the potassic metasomatism in the serpentinites (phlogopitites) are alkaline, undersaturated with respect to microcline and quartz of the intrusive aplopegmatite, and albitizing.

The occurrence of emeralds is considered to be due to the efficiency of the metasomatic trap rather than a significant magmatic pre-enrichment in Be. The occurrence of strong chemical gradients in the zones of preferential circulation of the solutions (zones 6 and 5 of the exo-phlogopitites, plagioclasites, and endo-phlogopitites) constitutes highly favorable conditions for the beryl formation.

Elevated temperature and alumni and silica activities are the main factors for emerald formation in the IA sub-type (Figure 17). At high temperatures (T > 600 °C) and medium to high silica activity, beryl, phenakite (Be_2SiO_4), and chrysoberyl ($BeAl_2O_4$) are stable [53]. Emerald occurs in quartz veins, such as at the Kafubu deposits in Zambia [50,70], or greisens (an association of quartz and white micas), such as at the deposits of Delbegetey in Kazakhstan [51] and Khaltaro in Pakistan [83]. At lower silicate activities, albite is always stable and emerald is replaced by alexandrite (Cr-chrysoberyl) + phenakite. This is the case for emerald-bearing desilicated pegmatites in M-UMR, such as in Russia and Brazil. At higher Al_2O_3 activities, corundum can crystallize, such as at the Poona emerald deposit in Australia [84].

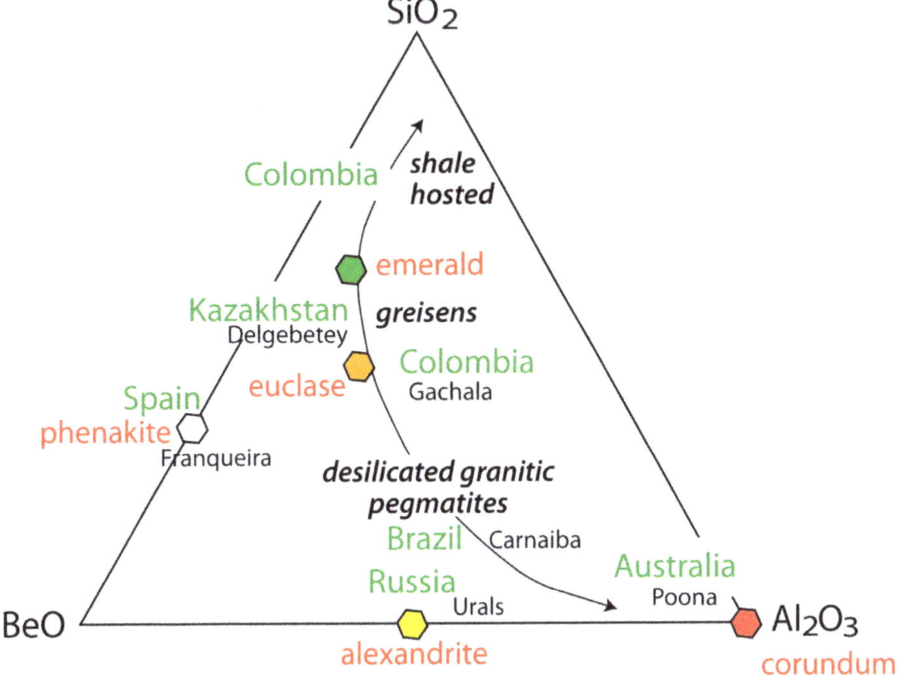

Figure 17. The ternary diagram of BeO-Al$_2$O$_3$-SiO$_2$ visualises the major types of emerald deposits as a function of their typical mineral assemblages, such as alexandrite, euclase, phenakite, and corundum ('BASH' system modified from [53]). Follow the discussion in the text.

The classic model of these granitic-metasomatic deposits presents much variability in terms of the geometry, chemical composition of the rocks, geological contacts, etc.:

1. At the Tsa da Gliza deposit in Canada, Be-bearing magmatic fluids from the neighboring granite reacted with the Cr- and V-bearing M-UMR [46]. Emeralds formed in aplite, pegmatite, and quartz veins surrounded by biotite schists. There is a continuum between the crystallization of granitic rocks, fluid rock-interaction, and emerald formation (Figure 18).

2. In the Kafubu area in Zambia (Figure 1c), emerald is found predominantly in metamorphosed M-UMR with phlogopite schist or in quartz-tourmaline veins adjacent to pegmatites.

3. At the Sandawana deposit in Zimbabwe, pegmatites intruded the M-UMR, but they are folded and fluid circulation in shear zones formed phlogopite schist. The fluid-rock interaction is coeval with the regional deformation [52]. Such phenomena are also found in the Carnaíba deposits in Brazil, where the dissolution of quartz from pegmatite is common [76].

4. At the Ianapera deposit in Madagascar, two coeval emerald deposits coexist (Figure 19; [11]): (1) A proximal one, formed at the contact between pegmatites and UMR; and (2) a distal one, hosted in biotite schist in fractures developed in mafic rocks with widespread fluid circulation affecting all the geological formations. Similarly, the Trecho Novo and Trecho Velho deposits at Carnaíba, Brazil [78] are also distal in nature.

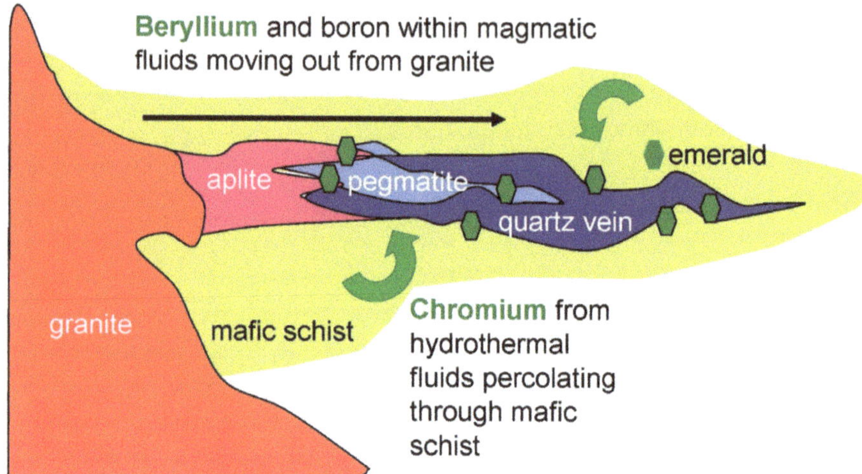

Figure 18. Genetic model proposed for the emerald mineralization at Tsa da Glisza, Yukon Territory, Canada.

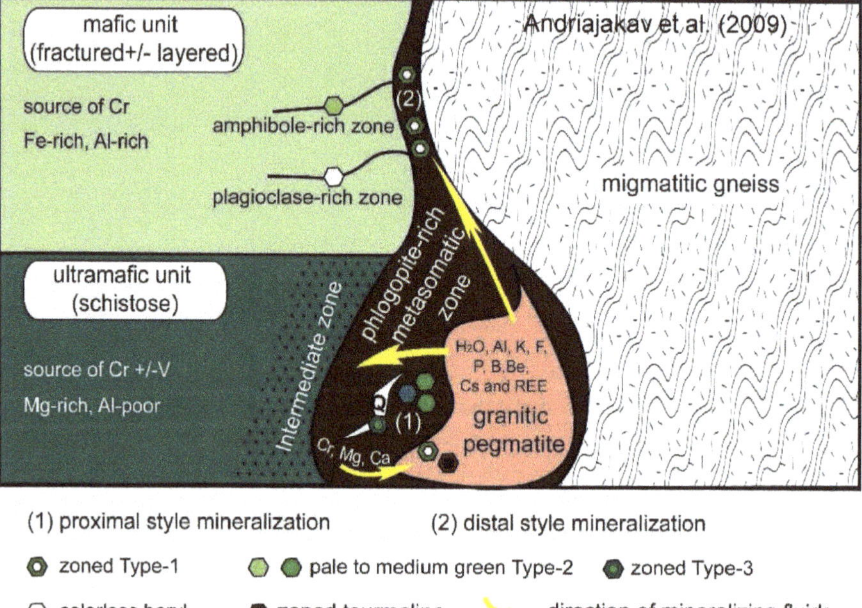

Figure 19. Schematic representation of the two styles of mineralization evidenced for the Ianapera emerald deposit, Madagascar: (**1**) Proximal mineralization occurs at the contact of pegmatite veins with ultramafic units. Emerald mineralization is hosted in metasomatic phlogopitite and desilicated pegmatites or quartz-tourmaline veins, at the contact between migmatitic gneiss and garnet amphibolite; (**2**) distal style is formed by phlogopite veins crosscutting mafic rocks. In addition to phlogopite, these veins contain Mg-amphibole, apatite, and dolomite, and minor quartz, calcite, zircon, plagioclase, and chlorite.

6.1.2. Sub-Type IB: Tectonic Magmatic-Related Emerald Deposits Hosted in (Meta)-Sedimentary Rocks (China, Canada, Kazakhstan, Norway, Australia; See Table 2)

These emerald occurrences are associated with granites that intrude sedimentary or meta-sedimentary lithologies. These deposits are generally sub-economic and several variants are described, depending on the nature of the host-rocks and the mineralization styles.

1. At Dyakou (China), the Lower Cretaceous porphyritic granite intruded biotite granofels, quartzite, gneiss, and plagioclase amphibolite of Lower and Upper Neoproterozoic formations. The intrusion formed skarns and dyke swarms of quartz veins, which are crosscut by pegmatites [85]. Emerald in quartz veins is less abundant than in the pegmatites, but of higher quality (Figure 1b). Some pegmatites show a local zoning with an outer zone enriched in K-feldspar and an inner zone of emerald and quartz.

2. At the Lened V-rich emerald occurrence (Canada), Be and other incompatible elements (i.e., W, Sn, Li, B, and F) in the emerald, vein minerals, and surrounding skarn were derived during the terminal stages of crystallization of the proximal Lened pluton [21,47]. Decarbonation during pyroxene-garnet skarn formation in the host carbonate rocks probably caused local overpressuring and fracturing that allowed ingress of magma-derived fluids and formation of quartz-calcite-beryl-scheelite-tourmaline-pyrite veins. The vein fluid was largely igneous in origin, but the dominant emerald chromophore V was mobilized by metasomatism of V-rich sedimentary rocks (avg. 2000 ppm V) that underlie the emerald occurrence [21].

3. At Delbegetey (Kazakhstan), the emerald mineralization is confined to the granite that hornfelsed carboniferous sandstones. Emerald is found in muscovite greisen formed in the wall-rocks of muscovite-tourmaline-fluorite-bearing quartz veins [51].

4. At Eidswoll (Norway), the emerald Byrud Gård mine is related to Permo-Triassic alkaline intrusions. The V-bearing emerald occurs in Middle Triassic pegmatite veins that intruded Cambrian Alum shales and quartz syenite sills [20]. Vanadium and Cr were probably leached from the alum shales by the mineralizing fluids [13].

5. At Emmaville-Torrington (Australia), emerald is located in pegmatite, aplite, and quartz veins associated with the Mole granite. The granite intrudes a Permian metasedimentary sequence consisting of meta-siltstones, slates, and quartzites [19]. The emerald-bearing pegmatite veins contain quartz, topaz, K-feldspar, and mica. Emerald is embedded in cavities and surrounded by dickite in the quartz-topaz veins. In the quartz lodes, emerald is associated with Sn-W-F minerals. At the Heffernan's Wolfram mine, emerald occurs with wolframite in vugs in the pegmatites [86].

6.1.3. Sub-Type IC: Tectonic Magmatic-Related Emerald Deposits Hosted in Peralkaline Granites (Nigeria)

These emerald occurrences are located in the Jurassic younger granite ring complexes of the anorogenic magmatism of Nigeria [68]. These granites crystallized in a volcanic-subvolcanic environment. They are generally peralkaline with perthitic K-feldspar, sodic amphiboles, and alkaline pyroxene. The roof zone of the intrusions is characterized by disseminated tin, tungsten, niobium, tantalum, and zinc mineralization related to sodic or potassic metasomatism, sheeted quartz vein systems, pegmatite pods and veinlets, replacement bodies, and fissure-filling veins (Figure 20). The emerald mineralization is located in sporadic pegmatitic pods with quartz and feldspar, as well as topaz and gem quality aquamarine. The source of chromium for the emerald is debated, but it could be the consequence of the mode of emplacement of the granites, which involved mechanisms of underground cauldron subsidence. The caldera produced ignimbrite, rhyolite, and thin Cr-bearing basic flows, which collapsed during doming or swelling and intrusion of the younger granites. Assimilation of the previous basic volcanic rocks or local fluid interaction between the pegmatites and the basic flows could have enriched the metasomatic fluids in Cr and Fe, resulting in the formation of emerald-aquamarine beryls.

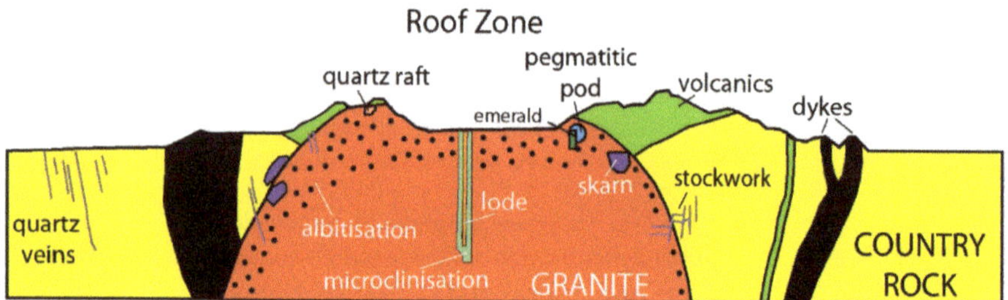

Figure 20. Ring complexes of Nigeria and emerald mineralization. Idealized cross-section showing structural setting and styles of mineralization.

6.2. Tectonic Metamorphic-Related (Type II)

Type II deposits are found in metamorphic environments with facies varying from high anchizone metamorphic illite crystallinity (Colombian black shales) to medium pressure conditions where the most exposed metamorphic rocks belong to the greenschist (ex: Swat talc-carbonate schist, Pakistan), amphibolite (ex: Habachtal metamorphic series, Austria), and up to granulite facies, with local partial melting of the rocks (migmatites of Hiddenite, USA). The majority of these deposits are characterized by the absence of granitic intrusions, and the deposits are linked to the circulation of fluids and metasomatism in thrusts, shear zones, or vein systems. Four sub-types are proposed in the new classification of emerald deposits (IIA, IIB, IIC, and IID).

6.2.1. Sub-Type IIA: Tectonic Metamorphic-Related Emerald Deposits Hosted in M-UMR (Brazil, Austria, Pakistan; See Table 1)

The Brazilian emerald deposits are partly associated with shear-zones cross-cutting M-UMR. The absence of a magmatic influence in their formation is constrained by field, geochemical, fluid inclusion, and stable isotope studies [15,76]. This is the case for the Santa Terezinha de Goiás deposit and the occurrence of Itaberaí. The economic importance of this deposit type is waning due to smaller emeralds, emerald dissemination, and higher artisanal mining cost and today represents only 4% of the total production of Brazil. The main Brazilian production is from type I deposits in Minas Gerais (74%) and Bahia (22%) [87]. Nevertheless, type IIA must be considered when prospecting in the Archean and Precambrian volcano-sedimentary series or greenstone belts.

1. The Santa Terezinha de Goiás deposit, located in central Goiás, produced 155 tons of emerald between 1981, date of the discovery, and 1988 [88]. The emerald grade was between 50 and 800 g/t. The infiltration of hydrothermal fluids is controlled by tectonic structures, such as the thrust and shear zones (Figure 21a). Pegmatite veins are absent and the mineralization is stratiform (Figure 21b). Emerald is disseminated within phlogopitites and phlogopitized carbonate-talc schists of the metavolcanic sedimentary sequence of Santa Terezinha [88–90]. Talc-schists provide the main sites for thrusting and the formation of sheath folds [91]. Emerald-rich zones are commonly found in the cores of sheath folds and along the foliation (Figure 21c). Two types of ore can be distinguished [88]: (i) A carbonate-rich ore composed of dolomite, talc, phlogopite, quartz, chlorite, tremolite, spinel, pyrite, and emerald; (ii) a phlogopite-rich ore composed of phlogopite, quartz, carbonates, chlorite, talc, pyrite, and emerald.

The distal São José two-micas granite, located 5 km from the emerald deposit, is a syntectonic foliated granite, which underwent a polyphase ductile deformation coeval to that observed in the emerald deposit [92]. C and S structures in the granite indicate shear deformation along a typical frontal thrust ramp and the granite overthrusted the Santa Terezinha sequence where the emerald deposit is located. D'el-Rey Silva and Barros Neto [92] suggested that the granite most probably was the source of Be for the formation of emerald in the Santa Terezinha de Goiás deposit.

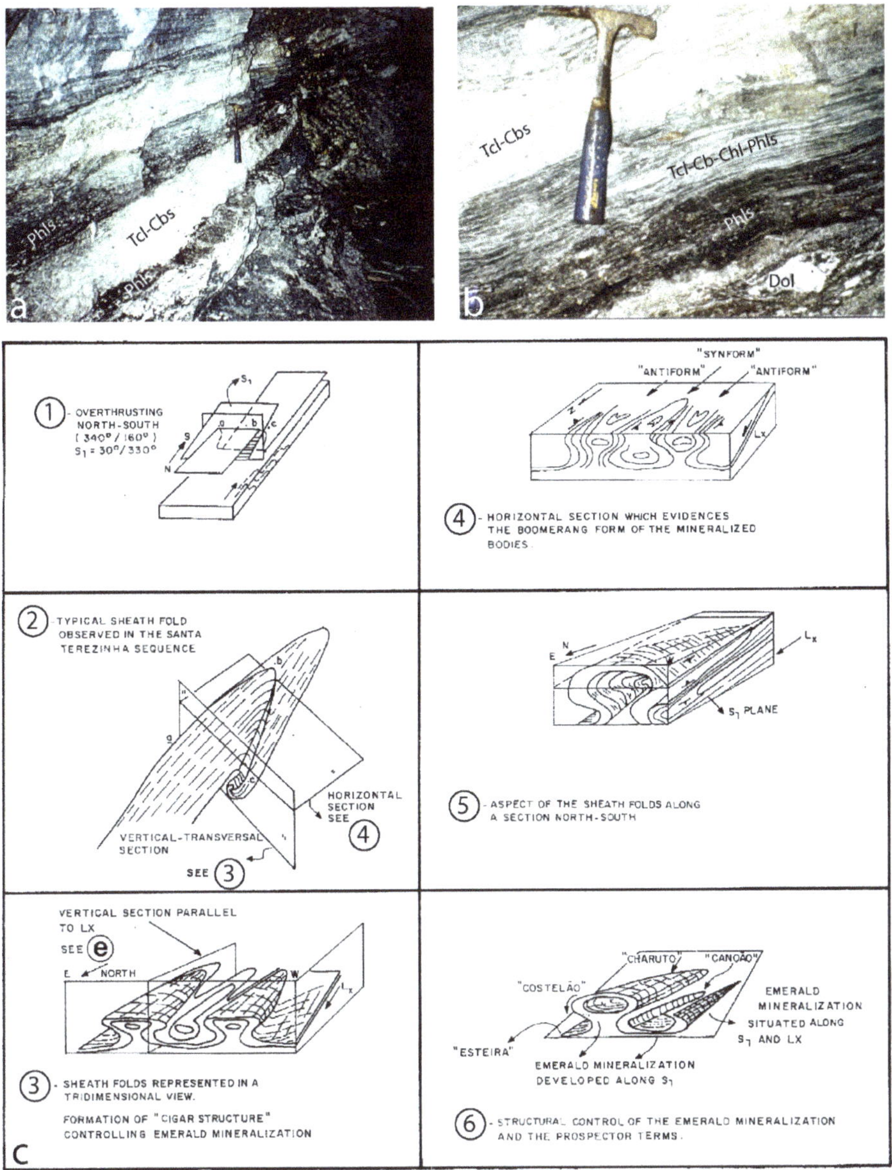

Figure 21. The Santa Terezinha de Goiás (Campos Verdes) emerald deposit, Goiás, Brazil: (**a**) Carbonated and phlogopite-rich emerald ores. The phlogopite schists (phls) underline the foliation of the talc-carbonate schists (Tcl-Cbs); (**b**) carbonates lenses (Dol) are observed within the talc-carbonate schists (Tcl-Cbs). The phlogopitisation affects the carbonated-talc-schists and the talc-carbonate-chlorite schist (Tcl-Cb-Chls) showing the "bed-by-bed" fluid injection along foliation planes; (**c**) structural evolution of the Santa Terezinha volcano-sedimentary sequence and the controls of the emerald mineralization on the basis of the structural study done on the Trecho Novo 167 underground mine (EMSA company property). Photographs: Gaston Giuliani.

2. The Habachtal deposit in the Austrian Alps has been studied in detail [77,93–97]. This alpine deposit is located in a contact zone, which overthrusts the volcano-sedimentary series of Habachtal (Habach Formation) on the ortho-augengneisses (central gneisses). The Paleozoic Habach formation is composed of a series of amphibolites, acid metavolcanics tranformed in muscovite schists, and black pelites with interlayered serpentinites and talc series. Two metamorphic events, one occurring before the Alpine event (P < 3 kb and T < 450 °C) and one occuring during the Alpine event (4.5 < P < 6 kb and 500 < T < 550 °C) were superimposed. The mineralized "blackwall zone", the equivalent of a phlogopitite, is a tectonic or shear zone 100 m wide, formed from UMR (serpentinites) pinched between orthogneisses and amphibolites. Emerald is disseminated in the "blackwall zone" phlogopitites, talc-actinolite, and chlorite schists. The metasomatic process involves fluid percolation that extracted Be from the muscovite schists (average Be content = 36 ppm) and Cr from the serpentinites (Cr content = 304 ppm) to facilitate the crystallization of emerald (Figure 22). Fluid inclusions trapped by emerald belong to the H_2O-CO_2-NaCl system [96] with two generations of fluid inclusions: An early generation (XCO_2 < 4 vol.%) and a late one (XCO_2 up to 11 vol.%). Emerald-metasomatic fluids were related to hydration phenomena due to the alpine metamorphism [96].

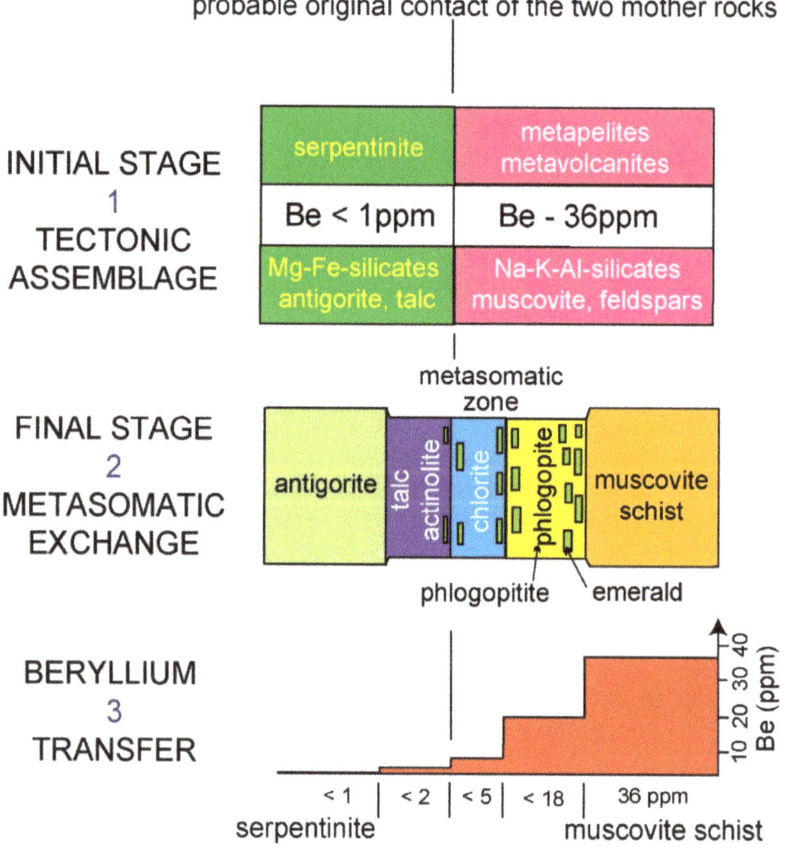

Figure 22. Model of formation of the Habachtal emerald deposit: (**1**) Initial stage showing the different lithologies with their respective Be contents; (**2**) final stage after regional metamorphism showing the final metasomatic rocks assemblages; (**3**) beryllium liberated from the muscovite schists is incorporated into the emerald crystals.

3.   The Swat-Mingora–Gujar Kili–Barang deposits are controlled by the Main Mantle Thrust [86,98–100]. The suture zone that marks the collision of the Indo-Pakistan plate with the Kohistan arc sequence is composed of a number of fault bounded rock melanges (blueschist, greenschist, and ophiolitic melanges).

The ophiolitic melange, which hosts the Pakistani emerald deposits, is composed mainly of altered ultramafic rocks with local cumulate, pillow lavas, and metasediments. Emerald occurs within hydrothermally altered serpentinites that show metasomatic zoning [100]: An outer zone composed of talc-magnesite $\pm$ chlorite $\pm$ micas; an intermediate zone consisting of talc-magnesite with dolomite veins; and an inner zone with dolomite-magnesite-talc schists and quartz-dolomite $\pm$ tourmaline $\pm$ fuschite veins. Emerald occurs disseminated in the inner and intermediate zones within or spatially associated with quartz-carbonate veins. Dilles et al. [100] obtained a $^{40}Ar/^{39}Ar$ Oligocene age of 23.7 $\pm$ 0.1 Ma for a fuchsite-quartz vein in the Swat emerald deposit.

Isotopic study of magnesite from the outer zone showed that the mineral association resulted from early metamorphic fluids [101], whereas the inner and intermediate zones resulted from the infiltration of hydrothermal fluids, which carried Si, Be, B, K, and Ca. Chromium came from the dissolution of chromite crystals in the serpentinites. Arif et al. [102] analysed Cr, Be, B, and other trace element contents in the ophiolitic rocks of the Indus suture zone in Swat. They showed that the Cr present in the Cr-bearing silicates (emerald, Cr-tourmaline, and Cr-muscovite) was derived from the original protolith. Beryllium and boron enrichments were found only in M-UMR affected by fractures and fluid circulation. In addition, analyses of small granitic dykes cutting granitic gneisses showed extreme B and Be enrichment. In consequence, Arif et al. [102] argued that the Be and B are sourced from a probable hidden leucogranite in depth.

6.2.2. Sub-Type IIB: Tectonic Metamorphic-Related Emerald Deposits Hosted in Sedimentary Rocks: Black Shales (Colombia, Canada, USA; Table 2)

Colombian emerald deposits are unique. Why?

The Colombian emerald deposits are located on both sides of the Eastern Cordillera sedimentary basin, with an eastern zone comprising the mining districts of Gachalá, Chivor, and Macanal, and a western zone, including the mines of La Glorieta-Yacopi, Muzo, Coscuez, La Pita, Cunas, and Peñas Blancas (Figure 23). Their distribution along both sides of the basin ensures abundant production of high-quality emerald. These deposits are unique because there is no connection with granites in their formation [103,104].

The emerald mineralization is hosted in Lower Cretaceous (135–116 Ma) sedimentary rocks composed of a thick succession of sandstone, limestone, black shale, and evaporites. The salt and sulphate rocks were necessary for the formation of emerald. These intercalations of evaporites are found both in the Guavio (Figure 24) and Rosablanca formations, in the eastern and western emerald zones, respectively. The evaporites are responsible for the high salinity of the basinal brines (~40 wt.% equivalent NaCl) and the circulation of H_2O-NaCl-CO_2-(Ca-K-Mg-Fe-Li-SO_4) fluids trapped by emerald during its growth ([105]; Figure 25).

The tectonic-sedimentary evolution of the Eastern Cordillera basin is unique (Figure 26). The formation of the emerald deposits is related to changes in the acceleration and convergence of the Nazca and South American plates that took place at: (1) 65 Ma, forming the emerald deposits on the eastern side of the basin [106]; and at (2) 38–32 Ma, generating the Muzo and Coscuez deposits on the western side [103]. Formation at different ages and conditions resulted in two drastically different styles of mineralization in the eastern and western emerald zones [107,108].

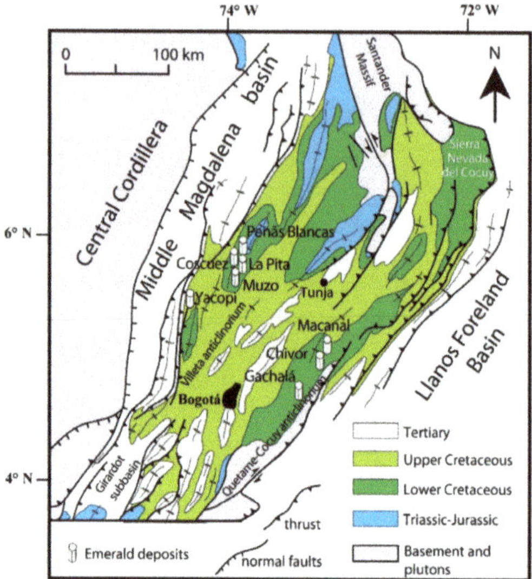

Figure 23. Simplified geological map of the basin from the Eastern Colombian Cordillera. The emerald deposits are hosted by Lower Cretaceous sedimentary rocks forming two mineralized zones located, respectively, on the eastern and western borders of the basin. On the western border, with the mining districts of La Glorieta-Yacopi, Muzo, Coscuez, La Pita, Cunas, and Peñas Blancas, and on the eastern border are Gachalá, Chivor, and Macanal.

Figure 24. Evaporites in the the Chivor mining area: (**a**) The gypsum deposits present decametric lenses of white gypsum/dolostone alternations hosted in a black matrix made of crushed and dismembered black shales similar to the main evaporitic regional breccia level in emerald deposits; (**b**) Dolomicritic beds are in alternation with nodulated beds of gypsum. Photographs: Yannick Branquet.

In the eastern emerald deposits, the Chivor mining district presents extensional structures extending from a brecciated regional evaporitic level (Figure 27), which acted as a local, gravity-driven detachment [107,108]. The brecciated rock unit in the Chivor area, which is in excess of 10 km long and 10 m thick (Figure 28a,b), is stratabound, i.e., parallel to the sedimentary strata, and dominantly composed of hydrothermal breccia (Figure 28c) made up of fragments of the hanging wall (carbonated carbon-rich BS, limestone, and whitish albitite (albitized black shale)) cemented by carbonates and pyrite (Figure 28d).

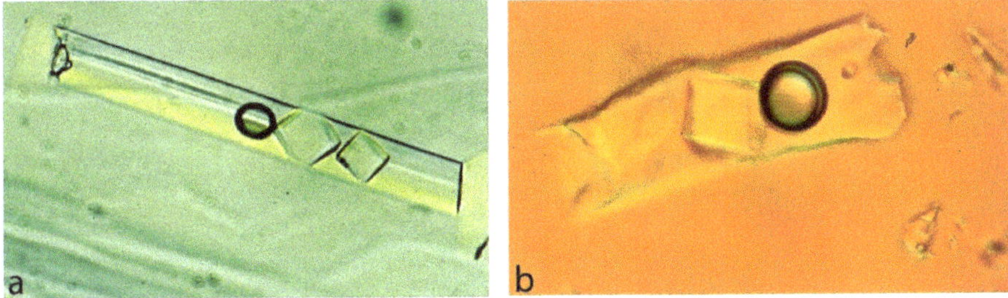

Figure 25. Highly saline basinal fluids trapped by fluid inclusions in Colombian emerald is strong evidence for the evaporitic origin of the mineralizing fluid: (**a**) Emerald from Oriente mine in Chivor (Eastern emerald zone). Primary multi-phase fluid inclusion trapped in an emerald. The cavity, 180 µm long, contains from the right to the left, two cubes of sodium chloride (halite), a rounded gas bubble, and two minute crystals of calcite all of which are wetted by a salty water occupying 75 vol.% of the cavity; (**b**) emerald from Coscuez (western emerald zone). The cavity, 40 µm long, contains 75 vol.% of salt water solution, 10 vol.% of gas corresponding to the vapour bubble, 15 vol.% of cubic halite crystal (NaCl), and a rounded crystal of carbonate (on the right of the cavity). The vapour phase is rimmed by liquid carbon dioxide. Photographs: Hervé Conge.

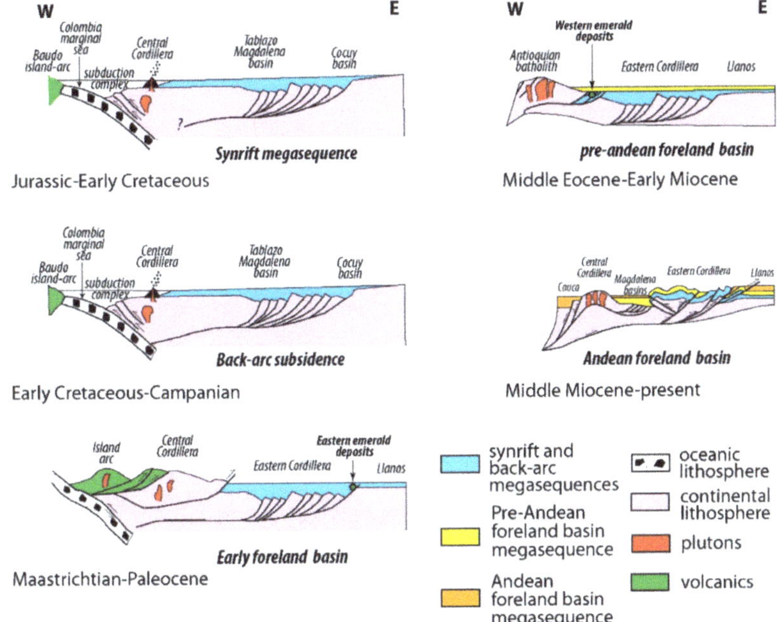

Figure 26. Basin development and tectonic history of the Llanos and Eastern Cordillera basins, and Middle Magdalena Valley in Colombia (modified from [109]). Four episodes of deformation have been recognized in the Tertiary of central Colombia: (**1**) Late Cretaceous–early Paleocene, with the formation at 65 Ma of the emerald deposits from the eastern zone; (**2**) middle Eocene with the creation of folds and thrusts in the Middle Magalena valley and western border of the Eastern Cordillera basin, and with the formation, between 38–32 Ma, of the emerald deposits from the western; (**3**) late Oligocene–early Miocene; and (**4**) late Miocene–Pliocene at 10.5 Ma where the Eastern Cordillera was uplifted and eroded, with the outcropping of the emerald deposits from the eastern and western emerald zones.

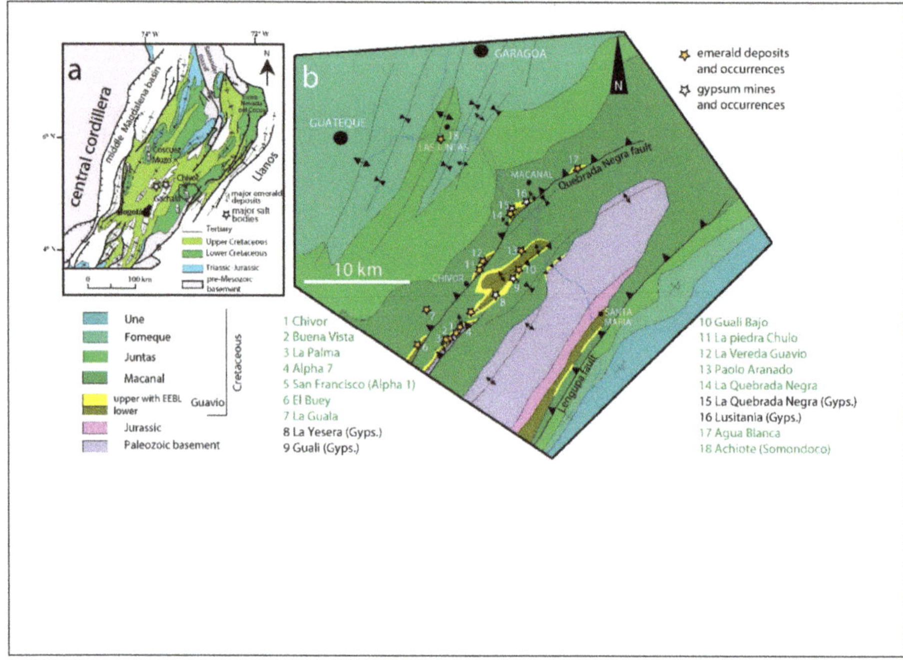

Figure 27. Deposits from the eastern emerald zone in Colombia:(**a**) Simplified geological map of the Eastern Cordillera with the location of the main emerald deposits. Inset is the location of Figure 27b; (**b**) geological map of the Chivor area. All emerald and gypsum deposits and occurrences are hosted within the Berriasian Upper Guavio Formation.

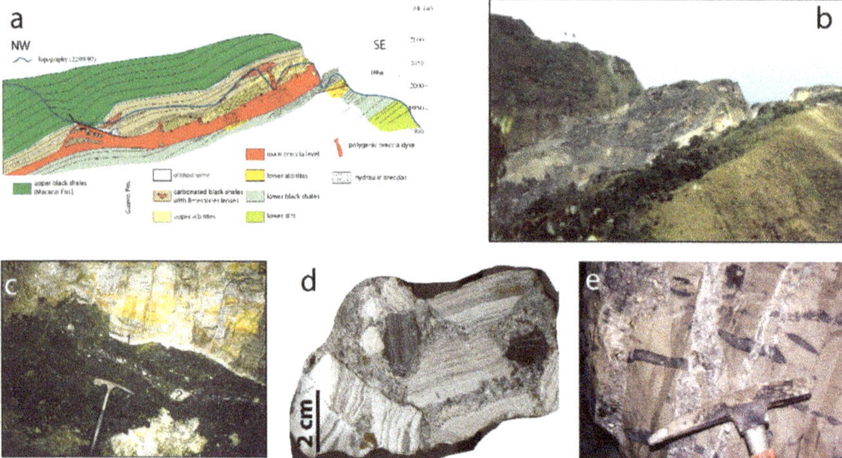

Figure 28. The Chivor mining area: (**a**) Geological cross-section through the Chivor emerald deposits; (**b**) south-eastern field view of the cross-section; (**c**) Chivor Klein pit. Upper contact of the main breccia level (in black) with albitites (1). The transport of clasts of albitite (2) within the breccia is marked by tails; (**d**) Oriente deposit. Polygenic breccia formed by clasts of albitite and black shales, cemented by pyrite, carbonates, and albite; (**e**) Oriente deposit. Carbonate-pyrite-emerald-bearing veins crosscutting albitite showing some remnants of black shale. Photographs **b** to **d**: Yannick Branquet; Photograph **e**: Gaston Giuliani.

All the mineralized structures, i.e., sub-vertical veins, extending from the roof of the brecciated level (Figure 28e), are mineralized listric faults that attest of the bulk extensional structures extending from the brecciated regional evaporitic level. At 65 Ma, to a small amount of horizontal stretching a large flow of hydrothermal fluids responsible for emerald deposition occurred through the regional evaporitic level. The observed structures of the breccia are diagnostic of hydraulic fracturing associated with the evaporite solution within the salt-bearing main breccia level. At the Cretaceous–Tertiary boundary, the overburden of the Guavio Formation in the Chivor area was about 5–6 km [110]. At that time, the area was slightly uplifted in an incipient foreland bulge [111]. This was [111] interpreted as extensional structures observed in the emerald deposits, resulting from flexural extension. Following this model, the emerald-related hydrothermal event recorded an abrupt change in the thermal and dynamic conditions of the Eastern Cordillera basin, which triggered regional-scale hot, deep, and over-pressured brines migration [110].

The western emerald deposits, such as Muzo (Figure 29a) and Coscuez (Figure 30), are characterized by compressive fold and thrust structures formed along tear faults. These tectonic structures are synchronous with the circulation of the hydrothermal fluids and emerald deposition. The deposits are hectometre-sized at most and display numerous folds, thrusts, and tear faults [112]. In the Muzo deposit, thrusts are marked by the calcareous BS, which overly siliceous BS (Figure 29b). All the tectonic contacts are marked by cm- to m-thick hydrothermal breccia called "cenicero", i.e., ashtray, by the local miners (Figure 29c,d). These white- or red-coloured breccias outline the thrust planes, which are associated with intense hydraulic fracturing due to overpressured fluids [113]. Multistage brecciation corresponds to successive fault-fluid flow pulses and dilatant sites resulting from shear-fracturing synchronous to the thrust fault propagation.

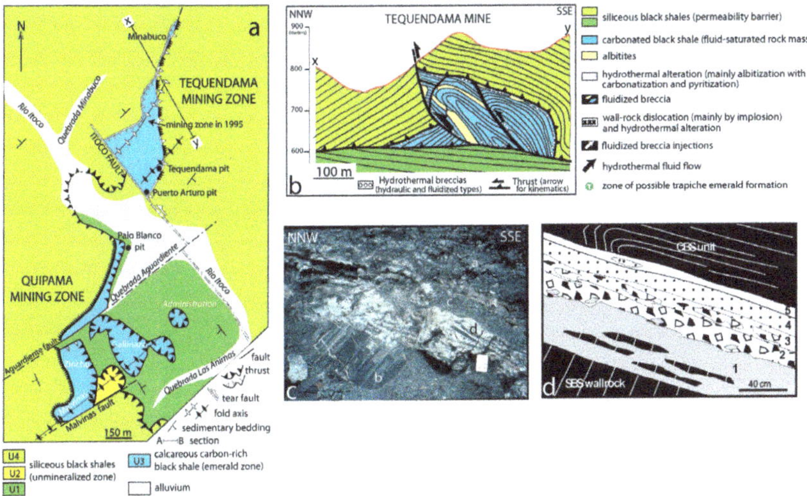

Figure 29. Tectonic style and lithologies of the Muzo emerald mining district, western emerald zone: (**a**) Geological map produced between 1994 and 1996. U1 through U4 represent the different tectonic units. The units, U1, U2, and U4, are comprised of barren siliceous black shales (cambiado), while U3 is composed of emerald-bearing calcareous carbon-rich black shale; (**b**) Tequendama mine cross-section along x-y (see Figure 29a). The cross-section shows the different lithologies and tectonic structures described by [112] and [107,113]. The thrusts and tear faults are associated with the presence of breccia, extensional veins, and related potential zones for trapiche emerald in the wall-rock dislocations and hydrothermal alteration [114]; (**c**) thrust verging NNW at Tequendama mine. The SSE dipping calcareous black shale unit (**a**) thrusts over the NNW dipping siliceous black shale unit (**b**). The siliceous beds are truncated at high angles by the thrust plane characterized by a hydrothermal breccia (**c**) and a thrust-parallel vein (**d**). Photograph: Gaston Giuliani; (**d**) texture and mineralogy of a thrust-parallel

and layered breccia called "cenicero" by the miners, outcrops of Puerto Arturo in 1994—95, Tequendama mine [107,113]. Layering of the breccia zoning: 1 = carbonate thrust vein; 2 = cemented breccia with pyrite and calcite crystals; 3 = tectonically reworked breccia with a red melange cemented by pyrite and ankerite; 4 = red layer totally cemented by pyrite and ankerite ("cenicero rojo"); 5 = pure pyrite layer ("cordón piritoso"). CBS = calcareous black shale; SBS = siliceous black shale.

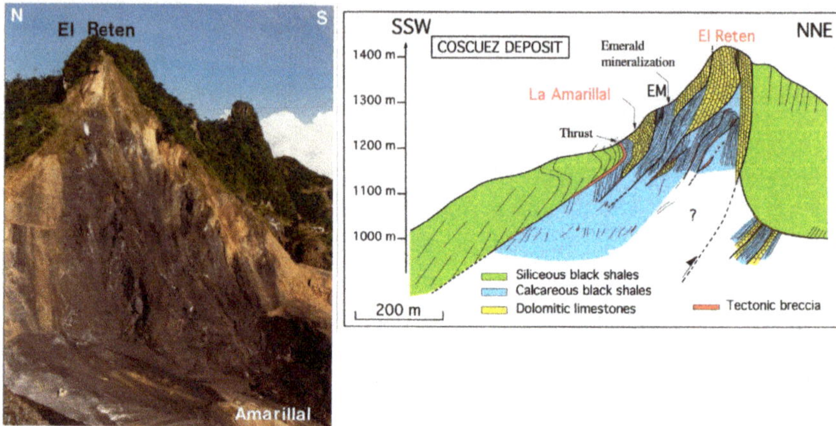

Figure 30. Tectonic style and lithologies of the Coscuez emerald mining district, western emerald zone: (**a**) General view of the Coscuez deposit in 1996. Photograph: Gaston Giuliani; (**b**) geological cross-section of the Coscuez deposit showing the lithostratigraphic column of the formations and the tectonic style marked by thrusts and faults. The emerald mineralization is linked to tectonic hydrothermal breccia zones, faults, thrusts, and stockwork veins [107,113].

The thermal-reduction of sulphate, at 300–330 °C, in the presence of organic matter in the black shale during the formation of the emerald-bearing veins is unique for Be-bearing mineralization [104,115]. Sulphates (SO_4^{2-}) in minerals of evaporitic origin are reduced by the organic matter of the BS to form hydrogen sulphide (H_2S) and hydrogen carbonate bounding (HCO_3^-), which are responsible for the precipitation of pyrite, carbonates, and bitumen in the veins via the following reactions:

$$(CH_2O)_2 + SO_4^{2-} \text{ (evaporitic origin)} \rightarrow Rb \text{ (pyrobitumen)} + 2HCO_3^- + H_2S \tag{3}$$

$$HCO_3^- + Ca^{2+} \rightarrow CaCO_3 \text{ (calcite)} + H^+ \tag{4}$$

$$2HCO_3^- + Ca^{2+} + Mg^{2+} \rightarrow CaMg(CO_3)_2 \text{ (dolomite)} + 2H^+ \tag{5}$$

$$7H_2S + 4Fe^{2+} + SO_4^{2-} \rightarrow 4\,FeS_2 \text{ (pyrite)} + 4H_2O + 6H^+ \tag{6}$$

The consequence of pyrite precipitation is the depletion of iron in the fluid prior to emerald precipitation. This impacted the gemmological features of the emerald as follows: (i) Iron-free absorption spectra with sharp Cr-V bands; (ii) iron-poor chemical fingerprinting; and (iii) optical properties that correlate with the low trace elements content [116].

Although, as noted above, the emerald deposits in Colombia are unique, two minor occurrences that show similar recipes for emerald formation have been reported:

1. An emerald occurrence was described near Mountain River in the northern Canadian Cordillera [117]. These emerald veins are hosted within siliciclastic strata in the hanging wall of the Shale Lake thrust fault. The emerald formed as a result of inorganic thermal-chemical sulphate reduction via the circulation of deep-seated hydrothermal carbonic brines through basinal

siliciclastic, carbonate, and evaporitic rocks (Figure 31). The deep-seated H_2O-NaCl-CO_2-N_2 brines, with a salinity up to 24 wt.% equivalent NaCl, were driven along deep basement structures and reactivated normal faults related to tectonic activity associated with the development of a back-arc basin during the Late Devonian to Middle Mississippian (385–329 Ma). The Mountain River emerald occurrence thus represents a similar and small-scale variation of the Colombian-type emerald deposit model [117].

2. Three emeralds were reported in the Uinta Mountains in Utah, USA [118,119]. The discovery was realized in the Neoproterozoic Red Pine shale, which is overlain by Paleozoic carbonate rocks. Based on the study of fibrous calcite hosted by the Mississippian carbonate units, subjacent to the hypothetic emerald-bearing shale, [120] proposed an amagmatic process for the formation of emeralds. The authors combined their chemical and isotope data on calcite, limestone, and the Red Pine shale using a model of formation similar to the Colombian type and involving thermal reduction of sulphates between 100 and 300 °C. Emeralds were not described [120] and one can question their existence in the Uinta Mountains. No one has described these emeralds since the discovery of [118].

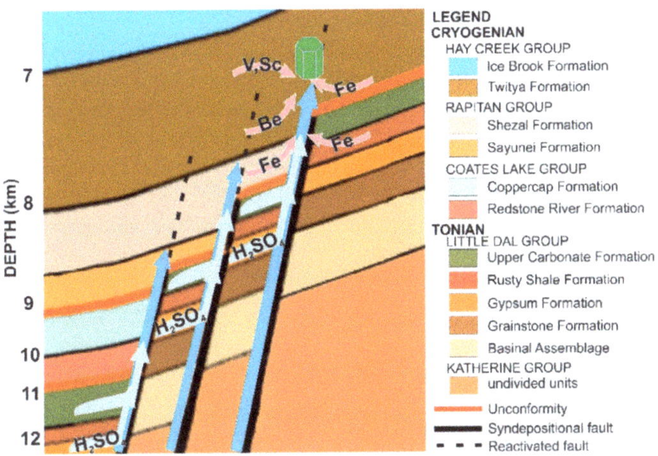

Figure 31. Idealized schematic model for the emerald mineralization of the Mountain River emerald occurrence located in Devonian to Mississippian platform sediments [117]. Emerald resulted from inorganic thermochemical sulphate reduction (light blue arrow) via the circulation of brines along a reactivated normal fault (dark blue arrow). The high-salinity brines result from the dissolution of evaporites lenses in the sediments. Be, V, Sc, and Fe were mobilized from the sedimentary formations (pinkish arrows).

6.2.3. Sub-Type IIC: Tectonic Metamorphic-Related Emerald Deposits Hosted in Metamorphic Rocks Other than M-UMR and Black Shales (Afghanistan, China, USA; Table 2)

This sub-type includes emerald-bearing quartz vein and veinlet deposits located in medium pressure metamorphic rocks from the greenschist to granulite facies:

1. The Panjsher emerald deposits in Afghanistan (Figure 1a) are located in the Herat-Panjsher suture zone along the Panjsher Valley. The suture zone, which marks the collision of the Indo-Pakistan plate with the Kohistan arc sequence, contains a number of faults, such as the Herat-Panjsher strike-slip fault, which was mainly active during the Oligocene-Miocene [121]. The emerald deposits lie southeast of the Herat-Panjsher Fault in the Khendj, Saifitchir, and Dest-e-rewat Valleys. The deposits are hosted in the Proterozoic metamorphic basement formed by migmatite, gneiss, schist, marble, and amphibolite. The basement is overlain to

the northwest by a Paleozoic metasedimentary sequence crosscut by Triassic granodiorite [86]. During the Oligocene, the Proterozoic rocks of the Panjsher valley were affected by the intrusion of granitoids [86] 20 km north and south of the emerald mining district [122]. The emerald deposits are hosted by metamorphic schists that have been affected by intense fracturing, fluid circulation, and hydrothermal alteration, resulting in intense albitization and muscovite-tourmaline replacements [122]. Emerald is found in vugs and quartz veins associated with muscovite, tourmaline, albite, pyrite, rutile, dolomite, and Cl-apatite [122]. Ar-Ar dating on a muscovite from the emerald-bearing quartz veins at the Khendj mine gave an Oligocene age of 23 ± 1 Ma [123]. At the moment, the sources of Cr and Be remain unclear.

2. The Davdar emerald deposit is located in the western part of Xinjiang Province, China. The deposit is formed by emerald-bearing quartz-carbonate veins associated with a major northwest-southeast trending fault zone [18]. The deposit is hosted by lower Permian meta-sedimentary rocks, including sandstone, dolomitic limestone, siltstone, and shale, which have been metamorphosed at upper greenschist conditions [124] to produce metasedimentary host rocks, which include quartzite, marble, schist, and phyllite, prior to the emplacement of the emerald-bearing veins. Basaltic dykes of an unknown age, which are up to 10 m wide and crop out along strike lengths of up to 200 m, are the only intrusive igneous rocks known in the area. The dykes are emplaced along the northwest-southeast fault zone; no visible contacts are exposed between the dykes and the emerald-bearing veins. The emerald-bearing veins, which are up to about 20 cm wide, contain epidote, K-feldspar, tourmaline group minerals, carbonates, and iron oxides. Quartz and emerald crystals up to a few centimeters long are found in the veins. Alteration haloes up to a few cm wide occur around the veins. In the sandstone and dolomitic limestone, the alteration halo is barely visible, but it is conspicuous as a bleached white halo in the phyllite. The alteration halo is generally enriched in fine-grained silica with variable amounts of quartz, biotite, muscovite, feldspar, carbonate, and tourmaline. It is representative of a retrograde metamorphic assemblage typical of greenschist facies minerals (epidote, plagioclase, potassic feldspar, quartz, biotite, and chlorite). Emerald typically occurs in the quartz veins and not in the host rocks or the alteration haloes.

3. Hiddenite emerald was discovered in North Carolina northeast of the community of Hiddenite in 1875. Since then, a number of notable samples have been discovered, primarily from the Rist and North American Gem mines [125]. Over 3500 carats of emerald were extracted from the latter in the 1980s, including the 858 ct (uncut) "Empress Caroline" crystal [126]. At the Rist property the emeralds occur in quartz veins and open cavities (50% of the veins) that occupy NE-trending sub-vertical fractures in folded metamorphic rocks [126,127]. The hiddenite area is underlain by Precambrian migmatitic schists, gneisses, and interlayered calc-silicate rocks, metamorphosed in the upper amphibolite facies. The area is locally intruded by the Rocky face leucogranite. The quartz veins range in size from 2 to 100 cm wide, 30 cm to 7 m long, and 10 cm to 5 m high. Most of the veins are not interconnected and represent tensional gash fractures that sharply crosscut the prominent metamorphic fabric of the host rocks, suggesting that they formed during late or post metamorphic brittle-ductile deformation [126]. Wise and Anderson [127] identified four cavity assemblages: (1) An emerald-bearing assemblage composed of albite, beryl, calcite, dolomite, siderite, muscovite, cryptocrystalline quartz, rutile, and sulfides with clays; (2) a Cr-spodumene-bearing assemblage, which includes calcite, muscovite, and quartz. The green Cr-bearing spodumene, locally referred to as "hiddenite", occurs in only minor amounts; (3) a calcite assemblage dominated by calcite and quartz; and (4) an amethyst assemblage characterized by amethystine quartz, calcite, muscovite, and chabazite. Emerald and spodumene rarely occur together in the same vein or cavity. Within the emerald-bearing cavities, beryl crystals up to 20 cm in length are closely associated with dolomite, muscovite, and quartz. The crystals are typically color-zoned with a pale green to colorless core and an emerald-green rim. Speer [126] described the veins and reported that the emeralds occur as free-standing crystals attached to

cavity walls and as individual collapsed fragments. The collapsed crystals that have fallen from walls of the cavities show cementation phenomena, while the attached crystals exhibit dissolution, re-growth, and over-growth. Bleached wall-rock alteration halos up to 9 cm wide and rich in silica and chlorite are commonly peripheral to veins and crystal cavities. Wise and Anderson [127] pointed out that the emerald and Cr-spodumene mineralization in quartz veins and cavities is similar to what is seen in alpine-type fissures. In the absence of a pegmatitic or granitic body, the source of Be and Li remains in question; the source of Cr and V is also uncertain, given that M-UMR are unknown in the area. Speer [126] specified that the veins originated as hydrothermal filling of tensional sites during the waning ductile/brittle stages of metamorphism. Apparently, the geological setting and genesis of the Hiddenite emerald occurrences are unique.

6.2.4. Sub-Type IID: Tectonic Metamorphosed or Remobilized Type IA Deposits, Tectonic Hidden Granitic Intrusion-Related Emerald Deposits, and Some Unclassified Deposits (Egypt, Australia, perhaps also Brazil, Austria, Pakistan, Zambia; Table 2)

This sub-type includes deposits probably genetically linked to hidden granitic intrusions (Swat valley) and those where metamorphism has blurred the distinction between metamorphic and magmatic origin (Habachtal, Eastern desert in Egypt). This sub-type permits reclassification and debate on the genesis of several deposits, including the following located in M-UMR:

1. Those for which the genesis is considered to be the consequence of regional metamorphism but with multi-stage emerald formation (Eastern desert of Egypt).
2. The sub-type IIA (Santa Terezinha de Goiás, Habachtal, and Swat Valley) where metamorphic-metasomatic deep crustal fluids circulated along faults or shear zones and interacted with M-UMR with apparently no magmatic intrusion.
3. The mineralization stages for the Poona deposit where emerald and ruby are associated.

It is also useful for considering deposits in meta-sedimentary rocks for which insufficient geological knowledge renders the genesis and classification uncertain (Panjsher Valley and Davdar deposits) and for the Hiddenite occurrences where the source of Be and origin of the mineralizing fluids are unknown.

The Egyptian emerald occurrences of Gebels Zabara, Wadi Umm Kabu, and Sikait occur in a N-W trending band circa 45 km long in the Nugrus thrust [128–130]. The deposits are located in a volcano-sedimentary sequence featuring an ophiolitic tectonic melange composed of metamorphosed M-UMR overlying biotite orthogneiss. Syntectonic intrusions of leucogranites and pegmatites occurred along the ductile shear-zone [43]. The study of [130] described three beryl-emerald generations that crystallized during magmatic, post-magmatic hydrothermal, and regional-metamorphic events. The genetic succession was based on chemical and microstructural studies. The original colorless Cr-poor beryl and phenakite of pegmatite origin has been partly replaced by the formation of Cr-rich beryl (emerald) through K-Mg metasomatism. At the Gebel Sikait, [130] described the occurrence of emerald in phlogopitites (i) at the contact between meta-pegmatite, meta-pelite, and meta-greisen veinlets of up to 10 cm in thickness and (ii) in folded quartz layers. At Gebel Zabara, the emerald is either in phlogopitites and talc-carbonate-chlorite-actinolite schists present in the serpentinite bodies, or in quartz veins. At Gebel Umm Kabo, emerald is within phlogopitites in contact with small lenses of quartz. The dating of phlogopite by K-Ar returned ages of 520 to 580 Ma [131] and by Rb/Sr returned ages of 591 ± 5.4 Ma, confirming the Panafrican orogen.

Based on microtextures, [128,129] suggested that emerald formation occurred during low-grade regional metamorphism. The emerald formation was controlled in detail by the local availability of Be present in the beryl-bearing meta-pegmatite and quartz veins, Cr in the meta-M-UMR, and the metamorphic fluids, all in the context of the late Pan-African tectonic-thermal event. This genetic model has been challenged [43], who pointed to the intrusions of syntectonic leucogranites with the presence of greisens and beryl-emerald-bearing pegmatites and quartz veins along the shear zones. Fluid inclusion studies have shown the presence of H_2O-NaCl-CO_2-CH_4 fluids with a salinity between 8 to 22 wt.% equivalent NaCl and a temperature of homogenization between 260 and 390 °C [43]. The oxygen isotope data for emeralds were consistent with both magmatic and metamorphic origins for the source of the mineralizing fluid [132]. Grundmann and Morteani [130] confirmed the existing $\delta^{18}O$ values with new isotopic data in the range of 9.9 to 10.7‰. They concluded that the complex interplay of magmatic and regional magmatic events during the genesis of the emeralds makes it impossible to relate their genesis to a particular event. The Pan-African regional metamorphic model is consistent with the remobilization of syntectonic Cr-poor beryl quartz veins and beryl-phenakite-bearing pegmatites.

The Egyptian emerald occurrence is a good example of the proposed sub-type IID emerald deposit: Magmatism, deformation, and remobilization by metamorphic-metasomatic fluids of the mixed Be and Cr reservoirs. The formation of emerald occurred during a regional tectonic event with the syntectonic intrusion of leucogranites and the injection of pegmatites, thrust and shear zone deformation accompanied by fluid circulation (with probable mixing of magmatic and metamorphic fluids), and reaction with rocks of different composition. The remobilization of Be, Cr, and V occurred at the same time through continuing regional tectonic activity. The oxygen isotopic composition of emerald with $\delta^{18}O$ values around 10‰ is similar to the oxygen magmatic signatures found for other worldwide type IA emerald deposits [132].

With the proposed sub-type IID, the genesis of the Santa Terezinha de Goiás, Habachtal, Swat Valley, and Poona deposits can be discussed and finally classified using the chemistry of the emerald and O-H stable isotopes.

At the Santa Terezinha de Goiás deposit, the ductile-fragile deformation coeval with the mineralization was strongly assisted by fluids migrating along shear planes under lithostatic fluid pressure at 500 °C [15]. The O-H isotopic composition of phlogopite and emerald is consistent with both magmatic (evolved crustal granites) and metamorphic fluids. This hypothesis was also proposed [133] based on fluid inclusion studies showing the mixing between carbonic and aqueous fluids. Nevertheless, considering the absence of granites and pegmatites in the underground mine, which reaches depths of up to 400 m [134], the low beryllium concentration in the Santa Terezinha volcano-sedimentary series (Be < 2 ppm), the lack of tourmaline in the metasomatic rocks, the control of the mineralization by shear zone structures, and the CO_2-H_2O-NaCl-($\pm N_2$) composition of the fluids, a metamorphic origin was preferred for the parental fluids of the emeralds [15,76]. The metamorphic hypothesis involves some input of Be-bearing metamorphic fluids released at the greenschist-amphibolite transition (T = 400–500 °C) or fluids generated at higher grades of metamorphism and channeled along transcrustal structures at the brittle-ductile transition. Such specific features are similar to those found for gold deposits in the vicinity of the emerald deposit in the Goiás metallogenetic province. The mineralizing fluids are channeled along lineaments and second order structures where CO_2 unmixing, wall-rock interaction, and concomitant ore precipitation are promoted during temperature and pressure fluctuations.

Chemical analyses of Santa Terezinha de Goiás emeralds led [65] and [23] to question the proposed metamorphic origin [76]. The emeralds have the highest Cs contents ever reported for emerald, with values between 907 and 980 ppm, and with Li contents between 142 and 155 ppm. The high content of Cs supports another hypothesis that magmatic fluids mixed with metamorphic fluids [23]. Following this genetic scheme, despite the absence of pegmatites and granites up to a 400 m depth in the mine and along the metamorphic strike, the influence of the magmatic fluids is evidenced by the chemistry of emerald. In such a ductile shear zone environment, the intrusion of felsic granitoids into the volcano-sedimentary sequences along thrusts is common [135] and could have happened at the Santa Terezinha deposit. This new chemical result confirms the hypothesis proposed by D'el-Rey Silva and Barros Neto [92] that the probable source of Be for emerald was the intrusion of syntectonic two-mica granite at São José do Alegre, located 5 km to the southwest of the emerald deposit. Whole rock Sm/Nd data from the granite and the volcano-sedimentary sequence at Santa Terezinha yielded ages of 510 ± 110 Ma (n = 6) and 556 ± 77 Ma, respectively [92,136]. $^{40}Ar/^{39}Ar$ ages on phlogopite grains from the emerald-bearing phlogopitites yielded ages of 550 ± 4 and 522 ± 1 Ma, respectively [137]. These ages show considerable overlap in a large window, which characterizes the Brasiliano orogenesis [138].

At the Habachtal deposit, the emerald-bearing phlogopitites called the "blackwall zone" are located at the tectonic contact between the orthogneisses and amphibolites of the Habach group. The metasomatic "blackwall zone" is formed in sheared melange zones surrounding tectonic lenses of the serpentinite-talc series at the contact with the tourmaline-garnet-mica-bearing metapelitic unit of the Habach Group.

Detailed textural studies on emerald-tourmaline and plagioclase porphyroblasts [77,93,95] recorded three metamorphic episodes and crystallization for these minerals. Deformation enhanced fluid circulation and metasomatic reaction and produced the emerald-tourmaline-bearing phlogopitites. Fluid inclusions in the emerald show similar characteristics to those in syn-metamorphic Alpine fissures in the Habach Formation. Grundmann and Morteani [77] proposed the genesis of emerald through syntectonic growth during regional metamorphism.

Trumbull et al. [97] used B isotopes of coexisting tourmaline in the metapelites and phlogopitites. The δ^{11}B isotope values suggest that two separate fluids were channelled and partially mixed in the shear zone during the formation of the metasomatic rocks. A regional metamorphic fluid carried isotopically light B as observed in the metapelite ($-14 < \delta^{11}$B < -10‰) and a fluid derived from the serpentinite association carried isotopically heavier B ($-9 < \delta^{11}$B < -5‰) typical for Middle Oceanic Ridge Basalt or an altered oceanic crust.

Grundmann and Morteani [77] pointed out that the source of Be was either the Be-rich garnet-mica schist series or the biotite-plagioclase gneisses (Be up to 36 ppm). Zwaan [52] was critical of this interpretation and warned that, in cases where pegmatitic sources of Be are not apparent, one must proceed with caution since fluids can travel far from granites and pegmatites. He pointed out that pegmatites do occur in the Zentral gneiss and the Habachtal emeralds contain up to 370 ppm of Cs [139], which suggest a pegmatitic source. However, the Cs data produced [23] are very different (79 < Cs < 157 ppm). The possibility of metamorphism of a pre-existing Be-bearing felsic rock occurrence cannot be excluded and must be considered when constructing a model in a medium- to high-grade metamorphic regime [16]. The question is similar to what has been reported for the origin of tungsten for the Felbertal scheelite deposit [139], which is located in the same metamorphic series as the Habachtal emeralds and is now considered to represent metamorphic remobilization of a Be-W-enriched Variscan granite [140]. This hypothesis is likely correct for emerald because (1) Hercynian aquamarine-bearing pegmatites were found in the Habach series [141] and (2) scheelite disseminations with chalcopyrite and molybdenite in the banded gneiss series of the Habachtal emerald deposit are drawn in the lithologic cross-section presented [77]. Following this hypothesis, the source of Be is magmatic.

The Swat-Mingora–Gujar Kili–Barang emerald deposits are thrust controlled and there is no magmatic or pegmatite intrusions visible in the field [86,101]. Emerald is either disseminated in carbonate-talc-fuchsite-tourmaline-quartz schists or in quartz veins and a network of fractures in magnesite rocks. The mean oxygen isotopic composition of emerald is remarkably uniform at δ^{18}O = 15.6 ± 0.4‰. The mean hydrogen isotopic composition of the channel waters is δD = -42.2 ± 6.6‰ and that of the fluid calculated from hydrous minerals, such as tourmaline and fuchsite, is δD = -47 ± 7.1‰. These O-H isotope data are consistent with both metamorphic and magmatic origins [101]. However, a magmatic origin is favored because the measured δD values of fuchsite and tourmaline are comparable to those found for muscovite and tourmaline from granites, such as the Makaland granitoid, exposed 45 km to the southwest of Mingora. The mineralization was probably caused by modified ^{18}O-enriched hydrothermal solutions derived from an S-type granitic magma [101].

The magmatic model proposed for Pakistani emerald deposits can be constrained by the ^{40}Ar/^{39}Ar ages obtained on the different rocks: 83.5 ± 2 Ma for the Shangla blue-schist melange, 22.8 ± 2.2 Ma for the tourmaline-beryl-fluorite-bearing Makaland Granite [142], and 23.7 ± 0.1 Ma for a fuschite mica from a quartz vein in the Swat emerald deposit [100]. However, the chemical data for Swat emerald presented [23] show low Cs contents (61 to 74 ppm), which are unusual for granite-related emeralds.

Finally, the genesis of the Swat emerald deposits can result from both metamorphic and magmatic contributions, with up to now a magmatic source not identified in the emerald mining districts.

At the Poona deposit, three styles of emerald mineralization have been identified in M-UMR from the Precambrian series of the northern Murchinson Domain [143]: (a) Emerald in phlogopitites formed at the contact of beryl-granite- muscovite-bearing pegmatites and M-UMR; (b) emerald with ruby-sapphire, topaz, and alexandrite in banded fluorite-margarite-beryl-bearing banded greisens in phlogopitites; and (c) quartz-margarite-topaz-bearing veins in phlogopitites.

A multi-stage mineralizing episode was proposed [143] of: First, the intrusion of granites (probably between 2724–2690 Ma) and circulation of fluids in the M-UMR produced greisens and quartz veins with topaz, beryl, quartz, and muscovite. The first episode was affected by regional metamorphism of greenschist to lower amphibolite facies. Metasomatic reactions occurred at the borders of the greisen zones with the formation of ruby, alexandrite, and emerald. The third episode corresponded to the retrograde phase of metamorphism (probably between 2710–2660 Ma), where corundum and alexandrite were partially or totally replaced by margarite, muscovite, and/or emerald.

Fluid inclusions in emerald combined with the O-H isotope compositions of both lattice and channel fluids of emerald confirmed multiple origins yielding both igneous and metamorphic signatures [16]. This emerald-ruby association is unique worldwide and merits more petrologic and geochemical studies before proposing a coherent genetic scheme.

The genesis of the Panjsher Valley [98] and Davdar [18] emerald deposits is not well understood due to difficulties with access and poor exposure. The similar geographic and geologic environments indicate that these two deposits may share a similar genetic model. Both deposits are hosted in layered meta-sedimentary rocks, with metamorphic facies ranging up to lower amphibolite. Emerald occurs in both veins and host rocks. The veins are predominantly composed of quartz and carbonate, with minor amounts of albitic plagioclase, phlogopite, tourmaline, scheelite, and pyrite. The host rocks vary from shale to carbonate and are thought to be Paleozoic. Proximal to the veins, hydrothermal alteration is dominated by quartz and calcite with lesser amounts of albite, phlogopite, tourmaline, and pyrite. In both localities, there are mafic to felsic intrusions as stocks, dykes, or sills. However, no clear relationship between the intrusive rocks and emerald mineralization has been established. The high-salinity fluids [18,56,144] and meta-sedimentary host rocks combined with the lack of observed igneous association could also be compatible with a tectonic metamorphic-related (type IIB) formational model, but more field work needs to be carried out on these deposits to map the local geology to prove or disprove an igneous link.

Emeralds from the Musakashi deposit in Zambia are of high quality with a bluish color very different from that observed for emerald originating from the Kafubu mining district [63]. Discovered in 2002, the deposit is located ~150 km west of the Kafubu mines. The geology of the deposits is unknown, however, small fragments of emerald were discovered in eluvium adjacent to quartz veins [50]. The internal features, chemical composition, solids, and three-phase fluid inclusions are quite different from those of Kafubu emeralds. Saeseasaw et al. [63] examined three-phase fluid inclusions (halite + vapor + liquid) in emerald from Musakashi, Panjsher Valley, Davdar, and Colombia. They conclude that Musakasi fluid inclusions look like those from Colombia. Nevertheless, the photographs show that they are polyphase and contain a cube of halite and a rounded salt, which is probably sylvite (KCl). Such multiphase inclusions with both halite and sylvite are found in both Panjsher and Davdar emeralds [62]. In addition, the chemistry of these emeralds overlaps with those from Colombian, Panjsher, and Davdar [63]. The geological setting is not precisely known, but the emeralds are very different from those for type IA deposits as shown by their very low Cs contents (3 < Cs < 10 ppm), which are similar to those found for Colombian and Davdar emeralds [63].

The genesis of the Hiddenite deposits remains obscure in terms of the sources of Be, Cr, and V, and the origin of the mineralizing fluids. Fluid inclusions are characterized as two populations of aqueous-carbonic fluids with two populations, those having high- and those having low-CO_2 contents, which underwent immiscibility between 230 and 290 °C, respectively [145]. The deposits are interpreted as late and low temperature metamorphic hydrothermal alpine-type quartz veins cutting meta-sedimentary and migmatitic biotite gneiss. The low temperature hydrothermal mineralization is confirmed by the association of quartz, carbonates, muscovite, and chabazite. The origin of the fluids must be clarified by the O-H isotopes' compositions of the different minerals associated with emerald, but a metamorphic origin has advanced [126,127].

7. Fluid Inclusions in Emerald

Emerald is one of the best hosts for fluid inclusions. It is commonly idiomorphic, well preserved, zoned, and displays growth zones optically, chemically, and especially via cathodoluminescence [20,85,117]. These growth zones facilitate the identification of primary vs. secondary fluid inclusions. Additionally, primary fluid inclusions often form during emerald precipitation and are elongated parallel to the host's *c* axis (Figure 32). The determination of fluid inclusion chemistry is generally limited to microthermometry [19,146,147], with more refined analyses performed via bulk leachate analyses or LA-ICP-MS or secondary ion mass spectrometry (SIMS) on quartz-hosted fluid inclusions petrographically determined as synchronous to emerald hosted inclusions [148,149]. In addition to fluid chemistry, fluid inclusions studies have proven most useful in determining the pressures and temperatures of emerald formation [47] and for determining if boiling is responsible for emerald colouration [19].

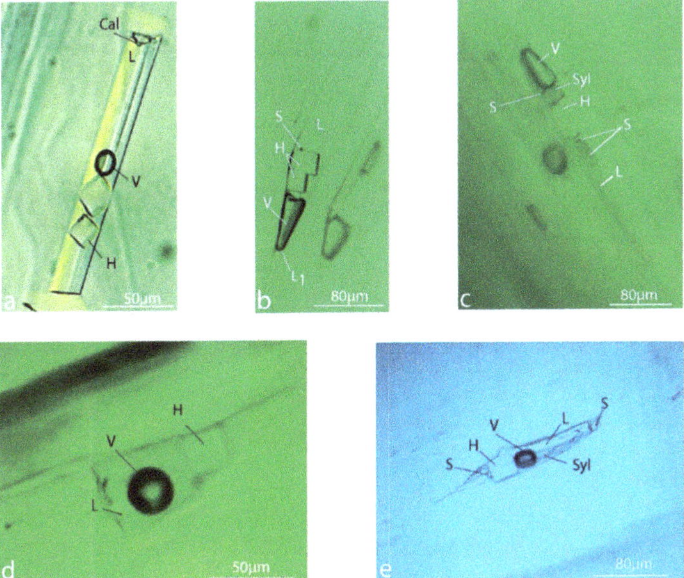

Figure 32. Primary fluid inclusions in emeralds: (**a**) Colombian fluid inclusion presenting a jagged and shredded outline. The cavity contains 75 vol.% of salted water solution (L), 10 vol.% of gas corresponding to the vapour bubble (V), 15 vol.% of halite (NaCl) daughter mineral, and a crystal of carbonate (Ca); (**b**) fluid inclusion in a Colombian emerald showing three cubes of halite (H), the liquid phase (L), the contracted vapour phase (V), a minute black phase (S), and a thin rim of liquid carbon dioxide (L₁) rim visible at the bottom part of the vapour phase; (**c**) multiphase fluid inclusions from Panjsher emerald (Afghanistan). They contain vapour and liquid phases (V + L), a cube of halite (H),

usually a primary sometimes rounded salt of sylvite (Syl), and aggregates of several anisotropic grains (S). The volume and the concentration of NaCl and KCl are different from those observed in Colombia. The overall salinity is estimated to 30 to 33 wt.% eq. NaCl (Vapnik and Moroz, 2001; (**d**) three-phase fluid inclusion in emerald from Nigeria emerald showing halite (H), liquid (L), and vapour (V) phases. Generally, the primary halite-bearing fluid inclusions are associated with coeval monophase or biphase (V + L) fluid inclusions; (**e**) multiphase fluid inclusion from the Davdar emeralds (China). The cavities contain liquid (L) and vapour (V) phases with daughter minerals, like halite (H) and sometimes sylvite (syl), and aggregates or multiple solid inclusions (S). The morphology of the cavities and the infilling looks like those found in emeralds from Afghanistan. Photographs: Gaston Giuliani.

Emerald-hosted fluid inclusions are, in general, aqueous dominant, with a wide range of salinities from dilute to salt saturated (Table 3). Numerous emerald deposits also have gaseous species contained within the fluid inclusions; the gaseous phases are generally dominated by CO_2, but other species, such as CH_4, N_2, and H_2S, have been identified via Raman spectroscopy in a number of emerald studies [47,150,151]. Raman spectroscopy can also be used to identify accidental and daughter inclusions within fluid inclusions [152]. Raman analyses of emerald hosted inclusions often prove challenging, as beryl/emerald is generally fluorescent and thus the inclusion spectrum is lost in the fluorescence from the host. However, different laser wavelengths and confocal Raman spectrometers can be used to limit the effects of host fluorescence; these applications to gases and solid inclusions contained within fluid inclusions were reviewed [153–155].

Bulk leachate analyses of fluid inclusions by thermal decrepitation or crushing are limited by contamination via the emerald host. However, leachate analyses of fluid inclusions hosted in quartz precipitated synchronously with emerald have proven successful for general chemistry and especially for halogen sourcing [105]. A further limitation of the bulk methods is the presence of multiple fluid inclusion generations and careful detailed petrographic studies should be undertaken to determine if specific quartz and emerald-hosted fluid inclusions are amenable to bulk techniques.

Hydrogen isotopes [156] are generally determined from trapped channel fluids, which are analogous to primary fluid inclusions. The advantage of channel fluid extraction is that channel fluids are normally three (or more) orders of magnitude more abundant than fluids trapped in fluid inclusions. The oxygen isotope signature of fluids responsible for emerald precipitation is generally determined via the measurement of structural oxygen within emerald/beryl or from other synchronously precipitated silicates. The hydrogen isotope signature of the fluids may also be inferred from the analyses of synchronously precipitated hydrous silicates, such as mica or tourmaline [101]. Additionally, the extraction of the channel fluids may also yield a quantitative determination of the weight percent of H_2O present in the emerald and this can be used to complement electron microprobe and LA-ICP-MS analyses to determine emerald chemistry.

Fluid fingerprinting to determine emerald provenance has proven incredibly useful. The pioneering studies [15,76] are consistently used to determine provenance and fluid sources, as done for the Biintal occurrence in Switzerland ([157]; Figure 33). Halogen bulk leachate analyses [105] can test for the presence of fluid interaction with evaporitic source rocks as well as providing more detailed information on the various dissolved salts present in the fluid inclusions. Although not yet routinely used, LA-ICP-MS, and to a lesser extent SIMS, can provide very detailed chemical analyses of formation fluids of emerald deposits, as they provide precise analyses of most of the elements of the periodic table and potentially their isotopes.

Table 3. Fluid inclusion data and oxygen isotope composition of several emerald deposits worldwide following the enhanced classification proposed for emerald deposits.

Type of Environment and Deposit	Tectonic Magmatic-Related Granitic Rocks in M-UM and SR (Type I)	Tectonic Metamorphic-Related	
		(Meta) Sedimentary Rocks (Type IIB)	Metamorphic Rocks (Types IIA- IIC-IID)
Temperature	300-680°C	300-330°C*	350-400°C [*1] 260-550°C [*2]
Pressure	0.5 to 7 kbar	3 to 4 kbar	1.6 kbar [*1D] 4-4 to 5 kbar [*2]
Salinity	2 to 45 wt.% eq. NaCl	40 wt.% eq. NaCl	30 to 33 wt.% eq. NaCl [*1P] 35 to 41 wt.% eq. NaCl [*1D] 2 to 38 wt.% eq.NaCl [*2]
Composition	H_2O-NaCl-($\pm CO_2$)-($\pm N_2$)-($\pm CH_4$)-(K,Be,F,B,Li,P,Cs)	H_2O-NaCl-($\pm CO_2$)-($\pm N_2$)-($\pm$hydrocarbon liquid)-F15(K,Mg,Fe,Li,SO4,Pb,Zn)	H_2O-NaCl-($\pm CO_2$) [*1D] H_2O-NaCl-KCl-$FeCl_2$-($\pm CO_2$) [*1P] H_2O-NaCl-($\pm CO_2$) [*2] CO_2 or CO_2-H_2O-NaCl-($\pm N_2$ ($\pm CH_4$) [*2] or H_2O-CH_4-CO_2-NaCl [*2]
Oxygen isotopes	$6.0 < \delta^{18}O < 15‰$	$16.2 < \delta^{18}O < 24.5‰$	Panjsher: $13.25 < \delta^{18}O < 13.9‰$ Davdar: $14.4 < \delta^{18}O < 15.8‰$ Sta Terezinha: $12.0 < \delta^{18}O < 12.4‰$ Gravelotte: $9.5 < \delta^{18}O < 9.7‰$ Gebel Sikait: $9.8 < \delta^{18}O < 10.7‰$ Habachtal: $6.5 < \delta^{18}O < 7.3‰$
Origin of the fluid	Metasomatic-Hydrothermal	Basinal brines that have dissolved evaporites	Metamorphic-Metasomatic

Salinity in wt.% eq. NaCl = weigth per cent equivalent NaCl; kbar = kilobars; $\delta^{18}O$ = ratio $^{18}O/^{16}O$ in per mil (‰). Type IIB deposits: * = Colombia. Some Types IIA, IIC, IID: [*1] = deposits of Panjsher ([P]) and Davdar ([D]); [*2] = deposits of Santa Terezinha de Goiás (Brazil), Gravelotte (South Africa), Gebel Sikait (Egypt), Habachtal (Austria)

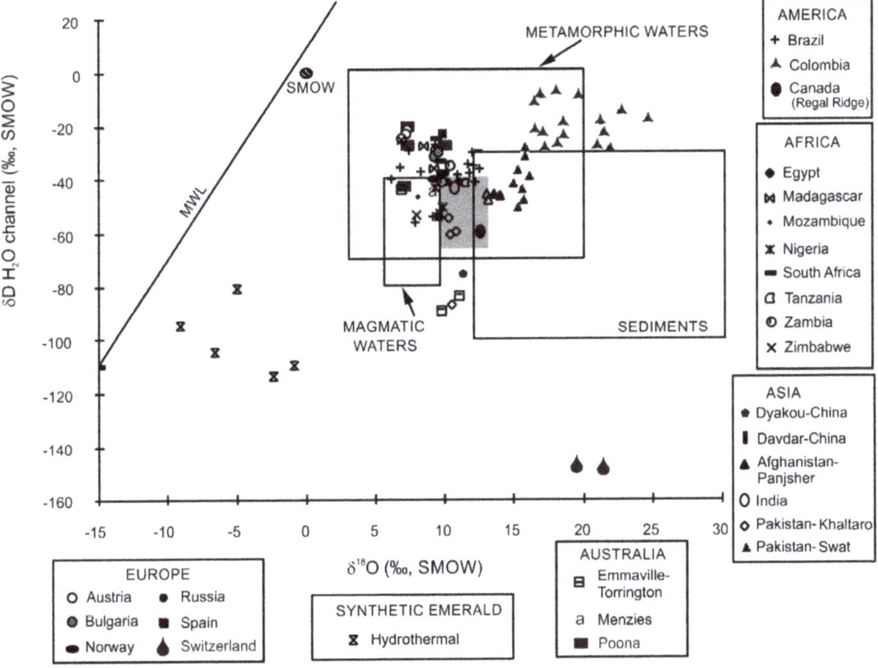

Figure 33. Channel δD H_2O versus $\delta^{18}O$ for emerald worldwide [15,76,157,158]. The isotopic compositional fields are from [159], including the extended (Cornubian) magmatic water box (grey). MWL = Meteoric Water Line, SMOW = standard mean ocean water.

The use of fluid inclusion petrography combined with microthermometry and halogen and cation bulk leachate analyses is useful for the discussion of the source of salts present in fluid inclusions [160]. The example of multi-phase halite-bearing fluid inclusions in emeralds (Figure 32) from Colombia (Eastern and Western emerald zones) and Afghanistan (Panjsher Valley) is very illustrative.

In Colombia, fluids trapped by emerald are commonly three-phase fluid inclusions (Figure 32a) characterized by the presence of a daughter mineral, i.e., halite (NaCl). At room temperature, the cavities contain 75 vol.% of salty water, i.e., aqueous brine (liquid H_2O), 10 vol.% of gas corresponding to the vapor bubble (V), and 15 vol.% of halite daughter mineral (H). However, some Colombian emeralds have multiphase fluid inclusions presenting a liquid carbonic phase (CO_2) forming up to 3 vol.% of the total cavity volume (Figures 25b and 32b), minute crystals of calcite (Figure 32a), very rare liquid and gaseous hydrocarbons [38], and sometimes two or three cubes of halite (Figure 32a,b), and sylvite (KCl).

The high Cl/Br ratio of the fluids (between 6300 and 18,900) indicates that the strong salinity of the brines is derived from the dissolution of halite of an evaporitic origin (Figure 34a; [105]). Cation exchanges, especially calcium, with the black shale host rocks are strong when compared to most basinal fluids (Figure 35; [161]) and are due to the relatively high temperature of the parent brines of emerald (T~300 °C). Indeed, these fluids are enriched in Ca (16,000 to 32,000 ppm), base metals (Fe ~5000 to 11,000 ppm; Pb ~125–230 ppm; Zn ~170–360 ppm), lithium (Li~400–4300 ppm), and sulfates (SO_4 ~400–500 ppm). In comparison, they have a composition and Fe/Cl and Cl/Br ratio similar to the fluids of the geothermal system of the Salton Sea in California (Figure 34b; [162]). The K/Na ratios confirm the Na-rich character of the fluids and the strong disequilibrium between K-feldspar and albite, as shown by the huge albitisation of the black shale (Figure 35).

In Afghanistan, primary multiphase halite-sylvite-bearing fluid inclusions (Figure 32c) are common for the Panjsher emeralds [15,86]. The fluids associated with emerald have total dissolved salts (TDS) between 300 and 370 g/L and the trapping temperature of the fluid is about 400 °C [144].

Crush-leach analyses of fluid inclusions indicate that the fluids are Cl-Na-rich and contain sulfates ($140 < SO_4 < 4300$ ppm) and lithium ($170 < Li < 260$ ppm), but very low to zero fluorine contents [160]. The K/Na ratio of the fluid inclusion confirms the disequilibrium, at ~400 °C, between K-feldspar and albite that drives the Na-metasomatism of the metamorphic schists and the deposition of albite in the veins (Figure 35). Crushing demonstrates that fluids are dominated by NaCl with Cl/Br ratios much greater than that of seawater (Figure 34a), indicating that the salinity was derived by the dissolution of halite. Thus, the high Cl/Br ratios are consistent with halite dissolution. The I/Cl versus Br/Cl ratios diagram (Figure 34b) also shows that the fluid inclusions have low I contents, which are also typical of brines derived from evaporite dissolution. They are comparable to the Hansonburg and contemporary fluids from the Salton Sea geothermal brines, both of which have dissolved evaporites [163,164].

Although it is not possible to unambiguously classify emerald deposits based solely on their fluid composition, there are some general observations that can prove useful: (1) The presence of salt cubes can limit possible modes of formation, as it is very unlikely that we can generate a salt saturated fluid in a purely metamorphic environment, hence the presence of salt cubes generally implies the input of igneous fluids or an interaction with evaporites and individual Ca/Na/K values may be specific to individual deposits; (2) compressible gases are seen in all types of emerald deposit, but gas ratios may also be used to identify individual deposits; and (3) specific daughter and accidental minerals contained within fluid inclusions cannot be used to identify specific deposit models within our classification system, but may again be used to fingerprint emeralds from specific deposits.

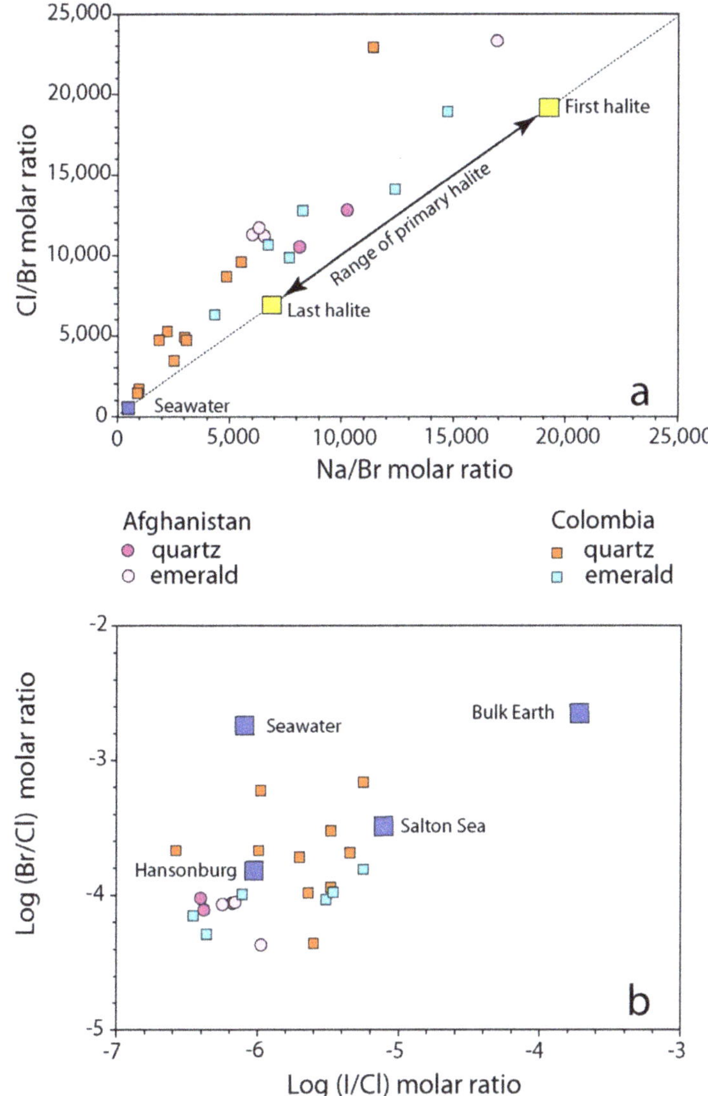

Figure 34. Origin of salinity in the emerald and quartz brines from Colombia and Afghanistan: (**a**) Analyses of the fluid inclusions from both emerald and quartz show a wide range of Na/Br and Cl/Br molar ratios that are much greater than those of primary halite and indicate a substantial loss of Br, typical of recrystallised halite for both emerald deposits; (**b**) log(I/Cl) versus log(Br/Cl) molar ratios of Afghan and Colombian fluid inclusions, which are depleted in both Br and I, indicative of evaporites contribution to the fluids in emerald and quartz. They are compared with the composition of fluids where evaporites are known to be involved, such as for the Salton Sea geothermal brines [163] and Hansonburg [164].

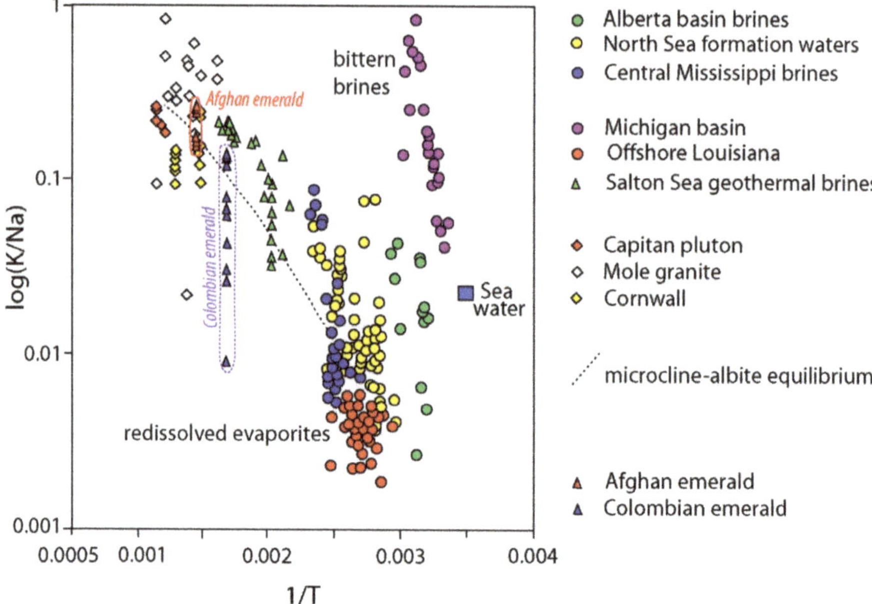

Figure 35. Diagram log (K/Na) molar ratio versus 1/T (°K) showing the evolution of the fluids associated with Colombian and Afghan emeralds relative to crustal fluids, including bittern brines, brines derived by dissolution of evaporites, and magmatic fluids. Sedimentary formation brines deviate significantly from the K-feldspar-albite equilibrium as well as for Afghan and Colombian brines, which are associated with a huge albitisation of their host-rock with the complete consumption of K-feldspar from, respectively, the schist and black shale.

8. Discussion of the Model of Formation of the Emerald Deposits Associated with M-UMR

The tectonic magmatic-related emerald deposits hosted in M-UMR (sub-type IA) occurred during orogenies from the end of the Archean (2.94 Ga) with the Gravelotte deposit in South Africa [165] to the Himalayan (9 Ma) with the Khaltaro deposit in Pakistan [83]. These deposits are related to plate tectonics over ~3.2 billion years with stable convection cells initializing continental drifts and subduction zones, and are related either to continental collisions or continent-to-continent rifting.

The classical model of sub-type IA relates granites with their dyke swarms of aplo-pegmatites and quartz veins in metamorphosed M-UMR, fluid-rock interaction, and infiltrational metasomatism [15,58,166,167]. An alternative genetic model for the formation of these emeralds is a regional tectonic and metamorphic model [77,130,168–170]. This genetic controversy involving sub-type IA deposits shows that (i) they are possibly genetically different; (ii) that alternative models are always dependent on a school of thought, which we absolutely want to apply for all deposits; (iii) these deposits share many common denominators in terms of geological setting, age of formation, nomenclature of rocks, source of the elements, and source of the fluids; and finally, (iv) each deposit belongs to the same family, but with a wide range of genres and uncountable typological varieties that allow us to follow the evolution of a mountain range marked by magmatic events accompanied by metamorphic remobilizations, which sometimes erase the primary geological features of the emerald deposit.

The following discussion examines the possible genetic links existing between the different sub-type IA and sub-types IIA and IID hosted in M-UMR within the geological and dynamic evolution of a continental crust.

8.1. Sub-Type IA Emerald

This is the fruit of the fluid-rock interaction producing metasomatism in both pegmatite and M-UMR, whatever the origin of the fluid. The classical sub-type IA results from the fluid circulation at the contact between these two geochemically contrasting rocks. The most representative deposits described in the literature are those from Carnaíba [76,78], Franqueira [171,172], Khaltaro [83], and Kafubu [70]:

1. The granite emplacement is related to a tectonic event, but the pegmatites are not deformed and metamorphosed. They are clearly intruding the M-UMR of the volcano-sedimentary series, and sometimes roof pendants on the granite as observed in Bode mine at Carnaíba (Figure 36) or at Franqueira;

2. At Carnaíba, the granite and pegmatite intrusions and fluid circulation are coeval [67], and the pegmatite is transformed into plagioclasite with disseminations of phlogopite (endo-plogopitite), and the M-UMR into phlogopitite (exo-phlogopitites). These exo-phlogopitites display clear zonation with a very sharp metasomatic front, i.e., metasomatic columns formed by infiltrational processes [173]. The metasomatic fronts where an additional phase appears in the mineral association correspond to the change of one of the determinant components from mobile to inert: Ca is displaced from the serpentinites to the center of the column for the formation of actinolite-tremolite, but also apatite at the border of the endo-phlogopitite. The occurrence of beryl is restricted to the most aluminous parts of the metasomatic zonation (plagioclasite, endo-phlogopitite, and exo-phlogopitites proximal to the plagioclasite). The inner part of the phlogopitite zonation plays the role of a filter for the Be-bearing fluids and constitutes a very efficient "metasomatic trap" where the mobile behavior of Cr favored the formation of emerald. At Khaltaro in Pakistan, non-deformed pegmatites and quartz veins crosscut amphibolites, which were metasomatized on 20 cm-wide selvages that are symmetrically zoned around the veins [83], as found at Carnaíba (see Figure 14). Mass-balance calculations on the metasomatic column (Figure 37) have shown that (a) in the inner and intermediate metasomatic zones, K, F, H_2O, B, Li, Rb, Cs, Be, Ta, Nb, As, Y, and Sr are gained and Si, Mg, Ca, Fe, Cr, V, and Sc are lost; and (b) in the outer zone, F, Li, Rb, Cs, and As are gained. The oxygen isotope composition of the hydrothermal minerals indicated the circulation of a single fluid of magmatic origin. At Kafubu, the regional metamorphic event pre-dates the emerald formation. The F-B-Li-rich phlogopitites are located at the contact between tourmaline veins and pegmatites with Mg-metabasites. The pegmatites of the Lithium-Cesium-Tantalum family are linked to hidden fertile B-F-Nb-Ta-Li-Cs-rich granite.

3. These deposits sometimes exhibit multi-stage Be-mineralization, as observed at the Carnaíba deposit: A second minor stage of metasomatism affected in some areas the emerald-bearing phlogopitites [15,90]. This stage is related to the intrusion of dyke swarms of quartz-muscovite veinlets with greisenisation of the granites, chloritisation, and muscovitisation of the phlogopitites, general silicification, and muscovitisation of the plagioclasites. This stage involves yellowish to whitish beryl, sometimes with molybdenite, scheelite, and schorlite.

4. These emerald mineralizations are interpreted to be due to the efficiency of the metasomatic trap rather than significant pre-enrichment in Be (5 to 11 ppm of Be in the Carnaíba granite). The occurrence of strong chemical gradients in the zone of preferential circulation of the solutions constitutes highly favorable conditions for the beryl crystallization.

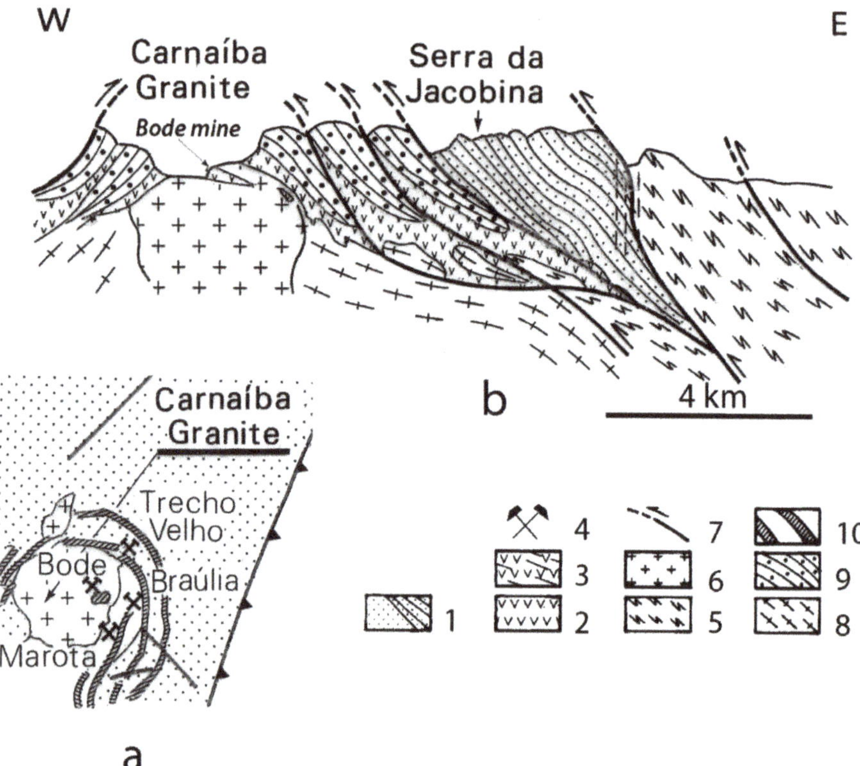

Figure 36. Schematic geological section of the emerald deposits of Carnaíba, Bahia state, Brazil: (**a**) The granite of Carnaíba and its emerald deposits cited in the present work. The Bode deposit is located in a roof-pendant of serpentinite present at the roof of the granite; (**b**) The Serra da Jacobina volcano-sedimentary sequence formed by intercalations of quartzite and serpentinite is crosscut by the Carnaíba granite. The pendants of serpentinite are present at the contact and on the roof of the granitic intrusion.

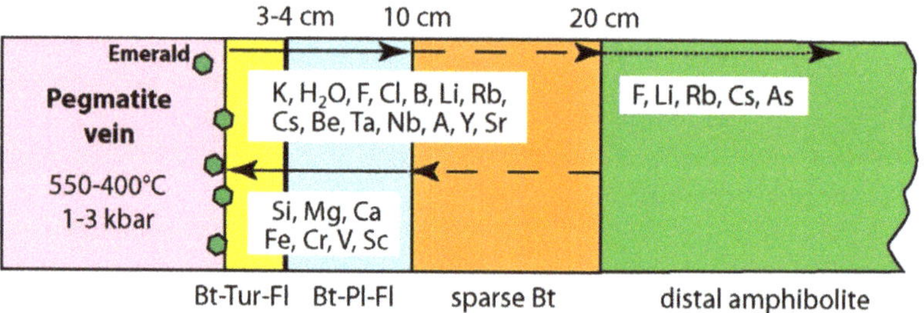

Figure 37. The Khaltaro emerald deposit, Nanga Parbat—Haramosh massif, Pakistan [83]. A symmetrically zoned metasomatic column is formed at the contact of the albitized pegmatites or hydrothermal quartz veins with amphibolite. The mass balance calculation in the metasomatized amphibolite indicates the gains and losses of components (see the transfers in the outer and inner zones). Emerald froms within the vein, near the contact with the altered amphibolite.

8.2. Multi-Stage Formation and Ages of Formation and Remobilization of Type IA Deposits

1. The Precambrian deposits located in the volcano-sedimentary series or greenstone belts are often folded and sheared, but metasomatic processes during emerald formation are generally coeval with the deformation, as in the deposits of Piteiras, Fazenda Bonfim, and Socotó (Brazil); Sumbawanga and Manyara (Tanzania); Kafubu (Zambia); and recently in the Gubaranda area from Eastern India [174].

2. The deposit at Sandawana in Zimbabwe [52,175] presents multi-stage formation; [169] advanced that the classical model of sub-Type IA cannot be applied. The Cs-Nb-Ta-bearing pegmatite veins that intruded UMR suffered the classical desilication with the formation of plagioclasite. During folding, shearing, and regional metamorphism, after the albitisation of the pegmatites, a reactive F-P-Be-Li-rich fluid of pegmatitic origin circulated in a shear zone, affecting the albitites and reacting with the UMR, to form emerald-bearing phlogopitites (Figure 38). Two generations of emerald are found: (i) Fine-grained crystals at the contact between albitite and phlogopitite and (ii) euhedral gem crystals formed later in phlogopitites either away from the albitite or in low-pressure zones next to the albitites. In that case, the albitites acted as incompetent levels, folded and sheared, and forming traps for euhedral emerald.

   The syntectonic pegmatites yielded an age of 2640 ± 40 Ma by U/Pb dating on monazite and an age of 2600 ± 100 Ma by the Pb-Pb method on microlite [176]. The $^{40}Ar/^{39}Ar$ dating on phlogopite and actinolite of the phlogopitites yielded a very disturbed age spectra and variable total gas ages between 2225 and 2447 Ma, with relative plateau ages of 1903 Ma for the phlogopite and of 1936 Ma for the amphibole [52]. Two ages were proposed for the formation of emerald: (i) An Archean age at 2640 Ma, which is the age of the intrusion of the pegmatites, or (ii) a Proterozoic age at around 2000 Ma, which corresponds to a major tectono-metamorphic episode that affected the Limpopo belt formed around the Zimbabwe craton. Zwaan [52] opted for the first hypothesis, considering that the deformation at circa 2000 Ma modified the isotopic argon clock of mica and amphibole, but these integrated Ar-ages between 2200 and 2500 Ma correlate with the Archean thermal event. The complexity of dating rocks that suffered deformational events and remobilization of material illustrates the complexity of classifying ore deposits. The Sandawana deposit belongs to sub-type IA and its genetic link with a magmatic source is obvious in terms of chemical elements, but it could be re-classified as sub-type IID if the age of the emerald is considered to be younger than 2400 Ma.

3. The possible genesis of sub-type IIA deposits, such as those of Swat Valley, Santa Terezinha de Goiás, and Habachtal [168], was discussed previously. The presence of meta-pegmatites can be suspected based on either chemical data or O-H isotope composition of emerald and associated minerals, but has not been found up to now due to the tectonic regime (thrust and shear zone) and the level of observation. These deposits are classified as sub-type IIC based on the geological environment and are considered to be metamorphic with probable mixing of magmatic fluids (high Cs content for the Santa Terezinha emeralds). This hypothesis is strengthened by the Na/Li vs. Cs/Ga chemical diagram presented by Schwarz [65]. Figure 39 shows that these emerald deposits are grouped in one chemical field very different from those of Colombia, Russia, and Nigeria. They are characterized by a high Cs/Ga ratio, indicating appreciable to high amounts of Cs (magmatic source), and a high Na/Li ratio. The high Na content of this emerald is correlated with a high mean H_2O content in the channels, as, determined for Santa Terezinha de Goiás (2.9 wt.%, n = 5), Habachtal (3.1 wt.%, n = 3), and Swat Valley (3.4 wt.%, n = 1) [15,16]. This is not just a coincidence, but is probably a genetic proxy Be-Cs source for emerald in these three deposits, i.e., magmatic sources with huge fluid circulation and metasomatism in a metamorphic environment.

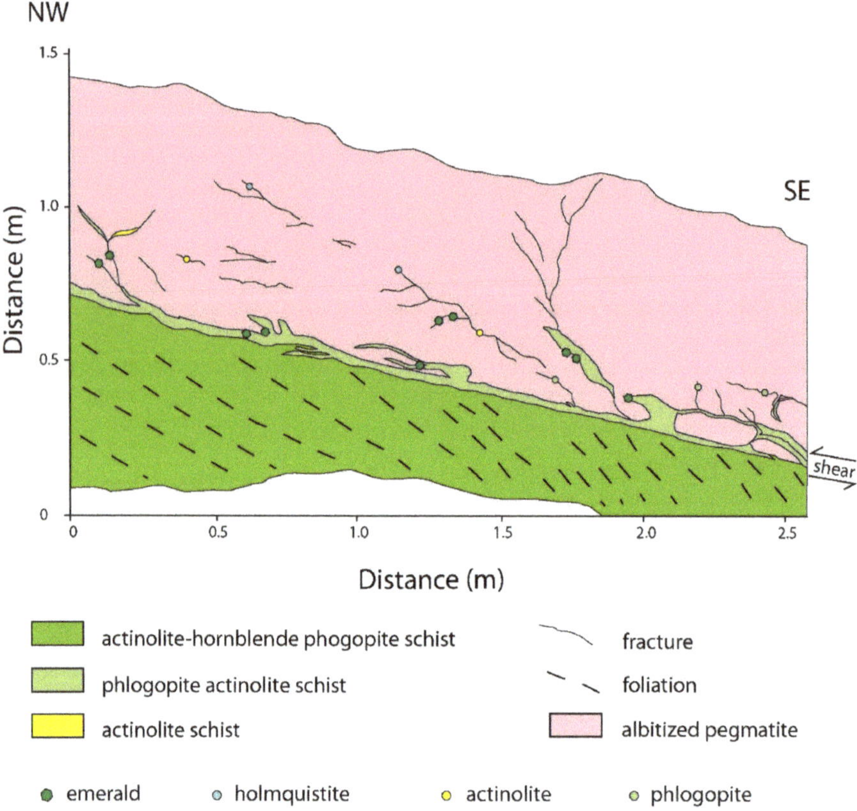

Figure 38. Cross-section of the mineralization at the Zeus underground mine, 200 ft. level, 26/28 stope, Sandawana deposit, Zimbabwe [167]. Fluids infiltrated along the schistosity of the actinolite-hornblende-phlogopite schist, and induced a metasomatic reaction with emerald formation at the foot wall of the pegmatite.

The Habachtal deposit is a complex deposit in terms of the genetic model and the previous discussion about the possible remobilization of Be-W enriched Hercynian pegmatites (possible sub-type IID deposit) by the regional metamorphism [168] opens the debate on the age of the formation of this deposit. The genesis of the emerald is metasomatic, but bound to the regional metamorphism of alpine age [77]. The K-Ar age obtained on phlogopite from the phlogopitites is 22 Ma, while the tracks of fission on apatite yielded ages of 9 Ma. The K-Ar age on muscovite from the muscovite schist is 27 Ma. The Rb/Sr dating realized on the zones of growth of garnets from the Schieferhülle formation, situated structurally above the Habach formation, indicated ages of crystallization between 62.0 and 30.2 Ma [177]. This age around 30 Ma is in agreement with the dates found for the end of the growth of garnet in the central Alps [178]. So, the best estimation established for the growth of the Habachtal emeralds would be situated around 30 Ma [179].

4. The deposits of the Eastern desert in Egpyt and Poona in Australia are good examples of sub-type IID, where sub-type IA deposits were remobilized by regional metamorphism with deformation and remobilization of older rocks, following the genetic model proposed by Grundmann and Morteani [77].

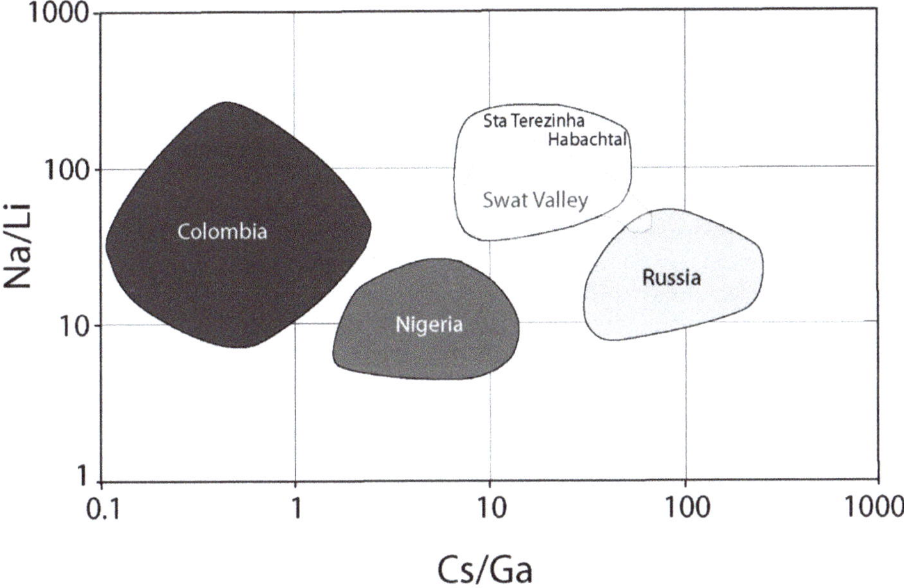

Figure 39. Cs/Ga versus Na/Li diagram of emeralds from Colombia, Nigeria, Russia, Pakistan, Austria, and Swat valley presented by Schwarz [65]. The diagram indicates that the chemical ratios of emerald from Santa Terezinha de Goiás (Brazil), Habachtal (Austria), and Swat valley (Pakistan) plot in the same population field. They are characterized by moderate to high Cs/Ga and very high Na/Li ratios. Discrepancies are observed for the concentrations of Cs and other elements when compared with the analysis presented by Aurisicchio et al. [23].

9. Exploration Now and in the Future

The majority of development of exploration methods for emerald has been for sub-type IIB deposits in Colombia. In Colombia, repeated washing of sediments is an effective and ubiquitous technique employed by local emerald-seeking "guaqueros" to reveal emerald in mineralized drainages. However, as pointed out by Lake [180], it is not effective to directly recover beryl by density separation because its specific gravity (~2.7) is similar to most rock-forming minerals. Geochemistry has also proven to be useful in Colombia. Escobar [181] studied the geology and geochemistry of the Gachalá area and found that Na enrichment and depletion of Li, K, Be, and Mo in the host rocks were very good indicators for locating mineralized areas. Beus [182] presented the results of a United Nations-sponsored geochemical survey of the streams draining emerald deposits in the Chivor and Muzo areas. The spatial distribution of areas with emerald mineralization was linked, on a regional scale, to intersections of the NNE- and NW-trending fault zones. The black shale units in those tectonic blocks that contain emerald mineralization were found to be enriched in CO_2, Ca, Mg, Mn, and Na, and depleted in K, Si, and Al [182]. The results of this study were tested with a stream sediment sampling program in the Muzo area. The results showed that samples collected from emerald-bearing tectonic blocks had anomalously low K/Na ratios. Beus [182] also suggested using a "composite" geochemical ratio that takes into account the albitization and leaching phenomena: $M = Na^3/(K \times Li \times Mo)$ (with Na and K in wt.% and Li and Mo in ppm) where M is an expression of the degree of metasomatic alteration. Other combinations, such as K, Li, Co or K, Li, Ba could be tried to determine the most contrasting value of M [182]. Subsequently, it was discovered that the Na content of the sediments was the best indicator of the mineralized zones in the drainage basins. Several new emerald occurrences were discovered by U.N. teams using the results of this study [183].

More recently, Ringsrud [183] reported that Colombian geologists were analyzing soil samples collected from altered tectonic blocks for Li, Na, and Pb to delineate emerald mineralization. Cheilletz et al. [103] showed that the Be content of black shale outside of the leached mineralized areas ranges from 2 to 6 ppm. Beryllium concentrations in the leached areas were found to range from 0.1 to 3.0 ppm [182]. Cheilletz and Giuliani [184] observed that the spatial coexistence of emerald districts and gypsum and anhydrite deposits could be used to prospect for new deposits.

Branquet et al. [107,113] observed that discovering new deposits will necessitate prospecting that is structurally oriented and focused on finding (1) the stratiform brecciated level in the eastern zone and (2) structural traps along regional tear faults in the western zone.

Geophysical techniques, such as induced polarization (IP) and magnetic surveys, have been used at Chivor and Muzo to delineate pyrite mineralization that is abundant and often associated with emerald-bearing veins [185]. Gutiérrez [186] tested several radiometric and magnetometric techniques at Chivor and Macanal deposits with some success.

Escobar [177] applied the criteria for emerald exploration in Colombia to stream sediment geochemical data for the Yukon and Northwest Territories in Canada. The criteria had to be adjusted for a number of factors, including the absence of Be in the Northwest Territories data and elevated Na due to plutonic alkali feldspar weathering into black shale drainages. Escobar [181] also noted that Colombian-type emerald deposits have relatively small footprints, and regional-scale geochemical surveys with data points 5–10 km apart may not be sufficient to show an emerald occurrence. Despite these constraints, Escobar [181] was able to identify several regions of interest.

With respect to other type II deposits, Arif et al. [187] observed that in the Swat valley in Pakistan, the emerald deposits occur in carbonate-altered ultramafic rocks, which also host Cr-rich dravite and "oxy-dravite". They suggested that the presence of high-Cr magnesian tourmaline, particularly in magnesite–talc-altered ultramafic rocks, can represent a criterion for further emerald exploration in the lower Swat region of Pakistan and in other ultramafic-hosted emerald-producing regions worldwide.

Wise [188] suggested that that the morphology of quartz, calcite, pyrite, and rutile crystals may serve as potential exploration guides for the discovery of hidden emerald deposits in the Hiddenite district of North Carolina. Emerald-bearing veins are characterized by: (1) quartz with multiple generations, including fine-grained doubly terminated crystals and very coarse-grained prismatic crystals; (2) calcite with largely rhombohedral habit, commonly accompanied by dolomite and siderite; (3) pyrite crystals dominated by octahedral faces; and (4) rutile that varies from single untwinned crystals to highly reticulated aggregates.

Some work has also been done in northwestern Canada to explore for type I deposits. Murphy et al. [189] plotted potential Be reservoirs and Cr and V reservoirs in the Yukon Territory and suggested that the best place to look is where the two come together. Lewis et al. [190] noted that all beryl occurrences in the Yukon Territory are intrusion-related, but for an intrusion to become enriched enough to reach Be-saturation to form beryl, it must be "ultrafractionated".

Future efforts will move towards the development of more effective exploration guidelines that consider the geological factors responsible for emerald formation. For type I deposits, this could involve exploring continental collision domains and considering the overlaps of Be and Cr-V reservoirs, mineralogical and geochemical anomalies (F, K, Li, P, B) linked to high intensity fluid/rock interaction, and deposit location, e.g., proximal or distal to the granitic intrusive. This last point is an important one as illustrated by the Ianapera deposit in Madagascar [11,135] where emerald occurs in two different, but coeval settings: (1) A proximal one, formed classically at the contact between pegmatites (and quartz veins) and M-UMR; and (2) a distal one formed outside the roof of the granite, with emerald-bearing phlogopitites in fractures and faults. The fluid percolation affected all the geological formations and different M-UMR, resulting in several types of zoned and un-zoned emerald habits with contrasting Cr-V chemistry.

10. Conclusions and Perspectives

The elaboration of a new classification proposed for emerald occurrences and deposits is the first step to building a new scheme for studying Be-mineralization from the field to the laboratory. The economic emerald deposits, in terms of volume and quality, are the Tectonic magmatic-related type hosted in M-UMR (Brazil, Zambia, Russia) or the Tectonic metamorphic-related type hosted in either low- (Colombia) to medium- (Afghanistan, China) temperature metamorphosed sedimentary rocks or medium-temperature metamorphosed M-UMR (Brazil, Pakistan).

The enhanced classification proposed in the present paper, based on objective geological criteria, takes into account the question of genesis of a number of deposits, which has been brought to the fore by research using new analytical facilities in the 21st century: The following topics were discussed: (i) Magmatic intrusives in M-UMR from classical intrusion to multi-stage formation through fluid circulation, metasomatism, and emerald formation synchronous with folding and shearing of regional metamorphic sequences; (ii) the presence of hidden intrusives that are probably the source of Be and Cs for emerald, such as in the Swat valley and Santa Terezinha de Goías deposits; and (ii) metamorphic remobilization of previous Be-magmatic mineralization over several orogenic episodes, such as for the Habachtal and Egyptian emerald occurrences. All these cases reflect the mixing of magmatic and metamorphic fluids.

The present work, which connects geological knowledge of the formation of gem deposits to studies of the properties and features of individual gems, is a step forward in the great challenge of geographic origin determination. The enhanced classification typology for emerald deposits opens a new framework for mineral exploration guidelines.

Author Contributions: G.G., L.A.G., D.M., A.E.F., and Y.B. contributed with their field and laboratory work, and experience to the realization of this description of the state-of-the-art of the geology and genesis of emerald deposits and the proposal of the enhanced classification typology.

Funding: The present paper was realized without any direct funding. The work is based on the knowledge and funding obtained in the course of more than 20 years work on the geology of gems.

Acknowledgments: The authors acknowledge the three reviewers for their constructive comments. We thank very much the guest editors of the special issue "Mineralogy and Geochemistry of Gems", namely Panagiotis Voudouris, Stefanos Karampelas, Vasilios Melfos, and Ian Graham for their invitation to present an extended review on emerald deposits. We acknowledge with thanks the editorial team of "Minerals" for their handling and the final layout. The authors want to thank Dietmar Schwarz, Vincent Pardieu, Victor Carillo, and Jean-Claude Michelou for their discussion and contribution of photographs. Gaston Giuliani wants to thank the Institut de Recherche pour le Développement (IRD, France) and the French CNRS and the Unversities of Lorraine and Paul Sabatier (CRPG/CNRS-Université de Lorraine and GET/IRD-Université Paul Sabatier) for academic and financial support for research on the geology of gems over the past thirty years. The authors want to thanks Mackenzie Parker for editing.

Conflicts of Interest: The authors declare no conflict of interest.

References

1. Groat, L.; Giuliani, G.; Marshall, D.; Turner, D. Emerald. In *Geology of Gem Deposits*; Raeside, E.R., Ed.; Mineralogical Association of Canada, Short Course Series 44: Tucson, AZ, USA, 2014; pp. 135–174.
2. In the News. Gemfields' October emerald auction: US$21.5 million in sales. *Incolor* **2018**, *37*, 22.
3. Schmidt, S. Une des plus grandes émeraudes au monde a été découverte dans une mine en Zambie. Trust My Sicence, Nature. 2018. Available online: https://trustmyscience.com (accessed on 11 February 2019).
4. Ibbetson, R. Miner finds giant 5,655-carat emeraldin Zambia worth up to £ 2m. *Daily Mail Online*, 30 October 2018.
5. Yager, Y.G.; Menzie, W.D.; Olson, D.W. *Weight of Production of Emeralds, Rubies, Sapphires, and Tanzanite from 1995 through 2005*; Open File Report 2008–1013; USGS Reston: Reston, VA, USA, 2008; 9p.
6. Wood, D.L.; Nassau, K. Characterization of beryl and emerald by visible and infrared absorption spectroscopy. *Am. Mineral.* **1968**, *53*, 777–800.
7. Damon, P.E.; Kulp, J.L. Excess helium and argon in beryl and other minerals. *Am. Mineral.* **1958**, *43*, 433–459.

8. Giuliani, G.; Marty, B.; Banks, D. Noble gases in fluid inclusions from emeralds: Implications for the origins of fluids and constraints on fluid-rock interactions. In Proceedings of the 18th Biennial Meeting of European Current Research on Fluid Inclusions, Siena, Italy, 4–5 July 2005. CD-ROM.

9. Charoy, B. Cristallochimie du béryl: L'état des connaissances. In *L'Émeraude*; Giard, D., Giuliani, G., Cheilletz, A., Fritch, E., Gonthier, E., Eds.; AFG-CNRS-ORSTOM Edition: Paris, France, 1998; pp. 47–54.

10. Schwarz, D.; Schmetzer, K. The definition of emerald: The green variety of beryl colored by chromium and/or vanadium. In *Emeralds of the World, (2002) ExtraLapis English No. 2: The Legendary Green Beryl*; Lapis International, LLC: East Hampton, CT, USA, 2002; pp. 74–78.

11. Andrianjakavah, P.R.; Salvi, S.; Béziat, D.; Rakotondrazafy, A.F.M.; Giuliani, G. Proximal and distal styles of pegmatite-related metasomatic emerald mineralization at Ianapera, southern Madagascar. *Miner. Deposita* **2009**, *44*, 817–835. [CrossRef]

12. Zwaan, J.C.; Jacob, D.E.; Häger, T.; Calvacanti Neto, M.T.O.; Kanis, J. Emeralds from the Fazenda Bonfim region, Rio Grande do Norte, Brazil. *Gems Gemol.* **2012**, *48*, 2–17. [CrossRef]

13. Rondeau, B.; Fritsch, E.; Peucat, J.J.; Nordru, F.S.; Groat, L.A. Characterization of emeralds from a historical deposit: Byrud (Eidsvoll), Norway. *Gems Gemol.* **2008**, *44*, 108–122. [CrossRef]

14. Artioli, G.; Rinaldi, R.; Stahl, K.; Zanazzi, P.F. Structure refinements of beryl by single-crystal neutron and X-ray diffraction. *Am. Mineral.* **1993**, *78*, 762–768.

15. Giuliani, G.; France-Lanord, C.; Zimmermann, J.-L.; Cheilletz, A.; Arboleda, C.; Charoy, B.; Coget, P.; Fontan, F.; Giard, D. Composition of fluids, δD of channel H_2O and $\delta^{18}O$ of lattice oxygen in beryls: Genetic implications for Brazilian, Colombian and Afghanistani emerald deposits. *Int. Geol. Rev.* **1997**, *39*, 400–424. [CrossRef]

16. Marshall, D.; Downes, P.J.; Ellis, S.; Greene, R.; Loughrey, L.; Jones, P. Pressure–temperature-fluid constraints for the Poona emerald deposits, Western Australia: Fluid inclusion and stable isotope studies. *Minerals* **2016**, *6*, 130. [CrossRef]

17. Brand, A.A.; Groat, L.A.; Linnen, R.L.; Garland, M.I.; Breaks, F.W.; Giuliani, G. Emerald mineralization associated with the Mavis Lake Pegmatite Group, near Dryden, Ontario. *Can. Mineral.* **2009**, *47*, 315–336. [CrossRef]

18. Marshall, D.; Pardieu, V.; Loughrey, L.; Jones, P.; Xue, G. Conditions for emerald formation at Davdar, China: Fluid inclusion, trace element and stable isotope studies. *Mineral. Mag.* **2012**, *76*, 213–226. [CrossRef]

19. Loughrey, L.; Marshall, D.; Jones, P.; Millsteed, P.; Main, A. Pressure-temperature-fluid constraints for the Emmaville-Torrington emerald deposit, New South Wales, Australia: Fluid inclusion and stable isotope studies. *Cent. Eur. J. Geosci.* **2012**, *4*, 287–299. [CrossRef]

20. Loughrey, L.; Marshall, D.; Ihlen, P.; Jones, P. Boiling as a mechanism for colour zonations observed at the Byrud emerald deposit, Eidsvoll, Norway: Fluid inclusion, stable isotope and Ar-Ar studies. *Geofluids* **2013**, *13*, 542–558. [CrossRef]

21. Lake, D.J.; Groat, L.; Falck, H.; Cempirek, J.; Kontak, D.; Marshall, D.; Giuliani, G.; Fayek, M. Genesis of emerald-bearing quartz veins associated with the Lened W-skarn mineralization, northwest territories, Canada. *Can. Mineral.* **2017**, *55*, 561–593. [CrossRef]

22. Santiago, J.S.; Souza, V.S.; Filgueiras de, B.C.; Jiménez, F.A.C. Emerald from the Fazenda Bonfim deposit, northeastern Brazil: Chemical, fluid inclusions and oxygen isotope data. *Braz. J. Geol.* **2018**, 1–16. [CrossRef]

23. Aurisicchio, C.; Conte, A.M.; Medeghini, L.; Ottolini, L.; De Vito, C. Major and trace element geochemistry of emerald from several deposits: Implications for genetic models and classification schemes. *Ore Geol. Rev.* **2018**, *94*, 351–366. [CrossRef]

24. Kovaloff, P. Geologist's report on Somerset emeralds. *S. Afr. Mining Eng. J.* **1928**, *39*, 101–103.

25. Zambonini, F.; Caglioto, V. Ricerche chimiche sulla rosterite di San Piero in Campo (Isola d'Elba) e sui berilli in generale. *Gazz. Chim. Ital.* **1928**, *58*, 131–152.

26. Leitmeier, H. Das Smaragdvorkommen in Habachtal in Salzburg und seine Mineralien. *Tscher. Miner. Petrog.* **1937**, *49*, 245–368. [CrossRef]

27. Otero Muñoz, G.; Barriga Villalba, A.M. *Esmeraldas de Colombia*; Banco de la Republica: Bogotá, Colombia, 1948.

28. Simpson, E.S. *Minerals of Western Australia*; Government Printer: Perth, Australia, 1948; Volume 1, pp. 195–207.

29. Gübelin, E.J. Emeralds from Sandawana. *J. Gemmol.* **1958**, *6*, 340–354. [CrossRef]

30. Vlasov, K.A.; Kutakova, E.I. *Izumrudnye Kopi*; Moscow Akademiya Nauk SSGR: Moscow, Russia, 1960; 252p.

31. Martin, H.J. Some observations on southern Rhodesian emeralds and chrysoberyls. *Chamb. Mines J.* **1962**, *4*, 34–38.

32. Petrusenko, S.; Arnaudov, V.; Kostov, I. Emerald pegmatite from the Urdini Lakes, Rila Mountains. *Annuaire de l'Université de Sofia Faculté de Géologie et Géographie* **1966**, *59*, 247–268.

33. Beus, A.A.; Mineev, D.A. *Some Geological and Geochemical Features of the Muzo-Coscuez Emerald Zone, Cordillera Oriental, Colombia*; Empresa Colombiana de Minas, Biblioteca: Bogotá, Colombia, 1972; 55p.

34. Hickman, A.C.J. The Miku emerald deposit. *Geol. Surv. Zambia Econ. Rep.* **1972**, *27*, 35.

35. Garstone, J.D. The geological setting and origin of emeralds from Menzies, Western Australia. *J. R. Soc. West. Aust.* **1981**, *64*, 53–64.

36. Hanni, H.A.; Klein, H.H. Ein Smaragdvorkommen in Madagaskar. *Zeitschrift der Deutschen Gemmologischen Gesellschaft* **1982**, *21*, 71–77.

37. Graziani, G.; Gübelin, E.; Lucchesi, S. The genesis of an emerald from the Kitwe District, Zambia. *Neues Jb. Miner. Monat.* **1983**, 175–186.

38. Kozlowski, A.; Metz, P.; Jaramillo, H.A.E. Emeralds from Sodomondoco, Colombia: Chemical composition, fluid inclusion and origin. *Neues Jb. Miner. Abh.* **1988**, *59*, 23–49.

39. Hammarstrom, J.M. Mineral chemistry of emeralds and some minerals from Pakistan and Afghanistan: An electron microprobe study. In *Emeralds of Pakistan*; Kazmi, A.H., Snee, L.W., Eds.; Van Nostrand Reinhold: New York, NY, USA, 1989; pp. 125–150.

40. Ottaway<sc>, T.L. The Geochemistry of the Muzo Emerald Deposit, Colombia. M.Sc. Thesis, University of Toronto, Toronto, ON, Canada, 1991.

41. Schwarz, D. Australian emeralds. *Austral. Gemmol.* **1991**, *17*, 488–497.

42. Schwarz, D.; Kanis, J.; Kinnaird, J. Emerald and green beryl from central Nigeria. *J. Gemmol.* **1996**, *25*, 117–141. [CrossRef]

43. Abdalla, H.M.; Mohamed, F.H. Mineralogical and geochemical investigations of beryl mineralizations, Pan-African belt of Egypt: Genetic and exploration aspects. *J. Afr. Earth Sci.* **1999**, *28*, 581–598. [CrossRef]

44. Gavrilenko, E.V.; Pérez, B.C. Characterisation of emeralds from the Delbegetey deposit, Kazakhstan. In *Mineral Deposits: Processes to Processing*; Stanley, C.J., Rankin, A.H., Bodnar, R.J., Naden, J., Yardley, B.W.D., Criddle, A.J., Hagni, R.D., Gize, A.P., Pasava, J.., Eds.; Balkema: Rotterdam, The Netherlands, 1999; pp. 1097–1100.

45. Alexandrov, P.; Giuliani, G.; Zimmerman, J.L. Mineralogy, age and fluid geochemistry of the Rila Emerald deposits, Bulgaria. *Econ. Geol.* **2001**, *96*, 1469–1476. [CrossRef]

46. Groat, L.A.; Marshall, D.D.; Giuliani, G.; Murphy, D.C.; Piercey, S.J.; Jambor, J.L.; Mortensen, J.K.; Ercit, T.S.; Gault, R.A.; Mattey, D.P.; et al. Mineralogical and geochemical study of the Regal Ridge showing emeralds, southeastern Yukon. *Can. Mineral.* **2002**, *40*, 1313–1338. [CrossRef]

47. Marshall, D.D.; Groat, L.A.; Falck, H.; Douglas, H.L.; Giuliani, G. Fluid inclusions from the Lened emerald occurrence; Northwest Territories, Canada: Implications for Northern Cordilleran Emeralds. *Can. Mineral.* **2004**, *42*, 1523–1539. [CrossRef]

48. Vapnik, Y.; Sabot, B.; Moroz, I. Fluid inclusions in Ianapera emerald, Southern Madagascar. *Int. Geol. Rev.* **2005**, *47*, 647–662. [CrossRef]

49. Vapnik, Y.; Moroz, I.; Eliezri, I. Formation of emeralds at pegmatite-ultramafic contacts based on fluid inclusions in Kianjavato emerald, Mananjary deposits, Madagascar. *Mineral. Mag.* **2006**, *70*, 141–158. [CrossRef]

50. Zwaan, J.C.; Seifert, A.V.; Vrána, S.; Laurs, B.M.; Anckar, B.; Simons, W.B.S.; Falster, A.U.; Lustenhouwer, W.J.; Muhlmeister, S.; Koivula, J.K.; et al. Emeralds from the Kafubu area, Zambia. *Gems Gemol.* **2005**, *41*, 116–148. [CrossRef]

51. Gavrilenko, E.V.; Calvo Pérez, B.; Castroviejo Bolibar, R.; Garcia Del Amo, D. Emeralds from the Delbegetey deposit (Kazakhstan): Mineralogical characteristics and fluid-inclusion study. *Mineral. Mag.* **2006**, *70*, 159–173. [CrossRef]

52. Zwaan, J.C. Gemmology, geology and origin of the Sandawana emerald deposits, Zimbabwe. *Scr. Geol.* **2006**, *131*, 211.

53. Barton, M.D.; Young, S. Nonpegmatitic deposits of beryllium: Mineralogy, geology, phase equilibria and origin. In *Beryllium: Mineralogy, Petrology, and Geochemistry*; Grew, E.S., Ed.; Mineralogical Society of America: Washington, DC, USA, 2002; Volume 50, pp. 591–691.

54. Giuliani, G.; Bourlès, D.; Massot, J.; Siame, L.L. Colombian emerald reserves inferred from leached beryllium of their host black shale. *Explor. Min. Geol.* **1999**, *8*, 109–116.

55. Giuliani, G.; Branquet, Y.; Fallick, A.E.; Groat, L.; Marshall, D. Emerald deposits around the world, their similarities and differences. *InColor* **2015**, *special issue*, 56–69.

56. Sabot, B. Classification des gisements d'émeraude: Apports des études pétrographiques, minéralogiques et géochimiques. Thèse de Doctorat, Université de Nancy, Nancy, France, 2002.

57. Dereppe, J.M.; Moreaux, C.; Chauvaux, B.; Schwarz, D. Classification of emeralds by artificial neural networks. *J. Gemmol.* **2000**, *27*, 93–105. [CrossRef]

58. Schwarz, D.; Giuliani, G. Emerald deposits—A review. *Austral. Gemmol.* **2001**, *21*, 17–23.

59. Schwarz, D.; Giuliani, G.; Grundmann, G.; Glas, M. Die Entstehung der Smaragde, ein vieldisskutiertes Thema. In *Smaragd, der Kostbarste Beryll, der Teuerste Edelstein*; Schwarz, D., Hochlitner, R., Eds.; ExtraLapis: Charleston, SC, USA, 2001; pp. 68–73.

60. Schwarz, D.; Giuliani, G.; Grundmann, G.; Glas, M. The origin of emerald—A controversial topic. *ExtraLapis Engl.* **2002**, *2*, 18–21.

61. Schwarz, D.; Klemm, L. Chemical signature of emerald. *Int. Geol. Congr. Abstr.* **2012**, *34*, 2812.

62. Giuliani, G.; Groat, L.; Ohnenstetter, D.; Fallick, A.E.; Feneyrol, J. The geology of gems and their geographic origin. In *Geology of Gem Deposits*; Raeside, E.R., Ed.; Mineralogical Association of Canada, Short Course Series 44: Tucson, AZ, USA, 2014; pp. 113–134.

63. Saeseasaw, S.; Pardieu, V.; Sangsawong, S. Three-phase inclusions in emerald and their impact on origin determination. *Gems Gemol.* **2014**, *50*, 114–132.

64. Ochoa, C.J.C.; Daza, M.J.H.; Fortaleche, D.; Jiménez, J.F. Progress on the study of parameters related to the origin of Colombian emeralds. *InColor* **2015**, *special issue*, 88–97.

65. Schwarz, D. The geographic origin determination of emeralds. *InColor* **2015**, *special issue*, 98–105.

66. Hainschwang, T.; Notari, F. Standards and protocols for emerald analysis in gem testing laboratories. *InColor* **2015**, *special issue*, 106–114.

67. Giuliani, G. La spirale du temps de l'émeraude. *Règne Minéral* **2011**, *98*, 31–44.

68. Kinnaird, J.A. Hydrothermal alteration and mineralization of the alkaline anorogenic ring complexes of Nigeria. *J. Afr. Earth Sci.* **1985**, *3*, 229–251. [CrossRef]

69. Larsen, B.T.; Olaussen, S.; Sundwoll, B.; Heeremans, M. The Permo-Carboniferous Oslo Rift through six stages and 65 million years. *Episodes* **2008**, *31*, 52–58.

70. Seifert, A.V.; Žaček, V.; Vrána, S.; Pecina, V.; Zachariáš, J.; Zwaan, J.C. Emerald mineralization in the Kafubu area, Zambia. *Bull. Geosci.* **2004**, *79*, 1–40.

71. Bastos, F.M. Emeralds from Itabira, Minas Gerais, Brazil. *Lapid. J.* **1981**, *35*, 1842–1848.

72. Hänni, H.A.; Schwarz, D.; Fischer, M. The emeralds of the Belmont mine, Minas Gerais, Brazil. *J. Gemmol.* **1987**, *20*, 446–456. [CrossRef]

73. de Souza, J.L.; Mendes, J.C.; da Silveira Bello, R.M.; Svisero, D.P.; Valarelli, J.V. Petrographic and microthermometrical studies of emeralds in the "Garimpo" of Capoeirana, Nova Era, Minas Gerais State, Brazil. *Miner. Deposita* **1992**, *27*, 161–168. [CrossRef]

74. Rondeau, B.; Notari, F.; Giuliani, G.; Michelou, J.-C.; Martins, S.; Fritsch, E.; Respinger, A. La mine de Piteiras, Minas Gerais, nouvelle source d'émeraude de belle qualité au Brésil. *Rev. Gemmol. AFG* **2003**, *148*, 9–25.

75. Walton, L. *Exploration Criteria for Colored Gemstone Deposits in Yukon*; Open file 2004-10; Geological Survey of Canada: Ottawa, ON, Canada, 2004; 184p.

76. Giuliani, G.; Cheilletz, A.; Zimmermann, J.-L.; Ribeiro-Althoff, A.M.; France-Lanord, C.; Féraud, G. Les gisements d'émeraude du Brésil: Genèse et typologie. *Chron. Rech. Min. BRGM* **1997**, *526*, 17–60.

77. Grundmann, G.; Morteani, G. Emerald mineralization during regional metamorphism: The Habachtal (Austria) and Leydsdorp (Transvaal, South Africa) deposits. *Econ. Geol.* **1989**, *84*, 1835–1849. [CrossRef]

78. Rudowski, L.; Giuliani, G.; Sabaté, P. Les phlogopitites à émeraude au voisinage des granites de Campo Formoso et Carnaíba (Bahia, Brésil): Un exemple de minéralisation protérozoïque à Be, Wo et W dans les ultrabasiques métasomatisées. *C. R. Acad. Sci.* **1987**, *304*, 1129–1134.

79. Rudowski, L. Pétrologie et géochimie des granites transamazoniens de Campo Formoso et Carnaíba (Bahia, Brésil) et des phlogopitites à émeraude associées. Thèse de Doctorat, Université Paris VI, Paris, France, 1989.

80. Muñoz, J.L. F-OH and Cl-OH exchange in micas with applications to hydrothermal ore deposits. In *Micas*; Bailey, S.W., Ed.; Mineralogical Society of America: Washington, DC, USA, 1984; Volume 13, pp. 469–493.

81. Guy, B. Contribution à l'étude des skarns de Costabonne (Pyrénées orientales, France) et à la théorie de la zonation métasomatique. Ph.D. Thesis, Université de Paris VI, Paris, France, 1979.

82. Korzhinskii, D.S. The theory of systems with perfectly mobile components and processes of mineral formation. *Am. J. Sci.* **1965**, *263*, 193–205. [CrossRef]

83. Laurs, B.M.; Dilles, J.H.; Snee, L.W. Emerald mineralization and metasomatism of amphibolite, Khaltaro granitic pegmatite hydrothermal vein system, Haramosh mountains, northern Pakistan. *Can. Mineral.* **1996**, *34*, 1253–1286.

84. Grundmann, G.; Morteani, G. Ein neues Vorkommen von Smaragd, Alexandrit, Rubin und Saphir in einem Topas-führenden Phlogopit-felds von Poona, Cue District, West Australien. *Zeitschrift der Deutschen Gemmologischen Gesellschaft* **1995**, *44*, 11–31.

85. Xue, G.; Marshall, D.; Zhang, S.; Ullrich, T.D.; Bishop, T.; Groat, L.A.; Thorkelson, D.J.; Giuliani, G.; Fallick, A.E. Conditions for early Cretaceous emerald formation at Dyakou, China: Fluid inclusion, Ar-Ar, and stable isotope studies. *Econ. Geol.* **2010**, *105*, 339–349. [CrossRef]

86. Kazmi, A.H.; Snee, L.W. Geology of the world emerald deposits: A brief review. In *Emeralds of Pakistan: Geology, Gemology and Genesis*; Kazmi, A.H., Snee, L.W., Eds.; Van Nostrand Reinhold Company: New York, NY, USA, 1989; pp. 165–228.

87. Martins, S. Brazilian emeralds. In Proceedings of the Oral Communication, II World Emerald Symposium, Bogotá, Colombia, 12–14 October 2018.

88. Biondi, J.C. Depósitos de esmeralda de Santa Terezinha (GO). *Rev. Bras. Geociências* **1990**, *20*, 7–24. [CrossRef]

89. Gusmão Costa, S.A. de Correlação da seqüência encaixante das esmeraldas de Santa Terezinha de Goiás com os terrenos do tipo greenstone belt de Crixás e tipologia dos depósitos. In Proceedings of the Boletim de Resumos Espandidos XXXVIII Congresso Brasileiro de Geologia Goiânia, Goiás, Brazil, 1986; Volume 2, pp. 597–614.

90. Giuliani, G.; D'El-Rey Silva, L.J.; Couto, P.A. Origin of emerald deposits of Brazil. *Miner. Deposita* **1990**, *25*, 57–64. [CrossRef]

91. D'el-Rey Silva, L.J.H.; Giuliani, G. Controle estrutural da jazida de esmeraldas de Santa Terezinha de Goiás: Implicações na gênese, na têctonica regional e no planajamento de lavra. In Proceedings of the Boletim de Resumos Espandidos XXXV Congresso Brasileiro de Geologia, Belém, PA, Brazil; 1988; Volume 1, pp. 413–427.

92. D'el-Rey Silva, L.J.H.; de Barros Neto, L.S. The Santa Terezinha-Campos Verdes emerald district, central Brazil: Structural and Sm-Nd data to constrain the tectonic evolution of the Neoproterozoic Brasília belt. *J. S. Am. Earth Sci.* **2002**, *15*, 693–708. [CrossRef]

93. Morteani, G.; Grundmann, G. The emerald porphyroblasts in the penninic rocks of the central Tauern Window. *Neues Jb. Miner. Monat.* **1977**, *11*, 509–516.

94. Grundmann, G. Polymetamorphose und Abschätzung der Bildungsb edin-gungen der smaragd-fürhrenden gesteinsserien der leckbachscharte, Habachtal, Österreich. *Fortschr. Mineral.* **1980**, *58*, 39–41.

95. Grundmann, G.; Morteani, G. Die geologie des smaragdvorkommens im Habachtal (Land Salzburg, Österreich). *Archiv. Lagerstättenforsch. Geol. Bundensanst* **1982**, *2*, 71–107.

96. Nwe, Y.Y.; Grundmann, G. Evolution of metamorphic fluids in shear zones: The record from the emeralds of Habachtal, Tauern window, Austria. *Lithos* **1990**, *25*, 281–304. [CrossRef]

97. Trumbull, R.B.; Krienitz, M.-S.; Grundmann, G.; Wiedenbeck, M. Tourmaline geochemistry and $\delta^{11}B$ variations as a guide to fluid-rock interaction in the Habachtal emerald deposit, Tauern window, Austria. *Contrib. Mineral. Petr.* **2009**, *157*, 411–427. [CrossRef]

98. Bowersox, G.; Snee, L.W.; Foord, E.E.; Seal, R.R., II. Emeralds of the Panjsher Valley, Afghanistan. *Gems Gemol.* **1991**, *27*, 26–39. [CrossRef]

99. Hussain, S.S.; Chaudhry, M.N.; Dawood, H. Emerald mineralization of Barang, Bajaur Agency, Pakistan. *J. Gemmol.* **1993**, *23*, 402–408. [CrossRef]

100. Dilles, J.H.; Snee, L.W.; Laurs, B.M. Geology, Ar-Ar age and stable isotopes geochemistry of suture-related emerald mineralization, Swat, Pakistan, Himalayas. In Proceedings of the Geological Society of America, Annual Meeting, Seattle, WA, USA, 1994; Abstracts. Volume 26, p. A-311.

101. Arif, M.; Fallick, A.E.; Moon, C.J. The genesis of emeralds and their host rocks from Swat, northwestern Pakistan: A stable-isotope investigation. *Miner. Deposita* **1996**, *31*, 255–268. [CrossRef]

102. Arif, M.; Henry, D.J.; Moon, C.J. Host rock characteristics and source of chromium and beryllium for emerald mineralization in the ophiolitic rocks of the Indus suture zone in Swat, NW Pakistan. *Ore Geol. Rev.* **2011**, *39*, 1–20. [CrossRef]

103. Cheilletz, A.; Féraud, G.; Giuliani, G.; Rodriguez, C.T. Time-pressure-temperature formation of Colombian emerald: An ^{40}Ar/^{39}Ar laser-probe and fluid inclusion-microthermometry contribution. *Econ. Geol.* **1994**, *89*, 361–380. [CrossRef]

104. Ottaway, T.L.; Wicks, F.J.; Bryndzia, L.T.; Kyser, T.K.; Spooner, E.T.C. Formation of the Muzo hydrothermal emerald deposit in Colombia. *Nature* **1994**, *369*, 552–554. [CrossRef]

105. Banks, D.; Giuliani, G.; Yardley, B.W.D.; Cheilletz, A. Emerald mineralisation in Colombia: Fluid chemistry and the role of brine mixing. *Miner. Deposita* **2000**, *35*, 699–713. [CrossRef]

106. Cheilletz, A.; Giuliani, G.; Branquet, Y.; Laumonier, B.; Sanchez, A.J.M.; Féraud, G.; Arhan, T. Datation K-Ar et ^{40}Ar/^{39}Ar à 65 ± 3 Ma des gisements d'émeraude du district de Chivor-Macanal: Argument en faveur d'une déformation précoce dans la Cordillère Orientale de Colombie. *C. R. Acad. Sci.* **1997**, *324*, 369–377.

107. Branquet, Y.; Cheilletz, A.; Giuliani, G.; Laumonier, B. Fluidized hydrothermal breccia in dilatant faults during thrusting: The Colombian emerald deposits case. In *Fractures, Fluid Flow and Mineralization*; McCaffrey, K.J.W., Lonergan, L., Wilkinson, J.J., Eds.; Geological Society London: London, UK, 1999; Volume 155, pp. 183–195.

108. Branquet, Y.; Giuliani, G.; Cheilletz, A.; Laumonier, B. Colombian emeralds and evaporites: Tectono-stratigraphic significance of a regional emerald-bearing evaporitic breccia level. In Proceedings of the 13th SGA Biennal Meeting, Nancy, France, 24–27 August 2015; Volume 4, pp. 1291–1294.

109. Cooper, M.A.; Addison, F.T.; Alvarez, R.; Coral, M.; Graham, R.; Hayward, A.B.; Howe, S.; Martinez, J.; Naar, J.; Pena, R.; et al. Basin development and tectonic history of the Llanos basin, Eastern Cordillera, and Middle Magdalena valley, Colombia. *AAPG Bull.* **1995**, *79*, 1421–1443.

110. Branquet, Y.; Cheilletz, A.; Cobbold, P.R.; Baby, P.; Laumonier, B.; Giuliani, G. Andean deformation and rift inversion, eastern edge of Cordillera Oriental (Guateque, Medina area), Colombia. *J. S. Am. Earth Sci.* **2002**, *15*, 391–407. [CrossRef]

111. Bayona, G.; Cortès, M.; Jaramillo, C.; Ojeda, G.; Aristizabal, J.J.; Harker, A.R. An integrated analysis of an orogen-sedimentary basin pair: Latest Cretaceous (Cenozoic evolution of the linked Eastern Cordillera orogen and the Llanos foreland basin of Colombia. *Geol. Soc. Am. Bull.* **2008**, *120*, 1171–1197. [CrossRef]

112. Laumonier, B.; Branquet, Y.; Lopès, B.; Cheilletz, A.; Giuliani, G.; Rueda, F. Mise en évidence d'une tectonique compressive Eocène-Oligocène dans l'Ouest de la Cordillère orientale de Colombie, d'après la structure en duplex des gisements d'émeraude de Muzo et de Coscuez. *C. R. Acad. Sci.* **1996**, *323*, 705–712.

113. Branquet, Y.; Laumonier, B.; Cheilletz, A.; Giuliani, G. Emeralds in the Eastern Cordillera of Colombia: Two tectonic settings for one mineralization. *Geology* **1999**, *27*, 597–600. [CrossRef]

114. Pignatelli, I.; Giuliani, G.; Ohnenstetter, D.; Agrosì, G.; Mathieu, S.; Morlot, C.; Branquet, Y. Colombian trapiche emeralds: Recent advances in understanding their formation. *Gems Gemol.* **2015**, *51*, 222–259. [CrossRef]

115. Giuliani, G.; France-Lanord, C.; Cheilletz, A.; Coget, P.; Branquet, Y.; Laumonier, B. Sulfate reduction by organic matter in Colombian emerald deposits: Chemical and stable isotope (C, O, H) evidence. *Econ. Geol.* **2000**, *95*, 1129–1153. [CrossRef]

116. Bosshart, G. Emeralds from Colombia. *J. Gemmol.* **1991**, *22*, 409–425. [CrossRef]

117. Hewton, M.L.; Marshall, D.D.; Ootes, L.; Loughrey, L.E.; Creaser, R.A. Colombian-style emerald mineralization in the northern Canadian Cordillera: Integration into a regional Paleozoic fluid flow regime. *Can. J. Earth Sci.* **2013**, *50*, 857–871. [CrossRef]

118. Keith, J.D.; Thompson, T.J.; Ivers, S. The Uinta emerald and the emerald-bearing potential of the Red Pine shale, Uinta Mountains, Utah. *Geol. Soc. Am. Abstr. Program* **1996**, *28*, 85.

119. Keith, J.D.; Nelson, S.T.; Thompson, T.J.; Dorais, M.J.; Olcott, J.; Duerichen, E.; Constenius, K.N. The genesis of fibrous calcite and shale-hosted emerald in a nonmagmatic hydrothermal system, Uinta mountains, Utah. *Geol. Soc. Am. Abstr. Program* **2002**, *34*, 55.

120. Nelson, S.T.; Keith, J.D.; Constenius, K.N.; Duerichen, E.; Tingey, D.G. Genesis of fibrous calcite and emerald by amagmatic processes in the southwestern Uinta Mountains, Utah. *Rocky Mt. Geol.* **2008**, *43*, 1–21. [CrossRef]

121. Tapponnier, P.; Mattauer, M.; Proust, F.; Cassaigneau, C. Mesozoic ophiolites, sutures, and large-scale tectonic movements in Afghanistan. *Earth Planet. Sci. Lett.* **1981**, *52*, 335–371. [CrossRef]

122. Sabot, B.; Cheilletz, A.; de Donato, P.; Banks, D.; Levresse, G.; Barrès, O. Afghan emeralds face Colombian cousins. *Chron. R. Min. BRGM* **2000**, *541*, 111–114.

123. Sabot, B.; Cheilletz, A.; de Donato, P.; Banks, D.; Levresse, G.; Barrès, O. The Panjsher-Afghanistan emerald deposit: New field and geochemical evidence for Colombian style mineralisation. In Proceedings of the European Union Geoscience XI, Strasbourg, France, 8–13 April 2001; Section OS 06. p. 548.

124. An, Y. The geological characteristics of the emerald deposit in Tashen Kuergan, Xinjiang. *Xinjiang Non-Ferrous Met.* **2009**, *29*, 9–10.

125. Wise, M.A. New finds in North Carolina. *ExtraLapis Engl.* **2002**, *2*, 64–65.

126. Speer, W.E. Hiddenite district, Alexander Co, North Carolina. In Fieldtrip Guidebook. In *Proceedings of the Southeastern Section 57th Annual Meeting*; Charlotte, NC, USA, 10–11 April 2008, The Geological Society of America: Boulder, CO, USA, 2008; 28p.

127. Wise, M.A.; Anderson, A.J. The emerald- and spodumene-bearing quartz veins of the Rist emerald mine, Hiddenite, North Carolina. *Can. Mineral.* **2006**, *44*, 1529–1541. [CrossRef]

128. Grundmann, G.; Morteani, G. "Smaragdminen der Cleopatra": Zabara, Sikait und Umm Kabo in Ägypten. *Lapis* **1993**, *7–8*, 27–39.

129. Grundmann, G.; Morteani, G. Emerald formation during regional metamorphism: The Zabara, Sikeit and Umm Kabo deposits (Eastern Desert, Egypt). In *Geoscientific Research in Northeast Africa*; Schandelmeier, T., Ed.; A.A. Balkema: Rotterdam, The Netherlands, 1993; pp. 495–498.

130. Grundmann, G.; Morteani, G. Multi-stage emerald formation during Pan-African regional metamorphism: The Zabara, Sikait, Umm Kabo deposits, South Eastern desert of Egypt. *J. Afr. Earth Sci.* **2008**, *50*, 168–187. [CrossRef]

131. Surour, A.A.; Takla, M.A.; Omar, S.A. EPR spectra and age determination of beryl from the Eastern Desert of Egypt. *Ann. Geol. Surv. Egypt* **2002**, *25*, 389–400.

132. Giuliani, G.; France-Lanord, C.; Coget, P.; Schwarz, D.; Cheilletz, A.; Branquet, Y.; Giard, D.; Pavel, A.; Martin-Izard, A.; Piat, D.H. Oxygen isotope systematics of emerald—Relevance for its origin and geological significance. *Miner. Deposita* **1998**, *33*, 513–519. [CrossRef]

133. Duarte, L.D.C.; Pulz, G.M.; D'el-Rey Silva, L.J.H.; Ronchi, L.H.; de Brun, T.M.M.; Juchem, P.L. Microtermometria das inclusões fluidas na esmeralda do distrito mineiro de Campos Verdes, Goiás. In *Caracterização e Modelamento de depósitos Minerais*; Ronchi, L.H., Althoff, F.J., Eds.; Eitoras Unisinos: São Leopoldo, RS, Brazil, 2003; pp. 267–292.

134. Olivera, J.A.P.; Ali, S.H. Gemstone mining as a develoment cluster: A study of Brazil's emerald mines. *Resour. Policy* **2011**, *36*, 132–141. [CrossRef]

135. Salvi, S.; Giuliani, G.; Andrianjakavah, P.R.; Moine, B.; Beziat, D.; Fallick, A.E. Fluid inclusion and stable isotope constraints on the formation of the Ianapera emerald deposit, Southern Madagascar. *Can. Mineral.* **2017**, *55*, 619–650. [CrossRef]

136. de Wit, M.J.; Hart, R.A.; Hart, R.J. The Jamestown ophiolite complex, Barberton mountain belt: A section through 3.5 Ga oceanic crust. *J. Afr. Earth Sci.* **1987**, *6*, 681–730. [CrossRef]

137. Biondi, J.C.; Poidevin, J.L. Idade da mineralização e da sequência Santa Terezinha (Goiás, Brasil). In Proceedings of the Boletim de Resumos Espandidos XXXVIII Congresso Brasileiro de Geologia, Camboriú, Brasil, 23–28 October 1994; pp. 302–304.

138. Ribeiro-Althoff, A.M.; Cheilletz, A.; Giuliani, G.; Féraud, G.; Barbosa Camacho, G.; Zimmermann, J.L. Evidences of two periods (2 Ga and 650-500 Ma) of emerald formation in Brazil by K-Ar and ^{40}Ar/^{39}Ar dating. *Int. Geol. Rev.* **1997**, *39*, 924–937. [CrossRef]

139. Calligaro, T.; Dran, J.-C.; Poirot, J.-P.; Querré, G.; Salomon, J.; Zwaan, J.C. PIXE/PIGE characterization of emeralds using an external micro-beam. *Nucl. Instrum. Meth. B* **2000**, *161–163*, 769–774. [CrossRef]

140. Höll, R.; Maucher, A.; Westenberger, H. Synsedimentary diagenetic ore fabrics in the strata- and time-bound scheelite deposits of Kleinarltal and Felbertal in the Eastern Alps. *Miner. Deposita* **1972**, *7*, 217–226. [CrossRef]
141. Raith, J.G.; Montanuiversität Leoben, Austria. Personal communication, 2010.
142. Kozlik, M.; Raith, J.G.; Gerdes, A. U-Pb, Lu-Hf and trace element characteristics of zircon from the Felbertal scheelite deposit (Austria): New constraints on timing and source of W mineralization. *Chem. Geol.* **2016**, *421*, 112–126. [CrossRef]
143. Maluski, H.; Matte, P. Ages of alpine tectonometamorphic events in the northwestern Himalaya (northern Pakistan) by ^{39}Ar/^{40}Ar method. *Tectonophysics* **1984**, *3*, 1–18.
144. Grundmann, G.; Morteani, G. Alexandrite, emerald, sapphire, ruby and topaz in a biotite-phlogopite fels from the Poona, Cue district, Western Australia. *Austral. Gemmol.* **1998**, *20*, 159–167.
145. Vapnik, Y.; Moroz, I. Fluid inclusions in Panjshir emerald (Afghanistan). In Proceedings of the 16th European Current Research on Fluid Inclusions, Faculdade de Ciências do Porto, Porto, Portugal, 17–20 July 2001; Noronha, F., Doria, A., Guedes, A., Eds.; Memoria n° 7: New York, NY, USA; pp. 451–454.
146. Lapointe, M.; Anderson, A.J.; Wise, M. Fluid inclusion constraints on the formation of emerald-bearing quartz veins at the Rist tract, Heddenite, North Carolina. *Atl. Geol. Abstr.* **2004**, *4*, 146.
147. Diamond, L.W. Introduction to gas-bearing aqueous fluid inclusions. In *Fluid Inclusions: Analysis and Interpretation*; Samson, I., Anderson, A., Marshall, D., Eds.; Mineralalogical Association of Canada: Quebec City, QC, Canada, 2003; Volume 32, pp. 101–158.
148. Bodnar, R.J. Introduction to aqueous-electrolyte fluid inclusions. In *Fluid Inclusions: Analysis and Interpretation*; Samson, I., Anderson, A., Marshall, D., Eds.; Mineralogical Association of Canada: Quebec City, QC, Canada, 2003; Volume 32, pp. 81–100.
149. Audétat, A.; Gunter, D.; Heinrich, C. Formation of a magmatic-hydrothermal ore deposit: Insights with LA-ICP-MS analysis of fluid inclusions. *Science* **2012**, *279*, 2091–2094.
150. Diamond, L.W.; Marshall, D.; Jackman, J.; Skippen, G.B. Elemental analysis of individual fluid inclusions in minerals by Secondary Ion Mass Spectrometry (SIMS): Application to cation ratios of fluid inclusions in an Archaean mesothermal gold-quartz vein. *Geochim. Cosmochim. Acta* **1990**, *54*, 545–552. [CrossRef]
151. Rosasco, G.; Roedder, E.; Simmons, J. Laser-excited Raman spectroscopy for non-destructive analysis of individual phases in fluid inclusions in minerals. *Science* **1975**, *190*, 557–560. [CrossRef]
152. Delhaye, M.; Dhamelincourt, P. Raman microprobe and microscope with Laser excitation. *J. Raman Spectrosc.* **1975**, *3*, 33–43. [CrossRef]
153. Dhamelincourt, P.; Schubnel, H. La microsonde moléculaire à laser et son application à la minéralogie et la gemmologie. *Rev. Gemmol. AFG* **1977**, *52*, 11–14.
154. Burruss, R.C. Raman spectroscopy of fluid inclusions. In *Fluid Inclusions: Analysis and Interpretation*; Samson, I., Anderson, A., Marshall, D., Eds.; Mineralogical Association of Canada: Quebec City, QC, Canada, 2003; Volume 32, pp. 279–290.
155. Burke, E.A.J. Raman microspectrometry of fluid inclusions. *Lithos* **2001**, *55*, 139–158. [CrossRef]
156. Taylor, R.; Fallick, A.; Breaks, F. Volatile evolution in Archean rare-element granitic pegmatites: Evidence from the hydrogen-isotopic composition of channel H_2O in beryl. *Can. Mineral.* **1992**, *30*, 877–893.
157. Marshall, D.; Meisser, N.; Ellis, S.; Jones, P.; Bussy, F.; Mumenthaler, T. Formational conditions for the Binntal emerald occurrence, Valais, Switzerland: Fluid inclusion, chemical composition and stable isotope studies. *Can. Mineral.* **2017**, *55*, 725–741. [CrossRef]
158. Groat, L.A.; Giuliani, G.; Marshall, D.D.; Turner, D. Emerald deposits and occurrences: A review. *Ore Geol. Rev.* **2008**, *34*, 87–112. [CrossRef]
159. Sheppard, S.M.F. Characterization and isotopic variations in natural waters. In *Stable Isotopes in High Temperature Geological Processes*; Valley, J.W., Taylor, H.P., O'Neil, J.R., Eds.; Mineralogical Association of America: Chantilly, VA, USA, 1986; Volume 16, pp. 165–183.
160. Giuliani, G.; Dubessy, J.; Ohnenstetter, D.; Banks, D.; Branquet, Y.; Feneyrol, J.; Fallick, A.E.; Martelat, E. The role of evaporites in the formation of gems during metamorphism of carbonate platforms: A review. *Miner. Deposita* **2017**, *53*, 1–20. [CrossRef]
161. Yardley, B.W.D.; Bodnar, R.J. Fluids in the Continental Crust. *Geochem. Perspect.* **2014**, *3*, 1–127. [CrossRef]
162. Yardley, B.W.D.; Cleverley, J.S. *The Role of Metamorphic Fluids in the Formation of Ore Deposits*; Geological Society: London, UK, 2013; pp. 1–393.

163. Williams, A.E.; McKibben, M.A. A brine interface in the Salton Sea Geothermal System, California: Fluid geochemical and isotopic characteristics. *Geochim. Cosmochim. Acta* **1989**, *53*, 1905–1920. [CrossRef]

164. Bohlke, J.K.; Irwin, J.J. Laser microprobe analyses of Cl, Br, I, and K in fluid inclusions: Implications for sources of salinity in some ancient hydrothermal fluids. *Geochim. Cosmochim. Acta* **1992**, *56*, 203–225. [CrossRef]

165. Poujol, M.; Robb, L.J.; Respaut, J.P. Origin of gold and emerald mineralization in the Murchinson greenstone belt, Kaapval craton, South Africa. *Mineral. Mag.* **1998**, *62A*, 1206–1207. [CrossRef]

166. Beus, A.A. *Geochemistry of Beryllium and Genetic Types of Beryllium Deposits*; W.H. Freeman: San Francisco, CA, USA; London, UK, 1966; 401p.

167. Smirnov, V.L. Deposits of beryllium. In *Ore deposits of the U.S.S.R.*; Pitman Publications: London, UK, 1977; Volume 3, pp. 320–371.

168. Franz, G.; Gilg, H.A.; Grundmann, G.; Morteani, G. Metasomatism at a granitic pegmatite-dunite contact in Galicia: The Franqueira occurrence of chrysoberyl (alexandrite), emerald, and phenakite: Discussion. *Can. Mineral.* **1996**, *34*, 1329–1331.

169. Zwaan, J.C.; Touret, J. Emeralds in greenstone belts: The case of Sandawana, Zimbabwe. *Münchener Geologische Hefte* **2000**, *28*, 245–258.

170. Franz, G.; Morteani, G. Be-minerals: Synthesis, stability, and occurrence in metamorphic rocks. In *Beryllium: Mineralogy, Petrology, and Geochemistry*; Grew, E.S., Ed.; Mineralogical Society of America: Washington, DC, USA, 2002; Volume 50, pp. 551–589.

171. Martin-Izard, A.; Paniagua, A.; Moreiras, D. Metasomatism at a granitic pegmatite-dunite contact in Galicia: The Franqueira occurrence of chrysoberyl (alexandrite), emerald and phenakite. *Can. Mineral.* **1995**, *33*, 775–792.

172. Martin-Izard, A.; Paniagua, A.; Moreiras, D.; Acevedo, R.D.; Marcos-Pascual, C. Metasomatism at a granitic pegmatite-dunite contact in Galicia: The Franqueira occurrence of chrysoberyl (alexandrite), emerald, and phenakite: Reply. *Can. Mineral.* **1996**, *34*, 1332–1336.

173. Korzhinskii, D.S. *Theory of Metasomatic Zoning*; Clarendon Press: Oxford, UK, 1970.

174. Sahu, S.S.; Singh, S.; Stapathy, J.S. Lithological and structural controls on the genesis of emerald occurrences and their exploration implications in and around Gubarabanda area, Singhbhum crustal province Eastern India. *J. Geol. Soc. India* **2018**, *92*, 291–297. [CrossRef]

175. Zwaan, J.C.; Kanis, J.; Petsch, J. Update on emeralds from the Sandawana mines, Zimbabwe. *Gems Gemol.* **1997**, *33*, 80–101. [CrossRef]

176. Holmes, A. The oldest dated minerals of the Rhodesian shield. *Nature* **1954**, *173*, 612–614. [CrossRef]

177. Christensen, J.N.; Selverstone, J.; Rosenfeld, J.L.; DePaolo, D.J. Correlation by Rb/Sr geochronology of garnet growth histories from different structural levels within the Tauern Window, Eastern Alps. *Contrib. Mineral. Petr.* **1994**, *118*, 1–12. [CrossRef]

178. Vance, D.; O'nions, R.K. Isotopic chronometry of zone garnets growth kinetics and metamorphic histories. *Earth Planet. Sci. Lett.* **1990**, *114*, 113–129. [CrossRef]

179. Gilg, A.; Technische Universität München, Germany. Personal comunication, 2009.

180. Lake, D.J. Are there Colombian-type emeralds in Canada's Northern Cordillera? Insights from regional silt geochemistry, and the genesis of emerald at Lened, NWT. M.Sc. Thesis, University of British Columbia, Vancouver, BC, Canada, 2017.

181. Escobar, R. Geology and geochemical expression of the Gachalá emerald district, Colombia. *Geol. Soc. Am. Abstr. Program* **1978**, *10*, 397.

182. Beus, A.A. Sodium—A geochemical indicator of emerald mineralization in the Cordillera Oriental, Colombia. *J. Geochem. Explor.* **1979**, *11*, 195–208. [CrossRef]

183. Ringsrud, R. The Coscuez mine: A major source of Colombian emeralds. *Gems Gemol.* **1986**, *22*, 67–79. [CrossRef]

184. Cheilletz, A.; Giuliani, G. The genesis of Colombian emeralds: A restatement. *Miner. Deposita* **1996**, *31*, 359–364. [CrossRef]

185. Rohtert, W.R.; Independant geologist. Personal comunication, 2017.

186. Gutiérrez, L.H.O. Evaluación magnetométrica, radiométrica y geoeléctrica de depósitos esmeraldíferos. *Geofísica Colombiana* **2003**, *7*, 13–18.

187. Arif, M.; Henry, D.J.; Moon, C.J. Cr-bearing tourmaline associated with emerald deposits from Swat, NW Pakistan: Genesis and its exploration significance. *Am. Mineral.* **2010**, *95*, 799–809. [CrossRef]
188. Wise, M.A. Crystal Morphology of quartz, calcite, pyrite and rutile; Potential tool for emerald exploration in the Hiddenite area (North Carolina). *Geol. Soc. Am. Abstr. Programs* **2012**, *44*, 25.
189. Murphy, D.C.; Lipovsky, P.S.; Stuart, A.; Fonseca, A.; Piercey, S.J.; Groat, L. What about those emeralds, eh? Geological setting of emeralds at Regal Ridge (SE Yukon) provides clues to their origin and to other places to explore. *Geol. Assoc. Canada Mineral. Assoc. Canada Abstr. Program* **2002**, *28*, 154.
190. Lewis, L.L.; Hart, C.J.R.; Murphy, D.C. Roll out the beryl. In *Yukon Geoscience Forum*; Publications du Gouvernement du Canada: Yukon Territory, NT, Canada, 2003.

Article

Emeralds from the Most Important Occurrences: Chemical and Spectroscopic Data

Stefanos Karampelas *, Bader Al-Shaybani, Fatima Mohamed, Supharart Sangsawong and Abeer Al-Alawi

Bahrain Institute for Pearls & Gemstones (DANAT), WTC East Tower, P.O. Box 17236 Manama, Bahrain;
Bader.Alshaybani@danat.bh (B.A.-S.); Fatima.Mohamed@danat.bh (F.M.);
Supharart.Sangsawong@danat.bh (S.S.); Abeer.Alalawi@danat.bh (A.A.-A.)
* Correspondence: Stefanos.Karampelas@danat.bh; Tel.: +973-1720-1333

Received: 1 August 2019; Accepted: 9 September 2019; Published: 19 September 2019

Abstract: The present study applied LA–ICP-MS on gem-quality emeralds from the most important sources (Afghanistan, Brazil, Colombia, Ethiopia, Madagascar, Russia, Zambia and Zimbabwe). It revealed that emeralds from Afghanistan, Brazil, Colombia and Madagascar have a relatively lower lithium content (^{7}Li < 200 ppmw) compared to emeralds from other places (^{7}Li > 250 ppmw). Alkali element contents as well as scandium, manganese, cobalt, nickel, zinc and gallium can further help us in obtaining accurate origin information for these emeralds. UV-Vis spectroscopy can aid in the separation of emeralds from Colombia and Afghanistan from these obtained from the other sources as the latter present pronounced iron-related bands. Intense Type-II water vibrations are observed in the infrared spectra of emeralds from Madagascar, Zambia and Zimbabwe, as well as in some samples from Afghanistan and Ethiopia, which contain higher alkali contents. A band at 2818 cm^{-1}, supposedly attributed to chlorine, was observed only in emeralds from Colombia and Afghanistan. Samples with medium to high alkalis from Ethiopia, Madagascar, Zambia and Zimbabwe can also be separated from the others by Raman spectroscopy based on the lower or equal relative intensity of the Type I water band at around 3608 cm^{-1} compared to the Type II water band at around 3598 cm^{-1} band (with some samples from Afghanistan, Brazil and Russia presenting equal relative intensities).

Keywords: emeralds; LA-ICP-MS; UV-Vis-NIR; FTIR; Raman; PL

---

1. Introduction

Emeralds, together with rubies and sapphires (red and blue corundum), as well as jadeite "jade" (jadeitite), diamonds and natural pearls, have been the most sought-after gems for several centuries. Emerald is the bluish-green to green to yellowish-green variety of beryl (with an ideal formula of Be$_3$Al$_2$SiO$_{18}$) coloured by chromium and/or vanadium (iron may also contribute to the colour, but to a lesser extent); beryl coloured solely by iron is green beryl (and not emerald) [1]. Transparent natural emeralds of homogenous vivid-green colour are the most researched. Most faceted gems are sold by carat (1 carat = 0.2 g), but their monetary value is not linearly correlated with their weight; bigger gems are rarer and can fetch higher prices. Absence or presence, type and degree of treatment are also important factors linked to gems' monetary value. In the case of emeralds, most gem-quality faceted stones contain surface-reaching fissures and, some of them, also cavities. In order to improve their clarity, the vast majority of emeralds are "filled" with a material having a refractive index similar to that of emerald (oil, resin or other) that reduces the visibility of the fissures and, sometimes, the cavities [2–4]. The degree of emerald clarity enhancement, which is not always directly linked with the amount of filling material, ranges from none to significant, with the former being very rare and more desirable [5]. Geographic origin is frequently requested from gemmological laboratories by customers as it is used by gem dealers as a brand name, is sometimes linked with history, exoticism, spirituality, etc., and might play an important role in the monetary value of a gem [6–23]. In parallel, over the last two decades, ethical issues related to gem mining have been in the spotlight, and end consumers demand transparency about the mine-to-market supply chain, in addition to detailed information on stones' provenance [23–26]. Origin determination for gems is also useful for archaeologists, curators, etc., as it can help them to better understand early trade routes [27–40].

Origin determination is based on gem characteristics linked to geological formation. However, gemmological laboratories are asked to issue reports mentioning a gem's geographical origin, which is related to politics rather than geology [12,15]. Geographic origin determination is getting more complicated, considering that gems can grow in similar geologic environments but in different countries; e.g., emeralds associated with granites-pegmatites and mafic-ultramafic rocks as in Kafubu, Zambia; Malyshevsk (the Ural Mountains), Russia; Mananjary, Madagascar; etc. In parallel, a gem can grow in more than one geological environment in the same country; e.g., emeralds occur in Zambia in both Kafubu near Kitwe (associated with granites-pegmatites and mafic-ultramafic rocks; Type IA occurrence—see classification below) and Musakashi near Solwezi (in eluvial lateritic soils adjacent to quartz veins; Type IID occurrence—see classification below) [16–19,21,31,41,42].

Emerald is a relatively rare mineral because it needs common elements such as silicon (Si), aluminium (Al) and oxygen (O), together with less common elements (that are rarely encountered together) such as beryllium (Be)—enriched in the crust, chromium (Cr)—typically enriched in mantle rocks, and/or vanadium (V), with iron (Fe) in limited concentrations; however, it can be found on all continents except Antarctica [16–18,21]. Emeralds of gem quality and economic importance are not always formed though; several parameters play an important role in gem formation [43,44]. It is important to have the right ingredients in just the right amounts (as previously mentioned): favourable "thermobarometric conditions", space to grow (with some exceptions), limited nucleation (i.e., few nuclei will evolve into a crystal) and stable growth conditions for a certain amount of time, but not for millions of years [44]. Importantly, post-growth phenomena that might damage the gem, such as mechanical fracturing, chemical etching, etc., should be absent [43].

A gemmological report with the origin for an emerald from a laboratory recognized by the international market can cost from ca. 100 up to 500 USD for gems <2 ct and >2000 USD for gems >50 ct, depending on the laboratory. Thus, gemmological laboratories receive principally medium- to high-quality "large" faceted (>0.5 ct and mostly >1 ct) emeralds "worth" an origin determination report. Nowadays these emeralds mostly come from Colombia (both the east and west side of central Cordillera), Zambia (Kafubu), Brazil (Itabira, Minas Gerais), Russia (Malyshevsk, the Ural mountains), Madagascar (Mananjary), Afghanistan (Panjsher Valley), Zimbabwe (Sandawana) and recently Ethiopia (Shakisso). In terms of monetary value, an emerald from Colombia fetches higher prices than an emerald of exactly the same size and quality from another country. Samples of similar size and quality other than Colombian fetch similar prices, but the traders still ask for an origin to be mentioned in the report.

The classification of emerald deposits is presented in several works, and recently an enhanced classification has been suggested [13,17,21]. According to this, the geological environment of the vast majority of occurrences producing gem-quality emeralds is classified as Type IA—tectonic-magmatic-related hosted in mafic–ultramafic rocks [21]. Only emeralds from Colombia are classified as Type IIB; tectonic-metamorphic-related hosted in sedimentary rock-black shale and emeralds from Afghanistan (Panjsher Valley) are classified as Type IIC; tectonic-metamorphic-related are hosted in metamorphic rocks [21].

Gemmological laboratories are issuing reports on emeralds (including origin determination), and coloured gems in general, after combining the results obtained by several methods [11,12,15,19,20]. The methods used should be non-destructive and rarely micro-destructive [45–48]. For the origin determination and characterization of emeralds, microscopy, FTIR (Fourier-Transform InfraRed) spectroscopy, UV-Vis-NIR (Ultraviolet-Visible-Near InfraRed) spectroscopy, chemistry such as EDXRF (Energy-Dispersive X-ray Spectroscopy), sometimes LA-ICP-MS (Laser Ablation–Inductively Coupled

Plasma-Mass Spectrometry) as well as LIBS (Laser-Induced Breakdown Spectroscopy) and, in same cases, Raman and PL (photoluminescence) spectroscopy are used [6,7,11,12,15,19,20,23,31,40–42,49–78]. Oxygen isotopes as well as fluid inclusions are also used for their study (see [21] for more information and further references); however, these methods are currently rarely used by gemmological laboratories. Gem-quality emeralds present some characteristics that might help gemmological laboratories build an accurate database to trace the origin of an unknown sample, in contrast with some other gems which might be more challenging (e.g., sapphires). For instance, the majority of gem-quality emeralds are found in primary deposits [76]; they contain inclusions that are associated with their geology and, due to their crystal structure and chemistry, several minor and trace elements directly linked to their growth environments could be present.

For the present work, 62 samples from eight countries' data were collected using LA-ICP-MS as well as UV-Vis-NIR, FTIR, Raman and PL spectroscopy. The data from the samples were compared, looking for potential differences linked to their geographical origin. This is the first study that combines all these methods on samples from the most important sources of gem-quality emeralds.

2. Materials and Methods

All 62 samples studied are listed in Table 1. Forty-three samples were rough, and small areas ("windows") were cut and polished in order to acquire better spectroscopic and chemical data, four of them were oriented (i.e., cut and polished parallel and/or perpendicular to the c-axis) and fifteen were faceted (all from Brazil). Most of the samples studied had a green or dark green homogenous colour. All samples from Russia were light green; some of the samples from Brazil and Colombia were also light green. All samples from Russia presented numerous inclusions.

Table 1. List of studied samples, along with their weight and colour ranges.

Locality	No. of Samples	Weight Range (ct)	Colour Range
Afghanistan (Panjsher Valley from Kherskanda)	9	0.27–0.68	Green
Brazil (Itabira)	18	0.17–1.88	Light green to green
Colombia (Coscuez)	8	0.11–2.42	Light green to green
Ethiopia (Shakisso)	4	0.43–1.69	Green
Madagascar (Mananjary from Irondro, Ambodivandrika and Morarano)	9	0.17–2.18	Green to dark green
Russia (Malyshevsk, Ural mountains)	5	0.33–0.80	Light green
Zambia (Kafubu)	6	0.47–2.50	Green
Zimbabwe (Sandawana)	3	0.24–0.56	Green

UV-Vis-NIR spectra were acquired using a Cary 5000 UV-Vis-NIR spectrometer (Varian Inc., Palo Alto, CA, USA) in the 250–1500 nm spectral range, with a spectral bandwidth and data interval of 0.7 nm and a scan rate of 60 nm/min for the UV-Vis region and a spectral bandwidth and data interval of 1.0 nm and a scan rate of 120 nm/min for the NIR region. Polarized spectra using a diffraction grating polarizer were acquired on the oriented samples.

FTIR spectra were acquired from 8000 to 300 cm^{-1} using a Nicolet iS5 spectrometer (Thermo Fischer Scientific, Waltham, MA, USA) with 4 cm^{-1} resolution and 500 scans (background spectra were collected using the same parameters). Most of the spectra were unpolarized (or partially oriented); i.e., acquired on randomly oriented samples; positioned to maximize the signal. The c-axis of unoriented samples was checked so that none of the spectra were acquired with the beam parallel to the axis.

The absorption coefficient (a) was plotted to all UV-Vis-NIR and FTIR spectra. This was calculated using the formula $a = 2.303\ A/d$, where A is the absorbance and d is the path length (or sample thickness for measurements on parallel polished windows) in cm.

Raman spectra were acquired using a Renishaw inVia spectrometer (Renishaw plc, Wotton-under-Edge, Gloucestershire, UK) from 100 to 2000 cm^{-1} and from 3300 to 3950 cm^{-1}, coupled with an optical microscope, 514 nm excitation wavelength (diode-pumped solid-state laser), 1800 grooves/mm grating, notch filter, 40-micron slit, a spectral resolution of around 2 cm^{-1} and calibrated using a diamond at 1331.8 cm^{-1}. For the 100–2000 cm^{-1} range, 40 mW laser power on the sample was used to acquire all Raman spectra (except for one sample from Colombia and one sample from Afghanistan, where a laser power of 8 mW was used to avoid spectra saturation linked to high sample luminescence), 50× short distance objective lens, an acquisition time of 20 s and five accumulations. As for the 3300–3950 cm^{-1} range, laser power of 0.8 mW on the sample was used to acquire all Raman spectra, 50× short distance objective lens, an acquisition time of 10 s (except for one sample from Colombia, where an acquisition time of 5 s was used) and 60 accumulations (except for the aforementioned sample from Afghanistan, where 40 accumulations were used).

Photoluminescence spectra from 550 to 900 nm were acquired on the samples using the Raman spectrometer with a 0.04 mW laser power on the sample, 50× short distance objective lens and an acquisition time of 10 s; different parameters were used for a sample from Afghanistan: 0.00008 mW laser power and an acquisition time of 20 s. Most spectra were acquired parallel and perpendicular to the samples' c-axis; the c-axis was positioned using a polariscope and a conoscope.

LA–ICP-MS chemical analysis was performed using an iCAP Q (Thermo Fisher Scientific; Waltham, MA, USA) Inductively Coupled Plasma-Mass Spectrometer (ICP-MS) coupled with a Q-switched Nd: YAG Laser Ablation (LA) device operating at a wavelength of 213 nm (Electro Scientific Industries, Fremont, CA, USA). A laser spot of 40 μm in diameter was used, along with a fluence of around 10 J/cm^2 and a 10 Hz repetition rate. The laser warmup/background time was 20 s, the dwell time was 30 s, and the washout time was 50 s. For the ICP-MS operations, the forward power was set at ~1550 W, the typical nebulizer gas (argon) flow was ~1.0 L/min and the carrier gas (helium) set at ~0.80 L/min. The criteria for the alignment and tuning sequence were to maximize the beryllium (Be) counts and keep the ThO/Th ration below 2%. NIST 610 and NIST 612 glasses were used for calibration standards. The time-resolved signal was processed in Qtegra ISDS software (version 2.10, Thermo Fisher Scientific; Waltham, MA, USA) using silicon (^{29}Si) as the internal standard, applying 31.35 wt % theoretical value for beryl. The limits of detection (LOD) and limits of quantification (LOQ) for each of the abovementioned elements are shown in Table 2. These limits differ from day to day (for every set of measurements), so they are presented as ranges, from the lowest to the highest. Three spots were analysed on every sample and five were analysed on samples from Ethiopia and Zimbabwe. The measured points were checked under a microscope (Nikon, Shinagawa, Tokyo, Japan) to make sure they are on green zones without inclusions (however, the presence of micro-inclusions cannot be completely ruled out).

Table 2. LA–ICP-MS detection limits and ranges in ppmw.

Limits	^{7}Li	^{23}Na	^{24}Mg	^{39}K	^{45}Sc
LOD	0.29–0.87	7.00–49.13	0.28–1.24	6.82–17.72	0.44–1.07
LOQ	0.87–2.42	21.00–147.38	0.83–3.73	20.45–53.16	1.32–3.20
Limits	^{51}V	^{52}Cr	^{55}Mn	^{56}Fe	^{59}Co
LOD	0.29–0.85	1.36–1.99	0.23–0.67	4.37–6.51	0.09–0.42
LOQ	0.86–2.54	4.07–5.96	0.70–2.01	13.12–19.52	0.28–1.25
Limits	^{60}Ni	^{66}Zn	^{69}Ga	^{85}Rb	^{133}Cs
LOD	2.71–6.11	0.65–2.28	0.13–0.45	0.15–0.28	0.02–0.07
LOQ	8.12–18.33	1.94–6.85	0.39–1.36	0.44–0.83	0.06–0.21

LOD: Limits of detection; LOQ: Limits of quantification.

The amount of water in emeralds is difficult to measure directly. However, it was found to be linked to Na_2O concentration [79] and can be calculated using the following equation: H_2O wt % = 0.5401 × ln(Na_2O wt %) + 2.1867 [80].

3. Results and Discussion

3.1. LA–ICP-MS

LA-ICP-MS data on the studied samples are presented in Table 3 and all acquired individual chemical analysis are presented in Tables S1–S8. ^{23}Na (sodium) and ^{24}Mg (magnesium) are the most abundant of the minor and trace elements measured in the studied samples, with all studied samples from Zimbabwe (Sandawana) having Na_2O > 2 wt %. It has been suggested that a Na_2O content of emeralds <1 wt % is considered low, 1 wt % < medium < 2 wt % and >2 wt % high [56]. The studied emeralds from Colombia had a relatively low Na_2O content. The samples from Afghanistan, Russia and Brazil had relatively low to medium Na_2O content, with those from Brazil having a medium sodium content as well as most of the studied samples from Afghanistan. Relatively medium to high, and fairly high, sodium quantities were presented by the samples from Ethiopia, Madagascar, Zambia and Zimbabwe, with the latter exhibiting the highest content amongst the studied samples. The calculated water content of the studied samples is listed in Table 3; the Colombian samples presented the lowest calculated water content (1.69–2.17%), followed by Russian samples (1.99–2.3%), samples from Brazil (2.17–2.47%) and samples from Afghanistan, with calculated water content ranging from low (1.93%) to relatively high (2.54%). The vast majority of the other studied samples presented a water content >2.5%. The plot of MgO vs. Na_2O presents a positive correlation (Figure 1), with a Na_2O/MgO ratio being below 1 for most of the studied samples; only the studied Russian samples presented a ratio >1, with a 1.33 median and a 1.34 average (see Table 3). Data of samples from Brazil, Russia, Zimbabwe, Zambia and Madagascar are similar to previously published data measured with a microprobe [52,55–57,77]; some of the Russian emeralds were found to contain higher MgO with a Na_2O/MgO ratio <1 [55].

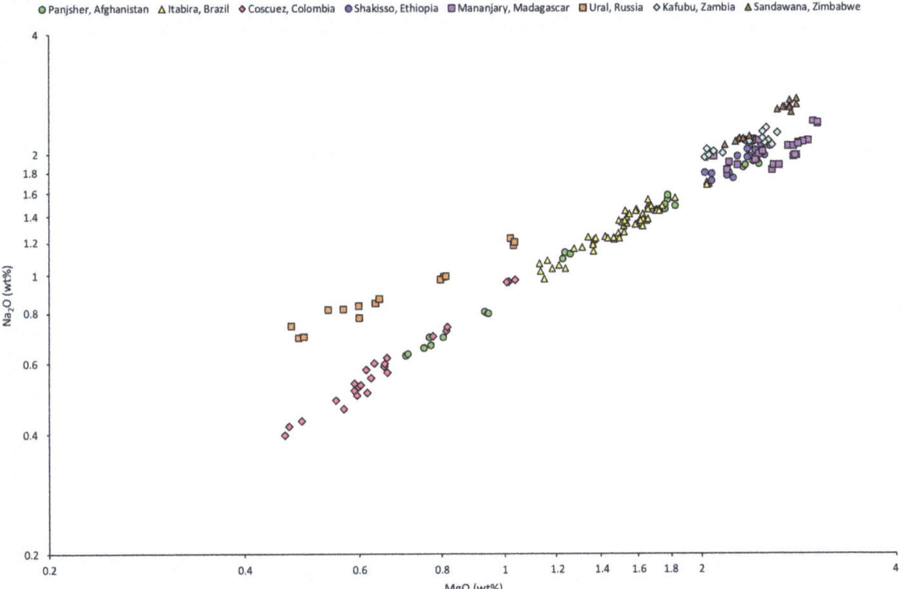

Figure 1. Binary plot of MgO (wt %) vs. Na_2O (wt %).

Table 3. LA–ICP-MS of the samples in ppmw.

Samples	Element	Min–Max	Average (SD)	Median
Afghanistan (Panjsher) 9 samples, 27 analysis	^{7}Li	84.9–162	115.81 (25.86)	108
	Na_2O (%)	0.63–1.91	1.27 (0.46)	1.44
	MgO (%)	0.71–2.45	1.51 (0.60)	1.69
	^{39}K	107–1540	713.63 (497.38)	710
	^{45}Sc	148–2390	669.07 (782.41)	256
	^{51}V	557–3130	1312.48 (766.18)	1100
	^{52}Cr	500–3840	1994.26 (1257.28)	2160
	^{55}Mn	BQL–3.09	0.78 (1.02)	BQL
	^{56}Fe	781–2530	1394.70 (550.92)	1270
	^{59}Co	BQL**	BQL	BQL
	^{60}Ni	BQL	BQL	BQL
	^{66}Zn	BQL	BQL	BQL
	^{69}Ga	10.1–28.7	17.17 (6.04)	14.8
	^{85}Rb	11–97.50	48.90 (30.84)	46.40
	^{133}Cs	22.1–75.9	40.11 (14.98)	41.6
	$^{52}Cr/^{51}V$	0.83–2.64	1.47 (0.59)	1.20
	Na_2O/MgO (%)	0.77–0.92	0.85 (0.04)	0.85
	Alkalis* (%)	0.49–1.61	1.03 (0.40)	1.16
	H_2O (%)	1.93–2.54	2.28 (0.22)	2.38
Brazil (Itabira) 18 samples, 54 analysis	^{7}Li	45.90–97.30	64.11 (14.60)	59.90
	Na_2O (%)	0.97–1.70	1.32 (0.17)	1.34
	MgO (%)	1.13–2.06	1.53 (0.21)	1.53
	^{39}K	152–385	246.83 (65.60)	239.50
	^{45}Sc	17.5–153	60.3 (31.47)	51.15
	^{51}V	52.50–177	116.52 (32.72)	117.50
	^{52}Cr	997–5700	2508.46 (1051.82)	2360
	^{55}Mn	4.49–24.10	14.10 (5.67)	13.40
	^{56}Fe	4540–8760	6407.59 (1122.38)	6220
	^{59}Co	1.94–2.96	2.47 (0.30)	2.47
	^{60}Ni	BQL	BQL	BQL
	^{66}Zn	28.40–87.40	55.31 (12.31)	57.20
	^{69}Ga	6.68–13.80	11.42 (1.73)	11.90
	^{85}Rb	19.10–52.60	32.73 (8.16)	31.25
	^{133}Cs	35.50–128	75.52 (23.10)	76.85
	$^{52}Cr/^{51}V$	8.30–39.45	21.93 (7.03)	21.05
	Na_2O/MgO (%)	0.81–0.94	0.87 (0.03)	0.87
	Alkalis* (%)	0.75–1.30	1.02 (0.13)	1.05
	H_2O (%)	2.17–2.47	2.33 (0.07)	2.35
Colombia (Coscuez) 8 samples, 24 analysis	^{7}Li	62–163	87.20 (35.04)	69.10
	Na_2O (%)	0.40–0.97	0.60 (0.16)	0.56
	MgO (%)	0.46–1.04	0.67 (0.16)	0.62
	^{39}K	BQL	BQL	BQL
	^{45}Sc	9.17–441	120.14 (135.30)	56.90
	^{51}V	879–6340	2530.92 (1724.94)	2175
	^{52}Cr	240–2820	800.71 (788.24)	432
	^{55}Mn	BQL	BQL	BQL
	^{56}Fe	507–1860	902.71 (493.08)	683
	^{59}Co	BQL	BQL	BQL
	^{60}Ni	BQL	BQL	BQL
	^{66}Zn	BQL	BQL	BQL
	^{69}Ga	25.7–58.5	39.30 (11.82)	33.95
	^{85}Rb	0.79–2.52	1.31 (0.48)	1.12
	^{133}Cs	4.83–12.1	8.54 (2.47)	8.31
	$^{52}Cr/^{51}V$	0.12–0.45	0.31 (0.09)	0.31
	Na_2O/MgO (%)	0.81–0.95	0.89 (0.04)	0.89
	Alkalis* (%)	0.30–0.73	0.46 (0.12)	0.43
	H_2O (%)	1.69–2.17	1.90 (0.13)	1.87

Table 3. *Cont.*

Samples	Element	Min–Max	Average (SD)	Median
Ethiopia (Shakisso) 4 samples, 20 analysis	^{7}Li	271–427	345.65 (51.78)	334.50
	Na_2O (%)	1.71–2.16	1.96 (0.14)	1.97
	MgO (%)	2.02–2.55	2.34 (0.15)	2.37
	^{39}K	290–444	374.45 (52.77)	371.50
	^{45}Sc	57–150	114.30 (34.50)	125.50
	^{51}V	96.30–123	112.70 (8.10)	114.50
	^{52}Cr	2000–5010	3655 (1060)	3795
	^{55}Mn	9–21.1	16.51 (4.22)	18.30
	^{56}Fe	3980–5390	4867 (507.64)	5050
	^{59}Co	1.27–2.20	1.65 (0.19)	1.65
	^{60}Ni	BQL–14.4	3.32 (5.91)	BQL
	^{66}Zn	32.20–44.80	37.10 (3.27)	36.40
	^{69}Ga	16.40–20.50	18.77 (1.18)	19.15
	^{85}Rb	50.60–64.70	57.47 (4.58)	57.95
	^{133}Cs	270–427	347.50 (58.82)	341.50
	^{52}Cr/^{51}V	19.80–43.66	32.08 (8.26)	32.95
	Na_2O/MgO (%)	0.78–0.92	0.84 (0.04)	0.83
	Alkalis* (%)	1.38–1.71	1.57 (0.10)	1.59
	H_2O (%)	2.48–2.60	2.55 (0.04)	2.55
Madagascar (Mananjary) 9 samples, 27 analysis	^{7}Li	57.50–128	98.45 (19.47)	104
	Na_2O (%)	1.82–2.41	2.05 (0.16)	2.01
	MgO (%)	2.07–3.02	2.58 (0.30)	2.59
	^{39}K	314–3150	1388.74 (808.89)	1090
	^{45}Sc	21–309	75.59 (86.47)	34.50
	^{51}V	102–386	224.85 (96.63)	229
	^{52}Cr	1490–3770	2380.37 (737.41)	2410
	^{55}Mn	8.65–28.50	15.22 (5.56)	15.30
	^{56}Fe	7310–11200	9824.81 (1131.24)	10,100
	^{59}Co	1.69–4.50	3.17 (0.70)	3
	^{60}Ni	18.10–38.60	28.94 (6.30)	30.60
	^{66}Zn	9.06–32	16.59 (6.65)	15.10
	^{69}Ga	6.13–13.10	8.15 (1.98)	7.72
	^{85}Rb	42.70–407	167.10 (106.29)	175
	^{133}Cs	105–1050	430 (297.51)	324
	^{52}Cr/^{51}V	4.49–22.54	12.45 (5.90)	10
	Na_2O/MgO (%)	0.71–0.97	0.80 (0.08)	0.80
	Alkalis* (%)	1.46–1.92	1.73 (0.13)	1.77
	H_2O (%)	2.51–2.66	2.57 (0.04)	2.56
Russia (Ural) 5 samples, 15 analysis	^{7}Li	736–911	826.93 (60.19)	831
	Na_2O (%)	0.70–1.23	0.91 (0.18)	0.85
	MgO (%)	0.47–1.04	0.70 (0.20)	0.64
	^{39}K	BQL–103	40.87 (41.97)	52
	^{45}Sc	19.50–72.90	43.52 (21.93)	44.20
	^{51}V	29.80–128	80.64 (37.30)	90.90
	^{52}Cr	318–1700	905.27 (446.14)	841
	^{55}Mn	13.50–22.60	19.16 (3.24)	20.60
	^{56}Fe	1210–1900	1600.67 (235.75)	1640
	^{59}Co	BQL–2.09	1.08 (0.73)	1.19
	^{60}Ni	BQL–23.20	12.90 (8.35)	16
	^{66}Zn	38.10–62.60	48.62 (7.97)	45.40
	^{69}Ga	6.28–19.50	13.69 (4.40)	13.70
	^{85}Rb	7.88–27.20	17.46 (7.61)	16.30
	^{133}Cs	252–568	350.53 (112.54)	308
	^{52}Cr/^{51}V	6.57–16.13	11.79 (3.27)	11.86
	Na_2O/MgO (%)	1.15–1.58	1.33 (0.13)	1.34
	Alkalis* (%)	0.65–1.04	0.80 (0.13)	0.75
	H_2O (%)	1.99–2.30	2.13 (0.10)	2.10

Table 3. *Cont.*

Samples	Element	Min–Max	Average (SD)	Median
	^{7}Li	492–741	639.17 (76.19)	661
	Na_2O (%)	1.95–2.32	2.12 (0.11)	2.14
	MgO (%)	2.02–2.62	2.32 (0.21)	2.35
	^{39}K	376–716	508.50 (121.79)	451
	^{45}Sc	19.80–63.40	41.63 (13.94)	44.25
	^{51}V	79.30–147	112.82 (24.87)	113.15
	^{52}Cr	349–2360	1430.94 (760.72)	1680
	^{55}Mn	11–32.30	22.52 (7.95)	24.70
Zambia (Kafubu)	^{56}Fe	6320–9590	8239.44 (932.06)	8440
6 samples, 18 analysis	^{59}Co	2.23–3.26	2.81 (0.31)	2.82
	^{60}Ni	BQL–28.40	20.02 (6.30)	19.95
	^{66}Zn	17.20–46.60	31.75 (10.54)	35.40
	^{69}Ga	11.90–17.10	14.84 (1.56)	14.90
	^{85}Rb	41.40–87.30	62.99 (16.16)	63.90
	^{133}Cs	941–1410	1201.72 (148.15)	1215
	^{52}Cr/^{51}V	3.82–25.38	13.20 (7.75)	13.92
	Na_2O/MgO (%)	0.82–1.00	0.92 (0.05)	0.92
	Alkalis* (%)	1.67–1.94	1.81 (0.08)	1.83
	H_2O (%)	2.55–2.64	2.59 (0.03)	2.60
	^{7}Li	512–1050	818.20 (214.17)	930
	Na_2O (%)	2.10–2.75	2.48 (0.24)	2.62
	MgO (%)	2.17–2.80	2.58 (0.23)	2.69
	^{39}K	230–434	354.73 (86.32)	407
	^{45}Sc	16.90–26.80	20.85 (3.92)	18.80
	^{51}V	185–280	219.27 (38.75)	198
	^{52}Cr	1430–2070	1790 (237.25)	1770
	^{55}Mn	47–93	62.85 (17.52)	53.40
Zimbabwe	^{56}Fe	4320–7050	6062.67 (1170.14)	6810
(Sandawana)	^{59}Co	1.98–2.55	2.26 (0.17)	2.25
3 samples, 15 analysis	^{60}Ni	BQL–19.70	14.67 (5.36)	16.60
	^{66}Zn	72.90–84.70	79.49 (3.49)	79.30
	^{69}Ga	24.80–33.70	28.37 (3.28)	27
	^{85}Rb	217–299	263.47 (32.29)	279
	^{133}Cs	274–756	589.07 (224.06)	739
	^{52}Cr/^{51}V	7.32–10.05	8.25 (0.85)	8.14
	Na_2O/MgO (%)	0.93–1.00	0.96 (0.02)	0.96
	Alkalis* (%)	1.68–2.29	2.04 (0.23)	2.18
	H_2O (%)	2.59–2.73	2.67 (0.05)	2.71

* Alkalis: ^{7}Li + ^{23}Na + ^{39}K + ^{85}Rb + ^{133}Cs; **BQL: Below Quantification Limits.

Chemical elements responsible for the colour of emeralds, such as Cr, V and Fe, can be seen in various amounts (e.g., ^{56}Fe up to 11200 ppmw for a sample from Madagascar). The ^{52}Cr/^{51}V ratio vs. ^{56}Fe is represented in Figure 2. The ^{52}Cr/^{51}V ratio is >3.8 for all studied Type IA samples. All studied samples from Colombia presented ^{52}Cr < ^{51}V, with a ^{52}Cr/^{51}V ratio ranging from 0.12 to 0.45. Only two of the nine studied samples from Afghanistan presented ^{52}Cr/^{51}V < 1, while the other studied samples had ratios reaching up to 2.64 (see Table 3 and Figure 2). The distribution and scattering of the analysed points in Figure 2 might be influenced by the colour of the analysed samples (e.g., all samples from Russia were light green in colour).

Alkali metals, other than ^{23}Na, are also present in emeralds in different concentrations (see ^{7}Li, ^{39}K, ^{85}Rb and ^{133}Cs in Table 3). The sum of the concentrations of all alkali metals (^{7}Li + ^{23}Na + ^{39}K + ^{85}Rb + ^{133}Cs) measured in the studied samples is divided here into low (sum < 1%), medium (1% < sum < 2%) and high (sum > 2%). Samples from Colombia presented a low sum of alkalis, ranging from 0.30% to 0.73% (average: 0.46%, median: 0.43%), from Russia 0.65% to 1.04% (average: 0.80%, median: 0.75%), from Brazil 0.75% to 1.30% (average: 1.02%, median: 1.05%), from Afghanistan 0.49% to 1.60% (average: 1.03%, median: 1.16%) and all the rest of the samples presented a medium to high sum of alkalis (2.29%

for a sample from Zimbabwe). Thus, the emeralds studied from Type IA occurrences can be separated into those with a low to medium sum of alkalis <1.30% (from Brazil and Russia) and those with a sum >1.35% (all the other emeralds of the same type; i.e., from Ethiopia, Madagascar, Zambia and Zimbabwe).

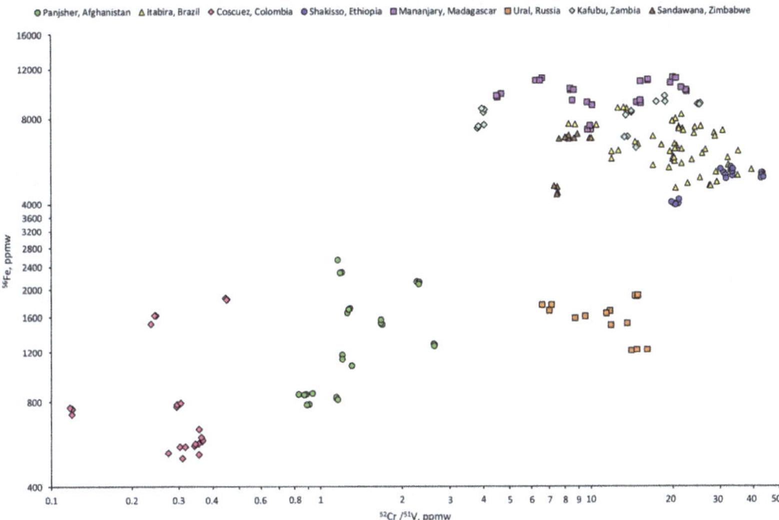

Figure 2. Binary plot of ^{52}Cr/^{51}V vs. ^{56}Fe.

The alkali metal concentration of emeralds from different localities can vary and their plots are useful for separating them [14,42,70,81]. The studied samples can be divided into those with relatively low ^{7}Li (<200 ppmw)—Afghanistan, Brazil, Colombia and Madagascar, and those with medium to high ^{7}Li (>250 ppmw)—Ethiopia, Russia, Zambia and Zimbabwe (see Table 3). In parallel, the samples from Colombia did not present a detectable amount of ^{39}K. The samples from Russia contained low amounts of potassium, with some measurements being below quantification limits (BQL). Samples from Madagascar showed the highest concentrations of potassium (up to 3150 ppmw, with 1389 ppmw average and 1090 ppmw median values) and the concentrations of emeralds from Afghanistan varied from low (107 ppmw) to high (1540 ppmw). In Figure 3, a ^{7}Li vs. ^{39}K binary plot of the studied samples is presented. Samples with low lithium (^{7}Li < 200 ppmw) can be separated from those with medium to high lithium (^{7}Li > 250 ppmw). Most of the samples from Russia are clustered separately from the other samples with medium to high lithium (Ethiopia, Zambia, Zimbabwe) as they contain relatively little potassium (^{39}K < 105 ppmw). The samples from Colombia are not plotted as they do not contain measurable potassium with LA-ICP-MS. In comparison with published data obtained with LA-ICP-MS [70,80,82], the studied samples from Brazil, Ethiopia, Madagascar, Zambia and Zimbabwe presented similar trends. Zambian samples presenting higher potassium and a similar content of lithium, as well as low lithium and high potassium, and samples from Zimbabwe, with relatively low lithium content, have also been presented in previous studies [42,81]. It is also mentioned in a previous publication that some samples from Russia could present higher potassium [55].

In Figure 4 a ^{39}K vs. ^{23}Na binary plot is presented. As mentioned previously, the samples from Brazil and Madagascar are the only ones studied that belong to the Type IA occurrence type and contain relatively little lithium (^{7}Li < 200 ppmw). It looks as samples from Madagascar can be further separated from samples from Brazil by using this plot, as the samples from Madagascar contain higher sodium and potassium than the Brazilian samples. Using the same plot, Russian samples can also be further separated from other samples of the Type IA occurrence type by those samples with medium to high lithium content (i.e., from Ethiopia, Zambia and Zimbabwe), as they contain low sodium and potassium.

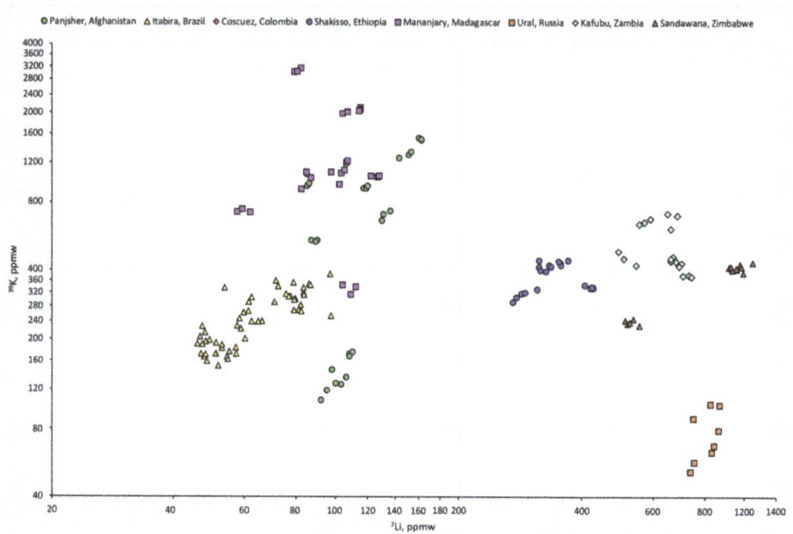

Figure 3. Binary plot of ^{7}Li vs. ^{39}K.

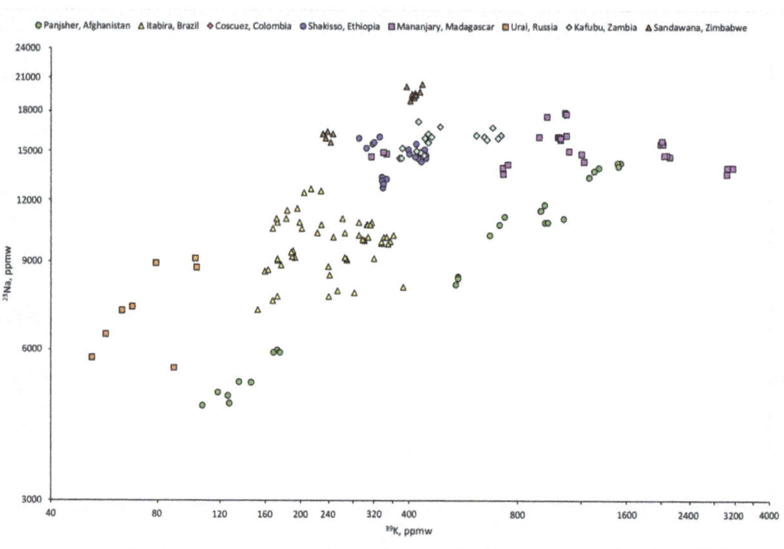

Figure 4. Binary plot of ^{39}K vs. ^{23}Na.

In Figures 5–8, ^{85}Rb vs. ^{133}Cs, ^{7}Li vs. ^{85}Rb, ^{7}Li vs. ^{133}Cs and ^{133}Cs vs. ^{23}Na binary plots are presented, respectively. All measured samples are plotted as they all presented concentrations of lithium, sodium, caesium and rubidium measurable with LA-ICP-MS (see Table 3). Apart for the samples from Brazil and Madagascar, which belong to Type IA occurrences, the samples from Colombia and Afghanistan present low ^{7}Li. Samples from Colombia also presented the lowest rubidium (^{85}Rb < 1.88 ppmw) and caesium (^{133}Cs < 12.1 ppmw), whereas samples from Zimbabwe presented the highest rubidium (167 ppmw < ^{85}Rb < 227 ppmw; 264 ppmw average and 279 ppmw median concentrations). Some of the samples from Madagascar also presented high rubidium (^{85}Rb > 100 ppmw, up to 317 ppmw). Samples from Zambia presented the highest caesium content (from 941 to 1410 ppmw, 1202 ppmw average and 1215 ppmw median values). According to the literature, some samples from Zambia (Kafubu) could have relatively low lithium [42,70], while some have lower caesium and rubidium [40]. Samples from Colombia are plotted separately in the three plots of Figures 5–7 as they contain the lowest rubidium, caesium and lithium. Samples from Zimbabwe and most samples from Madagascar contain high rubidium and caesium and so they are also plotted separately (see Figure 5). Samples from Madagascar and Zimbabwe can be separated as the former have a low lithium content and the latter have a high content (Figures 6 and 7). Samples from Ethiopia and Zambia are well separated in a ^{7}Li vs. ^{133}Cs binary plot (Figure 7), as samples from Zambia contain the highest caesium; however, a slight overlap was observed between the two in another publication [81]. Emeralds in close connection with highly evolved pegmatites present high lithium and caesium (see Figure 7; emeralds from Russia, Zambia and Zimbabwe) as most of the extremely fractionated rare-element granitic pegmatites of the complex LCT (lithium, caesium, tantalum) association are enriched in lithium and caesium [70,83]. Possible differences in the alkalis of emeralds of Type IA could be linked to the difference between their granites-pegmatites [83–87]. In comparison with published data obtained with LA-ICP-MS [42], the samples from Zambia and Zimbabwe presented similar trends; a few Zambian samples presented higher potassium and a similar content of lithium, and a few others presented lower lithium, lower caesium and higher potassium. The lithium and caesium content of the studied samples from Afghanistan did not vary much; however, potassium, sodium and rubidium exhibited a great variation (see Table 3). In Figure 8, in a ^{133}Cs vs. ^{23}Na binary plot, samples from Colombia, Brazil, Ethiopia, Russia and Zambia shows that individual measurements from each of these localities are clustered closely, whereas those obtained from Afghanistan, Madagascar and Zimbabwe present a wide variation.

^{45}Sc, ^{55}Mn, ^{59}Co, ^{60}Ni, ^{66}Zn and ^{69}Ga could also help with the determination of emeralds' origin. ^{59}Co, ^{60}Ni and ^{66}Zn were BQL for all the samples from Afghanistan and Colombia; all studied samples belonging to Type IA occurrence presented ^{59}Co and ^{66}Zn and, some of them, measurable ^{60}Ni with LA-ICP-MS (see Table 3). Nickel could be used to separate Malagasy from Brazilian samples (both with low lithium); it is detectable in the former (average: 28.94 ppmw) and BQL for the latter. Some samples from Afghanistan as well as most samples of Type IA occurrence presented detectable manganese, and the samples from Zimbabwe presented the highest zinc and manganese contents of the studied samples. Emeralds from Colombia presented the highest gallium content (27.5 to 58.5 ppmw), followed by emeralds from Zimbabwe (24.8 to 33.7 ppmw), while those from Brazil had the lowest (6.68 to 13.8 ppmw). Nevertheless, Colombian samples with low gallium are cited in the literature [42,64]. Scandium is also present in all studied samples, with emeralds from Afghanistan presenting a large variation and the highest content (148 to 2390 ppmw; high scandium in emeralds from Afghanistan was also noted in [40,42,64]). Also, Colombian samples presented a variation from relatively low to high content of scandium (9.17 to 441 ppmw), as well as the Malagasy samples (21 to 309 ppmw). In Figures 9 and 10, ^{69}Ga vs. ^{45}Sc and ^{7}Li vs. ^{45}Sc binary plots are presented, respectively. Samples from Colombia and Zimbabwe are separated from the rest in Figure 9 due to their higher gallium content as well as two out of nine of the studied samples due to their high gallium content. In Figure 10, the samples from Ethiopia are plotted separately in between the samples with low lithium and those with high lithium, as they also contain medium scandium concentrations.

In Figure 11, a binary plot of ^{133}Cs vs. ^{23}Na/^{7}Li ratio is presented; in this, the studied Ethiopian, Russian and Zambian emeralds (all with ^{7}Li > 250 ppmw) are well separated. The plotted points of Ethiopian samples overlap only with those from one sample from Zimbabwe and the plotted points from Zambia with the analysis of two out of the three studied samples from Zimbabwe. As mentioned above, the samples from Zimbabwe contain high rubidium and could be easily separated from the samples from Ethiopia and Zambia (see again Figures 5 and 6). Our results on the Russian emeralds present slightly lower ^{23}Na/^{7}Li ratios compared to those presented in [19].

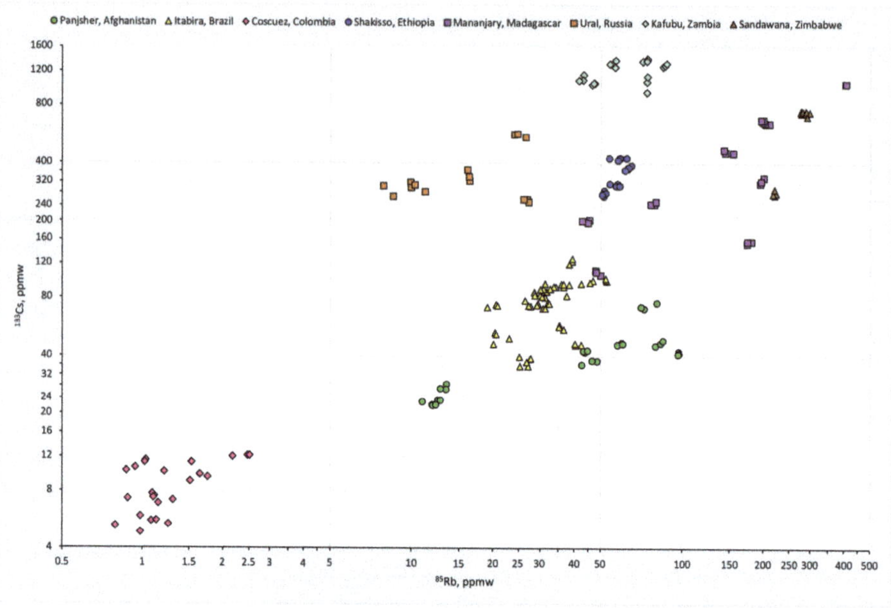

Figure 5. Binary plot of ^{85}Rb vs. ^{133}Cs.

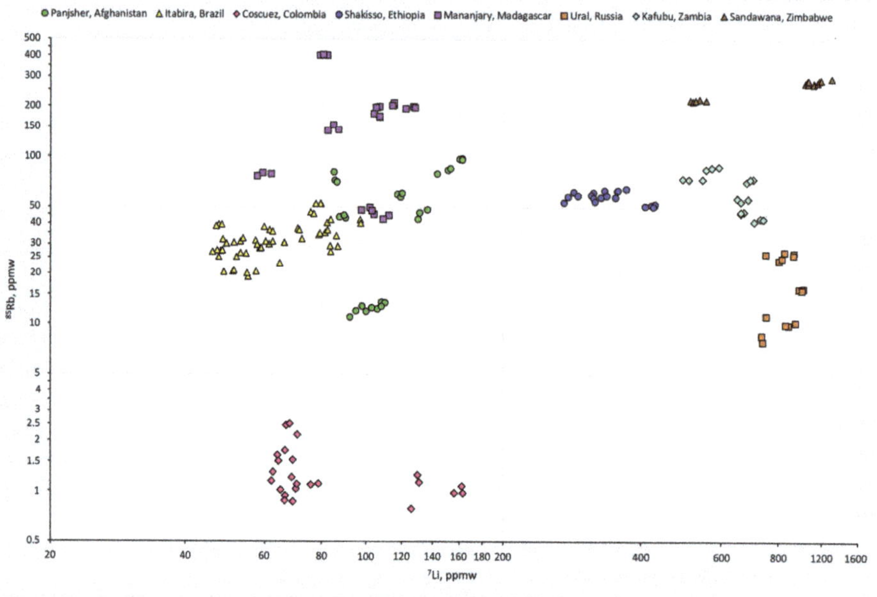

Figure 6. Binary plot of ^{7}Li vs. ^{85}Rb.

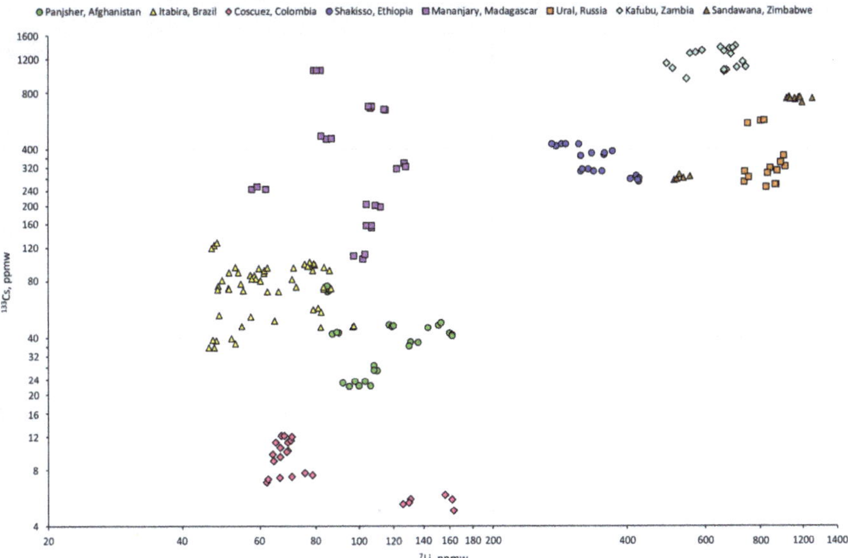

Figure 7. Binary plot of ^{7}Li vs. ^{133}Cs.

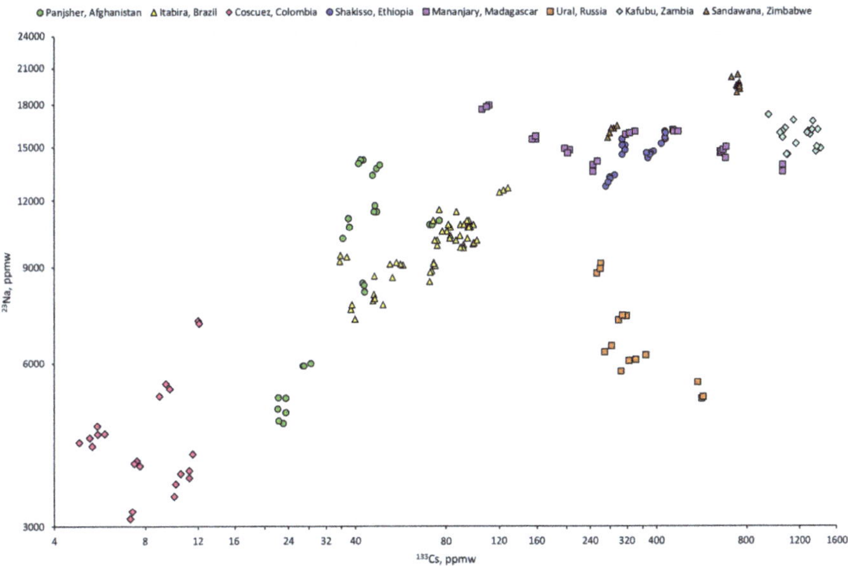

Figure 8. Binary plot of ^{133}Cs vs. ^{23}Na.

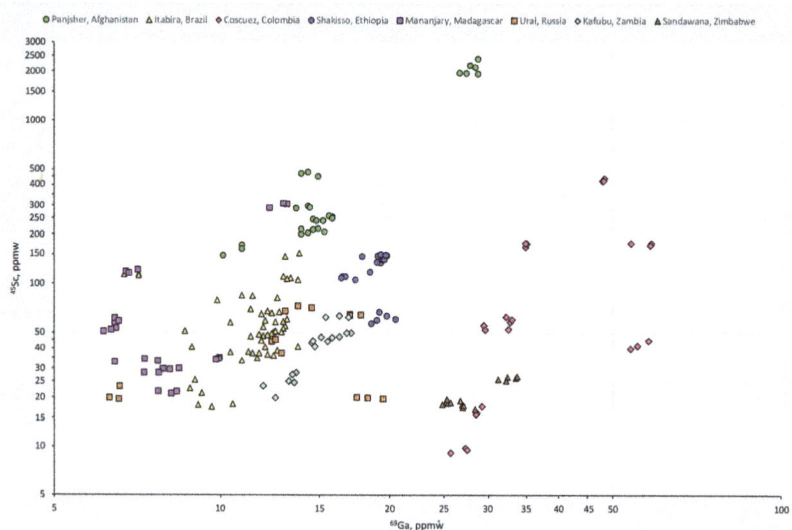

Figure 9. Binary plot of ^{69}Ga vs. ^{45}Sc.

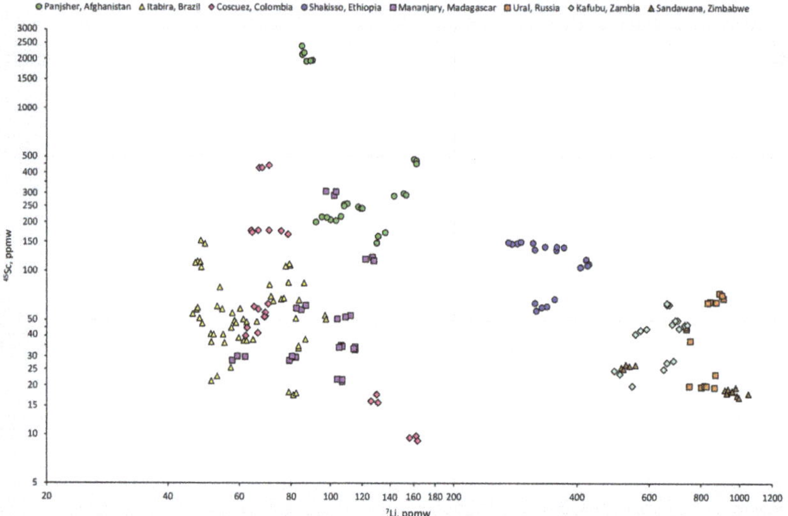

Figure 10. Binary plot of ^{7}Li vs. ^{45}Sc.

In Figure 12, a Ga vs. Zn vs. Li triplot is presented, with most samples plotted showing similarity to those presented in [64]; only the samples from Afghanistan differ. Samples belonging to Type II occurrences do not present any measurable zinc with LA-ICP-MS; thus, they are plotted separately from the samples from Type IA occurrences. From the latter, studied samples from Brazil are plotted separately from the studied samples from Madagascar (both present low lithium contents). Some points of the samples from Madagascar overlap with the other samples from Type IA occurrences, but they can be separated as they contain medium to high lithium.

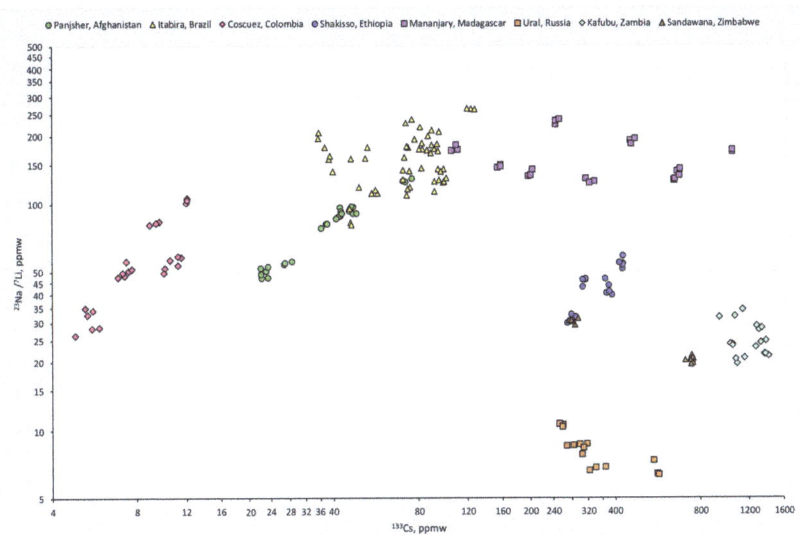

Figure 11. Binary plot of ratios ^{133}Cs vs. ^{23}Na/^{7}Li.

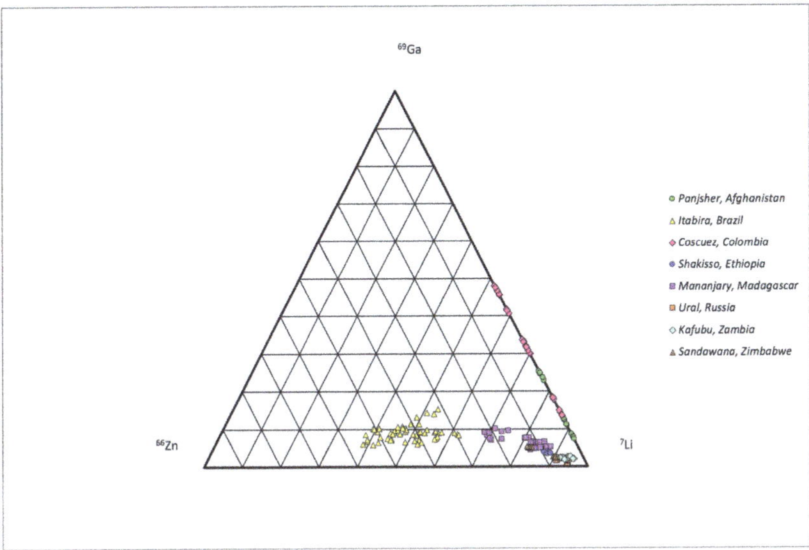

Figure 12. Ternary plot of ^{66}Zn vs. ^{69}Ga vs. ^{7}Li.

3.2. UV-Vis-NIR Spectroscopy

In Figures 13–15, UV-Vis polarised spectra with a spectral range of 250 to 900 nm of three green emeralds from Colombia, Afghanistan and Zambia, respectively, are presented. The light blue lines are for the ordinary-ray (o-ray) spectra and the orange lines are for the extraordinary-ray (e-ray) spectra. All emeralds from Colombia (see an example in Figure 13) presented absorptions due to Cr^{3+} and V^{3+} in the violet-blue part (around 430 nm) and absorption in the orange-red part (around 600 nm), with a shoulder at around 395 nm due to vanadium [34,42,53,62,88–92]. All spectra on the emerald from Afghanistan (Figure 14) show absorptions due to Cr^{3+} (some of the studied samples also presented the bands due to V^{3+}), with relatively low Fe^{2+}-linked absorption at around 830 nm and very weak absorption at around 370 nm (barely observed along the o-ray) due to Fe^{3+}. All studied emeralds classified as Type IA (i.e., from Brazil, Ethiopia, Madagascar, Russia, Zimbabwe and Zambia) presented absorptions linked to Cr^{3+}, Fe^{3+} and Fe^{2+} (see an example in Figure 15; Table 4). Some samples also presented additional weak bands in the red part of the electromagnetic spectrum, possibly linked to Fe^{3+}-Fe^{2+} charge transfer (e.g., from Zambia and Madagascar) [91].

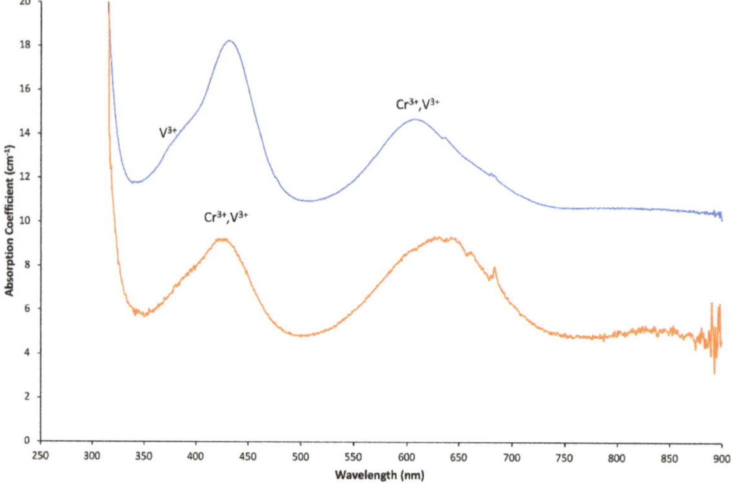

Figure 13. UV-Vis polarized spectra of an emerald from Colombia. The light blue and orange lines are for the o-ray and e-ray spectra, respectively. The upper spectrum has been vertically offset for clarity.

Table 4. UV-Vis absorptions linked with colouring elements of the studied samples.

Locality	Cr^{3+}	V^{3+}	Fe^{3+}	Fe^{2+}
Afghanistan	✓	✓	(✓)	(✓)
Brazil	✓		✓	✓
Colombia	✓	✓		(✓)
Ethiopia	✓		✓	✓
Madagascar	✓		✓	✓
Russia	✓		✓	✓
Zambia	✓		✓	✓
Zimbabwe	✓		✓	✓

✓: present; (✓): sometimes present with low intensity

In Figures 16–18, polarised spectra in the near infrared region (NIR) from 1300 to 1500 nm, of the samples shown in Figures 13–15, are presented. The bands in those figures are linked to water vibrations (overtone and combination). The more pronounced bands observed along the e-ray (orange line) are linked to Type I water (main band at around 1400 nm –around 7142 cm^{-1}) and those along the o-ray (light blue line) to Type II water (i.e., water linked with alkalis; main band at 1408 nm –around 7102 cm^{-1}) [88,93–95]. Consequently, Type II water bands are more intense in samples with a higher content of alkalis; thus, they are weaker in emeralds from Colombia (Figure 16). The bands are of medium intensity in samples from Brazil, Russia and seven out of nine samples from Afghanistan and one sample out of four from Ethiopia (Figure 17), and intense in all samples from Madagascar, Zambia and Zimbabwe, as well as three out of four from Ethiopia and two out of nine from Afghanistan (Figure 18).

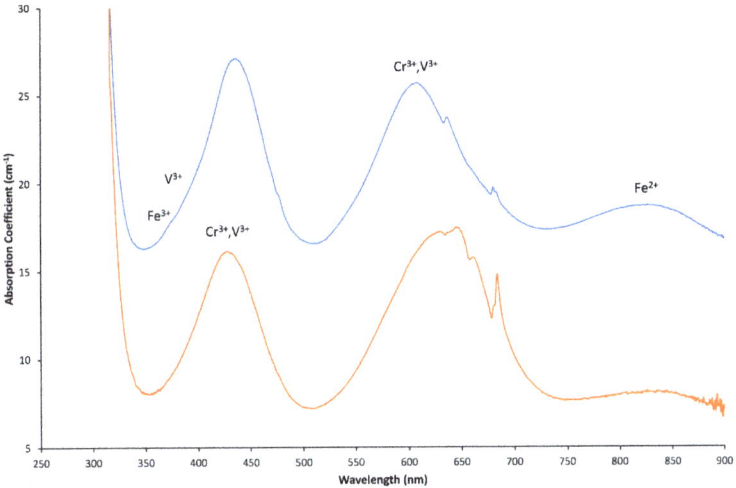

Figure 14. UV-Vis polarized spectra of an emerald from Afghanistan. The light blue and orange lines are for the o-ray and e-ray spectra, respectively. The upper spectrum has been vertically offset for clarity.

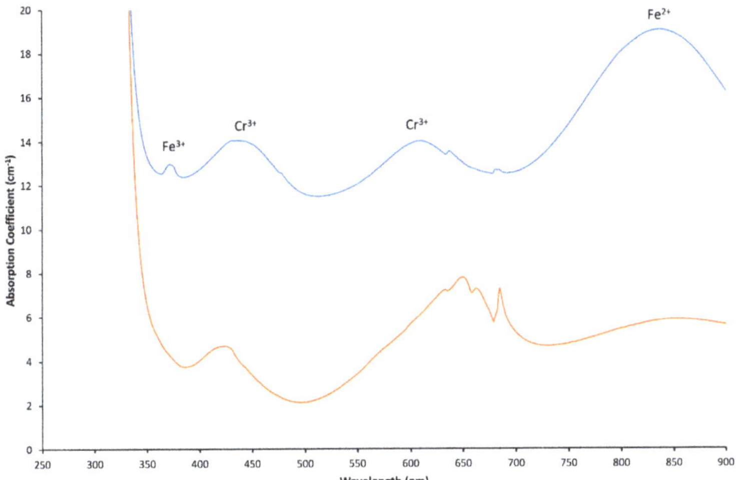

Figure 15. UV-Vis polarized spectra of an emerald from Zambia. The light blue and orange lines are for the o-ray and e-ray spectra, respectively. The upper spectrum has been vertically offset for clarity.

Thus, all emeralds from Type IA occurrences presented relatively important absorptions linked to Type II water; only the samples from Brazil, Russia and one (out of four) from Ethiopia showed medium to weak absorptions of Type II water and pronounced absorptions of Type I water. This is probably due to a lower content of alkalis in these samples compared to the other studied samples from Type IA occurrences. The samples from Colombia and most of the studied samples from Afghanistan (seven out of nine) show similar characteristics to those from Brazil, Russia as well as one from Ethiopia; those from Colombia exhibited the lowest intensity of Type II water bands. Two samples from Afghanistan (belonging to Type II occurrences) present relatively important absorptions linked to Type II water.

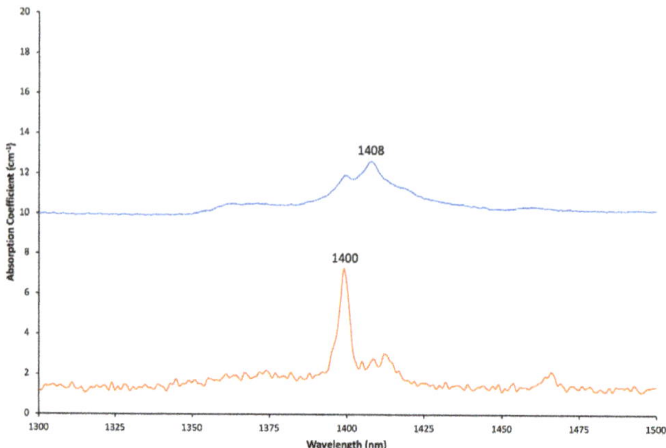

Figure 16. NIR polarized spectra of an emerald from Colombia. The light blue and orange lines are for the o-ray and e-ray spectra, respectively. The upper spectrum has been vertically offset for clarity.

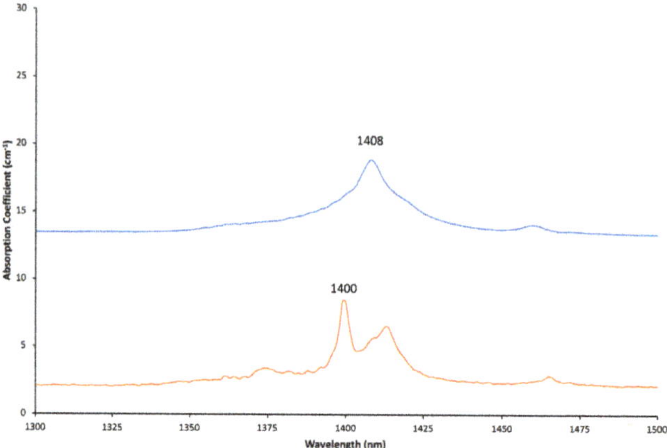

Figure 17. NIR polarized spectra of an emerald from Afghanistan. The light blue and orange lines are for the o-ray and e-ray spectra, respectively. The upper spectrum has been vertically offset for clarity.

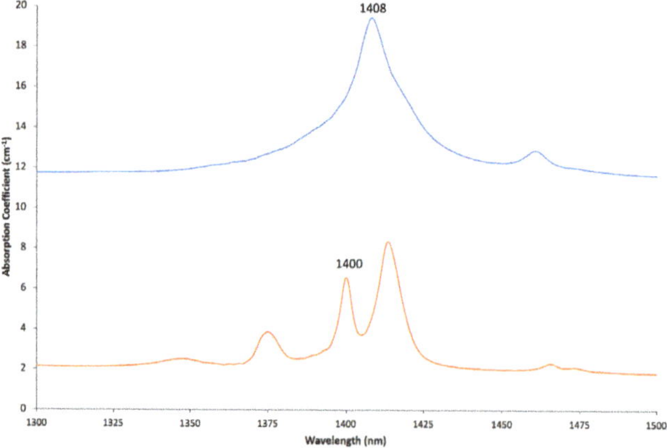

Figure 18. NIR polarized spectra of an emerald from Zambia. The light blue and orange lines are for the o-ray and e-ray spectra, respectively. The upper spectrum has been vertically offset for clarity.

3.3. FTIR Spectroscopy

FTIR spectra were acquired on randomly oriented samples (see an example in Figure 19). At around 3500 cm^{-1}, where the vibrations linked to water molecules' stretching are situated, all samples presented complete absorption due to their thickness. From 4500 to 6000 cm^{-1}, the combination water bands are situated with a series of bands at around 5270 cm^{-1} that are linked to Type I and Type II water. From 6500 cm^{-1} (1538.5 nm) to 7500 cm^{-1} (1333.3 nm), overtone and combination water bands are also observed, the same as was observed in the NIR region presented above [88,93,95].

At the region from 2200 to 2850 cm^{-1} (Figure 20), a series of bands linked to H_2O, D_2O, CO_2 and chlorine are present [66,76,95,96]. More precisely, the bands situated at 2470, 2640, 2670 and 2735 cm^{-1}, which are related to the stretching vibration of deuterated water, are observed in all the spectra of all the studied samples, with differing intensities [97]. An additional band at around 2290 cm^{-1}, is linked to water, is also observed in all studied samples; it is also found in some hydrothermal synthetic emeralds [14,98]. The series of bands from 2300 to 2400 cm^{-1} (with the main band situated around 2358 cm^{-1}) is attributed to CO_2 and vibrations linked to the presence of ^{13}C and ^{18}O isotopes [66,95]. The band at around 2818 cm^{-1} was found only in the samples from Colombia and Afghanistan and none of the others, but it was also observed in emeralds from Norway; this band is possibly linked to chlorine [65,66].

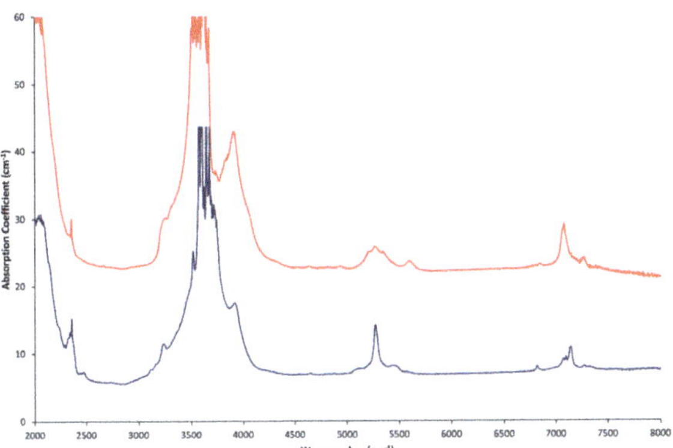

Figure 19. FTIR spectra of an emerald from Colombia (blue spectrum) and from Zimbabwe (red spectrum). The upper spectrum has been vertically offset for clarity.

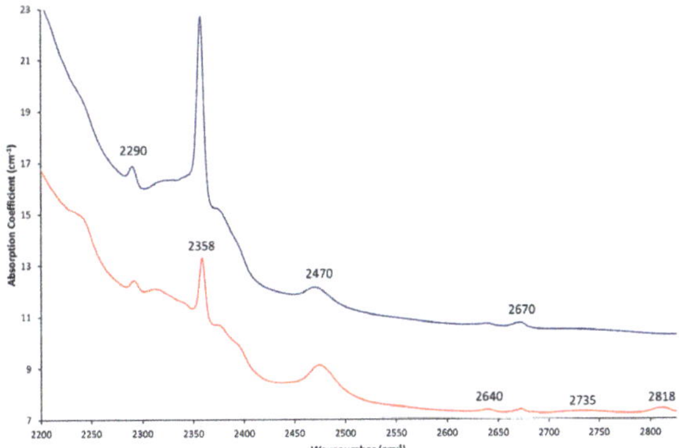

Figure 20. FTIR spectra of an emerald from Colombia (blue spectrum) and from Zimbabwe (red spectrum). The upper spectrum has been vertically offset for clarity.

Emerald FTIR spectra display strong polarisation phenomena [95,96]. For example, deuterated water bands are distinctly more pronounced in spectra acquired along the extraordinary-ray, and CO_2 bands are more pronounced along the ordinary-ray [95,96].

3.4. Raman Spectroscopy

Raman spectra ranging from 200 to 1300 cm^{-1} and from 3520 to 3680 cm^{-1} were acquired on different directions, with the laser beam perpendicular to the c-axis (spectra in black colour) and parallel to the c-axis (spectra in grey colour), without the use of a polariser. The results for an emerald from Colombia are shown in Figures 21 and 22, of an emerald from Ethiopia in Figure 23 and of an emerald from Zambia in Figure 24. Sometimes, it was challenging to acquire a proper spectrum with the 514 nm laser due to emeralds' chromium luminescence. Bands linked to Si_6O_{18} ring vibrations are situated below 600 cm^{-1}; the main band at around 686 cm^{-1} is due to Be-O stretching vibrations, and the main band at around 1070 cm^{-1} is due to Si-O and/or Be-O stretching [58,72,95,97,99,100]. The relative intensities of the Raman bands change following the different orientations; the band at around 686 cm^{-1} is more intense in spectra acquired with the laser parallel to the c-axis and the band at around 1070 cm^{-1} is more intense in spectra acquired with the laser perpendicular to the c-axis (Figure 21). The exact position and full width half maximum (FWHM) of the band at 1070 cm^{-1} was found to be useful for separating natural emeralds from their synthetic counterparts, as well as low-alkali from high-alkali emeralds, where the observed differences are due to silicon substitution with aluminium, beryllium and lithium along with sodium, potassium and caesium for charge compensation [72,73]. In the samples studied, the FWHM of the 1070 cm^{-1} band (measured from the spectra acquired with the laser perpendicular to the c-axis) in samples with higher alkali content is generally higher (FWHM < 22 cm^{-1} can be considered for those emeralds of low to medium alkali content), but the position of the band is not shifted towards higher Raman shifts as alkalis increase (see Table 5). It is worth noting that some of the samples studied showed different trends compared with [73]; the samples from Russia and Brazil (Itabira) presented a lower content of alkalis. Additionally, the FWHM depends on the spectral resolution; thus, it can be slightly different when using a different resolution.

The bands at around 3608 cm^{-1} and 3598 cm^{-1} are due to Type I water and Type II water, respectively, which also present polarisation phenomena (see again Figures 22–24) [67,69,99]. In the present study, the ratios of these Raman peaks' intensities (I_{3608}/I_{3598}) were greater for the spectra acquired with the laser parallel to the c-axis. The described relative intensities presented in Table 5 are for the spectra acquired with the laser perpendicular to the c-axis.

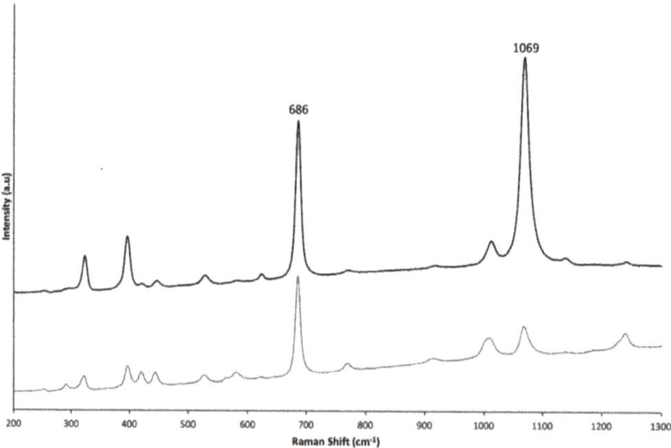

Figure 21. Raman spectra from 200 to 1300 cm^{-1} of an emerald from Colombia. The black and grey spectra are acquired with the laser beam perpendicular to the c-axis and parallel to the c-axis, respectively. The upper spectrum has been vertically offset for clarity.

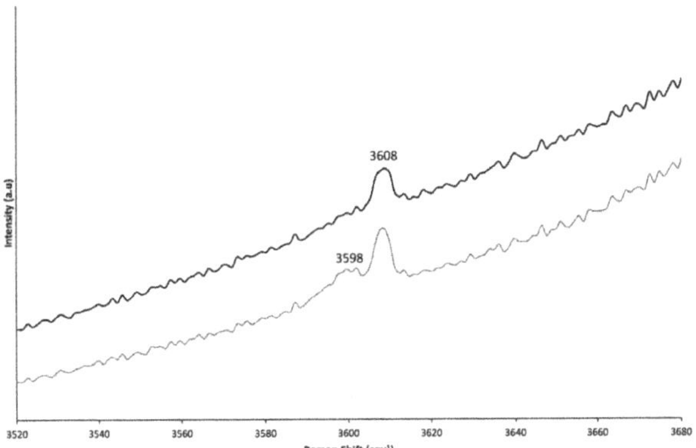

Figure 22. Raman spectra from 3520 to 3680 cm^{-1} of an emerald from Colombia. The black and grey spectra are acquired with the laser beam perpendicular to the c-axis and parallel to the c-axis, respectively. Note that the upper spectrum is shifted higher for clarity.

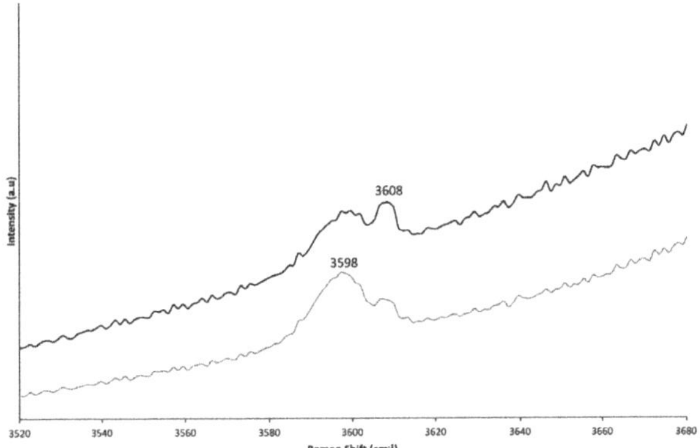

Figure 23. Raman spectra from 3520 to 3680 cm^{-1} of an emerald from Ethiopia. The black and grey spectra are acquired with the laser beam perpendicular to the c-axis and parallel to the c-axis, respectively. Note that the upper spectrum is shifted higher for clarity.

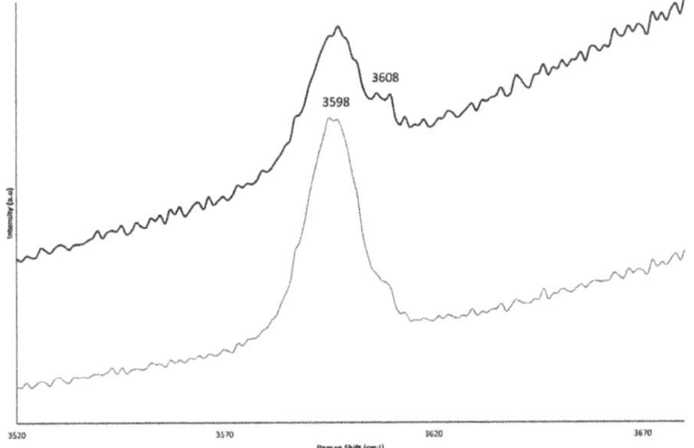

Figure 24. Raman spectra from 3520 to 3680 cm^{-1} of an emerald from Zambia. The black and grey spectra are acquired with the laser beam perpendicular to the c-axis and parallel to the c-axis, respectively. Note that the upper spectrum is shifted higher for clarity.

A clear separation between emeralds from Type IA and Type II occurrences cannot be made using Raman spectra (see again Table 5). This is due to the variation of alkali elements' content in the samples from the various geological environments, which might overlap. The Raman peak intensity of I_{3608} is higher than I_{3598} ($I_{3608} > I_{3598}$; see Figure 22) for all studied samples from Colombia, some samples from Afghanistan, some from Brazil and from Russia; it is more or less equal ($I_{3608} = I_{3598}$; see Figure 23) for most samples from Afghanistan, Brazil and Russia, as well as some from Ethiopia, and lower ($I_{3608} < I_{3598}$; see Figure 24) for all studied samples from Madagascar, Zambia and Zimbabwe as well as most from Ethiopia (see Table 5).

Table 5. Position and FWHM of the Raman band at around 1070 cm^{-1}, relative intensities of the Raman bands at 3598 cm^{-1} and 3608 cm^{-1} and position of the R1 photoluminescence bands for the samples from different localities. All observations were made using spectra acquired with a laser beam perpendicular to the c-axis.

Locality	Position & FWHM	I_{3608} & I_{3598} Intensities	R1 Position (PL)
Afghanistan	1068.16–1069.92 FWHM = 19.8–24.8	$I_{3608} = I_{3598}$ or $I_{3608} > I_{3598}$	683.7–684.2
Brazil	1068.38–1069.87 FWHM = 22.6–23.5	$I_{3608} = I_{3598}$ or $I_{3608} > I_{3598}$	683.9–684
Colombia	1069.01–1069.09 FWHM = 19.6–20.7	$I_{3608} > I_{3598}$	683.6–683.8
Ethiopia	1069.24–1070.15 FWHM = 23.6	$I_{3608} < I_{3598}$ or $I_{3608} = I_{3598}$	684–684.2
Madagascar	1069.44–1069.89 FWHM = 23.3	$I_{3608} < I_{3598}$	684.2–684.4
Russia	1068.81–1069.97 FWHM = 19.8	$I_{3608} = I_{3598}$ or $I_{3608} > I_{3598}$	683.7–683.8
Zambia	1069.11–1069.58 FWHM = 23.9	$I_{3608} < I_{3598}$	684.3-684.4
Zimbabwe	1069.91–1071.81 FWHM = 24.6–27.4	$I_{3608} < I_{3598}$	684.3-684.4

3.5. PL Spectroscopy

PL spectroscopy of emeralds was suggested as a useful tool to separate natural emeralds from synthetic ones, as well as to aid in emerald origin determination [58,61,72,74]. PL bands' intensities and positions vary slightly in different orientations relative to the c-axis [74]. In Figure 25, the photoluminescence spectra with a spectral range from 650 to 850 nm, acquired with the laser beam perpendicular to the c-axis, of emeralds from Colombia (upper spectrum) and Zimbabwe (bottom spectrum) are presented. Both samples presented two sharp bands at around 680 and 684 nm linked to Cr^{3+} (also known as R2 and R1 lines, respectively), as well as a broad band centred at around 720–740 nm, also linked with chromium [74,101]. It has been suggested that the exact position of the R1 band can give valuable clues on emeralds' natural vs. synthetic nature (no synthetics with R1 > 683.7 nm) as well as on their geologic origin (schist-origin emeralds with R1 < 683.9 nm [74]; see also Figure 26). Table 5 lists the exact position ranges of this band in the spectra of the studied samples. The studied samples show similar trends to those previously presented [74]; however, the emeralds from Brazil and Russia presented R1 bands shifted towards lower wavelengths and at 683.9–684 nm and 683.7–683.8 nm, respectively. Additionally, the R1 band of the studied samples from Afghanistan can vary in position, ranging from 683.6 to 684.2 nm.

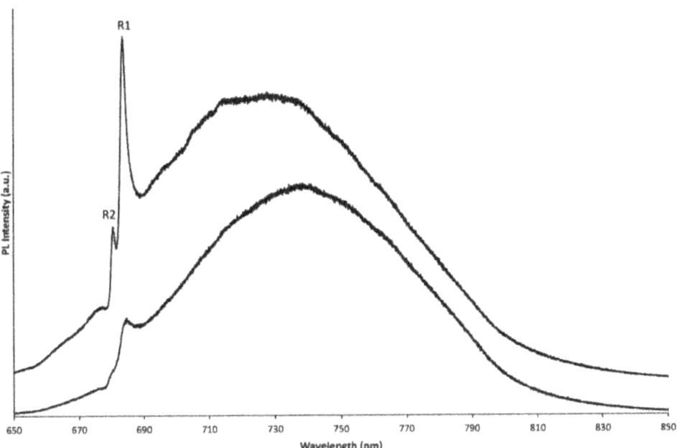

Figure 25. PL spectra from 650 to 850 nm of an emerald from Colombia (upper spectrum) and an emerald from Zimbabwe (bottom spectrum). Note that the upper spectrum is shifted higher for clarity.

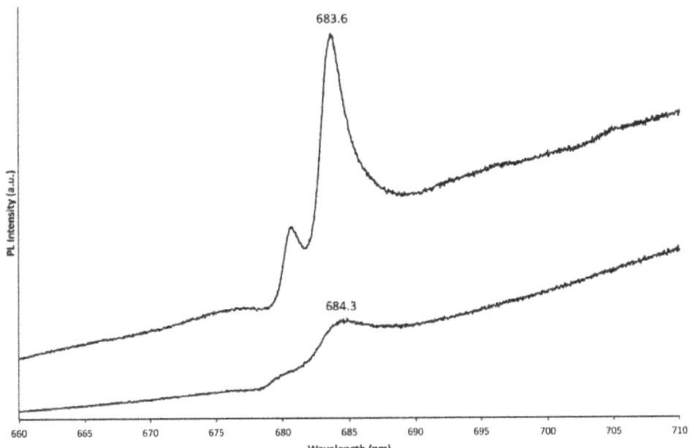

Figure 26. PL spectra from 660 to 710 nm of an emerald from Colombia (upper spectrum) and an emerald from Zimbabwe (bottom spectrum). Note that the upper spectrum is shifted higher for clarity.

4. Conclusions

LA-ICP-MS measurements on emerald samples from the eight most important sources reveal a relatively low lithium content (^{7}Li < 200 ppmw) for all studied emeralds from Type II occurrences (Afghanistan and Colombia). Additionally, certain emeralds belonging to Type IA occurrences (all samples from Brazil and Madagascar) can also present low lithium content (^{7}Li < 200 ppmw), whereas the emeralds from other Type IA occurrences (Ethiopia, Russia, Zambia and Zimbabwe) present medium to high lithium contents (^{7}Li > 250 ppmw). Measurements of the concentrations of a combination of alkali elements present in Type IA and Type II occurrences can help with the identification of emeralds from different mining areas. Scandium, manganese, cobalt, nickel, zinc and gallium can further aid with the separation. Origin determination cannot be performed by only studying one element or a single binary plot; it requires a combination of several.

UV-Vis spectra are useful in the separation of Type II emeralds from Type IA emeralds as the former contain iron-related bands of low intensity and the latter are of high intensity. Other spectroscopic data (FTIR, Raman and photoluminescence spectroscopy) on emeralds can further help to separate emeralds with low alkali element contents from those with high alkali element contents. However, there is an overlap between samples from different geological environments; separation using solely these methods should be undertaken with caution.

Apart from the microscopic characteristics, it seems that using the combination of spectroscopic and chemical characteristics (on green zones) presented in this work, a gem-quality emerald of "unknown" location and of relatively large size (>1 ct), similar to client stones submitted to gemmological laboratories, can be identified. First sorting could be done by using UV-Vis spectroscopy and the intensity of iron-related bands, as emeralds from Type IA occurrences (Brazil, Ethiopia, Madagascar, Russia, Zambia and Zimbabwe) contain intense iron-related bands and in emeralds from Type II occurrences (Afghanistan and Colombia) iron-related bands are absent or of low intensity. Emeralds from Type IA occurrences can be separated into those with low lithium (<200 ppmw for emeralds from Brazil and Madagascar) and those with medium to high lithium (>250 ppmw for emeralds from Ethiopia, Russia, Zambia and Zimbabwe). Low-lithium emeralds from Brazil contain less sodium, less potassium and no (BQL) nickel as well as more zinc compared to those from Madagascar, so plots combining these elements could help to separate these two. Medium- to high-lithium emeralds from Russia contain less sodium and potassium and rubidium than those from Ethiopia, Zambia and Zimbabwe. Samples from Zimbabwe present higher rubidium, gallium and manganese compared to those from Ethiopia and Zambia. Samples from Zambia contain more caesium and lithium and less scandium than the samples from Ethiopia, so they can be separated using plots combining these elements as well as a binary plot of the ^{133}Cs vs. ^{23}Na/^{7}Li ratio. Samples from Type II occurrences from Afghanistan and Colombia can be separated as the former contain higher concentrations of potassium, rubidium and caesium than the latter, and the scandium and gallium ratios differ as well.

The abovementioned scheme should be confirmed after studying a larger sample from these areas and taking into account the microscopic features of the samples. Studies on a larger number of gem-quality samples of green to dark green colour from various mining areas in the same region, along with a detailed study of their host rocks (especially of Type IA emeralds), should be performed in order to better understand the link between the trace elements and their geology and to better predict possible variation in the trace elements of emeralds in the same mining region. Statistical analyses (e.g., discriminant analysis) of the chemical data should be performed, as this might provide an additional tool for emeralds' geographic origin determination. Samples from less productive mining areas that have supplied gem-quality emeralds in the past, and might produce again in the future, should be also studied (e.g., Swat Valley, Pakistan; Mushakashi, Zambia; Bahia, Brazil). Additional measurements with a microprobe might help us to better understand the crystallochemistry of the samples. Standards for LA-ICP-MS made of doped emeralds with various elements might decrease any matrix effect and, in parallel, help with the measurement of trace elements with greater accuracy. Additionally, LA-ICP-MS with improved detection limit or analysis with secondary ion mass spectrometry (SIMS) may help further.

Polarised FTIR spectra on well-oriented samples should also be acquired in order to look for potential differences between samples from different mines. Polarized Raman spectra should also be collected in order to have a more accurate comparison between the vibrational data of different emeralds.

Supplementary Materials: The following are available online at http://www.mdpi.com/2075-163X/9/9/561/s1, Table S1: LA-ICP-MS analysis in ppmw of the studied emerald samples from Panjsher, Afghanistan; Table S2: LA-ICP-MS analysis in ppmw of the studied emerald samples from Itabira, Brazil; Table S3: LA-ICP-MS analysis in ppmw of the studied emerald samples from Coscuez, Colombia; Table S4: LA-ICP-MS analysis in ppmw of the studied emerald samples from Shakisso, Ethiopia; Table S5: LA-ICP-MS analysis in ppmw of the studied emerald samples from Mananjary, Madagascar; Table S6: LA-ICP-MS analysis in ppmw of the studied emerald samples from Ural, Russia; Table S7: LA-ICP-MS analysis in ppmw of the studied emerald samples from Kafubu, Zambia; Table S8: LA-ICP-MS analysis in ppmw of the studied emerald samples from Sandawana, Zimbabwe.

Author Contributions: S.K. formulated the paper, designed the experiments, participated in the data interpretation and wrote the manuscript. B.A.-S and F.M. prepared the experiments, did part of the LA-ICP-MS analysis, performed a data reduction, edited the manuscript and drew the figures and tables. S.S. designed, supervised and evaluated the LA-ICP-MS analysis. A.A.-A. selected the samples, designed the experiments and edited the manuscript.

Funding: This research received no external funding.

Acknowledgments: The authors wish to thank Dietmar Schwarz (ICA Gem Lab, Bangkok, Thailand) for the samples from Brazil and Russia, Christian Dunaigre (C. Dunaigre Consulting GmbH, Le Mont-Sur-Lausanne, Switzerland) for the samples from Brazil, Hanco Zwaan (Naturalis Biodiversity Center, Leiden, Netherlands) for the samples from Zimbabwe, and Vincent Pardieu (VP Consulting, Manama, Bahrain) for collecting the samples for DANAT directly from the mining areas (Madagascar, Zambia, Afghanistan and Colombia) or from trusted sources (Ethiopia).

Conflicts of Interest: The authors declare no conflict of interest.

References

1. Fritsch, E.; Rossman, G. An update on color in gems Part 3: Colors caused by band gaps and physical phenomena. *Gems Gemol.* **1988**, *24*, 81–103. [CrossRef]
2. Johnson, M.L.; Elen, S.; Muhlmeister, S. On the identification of various emerald filling substances. *Gems Gemol.* **1999**, *35*, 82–107. [CrossRef]
3. Kiefert, L.; Hänni, H.A.; Chalain, J.-P.; Weber, W. Identification of filler substances in emeralds by infrared and Raman spectroscopy. *J. Gemmol.* **1999**, *26*, 501–520. [CrossRef]
4. McClure, S.F.; Moses, T.M.; Tannous, M.; Koivula, J. Classifying emerald clarity at the GIA gem trade laboratory. *Gems Gemol.* **1999**, *35*, 176–185. [CrossRef]
5. Information Sheet #5 Standardised Gemmological Report Wording. Available online: http://www.lmhc-gemology.org/pdfs/IS5_20121209.pdf (accessed on 30 July 2019).
6. Hänni, H.A. Origin determination for gemstones: Possibilities, restrictions and reliability. *J. Gemmol.* **1994**, *24*, 139–148. [CrossRef]
7. Dereppe, J.M.; Moreaux, C.; Chauvaux, B.; Schwarz, D. Classification of emeralds by artificial neural networks. *J. Gemmol.* **2000**, *27*, 93–105. [CrossRef]
8. Schwarz, D.; Giuliani, G.; Grundmann, G.; Glas, M. The origin of emerald—A controversial topic. *Extralapis Engl.* **2002**, *2*, 18–21.
9. Gübelin Gem Lab. The roots of origin determination. *Jewel. News Asia* **2006**, *263*, 66–71.
10. Gübelin Gem Lab. The limitation of origin determination. *Jewel. News Asia* **2006**, *264*, 56–62.
11. Gübelin Gem Lab. A holistic method to determining gem origin. *Jewel. News Asia* **2006**, *264*, 118–126.
12. Krzemnicki, M.S. Determining the origin of gemstones: Challenges and perspectives. *InColor* **2007**, *4*, 6–11.
13. Groat, L.A.; Giuliani, G.; Marshall, D.D.; Turner, D. Emerald deposits and occurrences: A review. *Ore Geol. Rev.* **2008**, *34*, 87–112. [CrossRef]
14. Rossman, G.R. The geochemistry of gems and its relevance to gemology: Different traces, different prices. *Elements* **2009**, *5*, 159–162. [CrossRef]
15. Nguyen Bui, H.A.; Fritsch, E.; Rondeau, B. Geographical origin: Branding or Science? *InColor* **2012**, *19*, 30–39.
16. Giuliani, G.; Groat, L.; Ohnenstetter, D.; Fallick, A.E.; Feneyrol, J. The geology of gems and their geographic origin. In *Geology of Gem Deposits*; Raeside, E.R., Ed.; Mineralogical Association of Canada: Tucson, AZ, USA, 2014; Volume 44, pp. 113–134.
17. Groat, L.; Giuliani, G.; Marshall, D.; Turner, D. Emerald. In *Geology of Gem Deposits*; Raeside, E.R., Ed.; Mineralogical Association of Canada: Tucson, AZ, USA, 2014; Volume 44, pp. 135–174.
18. Giuliani, G.; Branquet, Y.; Fallick, A.E.; Groat, L.; Marshall, D. Emerald deposits around the world, their similarities and differences. *InColor* **2015**, 56–69.
19. Schwarz, D. The geographic origin determination of emeralds. *InColor* **2015**, 98–105.
20. Hainschwang, T.; Notari, F. Standards and protocols for emerald analysis in gem testing laboratories. *InColor* **2015**, 106–114.
21. Giuliani, G.; Groat, L.A.; Marshall, D.; Fallick, A.E.; Branquet, Y. Emerald Deposits: A Review and Enhanced Classification. *Minerals* **2019**, *9*, 105. [CrossRef]
22. Ogden, J.M. Rethinking laboratory reports for the geographical origin of gems. *J. Gemmol.* **2017**, *35*, 416–423. [CrossRef]
23. Cartier, L.E.; Ali, S.H.; Krzemnicki, M.S. Blockchain, chain of custody and trace elements: An overview of tracking and traceability opportunities in the gem. *J. Gemmol.* **2018**, *36*, 212–227. [CrossRef]
24. Archuleta, J.-L. The color of responsibility: Ethical issues and solutions in colored gemstones. *Gems Gemol.* **2016**, *52*, 144–160. [CrossRef]

25. Ali, S.H.; Giurco, D.; Arndt, N.; Nickless, E.; Brown, G.; Demetriades, A.; Durrheim, R.; Enriquez, M.A.; Kinnaird, J.; Littleboy, A.; et al. Mineral supply for sustainable development requires resource governance. *Nature* **2017**, *543*, 367–372. [CrossRef] [PubMed]

26. Makki, M.; Ali, S.H. Gemstone supply chains and development in Pakistan: Analyzing the post-Taliban emerald economy in the Swat Valley. *Geoforum* **2019**, *100*, 166–175. [CrossRef]

27. Gonthier, E. The symbolic representation of famous emeralds. In *L'Émeraude*; Giard, D., Giuliani, G., Cheilletz, A., Fritch, E., Gonthier, E., Eds.; AFG-CNRS-ORSTOM Edition: Paris, France, 1998; pp. 27–32.

28. Giuliani, G.; Chaussidon, M.; Schubnel, H.J.; Piat, D.; Rollion-Bard, C.; France-Lanord, C.; Giard, D.; de Narvaez, D.; Rondeau, B. Oxygen isotopes and emerald trade routes since antiquity. *Science* **2000**, *287*, 631–633. [CrossRef]

29. Calligaro, T.; Dran, J.-C.; Poirot, J.-P.; Querré, G.; Salomon, J.; Zwaan, J.C. PIXE/PIGE characterization of emeralds using an external micro-beam. *Nucl. Instrum. Meth.* **2000**, *161–163*, 769–774. [CrossRef]

30. Giuliani, G.; Chaussidon, M.; France-Lanord, C.; Guerra, H.S.; Chiappero, P.J.; Schubnel, H.J.; Gavrilenko, E.; Schwarz, D. L'exploitation des mines d'émeraude d'Autriche et de la Haute Egypte à l'époque Gallo-Romaine: Mythe ou réalité? *Rev. Gemmol* **2001**, *143*, 20–24. (In French)

31. Rondeau, B. Matériaux gemmes de référence du Museum National D'Histoire Naturelle: Exemples de valorisation scientifique d'une collection de minéralogie et gemmologie. Ph.D. Thesis, University of Nantes, Nantes, France, 2003.

32. Harrell, J.A. Archaeological geology of the world's first emerald mine. *Geosci. Can.* **2004**, *31*, 69–76.

33. Aurisicchio, C.; Corami, A.; Ehrman, S.; Graziani, G.; Cezaro, S.N. The emerald and gold necklace from Oplontis, Vesuvian Area, Naples, Italy. *J. Archaeol. Sci.* **2006**, *33*, 725–734. [CrossRef]

34. Schwarz, D.; Pardieu, V. Emeralds from the Silk Road countries. A comparison with emeralds from Colombia. *InColor* **2009**, *12*, 38–43.

35. Strack, E.; Kostov, R.I. Emeralds, sapphires, pearls and other gemmological materials from the Preslav gold treasure (X century) in Bulgaria. *Geochem. Mineral. Petrol.* **2010**, *48*, 103–123.

36. Karampelas, S.; Wörle, W.; Hunger, K.; Hanspeter, L.; Bersani, D.; Gübelin, S. Study of gems of a ciborium from Einsiedeln Abbey. *Gems Gemol.* **2010**, *46*, 291–295. [CrossRef]

37. Karampelas, S.; Wörle, W.; Hunger, K.; Hanspeter, L. Micro-Raman spectroscopy on two chalices from the Benedictine Abbey of Einsiedeln: Identification of gemstones. *J. Raman Spectrosc.* **2012**, *43*, 1833–1838. [CrossRef]

38. Farges, F.; Panczer, G.; Benbalagh, N.; Riondet, G. The Grand Sapphire of Louis XIV and The Ruspoli Sapphire. *Gems Gemol.* **2015**, *51*, 392–409.

39. Panczer, G.; Riondet, G.; Forest, L.; Krzemnicki, M.S.; Carole, D.; Faure, F. The Talisman of Charlemagne: New Historical and Gemological Discoveries. *Gems Gemol.* **2019**, *55*, 40–46. [CrossRef]

40. Aurisicchio, C.; Conte, A.M.; Medeghini, L.; Ottolini, L.; De Vito, C. Major and trace element geochemistry of emerald from several deposits: Implications for genetic models and classification schemes. *Ore Geol. Rev.* **2018**, *94*, 351–366. [CrossRef]

41. Zwaan, J.C.; Seifert, A.V.; Vrána, S.; Laurs, B.M.; Anckar, B.; Simons, W.B.S.; Falster, A.U.; Lustenhouwer, W.J.; Muhlmeister, S.; Koivula, J.K.; et al. Emeralds from the Kafubu area, Zambia. *Gems Gemol.* **2005**, *41*, 116–148. [CrossRef]

42. Saeseasaw, S.; Pardieu, V.; Sangsawong, S. Three-phase inclusions in emerald and their impact on origin determination. *Gems Gemol.* **2014**, *50*, 114–132.

43. Groat, L.A.; Laurs, B.M. Gem formation, production, and exploration: Why gem deposits are rare and what is being done to find them. *Elements* **2009**, *5*, 153–158. [CrossRef]

44. Fritsch, E.; Rondeau, B.; Devouard, B.; Pinsault, L.; Latouche, C. Why are some crystals gem quality? Crystal growth consideration on the "gem factor". *Can. Mineral.* **2017**, *55*, 521–533. [CrossRef]

45. Gramaccioli, C. Application of mineralogical techniques to gemmology. *Eur J Miner.* **1991**, *3*, 703–706. [CrossRef]

46. Fritsch, E.; Rondeau, B. Gemology: The developing science of gems. *Elements* **2009**, *5*, 147–152. [CrossRef]

47. Devouard, B.; Notari, F. Gemstones: From the naked eye to laboratory techniques. *Elements* **2009**, *5*, 163–168. [CrossRef]

48. Karampelas, S.; Kiefert, L. Gemstones and minerals. In *Analytical Archaeometry: Selected Topics*, 1st ed.; Edwards, H.G.M., Vandenabeele, P., Eds.; Royal Society of Chemistry: London, UK, 2012; pp. 291–317.

49. Gübelin, E.J.; Koivula, J.I. *Photoatlas of Inclusions in Gemstones*, 5th ed.; Opinio: Basel, Switzerland, 2004; Volume 1, p. 532.

50. Gübelin, E.J.; Koivula, J.I. *Photoatlas of Inclusions in Gemstones*; Opinio: Basel, Switzerland, 2005; Volume 2, p. 829.

51. Gübelin, E.J.; Koivula, J.I. *Photoatlas of Inclusions in Gemstones*; Opinio: Basel, Switzerland, 2008; Volume 3, p. 672.

52. Hänni, H.A.; Schwarz, D.; Fischer, M. The emeralds of the Belmont mine, Minas Gerais, Brazil. *J. Gemmol.* **1987**, *53*, 446–456. [CrossRef]

53. Bosshart, G. Emeralds from Colombia (Part 2). *J. Gemmol.* **1991**, *22*, 409–425. [CrossRef]

54. Bowersox, G.; Snee, L.W.; Foord, E.E.; Seal II, R.R. Emeralds of the Panjshir Valley, Afghanistan. *Gems Gemol.* **1991**, *27*, 26–39. [CrossRef]

55. Schmetzer, K.; Bernhardt, H.; Biehler, R. Emeralds from the Ural Mountains, USSR. *Gems Gemol.* **1991**, *27*, 86–99. [CrossRef]

56. Schwarz, D. Emeralds from the Mananjary region, Madagascar: Internal features. *Gems Gemol.* **1994**, *30*, 88–101. [CrossRef]

57. Zwaan, J.C.; Kanis, J.; Petsch, J. Update on emeralds from the Sandawana mines, Zimbabwe. *Gems Gemol.* **1997**, *33*, 80–101. [CrossRef]

58. Moroz, I.; Panczer, G.; Roth, M. Laser-induced luminescence of emeralds from different sources. *J. Gemmol.* **1998**, *26*, 316–320. [CrossRef]

59. Moroz, I.; Eliezri, I.Z. Mineral inclusions in emeralds from different sources. *J. Gemmol.* **1999**, *26*, 357–363. [CrossRef]

60. Moroz, I.; Roth, M.L.; Deich, V.B. The visible absorption spectroscopy of emeralds from different deposits. *Aust. Gemmol.* **1999**, *20*, 315–320.

61. Moroz, I.; Roth, M.; Boudeulle, M.; Panczer, G. Raman microspectroscopy and fluorescence of emeralds from various deposits. *J. Raman Spectrosc.* **2000**, *31*, 485–490. [CrossRef]

62. Fritsch, E.; Rondeau, B.; Notari, F.; Michelou, J.C.; Devouard, B.; Peucat, J.J.; Chalain, J.P.; Lulzac, Y.; de Narvaez, D.; Arboleda, C. Les nouvelles mines d'émeraude de La Pita (Colombie) 2ème partie. *Rev. De Gemmol.* **2002**, *144*, 13–21. (In French)

63. Zwaan, J.C. Gemmology, geology and origin of the Sandawana emerald deposits, Zimbabwe. *Scr. Geol.* **2006**, *131*, 211.

64. Abduriyim, A.; Kitawaki, H. Applications of laser ablation–inductively coupled plasma-mass spectrometry (LA-ICP-MS) to gemology. *Gems Gemol.* **2006**, *42*, 98–118. [CrossRef]

65. Ochoa, C.J.C.; Daza, M.J.H.; Fortaleche, D.; Jiménez, J.F. Progress on the study of parameters related to the origin of Colombian emeralds. *InColor* **2015**, 88–97.

66. Rondeau, B.; Fritsch, E.; Peucat, J.J.; Nordru, F.S.; Groat, L.A. Characterization of emeralds from a historical deposit: Byrud (Eidsvoll), Norway. *Gems Gemol.* **2008**, *44*, 108–122. [CrossRef]

67. Huong, L.T.T. Microscopic, Chemical and Spectroscopic Investigations on Emeralds of Various Origins. Ph.D. Thesis, University of Mainz, Mainz, Germany, 2008.

68. Smith, C.P.; Quinn, D.E. Inside emeralds. *Rapaport Diam. Rep.* **2009**, *32*, 139–148.

69. Huong, L.T.T.; Hager, T.; Hofmeister, W. Confocal micro-Raman spectroscopy: A powerful tool to identify natural and synthetic emeralds. *Gems Gemol.* **2010**, *46*, 139–148. [CrossRef]

70. Zwaan, J.C.; Jacob, D.E.; Häger, T.; Calvacanti Neto, M.T.O.; Kanis, J. Emeralds from the Fazenda Bonfim region, Rio Grande do Norte, Brazil. *Gems Gemol.* **2012**, *48*, 2–17. [CrossRef]

71. Cronin, D.P.; Rendle, A.M. Determining the geographical origins of natural emeralds through nondestructive chemical fingerprinting. *J. Gemmol* **2012**, *33*, 1–13. [CrossRef]

72. Bersani, D.; Azzi, G.; Lambruschi, E.; Barone, G.; Mazzoleni, P.; Raneri, S.; Longobardo, U.; Lottici, P.P. Characterization of emeralds by micro-Raman spectroscopy. *J. Raman Spectrosc.* **2014**, *45*, 1293–1300. [CrossRef]

73. Huong, L.T.T.; Hofmeister, W.; Hager, T.; Karampelas, S.; Kien, N.D.T. Identifying natural and synthetic emeralds by vibrational spectroscopy. *Gems Gemol.* **2014**, *50*, 287–292.

74. Thomson, D.B.; Kidd, J.D.; Astrom, M.; Scarani, A.; Smith, C.P. A comparison of *R*-line photoluminescence of emeralds from different origins. *J. Gemmol.* **2014**, *34*, 334–343. [CrossRef]

75. Renfro, N.D.; Koivula, J.I.; Muyal, J.; McClure, S.F.; Schumacher, K.; Shigley, J.E. Chart: Inclusions in Natural, Synthetic, and Treated Emerald. Available online: https://www.gia.edu/gems-gemology/winter-2016-inclusions-natural-synthetic-treated-emerald (accessed on 30 July 2019).

76. Kievlenko, E.Y. Beryl. In *Geology of Gems*, 1st ed.; Soregaroli, A., Ed.; Ocean Pictures Ltd.: Littleton, CO, USA, 2013; pp. 73–119.

77. Schwarz, D.; Henn, U. Emeralds from Madagascar. *J. Gemmol.* **1992**, *23*, 140–149. [CrossRef]

78. McManus, C.E.; Dowe, J.; McMillan, N.J. Quantagenetics® analysis of laser-induced breakdown spectroscopic data: Rapid and accurate authentication of materials. *Spectrochim. Acta B* **2018**, *145*, 79–85. [CrossRef]

79. Giuliani, G.; France-Lanord, C.; Zimmermann, J.-L.; Cheilletz, A.; Arboleda, C.; Charoy, B.; Coget, P.; Fontan, F.; Giard, D. Composition of fluids, δD of channel H_2O and $\delta^{18}O$ of lattice oxygen in beryls: Genetic implications for Brazilian, Colombian and Afghanistani emerald deposits. *Int. Geol. Rev.* **1997**, *39*, 400–424. [CrossRef]

80. Marshall, D.; Downes, P.J.; Ellis, S.; Greene, R.; Loughrey, L.; Jones, P. Pressure–temperature-fluid constraints for the Poona emerald deposits, Western Australia: Fluid inclusion and stable isotope studies. *Minerals* **2016**, *6*, 130. [CrossRef]

81. Schollenbruch, K.; Link, K.; Sintayehu, T. Gem quality emeralds from Southern Ethiopia. *InColor* **2017**, *35*, 48–54.

82. Vertiriest, W.; Wongrawang, P.A. gemological description of Ethiopian emeralds. *InColor* **2018**, *40*, 73–75.

83. Cerny, P.; Meintzer, R.E.; Anderson, A.J. Extreme fractionation in rare-element granitic pegmatites: Selected examples of data and mechanisms. *Can. Mineral.* **1985**, *23*, 381–421.

84. London, D.; Morgan VI, G.P. The pegmatite puzzle. *Elements* **2012**, *8*, 263–268. [CrossRef]

85. Linnen, R.L.; van Lichterverlde, M.; Cerny, P.H. Granitic pegmatites as sources of strategic metals. *Elements* **2012**, *8*, 275–280. [CrossRef]

86. Simmons, W.B.; Pezzotta, F.; Shigley, J.E.; Beurlen, H. Granitic pegmatites as reflections of their sources. *Elements* **2012**, *8*, 281–287. [CrossRef]

87. Cerny, P.; London, D.; Novak, M. Granitic pegmatites as reflections of their sources. *Elements* **2012**, *8*, 289–294. [CrossRef]

88. Wood, D.L.; Nassau, K. Characterization of beryl and emerald by visible and infrared absorption spectroscopy. *Am. Mineral.* **1968**, *53*, 777–800.

89. Platonov, A.N.; Taran, M.N.; Minko, O.E.; Polshyn, E.V. Optical absorption spectra and nature of color of iron-containing beryls. *Phys. Chem. Miner.* **1978**, *3*, 87–88.

90. Taran, M.N.T.; Rossman, G.R.R. Optical spectroscopic study and a re-examination of the beryl, cordierite, and osumilite spectra. *Am. Mineral.* **2001**, *86*, 973–980. [CrossRef]

91. Schmetzer, L. Letter: Comment on analysis of three-phase inclusions in emerald. *Gems Gemol.* **2014**, *50*, 316–319.

92. Spinolo, G.; Fontana, I.; Galli, A. Optical absorption spectra of Fe^{2+} and Fe^{3+} in beryl crystals Optical absorption spectra of Fe^{2+} and Fe^{3+} in beryl crystals. *Phys. Stat. Sol. (b)* **2007**, *244*, 4660–4668. [CrossRef]

93. Wood, D.L.; Nassau, K. Infrared spectra of foreign molecules in beryl. *J. Chem Phys.* **1967**, *47*, 2220–2228. [CrossRef]

94. Aurisicchio, C.; Grubessi, O.; Zecchini, P. Infrared spectroscopy and crystal chemistry of the beryl group. *Can. Mineral.* **1994**, *32*, 55–68.

95. Charoy, B.; Donato, D.P.; Barres, O.; Pinto-Coelho, C. Channel occupancy in an alkali-poor beryl from Serra Branca (Goias, Brazil): Spectroscopic characterization. *Am. Mineral.* **1996**, *81*, 395–403. [CrossRef]

96. Donato, D.P.; Cheilletz, A.; Barres, O.; Yvon, J. Infrared spectroscopy of OD vibrators in minerals at natural dilution: Hydroxyl groups in talc and kaolinite, and structural water in beryl and emerald. *Appl. Spectrosc.* **2004**, *58*, 521–527. [CrossRef]

97. Adams, D.M.; Gardner, I.R. Single-crystal vibrational spectra of beryl and dioptase. *J. Chem. Soc.* **1974**, *14*, 1502–1505. [CrossRef]

98. Duroc-Danner, J.M. The identification value of the 2293 cm^{-1} infrared absorption band in natural and hydrothermal synthetic emeralds. *J. Gemmol.* **2006**, *30*, 75–82. [CrossRef]

99. Hagemann, H.; Lucken, A.; Bill, H.; Gysler-Sanz, J.; Stalder, H.A. Polarized Raman spectra of beryl and bazzite. *Phys. Chem. Miner.* **1990**, *17*, 395–400. [CrossRef]

100. Kim, C.C.; Bell, M.I.; McKeown, D.A. Vibrational analysis of beryl ($Be_3Al_2Si_6O_{18}$) and its constituent ring (Si_6O_{18}). *Phys. B. Condens. Matter* **1995**, *205*, 193–208. [CrossRef]
101. Wood, D.L. Absorption, fluorescence, and Zeeman effect in emerald. *J. Chem. Phys.* **1965**, *42*, 3404–3410. [CrossRef]

Article

Diversity in Ruby Geochemistry and Its Inclusions: Intra- and Inter- Continental Comparisons from Myanmar and Eastern Australia

Frederick L. Sutherland [1,*], Khin Zaw [2], Sebastien Meffre [2], Jay Thompson [2], Karsten Goemann [3], Kyaw Thu [4], Than Than Nu [5], Mazlinfalina Mohd Zin [6] and Stephen J. Harris [7]

[1] Mineralogy & Petrology, Geosciences, Australian Museum, 1 William Street, Sydney, NSW 2010, Australia
[2] CODES Centre of Ore Deposit and Earth Sciences, University of Tasmania, Hobart, Tas 7001, Australia; khin.zaw@utas.edu.au (K.Z.); sebastien.meffre@utas.edu.au (S.M.); jay.thompson@utas.edu.au (J.T.)
[3] Central Science Laboratory, University of Tasmania, Hobart, Tas 7001, Australia; karsten.goemann@utas.edu.au
[4] Geology Department, Yangon University, Yangon 11041, Myanmar; macle45@gmail.com
[5] Geology Department, Mandalay University, Mandalay 05032, Myanmar; thanthannu83@gmail.com
[6] Geology Programme, Faculty of Science and Technology, The National University of Malaysia (UKM), Selangor 43600, Malaysia; farlinnzin@gmail.com
[7] School of Biological, Earth and Environmental Sciences, University of New South Wales, Sydney, NSW 2052, Australia; s.j.harris@student.unsw.edu.au
* Correspondence: linsutherland1@gmail.com; Tel.: +61-2-65826553

Received: 28 November 2018; Accepted: 25 December 2018; Published: 5 January 2019

Abstract: Ruby in diverse geological settings leaves petrogenetic clues, in its zoning, inclusions, trace elements and oxygen isotope values. Rock-hosted and isolated crystals are compared from Myanmar, SE Asia, and New South Wales, East Australia. Myanmar ruby typifies metasomatized and metamorphic settings, while East Australian ruby xenocrysts are derived from basalts that tapped underlying fold belts. The respective suites include homogeneous ruby; bi-colored inner (violet blue) and outer (red) zoned ruby; ruby-sapphirine-spinel composites; pink to red grains and multi-zoned crystals of red-pink-white-violet (core to rim). Ruby ages were determined by using U-Pb isotopes in titanite inclusions (Thurein Taung; 32.4 Ma) and zircon inclusions (Mong Hsu; 23.9 Ma) and basalt dating in NSW, >60–40 Ma. Trace element oxide plots suggest marble sources for Thurein Taung and Mong Hsu ruby and ultramafic-mafic sources for Mong Hsu (dark cores). NSW rubies suggest metasomatic (Barrington Tops), ultramafic to mafic (Macquarie River) and metasomatic-magmatic (New England) sources. A previous study showed that Cr/Ga vs. Fe/(V + Ti) plots separate Mong Hsu ruby from other ruby fields, but did not test Mogok ruby. Thurein Taung ruby, tested here, plotted separately to Mong Hsu ruby. A Fe-Ga/Mg diagram splits ruby suites into various fields (Ga/Mg < 3), except for magmatic input into rare Mogok and Australian ruby (Ga/Mg > 6). The diverse results emphasize ruby's potential for geographic typing.

Keywords: ruby; Mogok; Mong Hsu; New South Wales; trace elements; LA-ICP-MS analysis; inclusions; U–Pb age-dating; genetic diversity; geographic typing

---

1. Introduction

1.1. Background

Corundum is an aluminum oxide mineral in which trace element substitution of Al by Fe, Ti, V and Cr act as chromophores. Rubies and sapphires are varieties in which ruby develops a red color with sufficient entry of Cr^{3+}, whereas sapphire develops blue, green, yellow purple, violet, mauve and pink colors due to substitutions of the other chromophore elements in the presence of lesser Cr contents. Numerous gem corundum-bearing sites are known in SE Asia, where Mogok and Mong Hsu in Myanmar are well known examples of ruby deposits [1–3]. In contrast, East Australia is particularly noted for widespread placer sapphire deposits [2,3] and only scattered ruby-bearing deposits with rubies that are unusual in character [4]. Ruby occurs in diverse geological settings, with ages ranging from Neoarchean [5]-Proterozoic [3,6] to <5 Ma, and can form at wide ranges in temperature, pressure, and fluid activity and oxidation conditions [1–7]. The ruby deposits are primarily formed by regional metamorphism and/or from transportations from depth by volcanic processes, but the origins of the corundum can be multi-staged processes [3,4]. Eluvial and alluvial (placer) ruby deposits can form by weathering of the primary sources. Trace element studies together with O-isotope data and age dating can be used to discriminate between lithological sources [7–11]. Geochemical testing, initially leading to finger-printing [12], and further aided by O-isotope studies have become routinely used to enable geographic typing of the corundum suites [13–16]. This now achieves greater control through use of a wider range of minerals included in corundum for providing precise isotopic dating [17–19]. Quality ruby is an expensive gemstone and is mostly not favored for even micro-destructive study. Optical mineralogy combined with Raman spectroscopy remains a non-destructive technique for their testing, along with their solid/melt inclusions, for origin determination. [20,21].

In this study [22], we present new comparative trace element results and age data from Myanmar ruby fields, which typify metamorphosed and metasomatized carbonate and skarn settings at Mogok and Mong Hsu, and compare them with eastern Australia ruby fields, which typically carry ruby xenocrysts derived from basalt fields and found in placer deposits. This allows discussion of the extreme individual diversities and geochemical characteristics found within and between Myanmar ruby deposits and placer ruby sources in eastern Australia.

The regional tectonic settings of the Mogok and Mong Hsu, Myanmar and East Australian gem regions involve the western Pacific continental margins, associated with the Asian and Australian plates. The distribution of these zircon-corundum associations is shown in Figure 1, derived from references [23,24]

1.2. Local and Geological Settings, Myanmar

The corundum-bearing gem deposits in the Mogok area include both in situ and secondary deposits. They were described in detail [25,26] and only a relevant account is outlined here. A geological map, and associated gem workings of the Mogok area, is shown in Figure 2. The Mogok gem stone tract at the northern Mogok Metamorphic Belt (MMB) [27–32] is a source of world-class rubies, sapphires and other gemstones. Myanmar ruby genesis is associated with carbonates [33,34].

The Mogok area is characterized by high-grade metamorphic rocks; the dominant unit is banded gneiss with biotite, garnet, sillimanite and oligoclase. It is also interspersed with quartzite and bands and lenses of marble with ruby [25,26]. The metamorphic rocks are intruded by alkaline igneous rocks (mostly Oligocene-early Miocene sodic nepheline-syenite, and syenite-pegmatite and urtite suites) and early Oligocene leucogranites. Earlier Jurassic (?) to early Cretaceous mafic-ultramafic peridotites, norites minor dolerites and basalts are considered to represent layered cumulate intrusions rather than ophiolite sequences. Biotite granitoids are widely exposed at Kabaing, (16 Ma) and Thabeikkyin (130 Ma) near Mogok and some Oligocene-Miocene syenite pegmatites contain sapphires [25,26]. The Kabaing granitoids and metasedimentary rocks are commonly intruded by late-stage pegmatites and aplites [35].

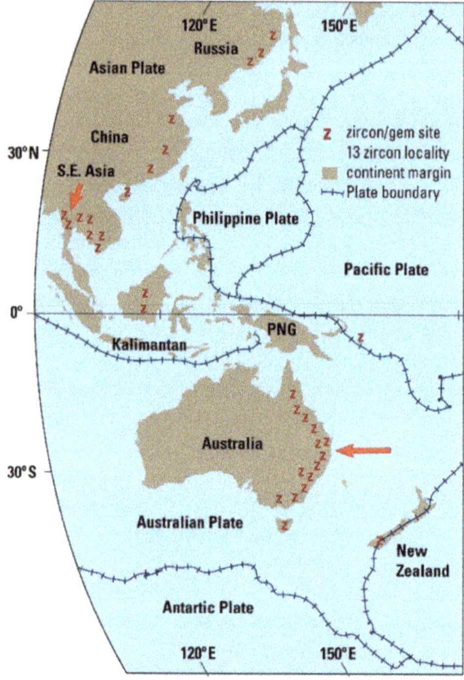

Figure 1. Regional setting and location of Myanmar and East Australian gem deposits, within the West Pacific continental margin zircon-corundum gem deposit zones (z), along the Asian and Australian plates [23,24]. Mogok and Mong Hsu, Myanmar, deposits (short arrow) are the western most sites [26] and New South Wales deposits (long arrow are among the eastern most group [9].

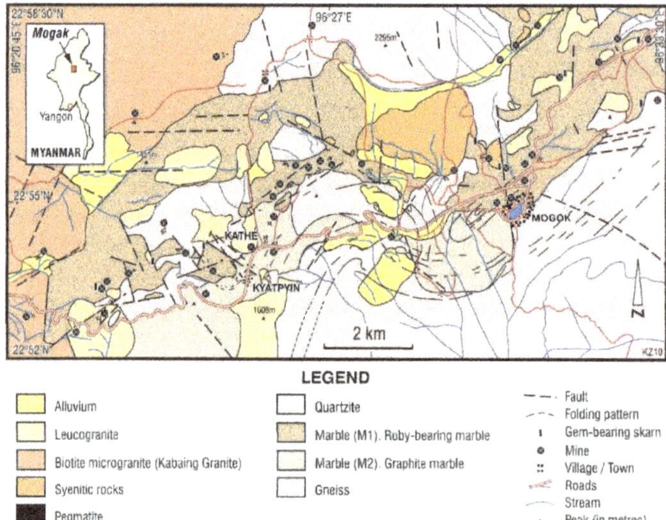

Figure 2. Geological map of Mogok Gem Tract, modified from reference [11], showing main lithological units and general distribution of gem deposits. The Thurein Taung workings lie ~4 km East of the 96°20'45" E longitude map border and 2.3 km North of the 22°52' N latitude map border.

The Thurein Taung study site is ~23 km W of Mogok Township at 96°22′ 20.7 E, 22°54′ 12.7 N. A prominent hill here has a core of steeply dipping white marble intruded by ijolite (Figure 3a,b). It is flanked by diopside-marble intruded by alkali syenite pegmatite on one side and by gneiss on the other flank. A skarn deposit in contact with leucogranite lies near the SW summit, where ruby and painite are found associated with titanite, anatase, rutile, baddelyite, axinite, elbaite, schorl, dravite and zircon [1]. The hill is covered in parts by alluvium and mine waste.

Figure 3. Thurein Taung ruby locality: (**a**) Hill exposing gem workings and scene for underlying geology section. Viewed from the east. Basal distance across view is ~600 m; (**b**) Sketch diagram across geological units (symbols and labels). V/H = 1, M1 = White Marble, M2 = Diopside marble, Gn = Gneiss, ε1 = ijolite dyke, ε2 = alkali syenite pegmatite dyke. Sapphire-bearing lithologies intrude the marbles (blue arrows), with sapphire deposits (s, inset). Ruby is associated with skarn near the top of the hill. Images; Kyaw Thu.

In comparison, the Mong Hsu area (Figure 4), the second-largest ruby deposit in Myanmar, is located in sedimentary and regional metamorphic rocks [17,27].

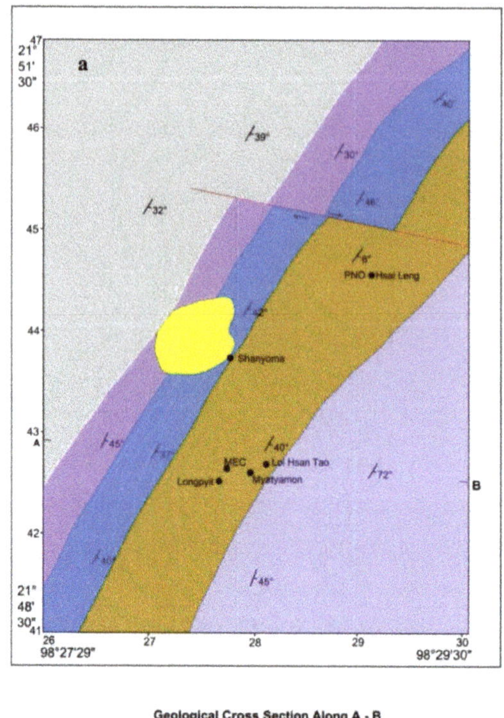

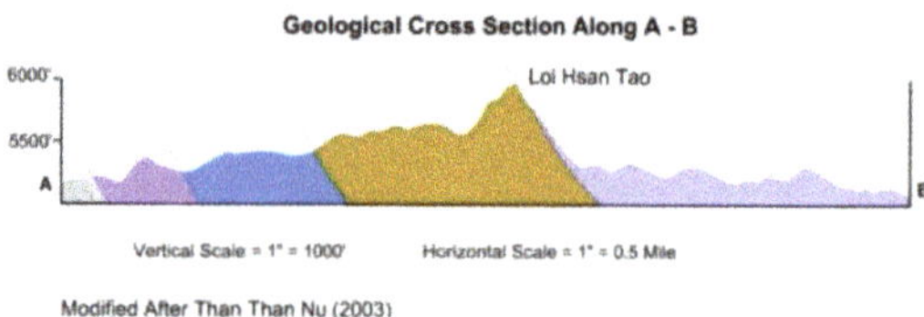

Figure 4. Geological map (**a**) and geological section (**A–B**) of Loi Hsan Tao Mountain, Mong Hsu area, Shan State, Myanmar, based on Than Than Nu in [36]. The ESE—dipping Lower Paleozoic fold belt sequence, younging from west to east includes Biotite Schist (grey), Calc-Silicates (purple), Dolomitic marble (brown), Phyllite and Quartzite (mauve). Alluvium (yellow) unconformably overlies the sequence. Structural features include strike directions (longer lines) with amount of dip (short lines normal to strike) and major fault (red line). The map north length represents 33.3 km. The ruby workings are indicated at around Lo Hasen Tao Mountain on map at cross section A–B, H/V = 2.64).

The regionally metamorphosed rocks (probably of Palaeozoic age) include biotite–almandine–staurolite schist, diopside calcsilicate rocks and biotite quartzite interbedded with biotite phyllite and ruby-bearing white marble (Figure 4). Some granitic pegmatites intruded the metamorphic rocks in the area and can be traced into the eastern and southeastern parts as far as Than Lwin River. In addition, massive Plateau Limestone (Devonian?) occupies the northern part of the area and shale, siltstone, minor limestone beds (Silurian) and bedded limestone (Ordovician) are present in the middle and southeastern parts of the Mong Hsu area.

Primary occurrences of ruby were discovered around Loi Hsan Tao Ridge (elevation 1750 m), about 3.6 km southeast of Mong Hsu town in Shan State of Myanmar [26,36,37]. In the Mong Hsu area, a series of medium grade regionally metamorphosed rocks comprised of schist, calc-silicate rock, quartzite, phyllite and ruby bearing dolomitic marble are exposed along Loi Hsan Tao ridge. The ruby occurs in a 330 m thick white fine-grained dolomitic marble within that rock sequence which dips southeast between 30–70°. The main mode of occurrence of Mong Hsu ruby is in interconnected weak planes or fissures which run parallel to the foliation of the host marble at upper levels but become steeper and cross cut the foliation at depth. Ruby grains, lacking orientation, occur mostly as aggregates together with vein calcite. Flat crystals and half-formed crystals are frequently found. At some work sites, zonal occurrences of mineral assemblages with depth are observed, showing dolomite, tremolite and talc in the upper zone, brucite and tourmaline in the middle zone and wollastonite in a lower zone.

1.3. Local and Geological Settings, East Australia

In East Australian gem fields, ruby is an accessory associate of other more abundant gem xenocrysts, largely zircon and sapphire, found within scattered alluvial placer deposits [4,7,38]; Figure 5. The xenocrysts are survivors from deeper seated source rocks breached by and transported to the surface by basaltic eruptions, before erosional release into placers. In this study, rubies from three New South Wales deposits were chosen for comparison, being from diverse settings within Paleozoic fold belts in Australia [39], and representing well-characterized suites from previous studies.

At Barrington Tops, ruby occurs as fragmented and corroded composites within alluvial deposits [40], which were derived from repeated basaltic events from the long-lived Barrington Tops volcano, which had erupted for over 55 Ma [41]. The ruby-sapphirine-spinel composites suggest a metamorphic origin at 780–940 °C [42,43]. The ruby distribution within the placers overlies three separate distinctive granitoid intrusions emplaced within the underlying fold belt sequence and age-dated by zircon SHRIMP U-Pb methods at 268, 273 and 277 Ma [44]. The granitoids include mafic granodiorites formed by interactions of mantle-derived melts with crustal rocks [45]. Southwest-facing faulted slivers of relict ophiolites to the northwest include tectonized harzbugites and intrusive mafic rocks [46]. This diverse basement provides complex metamorphic, metasomatic and contact metamorphic sources for potential ruby genesis. The Macquarie River ruby grades into pink sapphire and its source region is poorly constrained, as stones were recovered as accessories in gold dredging operations in the alluvial river bed. It accompanies diamonds and represents a distinctive highly Mg-rich, high temperature genetic type [47]. The New England ruby is a rare accessory in gem field workings within the dissected Maybole shield volcano sequences [48]. It has most unusual color zoning from ruby cores into sapphire rims and consistently high Ga content and Ga/Mg ratio compared to world-wide ruby. Rubies of diverse character other than those detailed in this study are found at Yarrowitch and Tumbarumba, NSW and ENE of Melbourne and are described in references [4,9].

The rubies compared with and discussed here from the central-north New South Wales, Australian region (Figure 5) are quite different in external features to the diverse Myanmar Mogok and Mong Hsu materials and their contained inclusion suites (Figure 6a,b and Figure 7). The NSW suites lie within a NE-SW rectangular zone ~220 km in length and up to 180 km in width. They show contrasting composite mineral growth (Barrington Tops; Figure 8), high pressure mineral inclusions in pink to red gradational sapphire-ruby suites (Macquarie River; Figure 9a,b, after [4]) and unusual ruby core to multiple zoned sapphire outer zones (New England; Figure 10).

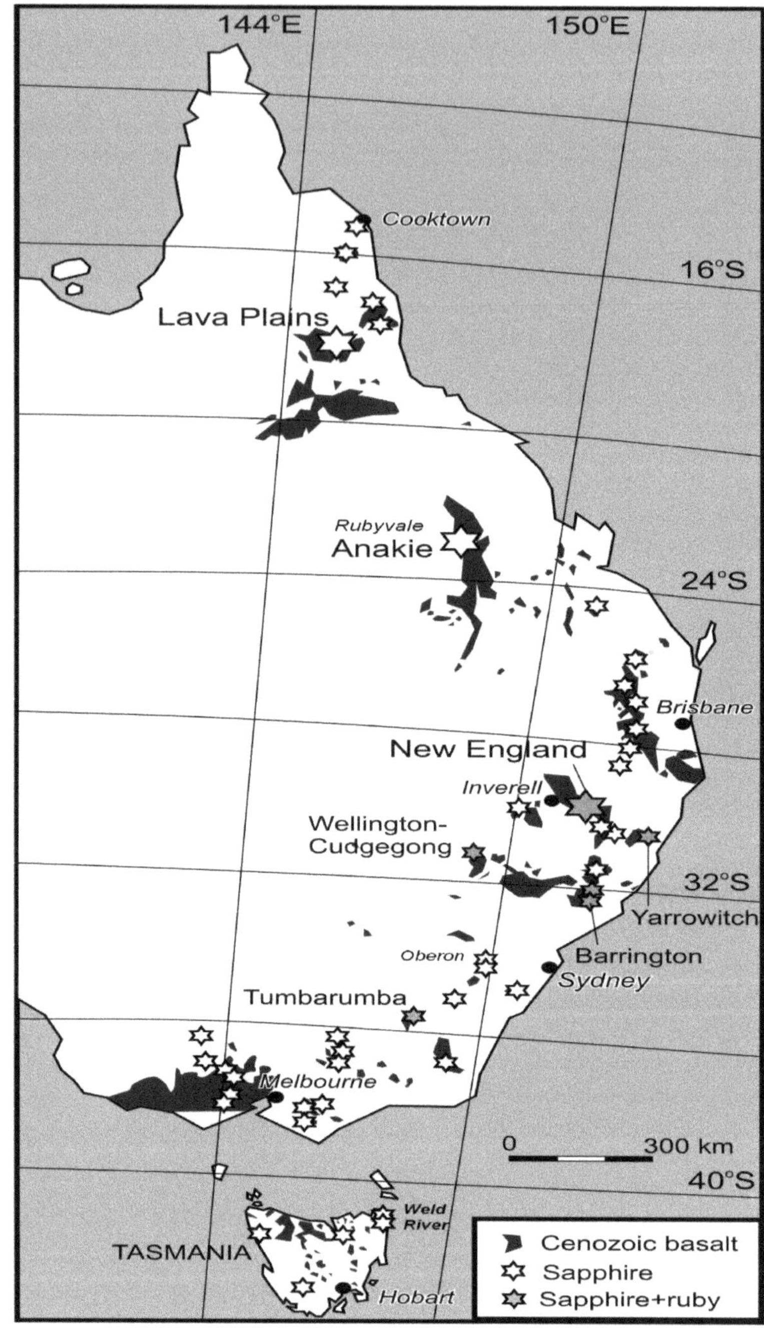

Figure 5. East Australian basalt-derived alluvial gem corundum deposits, showing ruby and locating Barrington Tops, Macquarie River (Wellington) and New England comparative sites. Map derived from reference [9].

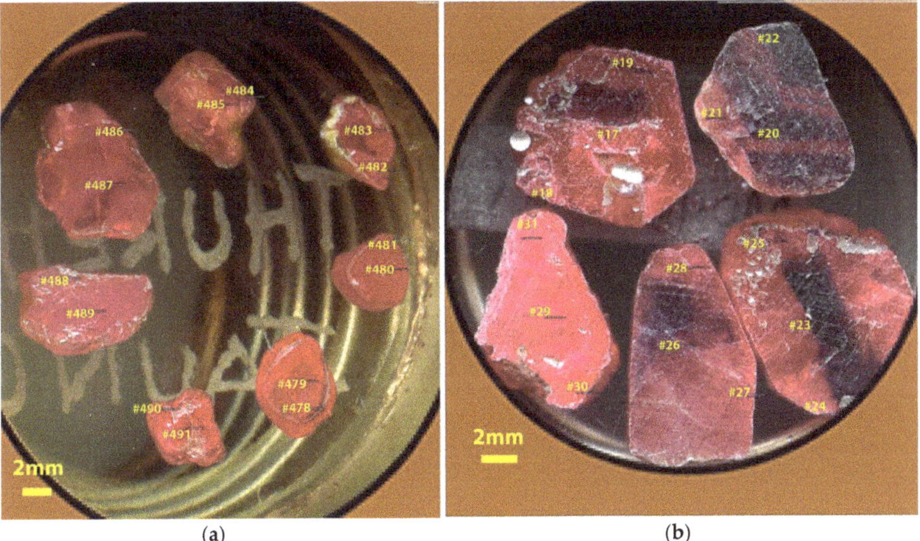

Figure 6. Photos showing mineralogical and textural characteristics of Mogok and Mong Hsu ruby samples. (**a**) Thurein Taung ruby crystals embedded in 25 mm diameter mount. (**b**) Mong Hsu ruby crystals embedded in 25 mm diameter mount. Yellow # nos. are analytical spots, listed in Tables A1 and A2.

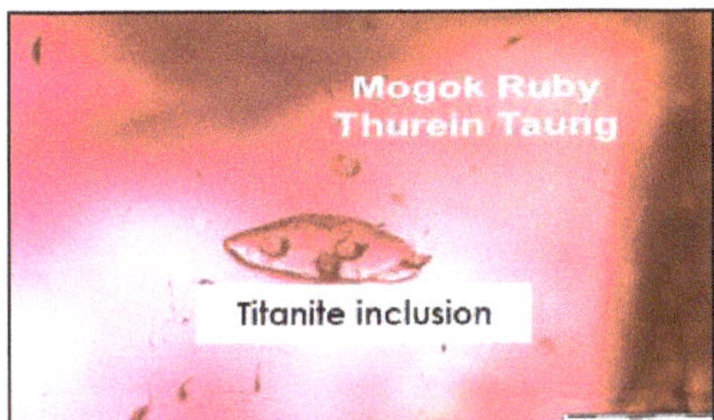

Figure 7. Thurein Taung ruby crystal hosting composite titanite inclusion 0.025 mm in length. The small circular structures represent ablation pits from initial analyses. Photo; Khin Zaw.

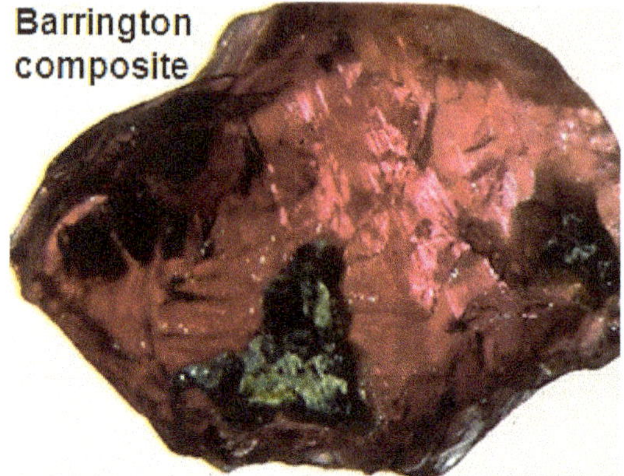

Figure 8. Barrington Tops ruby-sapphirine-spinel composite grain used for thermometric study [42,43]. Sample size ~5 × 7 mm. Ruby main mass (purple red); sapphirine (dark green), center bottom and center right; spinel (black), center left and outer fusion crust. Photographer G. Webb, Australian Museum.

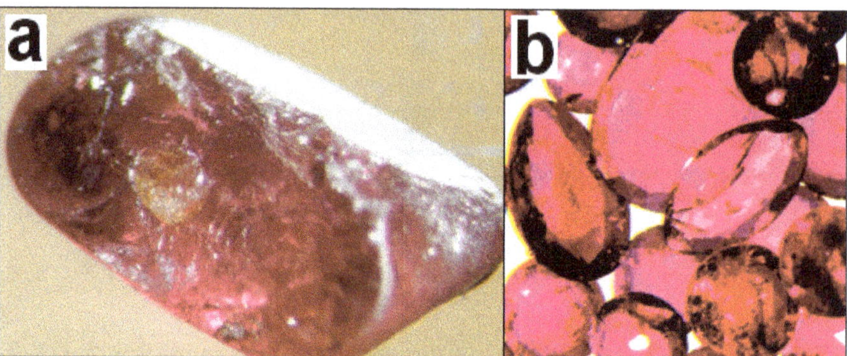

Figure 9. Photos of ruby features from Macquarie-Cudgegong River system alluvial deposits, NSW (**a**) ruby fragment, a few mm across with pyrope garnet inclusion (center). (**b**) Facetted stones of ruby and pink sapphire, up to 3 mm across [47]. Photographer G. Webb, Australian Museum.

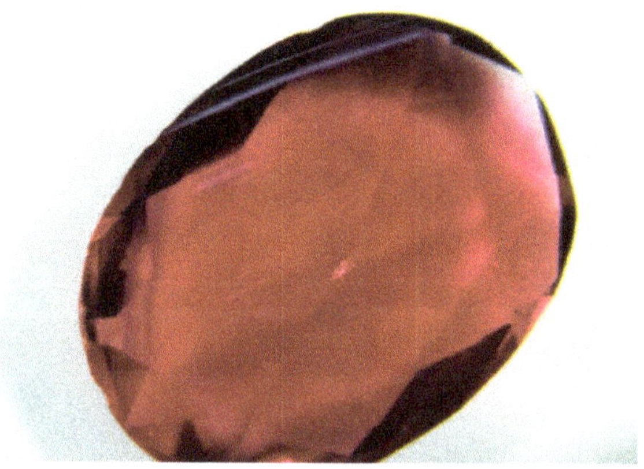

Figure 10. Facetted ruby (2.31c) from New England basaltic gem field [48]. Australian Museum Collection. Photo G. Webb, Australian Museum.

2. Materials and Methods

2.1. Materials

Examples of Myanmar ruby samples analyzed for trace elements by LA-ICP-MS and an inclusion used for U-Pb age-dating are shown in Figures 6 and 7. Examples of East Australian ruby samples with previous analytical trace element data are shown in Figures 8–10.

The Australian rubies were previously analyzed from Barrington Tops and Macquarie River [9] and New England [48], and results are used for comparative plots in variation diagrams in this study.

2.2. Analytical Methods

2.2.1. Analytical Methods, Trace Element and Isotopic Dating Analysis

The Myanmar ruby samples and mineral inclusions used for U-Pb age dating were analyzed for trace elements contents by LA-ICP-MS techniques, supplemented by further EDS analysis (see below) at the University of Tasmania. Previously determined trace elements in New South Wales, East Australian ruby samples, were used for the comparisons had been analyzed by similar LA-ICP-MS methods at University of Tasmania (Barrington Tops, Macquarie River) and GEMOC Key Centre facilities, Macquarie University, Sydney (New England gem field). They are detailed in other publications [9,48].

2.2.2. Trace Elements Analysis

Most trace element analyses were performed at CODES, University of Tasmania, using an ASI RESOLution S155 laser ablation system with Coherent Compex Pro 110 excimer laser operating at 193 nm wavelength and ~20 ns pulse width. The laser ablation system was coupled to an Agilent 7900 quadrupole inductively coupled plasma mass spectrometer (ICP-MS). Ablation was performed in pure helium flowing at 0.35 L/min and immediately mixed with argon flowing at 1.05 L/min after ablation. Depending on the session, the laser was pulsed at either 5 Hz or 10 Hz with either a 43 or a 51 µm spot size and a fluence of ~4 J·cm^{-2}. Each analysis consisted of 30 s of background gas, followed by 30 s of ablation time counting for 10 ms on each isotope. Primary calibration of trace elements was done with the NIST612 reference glass using GeoReM preferred values. A secondary standard correction was done using GSD-1g and BCR-2g reference glasses, again with GeoReM preferred values. Data reduction was done in an in-house macro-based Excel workbook, with aluminum as the internal standard element assuming stoichiometric proportions. Spot analyses were made at core and rim positions on each of the analyzed grains.

Backscattered electron (BSE) imaging and energy dispersive X-ray spectrometry (EDS) major elements analyses were performed at the Central Science Laboratory, University of Tasmania. This used a Hitachi SU-70 field emission scanning electron microscope (SEM, Hitachi High Technologies, Hitachinaka-shi, Japan) fitted with a Hitachi photo diode BSE detector and an Oxford AZtec XMax80 EDS system at an accelerating voltage of 15 kV and beam current of around 2 nA (Oxford Instruments Nanoanalsis, High Wycombe, UK). Elements were calibrated on a range of natural and synthetic mineral standard reference materials. Cobalt metal was used as beam measurement standard, i.e., to indirectly determine the relative change in beam current compared to the time of element calibration. The sample was coated with around 20 nm of carbon prior to analysis using a Ladd 40000 carbon evaporator (Ladd Research Industries, Williston, VT, USA).

2.2.3. U-Pb Isotopic Analysis

All U-Pb analyses were conducted on the same instrumentation described above, but with the addition of N$_2$ gas for higher sensitivity. Analyses were done with a 19 µm spot size at 5 Hz and 1.9 J cm^{-2} laser fluence. Each analysis consisted of 30 s of background gas, followed by 30 s of ablation time. Isotopes analyzed were ^{31}P, ^{49}Ti, ^{56}Fe, ^{89}Y ^{91}Zr, ^{93}Nb, ^{139}La, ^{140}Ce, ^{141}Pr, ^{146}Nd, ^{147}Sm, ^{151}Eu, ^{157}Gd, ^{163}Dy, ^{165}Ho, ^{166}Er, ^{169}Tm, ^{172}Yb, ^{175}Lu, ^{178}Hf, ^{181}Ta, ^{202}Hg, ^{204}Pb, ^{206}Pb, ^{207}Pb, ^{208}Pb, ^{232}Th, ^{235}U, and ^{238}U counting for 5 ms on each isotope, except for Pb isotopes and ^{238}U which had 25 and

15 ms counting times respectively. Each of the three minerals targeted for U-Pb analyses had a mineral specific primary standard for the U-Pb calibration: the 91500 zircon, for zircon analyses, Phalaborwa baddeleyite for baddeleyite analyses and an in-house titanite standard (19686) for titanite analyses. Calibration of the $^{207}Pb/^{206}Pb$ ratio was done using the NIST610 reference glass measured at the same conditions as the unknowns and using values from [49]. Calibration of select trace element data measured along with Pb/U ratios was done using the NIST610 reference glass using GeoReM preferred values and with Zr as the internal standard element for baddeleyite and zircon and Ti as the internal standard element for titanite; all assuming stoichiometric proportions in each mineral respectively. Data reduction for U-Pb and trace element concentrations is done in an in-house macro-based Excel workbook with details about the methodology described in reference [50].

Accuracy of the U-Pb ages was checked using a variety of reference minerals of known (ID-TIMS) age treated as unknowns. These include the Temora [51] and Plesovice [52] zircons, FC-1 baddeleyite [53], and the FCT-3 [54] and 100606 [55] titanites. All of these secondary reference materials were accurate within their uncertainties. Concordia plots and intercept ages are done using Isoplot 4 [56]. Where a common Pb estimate is lacking, the Stacey and Kramer [57] model Pb at the age of the mineral was used for the anchor on the Concordia intercept ages. Uncertainties are calculated using the method of Horstwood et al. [58], where systematic uncertainties are added after the Concordia intercept ages are calculated. Details of the geochronologic data are supplied in Supplementary Materials.

3. Results

3.1. Trace Element Variations

The main trace element values measured in Thurein Taung and Mong Hsu ruby samples are given in Table 1. This breaks the data up into rim and core measurements of the analyzed crystals for easier comparison, while individual results from each analytical spot are detailed in Tables A1 and A2. Some rubies appeared to indicate noticeable Si and Ca values above the relatively high detection limits for those elements, particularly in Thurein Taung ruby. Such levels that enter the corundum crystal structure invoke potential presence of nano inclusions of silica [59] or even particulate calcsilicate material [13]. Because of potential large variability in Si and Ca within corundum analyses, due to LA-ICP-MS analytical interreference affects [60], these element values were not reported in Tables 1, A1 and A2. Other trace elements are mostly at detection level or negligible (1 or <1 ppm) in concentration. A few Thurein Taung rubies contain 1–2 ppm B, while Mong Hsu rubies contain 1–3 ppm Cu and 1–9 ppm Zn.

Table 1. Trace element comparative ranges and averages (ppm), Myanmar ruby samples.

Sample	Mg	Ti	V	Cr	Fe	Ga
Thurein Taung						
Rims, range	27–204	38–1241	110–401	432–4599	38–483	20–170
n = 9, average	85	254	243	2468	143	79
Cores, range	35–189	53–283	102–421	618–4406	24–461	20–152
n = 9, average	78	139	225	2494	127	75
Mong Hsu 1						
Rims, range	27–75	336–3201	73–628	429–3545	10–32	48–78
n = 7, average	48	1410	391	2026	14	69
Cores, range	42–95	1168–2550	75–649	919–4625	<8–17	48–80
n = 4, average	70	1815	324	3199	<9	66
Mong Hsu 2						
Rims, range	20–147	42–1492	183–1012	959–16,388	11–51	68–105
n = 14, average	68	778	342	5600	29	83
Cores, range	41–226	100–2633	228–658	1194–27,386	15–54	81–103
n = 8, average	115	1210	417	7015	26	89

Data based on detailed analyses listed in Appendix A Tables A1 and A2. < = value bdl.

The main chromophore Cr is distinctly higher in Mong Hsu set 2 rubies (max. ~2.7 wt %, av. 0.65 wt %), than in Mong Hsu set 1 (max. 0.45 wt %, av. 0.23 wt %) and Turein Taung (max. 0.46 wt %, av. 0.25 wt %) rubies. One set 2 zoned crystal is significantly more enriched in Cr (max. 2.7 wt %, av. 1.9 wt %) than for 7 others in the set (max. 1 wt %, av. 0.44 wt %). The most variable chromophore is Ti, which is lowest in Thurein Taung rubies (max. 1250 ppm, av. 265 ppm) compared with Mong Hsu set 1 (max. 3200 ppm, av. 1362 ppm) and Mong Hsu set 2 (max. 2633 ppm, av. 935 ppm) rubies.

In Turein Taung rubies, average rim ppm values in Cr are less than average core values (− values), but are greater in Mg, Ti, V, Fe, Ga rim values (+ values). The rubies, however, are relatively homogenous with rim to core zones showing limited ppm variations (Mg +7; Ti +115, V +18; Cr −26; Fe +16; Ga +4). In contrast, wider average zonal spreads appear in ruby sets Mong Hsu 1 (Mg −22; Ti −405; V +67; Cr −1173; Fe + >9, Ga +3) and Mong Hsu 2 (Mg −47; Ti −432; V −75, Cr −1415; Fe +3; Ga − 6).

Thurein Taung V values at 102–421 ppm lie well below more extreme V values at 900–5500 ppm in nearby parts of the Mogok ruby field [11,13]. Thurein Taung Cr/V ratios, however, range up to 37 and suggest links with the Mogok V-rich ruby province. In comparison with Thurein Taung, the Mong Hsu ruby sets include higher V values up to ~1012 ppm and range into even higher Cr/V ratios (Mong Hsu 1, 1.2–37.1, av. 11.3; Mong Hsu 2, 4.0–59.5, av. 18.0).

3.2. U-Pb Ages

All zircon and titanite analyses were done in-situ, as small grains present as inclusions within corundum megacryst hosts. Baddeleyite was analyzed as large euhedral megacrysts. All age results and all secondary reference material analyses are presented in Supplementary Materials. Zircon and ilmenite inclusions were targeted for U-Pb analysis in the Mong Hsu rubies. These inclusions were generally >30 μm in size and euhedral in shape. Ilmenite analyses contained modest U contents (<70 ppm), however, they contained significant common Pb and so were useful as a common Pb anchor point for the zircon analyses. The zircons have significantly high U, up to 3000 ppm, and high U/Th ratios (up to 40). Analyses show significant variation in common Pb corrected age (207Pb correction), often with variable Pb/U ages within single analyses. A Concordia intercept age of 23.9 Ma (± 1.0/1.03 Ma, including systematic uncertainties respectively), calculated on the four youngest zircons formed a coherent population (MSWD 1.10, probability of fit 0.35) with ilmenite analyses included for use as a common Pb anchor (Figure 11).

Zircon and titanite inclusions were also targeted for U-Pb analysis in the Thurein Taung ruby. The zircons were generally <30 μm in size, while the titanite was a single grain ~250 by ~150 μm in size and euhedral in shape. Multiple analyses were done on this single grain to constrain the age as no other titanite grains were exposed at the surface. A Concordia intercept age of 32.34 Ma (± 0.97/1.03 Ma) was calculated on the 7 titanite analyses on this single grain with the age anchored on a ^{207}Pb/^{206}Pb ratio of 0.837 (Figure 12). Three zircons included in the same ruby megacryst as the titanite give a range of older ages from ~50 to ~100 Ma, despite being a few millimeters away from the titanite.

Two baddeleyite megacrysts from the Thurein Tuang, ~5 × 10 mm in size, were analyzed targeting both rims and cores. A wide range of ages were measured in the two megacrysts from ~40 Ma to ~110 Ma. Eight of the analyses formed a coherent age population at 103.3 Ma (± 2.2/2.45 Ma). There was no obvious correlation with the measured age and location (core vs. rim) of the baddeleyite, with some rims containing the ~103 Ma age population and some cores containing significantly younger ages. Likewise, there is no correlation with the trace element chemistry of the baddeleyite and the measured ages.

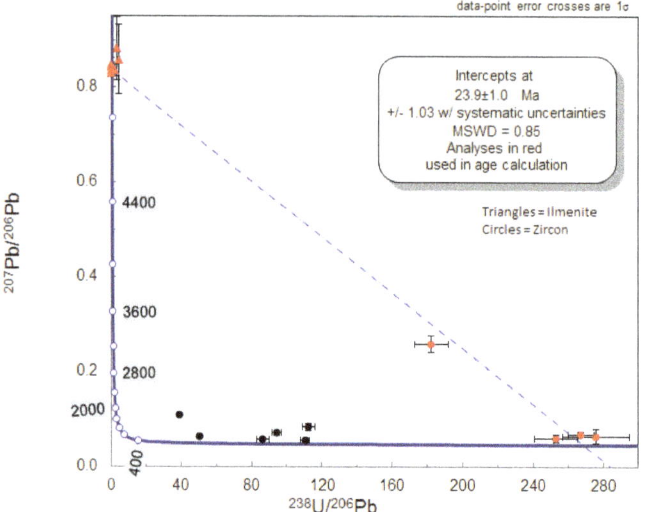

Figure 11. Concordia diagram showing zircon plots (red spots) used to constrain a concordia age of the four youngest zircons within the Mong Hsu host ruby. Age measurements on two baddeleyite megacrysts from the site (grey spots) are shown, but relationships to Mong Hsu ruby are uncertain.

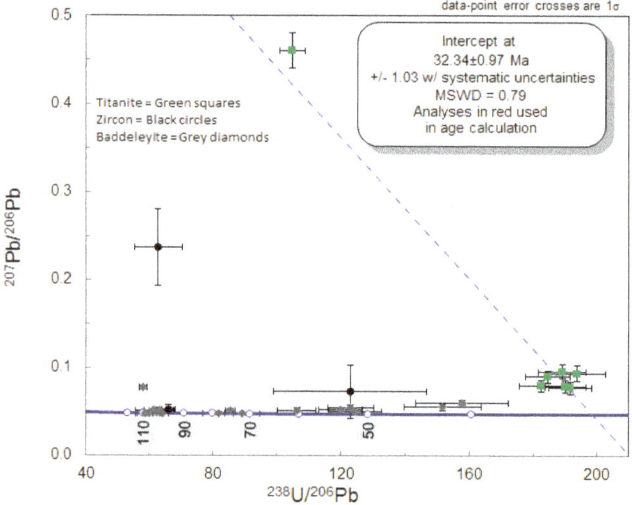

Figure 12. Concordia diagram showing titanite plots (green spots) and error bars constraining the intercept age of the titanite composite inclusion in Thurein Taung ruby. Comparative age plots are shown for older zircon included ruby (grey) and baddeleyite megacrysts (black) from Thurein Taung.

Trace elements in the young 23–27 Ma zircon in Mong Hsu ruby were analyzed by LA-ICP-MS during age-dating (Supplementary Materials). Three zircons showed consistent results, based on an assumed Zr content of 493,000 ppm, with Hf 17,334–18,659, U 2012–3407, Th 155–241, U/Th 13.6–14.1, Y 99–161 and very low LREE-MREE and enriched HREE. One zircon contains a Ti-rich inclusion, probably rutile that skewed results, giving excessive Ti and reducing other values. The ilmenite used as an anchor for common Pb ranged in Ti/Fe ratios between 0.47–3.73 and contained minor amounts of P, Zr, Nb and LREE.

3.3. Mineral Inclusion Analyses

Mineral phases distributed in the dated composite titanite inclusion and dated zircon xenocrysts within host Thurein Taung ruby were targeted for EDS analysis using BSE images (Figures 13 and 14; Table 2). The composite inclusion is an elongated subhedral, titanite crystal ~270 μm long, intergrown with two euhedral nepheline crystals, 30 and 65 μm across, on its margins that constitute ~4% of the inclusion. Mineral formulas given in Table 2 were based on O calculated by cation stoichiometry, with titanite standardized on Si.

Detailed LA-ICP-MS trace elements including REE were acquired for the Thurein Taung composite titanite and zircon inclusions (Supplementary Materials). Titanite analyses (n = 7), based on an assumed Ti content of 181,607 ppm, contain accessory ppm of P (300–837), Fe (134–171), Y (555–583), Zr (1100–1997), Nb (828–890), Ta (53–59), Th (651–733), U (248–3206), and have av. Th/U (2.6). Analyses are moderately elevated in LREE-HREE (La–Lu 1726–1906). Zircon analyses, using assumed Zr of 493,000 ppm, are enriched in Hf (16,890–19,504), show minor elevations in Ti, (27–61), Fe (<14–165), Y (57–66), Th (22–54), with av. Th/U (0.35). They are noticeably low in LREE-MREE but show mild enrichment in HREE. Baddeleyite analyses, using assumed Zr of 740,000 ppm, are enriched in Hf 16890–1950), have elevated Ti (1537–2813), Nb (308–1147), Ta (262–525), minor P (117), Pb (19–118) and negligible REE and Th.

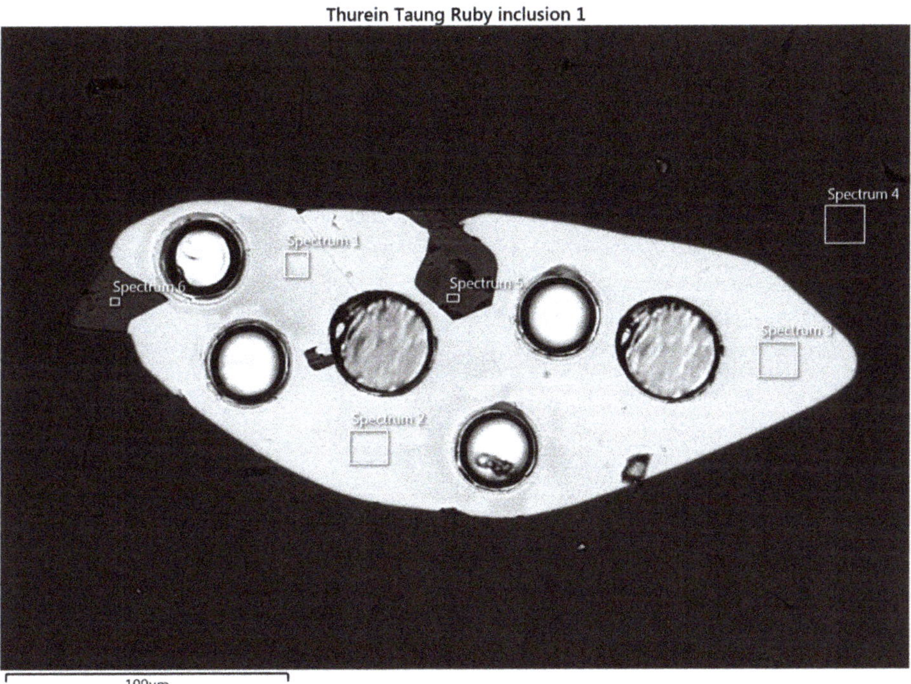

Figure 13. BSE image of titanite-nepheline inclusion (light grey) in Thurein Taung ruby (dark grey background). Darker euhedral indents in the inclusion (top and left) are nepheline. Circular pits mark age-dating sites, while EDS spectrum sites (rectangular boxes) represent titanite analyses, listed in Table 2).

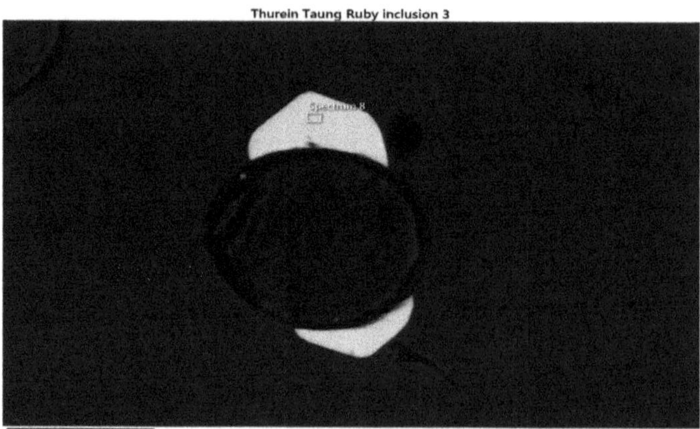

Figure 14. Zircon crystals in Thurein Taung ruby (background), showing circular age-dating pits and EDS spectrum sites (rectangular boxes) represented in analyses listed in Table 2.

Table 2. EDS element analyses (wt %), inclusion mineral phases, Thurein Taung ruby.

Titanite (Composite Inclusion) [1] $Ca_{4.00}$ $(Ti_{3.35}, Al_{0.68})_{4.03}$ Si_4 $(O_{19.72}, F_{0.55})_{20.27}$.					
Element (σ)	O (0.4)	Si (0.1)	Al (0.1)	Ca (0.1)	Ti (0.1)	F (0.1)
Av., n = 3	41.7	14.2	2.3	20.3	20.3	1.3

Nepheline (composite inclusion) [2] $(Na_{0.69}, K_{0.07}, Ca_{0.08})_{0.91}$ $(Si_{1.1} Al_{0.9})_{2.0}$ O_4.					
Element (σ)	O (0.2)	Si (0.1)	Al (0.1)	Ca ($<$ 0.1)	Na (0.1)	K ($<$ 0.1)
Av., n = 3	45.3	20.9	17.9	2.2	11.2	1.8

Zircon 1 (xenocryst) [3] $(Zr_{1.06}, Hf_{0.01})_{1.07}$ $Si_{01.05}$ O_4.				
Element (σ)	O (0.2)	Si (0.1)	Al (0.0)	Zr (0.3)	Hf (0.2)
Av., n = 1	32.4	14.9	bdl	48.8	1.2

Zircon 2 (xenocryst) [4] $(Zr_{1.00}, Hf_{0.01})_{1.01}$ $Si_{0.99}$ O_4.				
Element (σ)	O (0.1)	Si (0.1)	Al (0.0)	Zr (0.3)	Hf (0.2)
Av., n = 1	31.4	14.9	bdl	48.1	1.2

[1,2,3,4] Calculated mineral formula of analyzed mineral. bdl below detection limit.

4. Discussion

The gem mineralization associated with the Myanmar ruby deposits is inherited from the stratigraphy and dynamic tectonic events that built the geological framework within the gem tracts [59,61,62]. The initial discussion considers the age-dating results on inclusions within Myanmar rubies from the two studied deposits and examines their relationships to previous dating in ruby deposits in SE Asia. This is followed up by synthesizing the trace element results from the Myanmar sites and Australian ruby suite data within a range of genetic classification diagrams. With characterization, the suites will be examined for their geochemical traits that enable distinctions between these suites. Finally, comparative elemental 'fingerprinting' is considered in a broader context of geographic typing of rubies within a global perspective.

4.1. Mogok and Mong Hsu Ruby Ages

The U-Pb age of the Thurein Taung composite titanite inclusion at 32.4 ± 1 Ma, matches the age of small leucocratic granite bodies that intrude the area [25]. The near-euhedral crystal form suggests a likely syngenetic origin and a host ruby age linked to alkaline melts and skarn formation associated with that leucocratic granite event. The presence of subordinate nepheline in the titanite composite, however, needs consideration in respect to a potential under-saturated alkaline source. Nepheline-bearing rocks are known to intrude and be faulted against the local marbles in this vicinity [25]. These urtite-ijolite members and syenite rocks, however, seem younger than the composite inclusion age. This is based on a 25 Ma age for a foliated syenite and observations that the syenites themselves are intruded by the urtites and by pegmatites associated with the Kabaing granite, dated at 16.8 ± 0.5 Ma [25,61]. This, in conjunction with a zircon U-Pb age of 16.1 Ma in a painite overgrowth on ruby from Wetloo [25], suggests multiple periods of ruby growth may have taken place in the Mogok skarn environment [63]. This aspect is further supported by Ar-Ar age dating of phlogopite mica in syngenetic growth with ruby in the Mogok field at 17.1 ± 0.2 Ma–18.7 ± 0.2 Ma [33]. U-Pb dating of zircon inclusions is rare in constraining Mogok ruby ages, but has been used on syngenetic zircon to date quality Mogok sapphires [64]. Mogok sapphires have links with under-saturated alkaline intrusions [25]. Age dates obtained so far for host sapphires [64] gave 26.7 ± 4.2 and 27.5 ± 2.80), close to, but just outside error for the Thurein Taung composite inclusion age (this study).

Mong Hsu zoned ruby-sapphire crystals are found in host veins composed of calcite, a Mg-rich chlorite group member, Cr-bearing muscovite and opaque oxides, which traverse recrystallized dolomitic marble [37], or occur in disseminated grains in a dominant dolomite matrix with minor calcite and intergranular phlogopite [65]. Solid inclusions within the rubies themselves identified by laser Raman spectroscopy included Cr-bearing muscovite, paragonite-margarite solid solution, Mg-rich chlorite, rutile, quartz, dolomite and diaspore [66]. In addition, the present study identified zircon and ilmenite which enabled U-Pb age determination of the host ruby age. The Mong Hsu ruby age formation, based on the young zircon inclusions, at 23.9 ± 1.0 Ma is clearly younger than the Thurein Taung titanite inclusion age by some 8–9 Ma. The later age for Mong Hsu ruby generation fits in with a declining phase of regional metamorphism [36,61], within fluid-rich conditions which included H_2O, CO_2, F and other components [37]. An isochore estimation from fluid inclusion compositions within the ruby gave a P-T range of 0.20–0.25 GPa and 500–550 °C for the fluid activity, which may represent hydrothermal input from deeper magmatism [36]. This low P-T range for ruby growth was questioned from zircon-in-rutile thermometry on an inclusion in a ruby. The estimated T between 615–690 °C suggested upper amphibolite facies T in the 0.4–0.6 GPa range [66]. The present U-Pb zircon dating now yields a firmer early Miocene 24 Ma age for this metamorphic/magmatic event.

4.2. Myanmar and East Australian Ruby Trace Element Comparisons

Mong Hsu ruby crystal growth and trace element analysis was used to establish a geochemical identity [66]. The study recorded dark cores, which were elevated in Cr, V, Mg, Ta, W and Th values. This trend also was noted here, in a core significantly enriched (ppm) in Cr 10,036, Ti 1272, V 909, and Mg 1272, although Ta, W and Th remained low. The earlier study suggested a distinctive geochemical field for Mong Hsu ruby, compared with other ruby fields, when Cr/Ga was plotted against Fe/(V + Ti). Strangely, however, the compared fields did not include Mogok ruby. To further test the claim, ruby analyses from Mong Hsu, Thurein Taung, Mogok, from this study and data from New England, East Australia [48] are compared with the proposed Mong Hsu field in Figure 15.

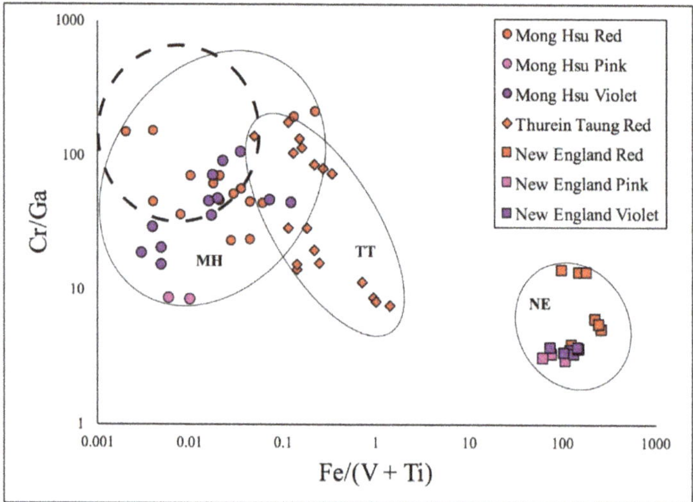

Figure 15. Ruby plots of Mong Hsu (MH), Thurein Taung (TT), Myanmar, and New England, East Australia (NE) are shown within oval enclosing field boundaries in a Cr/Ga vs. Fe/(V + Ti) diagram. The plots are color- and locality- coded to indicate multi-zoned crystals (top right legend). The Mong Hsu ruby field designated by Mittermayr et al. [66] is indicated by a dashed circle.

The Mong Hsu ruby plots (red zones) from the present study largely occupy the defined limits for the proposed Mong Hsu field. The violet-blue zones, however, mostly lie outside the field with lower Cr/Ga values, while pink variants show even lower Cr/Ga. Thurein Taung plots suggests some separation between two subset components. The diagram shows good separation between Mong Hsu and Thurein Taung ruby fields and extreme separation from the New England field. These trace element ratios thus appear effective in establishing a Mong Hsu ruby identity.

Further relationships between Myanmar and East Australian ruby suites are explored based on element oxide indices that classify corundum into genetic fields related to host rock type associations [3]. Comparative results are shown in Figures 16–18. In Mong Hsu rubies red zones mostly plot in the ruby-in marble field, dark violet-blue zones mostly fall in the mafic-ultramafics field and pink to white zones lie near the boundary region. These plots correspond to the R, V and I zones described by Peretti et al. [37], considered to mark repeated influx of fluids rich in F. The F interaction decreased the Ti chromophore activity through a complexing process and effectively raised Cr chromophore ratios to color red zones. The new Mong Hsu plots support multiple interplay between separate source components, involving marble host lithologies and invading fluids from deeper seated magmatic/metamorphic sources to generate multiple color zoned corundum. The Thurein Taung plots, in contrast, nearly all closely group in the ruby-in marble field, which suggests a simpler origin.

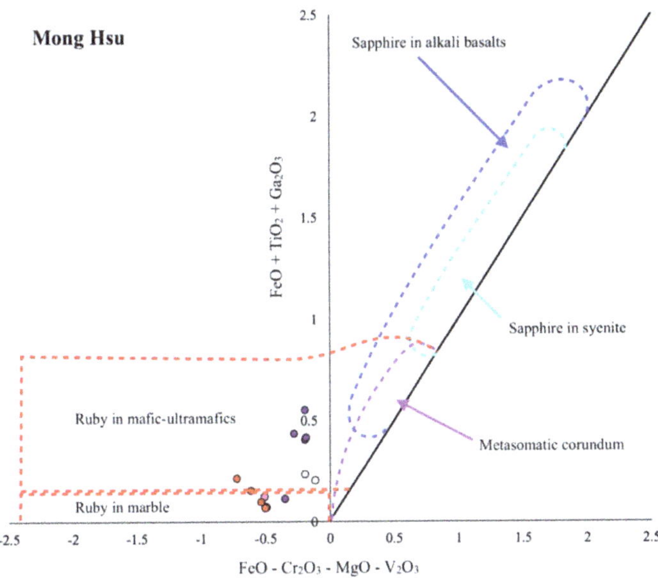

Figure 16. Discrimination metal oxide plots, with petrological fields after [3], for Mong Hsu ruby analyses. The plots are color-coded in relation to red, violet-blue-black and white zoning.

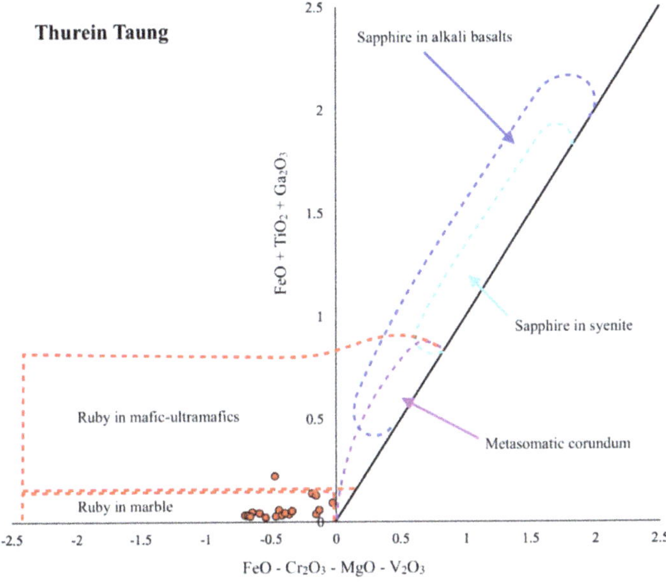

Figure 17. Discrimination metal oxide plots, with petrological fields after [3], for Thurein Taung ruby analyses. The plots are color-coded.

East Australian plots (Figure 18) present varied lithological affinities. Barrington Tops ruby has metasomatic corundum affinities. Macquarie River ruby in contrast forms a tight linear array near the base of the Ruby-in-mafic-ultramafics field, while the multiple color zone components of New England ruby spread along the Metasomatic corundum field to slightly overlap the Ruby-in-mafic-ultramafics field. Myanmar ruby fields have strong connections with carbonate sources, unlike East Australian ruby sources, although both groups include some mafic-ultramafic source inputs.

A trace element variation diagram, introduced as a tool to separate blue-colored sapphires of metamorphic, transitional and magmatic origin, by Peucat et al. [67], plotted Fe (ppm) content against Ga/Mg ratio. This tool received wider use being extended to sapphire of other colors and even ruby, as it is normally a metamorphic/metasomatic corundum and seemed to conform to the classification boundaries. [9,68,69]. Recent studies show that individual corundum crystals can show significant ranges in their Ga/Mg ratio [70], and that rare rubies from Myanmar and East Australia show high Ga/Mg ratios that lie in the magmatic field [13,48]. Furthermore, some sapphire/ruby suites with low Ga/Mg, previously thought metamorphic can contain melt inclusions suggesting a magmatic origin [71]. This makes the Ga/Mg ratios considered in isolation unreliable genetic indicators. Plots of Fe (or Fe/Mg) against Ga/Mg, nevertheless, remain useful in aiding distinctions between fields from different localities (Figure 19).

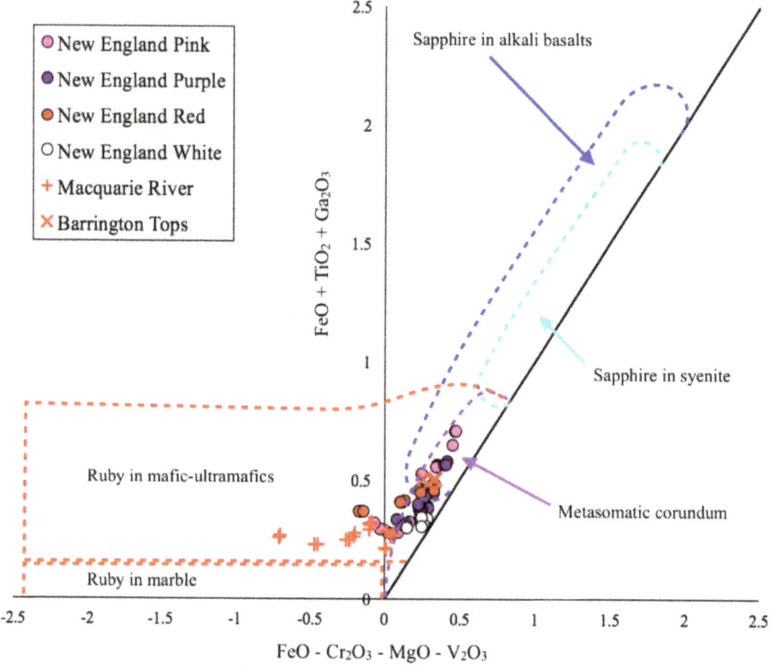

Figure 18. Discrimination diagram of metal oxide plots, within petrological fields after [3], for East Australian plots from Barrington Tops, Macquarie River and New England rubies. Plots are color coded to show the range of color and zoning. Symbols for different localities are shown in the legend.

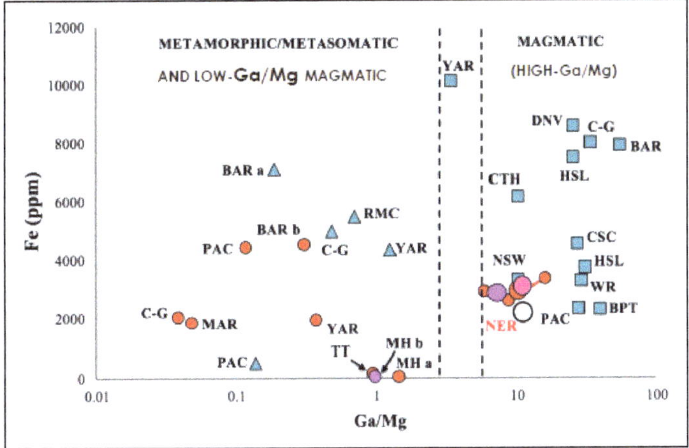

Figure 19. Fe-Ga/Mg diagram after [55], showing averaged plots from selected examples of ruby and sapphire suites from East Asian, East Australian and other localities, from data after [9,11,59,72]. Short dash lines mark a transitional zone (Ga/Mg 3–6) for Metamorphic/Magmatic fields [66]. Symbols include: Red-filled circles, ruby; Red-violet-white filled circles, zoned ruby-sapphire suites; Blue-filled triangles, low-Ga meta-sapphire suites; Blue-filled squares, higher-Ga sapphire suites. East Asian locality identity letters include: BT, Bo Phloi, Thailand, sapphire; CTH, Chanthanburi, Thailand sapphire; CSC, Changle, Shadong, China, sapphire; DNV, Dak Nong, Vietnam sapphire; MH, Mong Hsu, Myanmar; zoned ruby-sapphire suite; PAC, Pailin (west), Cambodia, ruby and sapphire suites; TT, Thurein Taung, Myanmar, ruby suite. East Australian locality identity letters are: BAR, Barrington Tops, magmatic sapphire, (a) meta-sapphire (b) ruby; C-G, Cudgegong-Gulgong, ruby, sapphire; MAR, Macquarie River, ruby, meta-sapphire, high-Ga sapphire; NSW, New South Wales, high-Ga sapphire; NER, New England, zoned ruby-sapphire; YAR, Yarrowitch, ruby, meta-sapphire, transitional sapphire, WR, Weld River, Tasmania, high Ga-sapphire. Other: RMC, River Mayo, Colombia, South America sapphire.

The studied Thurein Taung and Mong Hsu, Myanmar suites (Figure 19) show very low Fe and cluster at Ga/Mg ~1–2. Note that dark violet zones of Mong Hsu ruby-sapphire (MHb) separate into a slightly lower Ga/Mg group than for ruby zones. These carbonate- interacting Myanmar ruby suites are quite distinct in values from West Pailin, Cambodian ruby suite (PAC). East Australian ruby suites, in contrast, have a more extended separation and diversity than the plotted SE Asian ruby suites, extending in Fe values from ~1800–4200 ppm and Ga/Mg range from ~0.054 to 25.

Thurein Taung ruby is just one locality among many ruby deposits in the Mogok gem tract for which LA-ICP-MS analyses are available for comparison [11,13,59]. One Mogok site, Le-U, has ruby notably high in Si (1060–4290 ppm), trace B (10–35 ppm) and Sn (2–18 ppm) and $\delta^{18}O$ (20.4 mil %), and was attributed to a skarn origin [13]. This ruby is also exceptional in its high Ga contents (~370–720 ppm) and Ga/Mg ratios (46–521), exceeding those of the New England, Australian high-Ga ruby (compare Figures 19 and 20). In a further variation diagram, Le-U ruby with V + Cr (2270–31,890 ppm) plotted against Fe + Ti (4–12 ppm) falls in a distinct field outside other Mogok site plots (Figure 21). This is quite dissimilar to the Thurein Taung data (V + Cr 542–4744 ppm; Fe + Ti 82–2295, which extends across the general Mogok High-V metamorphic ruby field.

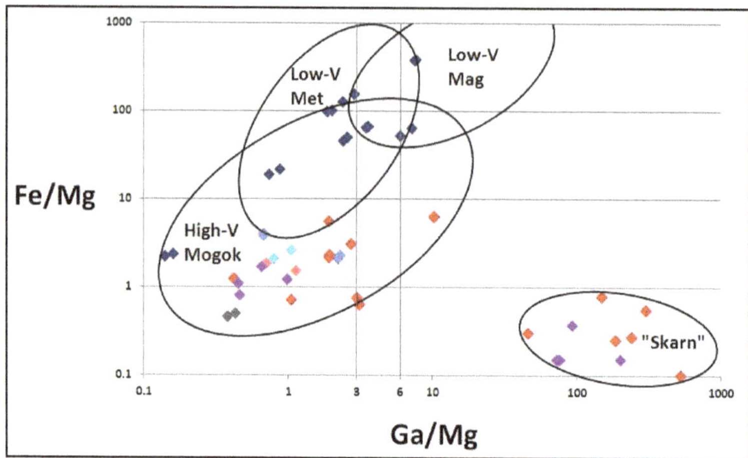

Figure 20. Fe/Mg vs. Ga/Mg plots Mogok corundum, as color coded plots within outlined fields for Mogok data [11,13]. Note extreme Ga/Mg ratios in a proposed "skarn" field for Le-U ruby (red diamonds) and Ohn-bin-ywe-htwet sapphire (mauve diamonds) [13], compared with Ga/Mg ratios plotted for Thurein Taung and Mong Hsu, Myanmar, Barrington Tops, Macquarie River and New England, East Australian ruby suites (Figure 18).

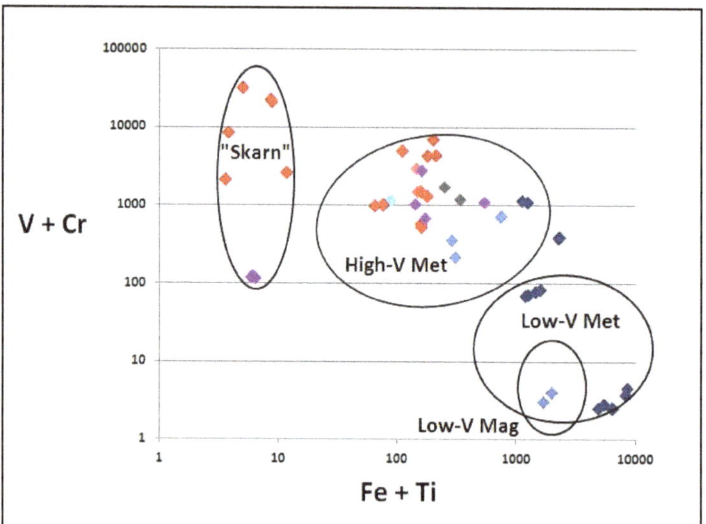

Figure 21. V + Cr vs. Fe + Ti diagram showing Mogok gem corundum, as color coded plots, as in Figure 19, within outlined fields for Mogok data from [11,13]. Note the extreme V + Cr enrichment and extreme low Fe + Ti for the suggested Mogok "Skarn field". Such low Fe + Ti is not shared by the other Myanmar or East Australian ruby data in this paper or in the cited literature.

4.3. Ruby Diversity and Geographic Typing

This study examined two ruby sites from Myanmar and three ruby sites from East Australia, using rubies derived from quite diverse geological settings. Some ruby suites came from known host rock settings and others from alluvial deposits as xenocrysts transported by basalts that erupted through hidden fold-belt settings. The color zoning and geochemistry of the different ruby suites and their age formation relationships show considerable diversity within in both inter- and intra-continental settings. The diversity described here is only a limited sample of the potential diversity of ruby deposits world-wide, as revealed in a selected range of major surveys [2,3,15,34,72–76].

Several ruby suites within this intra-continental comparison show unusual characteristics that seem ideal for geographic typing. The Mong Hsu zoned ruby-sapphire stands out, not only for its contrasting dark cores and red rims, but also for added complexity. The zoning also includes blue patches that differ from the dark cores in showing O–H stretching, when examined by FTIR absorption spectrometry [77]. Within some Mogok rubies, very high V levels, exceeding 30,000–50,000 ppm, with V/Cr > 1–26, as at Sin-Khwa, would be distinctive [11]. Furthermore, extreme Si-Ca-Ga-enriched ruby, as at Le-u seems unique [13]. The Le-u ruby was retrieved from an alluvial placer, where rubies were derived from marbles that had interacted with CO_2-rich fluids and nearby syenite intrusion [78]. Among East Australian rubies, Barrington Tops, Macquarie River and New England suites all show distinctive features. Barrington Tops ruby composites, although similar in mineralogy and trace element chemistry to West Pailin, Cambodian examples [79] differ from them in their $\delta^{18}O$ isotope ranges [9]. Macquarie River ruby shows extreme Mg enrichment relative to Fe and Ti [9] and in a Mg ($\times 100$)–Fe–Ti ($\times 10$) diagram all plots fell in the end-member Mg apex (Mg 90–100, Fe 0–10, Ti 0–10). New England color-zoned, Ga-enriched ruby, although approached in Ga values elsewhere, e. g., East Africa, differs in Cr-colored red cores trending to non-red rims [48], rather than the reverse trend [80].

A plethora of detailed data now exists on ruby characteristics available from many global sites, as exploration, mining and gem researchers, gem institutions and trade laboratories continue to open-up studies on this valuable gemstone [2,81–83]. This, with improved techniques of characterizing ruby through analytical advances [84], and more sophisticated statistical methods of identifying links/differences in analytical data from known sources [85–88], will enlighten future studies. Detailed mineral assemblage studies and P-T conditions of their formation within ruby-host lithology will become further refined to use as potential exploration vectors for new ruby deposits [5].

5. Conclusions

Several ruby deposits in two different tectonic terranes provide diverse ages and chemistry.

Myanmar ruby ages of ~32 Ma at Mogok and at 23 Ma at Mong Hsu reflect the post-Paleozoic collision history and felsic magmatic activity in Myanmar, while older diverse ruby generated within dismembered fold belts in East Australia were delivered to surface placers by post-rift basalts.

Comparisons of ruby trace element chemistry from the Myanmar and Australian suites show considerable geographic distinctions, with carbonate and mafic-ultramafic affinities for Myanmar suites, and metamorphic and metasomatic affinities dominating in Australia.

Both Myanmar and East Australia ruby suites include examples of unusual Ga-rich ruby which appear to indicate generation involving magmatic inputs.

Considering the diversity and chemical distinctions shown by the Myanmar and Australian comparisons, in the context of global ruby distribution and detailed data accumulating on many ruby suites, geographic typing of ruby origins seems to be a viable future proposition.

Supplementary Materials: The following are available online at http://www.mdpi.com/2075-163X/9/1/28/s1.

Author Contributions: F.L.S. assembled the East Australian gem data, compared it with Mogok, and Mong Hsu data, interpreted the results of the analyses and wrote the script. K.Z. drafted the manuscript, provided Myanmar expertise and funded analytical work. S.M., J.T., K.G. and M.M.Z. provided technical input and ran the LA-ICP-MS and SEM geo-chronological and chemical analytical work. T.T.N. provided geology background, preliminary data on Mong Hsu ruby and additional samples and technical input. K.T. provided samples for analysis and field data on Mogok localities. S.J.H. provided data on New England ruby and plotted Myanmar and Australian diagrams.

Acknowledgments: The Editor of *Minerals* and Guest Editors are thanked for organizing the special issue on "Mineralogy and Geochemistry of Gems" and extending an invitation to submit a contribution for consideration. This particularly includes the Australian guest editor, Ian Graham, School of Biological, Earth and Environmental Sciences, University of New South Wales, Sydney. The pre-submission draft manuscript was read by Ross Pogson, Geoscience, The Australian Museum, Sydney. Bruce Wyatt, Port Macquarie, New South Wales, Australia assisted in tabulation of analytical results. Paula Piilonen, Canadian Museum of Nature, Ottawa, allowed use of an Asian–Australian gem deposit map for modification as Figure 1 in this study. Three reviewers are thanked for constructive comments and editorial staff provided helpful handling of the manuscript for publication.

Conflicts of Interest: The authors declare no conflict of interest.

Appendix A

Table A1. LA-ICP-MS trace element values (ppm) [1], Thurein Taung ruby.

Crystal No. Spot	Mg	Ti	V	Cr	Fe	Ga
1.1 (478), rim	116	171	112	4191	60	50
1.2 (479), core	94	142	102	3841	65	48
2.1 (480), core	50	200	243	2883	79	101
2.2 (481), rim	49	329	250	2785	65	98
3.1 (482), rim	45	63	310	2306	122	32
3.2 (483), core	63	94	386	618	117	39
4.1 (484), rim	27	38	281	2183	44	155
4.2 (485), core	40	64	296	2340	49	152
5.1 (486), rim	104	1241	401	2731	80	20
5.2 (487), core	100	203	421	2917	91	22
6.1 (488), rim	204	339	190	1073	483	125
6.2 (489), core	189	283	185	1035	461	128
7.1 (490), rim	63	96	145	4599	38	41
7.2 (491), core	77	120	156	4406	35	43
8.1 (492), rim	69	99	110	432	45	22
8.2 (493), core	59	93	125	3505	24	20
9.1 (494), rim	88	131	386	1909	353	170
9.2 (495), core	35	53	110	902	225	119

[1] Values based on assumed Al content of 52,900 ppm. < = value bdl.

Table A2. LA-ICP-MS trace element values (ppm) [1], Mong Hsu ruby sample sets.

Crystal No. Spot	Mg	Ti	V	Cr	Fe	Ga
Sample set 1						
1. rim1 (red)	27	336	231	3454	13	75
1. core (blue grey)	76	1168	254	4625	14	66
1. rim2 (red)	62	487	238	3342	32	75
2. rim1 (red)	35	654	628	2815	10	78
2. core (red)	67	1397	649	4470	<8	71
2. rim2 (red)	34	580	595	2780	19	76
3. rim (violet)	75	3201	73	1393	22	76
3. core (violet)	42	2550	75	2780	17	80
4. rim1 (light red)	55	1329	304	446	9	53
4. core (light red)	95	2145	316	919	<9	48
4. rim2 (light red)	49	1137	352	429	15	52
Sample set 2a						
1. rim (red)	37	383	187	4476	38	81
1. core (blue black)	84	585	658	9698	54	94
2. rim (red)	73	1285	301	5524	18	81
2. core (blue black)	149	2633	348	6736	14	85
3. rim1 (red)	147	1367	1012	7771	51	88
3. core (blue black)	187	1358	398	8180	37	85
3. rim2 (red)	42	616	270	2033	20	86
4. rim1 (red)	131	2079	306	2705	37	78
4. core (blue black)	58	889	349	2011	31	88
4. rim2 (red)	29	324	317	1961	40	92

Table A2. *Cont.*

Crystal No. Spot	Mg	Ti	V	Cr	Fe	Ga
Sample set 2b						
1. rim1 (red), #18	29	42	320	16,388	31	88
1. core (blue black), #17	41	100	591	27,386	33	81
1. rim2 (red), #19	20	63	269	16,013	42	77
2. rim1 (red), #24	69	1050	352	2994	23	75
2. core (blue black), #23	226	2296	508	5871	44	85
2. rim2 (red), #25	146	1136	465	6919	34	75
3. rim1 (red), #27	47	794	240	959	11	92
3. core (blue black); #26	103	681	257	1194	17	103
3. rim2 (red), #28	47	77	368	3195	12	105
4. rim1 (light red), #30	24	189	183	4811	12	68
4. core (light red), #29	75	1136	228	3226	15	87
4. rim2 (light red), #31.	104	1492	195	2607	34	78

[1] Values based on assumed Al of 52,300 ppm. < = value bdl. # spot no., in Figure 6.

References

1. Themelis, T. *Gems and Mines of Mogok*; A & T Publishing: Los Angeles, CA, USA, 2008; p. 356.

2. Hughes, R.W. *Ruby & Sapphire—A Collector's Guide*; Gem & Jewelry Institute of Thailand: Bangkok, Thailand, 2014; p. 384.

3. Giuliani, G.; Ohnenstetter, D.; Fallick, A.E.; Groat, L.; Fagan, A.J. The geology and genesis of gem corundum Deposits. In *Geology of Gem Deposits*, 2nd ed.; Mineralogical Association of Canada Short Course: Quebec, QC, Canada, 2014; Volume 44, pp. 23–78.

4. Sutherland, L.; Graham, I.; Harris, S.; Zaw, K.; Meffre, S.; Coldham, T.; Coenraads, R.; Sutherland, G. Rubis australasiens. *Rev. Assoc. Fr. Gemmol.* **2016**, *197*, 13–20.

5. Yakymchuk, C.; Szilas, K. Corundum formation by metasomatic reactions in Archean metapelite, SW Greenland: Exploration vectors for ruby deposits within high-grade greenstone belts. *Geosci. Front.* **2018**, *9*, 727–749. [CrossRef]

6. Saul, J.M.A. *A Geologist Speculates*; Les 3 Colonnes: Paris, France, 2014; p. 159.

7. Giuliani, G.; Fallick, A.; Rakatondrazafy, M.; Ohnenstetter, D.; Andriamamonjy, A.; Ralantarison, T.H.; Offant, Y.; Garnier, V.; Dunbigre, C.H.; Schwarz, D.; et al. Oxygen isotope systematics of gem corundum deposits in Madagascar. *Miner. Depos.* **2007**, *42*, 251–270. [CrossRef]

8. Giuliani, G.; Fallick, A.E.; Ohnenstetter, D.; Pegre, G. Oxygen isotope composition of sapphire from the French Massif Central: Implications for the origin of gem corundum in basalt fields. *Miner. Depos.* **2009**, *44*, 221–231. [CrossRef]

9. Sutherland, F.L.; Zaw, K.; Meffre, S.; Giuliani, G.; Fallick, A.E.; Graham, I.T.; Webb, G.B. Gem corundum megacrysts from East Australia basalt fields: Trace elements, O isotopes and origins. *Aust. J. Earth Sci.* **2009**, *56*, 1003–1020. [CrossRef]

10. Zaw, K.; Sutherland, L.; Dellapasqua, F.; Ryan, C.G.; Yui, T.-Z.; Mernagh, T.P.; Duncan, D. Contrasts in gem corundum characteristics, eastern Australian basaltic fields: Trace elements, fluid/melt inclusions and oxygen isotopes. *Mineral. Mag.* **2006**, *70*, 669–687. [CrossRef]

11. Zaw, K.; Sutherland, L.; Yui, T.-F.; Meffre, S.; Thu, K. Vanadium-rich ruby and sapphire within Mogok Gemfield, Myanmar: Implications for gem color and genesis. *Miner. Depos.* **2014**, *50*. [CrossRef]

12. Muhleimster, S.; Fritsch, E.; Shigley, J.E.; Devourd, B.; Laurs, B.M. Seperating natural and synthetic rubies on the basis of trace-element geochemistry. *Gems Gemol.* **1998**, *34*, 80–101. [CrossRef]

13. Sutherland, L.; Zaw, K.; Meffre, S.; Yui, T.-F.; Thu, K. Advances in trace element "Fingerprinting" of gem corundum, ruby and sapphire, Mogok Area, Myanmar. *Minerals* **2015**, *5*, 61–79. [CrossRef]

14. Sutherland, F.L.; Abduriyim, A. geographic typing of gem corundum: A test case from Australia. *J. Gemmol.* **2009**, *31*, 203–210. [CrossRef]

15. Giuliani, G.; Ohnenstetter, D.; Fallick, A.E.; Feneyrol, J. Les Isotopes de l'oxygèn, un tracer des origins géologique et/ou géographique de gemmes. *Rev. Assoc. Fr. Gemmol.* **2012**, *179*, 11–16.

16. Giuliani, G.; Ohnenstetter, D.; Fallick, A.E.; Groat, L.A. Geographic origin of gems tied to their geologic history. *InColor* **2012**, *19*, 16–27.
17. Saul, J.M. *A Geologist Speculates*, 2nd ed.; Les 3 Colonnes: Paris, France, 2014; p. 160.
18. Link, K. Age determination of zircon inclusions in faceted sapphires. *J. Gemmol.* **2015**, *34*, 692–700. [CrossRef]
19. Sorokina, E.S.; Rösel, D.; Häger, T.; Mertz-Kraus, R. LA-ICP-MS U–Pb dating of rutile inclusions within corundum (ruby and sapphire): New constraints on the Mozambique Belt. *Miner. Depos.* **2017**, *52*, 641–649. [CrossRef]
20. Gübelin, E.J.; Koivula, J.I. *Photoatlas of Inclusions in Gemstones*; Opinio Verlag: Basel, Switzerland, 2008; Volume 3, p. 672.
21. Hughes, R.W. *Ruby & Sapphire: A Gemologists Guide*; RWH Publishing: Bangkok, Thailand, 2017; p. 816.
22. Sutherland, F.L.; Zaw, K.; Meffre, S.; Thompson, J.; Goemann, K.; Thu, K.; Nu, T.T.; Zin, M.M.; Harris, S. Diversity in ruby geochemistry and its inclusions, intra- and -inter comparisons from Myanmar and Eastern Australia. Abstract IMA2018-1068. In Proceedings of the International Mineralogical Association Conference (IMA), Melbourne, Australia, 13–17 August 2018.
23. Sutherland, L.; Graham, I.; Yaxley, G.; Armstrong, R.; Giuliani, G.; Hoskin, P.; Nechaev, V.; Woodhead, J. Major zircon suites of the Indo-Pacific lithosphere margin: "Zircon Zip" and its petrogenetic implications. *Mineral. Petrol.* **2016**, *110*, 399–420. [CrossRef]
24. Piilonen, P.C.; Sutherland, F.L.; Danišik, M.; Porier, G.; Valley, J.W.; Rowe, R. Zircon xenocrysts from Cenozoic alkaline basalts of the Ratanakiri Volcanic Province (Cambodia), South east Asia—Trace element geochemistry, O–Hf isotopic composition, U–Pb and U (U–Th)/He geochronology—Revelations into the underlying lithospheric mantle. *Minerals* **2018**, *8*, 566. [CrossRef]
25. Thu, K. The Igneous Rocks of the Mogok Stone Tract: Their Distribution, Petrography, Petrochemistry, Sequence, Geochronology and Economic Geology. Ph.D. Thesis, University of Yangon, Yangon, Myanmar, 2007.
26. Thu, K.; Zaw, K. Gem deposits of Myanmar. In *Myanmar: Geology, Resources and Tectonics*; Chapter 23; Barber, A.J., Zaw, K., Crow, M.J., Eds.; Geological Society London Memoirs: London, UK, 2017; Volume 48, pp. 497–529. [CrossRef]
27. Barley, M.E.; Pickard, A.L.; Zaw, K.; Rak, P.; Doyle, M.G. Jurassic to Miocene magmatism and metamorphism in the Mogok metamorphic belt and the India-Eurasia collision in Myanmar. *Tectonics* **2003**, *22*, 1–11. [CrossRef]
28. Mitchell, A.H.G.; Htay, M.T.; Htun, K.M.; Win, M.N.; Oo, T.; Hlaing, T. Rock relationships in the Mogok metamorphic belt, Tatkon to Mandalay, central Myanmar. *J. Asian Earth Sci.* **2007**, *29*, 891–910. [CrossRef]
29. Mitchell, A.H.G.; Chung, S.L.; Oo, T.; Lin, T.H.; Hung, C.H. Zircon U–Pb ages in Myanmar: Magmatic–metamorphic events and the closure of a neo-Tethys ocean? *J. Asian Earth Sci.* **2012**, *56*, 1–23. [CrossRef]
30. Searle, M.P.; Noble, S.R.; Cottle, J.M.; Waters, D.J.; Mitchell, A.H.G.; Hlaing, T.; Horstwood, M.S.A. Tectonic evolution of the Mogok Metamorphic Belt of Burma (Myanmar) constrained by U–Th–Pb dating of metamorphic and magmatic rocks. *Tectonics* **2007**, *26*, TC3014. [CrossRef]
31. Searle, M.P.; Morley, C.K.; Waters, D.J.; Gardiner, N.J.; Htun, U.K.; Nu, T.T.; Robb, L.J. Tectonic and metamorphic evolution of the Mogok Metamorphic and Jade Mines belts and ophiolitic terranes of Burma (Myanmar). In *Myanmar: Geology, Resources and Tectonics*; Chapter 12; Barber, A.J., Zaw, K., Crow, M.J., Eds.; Geological Society London Memoirs: London, UK, 2017; Volume 48, pp. 261–293. [CrossRef]
32. Zaw, K. Overview of mineralization styles and tectonic-metallogenic setting in Myanmar. In *Myanmar: Geology, Resources and Tectonics*; Chapter 24; Barber, A.J., Zaw, K., Crow, M.J., Eds.; Geological Society London Memoirs: London, UK, 2017; Volume 48, pp. 531–556. [CrossRef]
33. Garnier, V.; Maluski, H.; Giuliani, G.; Ohnenstetter, D.; Schwarz, D. Ar–Ar and U–Pb ages of marble-hosted ruby deposits from central and southeast Asia. *Can. J. Earth Sci.* **2006**, *43*, 509–532. [CrossRef]
34. Garnier, V.; Giuliani, G.; Ohnenstetter, D.; Fallick, A.E.; Dubessy, J.; Banks, D.; Vinh, H.Q.; Lhomme, T.; Maluski, H.; Pêcher, A.; et al. Marble-hosted ruby deposits from Central and Southeast Asia: Towards a new genetic model. *Ore Geol. Rev.* **2008**, *34*, 169–191. [CrossRef]
35. Zaw, K. Geological evolution of selected granitic pegmatites in Myanmar (Burma): Constraints from regional setting, lithology, and fluid-inclusion studies. *Int. Geol. Rev.* **1998**, *40*, 647–662. [CrossRef]

36. Nu, T.T. A Comparative Study of the Origin of Ruby and Sapphire in the Mogok, Pyinlon and Mong Hsu Areas. Ph.D. Thesis, University of Mandalay, Mandalay, Myanmar, 2003.
37. Peretti, A.; Mullis, J.; Mouawad, F. The role of fluorine in the formation of color zoning in rubies from Mong Hsu, Myanmar (Burma). *J. Gemmol.* **1996**, *25*, 3–19. [CrossRef]
38. Graham, I.; Sutherland, L.; Zaw, K.; Nechaev, V.; Khanchuk, A. Advances in our understanding of the gem corundum deposits of the west Pacific continental margin. *Ore Geol. Rev.* **2008**, *34*, 200–215. [CrossRef]
39. Glen, R.A. Refining accretionary orogeny models for the Tasmanides of eastern Australia. *Aust. J. Earth Sci.* **2013**, *60*, 315–370. [CrossRef]
40. Roberts, D.L.; Sutherland, F.L.; Hollis, J.B.; Kennewell, P.; Graham, I. Gemstone characteristics, North-East Barrington Plateau, NSW. *J. Proc. R. Soc. NSW* **2004**, *137*, 99–122.
41. Sutherland, F.L.; Fanning, C.M. Gem-bearing basaltic volcanism, Barrington, New South Wales: Cenozoic evolution based on basalt K–Ar ages and zircon fission track and U–Pb isotope dating. *Aust. J. Earth Sci.* **2001**, *48*, 221–237. [CrossRef]
42. Sutherland, F.L.; Coenraads, R.R. An unusual ruby-sapphirine-spinel assemblage from Barrington volcanic province, New South Wales, Australia. *Mineral. Mag.* **1996**, *60*, 623–638. [CrossRef]
43. Owen, J.V.; Greenough, J.D. An empirical sapphirine-spinel exchange thermometer and its application to high grade xenoliths in the Pope Harbour dyke, Nova Scotia, Canada. *Lithos* **1991**, 317–326. [CrossRef]
44. Waltenberg, K.; Blevin, P.L.; Bodarkos, S.; Cronin, D.E. New SHRIMP U–Pb zircon ages from the New England Orogen, New South Wales. Geological Survey NSW Reports GS 2015/1124; Geological Survey of New South Wales: Sydney, Austraila, 2015.
45. Eggins, S.; Henson, B.J. Evolution of mantle-derived augite-hypersthene granodiorite by crystal-liquid fractionation: Barrington Tops Batholith, Eastern Australia. *Lithos* **1987**, *20*, 295–310. [CrossRef]
46. Aitchison, J.C.; Ireland, T.R. Age profile of ophiolitic rocks across the Late Palaeozoic New England Orogen, New South Wales: Implications for tectonic models. *Aust. J. Earth Sci.* **1995**, *42*, 11–23. [CrossRef]
47. Sutherland, F.L.; Coenraads, R.R.; Schwarz, D.; Raynor, L.R.; Barron, B.J.; Webb, B.G. Al-rich diopside in alluvial ruby and corundum-bearing xenoliths, Australian and SE Asian fields. *Mineral. Mag.* **2003**, *67*, 717–732. [CrossRef]
48. Sutherland, F.L.; Graham, I.T.; Harris, S.J.; Coldham, T.; Powell, W.; Belousova, E.A.; Martin, L. Unusual ruby-sapphire transition in alluvial megacrysts, Cenozoic basaltic gem field, New England, New South Wales, Australia. *Lithos* **2017**, *278–279*, 347–360. [CrossRef]
49. Baker, J.; Peate, D.; Waight, T.; Meyzen, C. Pb isotopic analysis of standards and samples using a 207Pb-204Pb double spike and Thallium to correct for mass bias with a double-focusing MC-ICP-MS. *Chem. Geol.* **2004**, *211*, 275–303. [CrossRef]
50. Thompson, J.M.; Meffre, S.; Danyushevesky, L. Impact of air, laser pulse width and fluence on U–Pb dating of zircons by LA-ICPMS. *J. Anal. At. Spectrom.* **2018**, *33*, 221–230. [CrossRef]
51. Black, L.P.; Kamo, S.L.; Allen, C.M.; Aleinikff, J.N.; Davis, D.W.; Korsch, R.J.; Foudoulis, C. TEMORA 1: A new zircon standard for Phanerozoic U–Pb geochronology. *Chem. Geol.* **2004**, *200*, 155–170. [CrossRef]
52. Sláma, J.; Košler, J.; Condon, D.J.; Crowley, J.L.; Gerdes, A.; Hanchar, J.M.; Horstwood, M.S.A.; Morris, G.A.; Nasdala, L.; Norberg, N.; et al. Plešovice zircon—A new natural reference material for U–Pb and Hf isotopic microanalysis. *Chem. Geol.* **2008**, *249*, 1–35. [CrossRef]
53. Paces, J.B.; Miller, J.D. Precise U–Pb ages of the Duluth Complex and related mafic intrusions, northeastern Minnesota: Geochronological insights to physical, petrogenetic, paleomagnetic, and tectomagmatic processes associated with the 1.1 Ga Midcontinent Rift System. *J. Geophys. Res.* **1993**, *98*, 13997–14013. [CrossRef]
54. Dazé, A.; Lee, J.K.W.; Villeneuve, M. An intercalibration study of the Fish Canyon sanidine and biotite ^{40}Ar/^{39}Ar standards and some comments on the age of the Fish Canyon Tuff. *Chem. Geol.* **2003**, *199*, 111–127. [CrossRef]
55. Best, F.C. The Petrogenesis and Ni-Cu-PGE Potential of the Dido Batholith, North Queensland, Australia. Ph.D. Thesis, University of Tasmania, Hobart, Australia, 2012.
56. Ludwig, K.R. *A User's Manual for Isoplot 3.6: A Geochronological Toolkit for Microsoft Excel 9 revision of April 8, 2008*); Berkeley Geochronology Center Special Publication No.4; Berkeley Geochronology Center Special Publication: Berkeley, CA, USA, 2008; p. 77.
57. Stacey, J.S.; Kramers, J.D. Approximation of terrestrial lead isotope evolution by a two-stage model. *Earth Planet. Sci. Lett.* **1975**, *26*, 207–221. [CrossRef]

58. Horstwood, M.S.A.; Košler, J.; Gehrels, J.; Jackson, S.E.; Mclean, N.M.; Paton, C.; Pearson, N.J.; Sircombe, K.; Sylvester, P.; Vermeesch, P.; et al. Community Derived standards for LA-ICP-MS U- (Th-) Pb Geochronology— Uncertainty propagation, Age Interpretation and Data Reporting. *Geostand. Geoanal. Res.* **2016**, *40*, 311–332. [CrossRef]

59. Harlow, G.E.; Bender, W. A study of ruby (corundum) compositions from the Mogok, Myanmar: Searching for chemical fingerprinting. *Am. Mineral.* **2013**, *98*, 1120–2013. [CrossRef]

60. Stone-Sundeberg, J.; Thomas, T.; Sun, Z.; Guan, Y.; Cole, Z.; Equall, R.; Emmett, J.L. Accurate reporting of key trace elements in ruby and sapphire using matrix-matched standards. *Gems Gemol.* **2017**, *53*, 438–451. [CrossRef]

61. Gardiner, N.J.; Robb, L.J.; Morely, C.K.; Searle, M.P.; Cawood, P.A.; Whitehouse, M.J.; Kirkland, C.L.; Roberts, N.M.W.; Myint, T.A. The tectonic and metallogenic framework of Myanmar: A Tethyan mineral system. *Ore Geol. Rev.* **2016**, *79*, 26–45. [CrossRef]

62. Mitchell, A. *Geological Belts, Plate Boundaries and Mineral Deposits in Myanmar, Chapter 7, Mogok Metamorphic Belt*; Elsevier: Amsterdam, The Netherlands, 2017; pp. 201–247. ISBN 978-0-12-803382-1.

63. Nissimboim, A.; Harlow, G.E. A study of ruby on painite from the Mogok Stone Tract. Research Track, Gem Localities and Formation. *Gems Gemol.* **2011**, *47*, 140–144.

64. Link, K. New age data for blue sapphire from Mogok, Myanmar. *J. Gemmol.* **2016**, *35*, 107–109.

65. Li, Q.; Zu, E. Mineral properties of ruby deposit in Mongshu Myanmar. In *Proceedings of the 2015 International Conference on Material Engineering and Environmental Science, Changsha, China, 28–29 November 2015*; Xu, Q., Ed.; World Scientific Publishing Co.: Singapore, 2016; ISBN 978-1-60595-323-6.

66. Mittermayr, F.; Konzett, J.; Hauzenberger, C.; Kaindl, R.; Schmiderer, A. Trace element distribution, solid-and fluid inclusions in untreated Mong Hsu rubies. In *European Geosciences Union General Assembly*; EGU 2008-A-10706, 2008 S/Ref-ID: 1607-7962/gra/EGU2008-A-10706, and poster pdf; Geophysical Research Abstracts 10: Vienna, Austria, 2008.

67. Peucat, J.J.; Ruffault, P.; Fritsch, E.; Bouhnik-Le Coz, M.; Simonet, C.; Lasnier, B. Ga/Mg ratio as a new geochemical tool to differentiate magmatic from metamorphic blue sapphires. *Lithos* **2007**, *98*, 261–274. [CrossRef]

68. Uher, P.; Giuliani, G.; Szakáll, S.; Fallick, A.; Strunga, V.; Vaculovič, T.; Ondine, D.; Gregáňova, M. Sapphires related to alkali basalts from the Corová Highlands, Western Carpathians (southern Slovakia): Composition and origin. *Geol. Carpath.* **2012**, *63*, 71–82. [CrossRef]

69. Giuliani, G.; Pivin, M.; Fallick, A.E.; Ohnenstetter, D.; Song, Y.; Demaiffe, D. Geochemical and oxygen isotope signature of mantle corundum megacrysts from Mbuji-Mayi kimberlite, Democratic Republic of Congo. *C. R. Geosci.* **2015**, *347*, 24–34. [CrossRef]

70. Harris, S.J.; Graham, I.T.; Lay, A.; Powell, W.; Belousova, E.; Zappetini, E. Trace element geochemistry and metasomatic origin of alluvial sapphires from the Orosmayo region, Jujuy Province, Northwest Argentina. *Can. Mineral.* **2017**, *55*, 595–617. [CrossRef]

71. Palke, A.C.; Wong, J.; Verdel, C.; Ávila, J.N. A common origin for Thai/Cambodian rubies and blue and violet sapphires from Yogo Mountain Gulch, Montana, USA. *Am. Mineral.* **2018**, *103*, 469–479. [CrossRef]

72. Sutherland, F.L.; Duroc-Danner, J.M.; Meffre, S. Age and origin of gem corundum and zircon megacrysts from Mercardese-Rio Mayo area, South West Colombia, South America. *Ore Geol. Rev.* **2008**, *34*, 155–168. [CrossRef]

73. Giuliani, G.; Debussey, J.; Banks, D.A.; Lhomme, T.; Ohnennstetter, D. Fluid inclusions from Asian marble deposits, genetic implications. *Eur. J. Mineral.* **2015**, *27*, 393–404. [CrossRef]

74. Sansawong, S.; Vertriest, W.; Saeseaw, S.; Pardieu, V. A Study of Rubies from Cambodia and Thailand. GIA News from Research. *Gems Geol.* **2017**. Available online: http://www.gia.edu (accessed on 28 March 2018).

75. Vysotskiy, S.V.; Velivetskya, T.A.; Sutherland, F.L.; Agoshkov, A.I. Oxygen isotope composition as an indicator of ruby and sapphire origin: A review of Russian occurrences. *Ore Geol. Rev.* **2015**, *68*, 164–170. [CrossRef]

76. Kuelen, N.; Poulsen, M.B.; Salimi, R. *Report on the Activities of the Ruby Project*; Danmarks og Greenlands Geologishe Undersolgelse Rapport; GEUS: Copenhagen, Denmark, 2016; p. 20.

77. Maneerafanasam, P.; Wathanakul, P.; Kim, Y.C.; Choi, H.M.; Choi, B.G.; Shim, K.B. Characterization of dark core and blue patch in Mong Hsu ruby. *J. Korean Cryst. Growth Cryst. Technol.* **2011**, *21*, 55–59.

78. Yui, T.-F.; Zaw, K.; Wu, C.-M. A preliminary stable isotope study on Mogok ruby, Myanmar. *Ore Geol. Rev.* **2008**, *34*, 192–199. [CrossRef]

79. Sutherland, F.L.; Giuliani, G.; Fallick, A.E.; Garland, M.; Webb, G.B. Sapphire-ruby characteristics, West Pailin, Cambodia. *Aust. Gemmol.* **2008**, *23*, 329–368.
80. Sorokina, E.S. Morphological and chemical evolution of corundum (ruby and sapphire): Crystal ontogeny reconstructed by EMPA, LA-ICP-MS and Cr^{3+} Raman mapping. *Am. Mineral.* **2016**, *101*, 2716–2722. [CrossRef]
81. De Leon, S.W.D. Jewels of Responsibility from Mines to Markets; Comparative Case Analysis in Burma, Madagascar and Colombia. Master's Thesis, University of Vermont, Burlington, MA, USA, 2008.
82. Rossman, G.R. The geochemistry of gems and its relevance to gemology. *Elements* **2009**, *5*, 159–162. [CrossRef]
83. Smith, C.; Fagan, A.J.; Clark, B. Ruby and pink sapphire from Appaluttoq, Greenland. *J. Gemmol.* **2016**, *35*, 294–306. [CrossRef]
84. Wang, H.; Krezemnicki, M.S.; Chalain, J.-P.; Lefevre, P.; Zhou, W.; Cartier, L.A. Simultaneous high sensitivity trace-element and isotopic analysis of gemstones using laser ablation inductively coupled plasma, time of flight mass spectrometry. *J. Gemmol.* **2016**, *55*, 212–223. [CrossRef]
85. Pornwilard, M.-M.; Hansawek, R.; Shiowatana, J.; Siripinyanod, A. Geographic origin classification of gem corundum using elemental fingerprint analysis by laser ablation inductively coupled plasma mass spectrometry. *Int. J. Mass Spectrom.* **2011**, *306*, 57–62.
86. Kochelek, K.A.; McMillan, N.J.; McManus, C.F.; Daniel, D.E. Provenance determination of sapphires and rubies using laser-induced breakdown spectroscopy and multivariate analysis. *Am. Mineral.* **2015**, *100*, 1921–1931. [CrossRef]
87. Emori, K.; Kitawaki, H. Geographic origin of determination of ruby and blue sapphire based on LA-ICP-MS and 3D-plots. In Proceedings of the 34th International Gemmological Conference Program, Vilnius, Lithuania, 26–28 August 2015; pp. 48–50.
88. Wong, J.; Verdel, C. Tectonic environments of sapphire and ruby revealed by a global oxygen isotope compilation. *Int. Geol. Rev.* **2018**, *60*, 188–195. [CrossRef]

Article

Origin of Blue Sapphire in Newly Discovered Spinel–Chlorite–Muscovite Rocks within Meta-Ultramafites of Ilmen Mountains, South Urals of Russia: Evidence from Mineralogy, Geochemistry, Rb-Sr and Sm-Nd Isotopic Data

Elena S. Sorokina [1,2,*], Mikhail A. Rassomakhin [3], Sergey N. Nikandrov [3], Stefanos Karampelas [4], Nataliya N. Kononkova [1], Anatoliy G. Nikolaev [5], Maria O. Anosova [1], Alina V. Somsikova [1], Yuriy A. Kostitsyn [1] and Vasiliy A. Kotlyarov [6]

[1] Vernadsky Institute of Geochemistry and Analytical Chemistry Russian Academy of Sciences (GEOKHI RAS), 119991 Moscow, Russia; nnzond@geokhi.ru (N.N.K.); masha_anosova@mail.ru (M.O.A.); orlova@geokhi.ru (A.V.S.); kostitsyn@geokhi.ru (Y.A.K.)

[2] Institute for Geosciences, Johannes Gutenberg University Mainz, 55122 Mainz, Germany

[3] Ilmen State Reserve, Chelyabinsk Region, 456317 Miass, Russia; miha_rassomahin@mail.ru (M.A.R.); nik@ilmeny.ac.ru (S.N.N.)

[4] Bahrain Institute for Pearls & Gemstones (DANAT), WTC East Tower, P.O. Box 17236 Manama, Bahrain; stefanos.karampelas@danat.bh

[5] Institute of Geology and Petroleum Technologies, Kazan Federal University, Department of mineralogy and lithology, 420008 Kazan, Russia; anatolij-nikolaev@yandex.ru

[6] Institute of Mineralogy, Ural Branch Russian Academy of Sciences, Chelyabinsk Region, 456317 Miass, Russia; 100126@mineralogy.ru

* Correspondence: elensorokina@mail.ru; Tel.: +7-(499)-137-14-84

Received: 19 November 2018; Accepted: 4 January 2019; Published: 11 January 2019

Abstract: Blue sapphire of gem quality was recently discovered in spinel–chlorite–muscovite rock within meta-ultramafites near the Ilmenogorsky alkaline complex in the Ilmen Mountains of the South Urals. More than 20 minerals were found in the assemblage with the blue sapphire. These sapphire-bearing rocks are enriched in LREE and depleted in HREE (with the negative Eu anomalies) with REE distribution similar to those in miascites (nepheline syenite) of the Ilmenogorsky alkaline complex. $^{87}Sr/^{86}Sr$ ratios in the sapphire-bearing rocks varied from 0.7088 ± 0.000004 (2σ) to 0.7106 ± 0.000006 (2σ): epsilon notation εNd is -7.8. The Rb-Sr isochrone age of 289 ± 9 Ma was yielded for the sapphire-bearing rocks and associated muscovite. The blue sapphires are translucent to transparent and they have substantial colorless zones. They occur in a matrix of clinochlore-muscovite as concentric aggregates within spinel-gahnite coronas. Laser Ablation-Inductively Coupled Plasma-Mass Spectrometry (LA-ICP-MS) analyses showed values with trace elements typical for "metamorphic" blue sapphires, with Ga/Mg < 2.7, Fe/Mg < 74, Cr/Ga > 1.5 (when Cr is detectable), and Fe/Ti < 9. Sapphires overlap "metasomatic" at "sapphires in alkali basalts" field on the $FeO–Cr_2O_3–MgO–V_2O_3$ versus $FeO + TiO_2 + Ga_2O_3$ discriminant diagram. The sapphires formed together with the spinel-chlorite-muscovite rock during metasomatism at a contact of orthopyroxenites. Metasomatic fluids were enriched with Al, HSFE, and LILE and genetically linked to the miascite intrusions of Ilmenogorsky complex. The temperature required for the formation of sapphire–spinel–chlorite–muscovite rock was 700–750 °C and a pressure of 1.8–3.5 kbar.

Keywords: corundum; blue sapphire; meta-ultramafic rocks; LA-ICP-MS; Rb-Sr and Sm-Nd isotopy; Ilmenogorsky complex; Ural Mountains; metasomatism

1. Introduction

Sapphire is referred to the blue color variety of corundum (Al_2O_3) with Fe and Ti as main chromophores. Other colored sapphires are called fancy and need the color prefix (e.g., pink sapphire, colorless sapphire), apart from the red variety, which is called ruby.

Corundum, of various colors and qualities, occurs in numerous localities in the Ilmen Mountains (Figure 1a). However, gem quality sapphires are found in solely to three of them, all in situ within the primary bearing rocks (See Figure 1b, mines 298, 349, and 418) [1–7]. Occurrences are located inside of the state's Ilmen Nature Reserve, though their commercial exploitation is forbidden; samples can be collected just for research purposes in close collaboration with Nature Reserve's researchers. The most studied gem quality sapphire, of the three, is associated with alkaline syenite pegmatites (Figure 1b, mines 298 and 349), first reported by Barbot de Marni (1828) [1]. These sapphires are genetically linked with the intrusions of miascites (a leucocratic variety of nepheline syenite discovered close to Miass in the Ilmen Mountains and named after the place of discovery).

Gem-quality sapphire mineralization also occurs within the Kyshtymsky stratum of the Ilmenogorsky complex [4]. This stratum is included in the Saitovsky series [5], one of the structural units of the Vishnevogorsky–Ilmenogorsky polymetamorphic zone. This zone is a fragment of regional post-collision shear, in which the Ilmenogorsky complex is the southern unit (see more on Regional geology of Ilmenogorsky complex in [6]) (Figure 1a,b). The series is composed of meta-terrigenous quartzite-schist strata, which include garnet-feldspar-biotite, quartz-mica, and garnet-feldspar-amphibole schists; quartzites and quartzite-gneisses; and, units with mafic–ultramafic affinities. Lenses and lenticular bodies within this association are made up of olivine-enstatite, enstatite, talc-anthophyllite, talc-carbonate, tremolite-anthophyllite rocks, and various amphibolites [7]. SHRIMP U-Pb zircon dating from these rocks reflects the complex evolution of the series: ~1.3 Ga (relict cores of terrigenous zircon); from about 460 to ~420 Ma (stage of metamorphic evolution), ~320 Ma, and ~280 Ma (stages of collisional deformation), for more information see [5]. The dimensions of the mafic–ultramafic units, which host the sapphires, the thickness of the schists and quartzites, and their structural relationships vary widely (Figure 1c).

Gem quality sapphires were found as aggregates within chlorite-phlogopite; hosted by meta-ultramafic rocks. They were discovered at a depth of approximately 300 m in a drill core situated at the southern part of the Ilmen State Reserve [2]. In 2003, one of the authors (Sergey N. Nikandrov) found gem-quality sapphire in the similar micaceous (commonly muscovite) association, located between the lakes Bolshoy Tatkul and Bolshoe Miassovo (mine 418, see Figure 1b). Pink sapphires and rubies were previously found in mafic–ultramafic rocks, but this is the first occurrence of sapphires within such a spinel–chlorite–muscovite metasomatic rock.

Sapphires of gem quality are rare and they are principally found in secondary placer deposits; these are found seldom in situ and only in a few types of primary host rocks [3]. Research on gem quality sapphires found in situ in primary rocks, e.g., as those recently discovered in spinel–chlorite–muscovite rock within meta-ultramafites near the Ilmenogorsky alkaline complex in the Ilmen Mountains of the South Urals, may provide clues to the unknown petrogenesis of those from secondary placer occurrences. The present article is the first complete characterization of these sapphires, focusing on their genesis. For this, local geology and petrography were studied as well as the mineralogy, chemistry, and isotopy of corundum–spinel–chlorite–mica rocks. The results are also compared to those that were acquired on other gem-quality sapphires found in the Ilmen Mountains (syenite pegmatites) and checked their possible linkage as well as with those from other deposits for better understanding the gem corundum genesis.

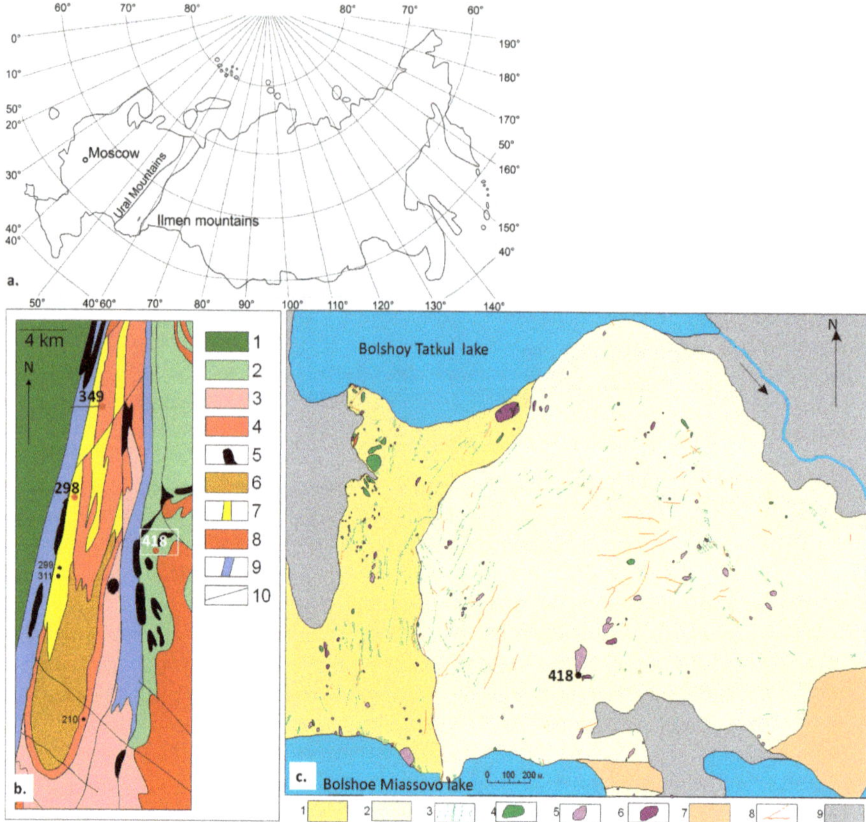

Figure 1. Location and geology of Ilmen sapphire deposits in meta-ultramafic host rocks: (**a**) General map of Russia showing location of Ilmen Mountains within Urals; (**b**) Regional geologic map of Ilmenogorsky complex after [4] with location of corundum mines within syenite pegmatites (mines 298 and 349) and within meta-ultramafic host rocks (mine 418): 1—sedimentary-volcanogenic metamorphosed rocks of the East Urals mega-zone (Upper Devonian-Lower Carboniferous), 2—volcanogenic-sedimentary metamorphosed strata of the Saitovsky series, 3—metamorphic rocks of the Ilmensky series, 4—metamorphic rocks of the Selyankinsky series, 5—meta-ultramafic rocks, 6—syenites and nepheline syenites, 7—fenites, 8—granitoids, 9—zones of blastomylonites, 10—faults, shear zones and other tectonic contacts. White box with 418 mine is shown in details in Figure 1b; (**c**) Regional geologic map after [8] with 418 mine of sapphire-spinel-chlorite-muscovite metasomatites within meta-ultramafic host rocks at 55.178434° N, 60.292494° E: 1—Lower-Kyshtymsky stratum, generally gneiss; 2—Upper-Kyshtymsky stratum, mostly quartzite; 3—traced horizons of various rocks (biotite and amphibole gneisses, diopside-scapolitic and kyanite schists, quartzites); 4—clastoliths and bodies of amphibolites; 5—tremolite-anthophyllite meta-ultramafites; 6—enstatite (±olivine) -antophyllite meta-ultramafites; 7—granite aplite; 8—granite pegmatites and granite dikes; 9—wetlands.

2. Materials and Methods

2.1. Sapphires

Five transparent to translucent colorless to blue corundum of gem quality and their solid inclusions from mine 418 were analyzed by the electron microprobe analyses (EMPA) at GEOKHI RAS using Cameca SX100 (CAMECA, Gennevilliers, France) and wavelength-dispersive (WDS) detection mode with 15 kV acceleration voltage, 30 nA beam current, and 3 μm beam size. Four studied sapphire samples present pronounced colorless and blue zones and one sample was completely colorless (Figure 2b,c and Table S1). A set of natural and synthetic reference materials was used for calibration and instrument-stability monitoring. An overlap correction was applied for V > Ti, Cr > V, Sc > Ta, and Nd > Ce. The detection limits for the measured elements varied for different analytical sessions and were less than 0.01 wt. %. Raman spectra were acquired to identify the inclusions inside the sapphires at the Gemological Institute of America laboratory (Carlsbad, CA, USA) using a Renishaw inVia Raman spectrometer combined with optical microscope. The system was coupled with Ar$^+$ laser emitting at 514 nm with a power of 10 mW on the sample. All of the measurements were from 200 to 2000 cm^{-1}, with a 60 s acquisition time, three cycles, at a resolution of about 1.5 cm^{-1}, and under 50× magnification at a room temperature. Rayleigh scattering was blocked using a holographic notch filter. Backscattered light was dispersed using a grating with 1800 grooves/mm. The spectrometer was calibrated at 520.7 cm^{-1} using Si as a reference.

Laser Ablation-Inductively Coupled Plasma-Mass Spectrometry (LA-ICP-MS) was performed on all five corundum samples that were previously analyzed by wavelength-dispersive electron microprobe analyses (WDS EMPA). Twenty-seven spots were measured on both blue and colorless zones (three spots on blue-colored zones of each four sapphire samples, one sapphire sample is completely colorless). Analysis took place in GEOKHI RAS using a New Wave Research UP-213 Nd:YAG laser (New Wave Research, Inc., Fremont, CA, USA) combined with Element-XR (Thermo Finnigan, Waltham, MA, USA) ICP-MS. Trace-element compositions were determined by monitoring ^{6}Li, ^{9}Be, ^{24}Mg, ^{27}Al, ^{44}Ca, ^{47}Ti, ^{51}V, ^{53}Cr, ^{55}Mn, ^{57}Fe, ^{59}Co, ^{60}Ni, ^{66}Zn, ^{71}Ga, ^{88}Sr, ^{89}Y, ^{91}Zr, ^{93}Nb, ^{181}Ta, ^{208}Pb, and ablating a material with a spot size of 55 μm at a repetition rate of 4 Hz and an energy density of about 21.17 J/cm^2. Warm up/background time was 90 s, dwell time was 100 s, and wash out time was 120 s. NIST SRM 610 and NIST SRM 612 glasses (National Institute of Standards and Technology; U.S. Department of Commerce: Gaithersburg, MD) were used as the reference materials and ATHO as an unknown for quality control (QCM). The time-resolved signal was processed in Glitter commercial software using ^{27}Al as the internal standard applying the theoretical value of 52.93 wt. % of Al in pure crystalline α-Al$_2$O$_3$ for analyzing the corundum unknowns. The measured concentrations of reference material and QCM agree for all elements within 10% and 15%, respectively, of preferable values by [9]. This larger discrepancy between the measured and preferable values can be attributed to isobaric interferences that cannot be resolved with the instrumentation used [10].

Absorption spectra on randomly oriented blue colored zones of sample No1 and No2 were recorded at Kazan Federal University at room temperature by using SHIMADZU UV-3600 spectrometer (Shimadzu Corp, Kyoto, Japan). Spectra were acquired from 185 to 3300 nm, with a data interval and spectral bandwidth of 1 nm and a scan rate of 300 nm/min.

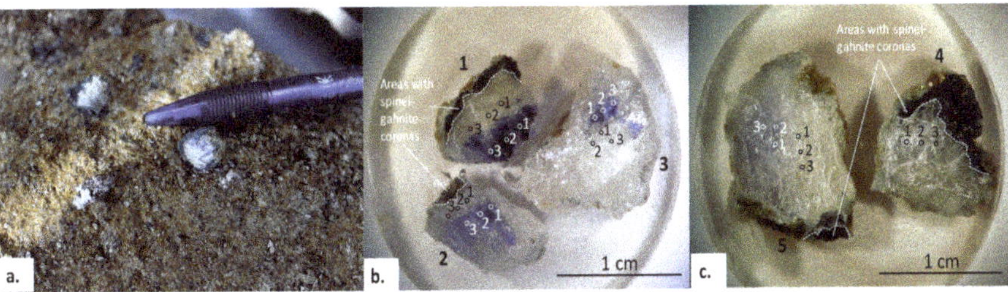

Figure 2. Photographs of corundum–spinel–chlorite–muscovite rock: (**a**) Corundum–spinel concentric aggregates embedded in a muscovite-chlorite matrix; (**b**) Corundum-blue sapphire–spinel samples No-s 1–3 with spot positions for Laser Ablation-Inductively Coupled Plasma-Mass Spectrometry (LA-ICP-MS) analyses from Table 1 mounted in the epoxy resin; (**c**) Corundum-blue sapphire-spinel samples No-s 4 and 5 with spots positions for LA-ICP-MS analyses from Table 1 mounted in the epoxy resin. Photographs by Elena S Sorokina.

Table 1. The trace-elements composition of sapphire colored zones by LA-ICP-MS [1].

No	Spot	Color	^{24}Mg	^{47}Ti	^{51}V	^{53}Cr	^{57}Fe	^{71}Ga	Ga/Mg	Fe/Mg	Cr/Ga	Fe/Ti	Mg × 100	Ti × 10
1	1	Cl	35.18	Bdl	1.05	56.17	1039.18	29.03	0.83	29.54	1.94	-	3518	-
	2	Cl	75.09	Bdl	Bdl	43.46	1121.67	32.05	0.43	14.94	1.36	-	7509	-
	3	Cl	46.83	Bdl	0.94	98.21	1072.46	35.79	0.76	22.9	2.74	-	4683	-
	1	Bl	13.09	279.22	1.23	81.71	973.14	28.99	2.21	74.34	2.82	3.49	1309	2792.2
	2	Bl	22.16	218.01	Bdl	Bdl	820	27.21	1.23	37.00	-	3.76	2216	2180.1
	3	Bl	21.93	189.74	Bdl	70.60	979.65	31.00	1.41	44.67	2.28	5.16	2193	1897.4
2	1	Cl	16.03	49.25	2.46	Bdl	793.43	29.13	1.82	49.50	-	16.11	1603	492.5
	2	Cl	17.68	267.84	1.11	40.05	915.65	31.11	1.76	51.79	1.29	3.42	1768	2678.4
	3	Cl	16.31	206.99	1.19	80.09	931.05	28.62	1.75	57.08	2.80	4.50	1631	2069.9
	1	Bl	22.01	167.50	0.76	520.29	981.77	30.45	1.38	44.61	-	5.86	2201	1675
	2	Bl	22.63	109.67	Bdl	63.15	931.31	28.85	1.28	41.15	2.19	8.49	2263	1096.7
	3	Bl	11.69	129.5	Bdl	58.41	788.16	31.82	2.72	67.42	1.83	6.09	1169	1295
3	1	Bl	15.22	215.82	1.34	87.16	1002.05	9.22	0.61	65.84	9.45	4.64	1522	2158.2
	2	Bl	19.98	174.27	Bdl	62.49	1021.17	27.56	1.38	51.11	2.27	5.86	1998	1742.7
	3	Bl	21.19	265.80	1.42	Bdl	853.12	26.88	1.27	40.27	-	3.21	2119	2658
	1	Cl	20.34	Bdl	0.92	95.93	1128.79	33.43	1.64	55.50	2.87	-	2034	-
	2	Cl	106.21	Bdl	1.56	114.49	767.40	32.59	0.31	7.23	3.51	-	10,621	-
	3	Cl	36.18	Bdl	1.19	Bdl	892.02	32.41	0.90	24.66	-	-	3618	-
4	1	Cl	16.86	185.12	1.32	Bdl	686.05	23.21	1.38	40.69	-	3.71	1686	1851.2
	2	Cl	15.53	127.12	1.31	Bdl	700.49	22.8	1.47	45.11	-	5.51	1553	1271.2
	3	Cl	11.91	90.24	Bdl	Bdl	801.94	24.81	2.08	67.33	-	8.88	1191	902.4
5	1	Cl	53.17	Bdl	Bdl	Bdl	949.86	26.94	0.51	17.86	-	-	5317	-
	2	Cl	34.99	Bdl	1.1	Bdl	934.3	30.5	0.87	26.70	-	-	3499	-
	3	Cl	19.22	Bdl	Bdl	Bdl	830.66	30.31	1.58	43.22	-	-	1922	-
	1	Bl	18.58	247.54	Bdl	39.46	800.69	26.99	1.45	43.09	1.46	3.23	1858	2475.4
	2	Bl	11.22	117.5	1.12	Bdl	764.53	27.66	2.47	68.14	-	6.5	1122	1175
	3	Bl	12.05	92.4	1.28	Bdl	732.68	27.34	2.27	60.80	-	7.93	1205	924

[1] Cl, BL—colorless and blue sapphire zones on a Figure 3b,c. ^{60}Ni was measured but was below the detection limit.

2.2. Host Rocks

Ten samples extracted from different rock types at mine 418, identified macroscopically, of about 100–150 g, were selected for host rock analyses (see Figure 3a and Table S2). Whole-rock major element analyses of these samples were performed at GEOKHI RAS using Energy Dispersive X-ray Fluorescence spectrometer (EDXRF) AXIOS Advanced (PANalytical B.V., Almelo, The Netherlands). The equipment provides the determination the quantitative concentrations of elements from oxygen to uranium from about 10^{-4} to 100 wt. %. The whole-rock trace element measurements were done in the Institute of Mineralogy Ural Brach Russian Academy of Sciences by using Inductively Coupled Plasma-Mass Spectrometer (ICP-MS) Agilent 7700×. Agilent and GSO provided reference multi-element solutions for ICP-MS on all measured trace elements; BCR-2 basalt glass was used as the internal standard. The values of the relative standard deviations during the measurement did not exceed the values established for this analytical equipment.

For Rb-Sr and Sm-Nd isotope studies, powder from two samples containing gem quality sapphire (K418-12 and K418-13, Table S2) was produced. The samples 12-1, 12-2, 13-1, and 13-2 were taken from the 100–150 g of samples 418-12 and 418-13 used previously for EDXRF and ICP-MS. Another 2 samples (rock1 and rock2) from a bigger portion of about 1 kg of sample 418-13 with sapphire, spinel, feldspar, and muscovite in minerals association, as well, as two muscovites (mica1 and mica2) extracted from appropriate samples rock1 and rock2 were chosen to constrain the Rb-Sr age. Sample dissolution was carried out in a mixture of hydrofluoric and nitric acids (5:1) on a shaker under incandescent lamps during three days. After evaporation, 1 mL of concentrated HCl acid was added three times to a dry residue. Rb, Sr, and REE were collected using fluoroplastic chromatographic columns with DowexW 50 × 8 synthetic ion-exchange resin. The collection was done by stepwise elution with 2.2 acid normality (n) of HCl (for Rb and Sr) and 4.0n HCl (for the Sm + Nd). From the REE mixture, Sm and Nd were collected by stepwise elution with 0.15n HCl, 0.3n HCl, and 0.7n HCl using polyethylene columns with synthetic ion-exchange resin Ln-spec. The isotopic studies took place on a Finnigan™ Triton multi-collector thermal ionization mass-spectrometer (TIMS) by ThermoFisher Scientific (Waltham, MA, USA) at GEOKHI RAS while using a two-tape (Re-Re) ion source for Rb, Sm, and Nd and a single-tape (Re) ion source for Sr. Measurements were performed in static mode with simultaneous recordings of ion currents for different isotopes. For the elimination of mass-dependent interference, normalization was carried out according to the exponential law for $^{86}Sr/^{88}Sr = 0.1194$ and $^{148}Nd/^{144}Nd = 0.241572$. The analyses of the international standards Sr-SRM-987 and Nd-LaJolla monitored the reproducibility and correctness of isotope measurements for Sr and Nd. The average value of $^{87}Sr/^{86}Sr = 0.710236 \pm 0.000009$ (2σ) and $^{143}Nd/^{144}Nd = 0.512095 \pm 0.000006$ (2σ). Concentrations of Rb, Sr, Sm, and Nd were determined by isotopic dilution using ^{85}Rb-^{84}Sr and ^{149}Sm-^{150}Nd tracers. The data-reduction was performed using in-lab software. Isoplot4.13 for Excel [11] was used for isochrone diagram with MSWD = 4.1.

Eight petrographic thin-sections of visually monolithic host rocks (samples 418-3, 418-4, 418-6, 418-7, 418-9. 418-10, 418-11, and 418-14), 24 randomly chosen corundum-spinel-chlorite-mica (sample 418-12) and one corundum-spinel (sample 418-13) wafers were studied for petrography (see Table S2). Ten of these wafers with different corundum morphology were used to perform the chemical composition of minerals found in association with corundum by EMPA at the Institute of Mineralogy Ural Branch Russian Academy of Sciences (Miass, Russia) using the electron microscope REMMA—202 M coupled to energy-dispersive spectrometer LZ-5. The operating conditions are 20–30 kV acceleration voltage and 4–6 nA beam current. The beam sizes varied from 1 to 2 µm. A set of natural and synthetic reference materials was used for calibration and instrument-stability monitoring.

Figure 3. Local geologic maps of 418 mine showing the locations of samples for this research at 55.178434° N, 60.292494° E: (**a**) Geological scheme of 418 mine: 1—meta-ultramafites; 2—ferruginated meta-ultramafites; 3—talc meta-ultramafites altering toward the center on a dense actinolite rock; 4—actinolite rock; 5—micaceous lenses with corundum; 6—lenses of actinolite-micaceous rock; 7—anthophyllite-asbestos veins; (**b**) The outcrop of 418 mine: 1—meta-ultramafic host rock of the main unit; 2—ferruginated and carbonated meta-ultramafites of the near-selvage zone; 3—talc zone in meta-ultramafic host rock; 4—the central zone. The numbers 418–3, 418–4, 418–6, 418–7, 418–9, 418–10, 418–11, 418–12, 418–13, and 418–14 indicate the numbers of samples in Table S1.

3. Results

3.1. Geology and Petrology of Sapphire-Bearing Rocks and Meta-Ultramafic Host Rocks

Sapphire occurrence is a plate-shaped symmetrical-zonal body with traces of considerable deformation, with a dip azimuth of 330°; dip angle is 40°; and, its thickness varies from 1.0 to 1.3 m (Figure 3a). The mine 418 with sapphire-spinel-chlorite-muscovite rock was opened from the surface by a trench measuring about 4 × 9 m (Figure 3b). The block of meta-ultramafites hosting corundum-spinel-chlorite-muscovite rock occupied the first tens of meters and up to about 35 m along the drainage ditch (Figure 3a). The meta-ultramafic host rocks consist of purely mono-mineral coarse-grained enstatite up to 4–5 cm in length with minor olivine and clinopyroxene; in some areas, they are replaced by talc and chrysotile-asbestos. Besides, there is a completely replaced pyrite by goethite in the rock. In the surface subsoil layer, there are chalcedony-quartz concretions up to 10 cm in length, while limonite and supergenic carbonate developed along the cracks of host rocks.

The following zones are distinguished by the structure of host and corundum-bearing rocks (from the selvage to the center, Figure 3b): 1. Meta-ultramafic host rock, which is macroscopically composed of orthopyroxene and clinopyroxene, and minor olivine, have clearly visible brittle deformations (sample 418-7, Table S2); 2. The ferruginated and carbonated meta-ultramafic host rock of the near-selvage zone has a common brownish-orange color; its minerals are almost completely replaced by quartz and carbonate; however, relics of olivine were still detectable (sample 418-6); 3. Talc zoning in the meta-ultramafic host rock (sample 418-4) alternates towards the center on a dense green-colored rock (sample 418-3). The transition to green-colored rock is smooth from almost mono-mineral talc to mono-mineral amphibole. In the amphibole zone, in addition to the monoclinic amphibole (actinolite-tremolite), rhombic anthophyllite was also detected; 4. The central zone is strongly affected by both weathering and deformation processes and it consists of lenticular lenses of various compositions (mica -sample 418-11-, actinolite, actinolite-mica, micaceous pockets with corundum, spinel pockets). In this zone, actinolite-tremolite rock is replaced by shallow rot; only small fragments remain in individual lenses (samples 418-9 and 418-10). The corundum, spinel, and other minerals that accompany them (Figure 2a–c) came from this zone during mineralogical studies described below (samples 418-12, 13, and 14).

3.2. Mineralogy of Sapphire-Hosted Micaceous Lenses

Two sapphire associations were identified, accompanied by spinel. In one, sapphire forms transparent to translucent crystals within the spinel corona. In the other, one lens is an assembly of small (sized about 1 cm) transparent to translucent corundum crystals with almost black spinel, not forming a corona around corundum crystals.

More than 20 major, minor, and accessory minerals are associated with sapphires. Some minerals are common (spinel, clinochlore etc.) and others represent rare or single findings (Ni-bearing sulfide mineralization and baddeleyite) (Table S3. Their texture relationships in regard of sapphire are shown in Table S4. Most minerals were identified under the optic microscope and by Raman Spectroscopy; chemical compositions of minerals were analyzed by EMPA and are presented in Tables S5–S12.

3.2.1. Major Minerals

Minerals of spinel- $MgAl_2O_4$ -gahnite $ZnAl_2O_4$ series were common detected in 3 mm thick dark green to black-colored corona around sapphire crystals, as well as in inclusions (Figure 4a). The dark-green to black color is likely due to Fe; Co traces (up to 0.28 wt. %) may also contribute to the coloration. Induction surfaces (i.e., surfaces of synchronous growth) that were detected between spinel and sapphire suggest their synchronous growth. However, spinel most likely started crystallizing at a slightly later stage and when sapphire was already growing in the system. Gahnite appears in spinel areas with higher ZnO contents (up to 22 wt. %). In zonal grains, ZnO values varied from 2.46 to 7.47 wt. % (Table S5). Muscovite $KAl_2(AlSi_3O_{10})(OH)_2$ formed micaceous lenses with embedded corundum, as well as micro-inclusions (up to 0.2 mm) within chlorite group minerals (Figure 4b). Muscovite is characterized by a higher MgO content from 0.5 to 2.5 wt. % in comparison to stoichiometric muscovite (Table S6). Clinochlore $Mg_5Al(AlSi_3O_{10})(OH)_8$ and amesite $Mg_2Al(AlSiO_5)(OH)_4$ form inclusions and rather large ingrowths (up to 5 mm) around spinel, inclusions within spinel, and in fractures along the border of corundum and spinel (Figure 4b). These minerals were also found in the rims around the uraninite-brockite and monazite-(Ce) (for the chemical composition, Table S6). Anorthite $CaAl_2Si_2O_8$ (An_{100}) is found in the "spinel lenses" forming crystals that are sized up to 3 mm in the intergrowth with amphibole and chlorite group minerals (see chemical composition in Table S7). All identified major minerals are syngenetic with sapphire.

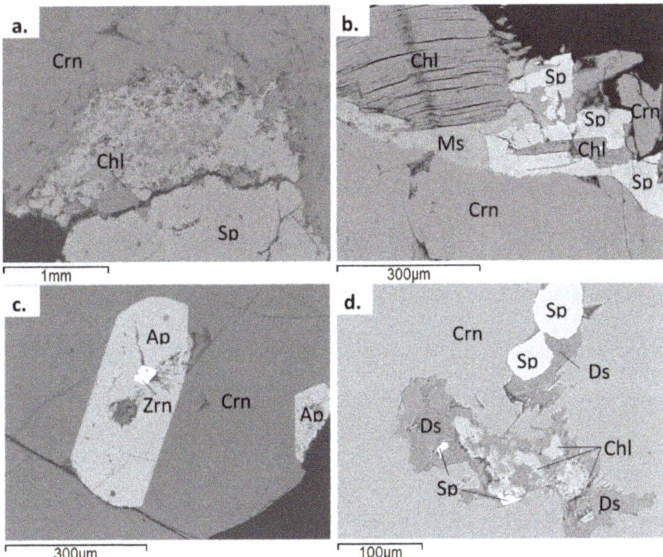

Figure 4. Backscattered (BSE) images of corundum-spinel-chlorite samples from the central zone: (**a**) Corundum-spinel intergrowth with inclusions of clinochlore; (**b**) The rim of corundum-spinel concentric aggregate with intergrowth of clinochlore and muscovite; (**c**) Inclusions of apatite and zircon within corundum; and, (**d**) Inclusions of spinel, clinochlore, and diaspore (Ds) within corundum. Mineral abbreviations are after [12]. BSE images by E.S. Sorokina.

3.2.2. Minor Minerals

Apatite group minerals $Ca_5(PO_4)_3(OH,F,Cl)$ were found within corundum fractures and chlorite group minerals as prismatic epigenetic inclusions up to 1 mm in length (Figure 4c). Its composition varies from chlorapatite to fluorapatite to hydroxylapatite (Table S8). Zircon $ZrSiO_4$ forms small zonal epigenetic inclusions (up to 0.1 mm in length) within the fractures of corundum, chlorite group minerals, and spinel. Besides zircon, the amphibole group minerals and apatite (Figure 4c) also fill out these fractures. Zircons contain up to 3.16–4.73 wt. % of HfO_2 (Table S9). Protogenetic diaspore $AlO(OH)$, along with clinochlore forms rims around inclusions of uraninite, intergrowth with clinochlore within corundum (Figure 4d), and platy aggregates in the fractures of barium feldspar. Diaspore contain up to 0.1 wt. % of SiO_2 and 0.34 wt. % of FeO_{tot}. Millerite NiS forms micro-inclusions (up to 15 µm) within corundum, spinel, and chlorite group minerals; the mineral is found in association with maucherite, heazlewoodite, and galena (Figure 5a). The FeO_{tot} content of millerite varies from 2 to 20 wt. % (Table S10). Maucherite $Ni_{11}As_8$ was detected in the intergrowth with heazlewoodite and millerite in crystals sized to 15 µm (for chemical composition, Table S10). Microlite group minerals $(Na,Ca)_2Ta_2O_6(O,OH,F)$ were found in small (up to 0.1 mm) grains and cube-octahedral crystals within chlorite group minerals (Figure 5b). Minerals characterized by the high UO_2 content (up to 11.34 wt. %); values of WO_3 and Nb_2O_5 both vary up to 3.15 wt. % and 3.57 wt. %; ThO_2 is 1.2 wt. % and Y_2O_3 1.61 wt. %. Nickeline $NiAs$ is found as inclusions within clinochlore and dravite; the size of mineral varied up to 20 µm (Figure 5c and Table S9). Uraninite UO_2 forms zonal micro-inclusions that are sized up to 150 µm (Figure 5d). The atomic number (pfu) is up to 0.41 in the mineral. Allanite-(Ce) $(CaCe)(Al_2Fe^{2+})(Si_2O_7)(SiO_4)O(OH)$—dissakisite-(Ce) $(CaCe)(Al_2Mg)(Si_2O_7)(SiO_4)O(OH)$ form zonal-sectorial and mosaic-like crystals up to 0.2 mm in length (Figure 5f); the oscillatory zonation is due to the different concentrations of CaO and REE (Table S11). Monazite-(Ce) $Ce(PO_4)$ has been found along the cracks of dissakisite-(Ce) grains, forming crystals of up to 20 µm in length (Figure 5f). CaO content ranges from 0.38 to 8.25 wt. % in the mineral,

Ce_2O_3 values vary from 23.11 to 37.53 wt. %, and La_2O_3 is 10–22 wt. %; in some analyses, additional SiO_2 and ThO_2 were detected (Table S9). Ni-bearing and REE-mineralization are likely syngenetic in origin.

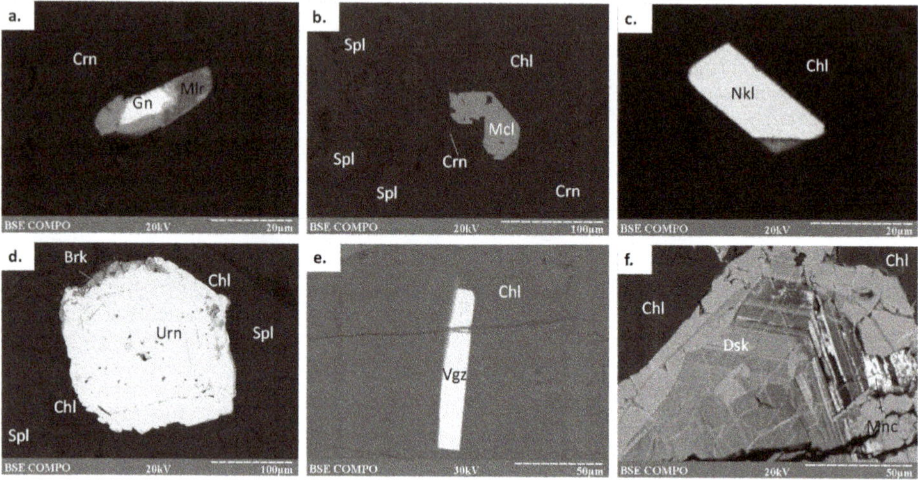

Figure 5. BSE images of corundum-spinel-chlorite samples from the central zone: (**a**) Millerite and galena inclusions within corundum; (**b**) Corundum-spinel-clinochlore intergrowth with inclusion of microlite; (**c**) Nickeline inclusion within clinochlore; (**d**) Uraninite-brokkite-clinochlore concentric aggregate within spinel; (**e**) Vigezzite inclusion within clinochlore; and, (**f**) Dissakisite-(Ce)-monazite-(Ce) aggregate within clinochlore. Mineral abbreviations are after [12]. BSE images by V.A. Kotlyarov.

3.2.3. Accessory Minerals

Vigezzite or fersmite $(Ca,Ce)(Nb,Ta,Ti)_2O_6$ forms likely syngenetic elongate-prismatic inclusions up to 70 µm long within clinochlore (Figure 5e). The content of Nd_2O_3 is about 3.5 wt. % in the mineral; La_2O_3 values vary up to 2.16 wt. %, Pr_2O_3 is up to 1.15 wt. %; and, FeO_{tot} up to 1 wt. %.

Heazlewoodite Ni_3S_2 is associated with corundum, spinel, clinochlore, and millerite. The size of the mineral is less than 0.3 mm. It contains up to 1.22 wt. % of FeO_{tot} [13]. Galena PbS is found in intergrowth with millerite in the form of crystals sized up to 10 µm (Figure 5a and Table S10). Gersdorffite NiAsS is identified within the inner part of a zonal aggregate composed of heazlewoodite and galena in contact between chlorite group minerals and corundum. CoO content in the mineral varies up to 13 wt. % (Table S10). The Ni- and Pb-bearing mineralization is likely syngenetic in origin. Dravite $Na(Mg_3)Al_6(Si_6O_{18})(BO_3)_3(OH)_3(OH)$ forms brownish-colored syngenetic or protogenetic crystals up to 3 cm in length. It lies in micaceous lenses in association with corundum and in "spinel lenses". The mineral contains lower FeO and CaO values in comparison with stoichiometric dravite; two values are up to 1.6 wt. % and 0.5 wt. %. Rutile TiO_2 forms syngenetic micro-inclusions within spinel sized up to 50 µm. The mineral is characterized by high tantalum content (up to 31.79 wt. % of Ta_2O_5) and Nb_2O_5 values of up to 1.46 wt. %. Hyalophane $(K,Ba)Al(Si,Al)_3O_8$ forms likely epigenetic micro-inclusions along the fractures within the chlorite group minerals; mineral size is up to 60 µm. BaO content in hyalophane is about 6 wt. % (Table S7). Celsian $Ba(Al_2Si_2O_8)$ forms lamellar epigenetic crystals and aggregates in fractures along a boundary between corundum and amesite; mineral size varies from 5 µm to 30 µm. BaO values in mineral are up to 33.67 wt. % (Table S7). Marialite $(Na,Ca)_4[Al_3Si_9O_{24}]Cl$ forms syngenetic inclusions within chlorite group minerals sized up to 20 µm; Cl content in the mineral is about 1.7 wt.%, CaO is 11 wt. %, Na_2O is about 7 wt. % (Table S7). Pargasite $NaCa_2(Mg_4Al)Si_6Al_2O_{22}(OH)_2$ and tschermakite $\square(Ca_2)(Mg_3Al_2)(Al_2Si_6O_{22})(OH)_2$ form syngeneitc micro-inclusions (up to 150 µm) within clinochlore of "spinel lenses" (Table S12). Baddeleyite ZrO_2 is found as an epigenetic micro-inclusion within zircon sized up to 5 µm.

3.3. Sapphire Mineralogy, Geochemistry, and UV-Vis-NIR Spectroscopy

Sapphires practically always form single crystals, with colorless translucent to blue transparent zones, sized up to 4 cm in length. Only small, less than 1ct, sapphires of gem quality can be faceted. A dark-green to black colored spinel-gahnite corona practically always surrounds sapphire crystals (Figure 2a–c). The shape of sapphire-spinel aggregates is often irregular and rounded due to numerous outgrowths of muscovite and clinoclore. Less common (in one or two lenses), there are short-prismatic and barrel-shaped sapphire crystals; however, without flat faces that are caused by intergrowth with the muscovite and spinel.

LA-ICP-MS measurements from different zones of sapphires are presented in Table 1. The position of analyzed spots is shown in Figure 3b,c. Mg, Ti, V, Cr, Fe, and Ga, elements commonly used to trace the geological origin were the only detected element of this method [14,15]. Si may also be present as a trace of sapphires; however, it cannot be measured by LA-ICP-MS due to high interference of $^{27}AlH^+$ with $^{28}Si^+$ and ^{28}SiH with $^{29}N-N$ [16]. In the blue-colored spots, Mg ranges from 11 to 23 parts per million weight (ppmw), Ti ranges from 92 to 279 ppmw, V from b.d.l. to 1 ppmw and Cr values ranges from b.d.l. to 87 ppmw (except one analyses detected 520 ppmw likely due to the ablation of diaspore inclusion; its spot was excluded from the plots provided below), relatively low Fe content (from 732 to 1021 ppmw), and fairly low Ga values (from 9 to 31 ppmw). The absorption spectra were comparable with those that were observed on sapphires of metamorphic or metasomatic origin and similar to those from Ilmen Mountains linked with syenite pegmatites [6,17]. Values of trace elements, such as Mg, Ti, Fe, Cr, Ga, and their ratios—Ga/Mg, Fe/Ti, Fe/Mg, and Cr/Ga have been used for the determination of the origin of "magmatic" versus "metamorphic" blue sapphires [14,15]. LA-ICP-MS trace element measurements of the studied sapphires showed Ga/Mg < 2.7, Fe/Mg < 74, Cr/Ga > 1.5, and Fe/Ti < 9, which is typical of sapphires of metamorphic origin [14,15]. On a Fe versus Ga/Mg diagram and a Fe-Ti × 10-Mg × 100 ternary plot [14], all measured corundum samples plotted in the overlapping field of "metamorphic" and "plumasitic" sapphires (Figures 6 and 7). Using the $FeO–Cr_2O_3–MgO–V_2O_3$ versus $FeO + TiO_2 + Ga_2O_3$ discriminant diagram of [18], the studied sapphires plot slightly off the bottom of the overlapping area with rubies found in "mafic–ultramafic rocks" and more in the field of "metasomatic" sapphires and sapphires in "alkali basalts" (Figure 8a,b). In the discriminant factor diagram by [18], the Ilmen sapphires within meta-ultramafites are plotted to the "plumasitic" field and they are close to those from skarns, which is mostly caused by their enrichment with Ti (Figure 9).

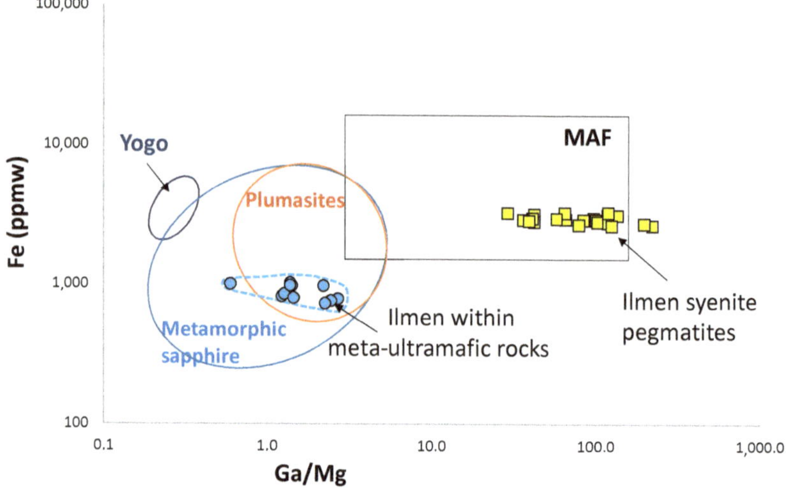

Figure 6. A plot of Ilmen sapphires within meta-ultramafic host rocks (bold blue circles) in a Fe versus Ga/Mg diagram showing boundaries for magmatic (MAF: Main Asian Field; magmatic sapphires in alkali basalt) and metamorphic sapphires are modified after [14,19]. Ilmen sapphires that plots from syenite pegmatites is from [6].

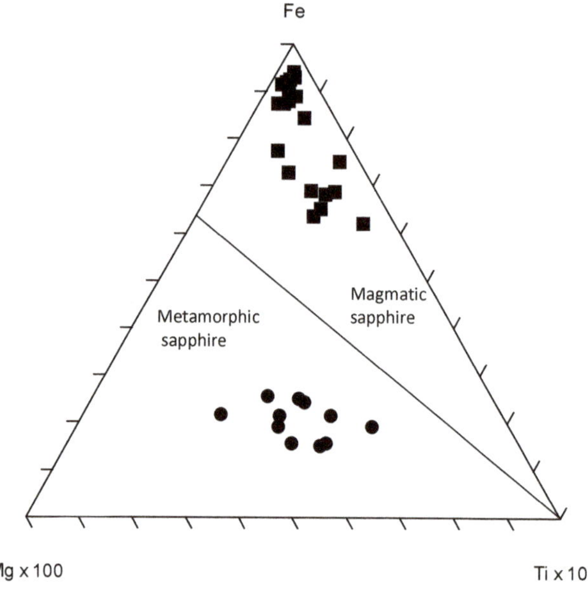

Figure 7. A Fe-Mg × 100-Ti × 10 ternary plot of Ilmen sapphires within ultramafic rocks (bold black circles) are modified after [14]. Values of Ilmen sapphires within syenite pegmatites (bold black squares) are from [6]).

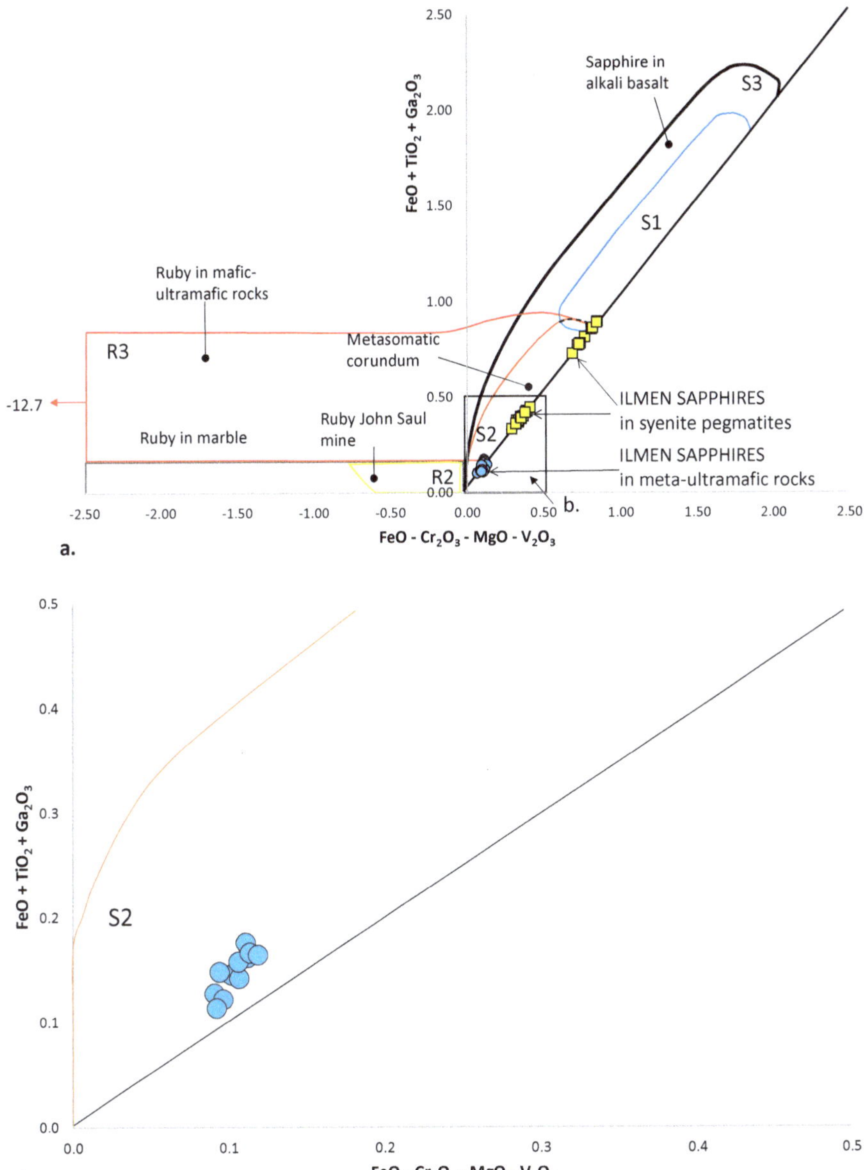

Figure 8. (**a**) A plot of Ilmen sapphires within meta-ultramafic rocks (bold blue circles) in a FeO–Cr$_2$O$_3$–MgO–V$_2$O$_3$ versus FeO–TiO$_2$–Ga$_2$O$_3$ diagram (in wt. %) used for the geological classification of corundum deposits are modified after [18]. The main fields defined for the different types of gem corundum deposit are: John Saul Ruby Mine (Kenya) type (R2) is for ruby; for sapphire, syenitic rocks (S1); metasomatites (S2); xenocrysts in alkali basalt and lamprophyre (S3). The domains of ruby in mafic–ultramafic rocks and S2 are overlapping. Values of Ilmen sapphires within syenite pegmatites are after [6]. (**b**) Focus on the distribution of Ilmen sapphires within metasomatites (S2) of the plot A).

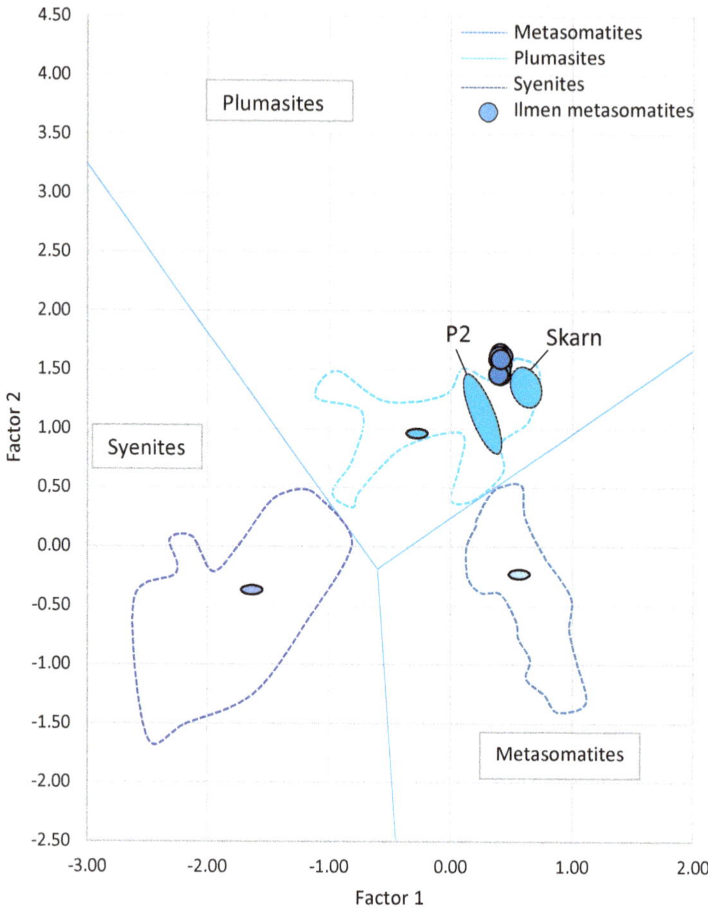

Figure 9. A plot of Ilmen sapphires within meta-ultramafic rocks (bold blue circles) in discriminant factors diagram is modified after [18]. Colored ovals are averages of syenites, plumasites, and metasomatites. P2 is an extension of metasomatic type according to [18].

3.4. Whole-Rock Geochemistry

EDXRF measurements of hosted meta-ultramafic and corundum-bearing rocks are presented in Table S13. According to the powder EDXRF values, meta-ultramafic host rocks have high SiO_2 content up to 50–61 wt. %, MgO was about 20–32 wt. %, CaO was 0.29–17.24 wt. %, Fe_2O_3 was 2.9–8.2 wt. %, Al_2O_3 was 0.11–3.97 wt. %, and low alkaline elements concentrations were also detected (in total, about 0.1–0.2 wt. %). Micaceous lenses with corundum and spinel, however, showed high MgO and Al_2O_3 contents (the two are about 16–25 wt. % and 16.26–38.52 wt. %, respectively), combined with variable SiO_2 (25–80 wt. %), 3.44–9.5 wt. % of Fe_2O_3, 0.26–1.39 wt. % of CaO, and low concentration of alkaline elements ($Na_2O + K_2O \leq 1$ wt. %) [20]. Thus, sapphire-bearing rocks show higher Al_2O_3 and K_2O, however, lower SiO_2 values than those detected in meta-ultramafites, but similar concentrations of Na_2O, MgO, CaO, and TiO_2 (Figure 10). As for numerous metasomatic deposits, there are difficulties in the calculation of mass balance for the newly discovered sapphire-spinel-chlorite-muscovite rock, its formation occurred with significant mass-transfer. The aluminum that is required for the formation of corundum is commonly passive in weakly acidic and neutral conditions; the transfer of aluminum is possible in alkaline and strongly acidic conditions [21]. The mobility of aluminum in a complex with sodium was previously discussed [22]. In the case of this sapphire occurrence, Al was particularly mobile and it was most likely transported by fluids that are linked to miascites (Table S13 and the later discussion). Additionally, sapphire-bearing rocks were enriched with LIL-elements, such as Ba (Ba-felspars), P (apatite), and Sr (as a trace-element in feldspar and muscovite), which came from miascite-generated fluids as well. However, the sources of Mg and some siderophyle elements (Ni) are linked likely to meta-ultramafic host rocks; their parts were re-worked by metasomatic fluids.

ICP-MS measurements are shown in Table S13. The distribution of REE in micaceous lenses has the general trend—an enrichment with LREE as compared to HREE (Figure 11a). This phenomenon is possibly due to the inclusions of LREE-bearing minerals, e.g., monazite-(Ce), allanite-(Ce) or others. In the meta-ultramafic host rocks, this trend is less pronounced or it disappears: REE contents normalized to chondrite are almost the same for both LREE and HREE. Nearly all of the studied rocks show negative Eu anomalies. The REE trend of these corundum-bearing rocks is close to that observed for Ilmen corundum-bearing syenite pegmatites (Sorokina et al., unpublished data) and to that of corundum syenite pegmatites hosted by ultramafic rocks in the French Pyrénées [23]. Enrichment in HSFE was also noted (mine 418); commonly observed for alkaline-ultramafic associations of the platform type (ultramafic rocks, syenites–carbonatites etc.; [24]), and similar to those of miaskite (nepheline syenite) of the Ilmenogorsky complex. On a trace element diagram normalized to primitive mantle (Figure 11b), all studied rocks present distinct K and Ti negative anomalies. However, the sapphire-bearing rocks also show positive Ta and Gd anomalies, but with different Sr anomalies.

3.5. Rb-Sr and Sm-Nd Isotope Measurements

Rb-Sr and Sm-Nd isotope measurements of sapphire-bearing rocks 418-12 and 418-13 previously analyzed by EDXRF and ICP-MS are presented in Table S14. As well, two samples of 418-13 rock containing feldspar and two samples of mica extracted from appropriated rock sample 418-13 associated with sapphire mineralization were analyzed in order to constrain the Rb-Sr age. Obtained values of Rb, Sr, Sm, and Nd by TIMS are closely similar or comparable to those that were received by ICP-MS (Tables S12 and S13). ^{87}Sr/^{86}Sr for samples 418-12 and 418-13 are varied from 0.708764 ± 0.000010 (2σ) to 0.710587 ± 0.000006 (2σ). For mica samples, these values are higher and vary from 0.711371 ± 0.000008 to 0.714131 ± 0.000016. The calculated Rb-Sr isochrone age for three rock samples and mica-2 sample is 289.3 ± 8.7 Ma (2σ), MSWD is 4.1 (Figure 12). Note, the calculated Rb-Sr age for another mica sample (mica-1) with three rock samples gave much higher error (352 ± 35 Ma) and higher MSWD (72), which was likely due to the secondary alteration of muscovite samples mica-1. $(^{87}$Sr/^{86}Sr$)_{289}$ of all samples recalculated on 289 Ma are about 0.707351 to 0.707919. ^{143}Nd/^{144}Nd for the samples 418-12 and 418-13 are 0.5122 ± 0.000008 (2σ) and 0.5122 ± 0.000012 (2σ), εNd$_{(289\ Ma)}$ is -7.8. The obtained high ^{87}Sr/^{86}Sr ratios and low εNd correspond to the formation of the newly discovered corundum-bearing rocks under the significant influence of crustal fluids.

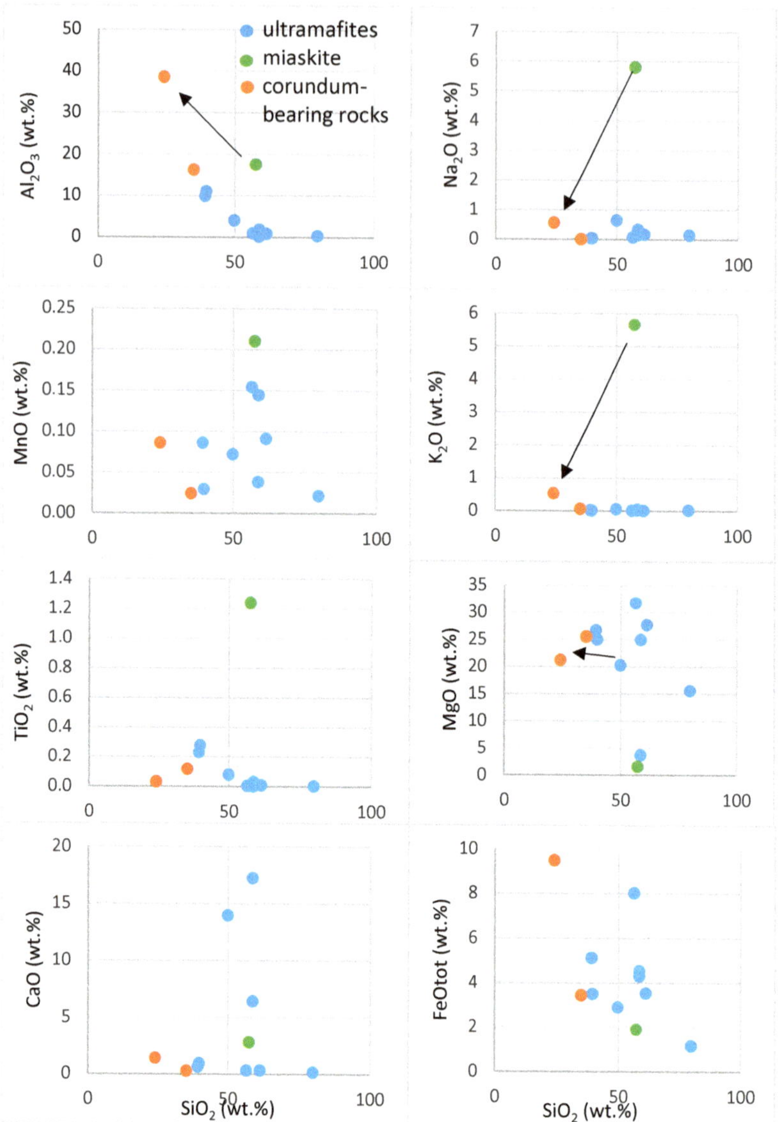

Figure 10. Variations of selected elements in whole-rock compositions from Table S13. Black arrows show the source of major elements of metasomatites. Chemistry of miascite is after [25].

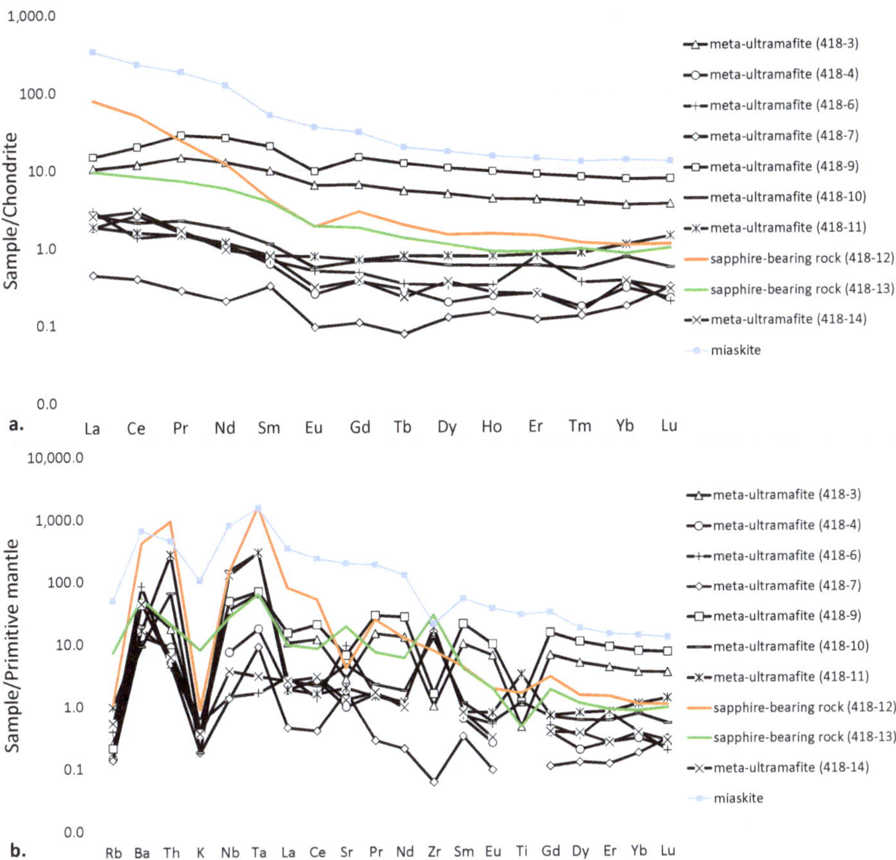

Figure 11. (**a**) Chondrite-normalized REE diagram of selected samples, the chondrite REE data are after [26]; (**b**) Trace elements diagram of selected samples normalized to primitive mantle, data on trace elements in primitive mantle are after [26]. Chemistry of miascite is after [25].

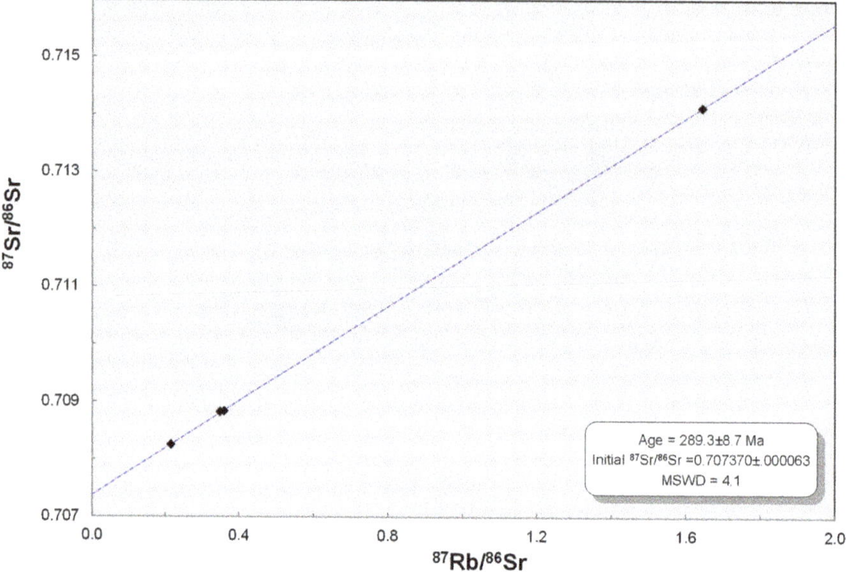

Figure 12. The Rb-Sr isochrone for three sapphire-bearing rocks and one muscovite samples associated with sapphire of 418 mine in Ilmen Mountains are with yielded age at 289.3 ± 8.7 Ma (2σ), MSWD is 4.1.

4. Discussion

4.1. Suggested Petrogenesis of Sapphire in the Meta-Ultramafic Host Rocks of the Ilmen Mountains

The studied blue sapphires surrounded by spinel coronas embedded in chlorite-muscovite matrix present metasomatites that are found within meta-ultramafic host rocks. These host rocks are most likely formed after metasomatic alteration (serpentinization) of orthopyroxenites; i.e., containing orthopyroxene with minor olivine and clinopyroxene. While the orthopyroxenites were formed at about 1.3 Ga and are related to the Saitovsky series of Kyshtimsky strata of the Ilmenogorsky complex, Rb-Sr and Sm-Nd isotopic data indicate their linkage with enriched mantle domains of types EM1 and EM2 [27].

The metasomatism required for the formation of these rocks with blue sapphires took place around 289.3 ± 8.7 Ma (measured by Rb-Sr dating of sapphire-bearing rocks and mono-mineral muscovite associated with sapphires). This age represents the cooling age after the metamorphic peak (transpression) that occurred in the Ilmenogorsky complex at about 330–320 Ma during the Uralian orogeny [28]. The metasomatic fluids formed after the partial melting of miascites occurred in the region (440–420 Ma). These fluids were also the source of HFSE, LILE, and alumina (remobilized under alkaline conditions [21,22]) needed for corundum formation, and transported these chemical elements at a minimum distance of about 4 km (calculated from the eastern flank of Ilmenogorsky complex) reaching the meta-ultramafic rocks. In Figure 11, similar REE trends between miascites and corundum-bearing rocks in mine 418 are presented (i.e., enrichment of LREE as compared to HREE), which is an important evidence of the metasomatic fluid source. Partial melting of miascites is associated with the formation of migmatites with more acidic composition from albitites to even granites. In the Ilmenogorsky miascite massif, alumina remobilization can also be associated with the formation of zeolitic veins and zeolitizated zones of pegmatites, filling late-forming zones with aluminum hydroxides (gibbsite, nordstrandite, boehmite, diaspore).

Desilication also took place during the contact of fluids with ultramafic rocks and sapphire formation. Desilication is commonly observed during corundum crystallization in numerous deposits worldwide ([1,29,30], etc.). The importance of buffering capacity of ultramafic host rocks with respect to silica in the Greenland ruby deposit has been previously shown [31]. In this deposit, the meta-ultramafic host rocks show similar buffering capacity in respect to the desilication of Al-bearing alkaline fluids that are linked to miascites. Initial ultramafic rocks are also most likely the source of Mg required for spinel and chlorite formation found in metasomatites linked with the corona around the sapphires, as well as the source of Ni for nickel-bearing minerals.

Sapphire crystallization was likely caused by the breakdown of diaspore possibly by the following reaction: 2diaspore $\rightarrow$ corundum + H_2O. Note, diaspore formed in meta-ultramafic host rock likely due to a previous metasomatic event that is linked with the partial melting of miascites. A previous metasomatic event was likely a significant time before the deposition of sapphire-bearing metasomatites. Chlorite formation occurred during the metasomatic reworking of meta-ultramafic host rocks. Meanwhile, the spinel corona most likely formed as a consequence of metasomatic fluid-chlorite interactions with the corundum as a reaction rim. Under decreasing pressure and temperature, muscovite crystallized during the breakdown of corundum: corundum + orthoclase + $H_2O \rightarrow$ muscovite; previously observed for a corundum-mica-tourmaline rock hosted by ultamafites of New Zealand [32].

Experiments have shown that blue sapphires (incorporation of Ti and Fe into the corundum lattice) can be formed at around 700 °C after muscovite and biotite, along with a peraluminous silicic melt at 2 kbar (water pressure) [33]. Corundum and hercynite spinel also coexist with orthoclase-rich feldspar and biotite at 750 °C [33]. Besides, other experiments have shown that muscovite can still be stable, slightly above 700 °C and at about 1.8 kbar [34] (Figure 13a). Thus, it seems that sapphire-spinel-clinochlore-muscovite metasomatites formed at a temperature around 700–750 °C and pressure about 1.8–3.5 kbar; at the eutectic point of corundum + orthoclase + $H_2O \leftrightarrow$ muscovite; where the three minerals are in equilibrium (Figure 13b).

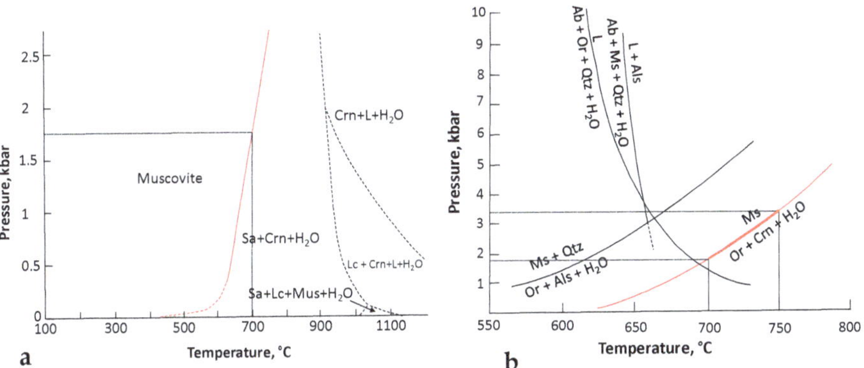

Figure 13. (**a**) The equilibrium curve (solid and dash) for the reaction Sa (sanidine) + Crn + $H_2O \leftrightarrows$ Mus, estimates of the equilibrium curve for the melting of sanidine and leucite—Lc (dash) in the muscovite composition, and possible relations between the assemblages Sa + Crn + H_2O and Mu + Sa + Lc + H_2O are modified after [34]. (**b**) Phase relations in peraluminous granitic systems in terms of pressure and temperature after [35]. The dashed line is a metastable extension of the reaction Ab (albite) + Ms (muscovite) + Qtz (quartz) + H_2O = melt + aluminum silicate to 2 kbar pressure. The H_2O-saturated haplogranite solidus (Ab + Or (orthoclase) + Qtz + H_2O = melt) and the subsolidus second sillimanite isograd (Ms + Qtz = Or + aluminum silicate + H_2O) were determined by [36] and [37]. The terminal muscovite reaction Ms = Or + Crn + H_2O comes from [35]. The diagram was modified after [33].

4.2. Comparison with Sapphire and Ruby from Metasomatites

To date, only pink sapphires to rubies (probably none of them of gem quality) were found in metasomatites within the ultramafic host rocks; e.g., they are known in the eastern Transvaal [38] and in the Barberton greenstone belt [39] of South Africa, in Zimbabwe [40], and in Westland, New Zealand [32]. Ruby-fuchsite rock, 'verdite' from Zimbabwe is enriched in Nias represented by gersdorfitte (NiAsS), which is also the case of the described occurrence in the Ilmen Mountains, in which andalusite, chlorite, margarite, tourmaline, diaspore, rutile, and native Bi are present as well. However, in Zimbabwe, 'verdite', which occurs in bodies a few kilometers length, was formed as the result of post-volcanic, pre-metamorphic exhalative alteration of komatiitic lavas [41]. Another approach to the genesis of 'verdite' suggests an alteration of ultramafic rocks in a high-temperature low-pH hydrothermal solution carrying LILE [42]. Ruby-bearing boulders of metasomatic origin from Westland in New Zealand, as is the case of the described occurrence, are found embedded in micaceous matrix associated, however, with fuchsite, margarite, and tourmaline [32]. In New Zealand, similarly to the described Ilmen sapphire occurrence, the crystallization of minerals in association with corundum required significant mass-transfer during the metasomatic process, possibly 50–90% [32]. Regionally metamorphosed pink sapphire to ruby deposits within an ultramafic protolith in Paranesti in Greece were also reported [43]. In common with the Ilmen sapphires, the Paranesti rubies are characterized by spinel-picotite inclusions. Spinel inclusions and Ni-bearing sulfides were also found within sapphires from some Australian occurrences [44]. However, the spinel inclusions in these sapphires are more Co-rich than Zn-rich, as is the case in the present Ilmen sapphires (Table S5). An inclusion of Zn-rich spinel within sapphire is also reported from a placer in Kedrovka, in the Russian Far East [45].

4.3. Comparison of Chemical Data with Sapphires from Other Geological Environments

In a Fe versus Ga/Mg diagram [14], all measured corundum samples were plotted to "metamorphic" and "plumasitic" population fields (Figure 6), clearly away from the "magmatic" blue sapphires that are found within syenite pegmatites of Ilmen Mountains (see again [6,17]). However, these trace-element diagrams should be used with the caution and possibly revised, since it was found that some of the sapphires from magmatic deposits could be plotted in the "metamorphic field" [46,47]. It is probably important to refine these plots using solely data of gem-quality samples and to group them by color (e.g., plot sapphire and rubies separately), taking into account that most of trace elements used for these are corundum chromophores (e.g., Fe, Ti, Mg, Cr, V). For instance, as shown in Figure S1, the population field is more scattered when plotting colorless opaque sapphires together with those of blue color.

On the $FeO-Cr_2O_3-MgO-V_2O_3$ versus $FeO + TiO_2 + Ga_2O_3$ discriminant diagram (Figure 8), the plot is slightly away from the area with rubies in "mafic–ultramafic rocks" and is mostly in the field with "metasomatic" sapphires and sapphires in "alkali basalts". The UV-Vis-NIR spectra of studied sapphires were comparable to those that were observed on sapphires of metamorphic or metasomatic origin.

5. Conclusions

Most gem-quality sapphires are mined from placers of unclear primary and possibly diverse origins. Studies of gem-quality sapphires that are found in situ within host rocks may provide clues to the petrogenesis of those found in secondary occurrences.

Gem-quality sapphire is known from syenite pegmatites in the Ilmen Mountains of the South Urals. An occurrence of gem-quality sapphires from these mountains was also discovered within meta-ultramafites where sapphires form concentric aggregates within spinel-gahnite coronas. Sapphires form irregular-shaped translucent to transparent crystals with colorless and blue zones. The concentric sapphire aggregates with spinel coronas are embedded in a clinochlore-muscovite matrix. In this occurrence, corundum(sapphire)–spinel(gahnite)–clinochlore-muscovite aggregates associate with anorthite (An$_{100}$), apatite, zircon, diaspore, microlite, allanite-dissakisite-(Ce), monazite-(Ce), dravite, Ba-feldspars, and various Ni-bearing minerals. The mantle-generated ultramafic host rocks (orthopyroxenites) that are composed of enstatite with minor olivine have undergone significant serpentinization. Altered meta-ultramafic host rocks have high contents of SiO$_2$ (approximately 50–61 wt. %) and MgO (approximately 20–32 wt. %), but low Al$_2$O$_3$ (0.11–3.97 wt. %), as well as alkaline elements concentrations (in total, about 0.1–0.2 wt. %), whereas the sapphire-bearing rocks showed high MgO (16–25 wt. %) and high Al$_2$O$_3$ (16.26–38.52 wt. %) contents combined with variable SiO$_2$ (25–80 wt. %), but low concentrations of alkaline element (Na$_2$O + K$_2$O $\leq$ 1 wt. %). Most of the sapphire-bearing rocks—in contrast with the meta-ultramafic host rocks—are enriched in LREE and depleted in HREE with negative Eu anomalies. Spider REE diagrams of the studied samples are similar to those of sapphires in the syenite pegmatites, linked to intrusions of miascites (nepheline syenite) in the Ilmen Mountains. Sapphire-spinel-clinochlore-muscovite rock formed after metasomatic alteration of orthopyroxenites at a temperature around 700–750 °C and pressure about 1.8–3.5 kbar. The metasomatic fluids enriched in HFSE, LILE, and alumina formed after the partial melting of miascites (nepheline syenites), while Mg and some siderophile elements (Ni) came from mantle-generated meta-ultramafic host rocks. Formation of this metasomatic rock occurred at about 289 $\pm$ 9 Ma (Rb-Sr dating of sapphire-bearing rocks and mono-mineral muscovite associated with sapphires), representing the cooling ages after the peak metamorphism at about 330–320 Ma during the Uralian orogeny.

Sapphire crystallized after the breakdown of diaspore by the reaction: 2diaspore $\rightarrow$ corundum + H$_2$O. Chlorite occurred during the metasomatic reworking of meta-ultramafic host rocks. However, the spinel corona surrounding sapphire aggregates most likely formed as a consequence of metasomatic fluid-chlorite interactions with the corundum. Under decreasing pressure and temperature, muscovite crystallized during the breakdown of corundum by the reaction: corundum + orthoclase + H$_2$O $\rightarrow$ muscovite.

Despite on obvious metasomatic genesis, LA-ICP-MS analyses of blue sapphires showed Ga/Mg < 2.7, Fe/Mg < 74, Cr/Ga > 1.5, and Fe/Ti < 9, in the range of "metamorphic" sapphires. The Fe versus Ga/Mg diagram and Fe-Ti × 10-Mg × 100 ternary plot should be used with the caution and possibly revised using solely data of gem-quality samples and to group them by color.

Supplementary Materials: The following are available online at http://www.mdpi.com/2075-163X/9/1/36/s1, Table S1: Description of sapphire-spinel samples used for WDS EMPA and LA-ICP-MS, Table S2: Mineral assemblage and mode of minerals in the thin-sections, Table S3: Minerals identified in association with the sapphire, Table S4: Texture relationship of minerals found in association with sapphire, Table S5: Chemical composition of spinel and gahnite, Table S6: Chemical composition of muscovite and clinochlore, Table S7: Chemical composition of feldspar group minerals and marialite, Table S8: Chemical composition of apatite group minerals, Table S9: Chemical composition of zircon and monazite-(Ce), Table S10: Chemical composition of sulphide minerals, Table S11: Chemical composition of allanite-(Ce) group minerals, Table S12: Chemical composition of amphibole group minerals, Table S13: EDXRF (wt. %) and ICP-MS (ppmw) analyses of host rocks from 418 mine, Table S14: Rb-Sr and Sm-Nd isotope measurements of corundum-bearing rocks. Figure S1: A plot of Ilmen sapphires within meta-ultramafic host rocks (blue-colored samples—bold blue circles, colorless samples—colorless circles) in a Fe versus Ga/Mg diagram showing boundaries for magmatic (MAF: Main Asian Field; magmatic sapphires in alkali basalt) and metamorphic sapphires, modified after [14,18].

Author Contributions: E.S.S. formulated the idea of a paper, collected the samples for studying, conducted a research on the Raman spectroscopy, LA-ICP-MS, and TIMS, and their following data-reduction process, designed EDXRF and WDS EMPA experiments, assembled some tables and figures, and wrote the manuscript; M.A.R. collected the samples for host rock studying, performed the petrography of corundum-bearing and host rocks, designed the EDS EMPA and ICP-MS experiments, assembled some tables and figures, provided the field data and mineralogy of host rocks; S.N.N. performed the petrography of host rocks and provided the geological setting data; S.K. helped with data interpretation and editing of the manuscript; N.N.K. performed WSD EMPA of selected samples; A.G.N. carried out UV-Vis-NIR spectroscopy of sapphires and provided its interpretation; M.O.A. assisted with LA-ICP-MS measurements and data-reduction process; A.V.S. assisted with both sample preparation process for TIMS and TIMS measurements; Y.A.K. created some ideas for TIMS measurements and interpretation of final TIMS data; V.A.K. performed the EDS EMPA measurements.

Funding: The research has been supported by the Council on grants of the President of the Russian Federation (project no. MK-4459.2018.5); the Raman measurements and field trip to Ilmen State Reserve in 2016 were supported by Gemological Institute of America through the R.T. Liddicoat Postdoctoral Research Fellowship for one author (ESS); UV-Vis-NIR measurements were supported by Russian Government Program of Competitive Growth of Kazan Federal University.

Acknowledgments: The work in the Ilmen Natural Reserve was conducted under an agreement on scientific collaboration between the GEOKHI RAS and the Ilmen Natural Reserve. The samples used in this study were transferred by the Ilmen State Reserve for scientific purposes. The authors thank the Editor of Minerals and Guest Editors for invitation to submit the manuscript for consideration and possible contribution in a special issue on "Mineralogy and geochemistry of gems". Pautov L.N. and Agakhanov A.A. (Fersman Mineralogical Museum RAS), Phillipova K.A. (Institute of Mineralogy Ural branch of RAS), and Medvedeva E.V. (Ilmen State Reserve Ural branch RAS) helped with electron-microprobe, ICP-MS measurements, and consultations on the topic of research. Kuz'mina T.G., Romashova T.V., Smirnov V.I., and Turkov V.A. (GEOKHI RAS) assisted in sample preparation process and EDXRF measurements. We are thankful to John Saul (Swala Gem Traders, Tanzania), four anonymous reviewers, and Guest Editor of this special issue Panagiotis Voudouris for improving the text of an earlier version of this article.

Conflicts of Interest: The authors declare no conflict of interest.

References

1. Barbot-de-Marni, P.N. About a new deposit of zircon, tantalite and corundum. *Min. J.* **1828**, *3*, 171–172. (In Russian)

2. Polyakov, V.O.; Bazhenov, A.G.; Petrov, V.I. Mineral associations of corundum of the Ilmen Mountains. In *New Data on the Mineralogy of Endogenous Deposits and Zones of Technogenesis in the Urals*; Ural Branch RAS: Sverdlovsk, Russia, 1991; pp. 15–21.

3. Giuliani, G.; Ohnenstetter, D.; Fallick, A.E.; Groat, L.; Fagan, A.G. The geology and genesis of gem corundum deposits. In *Geology of Gem Deposits*, 2nd ed.; Groat, L.A., Ed.; Mineralogical Association of Canada Short Course Series; Mineralogical Association of Canada: Tucson, AZ, USA, 2014; Volume 44, pp. 29–112.

4. Lennikh, V.I.; Valizer, P.M. To the geological scheme of the Ilmenogorsky complex. In *Geology and Mineralogy of the Ilmenogorsky Complex: Situation and Problems*; Ilmen State Reserve Ural Branch RAS: Miass, Russia, 2006; pp. 20–27. (In Russian)

5. Krasnobaev, A.A.; Puzhakov, B.A.; Petrov, V.I.; Busharina, S.V. Zirconology of metamorphites of the Kyshtym-Arakulian strata of the Sysert-Ilmenogorsky complex. *Proc. Zavaritsky Inst. Geol. Geochem. (Trudy Instituta Geologii i Geokhimii im. Akademika A.N. Zavaritskogo)* **2009**, *156*, 264–268.

6. Sorokina, E.S.; Karampelas, S.; Nishanbaev, T.P.; Nikandrov, S.N.; Semiannikov, B.S. Sapphire Megacrysts in Syenite Pegmatites from the Ilmen Mountains, South Urals, Russia: New Mineralogical Data. *Can Mineral.* **2017**, *55*, 823–843. [CrossRef]

7. Medvedeva, E.V.; Rusin, A.I.; Murdasova, N.M.; Kotlyarov, V.A. Mineralogy of andalusite-kyanite-sillimanite rocks of the Saitovsky series in Ilmeny Mountains, the South Urals. *Zapiski RMO* **2012**, *141*, 50–60. (In Russian)

8. Korinevsky, E.V. *Chaotic Formations of the Ilmenogorsky Metamorphic Complex of the Southern Urals and Their Nature*; Ural Branch RAS: Ekaterinburg, Russia, 2013; p. 112. (In Russian)

9. Jochum, K.P.; Weis, U.; Stoll, B.; Kuzmin, D.; Yang, Q.; Raczek, I.; Jacob, D.E.; Stracke, A.; Birbaum, K.; Frick, D.A.; et al. Determination of reference values for NIST SRM 610-617 glasses following ISO Guidelines. *Geostand. Geoanal. Res.* **2011**, *35*, 397–429. [CrossRef]

10. Jochum, K.P.; Scholz, D.; Stoll, B.; Weis, U.; Wilson, S.A.; Yang, Q.; Schwalb, A.; Börner, N.; Jacob, D.E.; Andreae, M.O. Accurate trace element analysis of speleothems and biogenic calcium carbonates by LA-ICP-MS. *Chem. Geol.* **2012**, *318–319*, 31–44. [CrossRef]

11. Ludwig, K.R. *Isoplot/Ex: A Geochronological Toolkit for Microsoft Excel*; Special Publication No. 1a (rev. 3); Berkeley Geochronology Center: Berkeley, CA, USA, 2004.

12. Whitney, D.L.; Evans, B.W. Abbreviations for names of rock-forming minerals. *Am Mineral.* **2010**, *95*, 185–187. [CrossRef]

13. Nikandrov, S.N.; Rassomakhin, M.A.; Nishanbaev, T.P. List of minerals of Ilmen Mountains (data for 2017). *Mineralogy* **2017**, *1*, 52–60. (In Russian)

14. Peucat, J.J.; Ruffault, P.; Fritch, E.; Bouhnik-Le-Coz, M.; Simonet, C.; Lasnier, B. Ga/Mg ratio as a new geochemical tool to differentiate magmatic from metamorphic blue sapphires. *Lithos* **2007**, *98*, 261–274. [CrossRef]

15. Sutherland, F.L.; Abduriyim, A. Geographic typing of gem corundum: A test case from Australia. *J. Gemmol.* **2009**, *31*, 203–210. [CrossRef]

16. Stone-Sundberg, J.; Thomas, T.; Sun, Z.; Guan, Y.; Cole, Z.; Equall, R.; Emmett, J. Accurate Reporting of Key Trace Elements in Ruby and Sapphire Using Matrix-Matched Standards. *Gems Gemol.* **2017**, *53*, 438–451. [CrossRef]

17. Sorokina, E.S.; Koivula, J.I.; Muyal, J.; Karampelas, S. Multiphase fluid inclusions in blue sapphires from the Ilmen Mountains, southern Urals. *Gems Gemol.* **2016**, *52*, 209–211.

18. Giuliani, G.; Caumon, G.; Rakotosamizanany, S.; Ohnenstetter, D.; Rakototondrazafy, M. Classification chimique des corindons par analyse factorielle discriminante: Application à la typologie des gisements de rubis et saphirs. *Revue Gemmol.* **2014**, *188*, 14–22.

19. Zwaan, J.C.; Buter, E.; Mertz-Kraus, R.; Kane, R.E. The origin of Montana's alluvial sapphires. *Gems Gemol.* **2015**, *51*, 370–391.

20. Sorokina, E.S.; Rassomakhin, M.A.; Nikandrov, S.N.; Anosova, M.O.; Kononkova, N.N.; Kuz'mina, T.G.; Romashova, T.V.; Philippova, K.A. Geochemistry of new blue corundum (sapphire) occurrence in Ilmen Mountains, South Urals of Russia: Clues to "metamorphic" sapphire petrogenesis in placer deposits. In Proceedings of the EGU General Assembly 2018, Vienna, Austria, 8–13 April 2018; Volume 20.

21. Azimov, P.Y.; Bushmin, S.A. Solubility of minerals of metamorphic and metasomatic rocks in hydrothermal solutions of varying acidity: Thermodynamic modeling at 400–800 °C and 1–5 kbar. *Geochem. Int.* **2007**, *45*, 1210–1234. [CrossRef]

22. Pokrovskii, V.A.; Helgeson, H.C. Thermodynamic properties of aqueous species and the solubility of minerals at high pressures and temperatures: The system Al_2O_3-H_2O-NaCI. *Chem Geol.* **1995**, *137*, 221–242. [CrossRef]

23. Pin, C.; Monchoux, P.; Paquette, J.-L.; Azambre, B.; Wang, R.C.; Martin, R.F. Igneous albitite dikes in orogenic lherzolites, western Pyrénées, France: A possible source for corundum and alkali feldspar xenocrysts in basaltic terranes. II. Geochemical and petrogenetic considerations. *Can. Mineral.* **2006**, *44*, 843–856. [CrossRef]

24. Rusin, A.I.; Valizer, P.M.; Krasnobaev, A.A.; Baneva, N.N.; Medvedeva, E.V.; Dubinina, E.V. Origin of the garnet-anortite-clinopyroxene-amphibole rocks from the Ilmenogorskii complex (Southern Urals). *Lithosphere* **2012**, *1*, 91–109. (In Russian)

25. Nedosekova, I.L.; Vladykin, N.V.; Pribavkin, S.V.; Bayanova, T.B. The Il'mensky-Vishnevogorsky Miaskite-Carbonatite Complex, the Urals, Russia: Origin, Ore Resource Potential, and Sources. *Geol. Ore Depos.* **2009**, *51*, 139–161. [CrossRef]

26. McDonough, W.F.; Sun, S.S. The composition of the Earth. *Chem. Geol.* **1995**, *120*, 223–253. [CrossRef]

27. Baneva, N.N.; Medvedeva, E.V.; Rusin, A.I. Geochemical characteristics of ultramafites of the Ilmenogorsky shear zone. *Proc. Zavaritsky Inst. Geol. Geochem. (Trudy Instituta Geologii i Geokhimii im. Akademika A.N. Zavaritskogo)* **2009**, *156*, 115–119. (In Russian)

28. Ivanov, K.S. The Main Features of the Geological History (1.6–0.2 Billion Years) and the Structure of the Urals. Unpublished Ph.D. Thesis, Urb RAS, Yekaterinbur, Russia, 1998.

29. Dzikowski, T.J.; Cempirek, J.; Groat, L.A.; Dipple, G.M.; Giuliani, G. Origin of gem corundum in calcite marble: The Revelstoke occurrence in the Canadian Cordillera of British Columbia. *Lithos* **2014**, *198–199*, 281–297. [CrossRef]

30. Zhang, Z.-M.; Ding, H.-X.; Dong, X.; Tian, Z.-L.; Mu, H.-C.; Li, M.-M.; Qin, S.-K.; Niu, Z.-X.; Zhang, N. The Eocene corundum-bearing rocks in the Gangdese arc, south Tibet: Implications for tectonic evolution of the Himalayan orogeny. *Geosci. Front.* **2018**, *9*, 1337–1354. [CrossRef]

31. Yakymchuk, C.; Szilas, K. Corundum formation by metasomatic reactions in Archean metapelite, SW Greenland: Exploration vectors for ruby deposits within high-grade greenstone belts. *Geosci. Front.* **2018**, *9*, 727–749. [CrossRef]

32. Grapes, R.; Palmer, K. (Ruby-sapphire)-chromian mica-tourmaline rocks from Westland, New Zealand. *J Petrol.* **1996**, *37*, 293–315. [CrossRef]

33. Icenhower, J.; London, D. An experimental study of element partitioning among biotite, muscovite, and coexisting peraluminous silicic melt at 200 MPa (H_2O). *Am. Mineral.* **1995**, *80*, 1229–1251. [CrossRef]

34. Yoder, H.S.; Eugster, H.P. Synthetic and natural muscovites. *Geochim. Cosmochim. Acta* **1955**, *8*, 225–280. [CrossRef]

35. Storre, B.; Karotke, E. An experimental determination of the upper stability limit of muscovite + quartz in the range 7–20 kb water pressure. *Neues Jahrb.Mineral. Monatsh.* **1971**, 237–240.

36. Merrill, R.B.; Robertson, J.K.; Wyllie, P.J. Melting reactions in the system $NaAlSi_3O_8$-$KAlSi_3O_8$-SiO_2-H_2O to 20 kilobars compared with results for the feldspar-quartz-H_2O and rock-H_2O systems. *J. Geol.* **1970**, *78*, 558–569. [CrossRef]

37. Althaus, E.; Karotke, E.; Nitsch, K.H.; Winkler, H.G.F. An experimental re-examination of the upper stability limit of muscovite plus quartz. *Neues Jahrb. Mineral. Monatsh* **1970**, *7*, 325–336.

38. Hall, A.L. On the marundites and allied corundum-bearing rocks in the Leysdorf district of the Eastern Transvaal. *Trans. Geol. Soc. S. Afr.* **1923**, *24*, 43–67.

39. Anhaesseur, C.R. On an Archean marundite occurrence (corundum margarite rock) in the Barberton Mountain Land, Eastern Transvaal. *S. Afr. J. Geol.* **1978**, *81*, 211–218.

40. Morrison, E.R. *Corundum in Rhodesia*; Rhodesia Geological Survey: Salisbury, Zimbabwe, 1972; 24p.

41. Schreyer, W.; Werding, G.; Abraham, K. Corundum-fuchsite rocks in greenstone belts of Southern Africa: Petrology, geochemistry and possible origin. *J. Petrol.* **1981**, *22*, 191–231. [CrossRef]

42. Kerrich, R.; Fyfe, W.S.; Barnett, R.L.; Blair, B.B.; Willmore, L.M. Corundum, Cr-muscovite rocks at O'Briens, Zimbabwe: The conjunction of hydrothermal desilicification and LIL-element enrichment–geochemical and isotopic evidence. *Contrib. Mineral. Petrol.* **1987**, *95*, 481–498. [CrossRef]

43. Wang, K.K.; Graham, I.T.; Lay, A.; Harris, S.J.; Cohen, D.R.; Voudouris, P.; Belousova, E.; Giuliani, G.; Fallick, A.E.; Greig, A. The Origin of a new Pargasite-Schist Hosted Ruby Deposit from Paranesti, Northern Greece. *Can. Mineral.* **2017**, *55*, 535–560. [CrossRef]

44. Guo, J.; O'Reilly, S.Y.; Griffin, W.L. Corundum from basaltic terrains: A mineral inclusion approach to the enigma. *Contrib. Mineral. Petrol.* **1996**, *122*, 368–386. [CrossRef]

45. Graham, I.; Sutherland, F.L.; Zaw, K.; Nechaev, V.; Khanchuk, A. Advances in our understanding of the gem corundum deposits of the West Pacific continental margins intraplate basaltic fields. *Ore Geol. Rev.* **2008**, *34*, 200–215. [CrossRef]

46. Palke, A.; Wong, J.; Verdel, C.; Ávila, J.N. A common origin for Thai/Cambodian rubies and blue and violet sapphires from Yogo Gulch, Montana, U.S.A.? *Am. Mineral.* **2018**, *103*, 469–479. [CrossRef]

47. Wong, J.; Verdel, C.; Allen, C. Trace-element compositions of sapphire and ruby from the eastern Australian gemstone belt. *Min. Mag.* **2017**, *81*, 1551–1576. [CrossRef]

Article

Corundum Anorthosites-Kyshtymites from the South Urals, Russia: A Combined Mineralogical, Geochemical, and U-Pb Zircon Geochronological Study

Maria I. Filina [1,*], Elena S. Sorokina [1,2], Roman Botcharnikov [2], Stefanos Karampelas [3], Mikhail A. Rassomakhin [4,5], Natalia N. Kononkova [1], Anatoly G. Nikolaev [6], Jasper Berndt [7] and Wolfgang Hofmeister [2]

[1] Vernadsky Institute of Geochemistry and Analytical Chemistry Russian Academy of Sciences (GEOKHI RAS), Kosygin str. 19, 119991 Moscow, Russia; elensorokina@mail.ru (E.S.S.); nnzond@geokhi.ru (N.N.K.)

[2] Institut für Geowissenschaften, Johannes Gutenberg Universität Mainz, J.-J.-Becher-Weg 21, 55128 Mainz, Germany; rbotchar@uni-mainz.de (R.B.); hofmeister@uni-mainz.de (W.H.)

[3] Bahrain Institute for Pearls & Gemstones (DANAT), WTC East Tower, P.O. Box 17236 Manama, Bahrain; stefanos.karampelas@gmail.com

[4] Institute of Mineralogy SU FRC MiG UB RAS, 456317 Miass, Chelyabinsk Region, Russia; miha_rassomahin@mail.ru

[5] Ilmen State Reserve SU FRC MiG UB RAS, 456317 Miass, Chelyabinsk Region, Russia

[6] Department of mineralogy and lithology, Institute of Geology and Petroleum Technologies, Kazan Federal University, 420008 Kazan, Russia; anatolij-nikolaev@yandex.ru

[7] Institut für Mineralogie, Westfälische Wilhelms Universität Münster, Corrensstrasse 24, 48149 Münster, Germany; jberndt@uni-muenster.de

* Correspondence: makimm@mail.ru; Tel.: +7(499)137-14-84

Received: 25 February 2019; Accepted: 11 April 2019; Published: 16 April 2019

Abstract: Kyshtymites are the unique corundum-blue sapphire-bearing variety of anorthosites of debatable geological origin found in the Ilmenogorsky-Vishnevogorsky complex (IVC) in the South Urals, Russia. Their mineral association includes corundum-sapphire, plagioclase (An_{61-93}), muscovite, clinochlore, and clinozoisite. Zircon, churchite-(Y), monazite-(Ce), and apatite group minerals are found as accessory phases. Besides, churchite-(Y) and zircon are also identified as syngenetic solid inclusions within the sapphires. In situ Laser Ablation Inductively Coupled Plasma Mass Spectrometry (LA-ICP-MS) U-Pb zircon geochronology showed the ages at about 290–330 Ma linked to the Hercynian orogeny in IVC. These ages are close to those of the syenitic and carbonatitic magmas of the IVC, pointing to their syngenetic origin, which is in agreement with the trace element geochemistry of the zircons demonstrating clear magmatic signature. However, the trace element composition of sapphires shows mostly metamorphic signature with metasomatic overprints in contrast to the geochemistry of zircons. The reason for this discrepancy can be the fact that the discrimination diagrams for sapphires are not as universal as assumed. Hence, they cannot provide an unambiguous determination of sapphire origin. If it is true and zircons can be used as traces of anorthosite genesis, then it can be suggested that kyshtymites are formed in a magmatic process at 440–420 Ma ago, most probably as plagioclase cumulates in a magma chamber. This cumulate rock was affected by a second magmatic event at 290–330 Ma as recorded in zircon and sapphire zoning. On the other hand, Ti-in-zircon thermometer indicates that processes operated at relatively lower temperature (<900 °C), which is not enough to re-melt the anorthosites. Hence, zircons in kyshtymites can be magmatic but inherited from another rock, which was re-worked during metamorphism. The most probable candidate for the anorthosite protolith is carbonatites assuming that metamorphic fluids could likely leave Al- and Si-rich residue, but removed Ca and CO_2. Further, Si is consumed by the silicification of ultramafic host rocks. However, kyshtymites do not show clear evidence of pronounced metasomatic zonation and evidence for large volume changes due to metamorphic alteration of carbonatites. Thus, the obtained data still do not allow for univocal reconstruction of the kyshtymite origin and further investigations are required.

Keywords: blue sapphire; anorthosites; kyshtymites; sapphire geochemistry; Ilmenogorsky-Vishnevogorsky complex; in situ LA-ICP-MS U-Pb zircon dating

---

1. Introduction

Corundum α-Al$_2$O$_3$ is the common mineral of many magmatic and metamorphic rocks. However, its blue gem-quality variety (i.e., sapphire) colored mainly by iron and titanium is rare and commonly found in secondary placers of debatable origin [1]. Recent studies of blue sapphires from different deposits worldwide demonstrate growing interest in their genesis due to findings of blue sapphires in situ within the primary rocks, for instance, those from alkali basaltic terrains [2–8].

Corundum in anorthosites is seldomly found worldwide and rarely of gem-quality, e.g., there is anorthosite occurrence with gem-quality pink corundum discovered in Fiskenaesset complex of W. Greenland. Pink corundum associated with coarse-grained, radial anthophyllite, green pargasite, green or red spinel, sapphirine, cordierite, and phlogopite [9]. Anorthosites with pink corundum are known from the Sittampundi Layered Complex, Tamil Nadu in India. These anorthosites belong to the rare group of metamorphosed Archean layered complexes, which are the part of oceanic crust formed in back-arc settings [10]. Anorthosites with colorless corundums were also found in Central Fiordland, New Zealand, where anorthositic complex is a part of the Tuyuan orogenic belt undergone a multiphase metamorphism of the amphibolite facies [11]. Another corundum anorthosite from the Chunky Gal Mountain (North Carolina, USA) [12] with pink-colored corundum is located in association with amphibolites and peridotites within alpine-type orogenic belt [13].

Corundum anorthosites (kyshtymites) are known for more than two centuries in South Urals of Russia; however, the genesis of these rocks remains enigmatic with the latest research results performed more than 50 year ago. For better understanding the origin of kyshtymites, their mineralogy, in situ LA-ICP-MS trace-element geochemistry and geochronology of zircon, and UV-Vis-NIR spectroscopy were studied by modern analytical techniques. The obtained data provide new insights into the origin of blue sapphires in anorthosites of Ilmenogorsky-Vishnevogorsky complex (IVC). Mineralogy and geochemistry of sapphires within kyshtymites were also compared to those found in other primary occurrences in IVC of South Urals (corundum-blue sapphire syenite pegmatites, and sapphire-bearing metasomatites within meta-ultramafic host rocks) and to those from secondary placer occurrences with similar geochemical and mineralogical features.

2. Geological Setting

The studied corundum deposit named "the 5th versta" was discovered by Karpinsky in 1883 [14]. Three kyshtymite veins were discovered during exploration of the deposit. Corundum was used mainly as an abrasive material, some of the crystals were gem-quality, however, these rough crystals did not exceed 1 carat. The exploration of the deposit was prosecuted until the 1930s and, currently, the occurrence is almost exhausted.

Blue sapphires in kyshtymites (the 5th versta deposit and the larger occurrence called Borzovsky deposit [15–17]) are located at the western flank of Vishnevogorsky nepheline syenite (miascite)-carbonatite alkaline complex of the South Urals with unique REE-mineralization [18] (Figure 1). These two deposits are accompanied by other primary sapphire occurrences in syenite pegmatites (mines 298 and 349) [2] and sapphires in metasomatites within meta-ultramafic host rocks (mine 418) (Figure 1) [3].

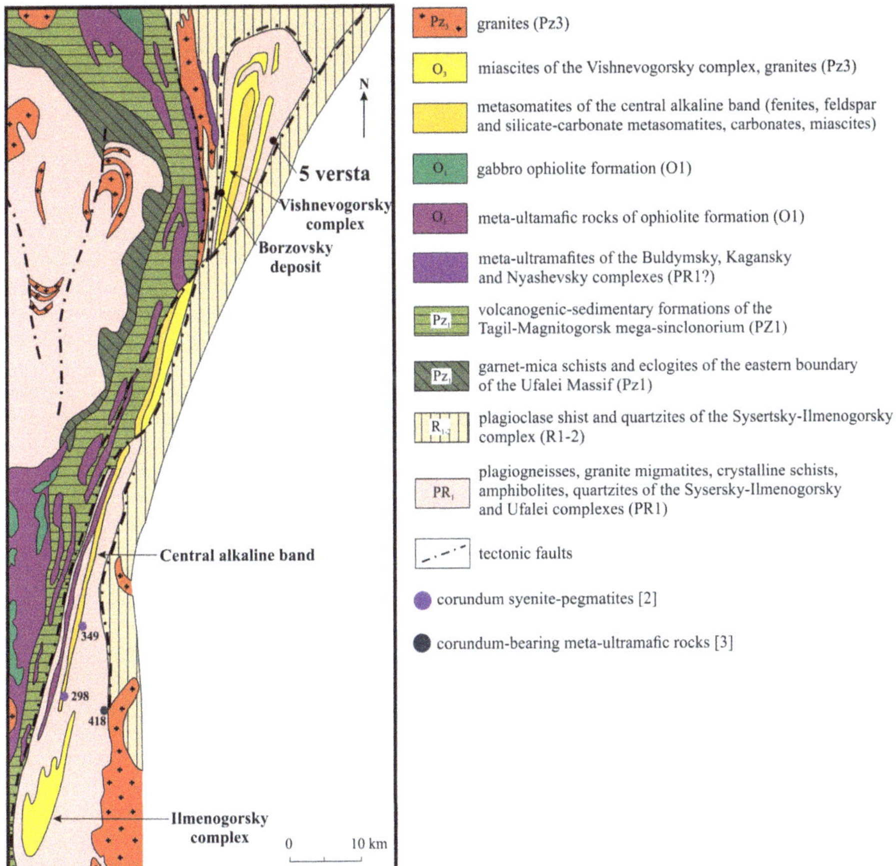

Figure 1. Geology of the Ilmenogorsky-Vishnevogorsky alkaline-carbonatite complex [18]. Coordinates of the 5th versta deposit are 55°54′03″N, 60°41′16″E.

The kyshtymites are found within the meta-ultramafic host rocks located among the quartzite shales of the meta-terrigenous Saitovsky series. Saitovsky series is one of the structural units within the Ilmenogorsky-Vishnevogorsky polymetamorphic zone, which is a deep fragment of the regional post-collisional shear [19]. The series with meta-ultramafites undergone the re-working during several thermal events (SHRIMP U-Pb zircon geochronology). The age of the mantle protolith was dated at ~1.3 Ga [20]. A stage of metamorphic evolution linked to the miascite intrusions was at about 450–420 Ma [21–23]. The stage of metamorphism and granite formation in the Sysertsky-Ilmenogorsky block linked to the Hercynian orogeny was at 360–320 Ma [24], whereas the ages of ~330–270 Ma corresponds to the collision processes [24].

At the beginning of the 20th century, the studied vein of kyshtymites of a lenticular body (Figure 2) was explored from the surface by quarry extended currently to the depth of about 3–4 m. The meta-ultramafic host rocks (initial orthopyroxenites) were composed mainly of enstatite, which undergone metasomatism (serpentinization). The reaction rim with a thickness of 10–25 cm consisting of chrysotile-asbestos was detected at the contact of the meta-ultramafic host rock with the kyshtymites [25].

Figure 2. Vein of corundum-blue sapphire anorthosites-kyshtymites: (**a**) the kyshtymite outcrop, thickness is about 3 m; and (**b**) A kyshtymite vein is at a contact with the meta-ultramafic host rock, their contact showed by a green-colored reaction rim consisting of chrysotile-asbestos.

Kyshtymites with meta-ultramafic host rocks are adjacent from the west by the miascites (nepheline syenites) of the Vishnevogorsky complex consisting of potassium feldspar (20–60 wt. %), nepheline (20–30 wt. %), lepidomelane (5–20% wt. %), amphibole (up to 20 wt. %), and plagioclase (up to 20 wt. %). Additionally, calcite (up to 3 wt. %), cancrinite, and sodalite were identified in miascites [26].

3. Materials and Methods

Fourteen kyshtymite samples with blue corundum-sapphires, one sample from the reaction rim, and one sample of meta-ultramafic host rock were investigated in this study. The list of research methods is provided in Table S1. Minerals of kyshtymites, meta-ultramafic host rock, and the reaction rim between them were identified by the optical microscopy in the petrographic thin-sections. Some of the samples were studied by the Raman spectroscopy at the Renishaw Moscow using the Renishaw inVia Raman spectrometer coupled with Ar^+ green Stellar-REN Modu-Laser (Renishaw plc, Gloucestershire, UK) with $\lambda = 514$ nm and 50× magnification at room temperature. The laser power was 10 mW on the sample with a 60 s acquisition time (three cycles), at a resolution of about 1.5 cm^{-1}. Rayleigh scattering was blocked using a holographic notch filter. Backscattered light was dispersed using a grating with 1800 grooves/mm. The spectrometer was calibrated at 520.7 cm^{-1} using Si as a reference.

The mineral chemistry was studied by the electron micro-probe analyses (EMPA) using Cameca SX 100 electron microprobe (CAMECA, Gennevilliers, France) in the wavelength-dispersive detection mode (WDS) at GEOKHI RAS, Moscow. The accelerating voltage was 15 kV, current was 30 nA, and beam size was from 3 to 5 μm. Both natural and synthetic reference materials were used for the instrument control: andradite for Si, jadeite for Na, orthoclase for K and Al, augite for Ca and Fe, olivine for Mg, rhodonite for Mn, TiO_2 for Ti, vanadinite for V, Cr_2O_3 for Cr, apatite for P, galena for Pb, $Rb_2Nb_4O_{11}$ for Nb, xenotime-(Y) for Y, metallic Gd for Gd, Pr_3PO_4 for Pr, Sm_3PO_4 for Sm, La_3PO_4 for La, and homogeneous glasses for Zr, Ta, Th, and U. The detection limits for almost all elements were less than 0.01 wt. %, the lower limits of the determined values were less about 0.03 wt. %. Correction coefficients are determined by PAP correlation (atomic number, fluorescence, and absorption correction).

Trace element composition of three representative sapphire crystals from samples K-8 and K-12 was determined using Laser Ablation–Inductively Coupled Plasma–Mass Spectrometry (LA-ICP-MS) at the Institute of Geology of Ore Deposits, Petrography, and Mineralogy RAS (IGEM RAS), Moscow. The analyses were conducted using New Wave Research UP-213 Nd:YAG laser (New Wave Research, Inc., Fremont, CA, USA) combined with the XSERIES 2 ICP-Mass Spectrometer (Thermo Scientific, Waltham, MA, USA). Trace-element concentrations were determined by the monitoring of ^{6}Li, ^{9}Be, ^{24}Mg, ^{27}Al, ^{44}Ca, ^{47}Ti, ^{51}V, ^{53}Cr, ^{57}Fe, ^{71}Ga, and ^{91}Zr and ablating a material with a spot size of 60 μm at a repetition rate of 10 Hz, and an energy density of about 14–15 J/cm^2. Warm up/background time was 15 s, dwell time was 40 s, and wash out time was 20 s. The NIST SRM 610 and NIST SRM 612 glasses were used as reference materials and the BHVO–2G glass as a quality control material (QCM). The time-resolved signal was processed in Igor (IOLITE) commercial software using ^{27}Al as the internal standard applying the theoretical value of 52.93 wt. % of Al in pure crystalline $\alpha–Al_2O_3$ for analyzing the corundum unknowns. The measured concentrations of reference material and QCM agree for all elements within 10% and 15%, respectively, of preferred values by Jochum et al. 2011 [27]. This larger discrepancy between measured and preferred values can be attributed to isobaric interferences that cannot be resolved with the instrumentation used [28].

UV-Vis-NIR (Ultraviolet-Visible-Near Infrared) absorption spectroscopy was acquired at Kazan Federal University on one sapphire (sample 12-K). Spectra were recorded at a room temperature using SHIMADZU UV-3600 spectrometer (Shimadzu Corp, Kyoto, Japan) with 2 light sources (one for UV and one for Vis-NIR) and 2 detectors (FEU R928 for UV-Vis and InGaAs for NIR) from 185 to 3300 nm range with a data interval and spectra bandwidth of 1 nm and a scan rate of 300 nm/min.

Chemical composition of six selected kyshtymite rock samples macroscopically representing different textures and containing from 30 to 50 wt. % of sapphires, one sample of meta-ultramafic host rocks, and one sample of reaction rim were used for whole-rock major element analyses at GEOKHI RAS, Moscow using Energy Dispersive X-ray Fluorescence spectrometer (EDXRF) AXIOS Advanced (PANalytical B.V., Almelo, The Netherlands). The equipment provides the determination of quantitative concentrations of elements from oxygen to uranium from about 10^{-4} to 100 wt. %.

Three selected kyshtymite rock samples containing from 50 to 70 wt.% of sapphires with gem-quality zones in crystals, one sample of meta-ultramafic host rocks, and one sample of the reaction rim between kyshtymites and host rocks previously analyzed by EDXRF were used for the whole-rock trace-element analyses at the Institute of Oceanology RAS, Moscow. Inductively Coupled Plasma–Mass Spectrometer (ICP-MS) Agilent 7500 (Santa Clara, Ca, USA) was applied to determine the contents of REE and trace-elements. Calibration of the sensitivity over the entire mass scale was carried out using 68-element solutions references (ICP-MS-68A, HPS, solutions A and B). The STM-2 standard was used as a quality control material. Indium was added to all sample solutions in the concentration of 10 ng/g to control signal stability. The detection limits of the elements were 0.1 ng/g for the heavy and medium elements, and 1 ng/g for the light elements. Analytical uncertainties were less than 1–3%.

Trace-element composition and U-Pb isotopic analysis of 6 zircon grains identified in the thin-section of samples 8-K were analyzed using Element2 ICP-Mass Spectrometer (ThermoFisher Scientific, Waltham, MA, USA) coupled with an Analyte G2 (Photon Machines Inc, Redmond, WA, USA) laser at the Westfälische Wilhelms Universität Münster. Before the analysis, the backscattered electron (BSE) and cathodoluminescence (CL) images, as well as maps in average weighted atomic numbers were taken at GEOKHI RAS, Moscow using the same equipment as for the EMPA WDS to identify the internal structures within the zircon grains. During the LA-ICP-MS measurements, gas flow rates were about 1.1 L/min for He, 0.9 L/min and 1.1 L/min for the Ar-auxiliary and sample gas, respectively. Cooling gas flow rate was set to 16 L/min. Trace-elements concentrations were determined by measuring of ^{29}Si, ^{43}Ca, ^{49}Ti, ^{51}V, ^{53}Cr, and REE with a spot size of 20 μm at a repetition rate of 10 Hz, and an energy density of about ~3–4 J/cm^2. Background time was 15 s, dwell time was 40 s, and wash out time was 20 s. NIST SRM 612 was used as the reference materials, and 91,500 zircon and BIR-1G glass as unknowns for quality control (QCM). The time-resolved signal was processed in GLITTER commercial software using Zr as the internal standard applying the stoichiometric value of Zr in pure crystalline ZrSiO$_4$ for analyzing the zircon unknowns. The measured concentrations of reference material and QCM agree for all elements within 10% and 15%, respectively, of preferred values by [29]. For U-Pb zircon dating ^{204}Pb, ^{206}Pb, ^{207}Pb, and ^{238}U were measured along, with ^{202}Hg to correct the interference of ^{204}Hg on ^{204}Pb, which is important to apply, if necessary, for a common Pb correction. Repetition rate was 10 Hz using an energy of ~3 J/cm^2 and a spot size of 25 μm. Ten unknowns were bracketed with three calibration standards GJ1 [29] to correct for instrumental mass bias. The data reduction was performed using in-house Excel spreadsheet [30]. Along with the unknowns, 91,500 reference zircon [31] was measured to monitor accuracy and precision of the analysis. Long term reproducibility of the reference zircon standard 91,500 [31] yielded a Concordia age of 1074.8 ± 8.8 Ma, which is indistinguishable to the age of 1065.4 ± 0.3 Ma for the 91,500 reference zircon determined by TIMS [31]. The Concordia diagrams and age calculations were made using Isoplot v. 4.13 (Ludwig, 2009) [32]. Weighted mean age calculations are given at 95% confidence level.

4. Results

4.1. Petrology and Mineralogy of Kyshtymites

Mineral composition and chemistry are shown in Tables S2–S4. Kyshtymite consists mainly of idiomorphic blue-colored transparent to translucent corundums–sapphires (up 50 wt. %) and plagioclase with composition varying from labradorite to anorthite An$_{61-93}$ (30–50 wt. %) (Table S3 and Figure 3).

Figure 3. Photography of the samples used in this study with dipyramidal-prismatic corundum-sapphires (Crn) crystals elongated along the *c* axis and fine-grained plagioclase (Pl): (**a**) sample 2-K; and (**b**) sample 7-K.

The rock has a porphyritic structure, where large corundum-sapphire crystals are located among the fine-grained plagioclase, muscovite, clinozoisite, and clinochlore (Figure 4). Other rock-forming minerals—muscovite, clinochlore, and clinozoisite—occupy up to 10 wt. %. Accessory minerals were detected as zircon, churchite-(Y), apatite, and monazite-(Ce) (Figure 5).

Table 1. Rim-to-rim LA-ICP-MS profiles of studied corundum-blue sapphire from anorthosites- kyshtymites (in µg/g).

Sample/No. of Spot		Color	Li	Be	Mg	Ti	V	Cr	Fe*	Ga	Ga/Mg	Fe/Ti	Cr/Ga	Fe/Mg
K-8-1	1	Wt	32.45	1.85	199.85	972.77	5.59	3.18	1166.20	36.40	0.18	1.20	0.09	5.84
	2	Bl	8.34	1.67	183.70	700.93	5.96	12.34	1477.19	38.57	0.21	2.11	0.32	8.04
	3	Bl	bdl	bdl	334.58	782.98	6.49	2.59	2099.17	36.90	0.11	2.68	0.07	6.27
	4	Wt	bdl	bdl	74.20	702.78	6.35	3.31	855.22	36.53	0.49	1.22	0.09	11.53
	5	Wt	bdl	2.20	69.22	231.35	7.54	2.59	1399.44	33.72	0.49	6.05	0.08	20.22
	6	Wt	bdl	3.12	67.05	115.41	8.52	1.61	1710.43	33.38	0.50	14.82	0.05	25.51
	7	Bl	bdl	2.51	66.78	101.64	9.08	2.44	2332.41	33.03	0.49	22.95	0.07	34.92
	8	Bl	bdl	bdl	291.17	458.20	5.35	3.86	1865.92	32.00	0.11	4.07	0.12	6.41
	9	Wt	bdl	bdl	388.58	297.79	4.39	7.07	1710.43	30.36	0.08	5.74	0.23	4.40
	10	Wt	bdl	bdl	741.16	585.52	4.74	6.51	855.22	30.76	0.04	1.46	0.21	1.15
K-8-2	1	Wt	bdl	bdl	131.29	986.27	5.53	6.04	1788.18	32.77	0.25	1.81	0.18	13.62
	2	Bl	bdl	bdl	59.29	782.19	6.27	1.89	2021.42	36.16	0.61	2.58	0.05	34.09
	3	Bl	bdl	bdl	46.67	706.48	6.38	1.60	2021.42	37.30	0.80	2.86	0.04	43.32
	4	Bl	bdl	bdl	296.46	559.58	6.17	2.40	1865.92	37.19	0.13	3.33	0.06	6.29
	5	Wt	bdl	3.47	127.59	164.38	8.63	2.52	1166.20	34.76	0.27	7.09	0.07	9.14
	6	Wt	bdl	2.67	94.50	122.82	8.55	2.90	1166.20	34.46	0.36	9.50	0.08	12.34
	7	Bl	bdl	2.33	92.62	133.67	8.60	1.86	1943.67	35.50	0.38	14.54	0.05	20.99
	8	Bl	bdl	bdl	77.21	583.13	6.59	0.92	2410.15	39.81	0.52	4.13	0.02	31.21
	9	Wt	6.83	bdl	95.29	656.99	5.98	0.73	1943.67	37.72	0.40	2.96	0.02	20.40
	10	Wt	bdl	bdl	91.85	711.78	5.17	0.93	1399.44	36.40	0.40	1.97	0.03	15.24
K-12-1	1	Bl	bdl	bdl	164.38	939.95	3.97	bdl	1010.71	42.99	0.26	1.08	-	6.15
	2	Wt	bdl	2.81	167.82	294.08	6.70	bdl	621.97	56.25	0.34	2.11	-	3.71
	3	Wt	bdl	1.96	188.47	210.70	6.22	bdl	1321.70	61.33	0.33	6.27	-	7.01
	4	Wt	bdl	3.60	147.68	216.79	6.49	bdl	4198.33	57.49	0.39	19.37	-	28.43
	5	Wt	bdl	3.81	94.97	125.20	6.11	bdl	855.22	48.65	0.51	6.83	-	9.00
	6	Wt	bdl	bdl	222.08	205.14	3.71	bdl	1554.94	50.95	0.23	7.58	-	7.00
	7	Wt	bdl	bdl	127.77	156.44	6.14	bdl	1321.70	46.38	0.36	8.45	-	10.34
	8	Bl	bdl	1.85	157.23	374.82	8.07	bdl	5209.04	52.41	0.33	13.90	-	33.13
	9	Bl	bdl	bdl	190.32	524.11	8.71	0.03	3265.37	54.58	0.29	6.23	-	17.16
	10	Bl	bdl	bdl	237.44	555.61	7.91	0.35	3109.87	50.90	0.21	5.60	0.01	13.10

Wt, Bl, white and blue colors of sapphire. * Calculated fromWDS EMPA; bdl, below the detection limit.

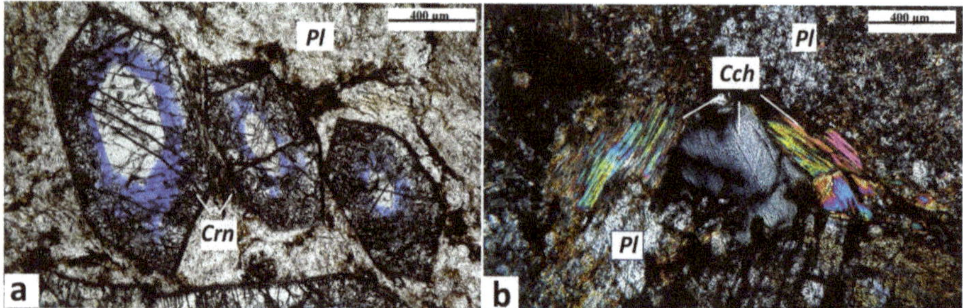

Figure 4. Photomicrographs of corundum-blue sapphire anorthosite-kyshtymite samples 8-K: (**a**) idiomorphic crystals of corundum-sapphire (Crn) with oscillatory white-blue zonation within the matrix of fine-grained plagioclase (Pl), parallel polarized light. The dotted arrows show the profiles measured by LA-ICP-MS (cf. Table 1). (**b**) Xenomorphic crystals of clinochlore (Cch) associated with plagioclase, crossed polarized light.

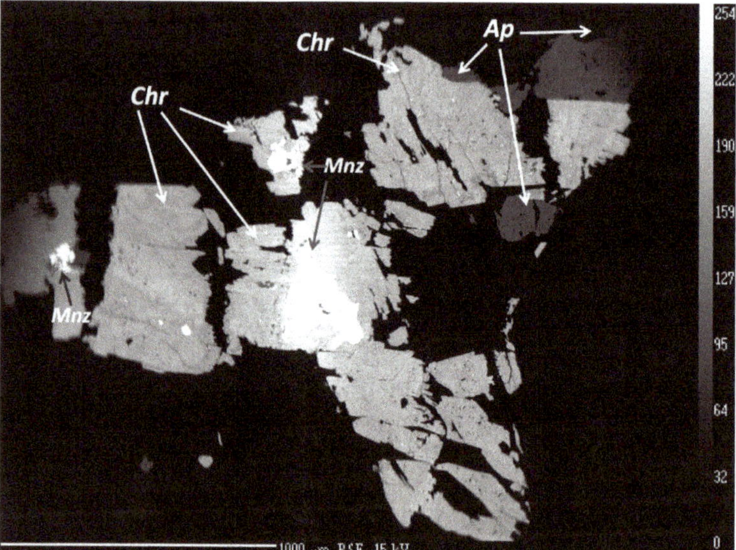

Figure 5. BSE image of churchite-(Y) (Chr), monazite-(Ce) (Mnz), and apatite group minerals (Ap) embedded in the plagioclase–muscovite–clinochlore matrix.

Muscovite occurs as plates of 0.1–0.2 mm in size common around plagioclase grains. The mineral is detected in association with clinochlore and clinozoisite, both epigenetic by nature. High MgO contents in muscovite (up to 1.71 wt. %) and K_2O in clinochlore (up to 7.11 wt. %) are associated with the replacement of muscovite by a clinochlore. Clinozoisite forms small rounded grains up to 0.1 mm in size replacing plagioclase. Clinozoisite contains up to FeO_{tot} of 2.07 wt. %.

Churchite-(Y), forms small syngenetic xenomorphic crystals of 30–70 μm in size commonly found as solid micro-inclusions within corundum-sapphire (Figure 5). Churchite-(Y) is also found in intergrowths with the monazite-(Ce) and likely apatite group minerals (Table S3). Zircon forms syngenetic and epigenetic prismatic or dipyramidal crystals with a size up to 40–100 μm, and contains up to 2.45 wt. % HfO_2 (Table S4).

4.2. Mineralogy, Geochemistry, Solid Inclusions, and UV-Vis-NIR-Spectroscopy of Corundum–Sapphire

Colorless to blue-colored translucent to transparent corundum-sapphires with the fractures passing through the entire crystals were found in kyshtymites. All sapphires show oscillatory zonation in the elongated dipyramidal-prismatic crystal sized up to 4 cm in length (Figure 3b). The most developed crystal faces are hexagonal prism (11$\bar{2}$0), pinacoid (0001), and hexagonal dipyramid (22$\bar{4}$3) (Figure 4a).

Sapphires from kyshtymites are almost inclusion-free except occasional finding of churchite-(Y) and zircon solid inclusions. Churchite-(Y) was found to be syngenetic with sapphires (Figure 5). The mineral contains traces of Gd_2O_3 (1.8–2.19 wt. %), Pr_2O_3 (0.10–0.17 wt. %), La_2O_3 (0.09–0.12 wt. %), Sm_2O_3 (0.91–1.05 wt. %), U_2O_3 (1.77–2.14 wt. %), and ThO_2 (0.55–72 wt. %). Inclusion of churchite-(Y) in sapphire has not been previously described. Therefore, it is likely the first identification and chemical analysis of churchite-(Y) solid inclusions within sapphires to the best of our knowledge.

Zircon solid inclusions in corundum-sapphire showed concentration of HfO_2 from 0.72 to 3.00 wt. % (one measurement showed 5.38 wt. %; see Table S5), which is almost the same as in zircon found in mineral association of kyshtymites (0.97–2.45 wt. %). Besides, the epigenetic zircon, muscovite, and clinochlore filling the sapphire fractures were also detected.

The UV-Vis-NIR spectra of the studied sapphires are comparable to those observed on sapphires of metamorphic or metasomatic origin [33] and of the other two occurrences in the region linked with syenitic pegmatite [2] and metasomatites within meta-ultramafic host rocks [3].

Corundum-sapphire trace element measurements by LA-ICP-MS are shown in Table 1. The Fe content in blue colored zones varies from 1010 to 5209 µg/g. The concentration of Mg (47–335 µg/g) and Ti (101–940 µg/g) in sapphire from kyshtymites are higher than that in blue sapphires from syenite pegmatites (mines 298 and 349) [2,34] and those in sapphires from metasomatites in meta-ultramafic host rocks (mine 418) [3] (Table S6), both located in Ilmen Mountains (see Figure 1). The Ga content remains low (30–61 µg/g) for all studied samples as in the case of sapphires in meta-ultramafites (mine 418), however, lower than in those of syenite pegmatites (Table S6). Cr is from b.d.l. to 12 µg/g and V from 4 to 9 µg/g. Detected Li and Be concentrations in the measured spots are presumably due to micro-inclusions within sapphires.

The 10,000 Ga/Al ratio is above 0.60–0.80, Ga/Mg ratio is 0.11–0.80, Fe/Mg is 6.15–43.32, Cr/Ga is 0.01–0.32, and Fe/Ti is 1.08–22.95. These ratios are common for metamorphic sapphires [35,36]. On the Fe versus Ga/Mg diagram, studied sapphires with blue color fall in the field of metamorphic sapphire, similar (but not overlapping) to those of meta-ultramafic host rocks of Ilmen Mountains. The results are partially overlapping with those of magmatic sapphires in Yogo Gulch in USA, Gortva in Slovakia, Baw Mar in Myanmar (filled symbols in Figure 6), metamorphic sapphires from Ratnapura in Sri-Lanka, and metasomatic sapphires from Kashmir in India (dotted lines in Figure 6).

On the Fe*Mg*100–Ti*10 ternary plot (Figure 7), the studied sapphires from kyshtymites also fall in the field of "metamorphic" sapphires overlapping those within Ilmen meta-ultramafic host rocks (mine 418). Meanwhile, blue sapphires from anorthosites-kyshtymites overlap most of the other known metamorphic (Ratnapura in Sri-Lanka—purple dotted lines in Figure 7), metasomatic (Kashmir in India—pink dotted lines in Figure 7), magmatic (Mogok in Myanmar, Yogo in USA, and Gortva, Slovakia—filled symbols in Figure 7), and placer sapphire occurrences (Balangoda in Sri Lanka, Pailin in Cambodia, and Ilakaka in Madagascar, and Montana in USA).

The sapphires from kyshtymites are also plotted to the "metamorphic" field on Fe/Mg vs. Ga/Mg diagram by Peucat et al. (2007) and Sutherland et al. (2009) [35,36] (Figure 8).

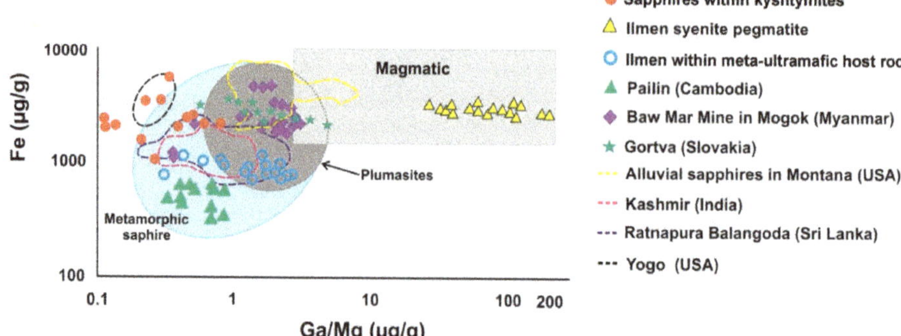

Figure 6. A Fe versus Ga/Mg diagram showing the boundaries for magmatic and metamorphic sapphires, modified after Peucat et al. (2007) [35] and Zwaan et al. (2015) [37] with blue sapphires within kyshtymites, Ilmen sapphires within syenite pegmatites [2], Ilmen sapphires within meta-ultramafic host rocks [3], and sapphire deposits from other regions after [35]: Ratnapura and Balangoda (Sri lanka), Kasmir (India), Yogo Gulch (USA), and Baw Mar Mine in Mogok. The plot of alluvial sapphires from Montana (USA) is after Zwaan et al. (2015) [37]; data on sapphires from the Hajacka, Gortva (Slovakia) are after Uher et al. (2012) [38].

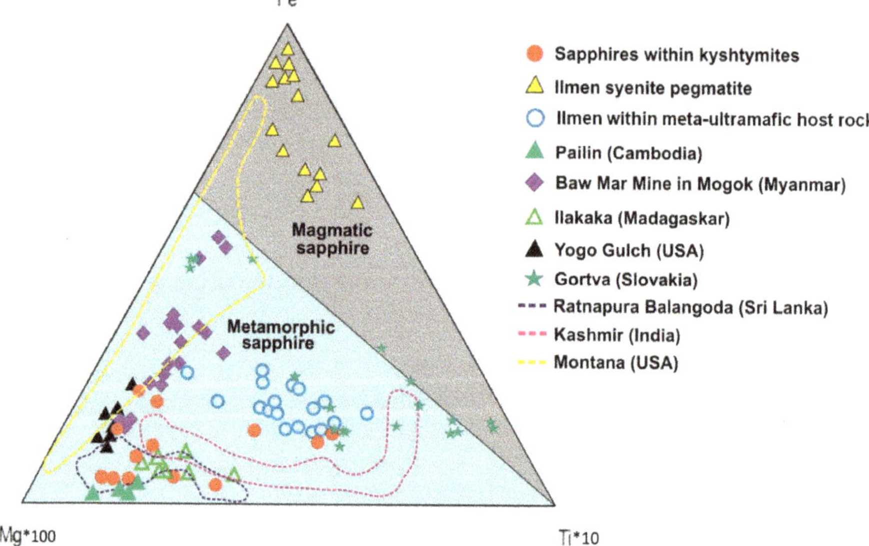

Figure 7. A Fe–Mg*100–Ti*10 ternary plot modified after Peucat et al. (2007) [35] of corundum-sapphire within kyshtymites, Ilmen sapphires within syenite pegmatites [2], Ilmen sapphires within the meta-ultramafic host rocks [3], and sapphire deposits from other regions after [35]: Pailin (Cambodia), Baw Mar Mine in Mogok (Myanmar), Ilakaka (Madaskar), Ratnapura and Balangoda (Sri lanka), Kasmir (India), and Yogo Gulch (USA). The plot of Gortva sapphire (Slovakia) is after Uher et al. (2012) [38] and Montana (USA) sapphires are after Zwaan et al. (2015) [37].

On the Fe–Cr*10–Ga*100 ternary plot (Figure 10), the sapphires from kyshtymites fall in the "magmatic field" similar to those "magmatic" sapphires from Australia [36] likely due to intermediate Fe and Ga contents, and absence of Cr values. However, they were plotted to "metasomatic" and "plumasitic" fields on discriminant factors diagram by Giuliani [40] (Figure 11), as in the case of those from Ilmen metasomatites within meta-ultramafic host rocks [3].

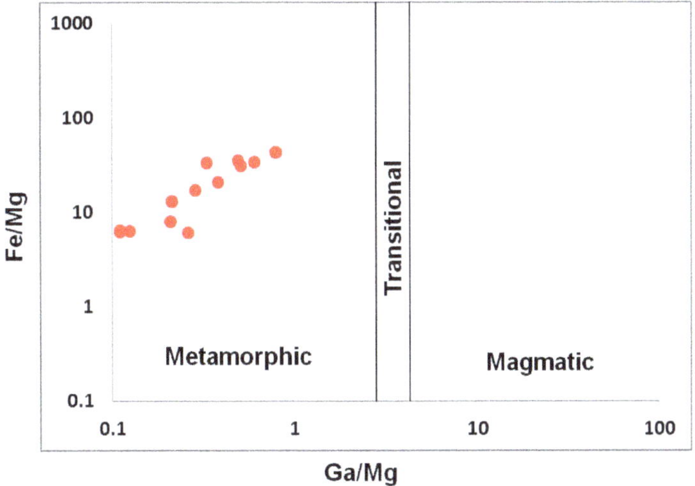

Figure 8. Fe/Mg vs. Ga/Mg discrimination diagram modified after [35,36].

Blue sapphires within kyshtymites overlap the "metasomatic" fields on FeO – Cr_2O_3 – MgO – V_2O_3 vs. FeO + TiO_2 + Ga_2O_3 discriminant diagram (Figure 9) by [35] as in case of those sapphires within Ilmen syenite pegmatites, Ilmen sapphires within meta-ultramafic rocks, sapphires from Gortva syenite xenoliths within alkali basalts [38], and sapphires from lamphrophiric dyke in Yogo Gulch [39].

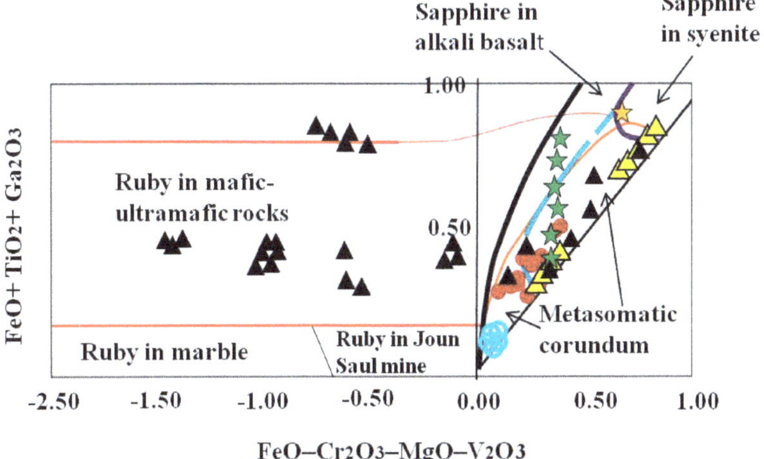

Figure 9. A FeO – Cr_2O_3 – MgO – V_2O_3 versus FeO + TiO_2 + Ga_2O_3 discriminant diagram (wt. %) is modified after Giuliani et al. (2014) [40] with extended "magmatic/syenitic" field with plotted corundum-sapphire within kyshtymites (red circles), Ilmen sapphires within syenite pegmatites (yellow triangles) [2], and within the meta-ultramafic host rocks (blue circles) [3]. The data on sapphires associated with xenoliths in alkali basalts (Gortva in Slovakia—green star; Loch Roag in Scotland—yellow star) were modified after Uher et al. (2012) [38]. The data on sapphires associated with ultramafic lamprophyre dike in Yogo Gulch (USA) (black triangles) are after [39]. Blue dotted line shows the extension of "magmatic/syenitic" sapphire field.

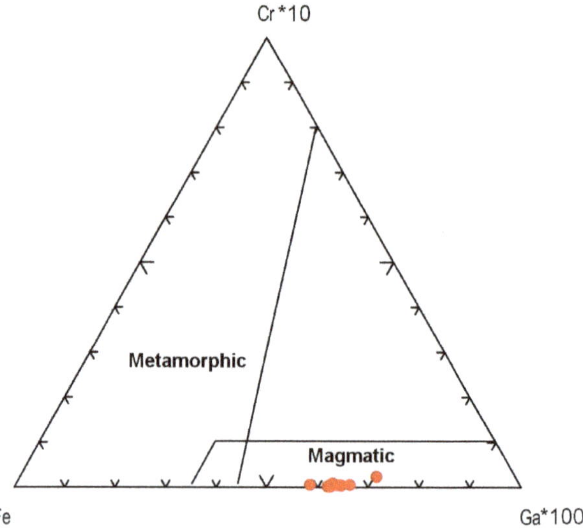

Figure 10. A Fe–Cr*10–Ga*100 ternary plot modified after Sutherland et al. (2009) [36].

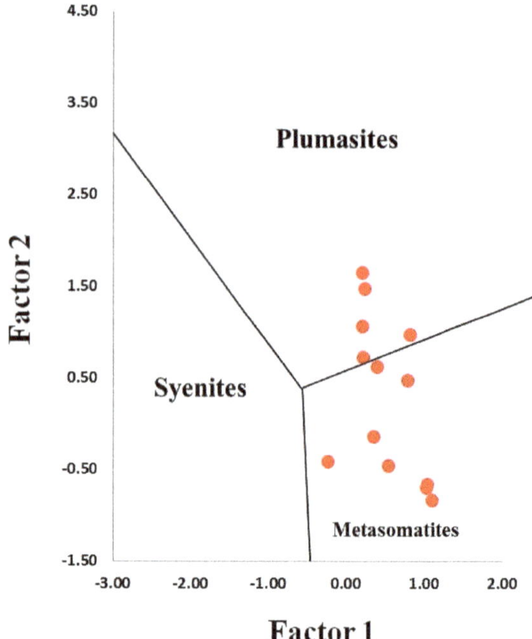

Figure 11. Discriminant factor diagram for identification of sapphire origin modified after Giuliani et al. (2014) [40].

4.3. Whole Rock Geochemistry of Kyshtymites and Elements Mobility

Major elements of kyshtymites measured by EDXRF are reported in Table S7. Kyshtymites have high Al_2O_3 (34.76–42.94 wt. %), Na_2O (0.82–4.01 wt. %), and K_2O (0.50–1.29 wt. %), but low SiO_2 content (40.84–42.72 wt. %) as well as variable CaO (5.89–15.79 wt. %) when compared with those of meta-ultramafic host rocks. Kyshtymites are characterized by low concentration of Fe_2O_3 (0.12–1.32 wt. %), MgO (0.60–2.86 wt. %), and TiO_2 (0.04–0.15 wt. %).

Meta-ultramafic host rocks are enriched in SiO_2 (69.72 wt. %), MgO (17.97 wt. %), and Fe_2O_3 (6.22 wt. %), while depleted in CaO (up to 0.25 wt. %), Al_2O_3 (up to 1.18 wt. %), and alkaline elements ($Na_2O + K_2O$ is 0.24 wt. %). The reaction rim between kyshtymites and meta-ultramafic host rocks also contains more MgO (28.08 wt. %), Fe_2O_3 (6.72 wt. %), and SiO_2 (45.09 wt. %) and less Al_2O_3 (4.29 wt. %), CaO (0.42 wt. %), and alkaline elements ($Na_2O + K_2$ is 0.14 wt. %) compared to those of kyshtymites. The concentration of TiO_2 was below the detection limit (Table S7, Figure 12).

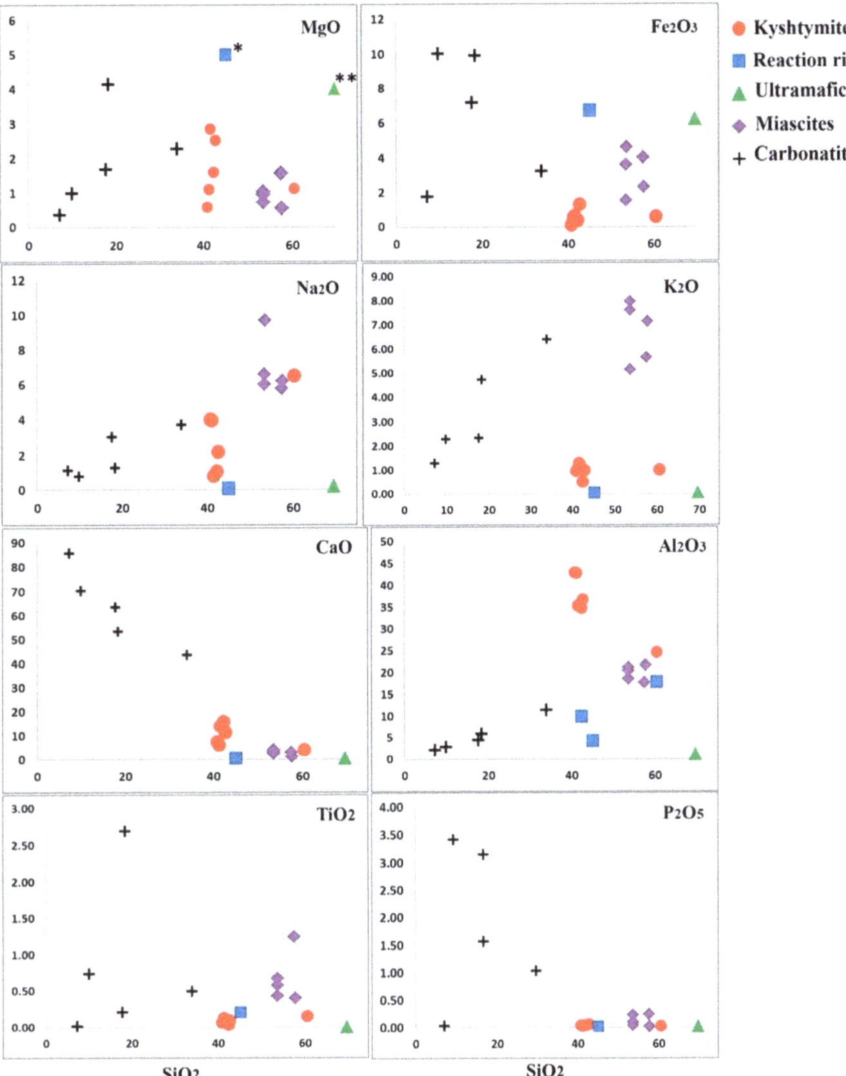

Figure 12. Harker diagrams for analyzed kyshtymites, reaction rim (* MgO, 28.08 wt. %) between kyshymites and meta-ultramafic host rocks, meta-ultramafic host rocks (** MgO, 17.97 wt. %), miascites, and carbonatites of Vishnevogorsky complex (chemistry of miascites and carbonatites is modified after Nedosekova et al. (2009) [18].

On the chondrite-normalized REE spider diagram, the main trend is the enrichment of the LREE comparing to the HREE (Figure 13). This trend is similar to those detected in nepheline syenites (miascites) and carbonatites of Vishnevogorsky complex [18]. However, kyshtymites are more enriched in REE than some miascites, and depleted in REE when compared to carbonatites (Figure 13). Some samples of kyshtymites show positive Eu patterns.

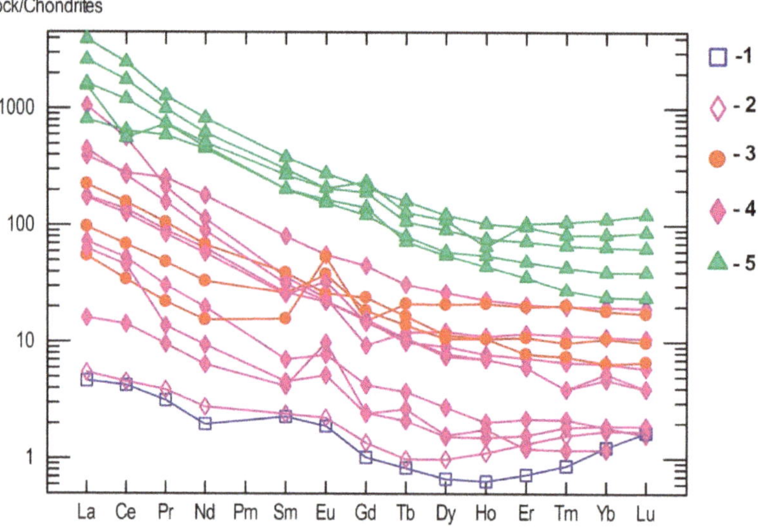

Figure 13. Chondrite-normalized REE spider diagram for the reaction rim between kyshymites and meta-ultramafic host rock (blue squares—1), meta-ultramafic host rocks (purple diamonds—2), kyshtymites (red circles—3), miascites (purple diamonds—4), and carbonatites (green triangles—5) of the Ilmenogorsky-Vishnevogorsky complex modified after [18] and Medvedeva E.V. (unpublished data); REE values in chondrite are after [41].

4.4. Trace Element Chemistry of Zircons and In Situ LA-ICP-MS U-Pb Zircon Geochronology

Nine zircon grains syngenetic with the sapphires from the kyshtymite sample 8-K were chosen for trace-elements measurements (Table S8) and in situ LA-ICP-MS U-Pb geochronological research (Table S9). The zircon grains ranged from about 50 μm × 50 μm to about 100 μm × 200 μm and showed common "magmatic" oscillatory zonation [42] visible in cathodoluminiscence images and maps in average weighted atomic numbers (Figure S1). Yellow and orange colors on the maps in Figure S1 correlate with the higher U and Th contents in zircons (see Table S8).

The REE spider diagram shows enrichment in HREE relative to LREE which is common for magmatic zircons [43] (Figure 14 and Figure S2). However, the zircons from kyshtymites show significantly higher values for most REE except for Tm, Yb and Lu. The U/Th ratio in studied zircons varies from 0.05 to 0.60 as in zircons from nepheline syenites (miascites) and carbonatites [43] with high Hf content of 6845–17482 μg/g (in the range of Hf values detected in zircon inclusions within sapphires, see Table S4), and Y concentration of 380–2370 μg/g. Zircons are also characterized by a positive Ce pattern Ce/Ce* = 1.39–27.36, whereas Eu pattern is absent (Eu/Eu* = 0.35–1.04). Titanium concentrations were below the detection limit except for 4 measurements with contents of 43-182 μg/g. One spot showed 2523 μg/g of Ti, most likely due to the ablation of inclusions. The kyshtymite crystallization temperature calculated by Ti-in-zircon thermometer $T(°C)_{zircon} = \frac{5080\pm30}{(6.01\pm0.03)-\log{(Ti,\mu g/g)}}$ − 273 [44] of three spots showed the temperatures of about 890–920 ± 30 °C.

Zircon/Chondrites

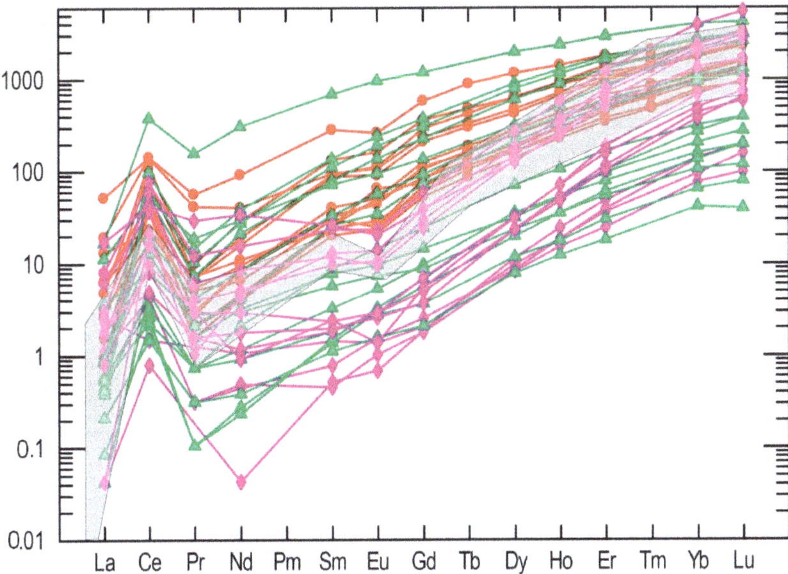

Figure 14. Chondrite-normalized concentration of REE in studied zircons from kyshtymites (samples 8-K—bold red circles), miascites (bold purple diamonds), and carbonatites (bold green triangles) of the Ilmenogorsky-Vishnevogorsky complex modified after (REE data on miascites and carbonatites are after [45]). Data on REE in chondrite are after [41]. Grey area is common REE distribution in magmatic zircons by Belousova et al. (2002) [43].

Twenty spots ablated on zircons during in situ LA-ICP-MS U-Pb geochronological measurements showed the stable ablation signal (cf. Table S8) indicating a homogeneous isotopic composition of the ablated volume. Eleven spots gave the Concordia age at 294 ± 6 Ma. Six spots showed the older Concordia age at 334 ± 10 Ma, which is common for the core areas of magmatic zircons (Figure S1). This older age was previously detected in zircons from carbonatites and nepheline syenites (miascites) of Ilmenogorsky-Vishnevogorsky complex [21–23,45] and is synchronous with the Hercynian metamorphism in this area [20,24] (Figure 15). Three spots were found to be in the Discordia, which is likely due to the later higher temperature event.

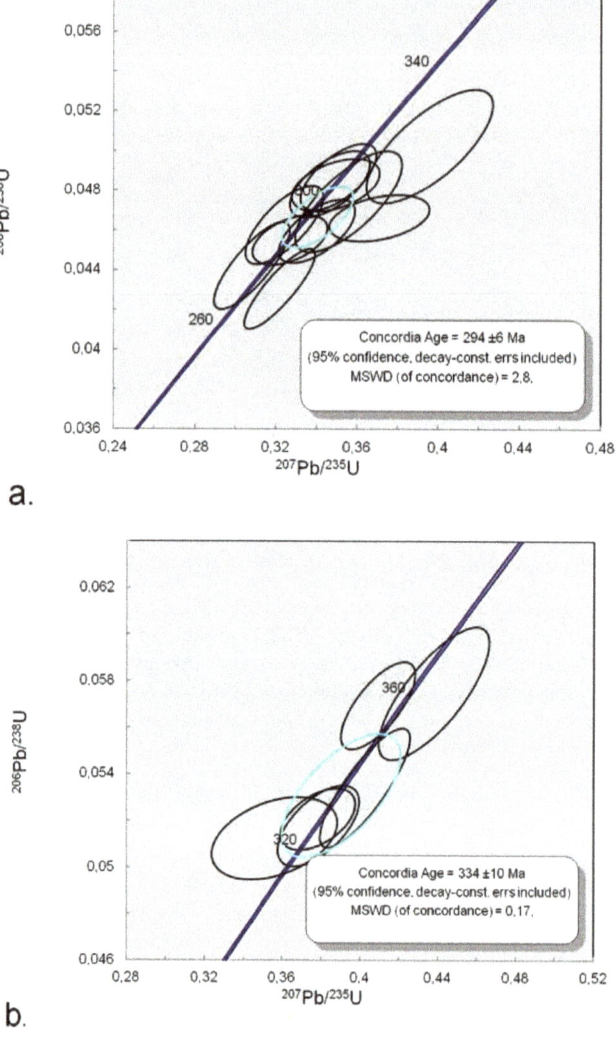

Figure 15. Concordia diagram for zircons from kyshtymite sample 8-K: (**a**) Concordia ages at 294 ± 6 Ma; and (**b**) 334 ± 10 Ma.

5. Discussion

5.1. Genetic Models of Corundum Anorthosites-Kyshtymites

There are several hypotheses about the origin of anorthosite-kyshtymites in South Urals, however, most of them are controversial. For instance, Fersman [46] considered kyshtymites as a product of granitic pegmatite desilication due to interaction with meta-ultramafic host rocks. According to Fersman's theory, an excess of alumina in melt crystalized in the form of corundum, and the reactive rims with biotite (phlogopite), actinolite, chlorite, and talc formed at a contact zone.

Fersman's hypothesis was criticized by Lodochnikov [47], pointing out the similar reaction rims at a contact of meta-ultramafites to the older sedimentary or igneous rocks with the absence of kyshtymites. According to Lodochnikov's idea, corundum anorthosites are hydrothermal formations, while active mineral-forming gas–liquid solutions are associated with the meta-ultramafites themselves. Later, Korzhinsky [48] put forward a hypothesis about the bimetasomatic origin of anorthositic veins with corundum under the influence of granite post-magmatic solutions intruded later than meta-ultramafic rocks. However, none of these theories have been confirmed during the exploration of deposit as there was no clear indication of any hydrothermal processes occurred here (e.g., the presence of dispersion halos, hydrothermal transformation of the host rocks, mineralization that determines the primary mineral zonation of hydrothermal deposits were not observed there).

The problem of the kyshtymite genesis was extensively described by Kolesnik for the Borzovsky kyshtymite occurrence (an analogue of the 5th versta deposit) [16,17]. Kolesnik suggested that the formation of kyshtymites is associated with metasomatic processes occurring during intrusion of granite dykes into ultramafic rocks. The development of corundum anorthosite replacing aplite-like granite dyke with sections of the pegmatoid structure at contact with the granite gneisses was considered as a possible formation mechanism [14]. However, this hypothesis was not supported during the exploration of deposits, because the source of aluminum, as well as calcium, required for the formation of corundum anorthosite -kyshtymites, is still debatable.

Our previous studies have shown a possible genetic link between kyshtymites (corundum anorthosites), miascites (nepheline syenites), and carbonatites of the Ilmenogorsky-Vishnevogorsky complex [25]: similar REE patterns, i.e., enrichment of LREE compared to HREE (Figure 13); anomalies in U, Nb, P;, Sr, and Ti (Figure S3); moderate and highly fractionated distributions of REE $(La/Yb)_N$ = 4.20–48.12, and a small Eu maximum $(Eu/Eu^* = 1.02–1.32)$. Both miascites and kyshtymites are extremely enriched in Al_2O_3 (up to 42.94 wt. % in kyshtymites and up to 22.76 wt. % in miascites [25]). Moreover, these rocks have similar accessory minerals (i.e., Y-bearing phases, apatite group minerals, and monazite-(Ce) [49]). One elder zircon age of kyshtymites and those determined in zircons from carbonatites show similar Concordia age of ca. 334 Ma [50] (Figure S4).

5.2. Magmatic vs. Metamorphic Origin of Kyshtymites

Assuming that sapphires in kyshtymites represent in situ minerals of the primary rock, their geochemical signatures can be used to decode the possible origin of the sapphire-bearing anorthosites. Several discrimination diagrams from the literature presented in Figures 6–11 provide constraints on the metamorphic vs. magmatic genesis of sapphires. According to the diagrams in Figures 6–8, sapphires from kyshtymites demonstrate metamorphic imprint, whereas plots in Figures 9–11 show that they have magmatic and metamorphic signature. Compositional profiles across sapphires (Table 1 and Figure 4) indicate existence of the core and rim zones and point to at least two events of sapphire formation. However, trace element compositions of both the rim and the core are within the variations between compositions of different sapphire grains. Hence, observed zonation does not provide any additional key to unravel sapphire crystallization environment.

The ambiguity in the interpretation of sapphire origin is most probably caused by the criteria used in discriminating between magmatic and metamorphic trace element signatures proposed previously (as in Figures 6–11). Recent studies have discovered transitional groups of sapphires having trace element compositions which are located between the proposed end-members on the discrimination diagrams and which are difficult to classify [35,38,51]. For instance, recently published data on geochemistry of sapphires with obvious magmatic origin like those from Gortva syenite xenoliths within alkali basalts (Slovakia) [38], sapphires from lamphrophiric dyke in Yogo Gulch (USA) [39], sapphires from syenite pegmatites of Ilmen Mountains (South Urals of Russia) [2] indicate that the compositions of "syenitic/magmatic" sapphires lie within the nominally "metasomatic" field (Figure 9). The magmatic origin of those rocks is also confirmed by the $\delta^{18}O$ data. The $\delta^{18}O$ value of the blue sapphire from xenolith of Gortva is 5.1 ± 0.1‰ [38] fits with the $\delta^{18}O$ range of sapphires associated with syenites/anorthoclasites [52]. The $\delta^{18}O$ values of sapphires from lamphrophiric dyke in Yogo Gulch (Montana, USA) showed 5.4–6.8‰ [39] overlapping the field defined for sapphires from lamprophyres (4.5–7.0‰) and sapphires from syenites (5.2–7.8‰) [52]. Sapphires from syenite pegmatite of the

Ilmen Mountains showed $\delta^{18}O$ about 4.3‰, i.e., in range defined for magmatic rocks (lamprophyre, basalt, and syenite) [53]. Magmatic origin of sapphires from Yogo Gulch, Gortva, and Ilmen syenite pegmatites is also confirmed by presence of syngenetic inclusions of primary magmatic minerals (Table S10). In fact, the presence of mainly Ca-plagioclase solid inclusions within sapphires from Yogo and in the lamprophyre xenoliths, as in case of kyshtymites, as well as trachytic melt inclusions indicates that a slab-related troctolitic or anorthositic protolith may be involved in their formation [39]. Plagioclase, as in case of kyshtymites, and alkali feldspar, as in case of Ilmen syenite pegmatites, were found in mineral association, while zircon, spinel, monazite-(Ce), ilmenite, and Y-REE phase were identified as solid inclusions within sapphires from Gortva. Columbite-(Fe), zircon, minerals of alkali feldspar group, monazite-(Ce), sub-micron grains of uraninite, muscovite, diaspore, and ilmenite were identified as syngenetic solid inclusions within the blue sapphire from syenite pegmatites of Ilmenogorsky complex (mines 298 and 349) [2,35]. Thus, the "magmatic/syenitic" field in Figure 9 could be extended toward the low boundary of plot with Ilmen sapphires within syenite pegmatites.

On the Fe vs. Ga/Mg diagram and Fe–Mg*100–Ti*10 ternary plot (Figures 6 and 7), sapphires from kyshtymites plotted in the "metamorphic" field overlapping those from lamphrophiric dyke of Yogo Gulch and Gortva syenite xenoliths, the igneous nature of which was shown above. Besides they are also overlap Pailin (Cambodia) placer sapphire xenocrysts trapped by alkali basalts. The $\delta^{18}O$ of last is 7.1–7.8‰, which is also in range defined for magmatic sapphire in syenite [53]. Moreover, sapphires from Pailin deposit [54] contain inclusions of pyrochlore, columbite-(Fe), goethite, zircon, monazite-(Ce), and rutile [55,56] as in case of Ilmen syenite pegmatites [2,35], while plagioclase identified there as well is also the case of kyshtymites. Sapphires from kyshtymites and Ilmen syenite pegmatites overlap with those from alluvial placer deposit in Montana (USA). Besides, anatase, Ca-rich plagioclase (Raman spectra of them match with anorthite and bytownite) and alkali feldspar (Raman spectra of them match with orthoclase) as in the case of Ilmen syenite pegmatites, along with rutile, ilmenite, monazite-(Ce), apatite group minerals, etc., were identified as solid inclusions within Montana sapphires (see Table S10).

These observations indicate that the existing discrimination diagrams do not provide a clear answer on the sapphire origin. In this sense, the "metamorphic" signature of the sapphires from kyshtymites could be only apparent and incorrect. Thus, based on the available classification of trace element compositions in sapphires, it is currently impossible to unambiguously determine the origin of kyshtymites.

Another possible genetic tool is the geochemistry of zircons from kyshtymites. The zircons demonstrate a clear REE pattern typical for magmatic zircons and similar to that found in zircons from syenites and carbonaties which are obviously magmatic rocks [45] (Figure 14 and Figure S2). Furthermore, both syenites and carbonatites contain also other types of zircons, which are interpreted as metamorphic, with significantly different trace element patterns (Figures 14 and 15). Since kyshtymites are syngenetic to magmatic syenites and carbonaties (see Figure 13), their REE signature of zircons can imply a magmatic origin of anorthosites. Further analysis of trace element compositions of zircons also shows magmatic origin within the continental crust (Figures 16 and 17).

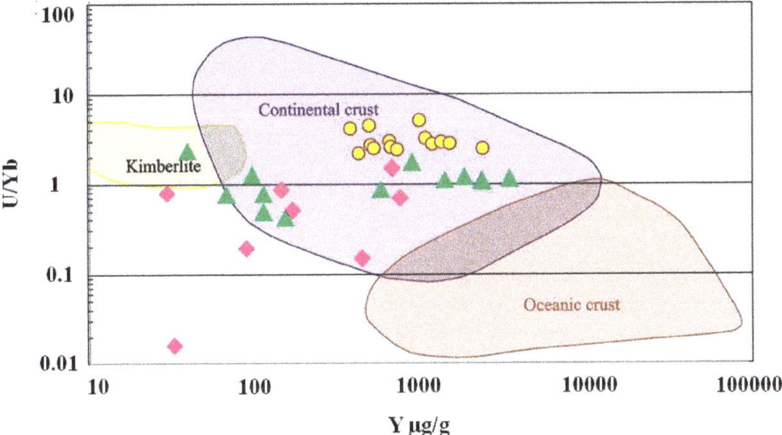

Figure 16. U/Nb vs. Y discriminant diagram of zircons from kyshtimtes (bold yellow circles), miascites (bold purple diamonds), and carbonatites (bold green triangles) of the Ilmenogorsky-Vishnevogorsky complex (the chemistry on miascites and carbonatites is modified after [45]). Fields for continental and oceanic crust, and kimberlite are from Grimes et al. (2007) [57].

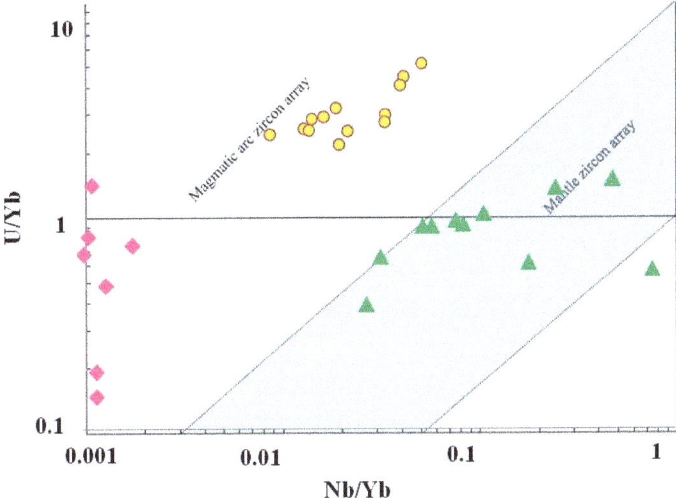

Figure 17. U/Nb vs. Nb/Yb discriminant diagram of zircons from kyshtimtes (bold yellow circles), miascites (bold purple diamonds), and carbonatites (bold green triangles) of the Ilmenogorsky-Vishnevogorsky complex (the chemistry on miascites and carbonatites is modified after [45]); fields for mantle and magmatic arc arrays are from Grimes et al. (2015) [58].

On the chondrite-normalized REE diagram for zircons (Figure 16), REE distributions show the main trend of enrichment in HREE compared to LREE with Ce pattern, which is similar to those observed in miascites and carbotanites of Ilmenogorsky-Vishnevogorsky complex. Positive patterns of Th, U, Hf, and Ta are also observed in chondrite-normalized REE diagrams (Figure S3).

Thus, formation of "magmatic" corundum-blue sapphire anorthosites could occur, likely during at least two stages. In the first stage, about 450–420 Ma [21–23,50], primary anorthosites formed as cumulates in the magmatic chamber, along with miascites and carbonatites of the Ilmenogorsky-Vishnevogorsky complex, with likely more Ca-rich plagioclase in mineral association. Formation of cumulates in magmatic chamber on the later stage in excess of H_2O was confirmed by experimental studies [59], whereas another study showed formation of cumulates at the top of magmatic chamber due to the lower density than the rest of magma [60]. Later, at 300–285 Ma, primary anorthositic cumulates probably could be re-melted during the collision process. Plagioclase recrystallized to more Ca-poor member, whereas the excess of aluminum crystallized in form of corundum, while sodium probably came from nepheline syenites (miascites).

However, even though zircons demonstrate clear evidence of magmatic origin, their signature may not represent the genesis of anothosites. Zircons are known as very stable and inert minerals and, hence, they could be inherited in anorthosites from another primary rock. Since both syngenetic syenites and carbonataites contain zircons, they can be a source of magmatic zircons in the system. On U/Nb vs. Y discriminant diagram (Figure 16), zircons from nepheline syenites (miascites) and carbonatites are plotted on the "continental crust" fields despite on their mantle origin [45] overlapping zircons from kyshtymites. On U/Nb vs. Nb/Yb discriminant diagram, zircons from miascites and kyshtymites are also plotted to "magmatic arc" field (Figure 17). In other words, the zircons in kyshtymites can be magmatic, but the host rock can be produced by a metamorphic process. Such a process must be a relatively low temperature process to preserve original zircons unaffected. The Ti-in-zircon thermomentry provides temperature estimates of about 890–920 °C, which is close to or below the closure temperature of the U-Pb geochronometer in zircons [61]. On the other hand, the U-Pb age determinations show only few discordant zircons (Figure 16) which could be affected by high-temperature events (either magmatic or metamorphic). Most zircons demonstrate concordant ages indicating that after their formation they were not exposed to high temperatures. Thus, compositions of zircons are also not an ultimate tool to reconstruct the origin of kyshtymites.

Typically, it is believed that anorthosites are formed by differentiation of mafic magmas when plagioclases become buoyant due to density contrast and accumulate at the roof of the magma chamber [58]. Such a process would require a relatively large volume of magma to produce plagioclase cumulates. The dimensions of kyshtymite anorthositic veins are quite small with the length from 15 to 70 m and thickness from 0.1 to 3 m, which are not realistic for anorthositic body of magmatic origin. Furthermore, formation of anorthosites by re-melting processes during orogenic events is also questionable as temperatures recorded by zircons are relatively low and obviously not enough to melt anorthite-rich substrate/rock.

The ages of 294 ± 6 Ma as determined by zircon geochronology correspond to the ages of Hercynian metamorphism during collision events in the area of Ilmenogorsky-Vishnevogorsky complex. Thus, formation of kyshtymites could also happen due to metamorphic processes. The kyshtymite vein shown in Figure 2 is located within the meta-ultramafic rocks separated by reaction rim consisting mostly of chrysotile-asbestos [62]. The ultra-mafic rocks originally having an orthopyroxenitic composition are strongly metamorphosed and enriched in silica (see Figure 12). This silica-enrichment process is likely linked to the Caledonian metamorphism [63] accompanied by the silicon metasomatism and partial removal of magnesium. Moreover, the excess of Si could also occur likely during the metamorphic re-working and desilication of carbonatites (see the discussion below). The importance of buffering capacity of ultramafic host rocks with respect to silica has been shown in ruby deposit from the Greenland [64]. The similar process observed on another corundum-blue sapphire-bearing deposit within meta-ultramafic host rocks (mine 418) in Ilmenogorsky complex [3].

However, the genesis of the contact zone is difficult to reconstruct since there is no clear evidence for a magmatic contact and there is no indication for any metasomatic zonation. Kyshtymites demonstrate Rb-Sr and Sm-Nd isotopic signatures ($^{87}Sr/^{86}Sr$ of 0.70637–0.706936 and εNd from −5.3 to −10.7) corresponding to the signature of the low-crustal material [25], whereas miascites and carbonatites show mantle-derived values [18]. The data lie on the miascites-carbonatites trend of the Ilmenogorsky-Vishnevogorsky complex and show increasing contribution from the crust (Figure S5). The age and isotopic data could be indicative of the late stage process of kyshtymite formation related to the Hercynian metamorphic event. The main question is the initial protolith to form anorthosites.

The geochemical features of syenites and carbonatites suggest that one of these rocks could serve as a substrate for metamorphic reactions to form anorthosites. Assuming that Al is mostly immobile major element and that carbonates can be decomposed by metamorphic fluids, it can be suggested that metamorphism of carbonatites can be responsible for the formation of Ca-Al-rich anorthosites. Carbonaties from Ilmenogorsky-Vishnevogorsky complex have in addition to CaO, about 5–30 wt. % of SiO_2 and 1–11 wt. % of Al_2O_3 on the CO_2-free basis [18]. If Ca and CO_2 are removed by the metamorphic fluids, the residue will be enriched in Al and Si. Assuming that host meta-ultramafic rocks can be effectively silicified and hence, consume Si, the concentration of Al in former carbonatite vein can reach very high values resulting in the formation of corundum and Ca-rich plagioclases. At the later stages with decreasing temperature, muscovite and clinozoisite, which particulary replace plagioclase, could form in kyshtymite veins as a lower-temperature phases. Clinochlore crystallized at the final stage. Its formation occurred due to replacement of muscovite and the introduction of Mg into the system from the meta-ultramaifc host rocks by metasomatic fluids. One evidence for such a process could be a UV-Vis-NIR spectra of sapphires in one kyshtymite sample which are similar to the spectra observed for metamorphic/metasomatic sapphires.

Such a scenario would explain the metamorphic signature of newly-formed sapphires and magmatic signature of zircons inherited from parental carbonatitic rock. On the other hand, such petrogenetic history of kyshtymites is not visible in the host rocks. It is expected that intensive metamorphism and decomposition of carbonates would result in the development of metamorphic/ metasomatic zonation around kyshtymite veins and will cause significant volume changes, which is not observed.

Thus, the new geochemical data presented in this study still do not provide a final and unambiguous answer on the origin of kyshtymites and further investigations are required, e.g., oxygen isotopy of corundum is promising for clarifying the genesis of the kyshtymites.

6. Conclusions

Sapphires were found in situ within primary rock—anorthosites (kyshtymites)—in the 20th century. Until now, the genesis of kyshtymites remains controversial. They are located at a boundary of the Ilmenogorsky-Vishnevogorsky alkaline complex of Russia's South Urals to the north of blue sapphires in syenite pegmatites (mines 298 and 349) and of metasomatites within meta-ultramafic host rocks (mine 418).

Measured trace elements are in the range for metamorphic sapphires: 10,000 Ga/Al > 0.60–0.80, Ga/Mg is 0.11–0.80, Fe/Mg is 6.15–43.32, Cr/Ga is 0.01–0.32, and Fe/Ti is 1.08–22.95. Sapphires are plotted to the "metamorphic" field in Fe versus Ga/Mg and Fe–Mg*100–Ti*10 diagrams. Besides, they also overlap "metasomatic" and extended "magmatic/syenitic" fields in $FeO – Cr_2O_3 – MgO – V_2O_3$ versus $FeO + TiO_2 + Ga_2O_3$ discriminant diagrams.

Two possible magmatic and metamorphic-metasomatic scenarios were proposed for the formation of anorthosites-kyshtymites. The magmatic formation of corundum anorthosites should take place during at least two stages. At about 450–420 Ma, primary anorthosites formed as cumulates in the magmatic chamber, along with miascites and carbonatites of the Ilmenogorsky-Vishnevogorsky complex. Further at 300–285 Ma, primary anorthositic cumulates could be re-melted during the collision process. Plagioclase recrystallized to more Ca-poor member, whereas the excess of aluminum crystallized in form of corundum. However, small sizes of anorthositic veins and obtained sub-solidus temperatures of 890–920 ± 30 °C by Ti-in-zircon thermometry do not support the hypothesis of magmatic origin.

The second possible scenario is the formation of kyshtymites through the metamorphic-metasomatic re-working of initial carbonatites. Calcium and CO_2 could be removed by metamorphic fluids, the residue would be enriched in Al and Si. While meta-ultramafic host rocks consume Si, the concentration of Al in former carbonatite vein can reach very high values resulting in the formation of corundum and Ca-rich plagioclases.

Trace-element chemistry and solid inclusions identified in sapphires from kyshtymites, along with those of syenite pegmatites of Ilmenogorsky complex, provide possible genetic links with sapphires from Montana (USA), Gortva (Slovakia), and Pailin (Cambodia). However, more research is required to unravel the nature of anorthositic-syenitic protoliths that could be involved in their formation.

Supplementary Materials: The following are available online at http://www.mdpi.com/2075-163X/9/4/234/s1, Table S1. The samples and methods used in the research; Table S2. Minerals identified in association with the sapphire; Table S3. Representative compositions (wt. %) of major minerals found in association with sapphire; Table S4. Representative compositions (wt. %) of minor minerals found in association with sapphire; Table S5. Representative compositions (wt. %) of the zircon inclusions in sapphire; Table S6. Chemical composition of sapphires from Ilmenogorsky-Vishnevogorsky complex and sapphire occurrences with possible anorthositic–syenitic origin. Table S7. Representative analyses of corundum-blue sapphire anorthosites-kyshtymites, meta-ultramafic host rocks, and reaction rim; Table S8. LA-ICP-MS trace-elements measurements of zircons from sample 8-K; Table S9 In situ LA-ICP-MS geochronology of zircons from sample 8-K; Table S10. Frequently detected solid inclusions and common minerals found in association with sapphires from Ilmenogorsky-Vishnevogorsky complex and sapphire occurrences with possible anorthositic-syenitic origin. Figure S1. Zircon CL images and maps in average weighted atomic numbers with the spots positions for trace-elements (No. of spots inside of circles) and U-Pb geochronological measurements (U-Pb ages inside of circles). Black, Concordia kyshtymite age; red, 6 Concordia elder ages; purple, 3 ages in Discordia. Figure S2. Chondrite-normalized concentration of REE and trace elements in studied zircons from kyshtymites (**a**), miascites (**b**), and carbonatites (**c**), of the Ilmenogorsky-Vishnevogorsky complex modified after [45]. Data on REE and trace elements in chondrite are after [41]. Figure S3. Trace-elements and REE distribution normalized to primitive mantle (the data on REE and trace-elements in primitive mantle are from [41]) in reaction rim (1) between kyshymites and meta-ultramafic host rock, meta-ultramafic host rocks (2), kyshtymites (3), miascites (4), and carbonatites (5) of the Ilmenogorsky-Vishnevogorsky complex modified after [18] and Medvedeva E.V (unpublished data). Figure S4. Concordia diagram for zircons from kyshtymite sample 8-K and carbonatites of Vishnevogorsky complex (sample 354 by Nedosekova [50]). Figure S5. Diagram εSr (T) vs. εNd (T) for kyshtimte and miascite of the Ilmenogorsky-Vishnevogorsky complex modified after [18,25], the diagram shows mantle reservoirs DMM, HIMU, EM1, EM2, MORB and OBI [65].

Author Contributions: M.I.F. formulated the idea of a paper, performed the petrography of kyshtymte and host rocks, designed experiments on EDXRF and WDS EMPA, assembled most of tables and figures, and wrote the manuscript; E.S.S. formulated the idea of a paper, designed experiments on Raman spectroscopy, WDS EMPA, ICP-MS, LA-ICP-MS sapphire and zircon geochemistry, and LA-ICP-MS U-Pb zircon geochronology, performed LA-ICP-MS sapphire geochemistry and its following data-reduction process, interpreted the zircon dating data and constructed Concordia diagrams, assembled some tables and figures, wrote the part of manuscript, and edited the final text of paper; R.B. supervised the LA-ICP-MS zircon geochemical and U-Pb zircon geochronological measurements, formulated the part of possible scenario for kyshtymites origin, and edited the manuscript; S.K. helped with data interpretation and editing of the manuscript; M.A.R. collected samples for research and provided the geological setting data; N.N.K. performed WSD EMPA, and produced zircon CL and BSE images and maps of average weighted atomic numbers; A.G.N. provided absorption spectra of sapphires and their interpretation; J.B. performed LA-ICP-MS zircon geochemistry and U-Pb zircon geochronology, and their following data-reduction process; and W.H. supervised the LA-ICP-MS zircon geochemical and U-Pb zircon geochronological measurements.

Funding: The research was supported by the Foundation of the President of Russian Federation (project No. MK-4459.2018.5). In situ LA-ICP-MS zircon trace-elements and geochronological measurements were supported by the German Academic Exchange Service (DAAD, No 57378441) and Stiftung zur Förderung der Edelsteinforschung of Johannes Gutenberg Universität Mainz (Germany).

Acknowledgments: The authors are grateful to Elena V. Medvedeva (Ilmen State Reserve, Russia) for the provided chemical analyses of miascites. The authors are grateful to the colleagues from GEOKHI RAS (T.G. Kuzmina, V.N. Ermolaeva, T.V. Romashova, V.A. Turkov, B.S. Semiannikov), Ya.V. Bychkova (Moscow State University, Russia), E.A. Minervina (IGEM RAS), K.Ponkratov (Renishaw Moscow), Delia Rösel (Technische Universität Bergakademie Freiberg, Germany), as well as Stephan Buhre and Tobias Häger (Johannes Gutenberg Universität Mainz, Germany) for their assistance in sample preparation and analytical studies of kyshmymites and miascites, as well as following data interpretation.

Conflicts of Interest: The authors declare no conflict of interest.

References

1. Giuliani, G.; Ohnenstetter, D.; Fallick, A.E.; Groat, L.; Fagan, A.G. The geology and genesis of gem corundum deposits. In *Geology of Gem Deposits*, 2nd ed.; Groat, L.A., Ed.; Mineralogical Association of Canada Short Course Series; Mineralogical Association of Canada: Tucson, AZ, USA, 2014; Volume 44, pp. 29–112.

2. Sorokina, E.S.; Karampelas, S.; Nishanbaev, T.P.; Nikandrov, S.N.; Semiannikov, B.S. Sapphire Megacrysts in Syenite Pegmatites from the Ilmen Mountains, South Urals, Russia: New Mineralogical Data. *Can. Mineral.* **2017**, *55*, 823–843. [CrossRef]

3. Sorokina, E.S.; Rassomakhin, M.A.; Nikandrov, S.N.; Karampelas, S.; Kononkova, N.N.; Nikolaev, A.G.; Anosova, M.O.; Orlova, A.V.; Kostitsyn, Y.A.; Kotlyarov, V.A. Origin of blue sapphire in newly discovered spinel–chlorite–muscovite rocks within meta-ultramafites of Ilmen Mountains, South Urals of Russia: Evidence from mineralogy, geochemistry, Rb-Sr and Sm-Nd isotopic data. *Minerals* **2019**, *9*, 36. [CrossRef]

4. Simonet, C.; Paquette, J.L.; Pin, C.; Lansner, B.; Fritsch, E. The Dusi (Garba Tula) sapphire deposit, Central Kenya–a unique Pan-African corundum-bearing monzonite. *J. Afr. Earth Sci.* **2004**, *38*, 401–410. [CrossRef]

5. Monchoux, P.; Fontan, F.; De Parseval, P.; Martin, R.F.; Wang, R.C. Igneous albitite dikes in orogenic lherzolites, western Pyrenees, France: a possible source for corundum and alkali feldspar xenocrysts in basaltic terranes. I. Mineralogical associations. *Can. Mineral.* **2006**, *44*, 817–842. [CrossRef]

6. Kan-Nyunt, H.P.; Karampelas, S.; Link, K.; Thu, K.; Kiefert, L.; Hardy, P. Blue sapphires from the Baw Mar Mine in Mogok. *Gems Gemol.* **2013**, *49*, 223–232. [CrossRef]

7. Khoi, N.N.; Hauzenberger, C.A.; Sutthirat, C.; Tuan, D.A.; Häger, T.; Van Nam, N. Corundum with Spinel Corona from the Tan Huong–Truc Lau Area in Northern Vietnam. *Gems Gemol.* **2018**, *54*, 4. [CrossRef]

8. Voudouris, P.; Mavrogonatos, C.; Graham, I.; Giuliani, G.; Melfos, V.; Karampelas, S.; Karantoni, V.; Wang, K.; Tarantola, A.; Zaw, K.; et al. Gem Corundum Deposits of Greece: Geology, Mineralogy and Genesis. *Minerals* **2018**, *9*, 49. [CrossRef]

9. Keulen, N.; Kalvig, P. Fingerprinting of corundum (ruby) from Fiskenæsset, West Greenland. *Geol. Surv. Denmark Greenland* **2013**, *25*, 53–56.

10. Karmakar, S.; Mukherjee, S.; Sanyal, S.; Sengupta, P. Origin of peraluminous minerals (corundum, spinel, and sapphirine) in a highly calcic anorthosite from the Sittampundi Layered Complex, Tamil Nadu, India. *Contrib. Mineral. Petrol.* **2017**, *172*, 67. [CrossRef]

11. Gibson, G.M. Margarite in Kyanite- and Corundum-Bearing Anorthosite, Amphibolite, and Hornblendite from Central Fiordland, New Zealand. *Contrib. Miner. Petrol.* **1979**, *68*, 171–179. [CrossRef]

12. Pratt, G.H. *Corundum and Its Occurrence and Distribution in the United States*; US Government Printing Office: Washington, DC, USA, 1906.

13. McElhaney, M.S.; McSween, H.Y. Petrology of the Chunky Gal Mountain mafic-ultramafic complex, North Carolina. *GSA Bull.* **1983**, *94*, 855–874. [CrossRef]

14. Koptev-Dvornikov, V.S.; Kuznetsov, E.A. *Borzovskoe Corundum Deposit: Petrological Study*; State Technical Publishing House: Moscow, Russia, 1931; p. 320. (In Russian)

15. Claire, M.O. Corundum and emery on the Urals. *Uralskiy Technik.* **1918**, *7*, 1–17. (In Russian)

16. Kolesnik, N.Y. *High-Temperature Metasomatism in Ultrabasic Massifs*; Science Publishing House: Novosibirsk, Russia, 1976; p. 240. (In Russian)

17. Kolesnik, N.Y.; Korolyuk, V.N.; Lavrent'ev, Y.G. Spinels and ore minerals of the Borzovsk deposit of corundum plagioclasites. *Notes Russian Mineral. Soc.* **1974**, *103*, 373–378. (In Russian)

18. Nedosekova, I.L.; Vladykin, N.V.; Pribakin, S.V.; Bayanova, T.B. The structure of the Ilmenogorsky-Vishnevogorsky Miaskite-Carbonatite Complex: Origin, ore-bearing. Sources of matter (Ural, Russia). *Geol. Ore Depos.* **2009**, *51*, 157–181. (In Russian) [CrossRef]

19. Rusin, A.I.; Krasnobaev, A.A.; Valizer, P.M. Geology of the Ilmen Mountains: situation and problems. In *Geology and Mineralogy of the Ilmenogorsky Complex: Situation and Problems*; IGZ UB RAS: Miass, Russia, 2006; pp. 3–19. (In Russian)

20. Krasnobaev, A.A.; Puzhakov, B.A.; Petrov, V.I.; Busharina, S.V. Zirconology of metamorphites of the Kyshtym-Arakulian strata of the Sysert-Ilmenogorsky complex. *Proc. Zavaritsky Inst. Geol. Geochem. (Trudy Instituta Geologii i Geokhimii im. Akademika A.N. Zavaritskogo)* **2009**, *156*, 264–268. (In Russian)

21. Kramm, U.; Blaxland, A.B.; Kononova, V.A.; Grauert, B. Origin of the Ilmenogorsk-Vishnevogorsk nepheline syenites, Urals, USSR, and their time of emplacement during the history of the Ural fold belt: A Rb-Sr study. *J. Geol.* **1983**, *91*, 427–435. [CrossRef]

22. Kramm, U.; Chernyshev, I.V.; Grauert, S. Zircon typology and U-Pb systematics: A case study of zircons from nefeline syenite of the Il'meny Mountains, Ural. *Petrology* **1993**, *1*, 474–485.

23. Chernyshev, I.V.; Kononova, V.A.; Kramm, U. Isotope geochronology of alkaline rocks of the Urals in the light of zircon uranium-lead data. *Geochemistry* **1987**, *3*, 323–338.

24. Ivanov, K.S.; Erokhin, Y.V. About the Age and nature of the metamorphic complexes of the Ilmenogorsk zone of the Urals. *Rep. Acad. Sci.* **2015**, *461*, 312–315.

25. Filina, M.I.; Sorokina, E.S.; Rassomakhin, M.A.; Kononkova, N.N.; Kostitsyn, Y.A.; Orlova, A.V. Genetic linkage of corundum plagioclazite-kyshtymite and miaskites of Ilmensky-Vishnevogorsky complex, Southern Urals, Russia: new data on Rb-Sr and Sm-Nd isotopic composition, geochemistry and mineralogy. *Geochem. Int.* **2019**. (accepted).

26. Arslanova, K.A.; Golubchina, M.N.; Iskanderova, A.D. *Geological Dictionary*; Nedra Publishing House: Moscow, Russia, 1978; Volume 2, p. 456. (In Russian)

27. Jochum, K.P.; Weis, U.; Stoll, B.; Kuzmin, D.; Yang, Q.; Raczek, I.; Jacob, D.E.; Stracke, A.; Birbaum, K.; Frick, D.A.; et al. Determination of reference values for NIST SRM 610-617 glasses following ISO Guidelines. *Geostand. Geoanal. Res.* **2011**, *35*, 397–429. [CrossRef]

28. Jochum, K.P.; Scholz, D.; Stoll, B.; Weis, U.; Wilson, S.A.; Yang, Q.; Schwalb, A.; Börner, N.; Jacob, D.E.; Andreae, M.O. Accurate trace element analysis of speleothems and biogenic calcium carbonates by LA-ICP-MS. *Chem. Geol.* **2012**, *318*, 31–44. [CrossRef]

29. Jackson, S.E.; Pearson, N.J.; Griffin, W.L.; Belousova, E.A. The application of laser ablation-inductively coupled plasma-mass spectrometry to in situ U–Pb zircon geochronology. *Chem. Geol.* **2004**, *211*, 47–69. [CrossRef]

30. Kooijman, E.; Berndt, J.; Mezger, K. U-Pb dating of zircon by laser ablation ICP-MS: Recent improvements and new insights. *Eur. J. Miner.* **2012**, *24*, 5–21. [CrossRef]

31. Wiedenbeck, M.; Alle, P.; Corfu, F.; Griffin, W.L.; Meier, M.; Oberli, F.; von Quadt, A.; Roddick, J.C.; Spiegel, W. Three natural zircon standards for U-Th-Pb, Lu-Hf, trace-element and REE analyses. *Geostand. Newsl.* **1995**, *19*, 1–23. [CrossRef]

32. Ludwig, K.R. *A User's Manual*; Barkeley Geochonology Center: Berkeley, CA, USA, 2009; p. 100.

33. Platonov, A.N.; Taran, M.N.; Balitsky, V.S. *The Nature of the Coloring of Gems*; Nedra Publishing House: Moscow, Russia, 1984; p. 196. (In Russian)

34. Sorokina, E.S.; Koivula, J.I.; Muyal, J.; Karampelas, S. Multiphase fluid inclusions in blue sapphires from the Ilmen Mountains, southern Urals. *Gems Gemol.* **2016**, *52*, 209–211.

35. Peucat, J.J.; Ruffault, P.; Fritch, E.; Bouhnik-Le Coz, M.; Simonet, C.; Lasnier, B. Ga/Mg ratio as a new geochemical tool to differentiate magmatic from metamorphic blue sapphires. *Lithos* **2007**, *98*, 261–274. [CrossRef]

36. Sutherland, F.L.; Zaw, K.; Meffre, S.; Giuliani, G.; Fallick, A.E.; Graham, I.T.; Webb, G.B. Gem-corundum megacrysts from east Australian basalt fields: trace elements, oxygen isotopes and origins. *Aust. J. Earth Sci.* **2009**, *56*, 1003–1022. [CrossRef]

37. Zwaan, J.C.; Buter, E.; Merty-Kraus, R.; Kane, R.E. The origin of Montana's alluvial sapphires. *Gems Gemol.* **2015**, *51*, 370–391.

38. Uher, P.; Giuliani, G.; Szaka, L.L.S.; Fallick, A.; Strunga, V.; Vaculovic, T.; Ozdin, D.; Greganova, M. Sapphires related to alkali basalts from the Cerov'a Highlands, Western Carpathians (southern Slovakia): Composition and origin. *Geologica Carpathica* **2012**, *63*, 71–82. [CrossRef]

39. Palke, A.C.; Wong, J.; Verdel, C.; Avila, J.N. A common origin for Thai/Cambodian rubies and blue and violet sapphires from Yogo Gulch, Montana, U.S.A.? *Am. Mineral.* **2018**, *103*, 469–479. [CrossRef]

40. Giuliani, G.; Caumon, G.; Rakotosamizanany, S.; Ohnenstetter, D.; Rakototondrazafy, M. Classification chimique des corindons par analyse factorielle discriminante: application a la typologie des gisements de rubis et saphirs. Chapter mineralogy, physical properties and geochemistry. *Revue Gemmol.* **2014**, *188*, 14–22.

41. Sun, S.S.; McDonough, W.F. Chemical and isotopic systematics of oceanic basalts: Implications for mantle composition and processes. *Geol. Soc. London Spec. Publ.* **1989**, *42*, 313–345. [CrossRef]

42. Fowler, A.; Prokoph, A.; Stenr, R.; Dupuis, C. Organization of oscillatory zoning in zircon: Analysis, scaling, geochemistry, and model of a zircon from Kipawa, Quebec, Canada. *Geochim. Cosmochim. Acta.* **2002**, *66*, 311–328. [CrossRef]

43. Belousova, E.A.; Griffin, W.L.; O'Reilly, S.Y.; Fisher, N.I. Igneous zircon: trace element compositon as an indicator of source rock type. *Contrib. Mineral. Petrol.* **2002**, *143*, 602–622. [CrossRef]

44. Watson, E.B.; Wark, D.A.; Thomas, J.B. Crystallization thermometers for zircon and rutile. *Contrib. Mineral. Petrol.* **2006**, *151*, 413–433. [CrossRef]

45. Nedosekova, I.L.; Belyatsky, B.V.; Belousova, E.A. Trace elements and Hf isotope composition as indicator of zircon genesis due to the evolution of alkaline-carbonatite magmatic system (Il'meny–Vishnevogorsky complex, Urals, Russia). *Geol. Geophys.* **2016**, *57*, 1135–1154. [CrossRef]

46. Fersman, L.E. *Pegmatites*; Publishing House of the Academy of Sciences of the USSR: Moscow, Russia, 1940; p. 712. (In Russian)

47. Lodochnikov, V.N. *Serpentines and Serpentinites Ilchirsk and Other Petrological Issues Associated with Them*; United Scientific and Technical Publishing: Moscow, Russia, 1936; p. 817. (In Russian)

48. Korzhinskiy, D.S. Essay on metasomatic processes. In *The Main Problems in the Theory of Magmatic Ore Deposits*; Publishing House of the Academy of Sciences of the USSR: Moscow, Russia, 1953; pp. 332–450. (In Russian)

49. Eskova, E.M.; Zhabin, A.G.; Mukhitdinov, G.N. *Mineralogy and Geochemistry of Rare Elements of the Vishnevogorsky Mountains*; Science Publishing House: Moscow, Russia, 1964; p. 318. (In Russian)

50. Nedosekova, I.L. U-Pb age and Lu-Hf isotopic systems of zircons Ilmenogorsky-Vishnevogorsky alkaline-carbonatitic complex, South Ural. *Lithosphere* **2014**, *5*, 19–31. (In Russian)

51. Sutherland, F.L.; Abduriym, A. Geographic typing of gem corundum: a test case from Australia. *J. Gemmol.* **2009**, *31*, 203–210. [CrossRef]

52. Giuliani, G.; Fallick, A.E.; Garnier, V.; France-Lanord, C.; Ohnenstetter, D.; Schwarz, D. Oxygen isotope composition as a tracer for the origins of rubies and sapphires. *Geology* **2005**, *33*, 249–252. [CrossRef]

53. Vysotsky, S.V.; Nechaev, V.P.; Kissin, A.Y.; Yakovlenko, V.V.; Velivetskaya, T.A.; Sutherland, F.L.; Agoshkov, A.I. Oxygen isotopic composition as an indicator of ruby and sapphire origin: A review of Russian occurrences. *Ore Geol. Rev.* **2015**, *68*, 164–170. [CrossRef]

54. Sutherland, F.L.; Giuliani, G.; Fallick, A.E.; Garland, M.; Webb, G. Sapphire-ruby characteristics, West Pailin, Cambodia: Clues to their origin based on trace element and O isotope analysis. *Aust. Gemmol.* **2008**, *23*, 329–368.

55. Sutherland, F.L.; Schwarz, D.; Jobbins, E.A.; Coenraads, R.R.; Webb, G. Distinctive gem corundum suites from discrete basalt fields: a comparative study of Barrington, Australia, and west Pailin, Cambodia, gemfields. *J. Gemmol.* **1998**, *26*, 65–85. [CrossRef]

56. Saeseaw, S.; Sangsawong, S.; Vertriest, W.; Atikarnsakul, U. *An In-Depth Study of Blue Sapphires from Pailin, Cambodia*; Gemological Institute of America report: Carlsbad, CA, USA, 2017; p. 45.

57. Grimes, C.B.; John, B.E.; Kelemen, P.B.; Mazdab, F.K.; Wooden, J.L.; Cheadle, M.J.; Hanghoj, K.; Schwartz, J.J. Trace element chemistry of zircons from oceanic crust: A method for distinguishing detrital zircon provenance. *Geology* **2007**, *35*, 643–646. [CrossRef]

58. Grimes, C.B.; Wooden, J.L.; Cheadle, M.J.; John, B.E. "Fingerprinting" tectono-magmatic provenance using trace elements in igneous zircon. *Contrib. Mineral. Petrol.* **2015**, *170*, 46. [CrossRef]

59. Botcharnikov, R.E.; Almeev, R.R.; Koepke, J.; Holtz, F. Phase relations and liquid lines of descent in hydrous ferrobasalt—Implications for the Skaergaard Intrusion and Columbia River flood basalts. *J. Petrol.* **2008**, *49*, 1687–1727. [CrossRef]

60. Arndt, N. The formation of massif anorthosite: Petrology in reverse. *Geosci. Front.* **2013**, *7*, 875–889. [CrossRef]

61. Leet, J.K.; Williams, I.S.; Ellis, D.J. Pb, U and Th diffusion in natural zircon. *Nature* **1997**, *390*, 159–162.

62. Biondi, J.C. Neoproterozoic Cana Brava chrysotile deposit (Goiás, Brazil): Geology and geochemistry of chrysotile vein formation. *Lithos* **2014**, *184*, 132–154. [CrossRef]

63. Varlakov, A.S.; Kuznetsov, G.P.; Korablev, G.G. *Hyperbasites of the Ilmenogorsky-Vishnevogorsky complex Complex (Southern Urals)*; Publishing House of the Institute of Mineralogy, Ural Branch RAS: Miass, Russia, 1998; p. 195. (In Russian)

64. Yakymchuk, C.; Kristoffer, S. Corundum formation by metasomatic reactions in Archean metapelite, SW Greenland: Exploration vectors for ruby deposits within high-grade greenstone belts. *Geosci. Front.* **2017**, *9*, 1–24. [CrossRef]

65. Hofmann, A.W. Mantle geochemistry: The message from oceanic volcanism. *Nature* **1997**, *385*, 219–229. [CrossRef]

 *minerals*

Article

Gem Corundum Deposits of Greece: Geology, Mineralogy and Genesis

Panagiotis Voudouris [1,*], Constantinos Mavrogonatos [1], Ian Graham [2], Gaston Giuliani [3], Vasilios Melfos [4], Stefanos Karampelas [5], Vilelmini Karantoni [4], Kandy Wang [2], Alexandre Tarantola [6], Khin Zaw [7], Sebastien Meffre [7], Stephan Klemme [8], Jasper Berndt [8], Stefanie Heidrich [9], Federica Zaccarini [10], Anthony Fallick [11], Maria Tsortanidis [1] and Andreas Lampridis [1]

[1] Faculty of Geology and Geoenvironment, National and Kapodistrian University of Athens, 15784 Athens, Greece; kmavrogon@geol.uoa.gr (C.M.); m.tsortanidi@hotmail.com (M.T.); andreaslampridis18@gmail.com (A.L.)

[2] School of Biological, Earth and Environmental Sciences, University of New South Wales, Sydney, NSW 2052, Australia; i.graham@unsw.edu.au (I.G.); kandy.wang@student.unsw.edu.au (K.W.)

[3] Université Paul Sabatier, GET/IRD et Université de Lorraine, C.R.P.G./C.N.R.S., 54501 Vandoeuvre, CEDEX, France; giuliani@crpg.cnrs-nancy.fr

[4] Faculty of Geology, Aristotle University of Thessaloniki, 54124 Thessaloniki, Greece; melfosv@geo.auth.gr (V.M.); wilelminikar@gmail.com (V.K.)

[5] Bahrain Institute for Pearls & Gemstones (DANAT), WTC East Tower, P.O. Box 17236 Manama, Bahrain; stefanos.karampelas@danat.bh

[6] UMR GeoResources, Faculté des Sciences et Technologies, Université de Lorraine, 54506 Vandoeuvre-lès-Nancy, France; alexandre.tarantola@univ-lorraine.fr

[7] CODES Centre of Ore Deposit and Earth Sciences, University of Tasmania, Tas 7001, Australia; khin.zaw@utas.edu.au (K.Z.); sebastien.meffre@utas.edu.au (S.M.)

[8] Institut für Mineralogie, Westfälische-Wilhelms Universität Münster, 48149 Münster, Germany; stephan.klemme@uni-muenster.de (S.K.); jberndt@uni-muenster.de (J.B.)

[9] Mineralogisch-Petrographisches Institut, Universität Hamburg, 20146 Hamburg, Germany; stefanie.heidrich@mineralogie.uni-hamburg.de

[10] Department of Applied Geosciences and Geophysics, University of Leoben, 8700 Leoben, Austria; federica.zaccarini@unileoben.ac.at

[11] Isotope Geosciences Unit, S.U.E.R.C., Glasgow G75 0QF, UK; anthony.fallick@glasgow.ac.uk

* Correspondence: voudouris@geol.uoa.gr; Tel.: +30-210-727-4129

Received: 13 December 2018; Accepted: 14 January 2019; Published: 15 January 2019

Abstract: Greece contains several gem corundum deposits set within diverse geological settings, mostly within the Rhodope (Xanthi and Drama areas) and Attico-Cycladic (Naxos and Ikaria islands) tectono-metamorphic units. In the Xanthi area, the sapphire (pink, blue to purple) deposits are stratiform, occurring within marble layers alternating with amphibolites. Deep red rubies in the Paranesti-Drama area are restricted to boudinaged lenses of Al-rich metapyroxenites alternating with amphibolites and gneisses. Both occurrences are oriented parallel to the ultra-high pressure/high pressure (UHP/HP) Nestos suture zone. On central Naxos Island, colored sapphires are associated with desilicated granite pegmatites intruding ultramafic lithologies (plumasites), occurring either within the pegmatites themselves or associated metasomatic reaction zones. In contrast, on southern Naxos and Ikaria Islands, blue sapphires occur in extensional fissures within Mesozoic metabauxites hosted in marbles. Mineral inclusions in corundums are in equilibrium and/or postdate corundum crystallization and comprise: spinel and pargasite (Paranesti), spinel, zircon (Xanthi), margarite, zircon, apatite, diaspore, phlogopite and chlorite (Naxos) and chloritoid, ilmenite, hematite, ulvospinel, rutile and zircon (Ikaria). The main chromophore elements within the Greek corundums show a wide range in concentration: the Fe contents vary from (average values) 1099 ppm in the blue sapphires of Xanthi, 424 ppm in the pink sapphires of Xanthi, 2654 ppm for Paranesti rubies, 4326 ppm for the Ikaria sapphires, 3706 for southern Naxos blue sapphires, 4777 for purple and 3301 for pink sapphire from Naxos plumasite, and finally 4677 to 1532 for blue to colorless sapphires from Naxos plumasites, respectively. The Ti concentrations (average values) are very low in rubies from Paranesti (41 ppm), with values of 2871 ppm and 509 in the blue and pink sapphires of Xanthi, respectively, of 1263 ppm for the Ikaria blue sapphires, and 520 ppm, 181 ppm in Naxos purple, pink sapphires, respectively. The blue to colorless sapphires from Naxos plumasites contain 1944 to 264 ppm Ti, respectively. The very high Ti contents of the Xanthi blue sapphires may reflect submicroscopic rutile inclusions. The Cr (average values) ranges from 4 to 691 ppm in the blue, purple and pink colored corundums from Naxos plumasite, is quite fixed (222 ppm) for Ikaria sapphires, ranges from 90 to 297 ppm in the blue and pink sapphires from Xanthi, reaches 9142 ppm in the corundums of Paranesti, with highest values of 15,347 ppm in deep red colored varieties. Each occurrence has both unique mineral assemblage and trace element chemistry (with variable Fe/Mg, Ga/Mg, Ga/Cr and Fe/Ti ratios). Additionally, oxygen isotope compositions confirm their geological typology, i.e., with, respectively $\delta^{18}O$ of $4.9 \pm 0.2\text{‰}$ for sapphire in plumasite, 20.5‰ for sapphire in marble and 1‰ for ruby in mafics. The fluid inclusions study evidenced water free CO_2 dominant fluids with traces of CH_4 or N_2, and low CO_2 densities (0.46 and 0.67 g/cm^3), which were probably trapped after the metamorphic peak. The Paranesti, Xanthi and central Naxos corundum deposits can be classified as metamorphic sensu stricto (s.s.) and metasomatic, respectively, those from southern Naxos and Ikaria display atypical magmatic signature indicating a hydrothermal origin. Greek corundums are characterized by wide color variation, homogeneity of the color hues, and transparency, and can be considered as potential gemstones.

Keywords: corundum megacrysts; ruby; sapphire; plumasite; metamorphic-metasomatic origin; Greece

1. Introduction

Rubies and sapphires, the two different varieties of the mineral corundum, are among the most popular gemstones used in jewelry. The color in corundum varies from brown, pink to pigeon-blood-red, orange, yellow, green, blue, violet etc. The main chromophore trace elements in corundum are Cr, Fe, Ti and V. Cr^{3+} produces pink and red, while $Fe^{2+}-Ti^{4+}$ pairs produce blue. Distinction of primary gem quality-rubies and sapphires in magmatic and metamorphic is mainly based on the classification schemes presented by Garnier et al. [1], Giuliani et al. [2] and Simonet et al. [3]. Corundum deposits were mainly formed in three periods, the Pan-African orogeny (750–450 Ma), the Himalayan orogeny (45–5 Ma) and the Cenozoic continental rifting with related alkali basalt volcanism (65–1 Ma) [2,4–10]. Magmatic and metamorphic ruby and sapphire deposits related to the Pan-African orogeny are present in southern Madagascar, East Africa, South India and Sri Lanka; those related to the Himalayan orogeny mostly occur in marbles from Central and Southeast Asia. Alkali basalt-related sapphires occur in Australia, northern and central Madagascar, Nigeria, Cameroon, French Massif Central and southeast Asia.

The use of trace element content of gem corundums, in association with their oxygen isotopic signature, is an effective tool in characterizing and interpreting their origin, and can also be used for the exploration, classification and comparison of gem corundum deposits especially those of disputed origin (e.g., placers, xenoliths, etc.) [2,7,11–17].

The presence of corundum in Greece (Xanthi, Naxos and Ikaria islands) has been known of since several years ago, but mostly from unpublished company reports [18–20]. It was only during the 80s, when detailed mineralogical, petrological studies investigated the geological environment of formation for the emery-hosted corundum on Naxos, Samos and Ikaria island [21], and the marble-hosted corundum in Xanthi area [22,23]. Iliopoulos and Katagas [24] described metamorphic conditions of formation for corundum-bearing metabauxites from Ikaria island. Fieldwork during

1998–2010 by the first two authors resulted in additional findings of corundum megacrysts in Greece, mostly within the Rhodope (Xanthi and Drama areas) and Attico-Cycladic (Naxos and Ikaria islands) tectono-metamorphic units [25–29]. According to these authors, both occurrences in the Rhodope are classified as metamorphic deposits related to meta-limestones (Xanthi deposit) and mafic granulites (Paranesti deposit), based on the scheme of Simonet et al. [3], and supported the theory of their formation during the high temperature-medium pressure retrograde metamorphic episode of carbonates and eclogitic amphibolites during the Cenozoic collision along the Nestos Suture Zone. The plumasite-hosted corundum deposits at Naxos island and those hosted in metabauxites from Ikaria island are classified as metasomatic and metamorphic, respectively [27,29]. Recently, Wang et al. [30] compiled multi-analytical geochemical, mineralogical and petrological studies on rubies from Paranesti Drama area, discussing their origin and comparing the deposit to those in other mafic-ultramafic complexes especially that of Winza in Tanzania.

In addition to the above localities, corundum has been also reported from the Koryfes Hill prospect, which is a telescoped porphyry-epithermal system hosted within Tertiary granitoids in the Kassiteres-Sapes area, on the south of the Rhodope massif (Figure 1). Hydrothermal corundum occurs within a transitional sericitic-sodic/potassic alteration of a quartz-feldspar porphyry, which forms the root zone of an advanced argillic lithocap comprising diaspore, topaz, pyrophyllite and alunite supergroup minerals [31]. Hydrothermal corundum forms up to 3-mm-large aggregates of deep blue-colored crystals that are separated from quartz (and K-feldspar/albite) by a sericite rim. Electron microprobe analyses indicate minor amounts of Fe (up to 0.34 wt. % FeO), Mg (up to 0.02 wt. % MgO) Ti (up to 0.13 wt. % TiO_2) and Cr (up to 0.12 wt. % Cr_2O_3) substituting for Al in the structure. The Koryfes Hill corundum occurrence belongs to those present in porphyry deposits elsewhere and is attributed to the magmatic class of corundum deposits, according to the classification scheme of Simonet et al. [3]. It was suggested by Voudouris [31] that the assemblage corundum-sericite at Koryfes Hill formed as a result of rapid cooling of ascending magmatic-hydrothermal solutions under pressures between 0.6 and 0.3 kb. However, due to its small grain size, which probably rather prevent its use as a gemstone this corundum occurrence will not be further discussed in the present paper.

The aim of this work is to present detailed information on the geology, mineralogy, geochemistry and fluid characteristics involved in the formation of rubies and sapphires in Greece, by expanding on the previous work of the Wang et al. [30], which was only focused on the Paranesti area. New laser ablation inductively coupled plasma mass spectrometry (LA-ICP-MS) and electron probe microanalysis (EPMA) data on corundum will help define their origin and compare them with other famous occurrences elsewhere.

2. Materials and Methods

For this study, the following 18 samples, collected by the first author, were investigated: Naxos samples (Nx1a, Nx1b, Nx2a, Nx2b, Nx3a, Nx3b, Nx4, Nx5a, Nx5b), Ikaria samples (Ik1a, Ik1b), Xanthi samples (Go1a, Go1b, Go5a, Go5b) and Drama samples (Dr1a, Dr1b, Dr2), ranging in color from colorless, to pink, purple, red and blue (Table 1). All occurrences are primary, which is rare for blue sapphire and relatively rare for ruby (S. Karampelas, pers. communication).

Thirty-five thin and ten thin-and-polished sections of corundum-bearing samples and host rocks were studied by optical and a JEOL JSM 5600 scanning electron microscope (SEM) equipped with back-scattered imaging capabilities, respectively, at the Department of Mineralogy and Petrology at the University of Athens (Athens, Greece). Quantitative analyses were carried out at the "Eugen F. Stumpfl" Laboratory installed at the University of Leoben, Austria, using a Superprobe Jeol JXA 8200 wavelength-dispersive electron microprobe (WDS), and at the Department of Mineralogy and Petrology, University of Hamburg (Hamburg, Germany) using a Cameca-SX 100 WDS. Analytical conditions were as follows: at Leoben, 20 kV accelerating voltage, 10 nA beam current, 1 μm beam diameter. Peak and backgrounds counting times were 20 and 10 s for major and 40 and 20 s for trace elements, respectively. The X-ray lines used were AlKα, SiKα, TiKα, GaKα, FeKα, CeLα, VKα, MgKα,

CrKα, and CaLα. The standards used were: corundum for Al, wollastonite for Si and Ca, rutile for Ti, GaAs for Ga, ilmenite for Fe, synthetic V for V, chromite for Mg and Cr. At Hamburg, accelerating voltage of 20 kV, a beam current of 20 nA and counting time of 20 s. The X-ray lines used were: AlKα, SiKα, TiKα, GaKα, FeKα, CeLα, VKα, MgKα, CrKb, and CaLα. The standards used were: andradite and vanadinite (for Si, Ca and V), and synthetic Al_2O_3 (for Al), TiO_2 (for Ti), Fe_2O_3 (for Fe), Cr_2O_3 (for Cr), Ga_2O_3 (for Ga) and MgO (for Mg). Corrections were applied using the PAP online program [32].

Table 1. Sample description of the studied Greek corundum crystals.

Sample	Variety	Color	Host Rock	Locality
Dr1a, Dr1b, Dr2	Ruby	Red	Pargasite-schist	Paranesti, Drama
Go1a, Go1b	Sapphire	Blue-colorless	Marble	Gorgona, Xanthi
Go5a	Sapphire	Pink-purple	Marble	Gorgona, Xanthi
Go5b	Sapphire	Pink	Marble	Gorgona, Xanthi
Nx1a, Nx2a, Nx2b, Nx3a, Nx3b, Nx4	Sapphire	Blue-colorless	Plumasite	Naxos (Kinidaros)
Nx1b	Sapphire	Purple	Plumasite	Naxos (Kinidaros)
Nx5a	Sapphire	Blue	Metabauxite	Naxos (Kavalaris)
Nx5b	Sapphire	Pink	Plumasite	Naxos (Kinidaros)
Ik1a, Ik1b	Sapphire	Blue	Metabauxite	Ikaria

LA-ICP-MS analyses were conducted at the CODES ARC Centre of Excellence in Ore Deposits of the University of Tasmania, Australia, and the Institute of Mineralogy, University of Münster, Germany. Analytical conditions were as follows: The LA-ICP-MS analysis at the CODES was performed under standard procedures using a New Wave UP-213 Nd: YAG Q-switched Laser Ablation System coupled with an Agilent HP 4500 Quadrupole ICP-MS. The international standard NIST 612 was used as the primary standard to calculate concentrations and correct for ablation depth, and the basaltic glass BCR-2 was used as the secondary standard. Analyses were standardized on the theoretical Al content for corundum at Al 529,227 ppm, with an error in analytical precision of 2–3%. Corundum sample ablation at the University of Münster was done with a pulsed 193 nm ArF excimer laser (Analyte G2, Photon Machines). A repetition rate of 10 Hz and an energy of ~4 J/cm^2 were used throughout the entire session. The beam spot diameter was set to 35 μm. Trace element analysis has been carried out with an Element 2 mass spectrometer (ThermoFisher). Forward power was 1250 W and reflected power <1 W, gas flow rates were 1.2 L/m for He carrier gas, 0.9 L/m and 1.2 L/m for the Ar-auxiliary and sample gas, respectively. Argon cooling gas flow rate was set to 16 L/min. Before starting analysis, the system was tuned on a NIST 612 reference glass measuring ^{139}La, ^{232}Th and ^{232}Th^{16}O to get stable signals and high sensitivity, as well as low oxide rates (^{232}Th^{16}O/^{232}Th < 0.1%) during ablation. A total of 30 elements were quantitatively analyzed using the NIST 612 glass as an external standard and ^{27}Al as internal standard, which had previously been determined by electron microprobe. Overall time of a single analysis was 60 s (20 s for background, 40 s for peak after switching laser on). Concentrations of measured elements were calculated using the Glitter software [33,34]. Standard reference glass BHVO-2G and BIR1-G were analyzed in order to monitor for precision and accuracy in the silicate phases in the course of this study. The obtained results match the published range of concentrations given in the GeoReM database (version 23).

Fluid inclusions studies were carried in a total of 14 double-polished sections prepared at the Department of Mineralogy, Petrology and Economic Geology of the Aristotle University of Thessaloniki (Thessaloniki, Greece). Routine heating and freezing runs were performed at a LINKAM THM-600/TMS 90 heating–freezing stage coupled to a Leitz SM-LUX-POL microscope. Calibration of the stage was achieved using organic standards with known melting points, and the H_2O phase transition from ice to liquid; the precision of the measurements, including reproducibility, was ± 0.2 °C. The SoWat program [35] was used to process fluid inclusion data.

Oxygen isotope analyses were performed at the Scottish Universities Environmental Research Centre, Glasgow, Scotland. The oxygen isotope composition of corundum was obtained using a modification of the CO_2 laser fluorination system similar to that described by Sharp [36], which was similar applied by Giuliani et al. [14]. The method involves the complete reaction of ~1 mg of ground corundum. This powder is heated by a CO_2 laser, with ClF_3 as the fluorine reagent. The released oxygen is passed through an in-line Hg-diffusion pump before conversion to CO_2 on platinized graphite. The yield is then measured by a capacitance manometer. The gas-handling vacuum line is connected to the inlet system of a dedicated VG PRISM 3 dual-inlet isotope-ratio mass spectrometer. Duplicate analyses of corundum samples suggest that precision and accuracy are $\pm 0.1\permil$ (1σ). All oxygen isotope ratios are shown in $\delta^{18}O$ ($\permil$) relative to Vienna standard mean ocean water (VSMOW).

3. Geological Setting

3.1. Regional Geology

The Hellenide orogen formed as a result of the collision between the African and Eurasian plates above the north-dipping Hellenic subduction zone from the Late Jurassic to the present [37–39]. From north to south, it consists of three continental blocks (Rhodopes, Pelagonia, and Adria-External Hellenides) and two oceanic domains (Vardar and Pindos Suture Zones). In the Aegean region, continuous subduction of both oceanic and continental lithosphere beneath the Eurasian plate since the Early Cretaceous resulted in a series of magmatic arcs from the north (Rhodope massif) to the south (Active South Aegean Volcanic Arc) [40]. The geodynamic evolution of Hellenides is characterized by a collisional phase taking place during the Mesozoic, followed by a continuous southward slab retreat in a back-arc setting, started at the Eocene but still ongoing, that triggered large-scale extension concomitant with thrusting at the southern part of the Hellenic domain [39,41]. The gem corundum deposits in Greece are spotted in two tectono-metamorphic units of the greater Hellenides Orogen: the Rhodope- and the Attico-Cycladic Massifs [29,30,42].

3.1.1. Rhodope Massif

The Rhodope Massif is considered part of the European continental margin [41,43]. It is a heterogeneous crustal body composed in its eastern and central parts of two sub-domains (Figure 1): the Northern Rhodope Domain and the Southern Rhodope Core Complex (both separated by the Nestos thrust fault and the Nestos Suture Zone) [41]. The Northern Rhodope Domain consists of a Lower unit of high-grade basement including orthogneisses derived from Permo-Carboniferous protoliths; this Unit includes four metamorphic core complexes (the Arda, Biala Reka-Kechros, and Kesebir-Kardamos migmatitic domes) which are considered to be equivalent to orthogneisses in the Southern Rhodope Core Complex (SRCC) [41,44]. The latter also consists of Triassic marbles with intercalated amphibolitic and metapelitic rocks [44,45].

The upper tectonic unit of the Northern Rhodope Domain includes high-grade basement rocks that have both continental and oceanic affinities, and with protoliths ranging in age from Neoproterozoic through Ordovician, and Permo-Carboniferous to Early Cretaceous [45]. The rocks of the Intermediate unit experienced high- to ultra-high-pressure metamorphism with subsequent high-grade amphibolite-facies overprint [46]. The Rhodope Massif has undergone a polycyclic alpine orogeny including UHP-eclogitic metamorphism, followed by HP granulite-facies, amphibolite-facies and finally by greenschist facies metamorphic events starting from Jurassic (~200–150 Ma) and lasting up to the Oligocene [47,48]. In the Rhodope region, syn-orogenic exhumation of the metamorphic pile initiated in the early Eocene (~55 Ma) and core complex extension followed the Cretaceous syn-metamorphic SW-directed thrusting [41,43,49,50]. The metamorphic core complexes were progressively exhumed along several ductile- to brittle shear zones, active from ~42–12 Ma [49,51].

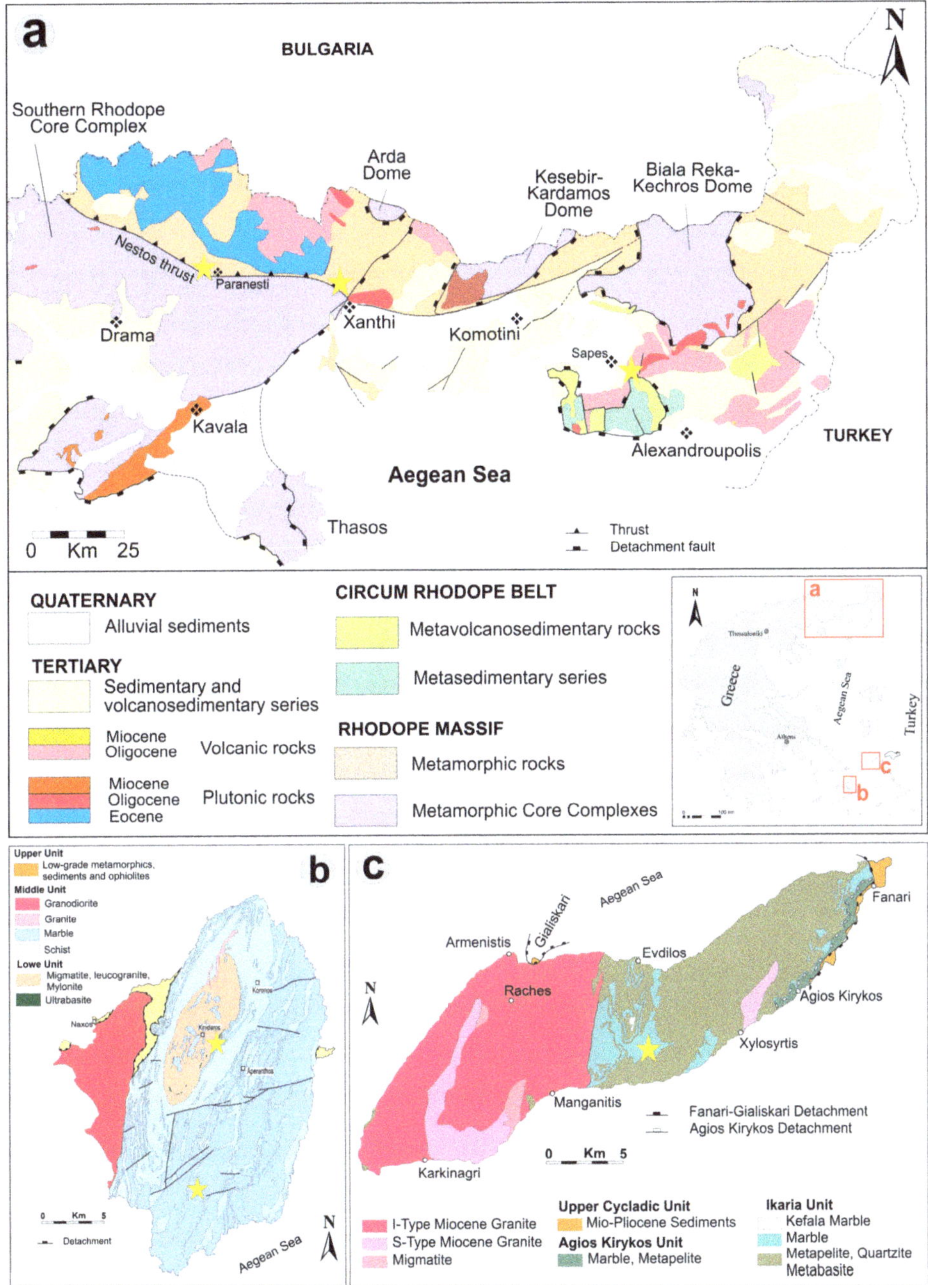

Figure 1. (**a**) Simplified geological map of eastern Rhodope massif, Greece showing the main tectonic zones and the distribution of Cenozoic igneous rocks (after Melfos and Voudouris [52] and references therein), showing location of corundums at Xanthi, Paranesti and Kassiteres/Sapes areas (stars); (**b**) Geological map of Naxos island showing locations of sapphires in plumasites at Kinidaros and meta-bauxites at Kavalaris Hill (stars) (after Duchêne et al. [53]; modified after Ottens and Voudouris [54]); (**c**) Geological map of Ikaria showing the location of blue sapphire in Atsida area (after Beaudoin et al. [55]; modified after Ottens and Voudouris [54]). Yellow stars mark the studied corundum occurrences.

Separating the upper and lower tectonic units in the Rhodope complex, the Nestos Suture Zone a NW trending shear zone was extensively studied by Papanikolaou and Panagopoulos [56], Krenn et al. [57], Nagel et al. [58] and Turpaud and Reischmann [45] and is composed from the bottom to the top of the following lithological types: (a) a lower 1-km-thick highly sheared "mélange" zone, consisting of amphibolites, garnet-kyanite schists, migmatites, orthogneisses and marbles; (b) a 1-km-thick sequence of augen-gneisses; (c) two layers of marbles intercalated with amphibolites and mylonitic amphibolites; (d) a layer of ortho-gneisses, characterized by biotite-gneisses with highly migmatized base (Sidironero). Gautier et al. [48] suggest a continuous thrusting tectonism along the Nestos Suture Zone until 33 Ma ago, and Nagel et al. [58] indicate that the Nestos Suture Zone constitutes the base of an Eocene thrusting wedge that also includes UHP units which were probably merged with the Nestos Zone during the thrusting event.

3.1.2. Attico-Cycladic Massif

The Cyclades consist of a lowermost Pre-Alpidic Basement Unit, the intermediate Cycladic Blueschist Unit, and the uppermost Pelagonian Unit [59]. The Blueschist Unit represents a polymetamorphic terrane which tectonically overlies the basement gneiss and consists of a metamorphosed volcano-sedimentary sequence of clastic metasedimentary rocks, marbles, calc-schists, and mafic and felsic meta-igneous rocks [38,39,60]. The Cyclades have been exhumed since the Eocene as metamorphic core complexes formed in low-, medium-, and/or high-temperature environments [50]. An Eocene (~52–53 Ma) high-pressure eclogite- to blueschist-facies metamorphism was followed by syn-orogenic exhumation of the blueschists in a cold retrograde path and then by early Oligocene amphibolite to greenschist facies, and finally, by high-temperature medium-pressure metamorphism and associated migmatites in the early Miocene [37,60–63]. The amphibolite to greenschist metamorphic event occurred during extension-related exhumation and was coeval with back-arc extension at the Rhodopes in northern Greece [64–66]. Exhumation of the Cycladic rocks as metamorphic core complexes was accommodated during the Oligocene–Miocene by several ductile to brittle detachment systems. The extensional event also allowed for various granitoids (granite, granodiorite, leucogranite) to be intruded throughout the Cyclades between 15 and 7 Ma [64].

On Naxos island, a migmatitic dome (pre-alpine basement), is overlain by the Blueschist unit containing alternating layers of marbles, schists and gneisses, and by the upper Pelagonian tectonic unit (Figure 1b). Blueschist relics at the SE part of the island indicate temperatures from 400 to 460 °C at minimum pressures of 7–9 kbar [21] for the high-pressure/low-temperature (HP/LT) Eocene metamorphic event. The metamorphic grade of the Oligocene–Miocene event increases from 400 °C at 6 kbar in marbles, schists, gneisses and amphibolites at the SE of the island, towards the core of the dome, where migmatites in the leucogneissic and amphibolite-facies sillimanitic schists rocks, suggest temperatures up to 700 °C at 6–8 kbar [21,67–70]. Jansen and Schuiling [67] distinguished six metamorphic zones (I–VI) at Naxos island based on mineral isograds. Maximum temperatures are 420 °C for zone I, inferred by the presence of diaspore; 420 to 500 °C (appearance of biotite) for zone II; 500 °C to 540–580 °C (disappearance of chloritoid) for zone III, 540–580 °C to 620–660 °C (kyanite-sillimanite transition) for zone IV, 620–660 °C to 660–690 °C (onset of melting) for zone V; and >690 °C for zone VI. The Naxos migmatitic core is surrounded by a discontinuous block of ultramafic (meta-peridotites) lenses, representing a thrust zone along which the metamorphic complex lies on top of the pre-alpine bedrock [71]. The peridotites underwent high-P metamorphism and then, after cooling amphibolite-facies conditions (T~600 °C), were finally re-heated and metamorphosed together with the country rocks. According to Jansen and Schuiling [67], pegmatites penetrating the ultramafic bodies in the sillimanite stability zone, are desilicated and composed of phlogopite, anorthite, corundum, chlorite, zoisite, tourmaline and beryl. Anorthite crystals are composed almost entirely of anorthite (98% An), while margarite can be found in places. According to Andriessen et al. [72] and Katzir et al. [71], aplites and pegmatites are of Early Miocene age (19–20 Ma), and were formed as a result of the crystallization of an in situ anatectic liquid, that was produced during the high-T metamorphic event. They represent a system that channeled fluids through the leucogneissic core to the lower metamorphic unit and resulted in the metasomatization of peridotites and in situ recrystallization of the peridotitic blackwalls. Corundum-bearing emery deposits occur in the metamorphic zones I–V following the increase in metamorphic grade from SE towards NW to the migmatite core [21,73].

Ikaria Island (Figure 1c), similarly to Naxos, represents a Miocene migmatite-cored metamorphic core complex, where peak-metamorphic temperatures range from 390 °C in the upper parts of the structure down to 625 °C in the core of the dome in the vicinity of migmatites and S-type granite [55,74]. Three main tectonic units are distinguished and are, from bottom to top, the Ikaria, Agios Kirykos and Fanari units, limited by two major shear zones [55,75,76]. The non-metamorphic Fanari unit consists of Miocene to Pliocene sandstones, conglomerates and ophiolitic molasses. The intermediate Agios Kirykos unit (previously called the Messaria unit) consists of alternating marble and metapelite layers, metamorphosed in greenschist-facies conditions. Finally, the Ikaria unit is composed of amphibolite-facies (ca. 6–8 kbar and 600–650 °C) metasediments including micaschist and marble layers and minor metabasites. Peak-metamorphic conditions were retrieved from the basal parts of the succession. Two synkinematic granitic bodies intrude the Ikaria unit: (a) an I-type Bt-granite (Raches granite) in the western part of the island, with K-Ar ages of 22.7 Ma [64]; and (b) an S-type Bt-Ms-granite (Xilosirtis granite) in the southern part of the island, with Rb-Sr ages of 18.1 Ma [64]. Liati and Skarpelis [77], Iliopoulos and Katagas [24], and Iliopoulos [75] studied the Ikaria metabauxitic rocks hosted in marbles of the Ikaria unit, and recorded a Jurassic age for the formation of bauxitic deposits, and upper greenschist to lower amphibolite-facies conditions for their metamorphism.

3.2. Local Geology

3.2.1. Xanthi

The Gorgona-Stirigma corundum occurrence is stratiform and distributed in marble layers (Figure 2a), reaching up to 50 m width, alternating with eclogitic amphibolites and gneisses [22,27,42]. The chemical composition of the marbles varies widely, highly locally impure marbles, rich in alumino-silicate minerals, are characterized by a mineralogical assemblage composed of calcite, dolomite, corundum, spinel, pyrophyllite, muscovite, paragonite, amphibole, zoisite, margarite, chlorite, olivine, serpentine, phlogopite, rutile, titanite, anorthite, garnet and Ni-tourmaline [22,23,27,42]. Spinel occurs as a rim around corundum crystals. The color of spinel ranges from blue to green and brown. Liati [22] defined the following contact assemblages: calcite-anorthite-zoisite, calcite-corundum-spinel, calcite-corundum-zoisite and margarite-zoisite-anorthite-chlorite. The corundum crystals occur within phyllosilicate-rich micro-shear zones along marble layers. Sapphire is of a pink, orange, purple to blue color, usually of tabular or barrel-shaped euhedral form and reaches sizes of up to 4 cm (Figure 3a–c). In some cases, blue corundum alters to green spinel. Red corundum has also been reported in this area, mainly from zoisite-bearing amphibolites [23], but could not be identified in the present study.

3.2.2. Paranesti

The Paranesti corundum occurrence is stratiform, oriented parallel to the main regional foliation, and distributed mainly in boudin-like lenses of amphibole schist (Figure 2b) [27,30,42]. The outcrops on the surface are spotted west of Perivlepto village at one hillside and one roadside location [30]. In the first site, the corundum-bearing amphibole schist lenses are surrounded by a narrow clinochlore schist zone and occur together with boudins of corundum-kyanite-amphibole schists. The schists are intruded by white-colored quartz-feldspar-mica-garnet pegmatitic veins, which do not present characteristics of desilication (Figure 2c). In the second site, the lenses are hosted by amphibolites, intercalated with kyanite-bearing gneisses and kyanite-amphibole-chlorite schists. Again, the transition from amphibolitic lithology towards corundum mineralization is made through a clinochlore-rich schist zone. Ruby crystals from Paranesti, range in size from <1 mm to 50 mm size (average size 5–10 mm, and are of pale pink to deep red color [30]. Their morphology is mainly flat tabular with basal planes paralleling the orientation of the main regional foliation [30], and less commonly prismatic and barrel-shaped (Figure 3d,e). They occur together with pargasite, which is the main amphibole of the assemblage, forming a similar image with Tanzanian rubies which are surrounded by zoisite. The ruby crystals coexist locally with kyanite (Figure 3f).

Figure 2. Field photographs demonstrating the geological setting of the ruby and sapphire occurrences in Greece. (**a**) Marble layers alternating with eclogitic amphibolites at Gorgona/Xanthi; (**b**) Boudinaged hornblende schists, intercalated within eclogitic amphibolites, marbles and metapegmatites at Paranesti/Drama; (**c**) The main ruby working at Paranesti. The pegmatite body at the lower part of the photograph shows no desilification; (**d**,**e**) Ultrabasic rocks between the migmatite dome-like core and the overlying kyanite-sillimanite grade schists and marbles in the two plumasite localities east of Kinidaros, Naxos island; (**f**) Corundum-bearing plumasite at the contact to ultrabasites at the locality shown in Figure 2e; (**g**) Panoramic view of the metabauxite bearing marbles at Kavalaris Hill, southern Naxos island; (**h**) Panoramic view for the metabauxite-bearing marbles of Ikaria unit; (**i**) Corundum-bearing metabauxite lenses within the Ikaria unit marbles.

3.2.3. Naxos

The corundums studied from Naxos island occur in two localities, representing different geological environments.

(a) In the central part of the island and about 2 km east and southeast of Kinidaros, two corundum-bearing plumasites (desilicated pegmatites) were studied (Figure 2d–f) [27,29,42]. The meta-peridotites occurred initially as lenses of few to several tens of meters size and are surrounded by phlogopitite, which formed at the contact between the metaperidotites and the intruding pegmatites. Today, most of the plumasites and metaperidotites in these two localities occur as loose fragments mainly due to erosion and agricultural activities in the area.

Figure 3. Field photographs and handspecimens demonstrating gem corundum crystals of Greece: (**a–c**) Pink, blue, to purple sapphires within Xanthi marbles; (**d,e**) Deep red-colored ruby with both tabular and barrel-shaped crystals within pargasite schist from Paranesti/Drama; (**f**) Ruby associated (replacing) kyanite, Paranesti/Drama; (**g**) Barrel-shaped blue and colorless sapphires hosted within anorthite-rich desilicated pegmatite (plumasite) and associated with phlogopite and tourmaline (schorlite) from the first Kinidaros locality, Naxos island; (**h**) Purple to pink sapphire within plumasite the later rimmed by phlogopite at the left part, second Kinidaros locality, Naxos island; (**i**) Blue sapphire in fissures within metabauxite at Kavalaris Hill, southern Naxos island; (**j,k**) Blue sapphires and margarite as free growing crystals of probably hydrothermal origin within fissure of metabauxites, Ikaria island.

The plumasites are surrounded by an up to 20-cm-wide zone of phlogopite and chlorite (blackwall) (Figure 3g). Within the plumasites, colorless to blue, purple and pink corundum may occur either as isolated crystals within the plagioclase matrix, and/or associated with tourmaline, and phlogopite (Figure 3g,h). Purple to pink sapphires have been also found entirely enclosed in phlogopitite in the blackwalls. The sapphire crystals are barrel-shaped, display macroscopic color zoning from colorless cores to blue rims or pink cores to purple rims and reach sizes up to 3 cm [27,42].

(b) In the southern part of the island, close to Kavalaris Hill (Figure 2g), corundum represents a single mineral rock termed "corundite" by Feenstra and Wunder [73], which was formed by the dissociation of former diasporites, in meta-carstic bauxites during prograde regional metamorphism. This first corundum occurrence in metabauxites also recorded as corundum-in isograd (T~420–450 °C and P~6–7 kbar, [73]), separates diasporites with the assemblage diaspore-chloritoid-muscovite-paragonite-calcite-hematite-rutile from emery characterized by corundum-chloritoid-muscovite-paragonite-margarite-hematite-rutile. Corundum and chloritoid intergrowths occur as a rim in metabauxites, while the corundum crystals occur mainly in the contact between marbles and metabauxites. The corundum from this locality does not occur in well-shaped crystals; it is, rather, massive, and of blue color (Figure 3i).

3.2.4. Ikaria

The metabauxite occurrence is located on mountain Atheras (Figure 2h) [20,75]. Already Iliopoulos [75], reported the presence of corundum as mineralogical constituent of meta-bauxite (emery) lenses occurring within dolomitic and calcic marble, which lies on top of gneiss, both being part of the Ikaria unit. In this area, four main layers of metabauxites, of up to 1 m width, were detected (Figure 2i). However, the corundum megacrysts described by Voudouris et al. [27,42] represent a second, late, generation of corundum formation within the emeries, since they fill together with margarite extensional fissures and networks of veins discordant to the metabauxite foliation (Figure 3j,k) [27,42]. The corundums are deep-blue in color, tabular to euhedral, reaching sizes up to 4 cm and are accompanied by Fe-chlorite, hematite, rutile and diaspore.

4. Mineralogy

The Paranesti rubies are dull to transparent, exhibiting clear parting and polysynthetic twinning. The rubies are associated with edenitic hornblende and tremolite, and rimmed by margarite, muscovite, chlorite and chromian spinel (Figure 4a,b and Figure 5a,b). The ruby crystals are surrounded by a margarite rim. Spinel appears as the main component of primary inclusions in the rubies. The association of ruby in this deposit is pargasite, chlorite (mainly clinochlore and rarely nimite), margarite, tremolite and/or monazite, while for the host rocks the assemblage is pargasite, anorthite, clinozoisite, chlorite and monazite. The Paranesti rubies are rimmed by Cr-bearing spinel and also include ulvospinel. Corundum is in equilibrium with amphibole, while in a later, retrograde stage margarite-muscovite, Cr-bearing spinel, orthoclase, chlorite, smectite and vermiculite, rim and/or replace corundum. Cr-bearing spinel is replaced by chlorite.

The corundums from Xanthi are transparent, with very clear parting and fine cracks. Polysynthetic twinning is common. Corundum may be isolated within the carbonate matrix (Figure 4c) or rimmed by amesite (an Al-rich serpentine member), margarite, titanite or by brown and green spinel. Blue sapphires are zoned, with alternating deep blue and colorless domains (Figure 4c). This zoning, or irregular color distribution in the sapphires, is attributed to different Fe and Ti contents in the crystals (see next paragraph). Ni-bearing brown tourmaline accompanying corundum in the paragenesis contains up to 4.4 wt. % NiO, much higher than the Ni content reported in tourmaline from Samos by Henry and Dutrow [78].

The sapphire crystals from Naxos present strong pleochroism with their color ranging from dark blue to colorless (Figure 4d–f). They are transparent, with smaller or bigger inner fractures. Hexagonal barrel-shaped corundum crystals exhibit blue color zoning. The blue color of corundum is observed either as a blue core surrounded by a white rim or as a blue-zoned outer rim surrounding a colorless core. Polysynthetic twinning is observed along the colorless parts of the crystals. Sapphires are associated with anorthite, phlogopite, zircon, margarite, muscovite, tourmaline and chlorite (Figures 4d–f and 5c–i). The sapphires are enveloped by an assemblage consisting of oligoclase/labradorite-orthoclase-muscovite and then by phlogopite, chlorite (Figure 5c–h). In sample Nx1, blue sapphire associated with phlogopite is rimmed by an intergrowth of oligoclase, anorthite, orthoclase and muscovite. Margarite, zircon, diaspore, phlogopite and chlorite are included in blue sapphire; however, it is considered that margarite, diaspore and chlorite are retrograde minerals postdating corundum formation (this study). In samples Nx4 and Nx4a, blue sapphire includes phlogopite, zircon and apatite and surrounded by anorthite, muscovite and tourmaline with minor amounts of monazite and chlorite. Phlogopite and anorthite are replaced by muscovite, margarite and chlorite. Tourmaline postdates anorthite and muscovite.

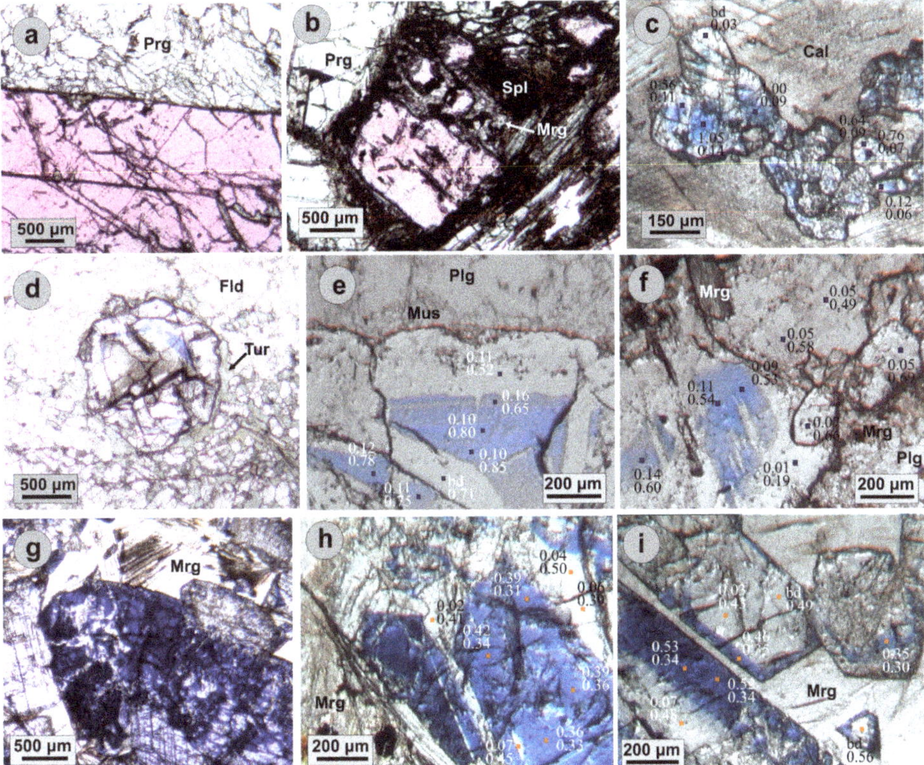

Figure 4. Microphotographs (plane polarized light) demonstrating mineralogical assemblages, color- and chemical zoning of corundum-bearing assemblages from Greece. Values in microphotographs (**c,e,f,h,i**) correspond to TiO_2 and Fe_2O_3 content in wt. % (upper and lower values, respectively) as obtained by electron probe microanalyses. (**a,b**) Ruby associated with pargasite (Prg) from Paranesti, Drama area. Note replacement of ruby by spinel (Spl) and margarite (Mrg) in Figure 4b (sample Dr1a); (**c**) blue to colorless sapphires included in calcite, Xanthi area (sample Go1b); (**d,e**) Zoned blue to colorless sapphires within anorthite (Plg) and rimmed by tourmaline (Tur) and muscovite (Mus), Kinidaros, Naxos island (sample Nx4); (**f**) Sapphire with blue core and colorless rim associated with margarite (Mrg) and anorthite (Plg), Kinidaros, Naxos island (sample Nx3); (**g**) Idiomorphic deep blue colored sapphire crystals surrounded by margarite, Ikaria island (sample Ik1b); (**h,i**) Zoned blue to colorless idiomorphic sapphires, surrounded by margarite, Ikaria island (sample Ik2).

On Ikaria Island, blue sapphires in the metabauxites are accompanied by the assemblage margarite + chloritoid + hematite/ilmenite + rutile (Figures 4g–i and 5j–l). Chloritoid fill fractures and also occurs as inclusions in margarite. The retrogression has continued to low temperatures as suggested by the presence of diaspore replacing corundum (Figure 4j–l) and chlorite replacing chloritoid. Corundum forms colorless cores and deep blue rims (Figure 4h,i). While transparent, their blue color is quite dark, with colorless spots appearing all over their surface. Corundum includes ilmenite, hematite, ulvospinel, rutile and zircon. In addition, a Zn-bearing green colored spinel, which forms solid solution between hercynite and gahnite, has been reported with chlorite and/or margarite by Iliopoulos [75] and Iliopoulos and Katagas [24].

In sample Ik1a, Ik1b, and Ik2 blue sapphire is surrounded by margarite, chlorite and green-spinel. Chlorite fills fractures and represents latest deposition. Chloritoid, ilmenite, ulvospinel, rutile, Ti-hematite and zircon occur as inclusions in sapphire. Sapphire is associated with hematite, zircon and chloritoid and is replaced by diaspore and margarite.

5. Mineral Chemistry

5.1. EPMA

Electron probe microanalyses (EPMA) for the Greece corundums are presented in Table 2. EPMA data from Paranesti/Drama rubies showed an extreme enrichment in Cr_2O_3, which reaches up to 2.65 wt. %, with the higher concentrations being consistent with intense red-colored grains. Beyond Cr_2O_3, the Paranesti rubies are characterized by moderate Fe_2O_3 vaues (up to 0.44 wt. %) and by traces of MgO, Ga_2O_3, SiO_2, TiO_2 and CaO, which range from below detection up to 0.17, 0.15, 0.09, 0.05 and 0.02 wt. %, respectively. Finally, V_2O_3 content is mostly below detection, but in some cases values up to 0.07 wt. % were detected.

Colored sapphires from Gorgona/Xanthi are characterized by elevated contents of TiO_2, MgO and Fe_2O_3. The blue color varieties show the highest TiO_2 content, which reaches up to 1.05 wt. %, while MgO and Fe_2O_3 reach up to 0.27 and 0.22 wt. %, respectively. In the pink-colored grains, TiO_2 and Fe_2O_3 contents are significantly lower (up to 0.17 and 0.06 wt. %, respectively) in comparison to the blue ones, but MgO content increases up to 0.36 wt. %. SiO_2 and CaO values are generally below 0.08 wt. %, but in one analysis SiO_2 is 0.51 wt. %, probably indicating a fine-grained silicate inclusion. V_2O_3 content tends to be higher in the pink varieties rather than the colorless to blue, with values ranging from below detection to 0.06 and 0.01 wt.%, respectively. Pink-colored grains contain more than double Ga_2O_3 content (up to 0.17 wt. %) compared to the blue ones (up to 0.07 wt. %). Finally, colorless grains diasplay very low TiO_2 content (from below detection up to 0.12 wt. %) and similar Fe_2O_3 contents to the blue domains.

In colored sapphires from the Naxos plumasite, Fe_2O_3 is the dominant impurity. In the blue vareties, the Fe_2O_3 content reaches up to 0.93 wt.%, and decreases significantly in the colorless (up to 0.69 wt. %), pink- (up to 0.50 wt. %) and the purple-colored grains (up to 0.34 wt. %). A similar decreasing trend is remarked for the Ga_2O_3 content, which reaches up to 0.18 wt. % in the white to blue sapphires, and decreases slightly in the pink and purple varieties, reaching up to 0.15 and 0.10 wt. %, respectively. On the other hand, TiO_2 and MgO seem both to follow a different trend: the highest values for both oxides characterize the purple-colored grains (up to 0.58 and 0.54 wt. %, respectively). Blue grains display intermediate values (up to 0.38 and 0.49 wt. %), followed by the pink grains (up to 0.27 and 0.35 wt. %), respectively. TiO_2 content in the colorless areas ranges from below detection to 0.12 wt. % similarly to the blue domains. The Cr_2O_3 content is up to 0.1 wt. % in the colorless to blue and purple varieties, and almost doubles in the case of pink sapphires (up to 0.18 wt. %). Traces of SiO_2, V_2O_3 and CaO were also detected in all samples (up to 0.13, 0.09 and 0.03 wt. %, respectively).

Blue sapphires from Ikaria island appear enriched in TiO_2 and Fe_2O_3 values, which reach up to 0.69 and 0.57 wt. %, respectively. SiO_2 shows a wide range from below detection to 0.44 wt. %, while MgO content is constantly low (up to 0.05 wt. %). The Cr_2O_3 and Ga_2O_3 content are almost fixed, reaching up to 0.14 wt. %. Traces of V_2O_3 and CaO were also detected (up to 0.04 and 0.02 wt. %, respectively). Colorless domains are characterized by very low TiO_2 values, which range from below detection up to 0.12 wt. %.

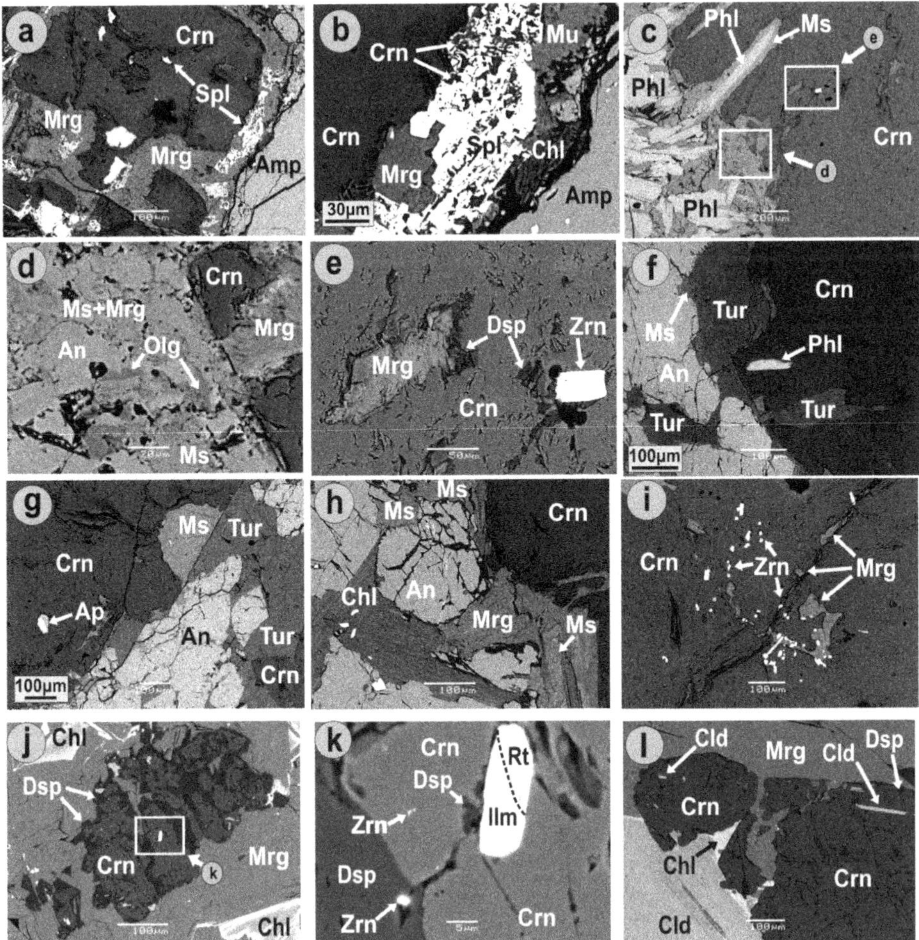

Figure 5. Microphotographs (SEM-BSE images) demonstrating mineralogical assemblages of corundum-bearing assemblages from Greece. (**a,b**) Ruby (Crn) including and rimmed by chromian spinel (Spl), margarite (Mrg), muscovite (Mu), chlorite (Chl) and pargasitic hornblende (Amp) (sample Dr1, Paranesti/Drama); (**c,d**) Blue sapphire (Crn) in association with anorthite (An) and phlogopite (Phl) replaced by muscovite (Mus), margarite (Mrg), and oligoclase (Olg) (Naxos island, sample Nx1); (**e**) Zircon (Zrn) is included in sapphire, diaspore (Dsp) and margarite (Mrg) fill fractures and are retrograde minerals. Naxos island, sample Nx1; (**f,g**) Blue sapphire including phlogopite and fluorite (Fl) is associated with anorthite (An) and replaced by tourmaline (Tur) and muscovite (Mus) (Naxos island, sample Nx4); (**h**) Blue sapphire and anorthite are rimmed by muscovite, margarite and chlorite (Chl) the late probably replacing phlogopite (Naxos island, sample Nx4a); (**i**) Blue sapphire including zircon is replaced by margarite (Naxos island, sample Nx4a); (**j–l**) Blue sapphire associated with margarite includes and replaces chloritoid (Cld) and are retrogressed to diaspore and chlorite, respectively. Zircon and Ilmenite (Ilm), rutile (Rt) are included in corundum (Ikaria island, sample Ik2).

Table 2. Chromophores and key trace elements electron probe microanalyses (EPMA) (wt. %) of the Greek corundum crystals.

Sample	Color	ppm	MgO	Al_2O_3	SiO_2	CaO	TiO_2	V_2O_3	Cr_2O_3	Fe_2O_3	Ga_2O_3	Total
Dr1a (n = 8)	red	aver	bd	96.96	0.03	0.01	0.01	0.01	1.82	0.39	0.02	99.23
		min	bd	96.52	0.03	bd	bd	bd	1.74	0.34	bd	98.80
		max	bd	97.39	0.04	0.02	0.01	0.01	1.99	0.43	0.09	99.66
Dr1b (n = 13)	red	aver	bd	96.98	0.04	0.01	0.01	0.01	2.14	0.37	0.02	99.55
		min	bd	96.13	0.02	bd	bd	bd	1.48	0.33	bd	98.95
		max	bd	97.58	0.09	0.01	0.01	0.01	2.65	0.44	0.07	100.10
Dr2 (n = 8)	red	aver	0.02	99.79	0.01	0.01	0.01	0.01	0.32	0.19	0.05	100.41
		min	bd	99.18	bd	bd	bd	bd	0.18	0.15	bd	99.78
		max	0.17	100.65	0.03	0.02	0.05	0.074	0.43	0.21	0.15	101.30
Go1a (n = 11)	blue-colorless	aver	0.07	98.02	0.09	0.01	0.59	0.01	0.04	0.11	0.01	98.94
		min	0.02	97.37	bd	bd	bd	bd	bd	0.07	bd	98.19
		max	0.27	98.61	0.51	0.01	0.94	0.01	0.08	0.14	0.04	99.36
Go1b (n = 15)	blue-colorless	aver	0.04	99.21	0.03	0.02	0.56	0.01	0.01	0.11	0.02	100.00
		min	bd	98.46	0.01	bd	bd	bd	bd	0.031	bd	99.70
		max	0.10	99.73	0.07	0.08	1.05	0.01	0.05	0.21	0.07	100.38
Go5b (n = 14)	pink	aver	0.05	100.93	0.01	0.01	0.06	0.01	0.03	0.04	0.05	101.18
		min	bd	100.22	bd	bd	bd	bd	bd	bd	bd	100.40
		max	0.36	101.58	0.03	0.02	0.17	0.06	0.06	0.06	0.17	101.99
Ik1a (n = 8)	blue	aver	bd	97.88	0.35	0.01	0.23	0.01	0.05	0.36	0.03	98.90
		min	bd	97.08	0.24	bd	bd	bd	bd	0.27	bd	98.16
		max	bd	98.45	0.44	0.016	0.53	0.04	0.13	0.50	0.12	99.36
Ik1b (n = 16)	blue	aver	0.01	99.13	0.04	0.01	0.25	0.01	0.03	0.43	0.03	99.91
		min	bd	98.44	0.02	bd	bd	bd	bd	0.14	bd	99.61
		max	0.01	99.68	0.06	0.01	0.68	0.04	0.10	0.57	0.14	100.21
Nx1a (n = 8)	blue-colorless	aver	0.01	98.25	0.09	0.01	0.22	bd	0.04	0.31	0.02	98.94
		min	bd	97.85	0.07	bd	bd	bd	bd	0.23	bd	98.77
		max	0.05	98.80	0.12	0.02	0.38	bd	0.12	0.45	0.08	99.42
Nx1b (n = 14)	purple	aver	0.12	99.80	0.02	0.01	0.27	0.01	0.01	0.27	0.04	100.54
		min	bd	98.08	bd	bd	0.08	bd	bd	0.20	bd	98.79
		max	0.54	100.57	0.07	0.02	0.58	0.06	0.04	0.34	0.11	101.28
Nx2b (n = 22)	blue-colorless	aver	0.10	100.26	0.02	0.01	0.07	0.01	0.01	0.45	0.05	100.98
		min	bd	93.73	bd	bd	bd	bd	bd	0.136	bd	99.21
		max	0.49	101.81	0.07	0.03	0.19	0.08	0.04	0.69	0.18	101.96
Nx3b (n = 15)	blue-colorless	aver	0.01	98.98	0.05	0.01	0.08	0.01	0.01	0.66	0.01	99.80
		min	bd	98.33	0.01	bd	0.01	bd	bd	0.19	bd	99.30
		max	0.02	99.47	0.13	0.01	0.16	0.01	0.04	0.93	0.05	100.23
Nx4 (n = 16)	blue-colorless	aver	0.01	98.09	0.09	0.01	0.10	0.01	0.01	0.72	0.01	99.02
		min	bd	97.69	0.05	bd	bd	bd	d	0.52	bd	98.67
		max	0.01	98.55	0.12	0.01	0.17	0.01	0.01	0.85	0.05	99.43
Nx5b (n= 12)	pink	aver	0.10	100.64	0.01	0.01	0.06	0.01	0.05	0.35	0.04	101.27
		min	bd	97.20	bd	bd	bd	bd	bd	0.16	db	98.91
		max	0.35	100.7	0.06	0.02	0.27	0.09	0.18	0.45	0.15	102.70

bd = below detection; aver = average; min = minimum; max = maximum.

5.2. LA-ICP-MS

The trace element results for averages, ranges and critical ratios are listed in Tables 3 and 4. Analyzed rubies from Paranesti/Drama are particularly enriched in Cr (up to 15,347 ppm), and less so in Fe (up to 4348 ppm), with the highest values for both elements coming from the most dark-colored grains. The Mg, Ga and V values are consistently low, (up to 31, 24 and 6 ppm, respectively). The low V contents result in consistent very low V/Cr ratios, typically <1. Calcium and Ti values display wide variations between 117–1444 ppm and 8–148 ppm, respectively, with higher values characterizing the most light-colored grains. Other trace elements are close or below detection limits.

Table 3. Chromophores and key trace elements (Mg and Fe) LA-ICP-MS analyses (ppm) of the Greek corundum crystals and their different chemical ratios.

Sample.	Color	ppm	Mg	Ti	V	Cr	Fe	Ga	V/Cr	V + Cr/Ga	Fe/Mg	Ga/Mg	Fe/Ti	Cr/Ga
Dr1a (*n* = 8)	red	aver	15	17	2.4	9814	2020	14	0.0003	718	162	1.11	133	718
		min	8	8	2	8326	1799	12	0.0002	602	86	0.53	75	602
		max	25	27	2.7	12,116	2181	16	0.0003	879	247	1.68	225	879
Dr1b (*n* = 13)	red	aver	17	39	3.79	14,322	3795	14	0.0003	998	295	1.13	109	998
		min	9	22	3.01	11,625	3352	12	0.0002	750	122	0.46	68	750
		max	31	58	5.71	15,347	4348	16	0.0004	1252	440	1.72	187	1252
Dr2 (*n* = 8)	red	aver	17	67	2.4	3289	2149	20	0.0008	162	157	1.48	58	161
		min	13	27	2.1	2431	1918	19	0.0006	122	91	0.95	16	122
		max	24	148	2.9	4056	2438	24	0.0011	190	209	2	83	189
Go1a (*n* = 11)	blue-colorless	aver	380	5007	90	97	1096	89	1.89	2.98	3.16	0.26	0.26	1.91
		min	227	1915	40	35	554	29	0.23	1.15	1.32	.07	0.10	0.29
		max	601	6462	159	251	1339	119	4.15	7.73	5.73	0.49	0.68	6.30
Go1b (*n* = 15)	blue-colorless	aver	264	3503	66	70	910	89	1.19	1.59	4.33	0.46	0.35	0.86
		min	78	1007	23	35	363	64	0.26	0.84	1.66	0.14	0.12	0.30
		max	536	6361	207	187	1309	121	2.96	3.71	9.25	0.94	1.20	2.92
Go5a (*n* = 8)	pink-purple	aver	28	102	20	104	1291	47	0.37	2.88	60	2.46	24	2.46
		min	13	30	5	4	962	9	0.04	0.34	21	0.43	4.98	0.43
		max	49	205	60	298	1782	100	1.36	8.13	130	7.45	56	7.45
Go5b (*n* = 14)	pink	aver	35	509	65	297	424	78	0.30	4.67	14	2.54	1.45	3.83
		min	17	39	47	105	294	72	0.04	2.13	8.17	4.42	0.51	1.27
		max	53	810	79	1082	494	87	0.68	14.68	24	6.57	9.19	14.08
Ik1a (*n* = 8)	blue	aver	13	1135	120	222	3237	85	0.57	4.04	445	12	3.10	2.65
		min	5	653	70	168	2710	72	0.22	3.30	62	1.50	1.93	1.97
		max	54	1942	154	313	3627	93	0.84	5.29	711	19	4.73	4.32
Ik1b (*n* = 16)	blue	aver	16	1390	111	224	5415	94	0.49	3.54	348	6.07	8.49	2.38
		min	13	267	31	103	3903	68	0.20	2.13	207	3.54	0.93	1.52
		max	24	4508	164	279	7324	114	0.67	5.24	491	9.76	20.3	3.21
Nx1a (*n* = 8)	blue-colorless	aver	47	368	16	691	1532	46	0.02	15	35	1.06	7.32	15
		min	32	105	14	504	1377	42	0.02	11	24	0.70	1.78	10
		max	71	943	19	851	1804	53	0.04	19	43	1.37	14	18

Table 3. *Cont.*

Sample.	Color	ppm	Mg	Ti	V	Cr	Fe	Ga	V/Cr	V + Cr/Ga	Fe/Mg	Ga/Mg	Fe/Ti	Cr/Ga
Nx1b (*n* = 14)	purple	aver	64	520	22	43	4677	63	0.52	1.03	82	1.12	12	0.48
		min	39	124	15	36	3324	55	0.34	0.82	27	0.37	4.16	0.67
		max	199	848	31	48	6670	76	0.86	1.14	123	1.57	44	0.83
Nx2a (*n* = 8)	blue-colorless	aver	45	489	24	3.6	3210	49	14	0.55	84	1.30	16	0.07
		min	24	40	19	0.9	2852	47	1.83	0.44	42	0.67	4.15	0.02
		max	82	761	36	11	3474	53	42	0.68	127	1.99	86	0.22
Nx2b (*n* = 22)	blue-colorless	aver	63	560	26	51	4241	60	0.50	2.23	94	1.23	2.14	1.65
		min	16	89	13	31	1758	48	0.09	0.91	24	0.31	0.97	0.51
		max	208	966	42	73	6678	72	1.01	6.01	217	3.01	7.29	4.42
Nx3a (*n* = 8)	blue-colorless	aver	50	264	23	115	3189	52	1.11	2.68	68	1.12	94	2.24
		min	35	10	17	5	2843	47	0.10	0.51	42	0.68	5.57	0.08
		max	79	598	30	247	3922	59	5.10	4.99	88	1.37	316	4.46
Nx3b (*n* = 15)	blue-colorless	aver	105	1944	28	71	2773	62	0.50	2.23	30	0.68	1.86	1.65
		min	98	454	25	23	2220	53	0.09	0.91	16	0.34	0.97	0.41
		max	198	3222	34	246	3317	78	1.01	6.01	50	0.93	7.29	4.42
Nx4 (*n* = 16)	blue-colorless	aver	51	377	21	45	3828	83	5.98	0.87	81	1.85	41	0.57
		min	16	18	9	1	1812	50	0.11	0.18	41	0.96	3.47	0.01
		max	117	846	32	111	6361	184	30	1.83	126	4.61	189	1.38
Nx5a (*n* = 8)	blue	aver	5	462	52	262	3706	87	0.20	3.64	1069	24	9.17	3.04
		min	2	238	43	227	3301	84	0.13	3.16	241	6.28	4.32	2.64
		max	14	784	65	339	4268	90	0.24	4.48	2309	47	18	3.97
Nx5b (*n* = 12)	pink	aver	52	181	20	428	3096	45	0.05	9.95	77	1.08	24	9.52
		min	17	60	14	274	2138	37	0.03	6	34	0.36	7.13	5.61
		max	133	348	43	548	4716	52	0.10	13	208	2.07	59	12

Table 4. Other trace elements LA-ICP-MS analyses (ppm) of the Greek corundum crystals.

Sample	ppm	Be	Na	Si	Ca	Mn	Ni	Zn	Rb	Sr	Zr	Nb	Mo	Sn	Ba	Ta	W	Pb	Th	U
Dr1a ($n = 8$)	aver	bd	4.5	294	175	0.19	0.26	0.04	0.05	0.13	0.02	0.03	0.01	0.18	0.45	0.01	0.04	0.02	0.01	0.01
	min	bd	1.0	185	117	0.14	0.10	0.03	0.00	0.00	0.02	0.02	0.01	0.07	0.01	0.01	0.02	0.01	0.01	0.01
	max	bd	20	520	282	0.20	0.47	0.08	0.37	0.59	0.03	0.04	0.01	0.47	3.15	0.01	0.12	0.02	0.01	0.01
Dr1b ($n = 13$)	aver	bd	236	1807	780	bd	57	bd	bd	bd	bd	0.07	0.36	bd	9.80	0.05	0.17	bd	0.02	0.01
	min	bd	152	bd	352	bd	32	bd	bd	bd	bd	0.03	0.09	bd	0.39	0.01	0.11	bd	0.01	0.01
	max	bd	323	2099	1202	bd	309	bd	bd	bd	bd	0.11	1.27	bd	47	0.14	0.29	bd	0.02	0.01
Dr2 ($n = 8$)	aver	bd	bd	bd	1414	bd	bd	bd	bd	bd	bd	0.15	bd	bd	bd	0.02	0.48	bd	0.01	0.01
	min	bd	bd	bd	1384	bd	bd	bd	bd	bd	bd	0.04	bd	bd	bd	0.02	0.23	bd	0.01	0.01
	max	bd	bd	bd	1444	bd	bd	bd	bd	bd	bd	0.21	bd	bd	bd	0.02	0.73	bd	0.01	0.01
Go1a ($n = 11$)	aver	bd	bd	bd	1748	bd	bd	bd	bd	bd	bd	28	bd	38	0.62	1.60	68	0.73	6.98	0.32
	min	bd	bd	bd	1482	bd	bd	bd	bd	bd	bd	0.46	bd	11	0.55	0.09	1.11	0.70	2.14	0.03
	max	bd	bd	bd	2014	bd	bd	bd	bd	bd	bd	90	1.46	79	0.69	4.01	311	0.76	21	0.72
Go1b ($n = 15$)	aver	bd	492	bd	2265	bd	bd	bd	bd	bd	bd	8.76	0.22	16	0.52	1.64	44	5.43	10	0.10
	min	bd	434	bd	1859	bd	bd	bd	bd	bd	bd	0.29	0.11	2.14	0.23	0.11	0.48	2.23	1.07	0.01
	max	bd	524	bd	2491	bd	bd	bd	bd	bd	bd	55	0.34	97	0.74	11	498	8.63	53	0.26
Go5a ($n = 8$)	aver	0.2	1.3	224	136	0.20	0.75	0.22	0.01	0.04	0.02	0.11	0.01	27	0.02	0.02	0.16	0.06	0.40	0.01
	min	0.1	bd	14	88	0.20	0.04	0.03	0.01	0.01	0.01	0.01	0.01	0.09	0.01	0.00	0.03	0.01	0.01	0.01
	max	0.5	3.3	369	217	0.20	2.04	0.44	0.01	0.24	0.05	0.52	0.01	213	0.06	0.04	0.46	0.34	1.17	0.03
Go5b ($n = 14$)	aver	bd	bd	bd	bd	bd	bd	bd	bd	bd	bd	0.16	0.45	bd	0.87	0.07	bd	bd	0.03	0.02
	min	bd	bd	bd	bd	bd	bd	bd	bd	bd	bd	0.16	0.33	bd	0.57	0.06	bd	bd	0.01	0.01
	max	bd	bd	bd	bd	bd	bd	bd	bd	bd	bd	0.16	0.61	bd	1.17	0.08	bd	bd	0.06	0.05
Ik1a ($n = 8$)	aver	0.45	116	2479	1700	0.44	0.56	2.25	0.16	1.45	0.07	2.56	0.01	4.97	0.35	0.96	1.19	2.95	1.74	0.12
	min	0.10	24	353	312	0.23	0.37	0.94	0.02	0.30	0.04	1.59	0.01	3.23	0.11	0.34	0.79	1.40	0.85	0.04
	max	0.70	313	6492	4158	1.05	0.81	4.30	0.48	3.39	0.11	4.03	0.01	7.66	0.71	1.73	1.89	4.85	2.45	0.23
Ik1b ($n = 16$)	aver	bd	bd	bd	bd	bd	bd	bd	bd	bd	bd	7.92	0.34	6.88	1.74	0.39	4.64	bd	4.34	0.04
	min	bd	bd	bd	bd	bd	bd	bd	bd	bd	bd	0.40	0.11	2.83	1.13	0.10	0.74	bd	0.04	0.01
	max	bd	bd	bd	bd	bd	bd	bd	bd	bd	bd	37	0.84	23	2.34	1.06	12	bd	31	0.22
Nx1a ($n = 8$)	aver	2.08	5.30	543	14	0.20	1.10	0.03	0.03	0.66	0.05	1.16	0.01	32	0.39	15	2.74	0.02	0.40	0.01
	min	0.18	1.00	399	14	0.20	0.79	0.03	0.00	0.01	0.00	0.11	0.01	5.61	0.01	3.82	0.16	0.02	0.01	0.01
	max	11.03	34	1110	14	0.20	1.65	0.03	0.18	4.75	0.15	3.89	0.01	108	3.01	46	8.74	0.02	2.45	0.02
Nx1b ($n = 14$)	aver	2.91	205	2303	1551	bd	bd	bd	bd	bd	bd	7.77	0.24	1040	0.11	0.39	43	bd	0.72	0.01
	min	1.31	141	2069	1351	bd	bd	bd	bd	bd	bd	0.41	0.12	166	0.10	0.16	4.04	bd	0.02	0.01
	max	4.50	253	2537	1874	bd	bd	bd	bd	bd	bd	19	0.49	1712	0.12	0.65	147	bd	2.57	0.02
Nx2a ($n = 8$)	aver	1.21	36	804	120	0.21	0.71	0.03	0.20	1.73	1.23	9.87	0.01	566	0.05	1.58	105	0.02	0.13	0.01
	min	0.10	1	466	14	0.11	0.35	0.03	0.01	0.01	0.76	0.23	0.01	197	0.01	0.18	0.63	0.02	0.01	0.01
	max	3.62	271	2209	648	0.35	1.39	0.03	0.04	11	2.25	36	0.01	887	0.17	7.14	480	0.02	0.48	0.03

Table 4. *Cont.*

Sample	ppm	Be	Na	Si	Ca	Mn	Ni	Zn	Rb	Sr	Zr	Nb	Mo	Sn	Ba	Ta	W	Pb	Th	U
Nx2b (*n* = 22)	aver	3.53	112	1998	2118	bd	bd	bd	0.92	0.88	bd	12	0.33	608	bd	3.12	147	0.53	1.48	0.08
	min	2.71	90	1848	1844	bd	bd	bd	0.66	0.68	bd	0.11	0.09	5	bd	0.18	1.08	0.51	0.01	0.01
	max	5.64	141	2252	2418	bd	bd	bd	1.25	1.15	bd	117	0.68	1842	bd	18	877	0.56	13	0.49
Nx3a (*n* = 8)	aver	2.79	1.58	501	14	0.20	1.06	0.03	0.02	0.22	0.83	7.71	0.01	340	0.27	2.16	100	0.02	0.07	0.01
	min	0.25	1.00	412	14	0.20	0.59	0.03	0.01	0.01	0.01	0.98	0.01	4.95	0.01	0.53	11	0.02	0.01	0.00
	max	17	5.65	693	14	0.20	1.67	0.03	0.11	1.56	1.64	36	0.01	574	2.05	6.02	424	0.02	0.21	0.01
Nx3b (*n* = 15)	aver	4.25	bd	bd	bd	bd	bd	bd	bd	bd	bd	9.87	0.23	560	0.17	4.85	49	bd	3.36	0.02
	min	3.28	bd	bd	bd	bd	bd	bd	bd	bd	bd	0.16	0.12	198	0.12	0.34	0.49	bd	0.04	0.01
	max	5.82	bd	bd	bd	bd	bd	bd	bd	bd	bd	48	0.34	947	0.22	13	245	bd	20	0.08
Nx4 (*n* = 16)	aver	7.05	1.69	500	14	0.23	0.83	0.03	0.01	0.02	1.80	112	0.01	244	0.06	111	34	0.03	0.83	0.06
	min	0.10	1.00	267	14	0.15	0.28	0.03	0.00	0.00	0.01	1.16	0.01	9.19	0.01	0.31	1.75	0.02	0.01	0.01
	max	38	12	920	14	0.61	1.92	0.03	0.17	0.12	7.56	769	0.01	684	0.38	549	157	0.10	4.69	0.55
Nx5a (*n* = 8)	aver	1.12	1.12	372	163	16	1.17	0.79	0.01	0.04	0.06	1.53	0.01	3.64	0.66	0.23	6.70	1.80	7.40	0.02
	min	0.10	1.00	328	98	0.20	0.70	0.22	0.01	0.01	0.02	1.14	0.01	2.31	0.07	0.11	3.80	0.18	4.20	0.01
	max	1.86	1.99	415	239	110	1.89	2.25	0.01	0.08	0.10	1.91	0.01	5.45	2.75	0.35	9.45	7.72	10	0.03
Nx5b (*n* = 12)	aver	bd	bd	bd	bd	bd	bd	bd	bd	bd	bd	1.49	0.63	4.30	bd	12	6.28	0.67	0.02	0.01
	min	bd	bd	bd	bd	bd	bd	bd	bd	bd	bd	0.10	0.28	3.22	bd	0.19	bd	0.49	0.01	0.01
	max	bd	bd	bd	bd	bd	bd	bd	bd	bd	bd	2.89	0.97	6.60	bd	96	13.1	1.03	0.02	0.01

Colored (colorless to blue, pink) sapphires from Gorgona/Xanthi display significant differences regarding the concentration of some chromophore elements. Colorless to blue sapphires are mostly characterized by significant variations of their Ti content which reflects their zoned coloration. Bluish areas display high Ti values, which are up to 6 times higher compared to colorless/white areas and reach up to 6462 ppm. Iron content is quite fixed in both colorless and blue areas, reaching up to 1339 ppm, but is slightly higher in the pink varieties (up to 1782 ppm). Pink sapphires are also characterized by much less Ti (up to 810 ppm) and significantly higher Cr (up to 1082 ppm) concentrations, compared to the colorless/blue areas where Cr values are generally less than 298 ppm. Vanadium and Ga content remains relatively fixed, regardless of the color, with maximum values of up to 227 and 121 ppm, respectively. Mg values are generally higher in the colorless to blue grains, with values up to 536 ppm, while in the pink grains is generally below 65 ppm. Si and Ca display values up to 369 and 217 ppm, respectively, while other elements, are generally close to or below the detection limit, with the exception of Sn and Ta, which reach up to 213 and 4 ppm, respectively, in the colorless to blue grains.

Iron is the most abundant trace element in analyzed sapphires form Ikaria island, with values reaching up to 7324 ppm. Ti displays high values as well, up to 4508 ppm. Cr, Mg and V values are relatively constant with maximum values reaching up to 313, 54 and 164 ppm, respectively. Gallium shows elevated concentrations, up to 114 ppm, resulting in very high values of the Ga/Mg critical ratio. Si and Ca display wide variations between 353–6492 ppm and 312–4158 ppm, with the highest values correlating and thus indicating submicroscopic silicate inclusions (possibly margarite). Other trace elements display generally low concentrations except for Sn, whose values reach up to 23 ppm.

Corundums from Naxos island are mostly characterized by high Fe concentrations, which reach up to 6678 ppm. Higher Fe values are related to areas with more intense blue coloration. Ti is considerably lower, reaching values up to 966 ppm, with the exception of one colorless to blue corundum, where higher values were detected, up to 3222 ppm. Chromium varies significantly, between 1 and 851 ppm. The higher Cr values characterize corundum grains with pink and purple hues. Mg and Ga values range between 2–208 ppm and 42-184 ppm, respectively. The higher Ga concentrations are remarked in the metabauxite-hosted sapphires, that accordingly display the highest Ga/Mg ratio values compared to any other Greek corundums (this study). Vanadium content is quite fixed in all samples and varies between 10 and 65 ppm, with the majority of the measured values being in the range of 20–40 ppm. Si and Ca reach values up to 2537 and 2418 ppm, respectively, probably reflecting the presence of submicroscopic inclusions. Sodium is generally low, but in some cases values up to 271 ppm were detected, probably due to submicroscopic inclusions of silicate minerals (plagioclase and/or paragonite). Almost all corundums contain traces of Be, which ranges between 0.10 and 38 ppm. Manganese is generally close to or below detection, but in the case of a corundum from metabauxites reaches up to 110 ppm. Most samples are characterized by high Sn values, which reach up to 1812 ppm, while Nb is also enriched, reaching up to 769 ppm. Finally, Ta reaches up to 549 ppm, while the rest trace elements are mostly below detection (Table 4).

6. Fluid Inclusions

The fluid inclusions (FI) in the studied corundums range in size between <1 μm and 55 μm. Only one type of two-phase vapor-liquid CO_2 (LcarVcar) FI is observed within all samples [79]. (Figure 6a–d,g). They often have an elongated negative crystal shape of the host corundum (Figure 6a,k). They occur as isolated, and based on their similarity in shape and the mode of occurrence, they are considered as primary FI (Figure 6a,c,f,j,k). However, the most significant criterion for identifying primary FI is when they occur along the growth zones of the corundum (Figure 6i), based on the criteria suggested by Bodnar [80].

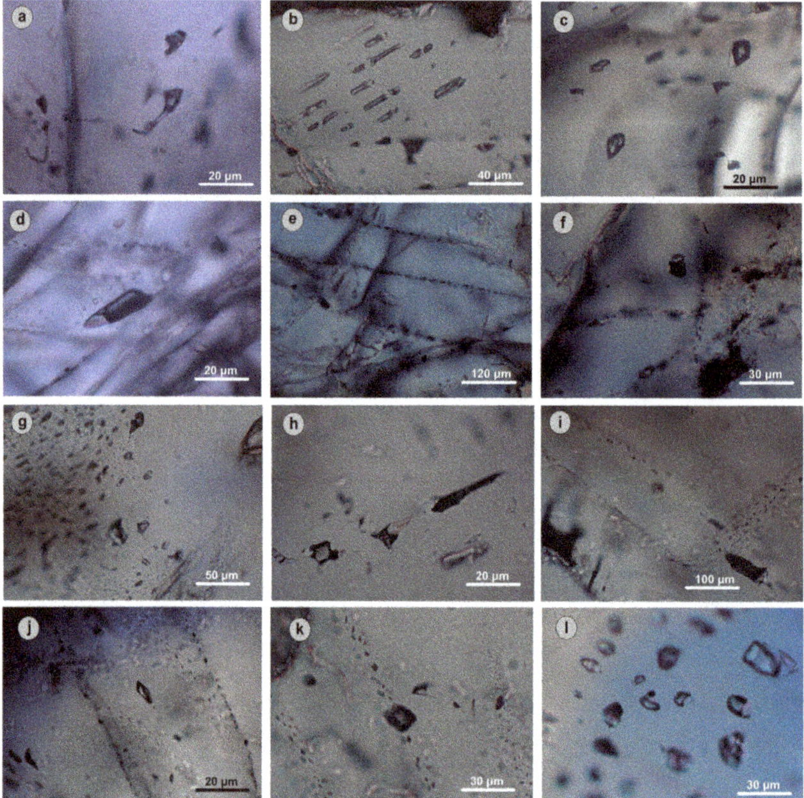

Figure 6. Photomicrographs of FI in corundums from the Rhodope massif of Greece, Paranesti and Gorgona (**a–f**) and from Naxos island the Attic-Cycladic massif (**g–l**). (**a**) Isolated primary two-phase vapor-liquid inclusions composed of $CO_2 \pm (CH_4$ and/or $N_2)$. The fluid inclusion at the upper part has a negative crystal shape, and the inclusion underneath shows necking-down deformation (Drama, thin-section Dr2). (**b**) Pseudosecondary two-phase carbonic FI along a fracture healed during crystal growth (Xanthi, thin-section Go5a). (**c**) Primary two-phase carbonic FI (Xanthi, thin-section Go5a). (**d**) Two-phase carbonic inclusion which shows evidence of post-entrapment modification (brittle crack and halo of neoformed inclusions) (Xanthi, thin-section Go5b). (**e**) Tiny FI along fracture plane trails, but due to their small size no phases can be identified (Xanthi, thin-section Go1b). (**f**) Isolated two-phase carbonic inclusion (Xanthi, thin-section Go1b); (**g**) Oriented cluster of pseudosecondary carbonic FI along fractures healed during crystal growth (thin-section, Nx1a). (**h**) Isolated primary two-phase carbonic inclusion with necking-down deformation (thin-section, Nx1a). (**i**) FI along growth zones of the corundum (thin-section, Nx1a). (**j,k**) Primary two-phase carbonic FI (thin-section, Nx2a). (**l**) Pseudosecondary two-phase carbonic FI along a fracture healed during crystal growth (thin-section, Nx2b).

Oriented clusters of $L_{car}-V_{car}$ FI along planes are very common in the studied samples. They are related to micro-cracks and subsequent fractures healed during crystal growth, and therefore are considered as pseudosecondary (Figure 6b,g,l). Numerous tiny FI (length <2 μm) along trails are usually observed, but due to their small size, the nature of the present phase(s) cannot be identified (Figure 6e). They are considered as secondary FI. The majority of the FI appears stretched and necked or empty due to leaking phenomena (Figure 6a,d,h). Microthermometric results were based on primary and pseudosecondary inclusions, without necking-down and post-entrapment modification evidence. However, many measurements were unsuccessful due to the black color of the carbonic fluid in the inclusions and the difficulty to observe phase changes [79].

Table 5 presents the microthermometric data of all studied FI from Drama, Xanthi, Ikaria and Naxos. The melting temperatures of CO_2 ($TmCO_2$) range between -57.3 and -56.6 °C, at or slightly lower than the triple point of CO_2 (-56.6 °C). This indicates that the fluid is dominated by CO_2 with very small quantities of CH_4 and/or N_2. Temperatures of Tm were not obtainable in the sample Dr1a from Paranesti (Drama). Clathrate nucleation was not observed in any measured fluid inclusion, demonstrating that liquid water (H_2O) was not incorporated in the whole process of the corundums formation. All FI homogenized to the liquid carbonic phase (LCO_2) at temperatures ($ThCO_2$) varying from 27.3 to 31.0 °C. This Th is close to the critical temperature of pure CO_2 and corresponds to densities of the source fluids from 0.46 to 0.67 g/cm^3. Figure 7 shows the histograms of the $ThCO_2$ of the FI from the four different studied corundum occurrences in Greece (Paranesti, Gorgona, Naxos, Ikaria).

Table 5. Microthermometric data of studied primary and pseudosecondary fluid inclusions from corundum crystals in Greece.

Sample	Locality	n	Tm CO_2 (°C)	Th CO_2 (°C)	Densities (g/cm^3)
Dr1a	Paranesti,	7	-57.2 to -56.7	28.7–31.1	0.46–0.64
Dr2	Drama	7	-57.1 to -56.6	27.4–30.4	0.57–0.67
Go1a		4	-57.0 to -56.8	29.2–30.2	0.58–0.62
Go1b	Gorgona,	3	-56.7 to -56.6	27.8–30.5	0.57–0.66
Go5a	Xanthi	6	-57.2 to -56.8	28.2–30.9	0.53–0.65
Go5b		6	-57.2 to -56.7	27.4–29.2	0.62–0.67
Ik1a	Ikaria	3	-56.9 to -56.6	30.1–30.8	0.54–0.59
Ik1b		6	-57.0 to -56.6	27.4–28.3	0.65–0.67
Nx1a		8	-57.2 to -56.6	27.3–31.0	0.51–0.67
Nx1b		10	-57.3 to -56.7	28.1–30.5	0.57–0.65
Nx2a	Naxos	4	-57.2 to -56.6	27.9–29.3	0.62–0.66
Nx2b		6	-57.1 to -56.8	28.9–31.0	0.51–0.63
Nx4b		4	-57.2 to -56.9	27.5–30.4	0.57–0.67

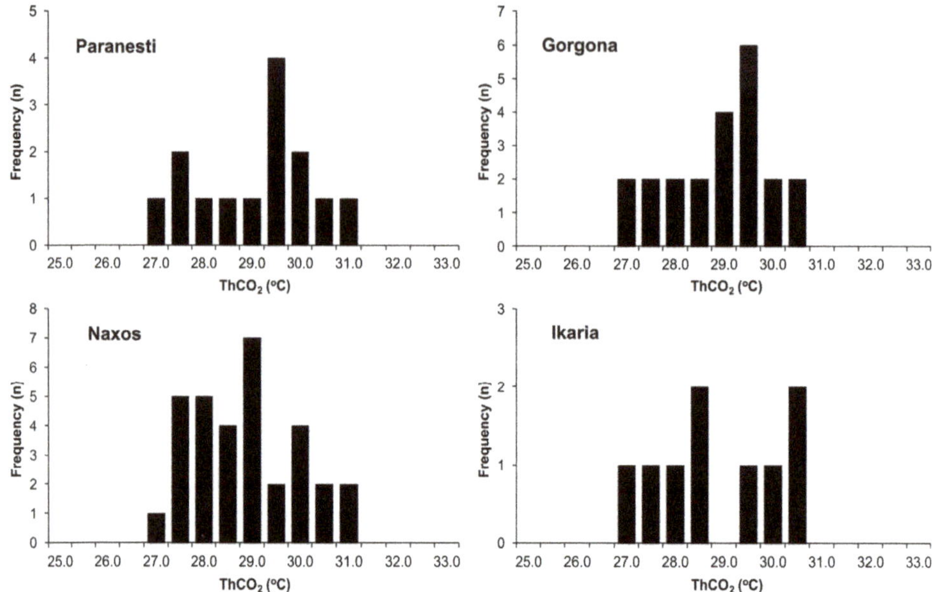

Figure 7. Histograms of homogenization temperatures of CO_2 ($ThCO_2$) to the liquid phase (LCO_2) in the fluid inclusion from the Rhodope massif (Paranetsi, Gorgona) and the Attic-Cycladic massif (Naxos, Ikaria) from Greece.

7. Oxygen Isotope Data

Oxygen isotope compositions (Table 6) confirm their geological typology, i.e., with, respectively, $\delta^{18}O$ of 4.9 ± 0.2‰ for sapphire in plumasite, 20.5‰ for sapphire in marble and 1‰ for ruby in mafics (Figure 8). The desilicated blue sapphire-bearing pegmatite from Naxos have similar oxygen isotope values to those from desilicated pegmatites in mafic host rocks from elsewhere [2,7,14,81,82]. The O-isotope composition of sapphire is buffered by the $\delta^{18}O$ value of the mafic host rock. The ruby in pargasite schists from Paranesti has a very low $\delta^{18}O$-value of 1‰ that can be interpreted in different ways: (i) inherited pre-metamophic reactions between sea-water and hot basic/ultrabasic rocks before subduction and metamorphism; (ii) syn-metamorphism depletion in ^{18}O related to hydration during amphibolite facies metamorphic conditions; (iii) post-amphibolite facies metamorphism with recrystallization under the effect of metasomatism of the metamorphosed mafic/ultramafic rocks with a high depletion in ^{18}O; and (iv) metamorphic/metasomatic conditions involving deeply penetrating meteoric waters along major crustal structures, see Wang et al. [83]. At the moment, the absence of more O-isotope data on this type of ruby precludes any of these possible hypotheses. The two $\delta^{18}O$-values of ruby in marble, between 20.5‰ and 22.1‰, are in agreement with the range of values found for this type of ruby worldwide [2,14].

Table 6. Oxygen isotope values of corundum from Greece. $\delta^{18}O$ corundum (‰, V-SMOW) (after Wang et al. [83]).

Sample	$\delta^{18}O$	Description
Nx2a	4.8	Blue to colorless sapphire in plumasite
Nx3a	5.1	Blue to colorless sapphire in plumasite
Dr1b	1.0	Red ruby in pargasite-schist
Go5a	22.1	Pink-purple sapphire in marble
Go5b	20.5	Pink sapphire in marble
Ik1a	22.4	Blue sapphire in metabauxite

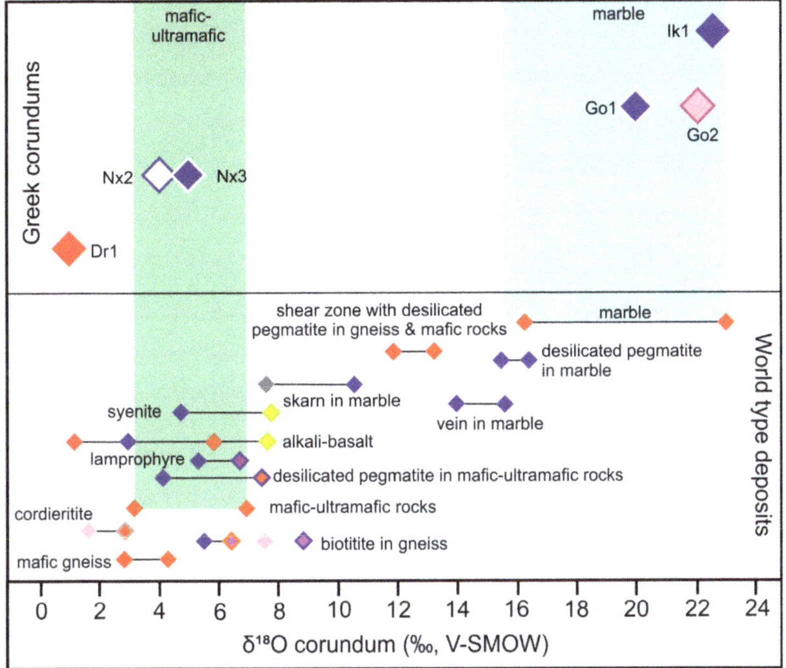

Figure 8. Oxygen isotopic composition of Greek corundum (after Wang et al. [83]) compared with the oxygen isotopic ranges from corundum deposits worldwide from the data of Giuliani et al. [2,7,14,81,82]. Color in diamonds represents the color of the studied corundums. Colorless sapphires are represented by white diamonds.

8. Discussion

8.1. Trace Elements Fingerpring: Metamorphic versus Magmatic Origin of Greek Corundum

Chromophore and genetic indicator elements (Fe, Cr, Ti, V, Ga and Mg) are commonly used to distinguish corundum from different primary sources using elemental diagrams [16,84–88]. In the (Cr + V)/Ga versus Fe/Ti diagram (Figure 9a, [84,85]), displaying the fields for metamorphic and magmatic corundums, the majority of the samples plot in the field of metamorphic corundum, exhibiting a large variation of Fe/Ti ratios. Some sapphires from Naxos island that plot in the magmatic field have very low Cr/Ga ratios. Rubies from Paranesti/Drama show high Cr/Ga and Fe/Ti ratios, followed by the pink and purple sapphires from Naxos island. Both colorless to blue and pink sapphires from Gorgona/Xanthi and blue sapphires from Ikaria island are characterized by relatively fixed Cr/Ga, but display variations with respect to their Fe/Ti ratios.

In the Fe – Cr × 10 – Ga × 100 discrimination diagram (Figure 9b; after Sutherland et al. [84]), the rubies from Paranesti are extremely rich in Cr_2O_3 and display a linear trend inside the metamorphic corundum field. Sapphires from Gorgona/Xanthi, Naxos and Ikaria islands are scattered in both the metamorphic and magmatic corundum fields. Colored sapphires from Gorgona/Xanthi are scattered in both the metamorphic and magmatic fields and display two trends, along the Ga–Fe and Ga–Cr lines, with the pink varieties plotting closer to the Cr_2O_3 edge, indicating a direct relationship between color and Cr_2O_3 content. Colorless to blue sapphires from Naxos island mostly plot along the Fe–Ga line, in the magmatic corundum field, while the pink and purple varieties display a linear trend towards the Cr_2O_3 edge, inside the metamorphic field, which is subparallel to the trend of pink sapphires from Gorgona/Xanthi. Blue sapphires from Ikaria island clearly plot in the metamorphic field.

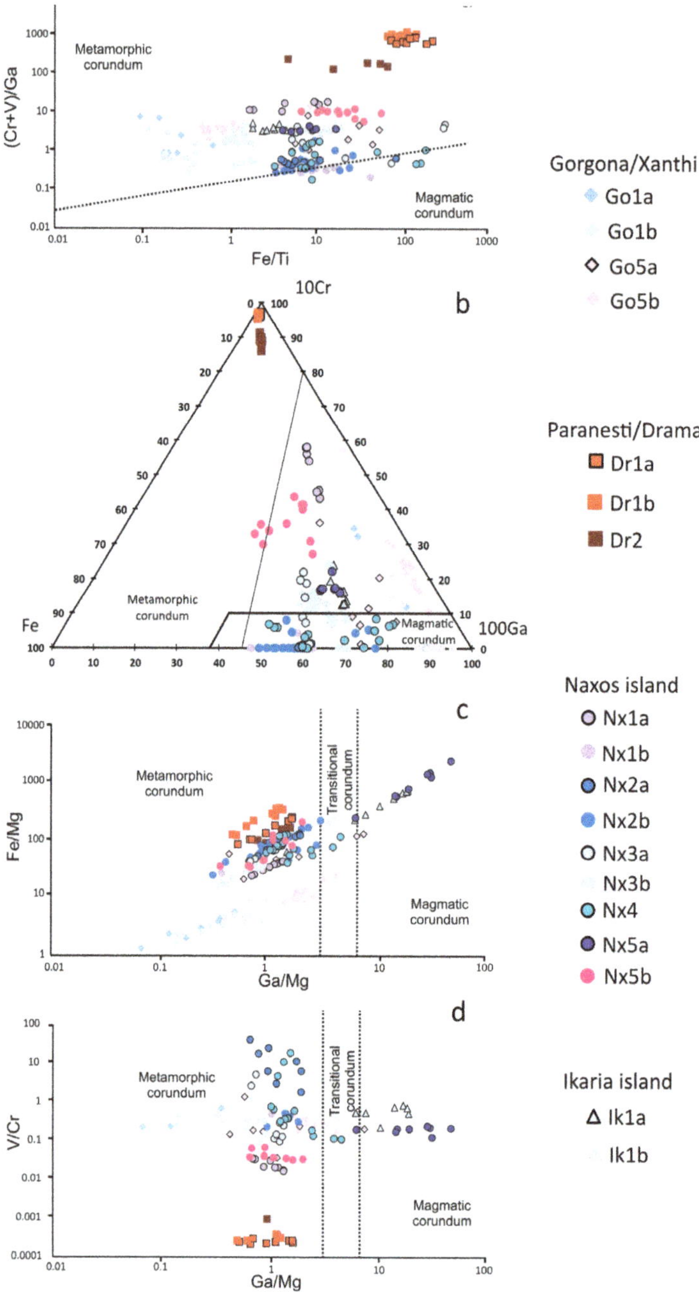

Figure 9. Greek corundum LA-ICP-MS analyses plotted on a (**a**) (Cr + V)/Ga versus Fe/Ti discrimination diagram separating the fields for magmatic and metamorphic corundums (after Sutherland et al. [84] and Harris et al. [85]); (**b**) trace element Fe – Cr × 10 – Ga × 100 discrimination diagram after Sutherland et al. [84]; (**c**) Fe/Mg versus Ga/Mg diagram discrimination separating the fields for magmatic, transitional and metamorphic corundums (after Peucat et al. [16], Sutherland et al. [84]; (**d**) V/Cr versus Ga/Mg discrimination diagram separating the fields for magmatic, transitional and metamorphic corundums (after Sutherland et al. [86]).

In the Fe/Mg versus Ga/Mg plot (Figure 9c), most samples plot in the metamorphic corundum field, except for sapphires from metabauxites of Naxos and Ikaria islands and two analyses of pink sapphires fom Gorgona/Xanthi. A few blue sapphires from Naxos island and pink sapphires from Gorgona/Xanthi plot in the in-between area of transitional corundum. Two distinct trends can be remarked in the metamorphic corundum field: Sapphires from Gorgona/Xanthi form a linear trend, with the pink varieties exhibiting higher both Fe and Ga concentrations. The rest of the analyzed corundums plot in a relatively small area, parallel to the previous trend, but are also characterized by higher Fe values, the highest of which characterize the rubies from Paranesti/Drama.

The V/Cr versus Ga/Mg diagram is useful for deciphering the genetic environments of corundum (Figure 9d). The majority of the analyzed samples plot in the metamorphic corundum field, with the exception of a few analyses that fit in the transitional field and the metabauxite-hosted sapphires from Ikaria and Naxos islands that have a magmatic signature (Figure 9d). In this diagram, rubies from Paranesti/Drama plot close to the Ga/Mg axis, as they are characterized by very low V content. On the other hand, colored sapphires from the other localities plot in two vertical trends. The first one, is characterized by fixed Ga/Mg content and refers to pink, purple and the majority of the colorless to blue sapphires from Naxos island, and expresses a decrease in the Cr content from the purple to the pink and finally the blue varieties and subsequent increase of their V content. The second trend is characterized by stable V/Cr ratio, but shows a significant variation regarding its Ga/Mg content. Gallium concentration increases by three orders from the blue sapphires of Gorgona/Xanthi towards the pink varieties of the same locality and finally to the metabauxite-hosted sapphires of Ikaria and Naxos islands which show the highest Ga values.

In the (FeO–Cr$_2$O$_3$–MgO–V$_2$O$_3$) versus (FeO + TiO$_2$ + Ga$_2$O$_3$) diagram after Giuliani et al. [87,88], which further differentiates the metamorphic environments (Figure 10), all rubies from Paranesti/ Drama clearly fall in the field of rubies with mafic/ultramafic origin. Pink sapphires from Gorgona/Xanthi, plot in an overlapping area between the marble-hosted rubies and the metasomatic corundum fields, while the rest of sapphires from this locality, which are colorless to blue in color, plot in a linear trend along the y axis of the diagram, indicating variable contents of Fe, Ti and Ga. The rest of the studied samples (Naxos and Ikaria sapphires) plot mostly in the field of metasomatic corundum, with some analyses plotting in the borders with the field of syenite-related sapphires. In general, the trends observed in Figures 9 and 10 reflect variations in chromophore element concentrations, even in samples from the same locality. This could be attributed to changes in chemistry and physicochemical conditions of the corundum-forming environment, perhaps due to metasomatic processes as described by Harris et al. [85].

8.2. Comparison with Corundum Deposits around the World

Various plots, based on trace element fingerprints, have been proposed as a useful tool for distinguishing the origin and genesis of gem corundum deposits (Figures 11 and 12; [4,16,89–93]).

In the Mg × 100 – Fe – Ti × 10 diagram (Figure 11a), rubies from Paranesti/Drama plot both in the magmatic and metamorphic fields, and mostly along the Fe–Mg line. Analyses that plot in the magmatic field are also of metamorphic origin and indicate the variation of the Fe/Mg ratio, thus suggesting a limitation of this chemical diagram. Sapphires from Gorgona/Xanthi and the majority of the Naxos island's sapphires plot in the metamorphic field. The blue and pink sapphires from Gorgona/Xanthi, as well as some blue sapphires from Naxos island, plot along the Mg–Ti line of low-Fe metamorphic sapphires (e.g., Ilakaka/Madagascar and Ratnapura Balangoda/Sri Lanka), subparallel to the so-called Kashmir trend. Other colored sapphires from Naxos plot subparallel to the Kashmir trend of metasomatic blue sapphires, but they are characterized by elevated contents of Fe in the purple- and Mg in the pink-colored varieties, and are scattered through the Bo Phloi/Thailand atypical blue sapphires. Metabauxite-hosted sapphires from Naxos and Ikaria islands display a Ti-rich magmatic trend and plot preferably along the Fe–Ti line and follow the Fe–Ti Pailin (Cambodia) of magmatic blue sapphires.

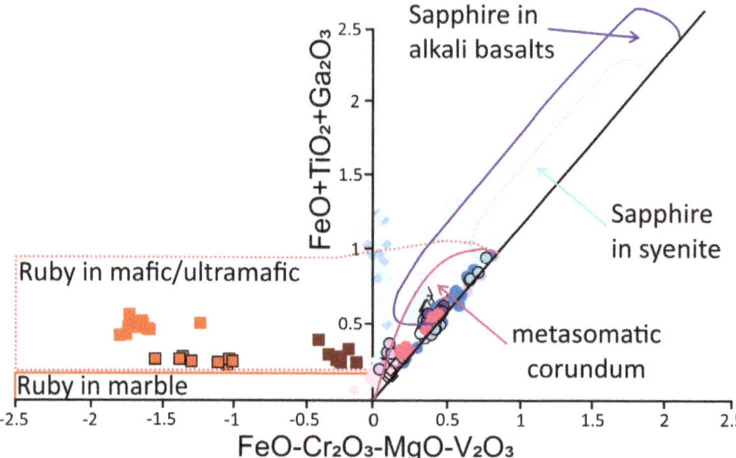

Figure 10. Greek corundum LA-ICP-MS analyses plotted on a FeO–Cr$_2$O$_3$–MgO–V$_2$O$_3$ versus FeO + TiO$_2$ + Ga$_2$O$_3$ discrimination diagram after Giuliani et al. [87,88]. Symbols as in Figure 9.

In the Fe × 0.1 − (Cr + V) × 10 − Ti plot, after Peucat et al. [16], rubies from Paranesti/Drama, as well as pink and purple sapphires plot close to the Cr + V edge, as they are enriched in Cr (Figure 11b). Moreover, pink sapphires from Gorgona/Xanthi, along with metabauxite-hosted sapphires from Ikaria and Naxos islands, plot close to the same edge, but due to their relative enrichment in V, rather than Cr. The blue sapphires from Gorgona/Xanthi, along with some blue sapphires from Naxos, plote close to the Ti edge, in a linear array parallel to the (Cr + V)–Ti line. Sapphires from Naxos are scarce in the diagram, mainly reflecting their variable content of Ti, with the weak-colored areas plotting close to or along the Fe–(Cr + V) line. The majority of the Naxos sapphires form a linear array, parallel to the Mogok marble-hosted trend, but they differ in respect to their lower Fe and higher Cr + V contents. Some Naxos sapphires plot close to the fields for Colombia and Umba blue sapphires hosted in desilicated pegmatites.

In the Fe versus Ga/Mg diagram (Figure 12a; after Peucat et al. [16], Zwaan et al. [89]), the majority of the analyzed corundums are scattered in the field of metamorphic sapphires, and specifically in the plumasitic sub-field. Some sapphires from Naxos island, with blue and pink color plot in the field of alluvial sapphires from Yogo Culch Montana, while blue sapphires from Gorgona/Xanthi are characterized by very low Ga/Mg ratios and thus plot slightly outside the metamorphic field. Metabauxite-hosted sapphires from Naxos and Ikaria islands plot along the Main Asian Field (MAF) of sapphires related to alkali basalt. They form a linear array indicating variations in their Ga/Mg ratio and their Fe content is slightly higher compared to sapphires from Pailin (Cambodia) and Ilmen (Russia). Finally, a few analyses of pink sapphires from Gorgona/Xanthi are plotted in the MAF field, and are characterized by slightly higher both Fe and Ga/Mg values, compared to the rest pink and colorless to blue sapphires of this location.

The Cr$_2$O$_3$ versus Fe$_2$O$_3$ diagram (Figure 12b), adapted from Schwarz et al. [90], shows different types of African deposits: marble-type from Mong Hsu and Mogok (Myanmar), desilicated pegmatite from Mangari (Kenya), amphibolitic-type from Chimwadzulu (Malawi), Songea and Winza (Tanzania) and basaltic-type from Thai (Thailand-Cambodia border region). The majority of the Paranesti/Drama rubies plot outside the fields of rubies from elsewhere, because they contain significantly more Cr$_2$O$_3$. Some samples, though, poorer in Cr$_2$O$_3$, fit well into the Winza (Tanzania) ruby field. Sapphire samples plot mostly along the Fe$_2$O$_3$ axis, with the majority of the compositions being comparable with colored sapphires from Winza.

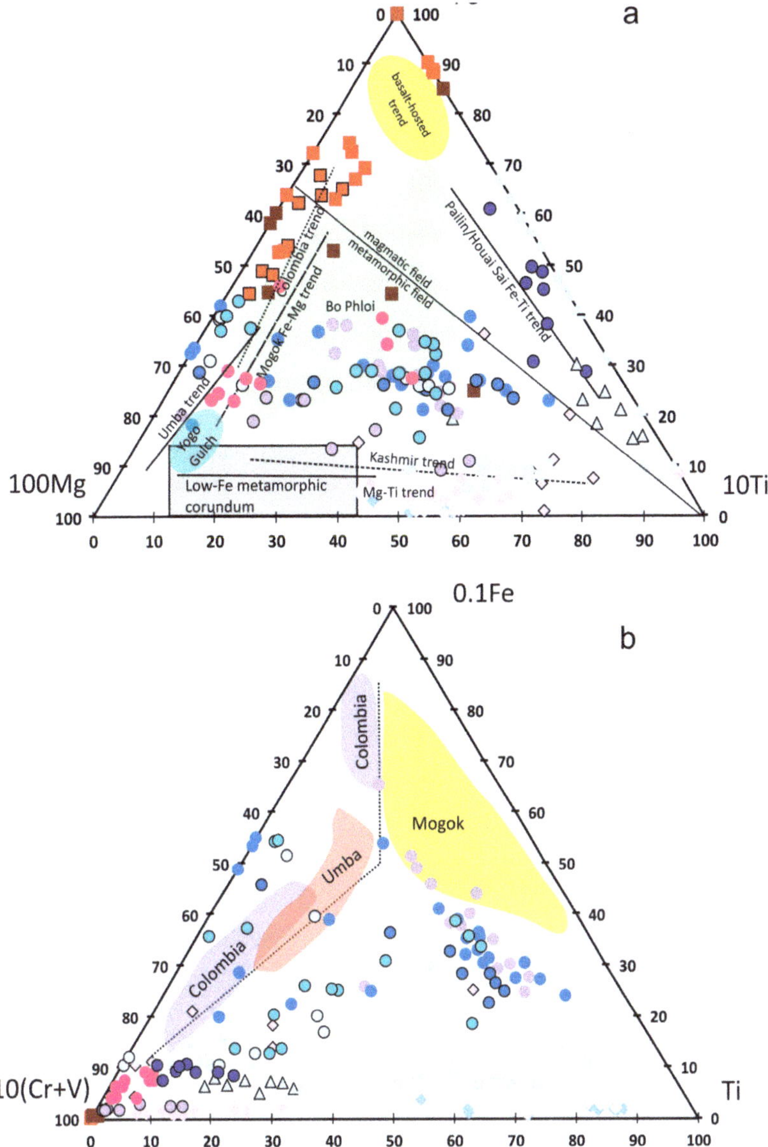

Figure 11. Greek corundum LA-ICP-MS analyses plotted on a (**a**) Mg × 100 − Fe − Ti × 10 discrimination diagram. Also shown are the metamorphic Fe–Mg and Fe–Ti trends of Mogok and low-Fe sapphires of Pailin, Ilakaka and Ratnapura Balangoda, the Umba and Kashmir trends of metasomatic blue sapphires, and the Bo Phloi scattering of atypical blue sapphires (modified after Peucat et al. [16]); (**b**) Fe × 0.1 − (Cr + V) × 10 − Ti diagram for Colombia, Mogok and Umba blue sapphires along with the plots from the Greek corundums, modified after Peucat et al. [16]. Symbols as in Figure 9.

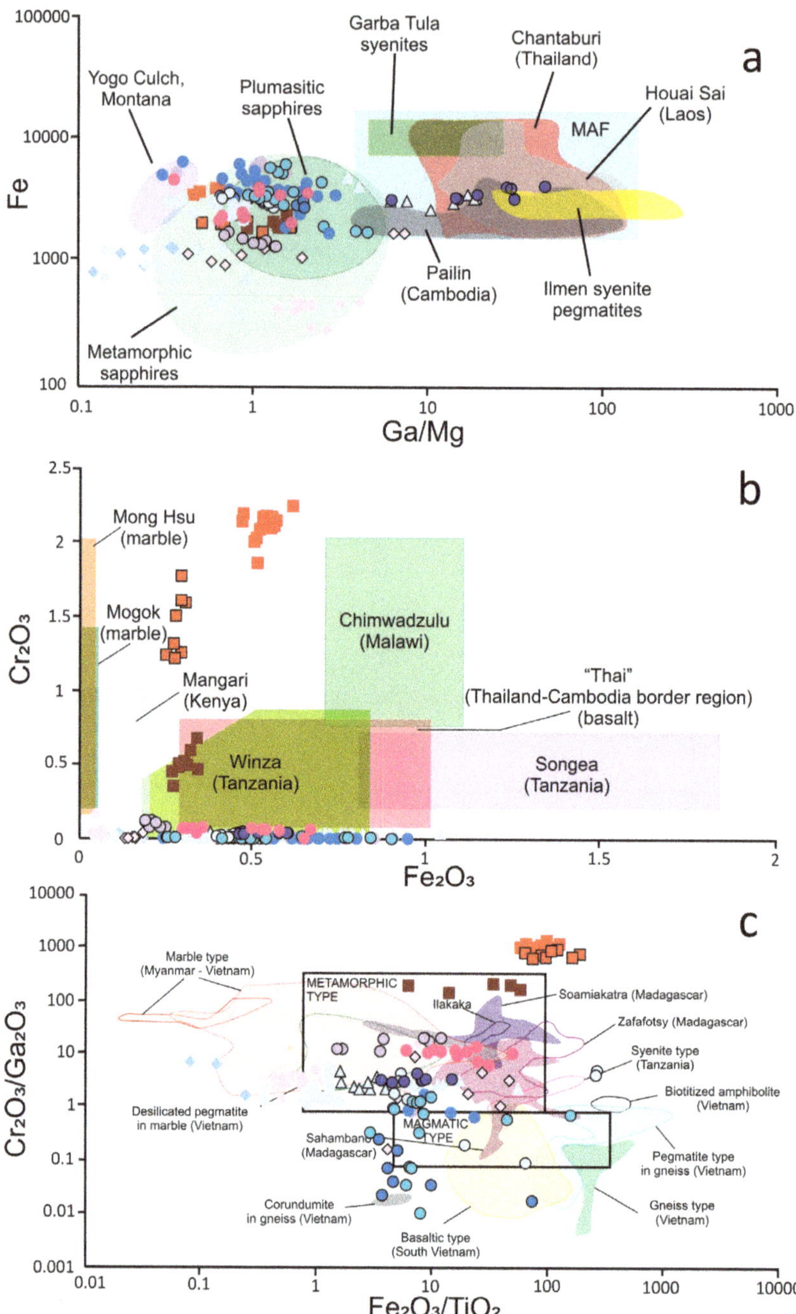

Figure 12. Greek corundum LA-ICP-MS analyses plotted on a (**a**) Fe versus Ga/Mg discrimination diagram (after Peucat et al. [16], Zwaan et al. [89]); (**b**) Cr_2O_3 versus Fe_2O_3 diagram demonstrating the fields of different African deposits (modified after Schwarz et al. [90]); (**c**) Cr_2O_3/Ga_2O_3 versus Fe_2O_3/TiO_2 plot demonstrating composition of diverse corundum deposits (modified after Rakotondrazafy et al. [91], Sutherland et al. [4], Pham Van et al. [92], Simonet et al. [93]. Symbols as in Figure 9.

The Cr_2O_3/Ga_2O_3 versus Fe_2O_3/TiO_2 diagram (Figure 12c; modified from Rakotondrazafy et al. [91]), displays the geochemical fingerprint of corundums from different types. The majority of analyzed corundums plot in the field of metamorphic origin. It is clear that rubies from Paranesti/Drama show higher Cr_2O_3 values, and thus cannot plot in fields of rubies from elsewhere.

Some rubies from Paranesti also fall close to the Soamiakatra rubies, which are hosted in clinopyroxenite enclaves of alkali basalts. Pink and purple sapphires from Naxos island display variation of their Fe_2O_3/TiO_2 ratio. The pink sapphires plot in a common field of the Soamiakatra and Sahambano sapphires (Madagascar), while the purple varieties are scattered in a sector overlapping both the marble- and the desilicated pegmatite in marble-types of Vietnam.

Colorless to blue sapphires from Naxos plumasites are scattered along the lower limit of the metamorphic field, indicating significant variation of their Fe_2O_3/TiO_2 values, while the Cr_2O_3/Ga_2O_3 values stay relatively fixed. Pink-colored sapphires from Gorgona/Xanthi plot partly inside the marble-hosted corundum from Vietnam, while the blue varieties plot in the extension of this trend, slightly outside the field, indicating minor Fe_2O_3 content. Metabauxite-hosted corundum from Naxos island plot slightly outside the magmatic corundum field, and their ratios are not comparable with any other corundum of the plot, except a few analyses, which plot close to the Vietnamese field of gneiss-hosted corundumite. Ikaria metabauxite-hosted samples plot well into the metamorphic corundum field and show a wide range of Fe_2O_3/TiO_2 values, covering the fields between marble-hosted and desilicated pegmatite in marbles (Vietnam), Sahambano and Zafafotsy (Magadascar) corundum deposits.

8.3. Fluid Characterization

The fluid inclusions study in the corundum (rubies and sapphires) from the four Greek occurrences revealed the presence of CO_2-dominated fluids with very small quantities of CH_4 and/or N_2, and relatively low densities, varying between 0.46 and 0.67 g/cm^3 [79]. Primary and pseudosecondary water-free carbonic fluid inclusions represent the main fluid, which was incorporated during the crystallization of rubies and sapphires. The absence of any significant change in the fluid composition and density of the primary and pseudosecondary inclusions possibly suggests that (i) the fluid was homogeneous and related to the same source; and (ii) the host rocks were not affected by the circulation of any external fluids. The CO_2-rich fluids are likely of metamorphic origin and probably derived from devolatilization of marble in most cases (Naxos and Ikaria).

Most of the corundum in different geological environments worldwide generally contains pure or nearly pure CO_2-bearing fluid inclusions. Previous studies have shown that in metamorphic complexes, CO_2-bearing fluids were incorporated in the corundum formation from granulite facies rocks in Sri Lanka [94], in the marble hosted ruby deposits from Luc Yen in North Vietnam [95] and in pegmatoids from the Nestos Shear Zone in Greece [57]. Corundum occurrences with pure or almost pure CO_2-bearing fluids, without any water included, were documented in sapphires from pegmatites in the Kerala district of India [96] and in a corundum bearing skarn from granulites in Southeast Madagascar [97]. In other corundum occurrences, such as in the Kashmir blue sapphires and in the Thailand sapphires, CO_2 is an important component of the source fluids [98,99]. The occurrence of high-density CO_2-rich fluid inclusions in granulite facies rocks shows that large amounts of CO_2 infiltrate the lower crust during the peak of metamorphism [100–102]. However, Hollister et al. [103] have shown that low-density CO_2 fluid inclusions must have been trapped after the peak of metamorphism. Two main sources for input of CO_2 fluids in the lower crust have been suggested [101,104,105]: from the mantle, or from the metamorphism of previously dehydrated crust. High-Al and low-Si protoliths in a high regional metamorphic grade can produce a pure supercritical CO_2 fluid and lead to the formation of corundum. A mantle-derived CO_2 has been suggested for Naxos by Schuiling and Kreulen [106].

In the present study, similar fluids containing almost pure CO_2 have been documented. Water was not identified in the fluid inclusions, neither by optical microscopy nor by any phase transition during microthermometry. This excludes the possibility of fluid immiscibility for the corundum formation and implies the presence of a primary water-free (or very poor) CO_2 dominated fluid. However, the presence of minor amounts of water in the paleofluid in all studied corundums cannot be excluded, due to the presence of hydrous mineral inclusions in the corundum. Already, Buick and Holland [107] have argued that the "primary" fluid inclusions in the metamorphic complex of Naxos have been compositionally modified by selective leakage of H_2O during uplift, and Krenn et al. [57] reported on recrystallization-induced leakage resulting in minor admixture of H_2O from former hydrous inclusions in corundum from pegmatites along the Nestos suture zone, Xanthi area.

It is very likely that the studied primary low-density fluid inclusions (d = 0.46–0.67 g/cm^3) were entrapped after the peak of metamorphism, suggesting that corundum formation took place during retrogression. The pseudosecondary, also low-density, carbonic fluid inclusions were entrapped in trails during the corundums formation process and are related to the evolution of the metamorphic events at a retrograde metamorphic regime of cooling and uplift and the subsequent exhumation, also after the peak of metamorphism. These variations in the density of the fluid inclusions can be interpreted as the result of pressure variation associated with successive localized microfracturing, a process which was suggested in the case of Luc Yen rubies of North Vietnam [95]. However, a continuous entrapment of the carbonic fluids during growth of corundum with pressure decreasing is not excluded.

8.4. Conditions of Greek Corundum Formation

The Xanthi and Paranesti corundums belong to metamorphic s.s. deposits, according to the classification of Giuliani et al. [2,7] and Simonet et al. [3], and more specifically, to those related to meta-limestones and mafic granulites, respectively. Both occurrences formed during the retrograde metamorphic path of high-temperature/medium-pressure metamorphism of platform carbonates and amphibolites during the Cenozoic collision that resulted in the Nestos Suture Zone. Wang et al. [30] concluded that Paranesti rubies were formed within an ultramafic precursor, most probably an aluminous clinopyroxenite, during amphibolite facies metamorphism. The estimated P–T conditions for their formation are 4 kbar < P < 7 kbar and 580 °C < T < 750 °C, and with subsequent retrogression.

The Paranesti (and Gorgona/Xanthi) corundum-bearing assemblages followed a nearly isothermal decompression within the amphibolite facies and then a further evolution towards the greenschist facies, along the path that was proposed by Krenn et al. [57] for the Nestos suture zone (Figure 13). This path records a transition from the kyanite to the sillimanite stability field during retrogression, as observed by former kyanite surrounded by fibrolitic sillimanite ([57,108] and this study). In the studied samples from Paranesti, corundum surrounds kyanite, and was probably formed after kyanite in the stability field of sillimanite during the retrogression. Mineralogical observations of the corundums from Paranesti ([30] and this study) indicate that the margarite occurs as reaction rims around ruby grains, suggesting that it was formed subsequently due to retrogression and according to the following reaction:

$$\text{Margarite} = \text{Anorthite} + \text{Corundum} + H_2O \tag{1}$$

In contrast to Paranesti, margarite in the Gorgona/Xanthi corundum-bearing marbles seems to be a prograde mineral ([22,109] and this study], indicating that corundum along with anorthite could have been formed by the breakdown of margarite (see Reaction (1)). In the studied samples, margarite is either in contact or is separated from corundum by a kaolinite rim, which probably represents earlier anorthite. As described by Storre and Nitsch [110] and Chatterjee [111,112], margarite breakdown to anorthite and corundum takes place at temperatures of 610, 625 and 650 °C for pressures of 6, 7 and 8 kbar, respectively. On the other hand, and under the assumption that margarite was formed together with corundum, a more complex reaction:

$$8\,\text{Diaspore} + \text{Pyrophyllite} + 2\,\text{Calcite} = 2\,\text{Margarite} + \text{Corundum} + 2\,CO_2 + 3\,H_2O \qquad (2)$$

as described by Okrusch et al. [113] and Haas and Holdaway [114] may explain this assemblage. This reaction requires the metastable persistence of diaspore + pyrophyllite, which should have otherwise reacted to form Al-silicate about 40–60 °C below the lower stability limit of corundum [113]. The assemblage corundum and chlorite within the Gorgona marbles can be formed according to the following reaction, as experimentally demonstrated by Seifert [115]:

$$\text{Mg-chlorite} + 6\,\text{Corundum} + 2\,\text{Spinel} = \text{sapphirine} + 4\,H_2O \qquad (3)$$

but the absence of sapphirine in the studied samples indicates that P–T did not exceed 6 kbar for temperatures between 620 and 720 °C [115]. The formation of the Xanthi corundum under dissociation of Mg-spinel + calcite into corundum + dolomite during the retrograde metamorphism of spinel-bearing dolomites according to the reaction:

$$\text{Spinel} + \text{Calcite} = \text{Corundum} + \text{Dolomite} + CO_2 \qquad (4)$$

was not observed, but cannot be ruled out. This reaction path has been described by Buick and Holland [116] in spinel-bearing dolomites from the leucocratic core of Naxos, where spinel crystals are separated from the calcite of the host rock by corundum + dolomite mantles, and they are often surrounded by Mg-chlorite + pargasite. The above authors suggested synchronous formation of Mg-chlorite + pargasite in place of spinel + calcite. Decomposition of spinel into corundum during the retrograde metamorphism in marbles as described by Reaction (4) was described as the major corundum-forming mechanism for the rubies at Jegdalek (Afghanistan), Hunza (Pakistan) and Luc Yen (Vietnam) [117].

In summary, conditions of corundum formation in the Gorgona marbles as suggested by Liati [22,109] and this study, are in accordance to those estimated by Wang et al. [30] for the Paranesti rubies. As an alternative hypothesis, corundum in the Gorgona marbles could have formed at lower pressures of about 3–4 kbar, during late stages of shear deformation from CO_2-rich fluids as proposed by Krenn et al. [57] for corundum–anorthite assemblages postdating kyanite in desilicated pegmatoid veins within gneisses and amphibolites of the Nestos suture zone, about 4 km south of the Gorgona locality.

In the absence of geochronological data for corundum-bearing assemblages within the Nestos suture zone, it remains difficult to state if they all represent the same metamorphic event, or can be attributed to different P–T conditions corresponding to different ages.

The plumasites from Kinidaros, Naxos contain tourmaline and beryl, in addition to corundum, anorthite and phlogopite, and originate from desilication of leucogranitic pegmatites. For these pegmatites, Matthews et al. [118] reported temperature variations from >700 °C to ~400 °C, based on oxygen isotope fractionation among quartz, tourmaline and garnet, and attributed their formation from anatectic melts during regional high-temperature metamorphism. According to these authors, crystallization of the pegmatitic magmas should initiate under water-undersaturated conditions, but with crystallization of anhydrous minerals and ascent, the magma should evolve to water-saturated conditions at 630 to 640 °C. Thus, some of the higher temperatures (T = 650 °C) given by the isotope thermometry could represent the crystallization at reduced water activity. In addition, Siebenaller [119] used fluid inclusion measurements in tourmaline, garnet and beryl from leucogranite pegmatites from Naxos island, to estimate P–T formation conditions for the pegmatites of between 5 kbar/600 °C and <2 kbar/450 °C, along the exhumation path of leucogranite (Figure 13). The absence of andalusite from the studied corundum-bearing assemblages constrains the lower limit of corundum formation along the retrograde path at about 3 kbar and 550 °C.

Finally, metabauxite occurrences on southern Naxos (Kavalaris Hill) correspond to the thermal dissociation of diaspore and the formation of corundum (α-AlOOH $\leftrightarrow$ Al$_2$O$_3$ + H$_2$O, Haas [120]), as described by Feenstra and Wunder [73]. The first corundum occurrence in Naxos metabauxites has been recorded as corundum-in isograd (T~420–450 °C and P~6 kbar, [73]). Similarly, the metabauxites from Ikaria were formed at T in the range 450–550 °C and P ~5–6kbar [75,77]. According to Iliopoulos [75], the latter correspond to corundum-chloritoid-bearing metabauxites from Naxos island (zones II and III of Feenstra [21]; Figure 13).

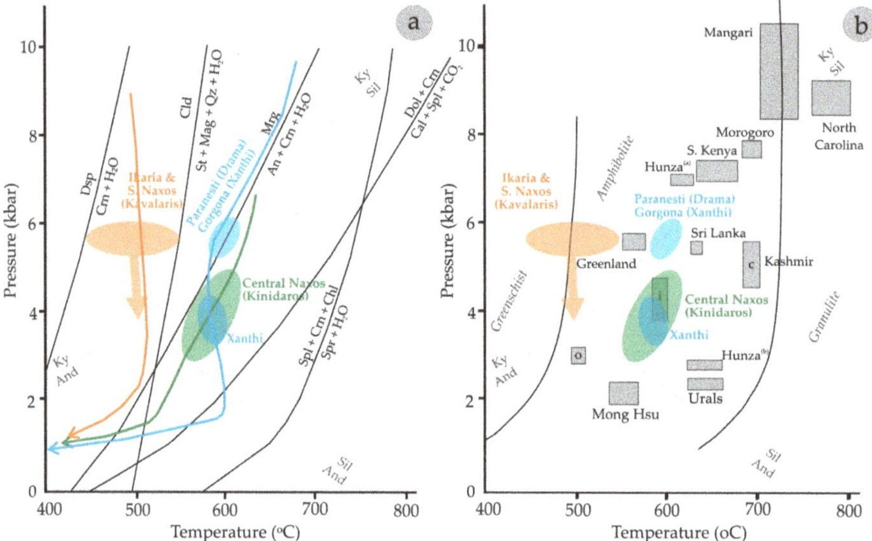

Figure 13. (**a**) P–T diagram showing mineral equilibria related to the formation of corundum in Greece (modified from Garnier et al. [117]). Upper blue area: Paranesti ruby and Gorgona sapphires ([30] and this study); Lower blue area: Alternative conditions for the Gorgona sapphires (for explanation see text); Orange area: Ikaria and southern Naxos metabauxites. The thick orange arrow indicates possible path for formation of vein-type sapphires from both localities; Green area: Central Naxos plumasites. Blue arrow: P–T–t path of high-grade rocks from the Nestos suture zone (from Krenn et al. [57]); Green arrow: P–T–t path of the tourmaline-garnet-beryl-bearing leucogranite from Naxos island (from Siebenaller [119]); Orange arrow: P–T–t path of the lower limit of metamorphic zone III (which hosts corundum-chloritoid-bearing metabauxites) from Naxos island (from Duchêne et al. [64]). The equilibrium curve "chloritoid + O$_2$ = staurolite + magnetite + quartz", which marks the disappearance of chloritoid, defines an upper limit for the studied chloritoid-bearing metabauxites at southern Naxos (Kavalaris) and Ikaria islands is from Feenstra [21]. Abbreviations: An = anorthite; And = andalusite; Cc = calcite; Cld = chloritoid; Clin = clinochlore; Co = corundum; Do = dolomite; Dsp = diaspore; Ky = kyanite; Ma = margarite; Mag = magnetite; Qz = quartz; Sill = sillimanite; Spr = sapphirine; Sp = spinel; St = staurolite; (**b**) Comparison between the hypothetical P–T conditions for the formation of corundum in Greece and in metamorphic deposits around the world (modified from Giuliani et al. [7], Simonet et al. [3]). Boxes indicate P–T fields of known deposits: North Carolina, Rubies from mafic granulites [121]; Morogoro, corundum-bearing anatexites [122]; Mangari, Southern Kenya, Metasomatic rubies [123–125]; Hunza, rubies in marbles [113]; Sri Lanka, sapphires from granulites [94]; Greenland, metasomatic rubies [126]; Kashmir metasomatic sapphires with three P–T boxes corresponding to the evolution of the fluids in the sapphire crystals from the center (c), to intermediate (i) and outer (o) zones [98]; Urals, rubies in marbles [127], and Mong Hsu, rubies in marbles [128].

However, the studied sapphires from both localities correspond to open space-filling material in extensional fissures and indicate a second, late stage of corundum formation. We interpret this corundum formation to be of metasomatic origin from a low-temperature CO_2-bearing metamorphic fluid, as already proposed for similar sapphire + margarite + tourmaline-filled veins from Naxos island [116]. This is in agreement with the findings of Tropper and Manning [129], who consider corundum-filled hydrofractures in the Naxos metabauxites to be products of retrograde cooling and decompression, consistent with kinematically late textures. This hypothesis is also supported by the fact that trace element fingerprints of Naxos and Ikaria metabauxite-hosted sapphire suggest magmatic affinities (this study). A hydrothermal origin of diaspore has been reported for the gem-quality diaspore crystals (var. zultanite) occurring in fissures of metabauxites in the İlbir Mountains, SW Turkey [130]. According to the above authors, the diaspore mineralization is caused by hydrothermal remobilization of primary bauxite components into crosscutting structures. Figure 13 displays an evolution (thick orange arrow) towards lower P–T conditions, along the P–T–t retrograde path of the metamorphic zone III (hosting corundum-chloritoid-bearing metabauxites) from Naxos island (according to Duchêne et al. [53]). Since the tectonometamorphic evolution of Ikaria is quite similar to that of Naxos [55,74], this path could probably reflect the formation of the vein-type sapphire assemblages in both Naxos and Ikaria islands.

8.5. Greece: A New Gem Corundum Province?

Greek corundum occurrences display a wide color variation, ranging from deep red, pink, purple, and blue to colorless, with crystal sizes of up to 5 cm. Among the studied occurrences, some corundums from Gorgona/Xanthi and the plumasite-hosted sapphires from Naxos display significant transparency and homogeneity of color and should be further examined for their suitability as potential cut gemstones. The rest of the studied corundums, especially the vivid-colored varieties, could be considered suitable for their use as cabochons. Although gem-quality corundums are considered to be absent in emery deposits [7], the Naxos and especially Ikaria blue sapphires are of gem (cabochon) quality, and are atypical for other metabauxite-hosted corundums. Future exploration is required in order to establish the potential for economic exploitation of the corundum-bearing areas in Greece, pointing towards a new gem corundum province in the world.

We highlight here the enrichment in Be of blue-colorless sapphires from Naxos plumasites, a feature which has only been reported from Ilakaka, Madagaskar, Sar-e-Sang, Afganistan and Weldborough, Tasmania, and was attributed to nano-inclusons of unidentified Be-rich phases [131]. The Naxos plumasitic sapphires, as well as those from the Xanthi marbles and Ikaria metabauxites, display very high values in Nb, Ta, W and Sn, much higher than those reported from the above occurrences. A further study of the concentration of these elements and especially of Be in the Greek corundums could be a useful tool for determining their origin [131].

9. Conclusions

Gem corundum in Greece covers in a variety of geological environments located within the Rhodope (Xanthi and Drama areas) and Attico-Cycladic (Naxos and Ikaria islands) tectono-metamorphic units. Pargasite-schist hosted a ruby deposit in the Paranesti/Drama area and marble hosted pink to blue sapphires in the Xanthi area, occurring along the UHP-HP Nestos suture zone. Plumasite-hosted sapphires from Naxos island display a wide color variation (blue to colorless and pink). Deep blue colored sapphires from Naxos and Ikaria islands are hosted within extensional fissures in metabauxite lenses within marbles. Various mineral inclusions in corundums are in equilibrium and/or postdate corundum crystallization, and reflect the surrounding mineralogical assemblages. Included in corundums are: spinel and pargasite (Paranesti), spinel, zircon (Xanthi), margarite, zircon, apatite, diaspore, phlogopite and chlorite (Naxos) and chloritoid, ilmenite, hematite, ulvospinel, rutile and zircon (Ikaria).

The chromophores of the studied corundums show a wide range in concentration and a unique trace element chemistry with variable critical ratios (Fe/Mg, Ga/Mg, Ga/Cr and Fe/Ti). Be, Nb, Sn, Ta and W are anomalously enriched in the plumasite-related sapphires from Naxos.

Based on the geological setting of formation and trace element fingerprints the Paranesti and Xanthi corundum occurrences can be classified as metamorphic s.s hosted mafics/ultramafics and marbles, respectively. Those from central Naxos are of metasomatic origin and are related to desilicated pegmatites crosscutting ultramafic rocks. Finally, blue sapphires from southern Naxos and Ikaria, hosted in metabauxites, display an atypical magmatic signature indicating a metasomatic origin by fluid-rock interaction. The study of fluid inclusions in corundum showed that they are CO_2-dominant with low density with very small quantities of CH_4 and/or N_2, and water-free. CO_2-rich fluids are probably of metamorphic origin and derived from devolatilization of carbonate formations. Greek corundums are characterized by a wide color variation, homogeneity of the color hues and transparency and could be considered as potential gemstones.

Author Contributions: P.V. and C.M. collected the studied samples. P.V. assisted by C.M., I.G., K.Z., S.M., S.K., K.W., S.H., St.K., J.B., F.Z., V.K., M.T. and G.L. obtained and evaluated the mineralogical data. G.G. and A.F. obtained and evaluated the oxygen isotopic data. V.M. and A.T. conducted the fluid inclusion measurements. P.V., C.M., G.G., V.M., A.T. and S.K. wrote the manuscript.

Funding: This research was partly funded by UNSW MREII Grants and Australian Research Council (ARC) Large Infrastructure and Equipment Funds (LIEF) grant LE0989067 and by LABEX ANR-10-LABX-21—Ressources21, Nancy, France.

Acknowledgments: The authors would like to thank Evangelos Michailidis for his kind help with the SEM in the University of Athens. Two anonymous reviewers and the Editor are especially thanked for their constructive comments that greatly improved the manuscript.

Conflicts of Interest: The authors declare no conflict of interest.

References

1. Garnier, V.; Ohnenstetter, D.; Giuliani, G.; Schwarz, D. Saphirs et rubis. Classification des gisements de corindon. *Règne Minéral* **2004**, *55*, 4–47.

2. Giuliani, G.; Ohnenstetter, D.; Garnier, V.; Fallick, A.E.; Rakotondrazafy, M.; Schwarz, D. The geology and genesis of gem corundum deposits. In *Geology of Gem Deposits*; Mineralogical Association of Canada Short Course Series; Groat, L., Ed.; Mineralogical Association of Canada: Quebec City, QC, Canada, 2007; Volume 37, pp. 23–78.

3. Simonet, C.; Fritsch, E.; Lasnier, B. A classification of gem corundum deposits aimed towards gem exploration. *Ore Geol. Rev.* **2008**, *34*, 127–133. [CrossRef]

4. Sutherland, F.L.; Schwarz, D.; Jobbins, E.A.; Coenraads, R.R.; Webb, G. Distinctive gem corundum from discrete basalt fields: A comparative study of Barrington, Australia, and west Pailin, Cambodia gem fields. *J. Gemmol.* **1998**, *26*, 65–85. [CrossRef]

5. Garnier, V.; Giuliani, G.; Ohnenstetter, D.; Schwarz, D.; Kausar, A.B. Les gisements de rubis associés aux marbres de l'Asie centrale et du Sud-est. *Règne Minéral* **2006**, *67*, 17–48.

6. Garnier, V.; Maluski, H.; Giuliani, G.; Ohnenstetter, D.; Schwarz, D. Ar–Ar and U–Pb ages of marble-hosted ruby deposits from central and southeast Asia. *Canad. J. Earth Sci.* **2006**, *43*, 509–532. [CrossRef]

7. Giuliani, G.; Ohnenstetter, D.; Fallick, A.E.; Groat, L.; Fagan, A.J. The geology and genesis of gem corundum deposits. In *Geology of Gem Deposits*, 2nd ed.; Mineralogical Association of Canada Short Course Series; Groat, L.A., Ed.; Mineralogical Association of Canada: Quebec City, QC, Canada, 2014; Volume 44, pp. 113–134. ISBN 9780921294375.

8. Graham, I.T.; Khin, Z.; Cook, N.J. The genesis of gem deposits. *Ore Geol. Rev.* **2008**, *34*, 1–2. [CrossRef]

9. Graham, I.; Sutherland, L.; Zaw, K.; Nechaev, V.; Khanchuk, A. Advances in our understanding of the gem corundum deposits of the West Pacific continental margins intraplate basaltic fields. *Ore Geol. Rev.* **2008**, *34*, 200–215. [CrossRef]

10. Sorokina, E.S.; Karampelas, S.; Nishanbaev, T.R.; Nikandrov, S.N.; Semiannikov, B.S. Sapphire megacrysts in syenite pegmatites from the Ilmen Mountains, South Urals, Russia: New mineralogical data. *Can. Miner.* **2017**, *55*, 823–843. [CrossRef]

11. Muhlmeister, S.; Fritsch, E.; Shigley, E.J.; Devouard, B.; Laurs, M.B. Separating natural and synthetic rubies on the basis of trace-element chemistry. *Gems Gemol.* **1998**, *34*, 80–101. [CrossRef]

12. Sutherland, F.L.; Schwarz, D. Origin of gem corundums from basaltic fields. *Aust. Gemmol.* **2001**, *21*, 30–33.

13. Saminpanya, S. Mineralogy and Origin of Gem Corundum Associated with Basalt in Thailand. Ph.D. Thesis, University of Manchester, Manchester, UK, 2000.

14. Giuliani, G.; Fallick, A.E.; Garnier, V.; France-Lanord, C.; Ohnenstetter, D.; Schwarz, D. Oxygen isotope composition as a tracer for the origins of rubies and sapphires. *Geology* **2005**, *33*, 249. [CrossRef]

15. Abduriyim, A.B.; Kitwaki, H. Determination of the origin of blue sapphire using laser ablation inductively coupled plasma mass spectrometry (LA-ICP-MS). *J. Gemmol.* **2006**, *30*, 23–26. [CrossRef]

16. Peucat, J.J.; Ruffault, P.; Fritsch, E.; Bouhnik-Le Coz, M.; Simonet, C.; Lasnier, B. Ga/Mg ratio as a new geochemical tool to differentiate magmatic from metamorphic blue sapphires. *Lithos* **2007**, *98*, 261–274. [CrossRef]

17. Sutherland, F.L.; Duroc-Danner, J.M.; Meffre, S. Age and origin of gem corundum and zircon megacrysts from the Mercaderes–Rio Mayo area, South-west Colombia, South America. *Ore Geol. Rev.* **2008**, *34*, 155–168. [CrossRef]

18. Papastamatiou, I. The Emery of Naxos. *Geol. Geoph. Res.* **1951**, *1*, 37–68.

19. Andronopoulos, V.; Tsoutrelis, C. Investigation of a Red Corundum Occurence in Xanthi, Greece. Unpublished Report; Greek Geol. Survey, Athens, Greece. 1964; 8p.

20. Ktenas, C. La géologie de l' île de Nikaria (rédigée des restes de l'auteur par G. Marinos). *Geol. Geoph. Res.* **1969**, *13*, 57–85.

21. Feenstra, A. Metamorphism of Bauxites on Naxos, Greece. Ph.D. Thesis, Instituut voor Aardwetenschappen, Rijksuniversiteit te Utrecht, Utrecht, The Netherlands, 1985.

22. Liati, A. Corundum-and zoisite-bearing marbles in the Rhodope Zone, Xanthi area (N. Greece): Estimation of the fluid phase composition. *Mineral. Petrol.* **1988**, *38*, 53–60. [CrossRef]

23. Zannas, I. Corundum occurrence at Stirigma-Gorgona area, northern Xanthi. Petrogenesis and color variations. *Miner. Wealth* **1995**, *97*, 9–18.

24. Iliopoulos, I.; Katagas, C. Corundum bearing metabauxites from Ikaria Island (Greece): Mineralogy and geochemistry. In Proceedings of the 10th International Congress of the Geological Society of Greece, Thessaloniki, Greece, 15–17 April 2004; pp. 423–424.

25. Voudouris, P. The minerals of Eastern Macedonia and Western Thrace: Geological framework and environment of formation. *Bull. Geol. Soc. Greece* **2005**, *37*, 62–77.

26. Voudouris, P.; Melfos, V.; Katerinopoulos, A. Precious stones in Greece: Mineralogy and geological environment of formation. Understanding the genesis of ore deposits to meet the demand of the 21st century. In Proceedings of the 12th Quadrennial IAGOD Symposium, Moscow, Russia, 21–24 August 2006.

27. Voudouris, P.; Graham, I.; Melfos, V.; Zaw, K.; Lin, S.; Giuliani, G.; Fallick, A.; Ionescu, M. Gem corundum deposits of Greece: Diversity, chemistry and origins. In Proceedings of the 13th Quadrennial IAGOD Symposium, Adelaide, Australia, 6–9 April 2010; Volume 69, pp. 429–430.

28. Voudouris, P.; Katerinopoulos, A.; Magganas, A. Mineralogical geotopes in Greece: Preservation and promotion of museum specimens of minerals and gemstones. Sofia Initiative "Mineral diversity preservation". In Proceedings of the IX International Symposium on Mineral Diversity Research and Reservation, Sofia, Bulgaria, 16–18 October 2017; pp. 149–159.

29. Graham, I.; Voudouris, P.; Melfos, V.; Zaw, K.; Meffre, S.; Sutherland, F.; Giuliani, G.; Fallick, A. Gem corundum deposits of Greece: A spectrum of compositions and origins. In Proceedings of the 34th IGC Conference, Brisbane, Australia, 5–10 August 2012.

30. Wang, K.K.; Graham, I.T.; Lay, A.; Harris, S.J.; Cohen, D.R.; Voudouris, P.; Belousova, E.; Giuliani, G.; Fallick, A.E.; Greig, A. The Origin of A New Pargasite-Schist Hosted Ruby Deposit From Paranesti, Northern Greece. *Can. Mineral.* **2017**, *55*, 535–560. [CrossRef]

31. Voudouris, P. Hydrothermal corundum, topaz, diaspore and alunite supergroup minerals in the advanced argillic alteration lithocap of the Kassiteres-Sapes porphyry-epithermal system, western Thrace, Greece. *Neues Jahrbuch für Mineralogie Abhandlungen* **2014**, *191*, 117–136. [CrossRef]

32. Pouchou, J.L.; Pichoir, F. Quantitative analysis of homogeneous or stratified microvolumes applying the model "PAP". In *Electron Probe Quantitation*; Heinrich, K.F.J., Newbury, D.E., Eds.; Plenum Press: New York, NY, USA, 1991; pp. 31–75.

33. Griffin, W.L.; Powell, W.J.; Pearson, N.J.; O'Reilly, S.Y. GLITTER: Data reduction software for laser ablation ICP-MS. In *Laser Ablation ICP-MS in the Earth Sciences: Current Practices and Outstanding Issues*; Mineralogical Association of Canada Short Course Series; Sylvester, P., Ed.; Mineralogical Association of Canada: Quebec City, QC, Canada, 2008; Volume 40, pp. 308–311.

34. Van Achterbergh, E.; Ryan, C.G.; Jackson, S.E.; Griffin, W.L. Data reduction software for LA-ICP-MS: Appendix. In *Laser Ablation-ICPMass Spectrometry in the Earth Sciences: Principles and Applications*; Mineralogical Association of Canada Short Course Series; Sylvester, P.J., Ed.; Mineralogical Association of Canada: St John's, NL, Canada, 2001; Volume 29, pp. 239–243.

35. Driesner, T.; Heinrich, C.A. The system H_2O-NaCl. I. Correlations for molar volume, enthalpy, and isobaric heat capacity from 0 to 1000 degrees C, 1 to 5000 bar, and 0 to 1 X-NaCl. *Geochim. Cosmochim. Acta* **2007**, *71*, 4880–4901. [CrossRef]

36. Sharp, Z.D. In situ laser microprobe techniques for stable isotope analysis. *Chem. Geol.* **1992**, *101*, 3–19. [CrossRef]

37. Jolivet, L.; Brun, J.P. Cenozoic geodynamic evolution of the Aegean region. *Int. J. Earth. Sci.* **2010**, *99*, 109–138. [CrossRef]

38. Ring, U.; Glodny, J.; Thomson, S. The Hellenic subduction system: High-pressure metamorphism, exhumation, normal faulting and large-scale extension. *Earth Plan. Sci.* **2010**, *38*, 45–76. [CrossRef]

39. Jolivet, L.; Faccenna, C.; Huet, B.; Labrousse, L.; Le Pourhiet, L.; Lacombe, O.; Lecomte, E.; Burov, E.; Denèle, Y.; Brun, J.-P.; et al. Aegean tectonics: Strain localization, slab tearing and trench retreat. *Tectonophysics* **2013**, *597*, 1–33. [CrossRef]

40. Fytikas, M.; Innocenti, F.; Manneti, P.; Mazzuoli, R.; Peccerillo, A.; Villari, L. Tertiary to Quaternary evolution of volcanism in the Aegean region. In *The Geological Evolution of the Eastern Mediterranean*; Dixon, J.E., Robertson, A.H.F., Eds.; Geological Society Special Publications: Oxford, UK, 1984; Volume 17, pp. 687–699.

41. Kydonakis, K.; Brun, J.-P.; Sokoutis, D. North Aegean core complexes, the gravity spreading of a thrust wedge. *J. Geophys. Res. Solid Earth* **2015**, *120*, 595–616. [CrossRef]

42. Voudouris, P. *Gemstones and Mineral-Megacrysts in Greece*; University New South Wales: Sydney, Australia, 2010.

43. Burg, J.-P. Rhodope: From Mesozoic convergence to Cenozoic extension. Review of petro-structural data in the geochronological frame. *J. Virtual Explor.* **2012**, *42*, 1–44. [CrossRef]

44. Brun, J.-P.; Sokoutis, D. Kinematics of the Southern Rhodope Core Complex (North Greece). *Int. J. Earth Sci.* **2007**, *96*, 1079–1099. [CrossRef]

45. Turpaud, P.; Reischmann, T. Characterization of igneous terranes by zircon dating: Implications for UHP occurrences and suture identification in the Central Rhodope, Northern Greece. *Int. J. Earth Sci.* **2010**, *99*, 567–591. [CrossRef]

46. Mposkos, E.; Kostopoulos, D. Diamond, former coesite and supersilicic garnet in metasedimentary rocks from the Greek Rhodope: A new ultrahigh-pressure metamorphic province established. *Earth Plan. Sci. Lett.* **2001**, *192*, 497–506. [CrossRef]

47. Liati, A. Identification of repeated Alpine (ultra) highpressure metamorphic events by U-Pb SHRIMP geochronology and REE geochemistry of zircon: The Rhodope zone of Northern Greece. *Contrib. Mineral. Petrol.* **2005**, *150*, 608–630. [CrossRef]

48. Gautier, P.; Bosse, V.; Cherneva, Z.; Didier, A.; Gerdjikov, I. Polycyclic alpine orogeny in the Rhodope metamorphic complex: The record in migmatites from the Nestos shear zone (N. Greece). *Bull. Geol. Soc. France* **2017**, *188*, 36. [CrossRef]

49. Bonev, N.; Burg, J.-P.; Ivanov, Z. Mesozoic–Tertiary structural evolution of an extensional gneiss dome—The Kesebir–Kardamos dome, eastern Rhodope (Bulgaria–Greece). *Int. J. Earth Sci.* **2006**, *95*, 318–340. [CrossRef]

50. Menant, A.; Jolivet, L.; Vrielynck, B. Kinematic reconstructions and magmatic evolution illuminating crustal and mantle dynamics of the eastern Mediterranean region since the late Cretaceous. *Tectonophysics* **2016**, *675*, 103–140. [CrossRef]

51. Wüthrich, E. Low Temperature Thermochronology of the Northern Aegean Rhodope Massif. Unpublished Ph.D. Thesis, ETH Zürich, Zürich, Switzerland, 2009.

52. Melfos, V.; Voudouris, P. Cenozoic metallogeny of Greece and potential for precious, critical and rare metals exploration. *Ore Geol. Rev.* **2017**, *59*, 1030–1057. [CrossRef]

53. Duchêne, S.; Aïssa, R.; Vanderhaeghe, O. Pressure-temperature-time evolution of metamorphic rocks from Naxos (Cyclades, Greece): Constraints from thermobarometry and Rb/Sr dating. *Geodin. Acta* **2006**, *19*, 301–321. [CrossRef]

54. Ottens, B.; Voudouris, P. *Griechenland: Mineralien-Fundorte-Lagerstätten*; Christian Weise Verlag: Munchen, Germany, 2018; 480p.

55. Beaudoin, A.; Augier, R.; Laurent, V.; Jolivet, L.; Lahfid, A.; Bosse, V.; Arbaret, L.; Rabillard, A.; Menant, A. The Ikaria hightemperature Metamorphic Core Complex (Cyclades, Greece): Geometry, kinematics and thermal structure. *J. Geodesy* **2015**, *92*, 18–41.

56. Papanikolaou, D.; Panagopoulos, G. On the structural style of Southern Rhodope (Greece). *Geol. Balc.* **1981**, *11*, 13–22.

57. Krenn, K.; Bauer, C.; Proyer, A.; Mposkos, E.; Hoinkes, G. Fluid entrapment and reequilibration during subduction and exhumation: A case study from the high-grade Nestos shear zone, Central Rhodope, Greece. *Lithos* **2008**, *104*, 33–53. [CrossRef]

58. Nagel, T.J.; Schmidt, S.; Janák, M.; Froitzheim, N.; Jahn-Awe, S.; Georgiev, N. The exposed base of a collapsing wedge: The Nestos Shear Zone (Rhodope Metamorphic Province, Greece): The Nestos shear zone. *Tectonics* **2011**, *30*. [CrossRef]

59. Bonneau, M. Correlation of the Hellenic nappes in the south-east Aegean and their tectonic reconstruction. *Geol. Soc. Lond. Spec. Publ.* **1984**, *17*, 517–527. [CrossRef]

60. Scheffer, C.; Vanderhaeghe, O.; Lanari, P.; Tarantola, A.; Ponthus, L.; Photiades, A.; France, L. Syn- to post-orogenic exhumation of metamorphic nappes: Structure and thermobarometry of the western Attic-Cycladic metamorphic complex (Lavrion, Greece). *J. Geodyn.* **2016**, *96*, 174–193. [CrossRef]

61. Grasemann, B.; Schneider, D.A.; Stöckli, D.F.; Iglseder, C. Miocene bivergent crustal extension in the Aegean: Evidence from the western Cyclades (Greece). *Lithosphere* **2012**, *4*, 23–39. [CrossRef]

62. Kruckenberg, S.C.; Vanderhaeghe, O.; Ferré, E.C.; Teyssier, C.; Whitney, D.L. Flow of partially molten crust and the internal dynamics of a migmatite dome, Naxos, Greece. *Tectonics* **2011**, *30*. [CrossRef]

63. Laurent, V.; Huet, B.; Labrousse, L.; Jolivet, L.; Monie, P.; Augier, R. Extraneous argon in high-pressure metamorphic rocks: Distribution, origin and transport in the Cycladic Blueschist Unit (Greece). *Lithos* **2017**, *272*, 315–335. [CrossRef]

64. Altherr, R.; Kreuzer, H.; Wendt, I.; Lenz, H.; Wagner, G.A.; Keller, J.; Harre, W.; Höhndorf, A. A late Oligocene/early Miocene high temperature belt in the Attic-Cycladic crystalline complex (SE Pelagonian, Greece). *Geol. Jahrb.* **1982**, *23*, 97–164.

65. Lister, G.S.; Banga, G.; Feenstra, A. Metamorphic core complexes of cordilleran type in the Cyclades, Aegean Sea, Greece. *Geology* **1984**, *12*, 221–225. [CrossRef]

66. Gautier, P.; Brun, J.P. Crustal-scale geometry and kinematics of late-orogenic extension in the central Aegean (Cyclades and Ewia Island). *Tectonophysics* **1994**, *38*, 399–424. [CrossRef]

67. Jansen, J.B.H.; Schuiling, R.D. Metamorphism on Naxos; petrology and geothermal gradients. *Am. J. Sci.* **1976**, *276*, 1225–1253. [CrossRef]

68. Wybrans, J.R.; McDougall, I. Metamorphic evolution of the Attic Cycladic Metamorphic Belt on Naxos (Cyclades, Greece) utilizing ^{40}Ar/^{39}Ar age spectrum measurements. *J. Metam. Geol.* **1988**, *6*, 571–594. [CrossRef]

69. Buick, I.S.; Holland, T.J.B. The P-T path associated with crustal extension, Naxos, Cyclades, Greece. In *Evolution on Metamorphic Belts*; Daly, J.S., Cliff, R.A., Yardley, B.W.D., Eds.; Geological Society Special Publications: Lomdon, UK, 1989; Volume 43, pp. 365–369.

70. Baker, J.; Matthews, A. The stable isotopic evolution of a metamorphic complex, Naxos, Greece. *Contrib. Mineral. Petrol.* **1995**, *120*, 391–403. [CrossRef]

71. Katzir, Y.; Valley, J.W.; Matthews, A.; Spicuzza, M.J. Tracking fluid flow during deep crustal anatexis: Metasomatism of peridotites (Naxos, Greece). *Contrib. Mineral. Petrol.* **2002**, *142*, 700–713. [CrossRef]

72. Andriessen, P.A.M.; Hebedaa, E.H.; Simonb, O.J.; Verschurea, R.H. Tourmaline K/Ar ages compared to other radiometric dating systems in Alpine anatectic leucosomes and metamorphic rocks (Cyclades and southern Spain). *Chem. Geol.* **1991**, *91*, 33–48. [CrossRef]

73. Feenstra, A.; Wunder, B. Dehydration of diasporite to corundite in nature and experiment. *Geology* **2002**, *30*, 119–122. [CrossRef]

74. Laurent, V.; Beaudoin, A.; Jolivet, L.; Arbaret, L.; Augier, R.; Rabillard, A.; Menant, A. Interrelations between extensional shear zones and synkinematic intrusions: The example of Ikaria Island (NE Cyclades, Greece). *Tectonophysics* **2015**, *651–652*, 152–171. [CrossRef]

75. Iliopoulos, I. Petrogenesis of Metamorphic Rocks from Ikaria Island. Ph.D. Thesis, University of Patras, Patra, Greece, 2005.

76. Ring, U. The Geology of Ikaria Island: The Messaria extensional shear zone, granites and the exotic Ikaria nappe. *J. Virtual Explor.* **2007**, *27*, 3.

77. Liati, A.; Skarpelis, N. The metabauxites of Ikaria island, Eastern Aegean, Greece. In Proceedings of the 5th International Symposium on eastern Mediterranean Geology, Thessaloniki, Greece, 14–20 April 2004; pp. 1427–1430.

78. Henry, D.J.; Dutrow, B.L. Compositional zoning and element partitioning in nickeloan tourmaline from metamorphosed karstbauxite from Samos, Greece. *Am. Mineral.* **2001**, *86*, 1130–1142. [CrossRef]

79. Karantoni, V.D. Gemological Study of Corundum, Variety of Ruby and Sapphire from Greece. Unpublished Master's Thesis, School of Geology, Aristotle University of Thessaloniki, Thessaloniki, Greece, 2018.

80. Bodnar, R.J. Introduction to fluid inclusions. In *Fluid Inclusions: Analysis and Interpretation*; Mineralogical Association of Canada Short Course Series; Samson, I.M., Anderson, A.J., Marshall, D.D., Eds.; Mineralogical Association of Canada: Quebec City, QC, Canada, 2003; Volume 32, pp. 1–8.

81. Giuliani, G.; Fallick, A.; Rakotondrazafy, M.; Ohnenstetter, D.; Andriamamonjy, A.; Ralantoarison, T.; Rakotosamizanany, S.; Razanatseheno, M.; Offant, Y.; Garnier, V.; et al. Oxygen isotope systematics of gem corundum deposits in Madagascar: Relevance for their geological origin. *Mineral. Depos.* **2007**, *42*, 251–270. [CrossRef]

82. Giuliani, G.; Fallick, A.E.; Ohnenstetter, D.; Pegre, G. Oxygen isotope composition of sapphire from the French Massif Central: Implications for the origin of gem corundum in basalt fields. *Mineral. Depos.* **2009**, *44*, 221–231. [CrossRef]

83. Wang, K.K.; Graham, I.T.; Martin, L.; Voudouris, P.; Giuliani, G.; Lay, A.; Harris, S.; Fallick, A.E. Fingerprinting Paranesti rubies through in-situ oxygen isotopes. *Minerals*. (under review).

84. Sutherland, F.L.; Zaw, K.; Meffre, S.; Giuliani, G.; Fallick, A.E.; Graham, I.T.; Webb, G.B. Gem corundum megacrysts from East Australia basalt fields: Trace elements, O isotopes and origins. *Aust. J. Earth Sci.* **2009**, *56*, 1003–1020.

85. Harris, S.J.; Graham, I.T.; Lay, A.; Powell, W.; Belousova, E.; Zappetini, E. Trace element geochemistry and metasomatic origin of alluvial sapphires from the Orosmayo region, Jujuy Province, Northwest Argentina. *Canad. Mineral.* **2017**, *55*, 595–617. [CrossRef]

86. Sutherland, F.; Zaw, K.; Meffre, S.; Yui, T.-F.; Thu, K. Advances in Trace Element "Fingerprinting" of Gem Corundum, Ruby and Sapphire, Mogok Area, Myanmar. *Minerals* **2015**, *5*, 61–79. [CrossRef]

87. Giuliani, G.; Lasnier, B.; Ohnenstetter, D.; Fallick, A.E.; Pégère, G. Les gisements de corindon de France. *Le Règne Minéral* **2010**, *93*, 5–22.

88. Giuliani, G.; Caumon, G.; Rakotozamisanny, S.; Ohnenstetter, D.; Rakotondrazafy, M. Classification chimique des corindons par analyse factorielle discriminante: Application à la typologie des gisements de rubis et saphirs. *Rev. Gemmol.* **2014**, *188*, 14–22.

89. Zwaan, J.H.; Buter, E.; Mertz-Kraus, R.; Kane, R.E. Alluvial sapphires from Montana: Inclusions, geochemistry and indications of a metasomatic origin. *Gems Gemmol.* **2015**, *51*, 370–391.

90. Schwarz, D.; Pardieu, V.; Saul, J.M.; Schmetzer, K.; Laus, B.M.; Giuliani, G.; Klemm, L.; Malsy, A.; Erel, E.; Hauzenberger, C.; et al. Rubies and Sapphires from Winza, Central Tanzania. *Gems Gemmol.* **2008**, *44*, 322–347. [CrossRef]

91. Rakotondrazafy, A.F.M.; Giuliani, G.; Ohnenstetter, D.; Fallick, A.E.; Rakotosamizanany, S.; Andriamamonjy, A.; Ralantoarison, T.; Razanatseheno, M.; Offant, Y.; Garnier, V.; et al. Gem corundum deposits of Madagascar: A review. *Ore Geol. Rev.* **2008**, *34*, 134–154. [CrossRef]

92. Pham Van, L.; Hoáng Quang, V.; Garnier, V.; Giuliani, G.; Ohnenstetter, D.; Lhomme, T.; Schwarz, D.; Fallick, A.E.; Dubessy, J.; Phan Trong, T. Gem corundum deposits in Vietnam. *J. Gemmol.* **2004**, *29*, 129–147.

93. Simonet, C.; Paquette, J.L.; Pin, C.; Lasnier, B.; Fritsch, E. The Dusi (Garba Tula) sapphire deposit, Central Kenya—A unique Pan-African corundum-bearing monzonite. *J. Afr. Earth Sci.* **2004**, *38*, 401–410. [CrossRef]

94. De Maesschalck, A.A.; Oen, I.S. Fluid and mineral inclusions in corundum from gem gravels in Sri Lanka. *Mineral. Mag.* **1989**, *53*, 539–545. [CrossRef]

95. Giuliani, G.; Dubessy, J.; Banks, D.; Quang, V.H.; Lhomme, T.; Pironon, J.; Garnier, V.; Trong Trinhd, P.; Van Longf, P.; Ohnenstetter, D.; et al. CO_2-H_2S-COS-S_8-AlO(OH)-bearing fluid inclusions in ruby from marble-hosted deposits in Luc Yen area, North Vietnam. *Chem. Geol.* **2003**, *194*, 167–185. [CrossRef]

96. Menon, R.D.; Santosh, M.; Yoshida, M. Gemstone mineralization in southern Kerala, India. *J. Geol. Soc. India* **1994**, *44*, 241–252.

97. Rakotondrazafy, A.F.M.; Moine, B.; Cuney, M. Mode of formation of hibonite ($CaAl_{12}O_{19}$) within the U-Th skarns from the granulites of S-E Madagascar. *Contrib. Mineral. Petrol.* **1996**, *123*, 190–201. [CrossRef]

98. Peretti, A.; Mullis, J.; Kündig, R. Die Kaschmir-Saphire und ihr geologisches Erinnerungsvermögen. *Neue Zürcher Zeitung* **1990**, *187*, 59.

99. Srithai, B.; Rankin, A.H. Fluid inclusion characteristics of sapphires from Thailand. In *Mineral Deposits: Processes to Processing. Proceedings of the 5th Biennial SGA Meeting and the 10th Quadrennial IAGOD Symposium, London, UK, 22–25 August 1999*; Stanley, C.J., Ed.; Balkema: Rotterdam, The Netherlands, 1999; pp. 107–110.

100. Touret, J.L.R. Le facies granulite en Norvege meridionale. II. Les inclusions fluids. *Lithos* **1971**, *4*, 423–436. [CrossRef]

101. Newton, R.C.; Smith, J.V.; Windley, B.F. Carbonic metamorphism, granulites and crustal growth. *Nature* **1980**, *288*, 45–50. [CrossRef]

102. Morrison, J.; Valley, J.W. Post-granulite facies fluid infiltration in the Adirondack Mountains. *Geology* **1988**, *16*, 513–516. [CrossRef]

103. Hollister, L.S.; Burruss, R.C.; Henry, D.L.; Hendel, E.M. Physical conditions during uplift of metamorphic terrains, as recorded by fluid inclusions. *Bull. Minéral.* **1979**, *102*, 555–561.

104. Lamb, W.M.; Valley, J.W. Metamorphism of reduced granulites in low-CO_2 vapor-free environment. *Nature* **1984**, *312*, 56–58. [CrossRef]

105. Newton, R.C. Fluids of granulite facies metamorphism. In *Fluid—Rock Interactions during Metamorphism*; Walther, J.V., Wood, B.J., Eds.; Springer: New York, NY, USA, 1986; Volume 5, pp. 36–59.

106. Schuiling, R.D.; Kreulen, R. Are thermal domes heated by CO_2-rich fluids from the mantle? *Earth Plan. Sci. Lett.* **1979**, *43*, 298–302. [CrossRef]

107. Buick, I.S.; Holland, T.J.B. The nature and distribution of fluids during amphibolite facies metamorphism, Naxos (Greece). *J. Metam. Geol.* **1991**, *9*, 301–314. [CrossRef]

108. Mposkos, E.; Krohe, A.; Baziotis, I. Alpine polyphase metamorphism in metapelites from Sidironero complex (Rhodope domain, NE Greece). In Proceedings of the XIX CBGA Congress, Thessaloniki, Greece, 23–26 September 2010; Volume 100, pp. 173–181.

109. Liati, A. Regional Metamorphism and Overprinting Contact Metamorphism of the Rhodope Zone, Near Xanthi (N. Greece): Petrology, Geochemistry, Geochronology. Unpublished Ph.D. Thesis, Technical University of Braunschweig, Braunschweig, Germany, 1986.

110. Storre, B.; Nitsch, K.-H. Zur Stabilität von Margarit im System CaO-Al_2O_3-SiO_2-H_2O. *Contrib. Mineral. Petrol.* **1974**, *43*, 1–24. [CrossRef]

111. Chatterjee, N. Margarite stability and compatibility relations in the system CaO-Al_2O_3-SiO_2-H_2O as a pressure-temperature indicator. *Am. Mineral.* **1976**, *61*, 699–709.

112. Chatterjee, N. The system CaO-Al_2O_3-SiO_2-H_2O: New phase equilibria data, some calculated phase relations, and their petrological applications. *Contrib. Mineral. Petrol.* **1984**, *88*, 1–13. [CrossRef]

113. Okrusch, M.; Bunch, T.E.; Bank, H. Paragenesis and petrogenesis of a corundum bearing marble at Hunza (Kashmir). *Mineral. Depos.* **1976**, *11*, 278–297. [CrossRef]

114. Haas, H.; Holdaway, M.J. Equilibria in the system Al_2O_3-SiO_2-H_2O involving the stability limits of pyrophyllite. *Am. J. Sci.* **1973**, *273*, 449–464. [CrossRef]

115. Seifert, F. Stability of sapphirine: A study of the aluminous part of the system MgO-Al_2O_3-SiO_2-H_2O. *J. Geol.* **1974**, *82*, 173–204. [CrossRef]

116. Markl, G.; Scharlau, A. Grosse Sapphire, Margarit und Tremolit aus den Smirgel-gruben auf Naxos, Griechenland. *Lapis Mineral. Mag.* **2008**, *33*, 26–35.

117. Garnier, V.; Giuliani, G.; Ohnenstetter, D.; Fallick, A.E.; Dubessy, J.; Banks, D.; Hoang Quang, V.; Lhomme, T.; Maluski, H.; Pêcher, A.; et al. Marble-hosted ruby deposits from Central and Southeast Asia: Towards a new genetic model. *Ore Geol. Rev.* **2008**, *34*, 169–191. [CrossRef]

118. Matthews, A.; Putlitz, B.; Hamiel, Y.; Hervig, R.L. Volatile transport during the crystallization of anatectic melts: Oxygen, boron and hydrogen stable isotope study on the metamorphic complex of Naxos, Greece. *Geochim. Cosmochim. Acta* **2003**, *67*, 3145–3163. [CrossRef]

119. Siebenaller, L. Fluid Circulations during Collapse of an Accretionary Prism: Example of the Naxos Island Metamorphic Core Complex (Cyclades, Greece). Ph.D. Thesis, University of Lorraine, Lorraine, France, 2008.

120. Haas, H. Diaspore-corundum equilibria determined by epitaxis of diaspore on corundum. *Am. Mineral.* **1972**, *57*, 1375–1385.

121. Tenthorey, E.A.; Ryan, J.G.; Snow, E.A. Petrogenesis of sapphirine-bearing metatroctolites from the Buck Creek ultramafic body, southern Appalachians. *J. Metam. Geol.* **1996**, *14*, 103–114.

122. Altherr, R.; Okrusch, M.; Bank, H. Corundum- and kyanite-bearing anatexites from the Precambrian of Tanzania. *Lithos* **1982**, *15*, 191–197. [CrossRef]

123. Mercier, A.; Debat, P.; Saul, J.M. Exotic origin of the ruby deposits of the Mangari area in SE Kenya. *Ore Geol. Rev.* **1999**, *14*, 83–104. [CrossRef]

124. Key, R.M.; Ochieng, J.O. The growth of rubies in south-east Kenya. *J. Gemmol.* **1991**, *22*, 484–496. [CrossRef]

125. Simonet, C. Géologie des Gisements de Saphir et de Rubis. L'exemple de la John Saul Mine, Mangari, Kenya. Ph.D. Thesis, Université de Nantes, Nantes, France, 2000.

126. Garde, A.; Marker, M. Corundum crystals with blue-red color zoning near Kangerdluarssuk, Sukkertoppen district, West Greenland. *Report Gronlands Geologiske Undersokelse* **1988**, *140*, 46–49.

127. Kissin, A.J. Ruby and sapphire from the Southern Ural Mountains, Russia. *Gems Gemmol.* **1994**, *30*, 243–252. [CrossRef]

128. Peretti, A.; Mullis, J.; Mouawad, F. The role of fluorine in the formation of color zoning in rubies from Mong Hsu, Myanmar (Burma). *J. Gemmol.* **1996**, *25*, 3–19. [CrossRef]

129. Tropper, P.; Manning, C.E. The solubility of corundum in H_2O at high pressure and temperature and its implications for Al mobility in the deep crust and upper mantle. *Chem. Geol.* **2007**, *240*, 54–60. [CrossRef]

130. Hatipoğlu, M.; Türk, N.; Chamberlain, S.; Akgün, M. Gem-quality transparent diaspore (zultanite) in bauxite deposits of the İbir Mountains, Menderes Massif, SW Turkey. *Mineral. Depos.* **2010**, *45*, 201–205. [CrossRef]

131. Pardieu, V. Blue sapphires and Beryllium: An unfinished world quest. *InColor* **2013**, *23*, 36–43.

Article

Fingerprinting Paranesti Rubies through Oxygen Isotopes

Kandy K. Wang [1,*], **Ian T. Graham** [1], **Laure Martin** [2], **Panagiotis Voudouris** [3], **Gaston Giuliani** [4], **Angela Lay** [1], **Stephen J. Harris** [1] and **Anthony Fallick** [5]

[1] PANGEA Research Centre, School of Biological, Earth and Environmental Sciences, University of NSW, 2052 Sydney, Australia; i.graham@unsw.edu.au (I.T.G.); angela.lay@unsw.edu.au (A.L.); s.j.harris@student.unsw.edu.au (S.J.H.)

[2] Centre for Microscopy Characterisation and Analysis, The University of Western Australia, 6009 Perth, Australia; laure.martin@uwa.edu.au

[3] Faculty of Geology and Geoenvironment, National and Kapodistrian University of Athens, 157 84 Athens, Greece; voudouris@geol.uoa.gr

[4] Université de Lorraine, IRD and CRPG UMR 7358 CNRS-UL, BP 20, 15 rue Notre-Dame-des-Pauvres, 54501 Vandœuvre-lès-Nancy, France; giuliani@crpg.cnrs-nancy.fr

[5] Isotope Geosciences Unit, S.U.E.R.C., Rankine Avenue, East Kilbride, Glasgow G75 0QF, UK; anthony.fallick@glasgow.ac.uk

* Correspondence: kandy.wang@student.unsw.edu.au; Tel.: +61-411418800

Received: 1 December 2018; Accepted: 30 January 2019; Published: 3 February 2019

Abstract: In this study, the oxygen isotope ($\delta^{18}O$) composition of pink to red gem-quality rubies from Paranesti, Greece was investigated using in-situ secondary ionization mass spectrometry (SIMS) and laser-fluorination techniques. Paranesti rubies have a narrow range of $\delta^{18}O$ values between ~0 and +1‰ and represent one of only a few cases worldwide where $\delta^{18}O$ signatures can be used to distinguish them from other localities. SIMS analyses from this study and previous work by the authors suggests that the rubies formed under metamorphic/metasomatic conditions involving deeply penetrating meteoric waters along major crustal structures associated with the Nestos Shear Zone. SIMS analyses also revealed slight variations in $\delta^{18}O$ composition for two outcrops located just ~500 m apart: PAR-1 with a mean value of 1.0‰ ± 0.42‰ and PAR-5 with a mean value of 0.14‰ ± 0.24‰. This work adds to the growing use of in-situ methods to determine the origin of gem-quality corundum and re-confirms its usefulness in geographic "fingerprinting".

Keywords: rubies; corundum; in-situ oxygen isotopes; Paranesti Greece; Nestos Shear Zone; Secondary ion mass spectrometry (SIMS)

---

1. Introduction

1.1. Oxygen Isotopic Studies in Corundums

Oxygen is an abundant element in the Earth's crust, mantle and fluids. Oxygen consists of three naturally-occurring stable isotopes: ^{16}O (99.76%), ^{17}O (0.04%) and ^{18}O (0.2%). $\delta^{18}O$ expressed as Vienna standard mean ocean water (VSMOW) in per mil is the standard for the oxygen isotopic composition which is a measure of the ratio of the stable isotopes oxygen-18 (^{18}O) and oxygen-16 (^{16}O). There are numerous applications of oxygen isotope geochemistry including paleoclimatology, urban forensics, geological genesis and many more [1–3]. Oxygen isotope fractionation is a function of the initial Rayleigh evaporation-precipitation cycle, temperature of the system and degree of water-rock interaction and therefore great care must be taken when interpreting oxygen isotope values [4–7].

Although worldwide corundum oxygen isotope values have been found in a wide range from -27‰ (Khitostrov, Russia) to +23‰ (Mong Hsu, Myanmar), most are in the range of +3‰ to +21‰ [8–10].

This criterion has often been used to determine the geological origin of coloured corundum and especially the gem corundums, rubies and sapphires. $\delta^{18}O$ has been particularly useful in determining the likely primary geological origin of placer corundums where the primary origin is uncertain [11]. As isotopic fractionation is a function of both temperature and geological processes, oxygen isotope data need to be treated with some degree of caution Thus, there are very few examples where oxygen isotopes have been used to "fingerprint" the geographic location [12].

1.2. Geological Setting and Sample Background

The Paranesti rubies are found within the Nestos Shear Zone (NSZ) of the Rhodope Mountain Complex (RMC) in north-eastern Greece (Figure 1). The tectonic and polymetamorphic record of this Northern Aegean region (including the RMC) reflects the Middle Jurassic to Neogene northeast dipping subduction and convergence of the African-Eurasian plates which resulted in the closure of the Tethys Ocean [13,14]. The NSZ is thought to be a one of the syn-metamorphic thrusts in the RMC that are responsible for regional metamorphic inversion, placing higher amphibolite-facies intermediate terranes onto upper-greenschist to lower amphibolite-facies rocks of the lower terrane [15,16].

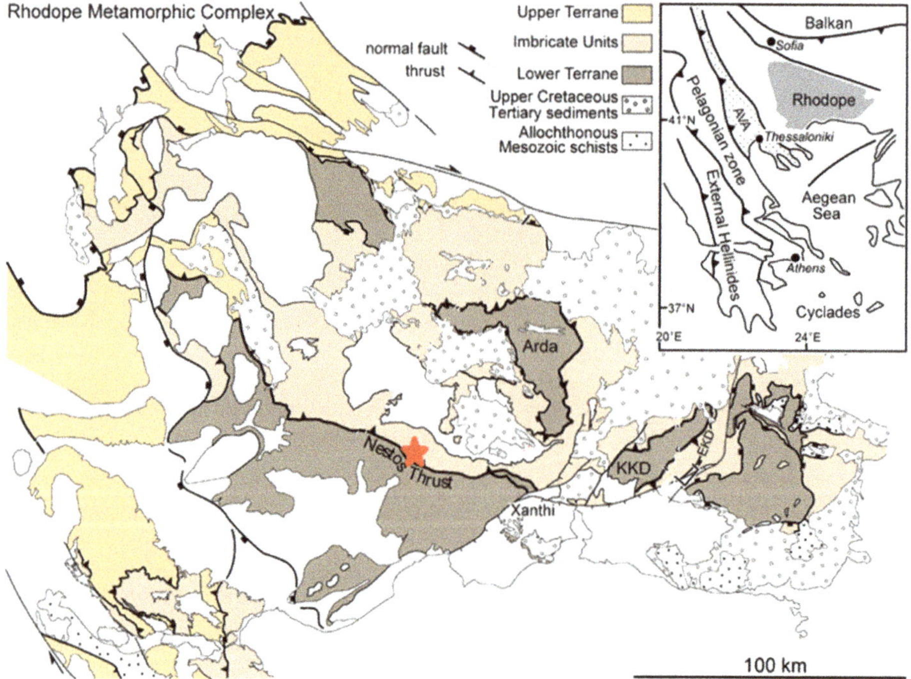

Figure 1. Geological map of the Rhodope Mountain Complex, with Paranesti located within the Nestos Shear Zone (red star) (Adapted from Moulas et al, 2017 [17]).

Based on an earlier systematic study on Paranesti [18], the ruby-bearing occurrences were found to be hosted in pargasite schist with a mafic/ultramafic protolith. The surrounding non-corundum-bearing chlorite schist was found to mainly be comprised of clinochlore. The ruby-bearing occurrence found on the hillside is referred to as PAR-1 (Figure 2a) and the road-side occurrence is termed PAR-5 (Figure 2b). Not all of the pargasite boudins nor the pegmatite intrusion found within the vicinity of the two sites contained corundum (Figure 2c,d).

Figure 2. Locality diagram of the ruby occurrences. (**a**) PAR-1 location on top of the hill. (**b**) PAR-5 location on the roadside. (**c**) Pargasite schist boudin found approx. 500 m north of PAR-5 without any corundum. (**d**) Pegmatite on top of the ruby-bearing pargasite schist at PAR-1.

A summary of the main findings from this previous study is listed in Table 1. Detailed LA-ICP-MS trace element analyses showed that the rubies are of metamorphic origin (Figure 3a) with minor partial metasomatic influences (Figure 3b). The high R^2 value based on the Fe/Mg vs Ga/Mg elemental discrimination diagram shows both PAR-1 and PAR-5 rubies to contain highly consistent trace element compositions (Figure 3a).

Table 1. Summary of prior Paranesti ruby results (Wang et al. 2017 [18]).

Attributes	PAR-1	PAR-5
Physical Characteristics		
Site	Hillside surface outcrop	Roadside surface outcrop—500 m east of PAR1
Grain-size	10 mm–20 mm	5 mm–10 mm
Colour	Deeper red than PAR-5 (generally)	Medium red
Inclusions	Spinels	None
Microscope view	More fractured, finer-grained	-
Host rock	Pargasite schist	Pargasite schist
EMPA Analyses (wt. %)		
Cr_2O_3	0.11–1.68	0.13–0.29
FeO	0.19–0.73	0.18–0.36
TiO_2	0–0.01	0–0.06
Ga_2O_3	0–0.04	0–0.04
LA-ICP-MS: Trace Element Analysis (ppm)		
Cr	360–2856	4–8627
Fe	1572–2664	1833–3822
V	1–3	2–5
Mg	7–42	8–376
Ti	6–184	10–190
Ga	14-23	13–29
Si	781–2456	837–2123
Ca	769–2119	653–1903

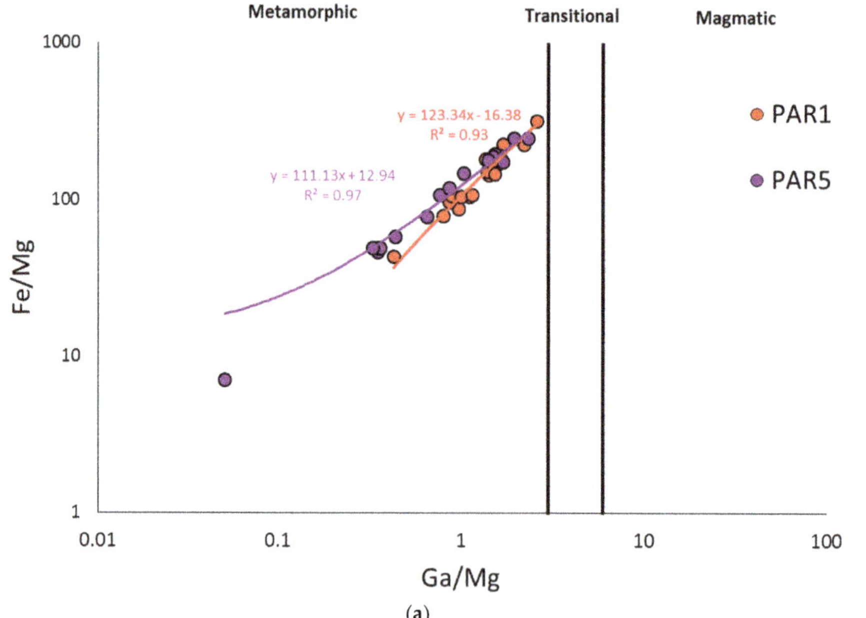

Figure 3. *Cont.*

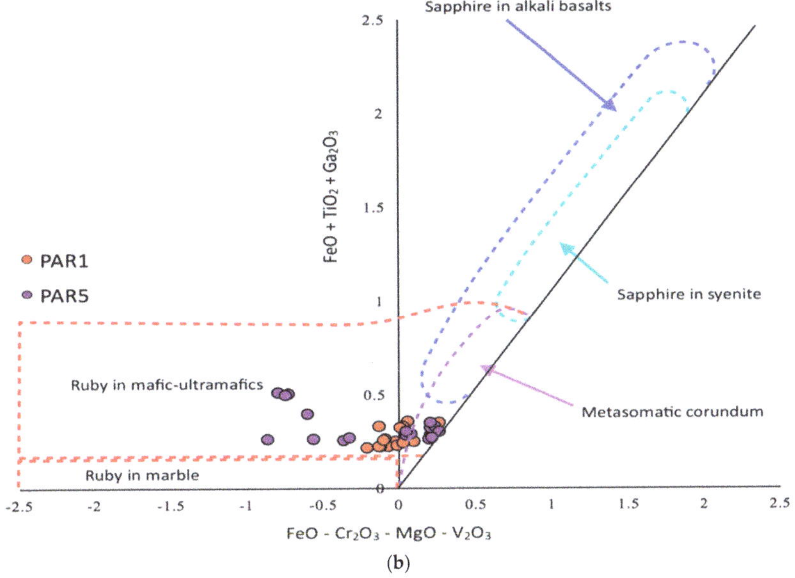

Figure 3. Trace element discrimination diagrams showing the fields for magmatic, metamorphic and metasomatic corundums, along with the plots for the Paranesti rubies. (**a**) FeMg vs. GaMg elemental diagram showing the metamorphic vs magmatic fields and SD lines with Paranesti ruby plots. Adapted with permission from Sutherland et al. 2014 [19]. (**b**) FeO + TiO$_2$ + Ga$_2$O$_3$ vs. FeO-Cr$_2$O$_3$-MgO-V$_2$O$_3$ elemental diagram showing a metasomatic origin as well as a mafic-ultramafic influence on the Paranesti rubies. Adapted with permission from Giuliani et al. 2014 [20].

2. Materials and Methods

Two different oxygen isotopic analytical methods have been used in this study in order to verify the oxygen isotope values of the Paranesti rubies. The rubies were mechanically extracted from the pargasite host matrix and carefully cleaned prior to being sent for analysis. In many samples, the ruby crystals occur in clusters of platy crystals where the grain sizes generally range between 0.5–1.5 cm (Figure 4a–c). Importantly, in the previous study [18] the ruby grains were generally found to be free of inclusions and thus amenable to in-situ analysis.

Figure 4. *Cont.*

(c)

Figure 4. Images of ruby samples from Paranesti. (**a**) Dark red ruby samples from PAR-1 in pargasite schist host rock 0.5–1.0 cm; (**b**) Cluster of pale ruby samples from PAR-5 in pargasite schist host rock 0.5–1.5 cm; (**c**) Clean PAR-1 ruby sample free from inclusions used for the 2009 Oxygen Isotope analysis.

2.1. Laser-Fluorination Method (2009)

In 2009, a reconnaissance study was conducted, whereby five individual grains, one each from different corundum localities/geological environments in Greece, were studied for their $\delta^{18}O$ composition. These included two colourless to blue sapphires in desilicified pegmatite from Naxos, one pink marble-hosted ruby from Kimi and one purple marble-hosted ruby from Xanthi. One medium red intensity ruby in pargasite schist from Paranesti (PAR-1) was also included. Oxygen isotope analyses were performed using a modification of the laser-fluorination technique described by Sharp [21] that was similar to that applied by Giuliani et al. in 2005 [11].

The method involves the complete reaction of ~1 mg of ground corundum. This powder is then heated by a CO_2 laser, with ClF_3 as the fluorine reagent. The released oxygen is passed through an in-line Hg-diffusion pump before conversion to CO_2 on platinized graphite. The yield is then measured by a capacitance manometer. The gas-handling vacuum line is connected to the inlet system of a dedicated VG PRISM 3 dual inlet isotope-ratio mass spectrometer. All oxygen isotope ratios are reported in $\delta^{18}O$ (‰) relative to Vienna standard mean ocean water (VSMOW). The secondary standard used for the laser-fluorination method was an internal quartz standard, NBS28 quartz that gave an average $\delta^{18}O$ value of 9.6‰. Oxygen yields differing significantly from the theoretical value of 14.07 µmol. per mg were taken as likely evidence of analytical artefact. Precision and accuracy on the internal quartz standard are ± 0.1‰ (1σ). Duplicate and triplicate analyses of sapphire and ruby suggested that this is appropriate for such materials.

2.2. Secondary Ion Mass Spectrometry (SIMS) Method (2017)

The 2017 analyses were performed exclusively on a range of coloured ruby grains from the two distinct Paranesti locations described in the previous study [13]. Unlike the 2009 analyses, these analyses were made using secondary ionisation mass spectrometry (SIMS) to analyse different areas of selected ruby grains in-situ to measure oxygen isotope ratios with less than one per mil (‰) level precision. Oxygen isotope ratios ($^{18}O/^{16}O$) in ruby were determined using a Cameca IMS 1280 multi-collector ion microprobe within the Centre for Microscopy, Characterisation and Analysis (CMCA), University of Western Australia (UWA). The materials examined in 2017 included 3 ruby grains from 3 different samples (57 analyses) from PAR-1 and 5 ruby grains from 5 different samples (44 analyses) from PAR-5. Each analysis point is shown in Figure 5a,b.

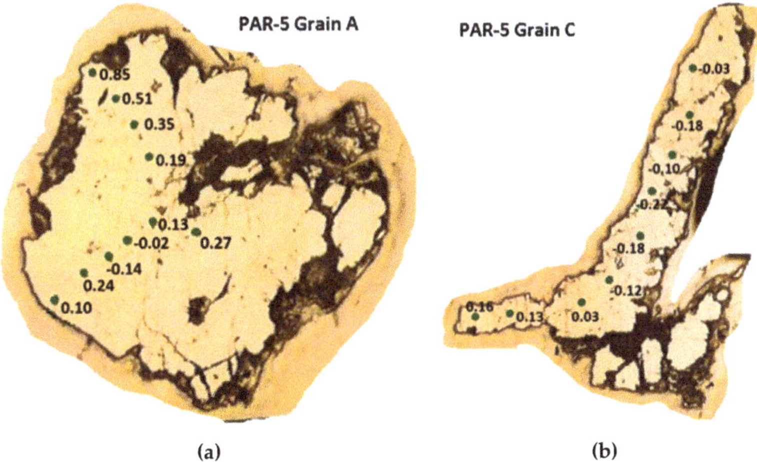

Figure 5. (**a**) SIMS in-situ analysis spot location individually marked on the ruby grain–PAR-5 Grain A; (**b**) SIMS in-situ analysis spot location individually marked on the ruby grain–PAR-5 Grain C.

The sample mounts were carefully cleaned with detergent, distilled water and ethanol in an ultrasonic bath and then coated with gold (30 nm in thickness) prior to SIMS O isotope analyses. For oxygen isotopic analyses, secondary ions were sputtered from the sample by bombarding its surface with a Gaussian Cs^+ beam and a total impact energy of 20 keV. The surface of the sample was rastered with a 2.5 nA primary beam over a 15 × 15 μm area. An electron gun was used to ensure charge compensation during the analyses. Secondary ions were admitted in the double focusing mass spectrometer within a 100 μm entrance slit and focused in the centre of a 4000 μm field aperture (×100 magnification). They were energy filtered using a 30 eV band pass with a 5 eV gap toward the high-energy side. ^{16}O and ^{18}O were collected simultaneously in multicollection mode in Faraday Cup detectors fitted with 10^{10} Ω and 10^{11} Ω, respectively. Each analysis includes a pre-sputtering over a 20 × 20 μm area during 30 s and the automatic centring of the secondary ions in the field aperture, contrast aperture and entrance slit and consisted of 20 four-second cycles which give an average internal precision of ~0.16‰ (2 SE).

External reproducibility during the analytical sessions was evaluated by repeating analyses in one single fragment of PAR-1. External reproducibility in this fragment was 0.3 and 0.4 per mil (2SD) during the two analytical sessions. In total, three large fragments of PAR-1 were analysed for their oxygen isotope composition, altogether yielding an average value of 0.9 ± 0.6 per mil (2SD, n = 57, Table 2). Raw oxygen isotope ratios were corrected for instrumental mass fractionation using the $\delta^{18}O$ composition of PAR-1, which oxygen isotope composition was obtained by laser fluorination method (from the 2009 study). Uncertainty on each $\delta^{18}O$ spot has been calculated by propagating the errors on instrumental mass fractionation determination, which include the standard deviation of the mean oxygen isotope ratio measured on the primary standard during the session and internal error on each sample data point. Corrected $\delta^{18}O$ (quoted with respect to Vienna Standard Mean Ocean Water or VSMOW) are presented in Supplementary Material Table S1.

Table 2. Oxygen isotope results from 2017 using the SIMS method.

Grain	δ^{18}O Min	δ^{18}O Max	δ^{18}O Mean	Number of Analyses
PAR-1a	0.64	1.62	1.00 ± 0.42	31
PAR-1b	0.44	1.17	0.67 ± 0.37	13
PAR-1c	0.77	1.68	1.27 ± 0.47	13
PAR-1 Total	0.44	1.68	1.00 ± 0.42	57
PAR-5central	-0.04	0.51	0.27	12
PAR-5a	-0.14	0.85	0.25	10
PAR-5b	-0.31	0.42	0.03	8
PAR-5c	-0.22	0.16	-0.06	9
PAR-5d	0.08	0.27	0.17	5
PAR-5 Total	-0.31	0.85	0.14 ± 0.24	44
Combined PAR-1 and PAR-5	-0.31	1.31	0.60	93

3. Results

3.1. Laser-Fluorination Results

Using the laser-fluorination method, the oxygen isotope ratio for the pargasite schist hosted PAR-1 ruby was found to be δ^{18}O 1.0‰. This analytical run also included a number of rubies and sapphires from different geological environments. Sapphires from desilicified pegmatites were found to range from 4.8‰ to 5.0‰ and rubies from marble-hosted deposits were found to range from 20‰ to 22‰ (Table 3).

Table 3. Oxygen isotope results from the 2009 reconnaissance using the laser-fluorination method, n = 1.

Sample	Location	Sample Type	Deposit Type	δ^{18}O
NAX2	Naxos, Greece	Colourless sapphire	Desilicified pegmatite	4.80
NAX3	Naxos, Greece	Colourless to blue sapphire	Desilicified pegmatite	5.05
PAR-1	Paranesti, Greece	Red ruby	Pargasite schist	1.00
KIM2	Kimi, Greece	Pink ruby	Marble-hosted	20.50
Xanthi	Xanthi, Greece	Purple-pink ruby	Marble-hosted	22.09

3.2. Secondary Ionisation Mass Spectrometry (SIMS) Results

The oxygen isotope ratios δ^{18}O (VSMOW) are presented in Table 2. PAR-5 results show values of -0.31‰ to 0.85‰ (0.14 ± 0.24), on average slightly lower compared to PAR-1 results 0.44‰ to 1.68‰ (1.00 ± 0.42) even though the two occurrences are only 500 m apart.

4. Discussion

4.1. Corundum Oxygen Isotopes as An Identifier for Geological Origin

A framework on the interpretation of the geological origin of gem corundums using the δ^{18}O ratio proposed by Giuliani et al is now widely adopted [11]. Based on this framework, rubies and pink sapphires can be classified into 5 categories based on their δ^{18}O value range.

1. Mafic gneiss hosted from 2.9‰ to 3.8‰;
2. Mafic-ultramafic rocks (amphibolite, serpentinite) from 3.2‰ to 6.8‰;
3. Desilicated pegmatites from 4.2‰ to 7.5‰;
4. Shear zones cross-cutting ultramafic lenses and pegmatites within sillimanite gneisses 11.9‰–13.1‰;
5. Marble-hosted rubies 16.3‰–23‰. This framework has been further validated by numerous subsequent corundum oxygen isotope studies [22–25].

The reconnaissance 2009 laser fluorination results on the sapphires and rubies from different geological environments very closely fits the oxygen isotope value ranges from the model of Giuliani et al. where over 200 corundum samples were analysed under the same method [1]. That is the marble-hosted value from 20‰ to 22‰ is within the range of 16.3‰ to 23‰ and the sapphires from the desilicified pegmatites with a value from 4.8‰ to 5‰ fits within the framework range from 4.2‰ to 7.5‰. Therefore, the $\delta^{18}O$ results obtained using the laser-fluorination method in 2009 are further validated as accurate measurements.

4.2. PAR-1 vs. PAR-5 Variations

The oxygen isotope values obtained using SIMS indicates that the Paranesti rubies have a narrow defined band of oxygen isotope signatures with a mean on +1‰ (ranging from −0.31‰ to 1.31‰). This is lower than any ratios based on the existing framework for rubies. There are further distinctive constrained values between PAR-1 (+0.65‰ to 1.31‰) and PAR-5 (−0.31‰ to 0.85‰). There is a slight overlap of the individual highest value in PAR-5 to the lowest value in PAR-1. The average for PAR-1 is +1‰ whilst the average for PAR-5 is +0.14‰.

There may be some differences between core-rim oxygen isotope values observed in the PAR-5 SIMS results where the core average (−0.02) is lower than the rim average (0.29). However, this is within the range of uncertainty when the errors are taken into account. This narrow range within individual localities and between the two localities that are 500 m apart is in stark contrast to the findings of Bindeman et al (2010) [7] who found variation within single 10 cm samples of up to 3‰ and variation within single ruby grains of up to 1.5‰. It also does not rule-out variances due to partitioning in individual crystals during growth. Therefore, a detailed cathodoluminescence analysis to determine the homogeneity or heterogeneity of the sample grains is suggested for future studies. As the traditional laser fluorination method consumes the entire grain, such subtle zoning would not be seen using this technique. Thus, the greater spatial resolution of the SIMS technique enables us to analyse discrete isotopic domains (i.e., rims, cores, sectors) within single corundum crystals.

4.3. Global Low to Ultra-Low Oxygen Isotope Corundum Comparison

$\delta^{18}O$ (SMOW) values for gem corundums below 1‰ are very rare and not shown on the original systematic framework by Giuliani et al 2005 [11]. Other than the Paranesti rubies shown above, the only negative value for corundums are from Karelia in north-western Russia and sapphire from a secondary deposit in Madagascar. Table 4 lists the global low to ultra-low oxygen isotope analyses for corundums.

The Madagascar sapphire deposit of Ilakaka with $\delta^{18}O$ of −0.3‰ to 16.5‰ is a consolidated placer formed in a sandstone environment. The geological origin of the different ranges of isotopic values found for the sapphires corresponds to at least five different geological environments [26]. The low $\delta^{18}O$ delta values for some sapphires correspond up to an unknown geological sapphire type.

The PAR-1 result of +1.0‰ is lower than most known primary corundum oxygen isotope values other than the unique ultra-low values of corundums from Karelia [28,30] and one instance of ruby from the Soamiakatra area of Madagascar [26]. The Karelia corundum formed under unique circumstances (see discussion below) and can be easily distinguished from the Paranesti rubies. The Madagascar rubies show much higher average $\delta^{18}O$ values and the minimum value obtained corresponds to the maximum value from Paranesti. Therefore, oxygen isotope analysis is a valuable tool that can be used to fingerprint the Paranesti rubies Figure 6 from other worldwide occurrences.

Table 4. Global comparison of corundums with low oxygen isotope values.

Country	District	$\delta^{18}O‰$ (Min)	$\delta^{18}O‰$ (Max)	Host Rock	Primary vs. Secondary	Corundum Type
Greece [1]	Paranesti-1 *	0.65	1.31	Pargasite schist	Primary	Ruby
Greece [1]	Paranesti-5 *	−0.31	0.85	Pargasite schist	Primary	Ruby
Madagascar [2]	Soamiakatra *	1.25	4.70	Pyroxenitic enclaves in basalt	Primary	Ruby
Madagascar [2]	Ilakaka *	−0.30	16.5	Placer in sandstone	Secondary	Sapphire
Madagascar [2]	Andilamena *	0.50	3.9	Placer in basalt	Secondary	Ruby
Russia [3,4,7]	Khitostrov ˆ	−26.3	−17.7	plagiogneiss	Primary	Corundum
Russia [4,7]	Khitostrov *	−26	-	Crn-St-Gt-Bi-Prg-Pl rocks with coarse grained Crn	Primary	Corundum
Russia [7]	Khitostrov *	−18.6	-	Ky-Crn-Pl, leucocratic	Primary	Corundum
Russia [3]	Varastskoye ˆ	−19.2	−11.3	plagiogneiss	Primary	Corundum
Russia [4]	Varastskoye #	−17.3	-	Crn-Cam rock, coarse grained	Primary	Corundum
	Varastskoye #	−19.2	-	Crn and Crn-St-Pl substituting Ky	Primary	Corundum
Russia [4]	Dyadina #	0.49	-	Inclusion of Cam-Crn in giant Gt	Inclusion	Corundum
Russia [4]	Dyadina #	0.10	-	Crn-Cam rock, coarse grained	Primary	Corundum
Russia [5]	Dyadina *	0.4	0.8	Corundum amphibolite	Primary	Corundum
Russia [4]	Kulezhma #	0.31	-	Cam-Crn rock	Primary	Corundum
Russia [4]	Pulonga #	0.67	-	Crn-Gt-Ged rock	Primary	Corundum
Russia [4]	Perusel'ka *	0.26	3.45	Crn-Cam rock, coarse grained	Primary	Corundum
Russia [5]	Perusel'ka #	0.6	-	Corundum-kyanite amphibolite	Primary	Corundum
Russia [5]	Perusel'ka #	1.5	-	Corundum amphibolite	Primary	Corundum
Russia [5]	Notozero *	−1.7	−1.5	Ged-Gt rocks with Crn and St	Primary	Corundum
Russia [4]	Mironova Guba ˆ	(2.34)	-	Cam-Crn rock	Primary	Corundum
Thailand [6]	Bo Rai *	1.30	4.20	Placer in basalt	Secondary	Ruby

* Individual grain analysis ˆ Whole-rock analysis # Only one analysis result, no range. Bi—biotite, Cam—Ca-amphibole, Crn—corundum, Ged—gedrite amphibole, Gt—garnet, Ky—kyanite, Pl—plagioclase, Prg—pargasitic amphibole, St—staurolite. [1] Wang et al. (2017) [18]; [2] Giuliani et al. (2007) [26]; [3] Vysotskiy et al. (2015) [27]; [4] Bindeman and Serebryakov (2011) [28]; [5] Vysotskiy et al. (2014) [12]; [6] Yui et al. (2006) [29]; [7] Bindeman et al. (2010) [7].

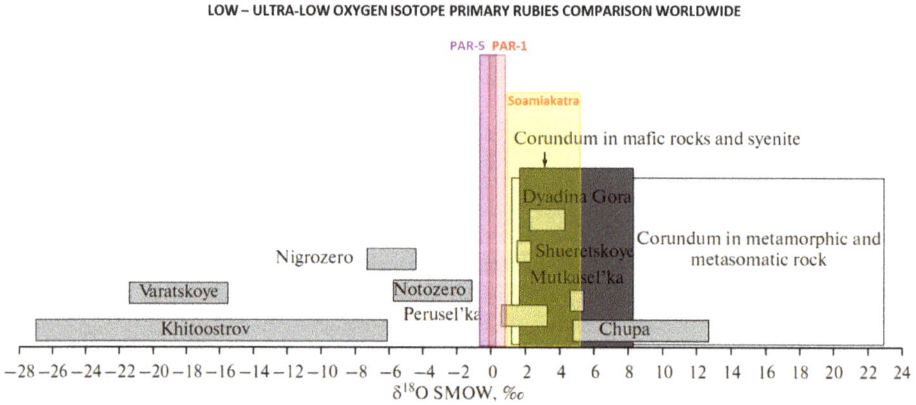

Figure 6. Oxygen isotopic comparison of rubies from Paranesti occurrences compared with low $\delta^{18}O$ corundums from Karelia in northwestern Russia and Soamiakatra in Madagascar.

4.4. Possible Causes for Low Oxygen Isotope Corundum Formation

There are several current hypotheses on how corundums can form with low $\delta^{18}O$ isotope ratios. These range from hydrothermal alteration of deeply penetrating surface meteoric waters to isotope separation by thermal diffusion during endogenous fluid flow [31,32].

4.4.1. Kinetic Isotope Fractionation

Kinetic isotope fractionation occurs when rapid thermal decomposition of hydrous phases results in isotope disproportionation into a high-$\delta^{18}O$ residue and a low-$\delta^{18}O$ fluid [33,34]. However, the high-δ^{18}O residue material should also be found within proximity of the studied samples for this hypothesis to apply. The water-rock interaction is kinetically restricted in supracrustal rocks and isotope fractionation factors are large at low temperatures, favouring higher-δ^{18}O solids [35]. In contrast, isotopic exchange is more rapid within a hydrothermal system. As the Paranesti rubies formed under amphibolite-facies conditions above 600 °C [18], significant kinetic isotope fractionation is highly unlikely and therefore rules out this hypothesis.

4.4.2. Thermal Diffusion

For thermal diffusion, the oxygen in a temperature gradient is redistributed with low δ^{18}O at the hotter end and high δ^{18}O at the colder end of melt or hydrous solution [31]. Akimova (2015) [32] has proposed a model of cascading thermo-diffusion within shear zones to explain the Karelian ultra-low δ^{18}O corundums. This scenario would require several individual thermal cells to align in the correct position simultaneously with a similar convection rate and timing. Given that only two locations have shown ruby-bearing pargasite schist with other very similar pargasite boudins nearby being ruby absent, this model appears to be less likely than the hydrothermal scenario.

4.4.3. Other Ultra-Low δ^{18}O Protoliths

Ultra-low δ^{18}O protoliths could potentially provide the low δ^{18}O during corundum syn-metamorphic formation [35]. However, a source for the ultra-low δ^{18}O protolith would be needed under this scenario such as a low δ^{18}O mantle reservoir or previously surface-exposed and then rapidly buried metamorphic rocks. Neither were observed at Paranesti. As oxygen isotope analyses were not performed on the whole-rock and associated mineral phases for the Paranesti occurrences, this hypothesis cannot be ruled out.

4.4.4. Hydrothermal Alteration Model

This hypothesis involves the conservation of the initial isotopic ratios of the protolith in the corundum-bearing rocks and then isotopic exchange between these rocks and meteoric waters before metamorphism [10]. Wang et al. [18] demonstrated that the rubies from Paranesti were syn-metamorphic and were largely free of inclusions. However, as shown by Bindeman et al (2010) [7], this does not preclude preservation of the initial ratios within the protolith for the Paranesti occurrence. Therefore, it is unlikely that the low isotopic ratios observed within the Paranesti rubies were due to preservation of the initial ratios within the protolith.

For granulite facies metamorphism, Wilson and Banksi (1983) [36] proposed three processes that could produce a low oxygen isotope value. These are (1) pre-granulite reaction between heated seawater and hot basic intrusives or an initial protolith such as a palaeosol for the sapphirine–spinel–(cordierite) assemblages; (2) syn-granulite depletion in ^{18}O related to dehydration during granulite metamorphism and removal of the resultant products of partial melting with a depletion in ^{18}O by up to 2‰ or 3‰ for the restite; and (3) post-granulite facies metamorphism with recrystallization under the effect of biotite and/or amphibole-metasomatism with depletion in δ^{18}O up to 4‰. Based on the previous study [18], the Paranesti rubies were found to have formed under amphibolite facies conditions and there is no evidence that they ever reached granulite facies within the specified zone. However, there are other locations within the Rhodope Mountain Complex (RMC) where regional metamorphism reached granulite facies conditions, though these are some distance away from Paranesti and no rubies are known from these locations.

The glacial meltwater influence during formation of corundums was proposed to explain the ultra-low δ^{18}O isotopic ratios observed for corundums from Karelia in north-western Russia [10,12,27]. However, there is no evidence suggesting the existence of glaciers in the Mediterranean region based on the reconstruction of the tectonic evolution of the East Mediterranean region since the late Cretaceous [37]. Could the RMC be a higher mountain with glaciers that have melted during ruby genesis? There is no evidence in the literature to suggest such and this would only be a remote possibility. Therefore, it is unlikely that a glacial meltwater source played a role during ruby formation at Paranesti. However, it is likely that meteoric water (but not glacial melt) interaction caused by downward flow of surface waters along deep crustal fractures/structures during the formation of the corundum would contribute in producing low $\delta^{18}O$ values for the Paranesti rubies.

5. Conclusions

The in-situ SIMS oxygen isotope analyses on the Paranesti rubies is the first time that a primary (and exclusively) ruby occurrence was found to have ~+1‰ for its $\delta^{18}O$-isotope composition. Based on the low $\delta^{18}O$ value and the local geology, it is most likely that the Paranesti rubies formed under metamorphic/metasomatic conditions involving deeply penetrating meteoric waters along major crustal structures related to the Nestos Shear Zone. PAR-5 is potentially closer to the source of the hydrothermal influence during ruby formation compared to PAR-1 and thus has a lower $\delta^{18}O$. Importantly, this study shows that in-situ gem corundum oxygen isotope analysis using the SIMS method may be used to determine the likely geographic origin for corundums lacking any provenance details. Importantly, with the SIMS method being only minimally destructive, with the analysis spot (15 × 15 μm) amenable to repolishing, a wider adoption of this technique has important applications/implications for the international gem and jewellery industry. A future area of research would be to apply this methodology for more gem mineral varieties other than corundum and emeralds. The aim of such future work would be to determine if in-situ SIMS oxygen isotope analysis can be used to both better understand gem formation and to see if it can be used to clearly separate the same gem mineral from different geographic locations.

Supplementary Materials: The following are available online at http://www.mdpi.com/2075-163X/9/2/91/s1, Table S1: Secondary Ionisation Mass Spectrometry (SIMS) Results.

Author Contributions: K.K.W. wrote the manuscript and interpreted the results of the analyses. I.T.G. collected the samples, provided technical input, funding and supervised the project. L.M. provided technical input and ran the SIMS analyses. P.V. collected the samples and provided geological expertise on the Paranesti region. G.G. and A.F. coordinated the laser fluorination analyses and provided technical input. A.L. and S.J.H. provided technical input.

Funding: This research received no external funding.

Acknowledgments: The authors would like to thank Joanne Wilde, formerly of the School of Biological, Earth and Environmental Sciences, UNSW Sydney, for making the polished mounts required for SIMS analysis. The authors acknowledge the facilities and the scientific and technical assistance of the Australian Microscopy & Microanalysis Research Facility at the Centre for Microscopy, Characterisation & Analysis, The University of Western Australia, a facility funded by the University, State and Commonwealth Governments.

Conflicts of Interest: The authors declare no conflict of interest.

References

1. Hoefs, J. *Stable Isotope Geochemistry*; Springer: Berlin/Heidelberg, Germany, 1997; ISBN 9783662033791.
2. Bowen, G.J.; Wassenaar, L.I.; Hobson, K.A. Global application of stable hydrogen and oxygen isotopes to wildlife forensics. *Oecologia* **2005**, *143*, 337–348. [CrossRef] [PubMed]
3. Miller, K.G.; Fairbanks, R.G.; Mountain, G.S. Tertiary oxygen isotope synthesis, sea level history and continental margin erosion. *Paleoceanography* **1987**, *2*, 1–19. [CrossRef]
4. Criss, R.E.; Taylor, H.P., Jr. Meteoric-hydrothermal systems. *Rev. Mineral.* **1986**, *16*, 373–424.
5. Valley, J.W. Stable Isotope Thermometry at High Temperatures. *Rev. Mineral. Geochem.* **2001**, *43*, 365–413. [CrossRef]
6. Bindeman, I. Oxygen Isotopes in Mantle and Crustal Magmas as Revealed by Single Crystal Analysis. *Rev. Mineral. Geochem.* **2008**, *69*, 445–478. [CrossRef]
7. Bindeman, I.N.; Schmitt, A.K.; Evans, D.A.D. Limits of hydrosphere-lithosphere interaction: Origin of the lowest-known $\delta^{18}O$ silicate rock on Earth in the Paleoproterozoic Karelian rift. *Geology* **2010**, *38*, 631–634. [CrossRef]

8. Giuliani, G.; Ohnenstetter, D.; Garnier, V.; Fallick, A.E.; Rakotondrazafy, M.; Schwarz, D. The geology and genesis of gem corundum deposits. In *Geology of Gem Deposits*; Groat, L.A., Ed.; Mineralogical Association of Canada: Québec, QC, Canada, 2007; Volume 37, pp. 23–78. ISBN 9780921294375.

9. Yui, T.-F.; Khin, Z.; Limtrakun, P. Oxygen isotope composition of the Denchai sapphire, Thailand: A clue to its enigmatic origin. *Lithos* **2003**, *67*, 153–161. [CrossRef]

10. Krylov, D.P.; Glebovitsky, V.A. Oxygen isotopic composition and nature of fluid during the formation of high-Al corundum-bearing rocks of Mt. Dyadina, northern Karelia. *Doklady Earth Sci.* **2007**, *413*, 210–212. [CrossRef]

11. Giuliani, G.; Fallick, A.E.; Garnier, V.; France-Lanord, C.; Ohnenstetter, D.; Schwarz, D. Oxygen isotope composition as a tracer for the origins of rubies and sapphires. *Geology* **2005**, *33*, 249. [CrossRef]

12. Vysotskiy, S.V.; Ignat'ev, A.V.; Levitskii, V.I.; Nechaev, V.P.; Velivetskaya, T.A.; Yakovenko, V.V. Geochemistry of stable oxygen and hydrogen isotopes in minerals and corundum-bearing rocks in northern Karelia as an indicator of their unusual genesis. *Geochem. Int.* **2014**, *52*, 773–782. [CrossRef]

13. Bonev, N.; Burg, J.-P.; Ivanov, Z. Mesozoic–Tertiary structural evolution of an extensional gneiss dome—The Kesebir–Kardamos dome, eastern Rhodope (Bulgaria–Greece). *Int. J. Earth Sci.* **2006**, *95*, 318–340. [CrossRef]

14. Krenn, K.; Bauer, C.; Proyer, A.; Mposkos, E.; Hoinkes, G. Fluid entrapment and reequilibration during subduction and exhumation: A case study from the high-grade Nestos shear zone, Central Rhodope, Greece. *Lithos* **2008**, *104*, 33–53. [CrossRef]

15. Barr, S.R.; Temperley, S.; Tarney, J. Lateral growth of the continental crust through deep level subduction–accretion: A re-evaluation of central Greek Rhodope. *Lithos* **1999**, *46*, 69–94. [CrossRef]

16. Nagel, T.J.; Schmidt, S.; Janák, M.; Froitzheim, N.; Jahn-Awe, S.; Georgiev, N. The exposed base of a collapsing wedge: The Nestos Shear Zone (Rhodope Metamorphic Province, Greece): The Nestos Shear Zone. *Tectonics* **2011**, *30*. [CrossRef]

17. Moulas, E.; Schenker, F.L.; Burg, J.-P.; Kostopoulos, D. Metamorphic conditions and structural evolution of the Kesebir-Kardamos dome: Rhodope metamorphic complex (Greece-Bulgaria). *Int. J. Earth Sci.* **2017**, *106*, 2667–2685. [CrossRef]

18. Wang, K.K.; Graham, I.T.; Lay, A.; Harris, S.J.; Cohen, D.R.; Voudouris, P.; Belousova, E.; Giuliani, G.; Fallick, A.E.; Greig, A. The Origin of a New Pargasite-Schist Hosted Ruby Deposit from Paranesti, Northern Greece. *Can. Mineral.* **2017**, *55*, 535–560. [CrossRef]

19. Sutherland, F.; Zaw, K.; Meffre, S.; Yui, T.-F.; Thu, K. Advances in Trace Element "Fingerprinting" of Gem Corundum, Ruby and Sapphire, Mogok Area, Myanmar. *Minerals* **2014**, *5*, 61–79. [CrossRef]

20. Giuliani, G.; Ohnenstetter, D.; Fallick, A.E. The geology and genesis of gem corundum deposits. In *Geology of Gem Deposits. Series: Mineralogical Association of Canada Short Course Series, 44*; Groat, L.A., Ed.; Mineralogical Association of Canada: Québec, QC, Canada, 2014; pp. 35–112. ISBN 9780921294542.

21. Sharp, Z.D. In situ laser microprobe techniques for stable isotope analysis. *Chem. Geol.* **1992**, *101*, 3–19. [CrossRef]

22. Garnier, V.; Giuliani, G.; Ohnenstetter, D.; Fallick, A.E.; Dubessy, J.; Banks, D.; Vinh, H.Q.; Lhomme, T.; Maluski, H.; Pêcher, A.; et al. Marble-hosted ruby deposits from Central and Southeast Asia: Towards a new genetic model. *Ore Geol. Rev.* **2008**, *34*, 169–191. [CrossRef]

23. Zaw, K.; Sutherland, L.; Yui, T.-F.; Meffre, S.; Thu, K. Vanadium-rich ruby and sapphire within Mogok Gemfield, Myanmar: Implications for gem color and genesis. *Miner. Deposita* **2015**, *50*, 25–39. [CrossRef]

24. Graham, I.; Sutherland, L.; Zaw, K.; Nechaev, V.; Khanchuk, A. Advances in our understanding of the gem corundum deposits of the West Pacific continental margins intraplate basaltic fields. *Ore Geol. Rev.* **2008**, *34*, 200–215. [CrossRef]

25. Sutherland, F.L.; Duroc-Danner, J.M.; Meffre, S. Age and origin of gem corundum and zircon megacrysts from the Mercaderes–Rio Mayo area, South-west Colombia, South America. *Ore Geol. Rev.* **2008**, *34*, 155–168. [CrossRef]

26. Giuliani, G.; Fallick, A.; Rakotondrazafy, M.; Ohnenstetter, D.; Andriamamonjy, A.; Ralantoarison, T.; Rakotosamizanany, S.; Razanatseheno, M.; Offant, Y.; Garnier, V.; et al. Oxygen isotope systematics of gem corundum deposits in Madagascar: Relevance for their geological origin. *Miner. Deposita* **2007**, *42*, 251–270. [CrossRef]

27. Vysotskiy, S.V.; Nechaev, V.P.; Kissin, A.Y.; Yakovenko, V.V.; Ignat'ev, A.V.; Velivetskaya, T.A.; Sutherland, F.L.; Agoshkov, A.I. Oxygen isotopic composition as an indicator of ruby and sapphire origin: A review of Russian occurrences. *Ore Geol. Rev.* **2015**, *68*, 164–170. [CrossRef]

28. Bindeman, I.N.; Serebryakov, N.S. Geology, Petrology and O and H isotope geochemistry of remarkably [18]O depleted Paleoproterozoic rocks of the Belomorian Belt, Karelia, Russia, attributed to global glaciation 2.4 Ga. *Earth Planet. Sci. Lett.* **2011**, *306*, 163–174. [CrossRef]

29. Yui, T.-F.; Wu, C.-M.; Limtrakun, P.; Sricharn, W.; Boonsoong, A. Oxygen isotope studies on placer sapphire and ruby in the Chanthaburi-Trat alkali basaltic gemfield, Thailand. *Lithos* **2006**, *86*, 197–211. [CrossRef]

30. Krylov, D.P. Anomalous $^{18}O/^{16}O$ ratios in the corundum-bearing rocks of Khitostrov, northern Karelia. *Doklady Earth Sci.* **2008**, *419*, 453–456. [CrossRef]

31. Bindeman, I.N.; Lundstrom, C.C.; Bopp, C.; Huang, F. Stable isotope fractionation by thermal diffusion through partially molten wet and dry silicate rocks. *Earth Planet. Sci. Lett.* **2013**, *365*, 51–62. [CrossRef]

32. Akimova, E.Y.; Lokhov, K.I. Ultralight Oxygen in Corundum-Bearing Rocks of North Karelia, Russia, as a Result of Isotope Separation by Thermal Diffusion (Soret Effect) in Endogenous Fluid Flow. *J. Mater. Sci. Chem. Eng.* **2015**, *3*, 42–47. [CrossRef]

33. Clayton, R.N.; Mayeda, T.K. Kinetic Isotope Effects in Oxygen in the Laboratory Dehydration of Magnesian Minerals. *J. Phys. Chem.* **2009**, *113*, 2212–2217. [CrossRef]

34. Mendybayev, R.A.; Richter, F.M.; Spicuzza, M.J.; Davis, A.M. Oxygen isotope fractionation during evaporation of Mg- and Si- rich CMAS-liquid in vacuum. *Lunar Planet. Inst. Sci. Conf. Abstr.* **2010**, *41*, 2725.

35. Bindeman, I.N.; Serebryakov, N.S.; Schmitt, A.K.; Vazquez, J.A.; Guan, Y.; Azimov, P.Y.; Astafiev, B.Y.; Palandri, J.; Dobrzhinetskaya, L. Field and microanalytical isotopic investigation of ultradepleted in ^{18}O Paleoproterozoic "Slushball Earth" rocks from Karelia, Russia. *Geosphere* **2014**, *10*, 308–339. [CrossRef]

36. Wilson, A.F.; Baksi, A.K. Widespread ^{18}O depletion in some precambrian granulites of Australia. *Precambrian Res.* **1983**, *23*, 33–56. [CrossRef]

37. Menant, A.; Jolivet, L.; Vrielynck, B. Kinematic reconstructions and magmatic evolution illuminating crustal and mantle dynamics of the eastern Mediterranean region since the late Cretaceous. *Tectonophysics* **2016**, *675*, 103–140. [CrossRef]

Article

Enigmatic Alluvial Sapphires from the Orosmayo Region, Jujuy Province, Northwest Argentina: Insights into Their Origin from in situ Oxygen Isotopes

Ian T. Graham [1,*], Stephen J. Harris [1], Laure Martin [2], Angela Lay [1] and Eduardo Zappettini [3]

[1] PANGEA Research Centre, School of Biological, Earth and Environmental Sciences, University of NSW, Sydney 2052, Australia
[2] Centre for Microscopy Characterisation and Analysis, The University of Western Australia, Perth 6009, Australia
[3] Argentine Geological-Mining Survey (SEGEMAR), Buenos Aires C1067ABB, Argentina
*   Correspondence: i.graham@unsw.edu.au; Tel.: +61-435-4296-445

Received: 21 May 2019; Accepted: 25 June 2019; Published: 27 June 2019

Abstract: This study sought to investigate in situ oxygen isotopes ($\delta^{18}O$) within alluvial colorless-white to blue sapphires from the Orosmayo region, Jujuy Province, NW Argentina, in order to provide additional constraints on their origin and most likely primary geological environment. Analyses were conducted using the in situ SIMS oxygen isotope technique on the same grains that were analyzed for their mineral inclusions and major and trace element geochemistry using EMPA and LA–ICP–MS methods in our previous study. Results show a significant range in $\delta^{18}O$ across the suite, from +4.1‰ to +11.2‰. Additionally, akin to their trace element chemistry, there is significant variation in $\delta^{18}O$ within individual grains, reaching a maximum of 1.6‰. Both the previous analyses and $\delta^{18}O$ results from this study suggest that these sapphires crystallized within the lower crust regime, involving a complex interplay of mantle-derived lamprophyres and carbonatites with crustal felsic rocks and both mantle- and crustal-derived metasomatic fluids. This study reinforces the importance of the in situ analysis of gem corundums, due to potential significant variation in major and trace element chemistry and ratios and even oxygen isotope ratios within discrete zones in individual grains.

Keywords: sapphires; corundum; in situ oxygen isotopes; Orosmayo Argentina; lamprophyre; secondary ion mass spectrometry (SIMS); carbonatite

1. Introduction

Since the seminal paper of Giuliani et al. (2005) [1], oxygen isotopes have been widely used in order to help constrain the likely geological environment of formation for gem sapphires and rubies, especially those from alluvial deposits whose primary formation is unknown [1–5], such as in the case of the sapphires from this study.

Although worldwide corundum oxygen isotope values cover a wide range from −27‰ (Khitostrov, Russia) to +23‰ (Mong Hsu, Myanmar), most are in the range of +3‰ to +21‰ [1,3,6–9]. This criterion has often been used to determine the geological origin of colored corundum, and especially the gem corundums, rubies and sapphires. However, there are only a few studies on in situ oxygen isotope analysis of gem corundums [10,11]. The next research frontier is the use of in situ oxygen isotopic compositions combined with detailed LA–ICP–MS and multivariate statistical methods to help "fingerprint" the geographic location (i.e., geographic typing) of gem corundums.

As there are a number of possible geological environments for gem corundum formation [3], any information which can help provide constraints on their origin is of great use. Trace element geochemistry of gem corundums has been widely used for this in the past, especially when distinct and/or unusual chemistry can provide even site-specific signatures useful for geographic typing [10–13].

In this study, we investigated the primary origin of the little-known alluvial colorless-white to blue sapphires from the Orosmayo region, Jujuy Province, NW Argentina using in situ oxygen isotope analysis. The discovery of alluvial sapphires in this region of NW Argentina dates back to 1991 during the exploration for alluvial gold in the Orosmayo region. These sapphires were first described by Zappettini and Mutti (1997) [14], and then in a more detailed study on the sapphires themselves by Harris et al. (2017) [15] using SEM, EMPA, and LA–ICP–MS techniques. A summary of the main findings of Harris et al. (2017) [15] is given in Table 1.

Table 1. Summary of main findings of Harris et al. (2017) [15] for the Orosmayo sapphire suite (bdl= below detection limit).

Color range	white, pale to deep blue, pale lilac, pale pink
Color distribution	oscillatory, sector (most common), star sapphires
Surface features	percussion marks, corrosion features
Mineral Inclusions	apatite, rutile, magnetite, phlogopite, trikalsilite, pyrite, baryte, monazite- (La), xenotime-(Y), zircon
Ga/Mg	0.3–10.4
Fe/Mg	4–117
Ti	47–8670
V	15–399
Cr	bdl-506
Ga	38–239
Mg	11–253
Elevated trace elements	Si, P, Ca, Y, La, Ce, Nd, Zn, Pb, Be, Nb, Ta, Sn, Th, U

Orosmayo (located at approximately 22°15′ S–66°25′ W) is located on the western slope of the Sierra de Rinconada, Jujuy Province, within the San Juan de Oro Basin of northwest Argentina (Figure 1; [14,15]). The area forms part of the *Puna* of northwestern Argentina which connects with the Altiplano of Bolivia to form a plateau [14,16]. The sapphires were sampled from stream sediments located at approximately 22°30′ S and 66°30′ W (see Harris et al. 2017; [15]).

2. Materials and Methods

The same exact grain mounts used in the previous study of Harris et al. (2017) [15] were also used in this study with the addition of a standard (PAR-1; see Wang et al. 2019 [11]), which was embedded into each polished mount before they were then repolished. This study analyzed 16 sapphire grains (AC-1 through to AC-16) obtained from the 37 original grains analyzed by Zappettini and Mutti (1997) [14]. The sapphires from this region range in size from a few mm up to 10 mm and all represent a single group in terms of their overall morphology and colors. However, for this study, the smaller grains were used in order to analyze as many grains as possible. The selected grains were mounted in two epoxy discs, with eight grains on each disc, and polished prior to analysis.

Secondary Ionization Mass Spectrometry (SIMS)

The oxygen isotope analyses were made using secondary ion mass spectrometry (SIMS) for the analysis of different areas of selected sapphire grains in situ to measure oxygen isotope ratios with per mil (‰) level precision. For this technique, oxygen ($^{18}O/^{16}O$) isotope ratios were analyzed using a Cameca IMS 1280 multi-collector ion microprobe within the Centre for Microscopy, Characterisation and Analysis (CMCA), University of Western Australia (UWA). The materials examined included the 16 variously colored and zoned sapphire grains (AC-1 to AC-16), which were embedded in the middle of one-inch-diameter resin mounts and polished flat. Results were calibrated using the standard PAR-1 of Wang et al. 2019 [11] embedded in a separate mount and measured before and after unknown mounts. Each analysis point is clearly marked on the sample images (Figure 2a,b).

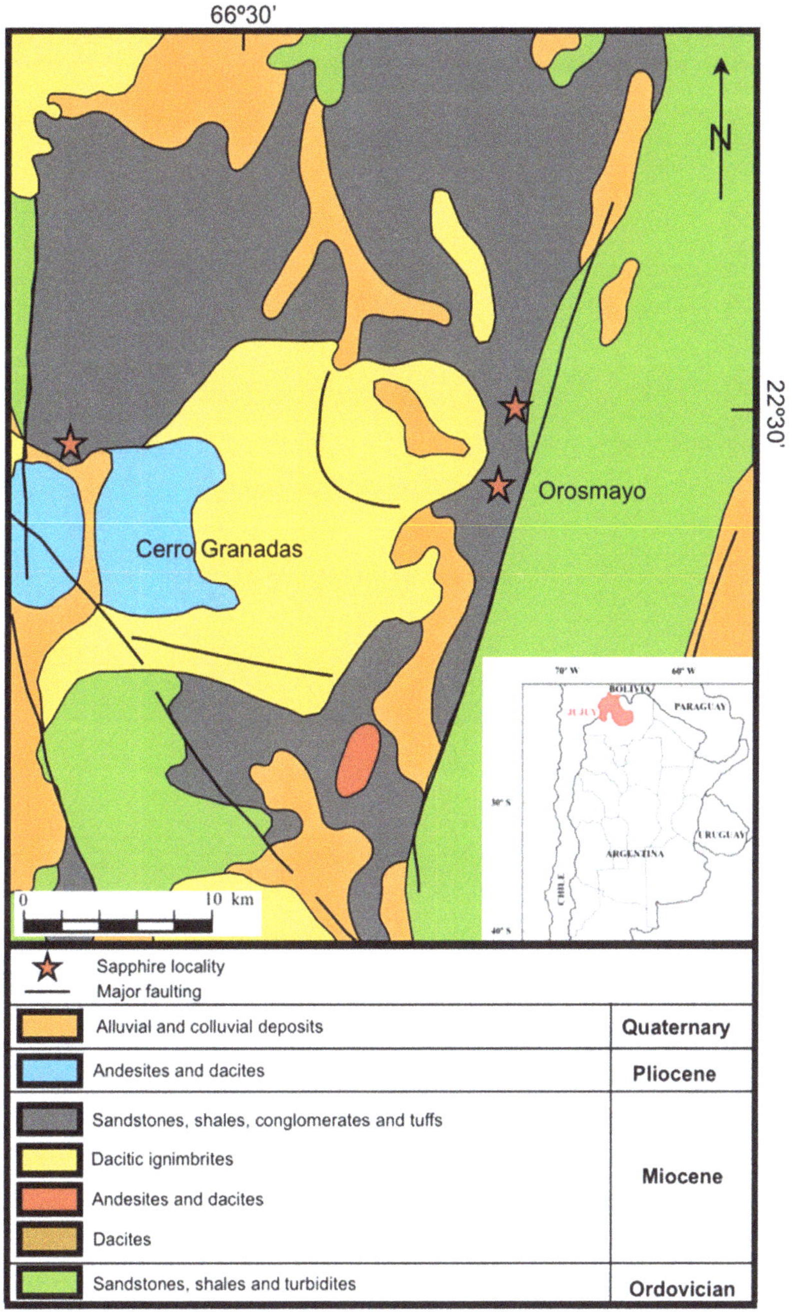

Figure 1. Map showing the location of Orosmayo in northwest Argentina (adapted from Harris et al. 2017) [15].

The sample mounts were carefully cleaned with detergent, distilled water, and ethanol in an ultrasonic bath and then coated with gold (30 nm in thickness) prior to SIMS O isotope analysis. For the oxygen isotopic analyses, secondary ions were sputtered from the sample by bombarding its surface with a Gaussian Cs^+ beam and a total impact energy of 20 keV. The surface of the sample was rastered with a 2.5 nA primary beam over a 15×15 µm area. An electron gun was used to ensure charge compensation during the analyses. Secondary ions were admitted in the double focusing mass spectrometer within a 100 µm entrance slit and focused in the center of a 4000 µm field aperture (100× magnification). They were energy-filtered using a 30 eV band pass with a 5 eV gap towards the high-energy side. ^{16}O and ^{18}O were collected simultaneously in multicollection mode in Faraday Cup detectors fitted with 10^{10} Ω and 10^{11} Ω, respectively. Each analysis included pre-sputtering over a 20 µm × 20 µm area for 30 s, and the automatic centering of the secondary ions in the field aperture, contrast aperture, and entrance slit, and consisted of 20 four-second cycles which give an average internal precision of ~0.16‰ (2 standard error (SE)).

External reproducibility during the analytical sessions was evaluated by repeating analyses in one single fragment of PAR-1. External reproducibility in this fragment was 0.3‰ and 0.4‰ (2σ) during the analytical sessions.

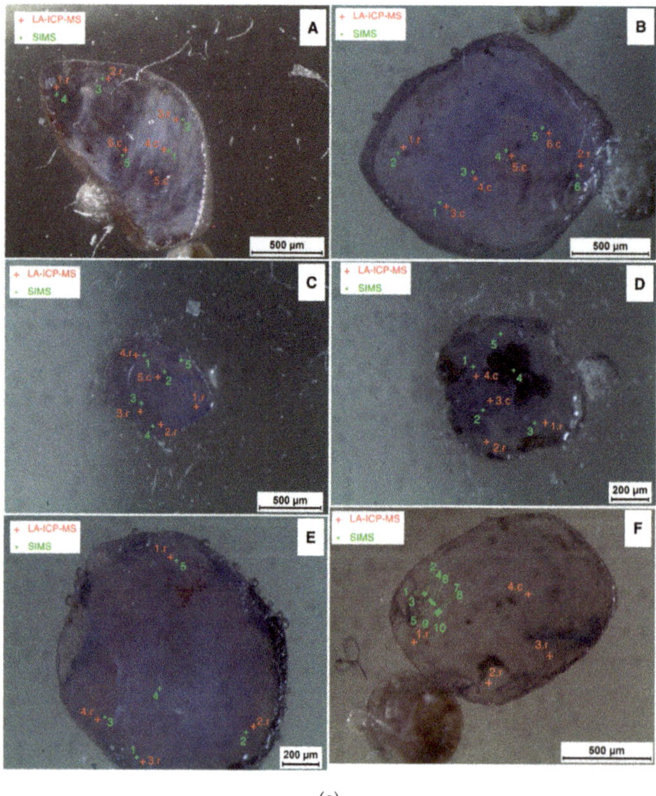

(a)

Figure 2. *Cont.*

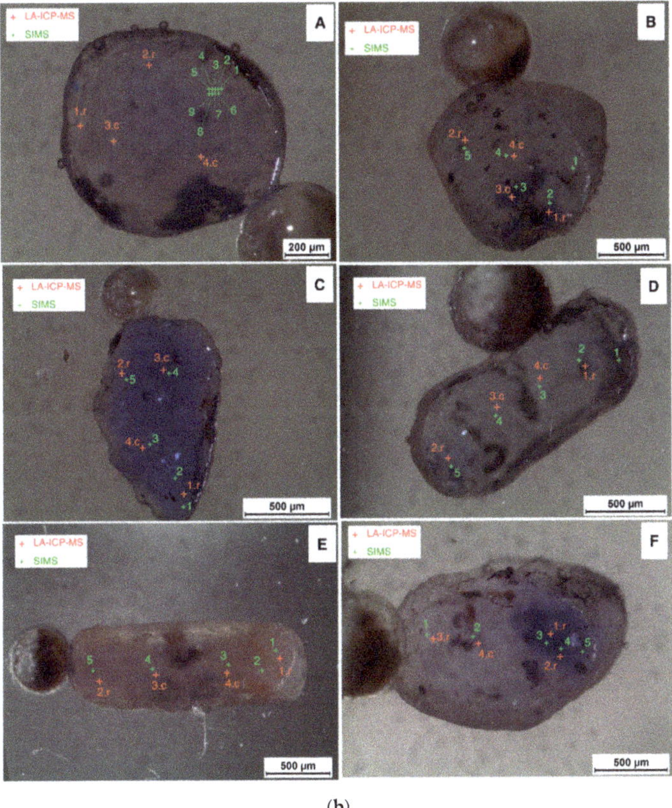

Figure 2. Images of the analyzed sapphires with both LA–ICP–MS and SIMS analysis spots indicated (**a**). A = AC-1, B = AC-2, C = AC-3, D = AC-4, E = AC-5 and F = AC-7. (**b**) A = AC-10, B = AC-11, C = AC-13, D = AC-14, E = AC-15 and F = AC-16).

3. Results

The high precision of oxygen isotope measurements using the SIMS method resulted in spot-to-spot reproducibility of 0.3‰ (2σ) during the analytical run, and individual spot analyses typically had errors of 0.5‰ (2σ). The oxygen isotope ratios $\delta^{18}O$ (reported relative to Vienna standard mean ocean water or VSMOW) are presented in Table 2. A summary of the results for each grain, along with characteristics of that grain are shown in Table 3. The complete SIMS oxygen isotope analytical results for the Orosmayo sapphires and PAR-1 ruby are given in Supplementary Materials Table S1. There was large variation between the different grains analyzed, overall from +4.1 to +11.2 $\delta^{18}O$‰, and a measurable variation within sample AC-3 up to 1.6‰ (Table 3). Although this variation is caused by the one outlier point (spot 3, also see Figure 3), sample AC-1 also showed a variation of 1.5‰ without any clear outliers (Table 3). There was no consistent zonation in terms of core to rim variation amongst the sapphires analyzed. This would suggest much more complex growth mechanisms than simply from core to rim, which is also suggested by the occurrence of complex sector-zoned sapphires and these have the highest oxygen isotope values. Other methods, such as a detailed cathodoluminescence study, would help in understanding the growth zonation and possible growth mechanisms. The lowest values correspond to apparently unzoned blue sapphires and the highest values mostly occur within sector zoned sapphires.

Table 2. $\delta^{18}O$ oxygen isotope results for grains AC-1 to AC-5, and AC-7, AC-10 to AC-11, AC-13 to AC-16.

Sapphire	Analysis Point	$\delta^{18}O$	2σ abs	Part of Crystal
AC-1	1	8.66	0.49	outer core
	2	7.52	0.49	outer rim
	3	7.66	0.47	outer rim
	4	8.08	0.48	outer rim
	5	8.97	0.46	inner rim
AC-2	1	5.24	0.48	outer rim
	2	5.68	0.50	outer rim
	3	5.76	0.46	outer core
	4	5.41	0.48	inner core
	5	5.43	0.46	outer core
	6	5.5	0.48	outer rim
AC-3	1	5.77	0.48	inner rim
	2	5.28	0.58	outer core
	3	4.12	0.50	inner rim
	4	5.61	0.48	outer rim
	5	5.4	0.46	outer rim
AC-4	1	5.69	0.51	inner rim
	2	5.33	0.49	inner rim
	3	4.97	0.48	outer rim
	4	5.89	0.49	outer core
	5	5.92	0.48	outer rim
AC-5	1	7.8	0.45	outer rim
	2	8.42	0.45	outer rim
	3	8.05	0.46	outer rim
	4	7.52	0.47	outer core
	5	8.71	0.47	outer rim
AC-7	1	6.09	0.45	outer rim
	2	5.42	0.46	outer rim
	3	6.14	0.50	inner rim
	4	6.02	0.47	inner rim
	5	6.81	0.47	inner rim
	6	5.42	0.49	inner rim
	7	6.59	0.46	inner rim
	8	6.66	0.47	inner rim
	9	5.56	0.51	inner rim
	10	5.47	0.48	inner rim
AC-10	1	9.28	0.46	inner rim
	2	9.42	0.46	inner rim
	3	9.44	0.45	inner rim
	4	9.02	0.46	inner rim
	5	9.28	0.45	inner rim
	6	8.95	0.47	inner rim
	7	9.27	0.45	inner rim
	8	9.21	0.46	inner rim
	9	9.02	0.45	inner rim
AC-11	1	11.22	0.46	inner rim
	2	10.94	0.45	outer rim
	3	10.88	0.46	outer core
	4	11.21	0.45	inner core
	5	10.65	0.46	inner rim
AC-13	1	5.68	0.44	outer rim
	2	5.46	0.48	inner rim
	3	5.97	0.46	outer core
	4	5.91	0.45	outer core
	5	5.96	0.45	outer rim
AC-14	1	6.18	0.47	outer rim
	2	6.23	0.47	outer core
	3	6.51	0.44	core
	4	6.68	0.48	core
	5	6.29	0.44	outer rim
AC-15	1	9.38	0.45	outer rim
	2	9.57	0.47	inner rim
	3	9.36	0.46	outer core
	4	8.91	0.45	outer core
	5	9.36	0.43	outer rim
AC-16	1	9.18	0.48	outer rim
	2	8.84	0.45	outer core
	3	8.65	0.45	outer core
	4	8.94	0.46	inner rim
	5	9.00	0.44	outer rim

Table 3. Summary of $\delta^{18}O$ results for the 12 samples analyzed.

Sapphire	Range	Range Variation	Average	Characteristics
AC-1	7.52–8.97	1.45	8.18	star sapphire
AC-2	5.24–5.76	0.52	5.50	magmatic zoning
AC-3	4.12–5.77	1.65	5.24	blue sapphire
AC-4	4.97–5.92	0.95	5.56	sector zoning
AC-5	7.52–8.71	1.19	8.10	blue-lilac
AC-7	5.42–6.81	1.39	6.02	pale lilac
AC-10	8.95–9.44	0.49	9.21	blue-lilac
AC-11	10.65–11.22	0.57	10.98	sector zoning
AC-13	5.46–5.97	0.51	5.80	deep blue
AC-14	6.18–6.68	0.50	6.38	sector zoning
AC-15	8.91–9.57	0.66	9.32	star sapphire
AC-16	8.65–9.18	0.53	8.92	sector zoning

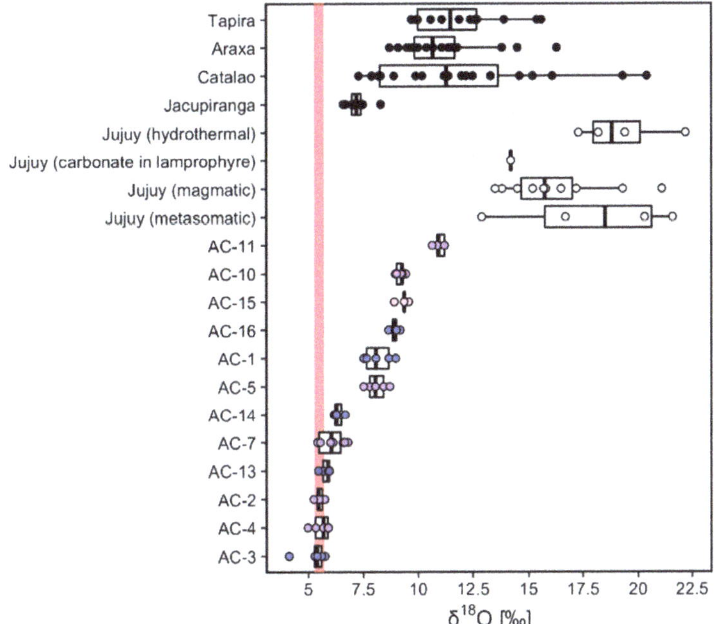

Figure 3. Diagram showing the range in oxygen isotope values for Jujuy sapphires (this study) in comparison to carbonatites and lamprophyres from the Jujuy region [17] and those of carbonatites from Brazil [18] and mantle $\delta^{18}O$ value (red band).

4. Discussion

4.1. Oxygen Isotope Groupings

Although there is a continuous spectrum in terms of trace element chemistry and $\delta^{18}O$ values, five main groupings are apparent: Group 1 ($n = 4$) comprises the blue sapphires and has $\delta^{18}O$‰ 4.1–5.9; Group 2 ($n = 2$) comprises pale lilac-blue sapphires with $\delta^{18}O$‰ 5.4–6.7; Group 3 ($n = 2$) consists of one star sapphire (AC-1) and a blue-lilac sapphire (AC-5) with $\delta^{18}O$‰ 7.5–8.9; Group 4 ($n = 3$) comprises a pale blue sapphire (AC-10), a star sapphire (AC-15) and a sector zone colorless-blue sapphire (AC-16) with $\delta^{18}O$‰ 8.6–9.6 and Group 5 ($n = 1$) is a sector zoned colorless-blue sapphire (AC-11) with $\delta^{18}O$‰ 10.6–11.2. These groupings, along with a summary of their trace element chemistry based on

Harris et al. (2017) [15] are shown in Table 4. Collectively, there is a range in $\delta^{18}O$ from 4.1‰ to 11.2‰ and a range in Ga/Mg of 0.3–10.4 (average range of 0.73–7.7) but no correlation or apparent relationship between Ga/Mg and the oxygen isotope values. Importantly, Baldwin et al. (2017) [19] found that sapphire megacrysts from the Siebenbirge Volcanic Field of Germany had a wide Ga/Mg of 1.0–23.4 and ascribed this to the influence of a carbonatitic melt in the crystallization of the sapphires. The range in Ga/Mg values for Jujuy Province sapphires at 0.30–10.40 fits well within this range and, based on this alone, would tentatively suggest a similar mechanism of formation. However, interpretation of Ga/Mg ratios in sapphires must be treated with caution as Sutherland et al. (2009) [20] found an even larger range in Ga/Mg for Australian basalt-hosted blue-green-yellow (BGY suite) sapphires with 0.24–31.60 for the Yarrowitch field and 0.17–193 for the Barrington field. This collectively lends support to the views of Baldwin et al. (2017) [19] and Palke et al. (2017) [21] who suggested that Ga/Mg ratios are not good discriminators for sapphire origin.

Table 4. Groupings based on $\delta^{18}O$‰ compositions with their main geochemical characteristics from Harris et al. (2017) [15].

Sapphire	Mg	Ti	V	Cr	Fe	Ga	Fe/Mg	Ga/Mg	Number of analyses
Group 1 (Blue Sapphires) with $\delta^{18}O$‰ from 4.12 to 5.92									
AC-2	38	73	64	456	789	140	22.3	4	6
AC-3	119	1987	154	359	1024	118	9.8	1.2	5
AC-4	103	1771	230	160	843	126	9.5	1.4	4
AC-13	162	428	155	91	1481	119	9.3	0.73	4
Group 2 (Pale Blue-Lilac Sapphires) with $\delta^{18}O$‰ from 5.42 to 6.68									
AC-7	87	1044	113	224	621	116	7.8	1.5	4
AC-14	20	190	16	27	1348	121	86	7.7	4
Group 3 (Star Sapphire AC-1; Blue-Lilac Sapphire AC-5) with $\delta^{18}O$‰ from 7.52 to 8.97									
AC-1	259	6619	187	45	927	146	4.3	0.8	6
AC-5	42	1077	67	209	677	99	17.3	2.5	4
Group 4 (Pale Blue Sapphire AC-10; Star Sapphire AC-15; Sector Zoned Colorless to Blue Sapphire AC-16) with $\delta^{18}O$‰ from 8.65 to 9.57									
AC-10	71	1037	76	89	777	105	11	1.5	4
AC-15	200	6831	163	309	1278	160	6.5	0.8	4
AC-16	38	348	63	202	581	99	17	2.9	4
Group 5 (Sector Zoned Colorless to Blue Sapphire) with $\delta^{18}O°/_{oo}$ from 10.65 to 11.22									
AC-11	106	1741	230	16	517	128	5	1.3	4

4.2. What Do the Oxygen Isotopes Tell Us about the Primary Source for Jujuy Sapphires?

A framework on the interpretation of the geological origin of gem corundums using the $^{18}O/^{16}O$ ratio first proposed by Giuliani et al. (2005) [1] is now widely adopted. Based on this framework, rubies and pink sapphires can be classified into 5 categories based on their $\delta^{18}O$ value range. 1. Mafic gneiss hosted 2.9–3.8‰; 2. Mafic-ultramafic rocks (amphibolite, serpentinite) 3.2–6.8‰; 3. Desilicated pegmatites 4.2–7.5‰; 4. Shear zones cross-cutting ultramafic lenses and pegmatites within sillimanite gneisses 11.9–13.1‰; 5. Marble-hosted rubies 16.3–23‰. This framework has been further validated by numerous subsequent corundum oxygen isotope studies [3–5,11,13,20,22–32]. A summary of worldwide oxygen isotope values for gem sapphires, along with their host rock and Ga/Mg ratios, is presented in Table 5. In a study of sapphires from the French Massif Central basaltic field, Giuliani et al. (2009) [25] found two distinctive groups. The first group with $\delta^{18}O$ from 4.4‰ to 6.8‰ were believed to have crystallized from felsic magmas in the upper mantle while the second group with higher $\delta^{18}O$ from 7.6‰ to 13.9‰ were believed to have been derived from lower crustal granulites. However, the oxygen isotope range for Jujuy sapphires spans across both of these groupings. Critically, the study of Harris et al. (2017) [15] showed that for Jujuy sapphires, there is a continuous spectrum in composition with significant trace element variation within single grains. This strongly suggests crystallization from the same source but under variable metasomatic conditions.

Table 5. Summary of worldwide oxygen isotope ranges and Ga/Mg for some sapphires (n.d = not determined).

Color	Host Rock	d^{18}O‰	Ga/Mg	Reference
Blue, yellow, brown	alkali basalts	4.6–5.2	5.7–11.3	[31]
Pink	kimberlites	4.3–5.4	1.9–3.9	[31]
Blue	lamprophyre	5.4–6.8	0.3–0.5	[21]
Blue, yellow, green, colorless, brown	alkali basalts	2.7–6.9	n.d	[29]
Blue, green, yellow	alkali basalts	4.5–5.6	3–18	[13]
Blue, grey-pink, pink	alkali basalts	3.8–5.9	0.9–4.8	[26]
Blue, violet	lamprophyre	4.5–7.7	0.2–0.6	[33]
Blue, green, pink, colorless	alkali basalts	4.4–6.8	n.d	[25]
Blue, green, grey, white	alkali basalts	7.6–13.9	n.d	[25]
Blue, grey, brown	alkali basalts	n.d	1.0–23.4	[19]
Blue, violet	alluvial	2.6–6.8	0.6–9.0	[34]
Blue, green	alkali basalts	4.3–6.4	n.d	[5]
Blue, grey, violet, red	syenite pegmatites	2.0–10.6	n.d	[5]
Blue, green, yellow	alkali basalts	1.34–39.29	5.3–5.4	[20]
Blue, green, yellow	alkali basalts	6.6–193.0	5.2–5.8	[20]
Blue, green, yellow, pink	alkali basalts	4.3–8.4	n.d	[1]
Blue, grey	amphibolite	4.2–4.9	n.d	[1]
Blue	desilicated pegmatite in mafic rocks	10.9–11.2	n.d	[1]
Yellow, green, colorless, brown	syenite/anorthoclasite	5.2–7.8	n.d	[1]
Blue, grey, colorless	skarn	7.7–10.7	n.d	[1]

In a study of carbonatites from Brazilian alkaline complexes, Santos and Clayton (1995) [18] found that the non-fenitized carbonatites from Jacupiranga had significantly lower oxygen isotope ratios (6.6–7.3) compared to the fenitized (i.e., K and Na metasomatism; Elliott et al. 2018, [35]) carbonatites of Araxa (9.7–16.3), Catalao (7.3–19.3), and Tapira (9.7–15.4). Zappettini et al. (1998) [17] analyzed a range of carbonatites from Jujuy Province and found that oxygen isotope values for magmatic carbonatites (13.5–21.1), metasomatic carbonatites (12.9–21.6), and hydrothermally altered carbonatites (17.3–22.2) overlapped each other but that magmatic carbonatites had lower values in general. They also analyzed one carbonatized lamprophyre sample which gave a value of 14.2 within the field of the magmatic carbonatites. The range in oxygen isotope compositions for Jujuy sapphires is much lower than all of these analyzed carbonatites and the lamprophyre from this region [17]. However, Harmon et al. (1981) [36] analyzed volcanic rocks from Cerro Galan in Catamarca Province, 400 km to the south of Jujuy, and found that these calc-alkaline volcanics had a range in oxygen isotope values from 8.3 to 10.7, and had formed directly from mantle-derived magmas but with likely partial crustal assimilation during ascent. This range is well within that of Groups 3–5 from the Jujuy suite. Additionally, the Brazilian fenitized carbonatites have a very similar range and would tentatively suggest that there was significant metasomatic influence in the Jujuy sapphire-forming environment, especially for the sapphires with the highest oxygen isotope ratios (Figure 3). It also suggests influence from an additional underlying deep source that had lower values than those for the carbonatites previously analyzed from Jujuy Province. This metasomatic influence would have been highly variable, as suggested by the wide range in oxygen isotope ratios and trace element contents and ratios for the Jujuy sapphires as shown by Harris et al. (2017) [15].

As suggested by Baldwin et al. (2017) [19], a possible origin of sapphires from carbonatitic melts could be identified by undertaking oxygen isotope measurements on the sapphires as, in general, carbonate-associated sapphires generally have higher isotopic ratios [1], as seen for the higher range amongst the Jujuy sapphires. However, interpreting such data would be problematic as there are no data available on oxygen isotope fractionation between corundum and co-existing carbonatitic melt [19].

4.3. Origin of Jujuy Sapphires

Any model proposed for the genesis of the alluvial Jujuy sapphires needs to account for (1) a predominant metamorphic source signature, and a less prominent mafic-ultramafic association and contrasting felsic/acidic association; (2) elevated Si contents, and unusually high Ga content for metamorphic sapphires (i.e., typically >100); (3) enrichments in Be, Nb, Ta, Sn, Th, and U within syngenetic rutile growths; (4) elevated light rare earth element (LREE) concentrations; (5) a bimodal mineral inclusion suite, including sulfides, feldspar, phosphates, and likely LREE-bearing minerals; (6) the range in Ga/Mg from 0.88 to 7.33; (7) significant variation in trace element contents, ratios, and oxygen isotope ratios within single grains; and (8) a wide but continuous range in oxygen isotope values from 4.1‰ to 11.2‰.

Given the presence of LREE-bearing inclusions and elevated LREE concentrations, it is possible that these elevated contents were the result of a carbonatitic parental melt. This is possible given reports of REE-bearing carbonatites within Jujuy Province. Based on $\delta^{13}C$ and $\delta^{18}O$ isotopic studies, Zappettini et al. (1998) [17] suggested that these carbonatites formed by the crystallization of carbonate magma with subsequent formation of metasomatic and hydrothermal carbonatitic veins. This is akin to the variably fenitized carbonatites from Brazil as shown above. Similarly, nearby Rio Grande monchiquites (163 ± 9 Ma) were shown to have been derived from heterogeneous enriched lithospheric mantle metasomatized by carbonatite fluids involving significant magma mixing (Hauser et al. 2010) [37]. Similar processes, involving exchanges between a carbonatite (Si-poor) and mantle-derived acidic magma (Si- and Ga-rich) could, in turn, be responsible for sapphire-bearing lamprophyre formation (e.g., the plumasite model; Peucat et al. 2007) [38] within Jujuy Province (e.g., Zappettini 1989) [39]. Factor analysis on Jujuy sapphires by Harris et al. (2017) [15] showed a largely continuous linear grouping from the metasomatic field of Giuliani et al. (2014) [4] into their plumasite field. The one exception to this trend was the star sapphires (i.e., the sapphires with the highest oxygen isotope ratios) which formed a distinctly separate grouping within the plumasite field of Giuliani et al. (2014) [4]. This once again provides supporting evidence for a sapphire-crystallization environment involving highly variable metasomatic interaction.

Collectively, the in situ study of Harris et al. (2017) [15] and this in situ study provide crucial evidence for understanding the origin of Jujuy sapphires. Based on their study of surface features, mineral inclusions, and major and trace element chemistry, Harris et al. (2017) [15] concluded that the Jujuy sapphires crystallized via metasomatic fluid exchange between an evolved crustal felsic system and intruding coeval mantle-derived carbonatized lamprophyres as part of a regional-wide alkaline magmatic event within NW Argentina during the Lower Cretaceous. Oxygen isotope ratios for the Jujuy sapphires largely support this model but allow us to go one step further and suggest that the influence of carbonatites on sapphire crystallization was greater than previously thought, and that the localized sapphire-former crustal environment was more complex, involving episodic interactions between mantle-derived lamprophyres, mantle-derived carbonatites, lower crustal felsic magmas and, most likely, both mantle and lower crustal metasomatic fluids. As stated by Harris et al. (2017) [15], such a complex interplay of widely diverse magma and fluid compositions would induce localized desilicification reactions that would lead to localized sapphire crystallization. Based on all of the available evidence, we suggest that most of these interactions would have occurred at the crust/mantle boundary, with initial crystallization within the upper mantle (i.e., sapphires with lower oxygen isotope signatures of 4.12–6.50) then continuing crystallization within the lower crust (i.e., sapphires with oxygen isotope signatures >6.50).

The wide range in chemistry and oxygen isotope compositions but within a single comagmatic group is explained by the degree and type of interaction (i.e., carbonatite–lamprophyre vs. carbonatite–granite) between these magmas/fluids, with some crystallizing at lower crustal levels as suggested by Palke et al. [21,34] for sapphires from Montana, USA. The large within-grain variation seen in trace element contents, trace element ratios, and oxygen isotope ratios is highly suggestive of a geochemically dynamic environment with rapidly changing degrees of interaction between the various diverse fluid and magmatic components, as described above.

This model accounts for the dominant metasomatic signature, while also accounting for the mafic–ultramafic (i.e., carbonatitic and lamprophyric magmas) and felsic (i.e., mantle-derived enriched acidic magmas and/or granitic anatectic melts) associations. It also accounts for the bimodal mineral inclusion suite, enrichment in LREE, Ga/Mg ratios, and range in trace element and oxygen isotope compositions within a single sapphire group that crystallized at the same time but across the crust–mantle boundary. The oxygen isotope values and range provide additional support to the initial hypothesis of Zappettini et al. (1997) [40], who proposed that the Jujuy sapphires formed via metasomatic exchanges between REE-bearing carbonatites and an evolved felsic magma during the Cretaceous (150–110 Ma), and were then subsequently incorporated as xenocrysts within lamprophyre dykes.

5. Conclusions

In situ SIMS oxygen isotope analyses on the Orosmayo sapphire suite revealed that they have values spanning a continuous range from +4.1 to +11.2 $\delta^{18}O$‰, with measurable variation within individual sapphires. Based on their trace element geochemistry, geochemical ratios, and oxygen isotopes, although there is a continuous spectrum of compositions, the sapphires can be subdivided into five suites reflecting different degrees of interaction between the various fluid/magmatic components involved in sapphire formation. Both the previous geochemical work and this study suggest that the Jujuy sapphires crystallized through the crust–mantle boundary in a localized environment involving episodic interactions between mantle-derived lamprophyres, mantle-derived carbonatites, lower crustal felsic magmas and, most likely, both mantle and crustal metasomatic fluids. This complex interplay of widely diverse magma and fluid compositions would induce localized desilicification reactions, leading to localized sapphire crystallization. This study reinforces the importance of carbonatites on sapphire crystallization. The significant range in trace element chemistry and oxygen isotope ratios within individual grains stresses the critical importance of undertaking in situ analysis on gem corundums in order to fully understand their environment of formation and to aid in their geographic typing.

Supplementary Materials: The following are available online at http://www.mdpi.com/2075-163X/9/7/390/s1, Table S1: Secondary Ion Mass Spectrometry (SIMS) Results.

Author Contributions: I.T.G. wrote the manuscript, was responsible for the conception of this project and interpreted the results of the analyses. S.J.H. provided input into the discussion and constructed some of the figures. L.M. Provided technical input and ran the SIMS analyses. A.L. provided technical input. E.Z. collected the samples and provided geological expertise on the Jujuy region of NW Argentina.

Funding: This research received no external funding.

Acknowledgments: The authors would like to thank Joanne Wilde, formerly of the School of BEES, for making the polished mounts required for SIMS analysis. The authors acknowledge the facilities and the scientific and technical assistance of the Australian Microscopy & Microanalysis Research Facility at the Centre for Microscopy, Characterisation & Analysis, The University of Western Australia, a facility funded by the University, State and Commonwealth Governments.

Conflicts of Interest: The authors declare no conflict of interest.

References

1. Giuliani, G.; Fallick, A.E.; Garnier, V.; France-Lanord, C.; Ohnenstetter, D.; Schwarz, D. Oxygen isotope composition as a tracer for the origins of rubies and sapphires. *Geology* **2005**, *33*, 249–252. [CrossRef]
2. Yui, T.-F.; Wu, C.-M.; Limtrakun, P.; Sricharin, P.; Boonsoong, A. Oxygen isotope studies on placer sapphires and ruby in the Chanthaburi-Trat alkali basalt gem field, Thailand. *Lithos* **2006**, *86*, 197–211. [CrossRef]
3. Giuliani, G.; Ohnenstetter, D.; Garnier, V.; Fallick, A.E.; Rakotondrazafy, M.; Schwarz, D. The geology and genesis of gem corundum deposits. In *Geology of Gem Deposits*; Groat, L.A., Ed.; Mineralogical Association of Canada: Quebec City, QC, Canada, 2007; Volume 37, pp. 23–78. ISBN 978092194375.

4. Giuliani, G.; Ohnenstetter, D.; Fallick, A.E.; Groat, L.; Fagan, J. The geology and genesis of gem corundum deposits. In *Geology of Gem Deposits (2nd edition)*; Groat, L.A., Ed.; Mineralogical Association of Canada: Quebec City, QC, Canada, 2014; Volume 44, pp. 29–112. ISBN 9780921294542.

5. Vysotskiy, S.V.; Nechaev, V.P.; Kissin, A.Y.; Yakovenko, V.V.; Ignat'ev, A.V.; Velivetskaya, T.A.; Sutherland, F.L.; Agoshkov, A.I. Oxygen isotopic composition as an indicator of ruby and sapphire origin: A review of Russian occurrences. *Ore Geol. Rev.* **2015**, *68*, 164–170. [CrossRef]

6. Yui, T.F.; Zaw, K.; Limtrakun, P. Oxygen isotope composition of the Denchai sapphire, Thailand: A clue to its enigmatic origin. *Lithos* **2003**, *67*, 153–161. [CrossRef]

7. Bindeman, I.N.; Schmitt, A.K.; Evans, D.A.D. Limits of hydrosphere-lithosphere interaction: Origin of the lowest-known o18O silicate rock on Earth in the Paleoproterozoic Karelian rift. *Geology* **2010**, *38*, 631–634. [CrossRef]

8. Bindeman, I.N.; Serebryakov, N.S. Geology, Petrology and O and H isotope geochemistry of remarkably ^{18}O depleted Paleoproterozoic rocks of the Belomorian Belt, Karelia, Russia, attributed to global glaciation 2.4 Ga. *Earth Planet. Sci. Lett.* **2011**, *306*, 163–174. [CrossRef]

9. Krylov, D.P. ^{18}O/^{16}O ratios in the corundum-bearing rocks of Khitostrov, northern Karelia. *Dokl. Earth Sci.* **2008**, *419*, 453–456. [CrossRef]

10. Sutherland, F.L.; Graham, I.T.; Harris, S.J.; Coldham, T.; Powell, W.; Belousova, E.A.; Martin, L. Unusual ruby-sapphire transition in alluvial megacrysts, Cenozoic basaltic gem field, New England, New South Wales, Australia. *Lithos* **2017**, *278*, 347–360. [CrossRef]

11. Wang, K.K.; Graham, I.T.; Martin, L.; Voudouris, P.; Giuliani, G.; Lay, A.; Harris, S.J.; Fallick, A.E. Fingerprinting Paranesti rubies through oxygen isotopes. *Minerals* **2019**, *9*, 91. [CrossRef]

12. Sutherland, F.L.; Abduriyim, A. Geographic typing of gem corundum: A test case from Australia. *J. Gemmol.* **2009**, *31*, 203–210. [CrossRef]

13. Sutherland, F.L.; Coenraads, R.R.; Abduriyim, A.; Meffre, S.; Hoskin, P.W.O.; Giuliani, G.; Beattie, R.; Wuhrer, R.; Sutherland, G.B. Corundum (sapphire) and zircon relationships, Lava Plains gem fields, NE Australia: Integrated mineralogy, geochemistry, age determination, genesis and geographical typing. *Mineral. Mag.* **2015**, *79*, 545–581. [CrossRef]

14. Zappettini, E.; Mutti, D. Zafiros de la Puna Argentina: Un Potencial recurso minero. *Recur. Miner.* **1997**, *2*, 33–62.

15. Harris, S.J.; Graham, I.T.; Lay, A.; Powell, W.; Belousova, E.; Zappettini, E. Trace element geochemistry and metasomatic origin of alluvial sapphires from the Orosmayo region, Jujuy Province, northwest Argentina. *Can. Mineral.* **2017**, *55*, 595–617. [CrossRef]

16. Chernicoff, C.J.; Richards, J.P.; Zappettini, E.O. Crustal lineament control on magmatism and mineralization in northwestern Argentina: Geological, geophysical, and remote sensing evidence. *Ore Geol. Rev.* **2002**, *21*, 127–155. [CrossRef]

17. Zappettini, E.O.; Rubiolo, D.; Hubberten, H.W. *Carbon and Oxygen Isotopes in Carbonatites from Puna, Jujuy and Salta, Argentina*; Congreso Nacional de Geologia Economica: Buenos Aires, Argentina, 1998; Volume 2. (In Spanish)

18. Santos, R.V.; Clayton, R.N. Variations of oxygen and carbon isotopes in carbonatites: A study of Brazilian alkaline complexes. *Geochim. Cosmochim. Acta* **1995**, *59*, 1339–1352. [CrossRef]

19. Baldwin, L.C.; Tomaschek, F.; Ballhaus, C.; Gerdes, A.; Fonseca, R.O.C.; Wirth, R.; Geisler, T.; Nagel, T. Petrogenesis of alkaline basalt-hosted sapphire megacrysts: Petrological and geochemical investigations of in situ sapphire occurrences from the Siebengebirge Volcanic Field, Germany. *Contrib. Mineral. Petrol.* **2017**, *172*, 1–27. [CrossRef]

20. Sutherland, F.L.; Zaw, K.; Meffre, S.; Giuliani, G.; Fallick, A.E.; Graham, I.T.; Webb, G.B. Gem-corundum megacrysts from east Australian basalt fields: Trace elements, oxygen isotopes and origins. *Aust. J. Earth Sci.* **2009**, *56*, 1003–1022. [CrossRef]

21. Palke, A.; Renfro, N.D.; Berg, R.B. Melt inclusions in alluvial sapphires from Montana, USA: Formation of sapphires as a restitic component of lower crustal melting? *Lithos* **2017**, *278*, 43–53. [CrossRef]

22. Giuliani, G.; Fallick, A.E.; Rakotondrazafy, M.; Ohnenstetter, D.; Andriamamonjy, A.; Rakotosamizanany, S.; Ralantoarison, T.; Razanatseheno, M.; Dunaigre, C.; Schwarz, D. Oxygen isotope systematics of gem corundum deposits in Madagascar: Relevance for their geological origin. *Miner. Depos.* **2007**, *42*, 251–270. [CrossRef]

23. Sutherland, F.L.; Duroc-Danner, J.M.; Meffre, S. Age and origin of gem corundum and zircon megacrysts from the Mercaderes–Rio Mayo area, South-west Colombia, South America. *Ore Geol. Rev.* **2008**, *34*, 155–168. [CrossRef]

24. Garnier, V.; Giuliani, G.; Ohnenstetter, D.; Fallick, A.E.; Dubessy, J.; Banks, D.; Vinh, H.Q.; Lhomme, T.; Maluski, H.; Pecher, A.; et al. Marble-hosted ruby deposits from central and southeast Asia: Towards a new genetic model. *Ore Geol. Rev.* **2008**, *34*, 169–191. [CrossRef]

25. Giuliani, G.; Fallick, A.; Ohnenstetter, D.; Pegere, G. Oxygen isotopes compositions of sapphires from the French Massif Central: Implications for the origin of gem corundum in basaltic fields. *Miner. Depos.* **2009**, *44*, 221–231. [CrossRef]

26. Uher, P.; Giuliani, G.; Szakall, S.; Fallick, A.; Strunga, V.; Vaculovic, T.; Ozdin, D.; Greganova, M. Sapphires related to alkali basalts from the Cerova Highlands, Western Carpathians (southern Slovakia): Composition and origin. *Geol. Carpathica* **2012**, *63*, 71–82. [CrossRef]

27. Vysotskiy, S.V.; Ignat'ev, A.V.; Levitskii, V.I.; Nechaev, V.P.; Velvetskaya, T.A.; Yakovenko, V.V. Geochemistry of stable oxygen and hydrogen isotopes in minerals and corundum-bearing rocks in Northern Karelia as an indicator of their unusual genesis. *Geochem. Int.* **2014**, *52*, 773–782. [CrossRef]

28. Sutherland, F.L.; Zaw, K.; Meffre, S.; Yui, T.F.; Thu, K. Advances in trace element "fingerprinting" of gem corundum, ruby and sapphire, Mogok Area, Myanmar. *Minerals* **2014**, *5*, 61–79. [CrossRef]

29. Rakotosamizanany, S.; Giuliani, G.; Ohnenstetter, D.; Rakotondrazafy, A.F.M.; Fallick, A.E.; Paquette, J.-L.; Tiepolo, M. Chemical and oxygen isotopic compositions, age and origin of gem corundums in Madagascar alkali basalts. *J. Afr. Earth Sci.* **2014**, *94*, 156–170. [CrossRef]

30. Zaw, K.; Sutherland, L.; Yui, T.F.; Meffre, S.; Thu, K. Vanadium-rich ruby and sapphire within Mogok Gemfield, Myanmar: Implications for gem color and genesis. *Miner. Depos.* **2015**, *50*, 25–39. [CrossRef]

31. Giuliani, G.; Pivin, M.; Fallick, A.E.; Ohnenstetter, D.; Song, Y.; Demaiffe, D. Geochemical and oxygen isotope signatures of mantle corundum megacrysts from the Mbuji-Mayi kimberlite, democratic Republic of Congo, and the Changle alkali basalt, China. *Comptes Rendus Geosci.* **2015**, *347*, 24–34. [CrossRef]

32. Wong, J.; Verdel, C. Tectonic environments of sapphire and ruby revealed by a global oxygen isotope compilation. *Int. Geol. Rev.* **2018**, *60*, 188–195. [CrossRef]

33. Palke, A.C.; Wong, J.; Verdel, C.; Avila, J.N. A common origin for Thai/Cambodian rubies and blue and violet sapphires from Yogo Gulch, Montana, U.S.A.? *Am. Mineral.* **2018**, *103*, 469–479. [CrossRef]

34. Palke, A.C.; Renfro, N.D.; Berg, R.B. Origin of sapphires from a lamprophyre dike at Yogo Gulch, Montana, USA: Clues from their melt inclusions. *Lithos* **2016**, *260*, 339–344. [CrossRef]

35. Elliott, H.A.L.; Wall, F.; Chakhmouradian, A.R.; Siegfried, P.R.; Dahlgren, S.; Weatherley, S.; Finch, A.A.; Marks, M.A.W.; Dowman, E.; Deady, E. Fenites associated with carbonatite complexes: A review. *Ore Geol. Rev.* **2018**, *93*, 38–59. [CrossRef]

36. Harmon, R.S.; Thorpe, R.S.; Francis, P.W. Petrogenesis of Andean andesites from combined O-Si isotope relationships. *Nature* **1981**, *290*, 396–399. [CrossRef]

37. Hauser, N.; Matteini, M.; Omarini, R.H.; Pimentel, M.M. Constraints on metasomatised mantle under Central South America: Evidence from Jurassic alkaline lamprophyre dykes from the Eastern Cordillera, NM Argentina. *Mineral. Petrol.* **2010**, *100*, 153–184. [CrossRef]

38. Peucat, J.J.; Ruffault, P.; Fritsch, E.; Bouhnik-La Coz, M.; Simonet, C.; Lasnier, B. Ga/Mg ratios as a new geochemical tool to differentiate magmatic from metamorphic blue sapphire. *Lithos* **2007**, *98*, 261–274. [CrossRef]

39. Zappettini, E. Geologia y Metalogenesis de la Region Comprendida Entre las Localidades de Santa Ana y Cobres, Provincias de Jujuy y Salta, Republica Argentina. Ph.D. Thesis, Facultad de Ciencias Exactas y Naturales, Universidad de Buenos Aires, Buenos Aires, Argentina, 1989.

40. Zappettini, E.; Mutti, D.; Bernhardt, H.J. Genesis de Zafiro y hercinita de la puna argentina. *Actas* **1997**, *2*, 1598–1602.

 *minerals*

Article

Chemical Characteristics of Freshwater and Saltwater Natural and Cultured Pearls from Different Bivalves

Stefanos Karampelas *, Fatima Mohamed, Hasan Abdulla, Fatema Almahmood, Latifa Flamarzi, Supharart Sangsawong and Abeer Alalawi

Bahrain Institute for Pearls & Gemstones (DANAT), WTC East Tower, P.O. Box 17236 Manama, Bahrain; Fatima.Mohamed@danat.bh (F.M.); Hasan.Abdulla@danat.bh (H.A.); Fatema.Almahmood@danat.bh (F.A.); Latifa.Flamarzi@danat.bh (L.F.); Supharart.Sangsawong@danat.bh (S.S.); Abeer.Alalawi@danat.bh (A.A.)
* Correspondence: Stefanos.Karampelas@danat.bh

Received: 19 April 2019; Accepted: 5 June 2019; Published: 12 June 2019

Abstract: The present study applied Laser Ablation-Inductively Coupled Plasma-Mass Spectrometry (LA-ICP-MS) on a large number of natural and cultured pearls from saltwater and freshwater environments, which revealed that freshwater (natural and cultured) pearls contain relatively higher quantities of manganese (Mn) and barium (Ba) and lower sodium (Na), magnesium (Mg) and strontium (Sr) than saltwater (natural and cultured) pearls. A few correlations between the host animal's species and chemical elements were found; some samples from *Pinctada maxima* (*P. maxima*) are the only studied saltwater samples with ^{55}Mn >20 ppmw, while some *P. radiata* are the only studied saltwater samples with ^{24}Mg <65 ppmw and some of the *P. imbricata* are the only studied saltwater samples with ^{137}Ba >4.5 ppmw. X-ray luminescence reactions of the studied samples has confirmed a correlation between its yellow-green intensity and manganese content in aragonite, where the higher Mn^{2+} content, the more intense the yellow-green luminescence becomes. Luminescence intensity in some cases is lower even if manganese increases, either because of pigments or because of manganese self-quenching. X-ray luminescence can be applied in most cases to separate saltwater from freshwater samples; only samples with low manganese content (^{55}Mn <50 ppmw) might be challenging to identify. One of the studied natural freshwater pearls contained vaterite sections which react by turning orange under X-ray due to a different coordination of Mn^{2+} in vaterite than that in aragonite.

Keywords: pearls; freshwater; saltwater; LA-ICP-MS; X-ray luminescence

1. Introduction

Pearls are probably the most appreciated organogenic gems and they are either natural or cultured. Natural pearls (NPs) are secreted accidentally, without human intervention within naturally formed sacs (cysts made of epithelium cells), by molluscs such as bivalves or gastropods and very rarely also by cephalopods. Cultured pearls (CPs) are formed by molluscs within a pearl sac (cyst) produced with human intervention; e.g., after transplantation of epithelial cells cut from the mantle—a.k.a. tissue—(with or without the implantation of a bead) by human. Pearls are also classified, following their external appearance, into nacreous and non-nacreous forms. Under an optical microscope, nacreous (natural and cultured) pearls show terrace like structures (sometimes looking like fingerprints) composed of aragonite and organic matter (mixture of beta-chitin and acidic glycoproteins) stacked in a "brick-wall" pattern (i.e., sheet nacre). All pearls without nacreous appearance are considered non-nacreous. The vast majority of natural and cultured pearls used in jewellery are found in bivalves and they have a nacreous surface which is entirely made out of aragonite [1].

Natural and cultured pearls may be separated following the growth environment of their host mollusc; the environment is either freshwater (i.e., living in rivers or lakes; FW) or saltwater (i.e., living in sea; SW). Usually, the implanted bead in cultured pearls (both SW and FW) is cut from a freshwater bivalve shell [2].

Freshwater cultured and natural pearls contain more manganese (Mn) than their saltwater counterparts. As a consequence, eye visible (nowadays captured with a digital camera in most cases) reactions under X-rays (a.k.a. X-ray luminescence) are different for samples found in different water environments. Freshwater (natural and cultured) pearls luminate a green-yellow colour and saltwater pearls (natural and cultured without bead) remain inert [3–5]. However, some cultured saltwater pearls with a bead (made out of freshwater shell) can also give a green–yellow form of luminescence under X-rays. This reaction is due to the manganese content of freshwater bead and the relatively thin nacre (i.e., thickness of nacre material covering the bead) [5]. Cathodoluminescence (CL) microscopy and spectroscopy were also used for pearl characterization, and it was suggested that Mn^{2+} differs from cultured freshwater to natural freshwater and to natural saltwater pearls [6].

Energy dispersive X–ray fluorescence (EDXRF), a non-destructive method commonly used on gems, is applied to separate freshwater from saltwater samples [7–9]. Saltwater samples contain more Sr and less Mn than freshwater samples. The SrO/MnO ratio was suggested to be used for freshwater and saltwater pearls separation, as it is >12 for saltwater pearls and <12 for freshwater pearls [8]. EDXRF is also used to detect treatments used on natural and cultured pearls to improve their colour—with inorganic substances such as silver, iodine, bromine etc. [10,11].

Few examples of Laser Ablation-Inductively Coupled Plasma-Mass Spectrometry (LA-ICP-MS) analysis on cultured and natural freshwater and saltwater pearls from different regions can be found in literature [12–16]. However, this method is widely applied on other biogenic calcium carbonates (of freshwater and saltwater) such as molluscs shells and corals, as they are considered to be valuable environmental monitors and archives of paleoclimates [17–22]. This is because of their immobility and their trace elements content, which could be linked with the water conditions they grew in, making them a useful indicator of climate pollution and ecosystem changes. Biological factors such as growth rate, age etc. can also influence biogenic carbonate chemistry, making the interpretation of chemical elements incorporation in calcium carbonate challenging [23].

For the present work, 1113 natural and cultured nacreous pearls were studied using X-ray luminescence and LA-ICP-MS. The studied samples were collected from various bivalves and different geographic areas. This study was carried out in order to better study natural and cultured pearls' chemical characteristics and look for potential differences linked to their origin (environment, host animal and/or geographic). This is the first study that combines these methods on such a large group of (natural and cultured) pearls.

2. Materials and Methods

All studied 1113 samples are listed in Table 1; 999 samples were natural saltwater pearls found in three different bivalves and areas. Two hundred forty-eight were natural saltwater pearls from *P. radiata*, fished in the early 1980s in the Arabian Gulf, around 15 km NNE off Bahrain ("main heirats") for a project conducted by the Bahrain Centre for Studies and Research, headed by Dr. Hashim Al-Sayed (ex-dean of College of Science, University of Bahrain), principally for environmental studies. The samples currently belong to the Bahrain National Museum. The pearls have various shapes, sized from 2.3 to 8.8 mm and weigh from 0.2 to 4.3 carats (1 carat = 0.2 grams). Their colours vary from light cream to yellow.

Eighty natural saltwater pearls were from *P. margaritifera*, fished early 1990s from the Red Sea, off Hurghada (Egypt) were also studied. All samples were reportedly found in the same animal. The samples have various shapes, are sized from 1.5 to 5.4 mm, weigh from 0.1 to 1.1 carats and their colours vary from white to light grey.

The rest of the studied natural saltwater pearls were six hundred seventy-one samples from Venezuela, part of a private collection. These samples were reportedly collected in the Pre to Early-Colombian Era from *P. imbricata* and kept in a jar for several years [9]. The pearls also have various shapes, sized from 2.9 to 19.5 mm, weigh from 0.2 to 47.5 carats and their colours vary from white to cream and sometimes light grey and light yellow.

Table 1. List of studied samples.

Environment	Bivalve	Area	No. of Samples	
Saltwater	*P. radiata* *	Arabian Gulf (Heirats, Bahrain)	248 ****	
	P. margaritifera *	Red Sea (Hurghada, Egypt)	80 (****)	
	P. imbricata *	Venezuela	671 ****	1058
	P. maxima **/***	Indonesia	53 (****)	
	P. maxima ***	Burma	3	
	P. fucata ***	Vietnam (Halong Bay)	3	
Freshwater	*Margaritifera margaritifera* *	Scotland (Spey river)	12	
	Unionidae indet. *	North American rivers	26	55
	Hyriopsis sp. ***	Chinese rivers and lakes	17	

* Natural pearls; ** Cultured pearls without bead; *** Cultured pearls with bead; **** 1 spot was analysed using LA-ICP-MS, (****) one spot was analysed using LA-ICP-MS in most pearls.

Fifty-nine cultured saltwater pearls from two different animals and three different areas were also studied. These are of *P. maxima*; where 21 samples with bead and 32 samples without bead are from different farms off Indonesia and 3 samples with bead from a farm off Burma as well as 3 samples (with bead) of *P. fucata* cultivated off Vietnam (Halong Bay). The samples have various shapes, from 3.5 to 13.3 mm in size, weigh from 0.3 to 28.2 carats and their colours vary from light cream to light yellow. Thus, a total number of 1058 of saltwater pearls (natural and cultured) were studied for the present work.

Twelve natural freshwater pearls found in *Margaritifera margaritifera* in the Spey river (Scotland) during two different expeditions at the same season in the late 1980s were also studied. The samples have various shapes, are 4.6–9.1 mm in size, weigh from 0.4 to 3.8 carats and their colours vary from white to light grey to light yellow.

Twenty-six natural freshwater pearls reportedly found in USA freshwaters and from various animals belonging to Unionidae (Unionidae indet.) were studied. The samples have various shapes, sized from 2.8 to 8.9 mm, weigh from 0.2 to 4.9 carats and their colours vary from white to light yellow to light grey to light purple to brown.

Seventeen cultured freshwater pearls, without bead, were all reportedly cultivated in Chinese freshwaters into animals from *Hyriopsis* sp.; which today dominate the gem market. The samples have various shapes, sized from 4.4 to 10.6 mm, weigh from 0.4 to 8.3 carats and their colours vary from white to cream to purple and to grey. A total number of 55 freshwater pearls (natural and cultured) were studied for the present work.

X-ray luminescence was studied with a PXI GenX-100 (Pacific X-ray Imaging, San Diego, CA, USA) under 100 kV and 5 mA (500 W), the samples were placed around 20 cm from the X–ray tube. A Nikon D850 camera (Nikon, Tokyo, Japan) was used with an AF–S Micro-Nikkor 105 mm lens (Nikon, Tokyo, Japan), utilizing an exposure time of 8 seconds, F6.3 aperture and ISO Hi 0.7. Natural and cultured pearls may change their colour after exposure to X-ray irradiation, but no alteration of samples surface was noticed after the performed measurement. The samples used for more than 100 times as references for X-ray luminescence may turn darker in colour.

Laser Ablation-Induced Coupled Plasma-Mass Spectrometer (LA-ICP-MS) chemical analysis were performed using a iCAP Q (Thermo Fisher Scientific; Waltham, MA, USA) Induced Coupled Plasma-Mass Spectrometer (ICP-MS) coupled with a Q-switched Nd:YAG Laser Ablation (LA) device operating at a wavelength of 213 nm (Electro Scientific Industries/New Wave Research, Fremont, CA, USA/San Diego, CA, USA). A laser spot of 40 µm in diameter was used, along with a fluence of around 5 J/cm^2 and a 10 Hz repetition rate. Laser warm up/background time was 20 s, its dwell time was 30 s, and its wash out time was 50 s. For the ICP-MS operations, the forward power was set at ~1550 W and the typical nebulizer gas (argon) flow was ~1.0 L/min and the carrier gas (helium) set at ~0.80 L/min. The criteria for the alignment and tuning sequence were to maximize Cobalt (Co), Lanthanum (La), Thorium (Th), and Uranium (U) counts and keep the ThO/Th ratio below 2%. A MACS-3 standard synthetic calcium carbonate ($CaCO_3$) pellet was used to minimize matrix effects [24]. The time-resolved signal was processed in Qtegra ISDS 2.10 software using calcium (^{43}Ca) as the internal standard applying 40.04 wt % theoretical value—calculated from pure aragonite ($CaCO_3$). Several isotopes were measured, but in this study only sodium (^{23}Na), strontium (^{88}Sr), barium (^{137}Ba) and lead (^{208}Pb) as well as manganese (^{55}Mn) and magnesium (^{24}Mg) are presented. These isotopes were selected as they present low to no matrix and gas blank related interferences (Mn and Mg are relatively high) [16,24]. 1022 samples were analysed with one spot only (all natural saltwater pearls from Bahrain and Venezuela, 59 out of 80 natural saltwater pearls from Egypt and 44 out of 53 cultured saltwater pearls from Indonesia) and in 91 samples three spots were analysed (Table 1). Chemical elements can be incorporated differently in various calcium carbonate polymorphs. All chemical analyses were acquired on spots made of aragonite (checked with Raman spectroscopy) with nacreous microstructure (checked with optical microscope). Limits of detection (LOD) and limits of quantification (LOQ) for each of the abovementioned elements are shown in Table 2. These limits differ from day to day (for every set of measurements) so they are presented as ranges from the lowest to the highest.

Table 2. Laser Ablation-Induced Coupled Plasma-Mass Spectrometer (LA-ICP-MS) detection limits and ranges in ppmw.

Limits	^{23}Na (ppmw)	^{24}Mg (ppmw)	^{55}Mn (ppmw)	^{88}Sr (ppmw)	^{137}Ba (ppmw)	^{208}Pb (ppmw)
LOD	1.62–70.84	0.02–0.57	0.09–0.65	0.01–0.05	0.01–0.39	0.01–0.02
LOQ	5.35–233.77	0.66–1.88	0.21–1.18	0.03-0.16	0.17–1.18	0.02–0.07

LOD: Limits of detection; LOQ: Limits of quantification.

Raman spectra were acquired in a Renishaw inVia spectrometer from 100 to 2000 cm^{-1}, coupled with an optical microscope, 514 nm excitation wavelength (diode-pumped solid-state laser), 1800 grooves/mm grating, notch filter, 40 microns slit, a spectral resolution of around 2 cm^{-1} and calibrated using a diamond at 1331.8 cm^{-1}. A laser power of 5 mW on the sample (to avoid any destruction of fragile organic matter) was used to acquire all Raman spectra, 50× long distance objective lens, an acquisition time was 30 seconds and 7 accumulations.

3. Results and Discussion

3.1. LA-ICP-MS Results

LA-ICP-MS data are presented in Table 3 where the results of saltwater and freshwater samples are listed together and in Table 4 the results by bivalve are listed. All acquired individual data-points are presented in Tables S1–S9. Measured freshwater samples contain higher ^{55}Mn than measured saltwater samples (see again Table 3). Manganese content of water is considered one of the important factors affecting molluscs and shells ^{55}Mn content. Rivers and lakes contain a higher content of ^{55}Mn than sea water [6,7,25].

Table 3. LA-ICP-MS analysis of saltwater and freshwater samples.

Samples	Element	Min–Max (ppmw)	Average (SD) (ppmw)	Median (ppmw)
Saltwater	^{23}Na	2130–7270	4711.27 (816.6)	4720
	^{24}Mg	29.5–950	241.29 (111.13)	240
	^{55}Mn	BQL–45.7	1.84 (5.01)	BQL
	^{88}Sr	518–1860	941.2 (197.31)	912
	^{137}Ba	BQL–11	1.18 (1.17)	0.84
	^{208}Pb	BQL–177	5.18 (14.88)	0.95
Freshwater	^{23}Na	1030–2450	1672.81 (303.89)	1680
	^{24}Mg	5.58–81.3	30.87 (16.45)	28.3
	^{55}Mn	17.4–1440	504.95 (386.34)	473
	^{88}Sr	70.6–2700	431.84 (389.41)	350
	^{137}Ba	13.2–249	72.76 (48.25)	61.6
	^{208}Pb	BQL–11.6	0.24 (1.24)	BQL

BQL: Below quantification limit; SD: Standard deviation.

Table 4. LA-ICP-MS analysis of the studied samples by mollusc.

Samples	Element	Min–Max (ppmw)	Average (SD) (ppmw)	Median (ppmw)
P. radiata (Heirats, Bahrain) Natural	^{23}Na	3340–7270	5249.52 (800.85)	5250
	^{24}Mg	29.5–477	147.45 (88.18)	131
	^{55}Mn	BQL–7.62	0.79 (1.38)	BQL
	^{88}Sr	518–1650	943.9 (219.42)	902
	^{137}Ba	BQL–4.32	0.78 (0.6)	0.58
	^{208}Pb	BQL–3.91	0.17 (0.4)	0.08
P. margaritifera (Red Sea, Egypt) Natural	^{23}Na	2630–7240	4442.05 (1204.07)	4595
	^{24}Mg	67.6–675	362.72 (109.6)	365
	^{55}Mn	BQL–5.5	2.24 (1)	2.31
	^{88}Sr	663–1560	1052.57 (180.98)	1030
	^{137}Ba	BQL–2.02	0.26 (0.26)	0.28
	^{208}Pb	BQL–1.74	0.3 (0.36)	0.17
P. imbricata (Venezuela) Natural	^{23}Na	3200–6240	4621.64 (564.29)	4600
	^{24}Mg	108–531	264.34 (72.01)	258
	^{55}Mn	BQL–10.3	0.33 (0.95)	BQL
	^{88}Sr	572–1620	898.07 (164.56)	880
	^{137}Ba	BQL–11	1.56 (1.32)	1.18
	^{208}Pb	0.35–177	8.58 (18.68)	2.42
P. maxima (Indonesia) Cultured	^{23}Na	2150–5930	4430.28 (860.33)	4460
	^{24}Mg	71.1–279	125.76 (55.38)	110
	^{55}Mn	2.6–37.4	15.37 (8.65)	13.5
	^{88}Sr	791–1540	1061.54 (151.92)	1050
	^{137}Ba	0.09–1.85	0.53 (0.38)	0.38
	^{208}Pb	BQL–1.27	0.17 (0.22)	0.11
P. maxima (Burma) Cultured	^{23}Na	2130–5000	3317.78 (1170.44)	2790
	^{24}Mg	107–172	139.67 (23.58)	139
	^{55}Mn	7.04–45.7	25.82 (14.42)	30.90
	^{88}Sr	840–1700	1279.78 (364.37)	1380
	^{137}Ba	0.86–2.39	1.58 (0.47)	1.55
	^{208}Pb	0.2–1.16	0.43 (0.32)	0.3
P. fucata (Halong Bay, Vietnam) Cultured	^{23}Na	2370–4950	3821.11 (1092.03)	4150
	^{24}Mg	237–950	476.11 (298.01)	306
	^{55}Mn	1.29–15.9	7.2 (6.31)	4.38
	^{88}Sr	888–1860	1284.22 (403.18)	1020
	^{137}Ba	0.36–1.47	0.88 (0.41)	0.78
	^{208}Pb	0.35–1.08	0.7 (0.24)	0.70
Margaritifera margaritifera (Spey river, Scotland) Natural	^{23}Na	1390–2220	1733.61 (202.26)	1725
	^{24}Mg	11.1–60.6	31.37 (11.79)	29.15
	^{55}Mn	58–896	355.25 (234.92)	343.5
	^{88}Sr	175–1030	559.94 (244.16)	599
	^{137}Ba	15.30–233	85.96 (56.2)	90.45
	^{208}Pb	BQL–0.39	0.08 (0.12)	BQL
Unionidae indet. (North American rivers and lakes) Natural	^{23}Na	1030–2300	1519.48 (BQL)	1510
	^{24}Mg	9.21–81.3	34.06 (16.96)	33.45
	^{55}Mn	17.40–1020	382.07 (298.5)	427.5
	^{88}Sr	70.6–2700	389.12 (527.14)	260.5
	^{137}Ba	13.2–249	77.04 (48.49)	65.9
	^{208}Pb	BQL–11.6	0.44 (1.79)	BQL
Hyriopsis sp. (Chinese rivers and lakes) Cultured	^{23}Na	1260–2450	1880.39 (BQL)	1870
	^{24}Mg	5.58–76.9	25.65 (17.51)	19.20
	^{55}Mn	45–1440	798.54 (431.41)	898
	^{88}Sr	241–690	406.75 (111.39)	378
	^{137}Ba	13.6–169	56.88 (37.33)	52
	^{208}Pb	BQL–0.35	0.04 (0.09)	BQL

BQL: Below quantification limit; SD: Standard deviation.

Most (40 out of 55) of the studied freshwater samples present ^{55}Mn >200 ppmw. In some samples ^{55}Mn is higher than 1000 ppmw and up to 1440 ppmw. Large amount of Mn^{2+} incorporation into biogenic aragonite has been attributed to local crystallographic alterations [26]. On the other hand, some freshwater samples (15 out of 55) present relatively low ^{55}Mn concentrations (<150 ppmw). It has been suggested that ^{55}Mn found in nacreous aragonitic freshwater mollusc shells is linked with the availability of Mn^{2+} in the sediment-water interface [27]. Manganese variations of shells were also linked to local geology, anthropogenic factors, animal's age and growth rates as well as phytoplankton blooms, water temperature and pH as well as others [28,29]. However, measured natural pearls from *Margaritifera margaritifera* bivalve fished off the same season (two consecutive years) at the same location are presenting great variability of ^{55}Mn (58–896 ppmw; Table 4).

All studied saltwater cultured pearls, as well as around half of the natural saltwater pearls present detectable ^{55}Mn using LA-ICP-MS reaching up to 45.7 ppmw (see again Tables 3 and 4). Noteworthy, 1/3 of the cultured saltwater samples present ^{55}Mn >15 ppmw and all studied natural saltwater pearls contained ^{55}Mn <15 ppmw. Half of the studied natural saltwater pearls did not present any detectable amounts of ^{55}Mn. Moreover, three natural freshwater pearls from US presented amounts of ^{55}Mn similar to those presented to some cultured saltwater pearls from *P. maxima* bivalve from Indonesia and Burma (Table 4).

^{88}Sr in studied saltwater samples is relatively higher than in freshwater samples (Table 3). Most studied freshwater samples present concentrations from 70.6 ppmw to 800 ppmw, with only four natural freshwater pearls presenting ^{88}Sr >800 ppmw. There are two natural pearls from *Margaritifera margaritifera* presenting ^{88}Sr values of 874 and 1150 ppmw and two natural samples from Unionidae indet. with about 1487 ppmw and 2663 ppmw. The other 24 studied natural samples from Unionidae indet. presented ^{88}Sr <470 ppmw. It was suggested that the variation of ^{88}Sr in aragonitic (nacreous and non-nacreous) bivalves is influenced by the animal's growth rates (which is linked with water temperature as well as other parameters) and physiological processes; ^{88}Sr variations are not under environmental control [30–32]. Strontium content measured in nacreous aragonitic freshwater bivalve inner shells has been recently suggested as a good proxy of water's ^{88}Sr [27].

Figure 1 is a binary plot of manganese oxide (MnO) and strontium oxide (SrO). This plot is similar to those previously presented using EDXRF chemical analysis [7–9], taking into account that the limit of detection for MnO is higher for EDXRF than for LA-ICP-MS. It was formerly published that the SrO/MnO ratio, measured with EDXRF, is separating freshwater (here presented as filled squares in the figure) from saltwater pearls, as it is >12 for saltwater pearls and <12 for freshwater pearls [8]. Most of saltwater samples can be separated using this ratio as they present higher strontium and lower manganese with a SrO/MnO ratio of >100. However, for the freshwater samples, most present lower strontium and higher manganese samples with a ratio of <1. Chemical analysis of some freshwater samples might lead to wrong conclusions if the previously published plots and ratio are used [8,9], these are the freshwater (natural and cultured) pearls with ^{55}Mn <150 ppmw (i.e., MnO <ca. 194 ppmw) or ^{88}Sr >950 ppmw (i.e., SrO >ca. 1124 ppmw). There are two natural freshwater pearls presenting SrO/MnO ratio >12. The SrO/MnO ratio of the studied samples is still not overlapping; all the studied saltwater samples present SrO/MnO ratio >17 and freshwater samples <16.

Interestingly, all studied freshwater samples presented higher ^{137}Ba concentration than the saltwater samples; no overlapping values between the studied freshwater and saltwater samples were observed (Table 3). Barium is present in various concentrations in freshwater and saltwater [33] and in some studies barium uptake of a nacreous and non-nacreous shell was principally related to water's barium content [34–36]. A strong inverse correlation was also found between salinity and barium in an aragonitic non-nacreous shell [30]. Other factors are also playing a role as the content is highly variable from a freshwater sample to another; even for those collected at the same region. For instance, measured ^{137}Ba of natural pearls from *Margaritifera margaritifera* bivalves collected from the Spey river (Scotland) varied from 15.3 to 233 ppmw (Table 4). Food supply and shell growth have also been suggested as influencing factors [37,38] along with another yet undetermined environmental factor [35].

Animal age might also play a role. The vast majority of saltwater samples present a detectable ^{137}Ba, while only 30 samples out of 1058 sample present undetectable samples. Also, only 120 samples out of the studied 1058 samples present ^{137}Ba above 2.5 ppmw. All these samples are natural pearls; 116 sample found in *P. imbricata*. Few of these are samples with ^{137}Ba >4.5 ppmw.

Figure 2 is the binary plot of ^{55}Mn and ^{137}Ba of the studied samples. All freshwater samples present higher ^{55}Mn and ^{137}Ba (shown as coloured filled squares in the plot) than the studied saltwater samples (shown as coloured empty circles in the figure). The population fields of these two group of samples are well apart, and the plot can be used to efficiently separate saltwater and freshwater (natural and cultured) pearls. In Figure 3, freshwater and saltwater samples are well separated. Moreover, freshwater samples present a somehow linear trend for ^{137}Ba vs. ^{88}Sr plots. This is not observed for saltwater samples and it further supports previous studies mentioning that ^{88}Sr incorporation differs in freshwater and saltwater bivalves [27]. This should be further studied on samples from known molluscs and areas like the studied natural pearls founded in *Margaritifera margaritifera* bivalve from the river Spey (Scotland).

^{23}Na is the most abundant element present in the studied samples in both saltwater and freshwater (Table 3) with higher concentrations for the first than for the latter. Sodium concentration of biogenic carbonate is linked with water salinity and pH [25,36]. All studied samples present diverse concentrations, not linked to specific bivalves or regions. Even the natural saltwater pearls found into the same *P. margaritifera* bivalve present a great variability of ^{23}Na from 2630 to 7240 ppmw (Table 4) and the natural freshwater pearls from Spey river with ^{23}Na from 1030 to 2450 ppmw. This may indicate that other factors can also influence cultured and natural pearls' sodium concentration, such as the age of the pearl sac's epithelial cells where it was found into. In the ^{23}Na and ^{137}Ba diagram presented in Figure 4, saltwater samples are more highly separated than the freshwater samples (filled squares).

Measured ^{24}Mg content was also lower on the studied freshwater samples than on the saltwater (Table 3). However, some saltwater samples present relatively low manganese content, similar to the content of freshwater samples. The lowest content on saltwater samples was measured for *P. radiata*, where 45 samples out of 284 sample has ^{24}Mg <65 ppmw (Table 3). Samples found in the same animal (see again Table 4 for *P. margaritifera*) have a great variation of ^{24}Mg (from 67.6 to 675 ppmw). It has been suggested that Mg content is not (or is little) related to water conditions (e.g., temperature) as previously believed, but instead is related to physiological processes [30,34,39].

^{208}Pb content was relatively low on all the studied samples, sometimes including BQL. All studied freshwater samples present below 0.4 ppmw of ^{208}Pb. Solely two freshwater samples, both natural from Unionidae indet., presented concentrations of >0.4 pppmw, one presented around 1.5 ppmw and the other around 9 ppmw. Most of the studied saltwater samples presented below 1.5 ppmw ^{208}Pb. Only five samples from *P. radiata* containing above 1.5 ppmw ^{208}Pb and up to 3.91 ppmw (Table 4). However, around 2/3 of the studied *P. imbricata* samples contained above 1.5 ppmw ^{208}Pb, with around 1/3 above 3.91 ppmw and up to 177 ppmw (see also Figure 5). It has been suggested that ^{208}Pb measured in aragonite bivalve shells can be linked with anthropogenic lead pollution [40]. Anthropogenic small-scale lead pollution might be the reason of ^{208}Pb >1 ppmw concentration in eight natural freshwater samples from *P. radiata*. Relatively high concentrations of ^{208}Pb observed in *P. imbricata* natural pearls are probably superficial due to samples storage, as some natural pearls fished near Central America around 16th century were stored in lead boxes (e.g., natural pearls of Santa Margarita shipwreck treasures were found [41]). However, a possible link with polluted water caused by human and lead content cannot be excluded.

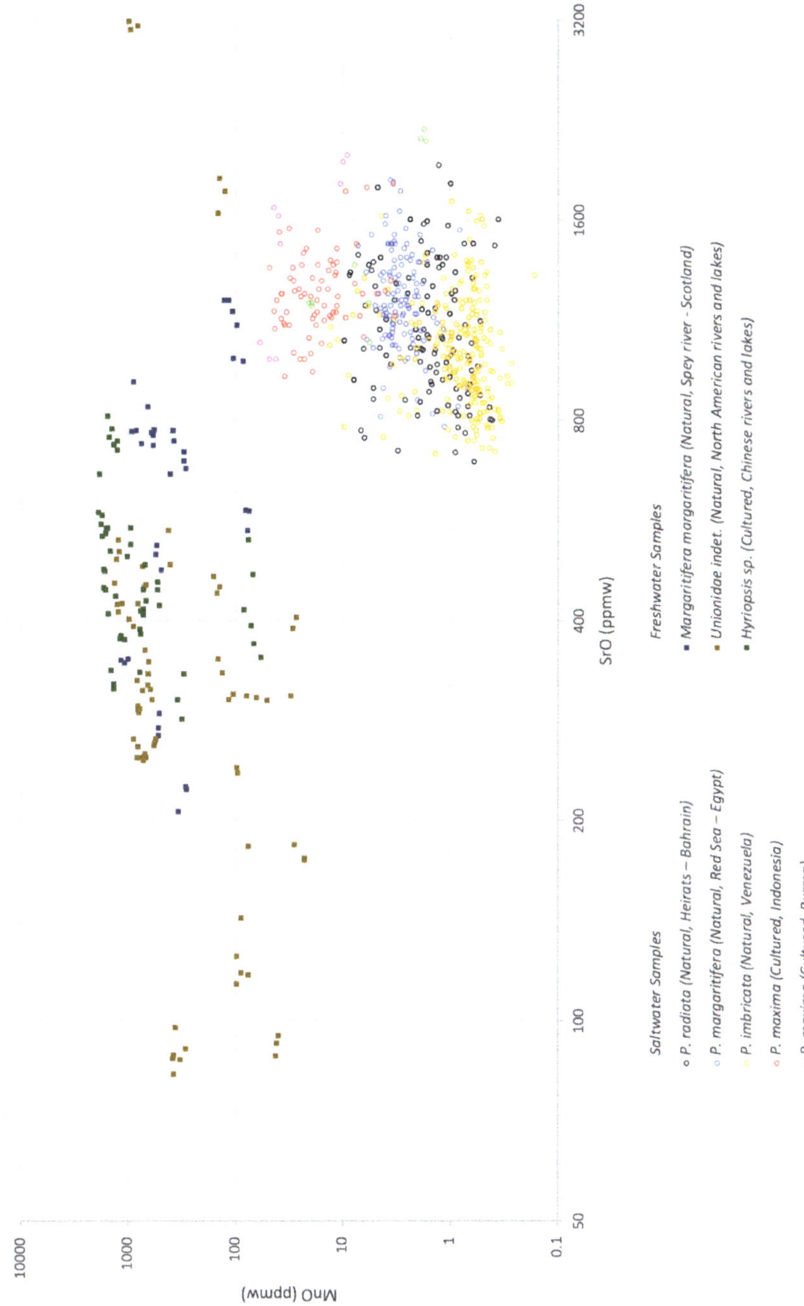

Figure 1. Binary plot of SrO vs. MnO.

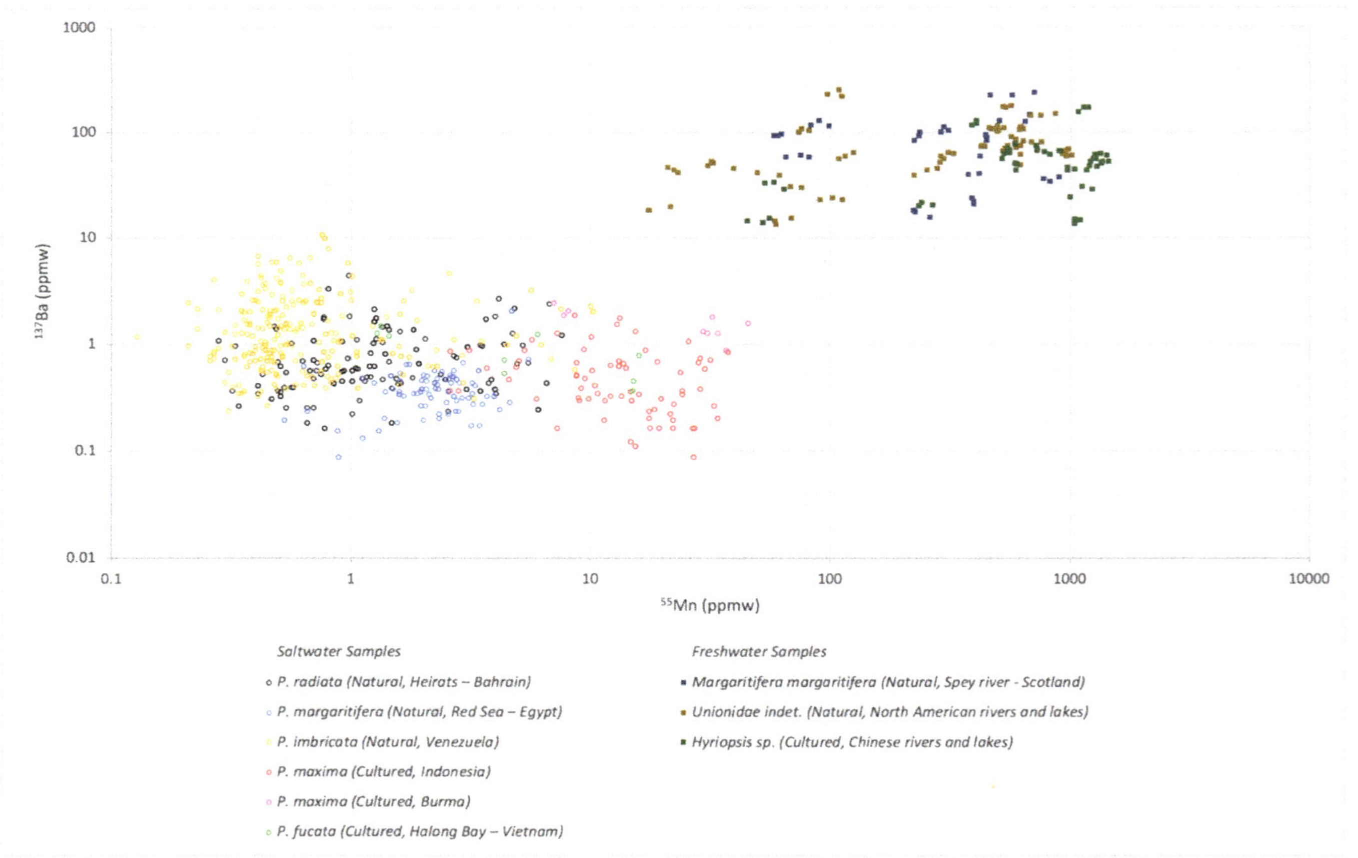

Figure 2. Binary plot of ^{55}Mn vs. ^{137}Ba of the studied samples.

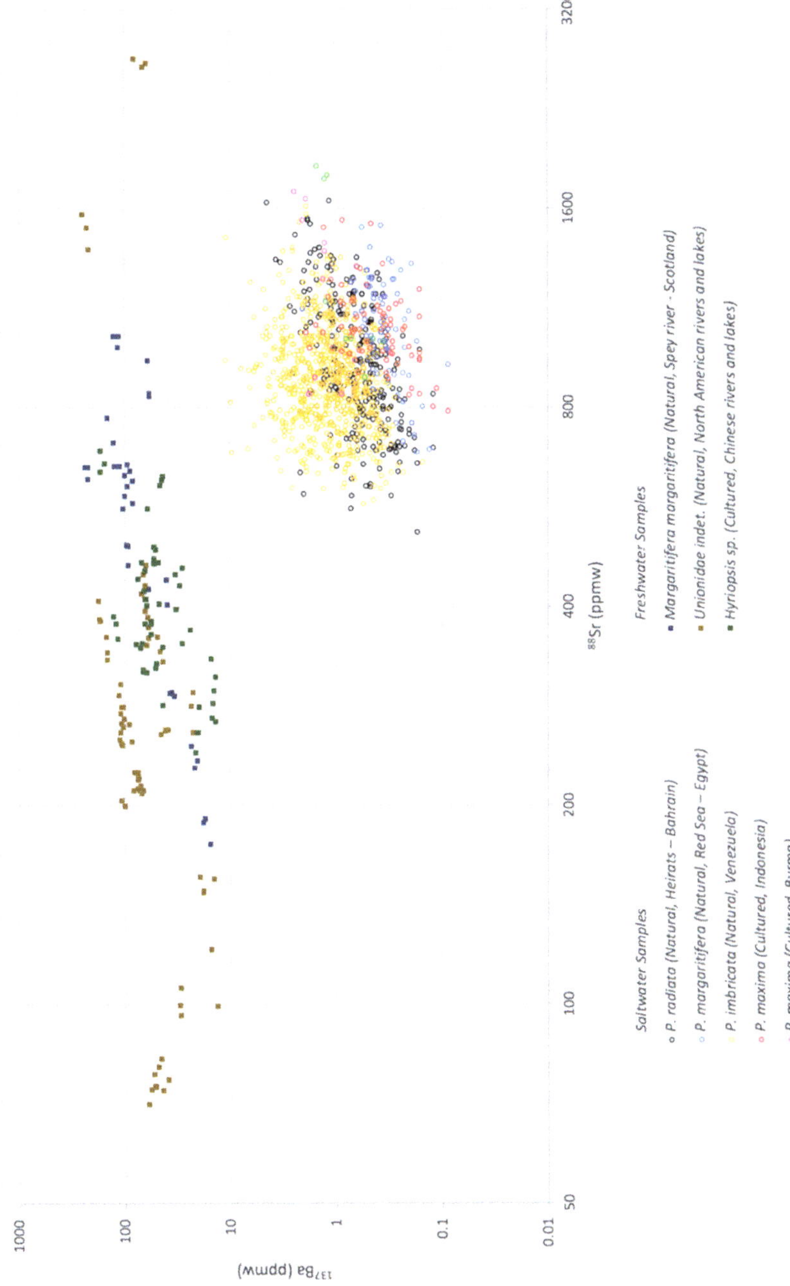

Figure 3. Binary plot of ^{88}Sr vs. ^{137}Ba of the studied samples.

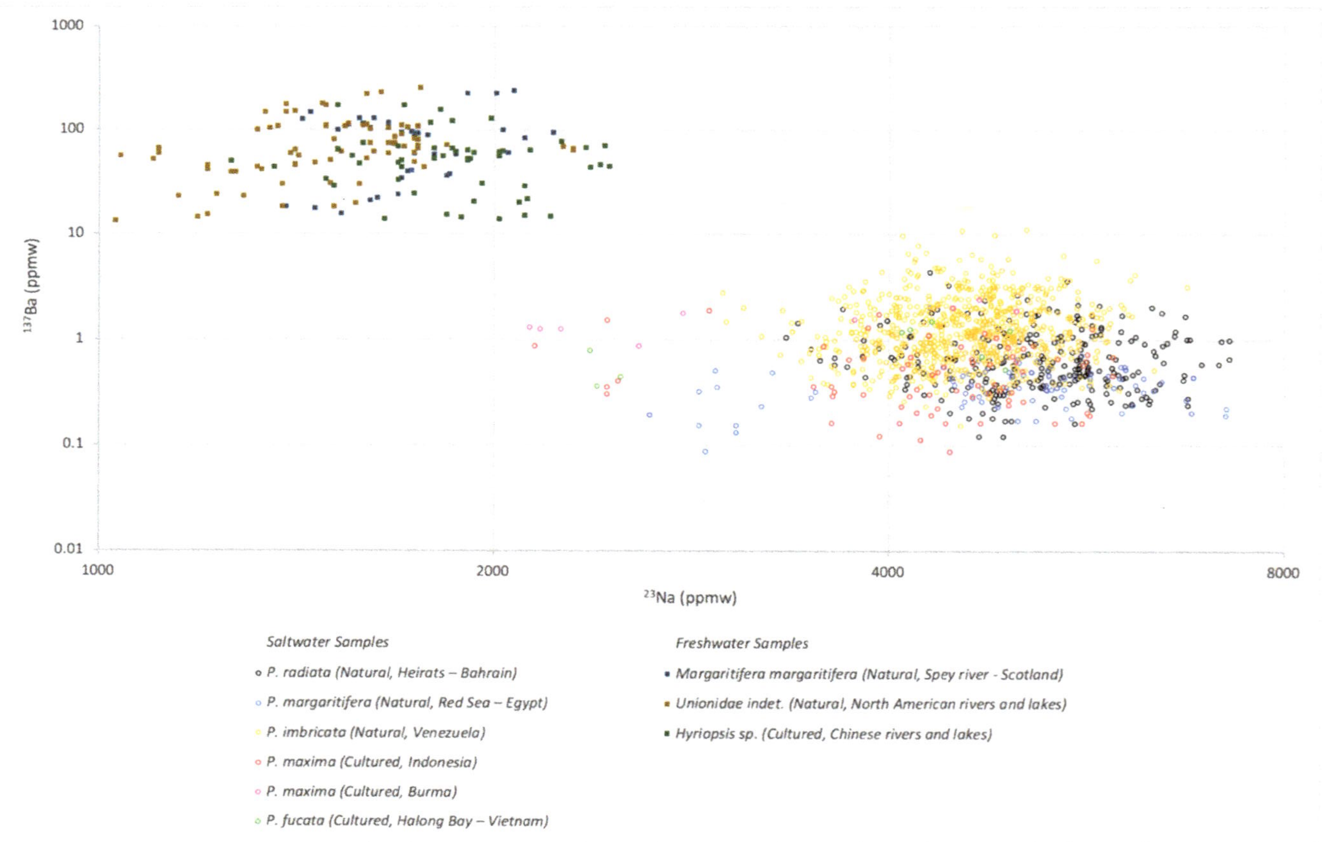

Figure 4. Binary plot of ^{137}Ba vs. ^{23}Na of the studied samples.

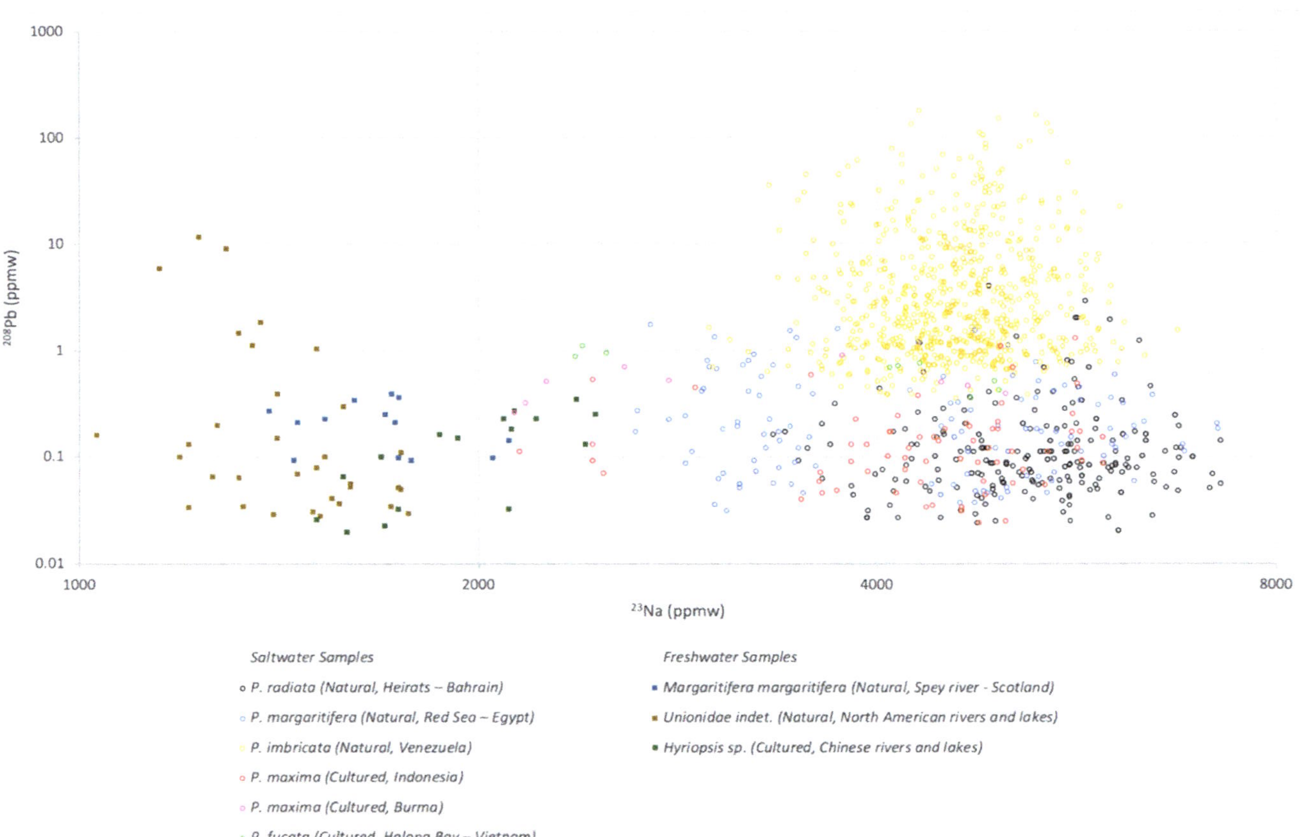

Figure 5. Binary plot of ^{23}Na vs. ^{208}Pb of the studied samples.

3.2. X-ray Luminescence and Raman Spectroscopy

Under X-rays, most natural and cultured (without bead) saltwater pearls remain inert and most natural and cultured freshwater pearls presented yellow-green luminescence of medium to high intensity (Figures 6–9). This luminescence is attributed to 9-fold coordination of oxygen around Mn^{2+} in aragonite [42]. Luminescence intensity is correlated with an Mn^{2+} content, as the higher Mn^{2+} content, the more intense the yellow-green luminescence appears to be [42,43] (Figure 6). Meanwhile, it has been noticed that a higher Mn^{2+} content could cause lower luminescence (see again Figure 6). Similar to what was previously observed with CL spectroscopy (and X-ray induced luminescence) of biogenic aragonites, this is probably due to self-quenching of manganese. The presence of Fe^{2+} (as well as Cu^{2+}, Co^{2+} and Ni^{2+}) could quench Mn^{2+} luminescence as well [42,43].

Some coloured natural and cultured pearls present photoluminescence phenomena [44]. None of the studied samples presented any reaction under X-rays related to pigments. It was previously mentioned that natural (and artificial) pigments of freshwater samples suppress luminescence under X-rays [5]. The studied freshwater samples confirm that coloured samples with higher [55]Mn (measured with LA-ICP-MS) present lower luminescence than less coloured samples with lower manganese (measured with LA-ICP-MS); see Figure 7.

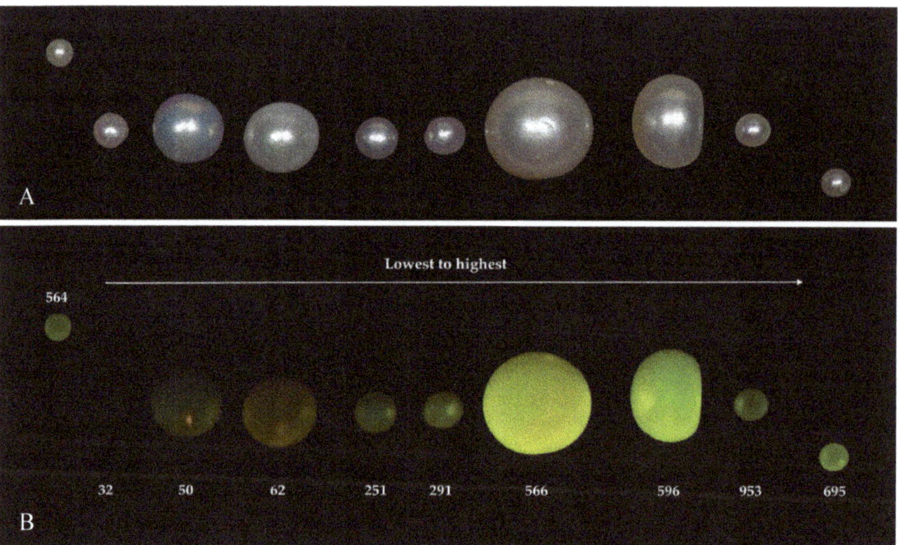

Figure 6. Freshwater samples under daylight (**A**) and under X-rays (**B**). The two freshwater samples upper left (diameter: 2.85 mm) and bottom right (diameter: 2.95 mm) used as reference. The concentrations of [55]Mn in ppmw are mentioned below for the studied samples and above for the reference samples. For the samples 3 analyses with LA-ICP-MS were acquired, the average values are presented (rounded to the nearest one).

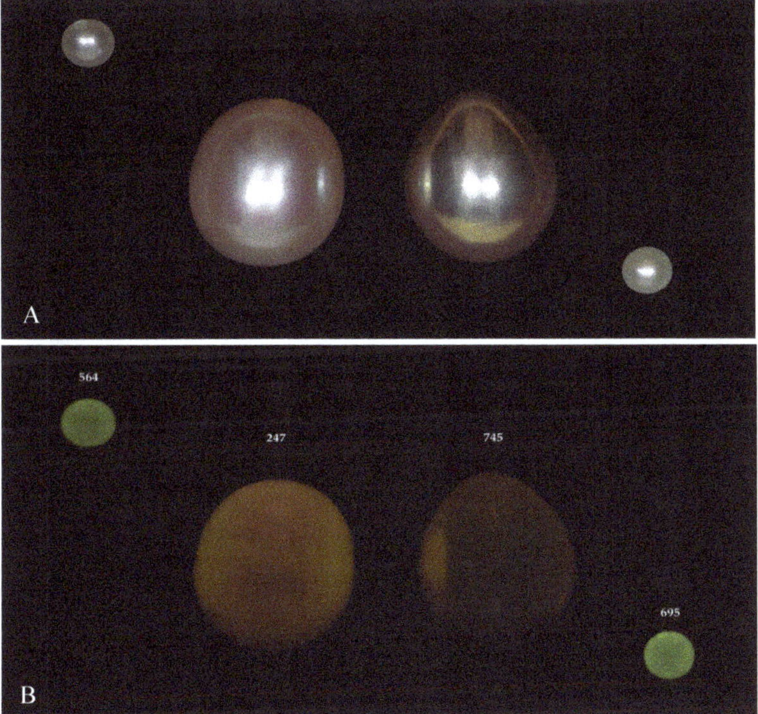

Figure 7. Freshwater samples under daylight (**A**) and under X-rays (**B**). Two natural-coloured freshwater samples (left sample of pink colour and right samples of grey colour) with the amount ^{55}Mn in ppmw. The two freshwater samples upper left (diameter: 2.85 mm) and bottom right (diameter: 2.95 mm) used as reference. For the samples 3 analyses with LA-ICP-MS were acquired, the average values are presented (rounded to the nearest one).

Some saltwater cultured pearls, with freshwater bead, also present intense yellow-green luminescence (Figure 8). The amount of ^{55}Mn measured on these samples is from 8 to 38 ppmw and it is not directly linked to the luminescence. The reaction under X-rays is due to the content of Mn^{2+} of the bead of these saltwater cultured samples and the relatively thin nacre thickness (i.e., thickness of the layer of saltwater nacre deposited around the freshwater bead) of the studied samples [5]. In Figure 8, the samples with the highest fluorescence has the thinnest nacre thickness (0.65 mm; measured using X-ray microradiography) and the other two samples have similar relatively thick nacre thickness (1.8 mm).

Figure 8. Saltwater cultured pearls (with freshwater bead) under daylight (**A**) and under X-rays (**B**). The first and second pearls from the left has a similar nacre thickness that equals 1.8 mm, while the third one has a nacre thickness of 0.65 mm (measured using X-ray microradiography). The two freshwater samples upper left (diameter: 2.85 mm) and bottom right (diameter: 2.95 mm) used as reference.

As mentioned above, some of the studied freshwater samples present ^{55}Mn <150 ppmw. This samples present luminescence of low to medium intensity under X-rays (see examples in Figures 6 and 9). On the other hand, some of the studied saltwater samples present detectable manganese measured with LA-ICP-MS. On our samples, X-ray luminescence appeared with low intensity at the studied samples only when ^{55}Mn was higher than 10 ppmw. However, this will be different, if different parameters (e.g., less X-ray power) are used to acquire luminescence images. All the studied freshwater samples presented luminescence as well as some of the studied saltwater samples of low intensity. The latter were sometimes difficult to separate them from the freshwater samples containing low manganese (Figure 9). The amount of manganese in freshwater samples is low and below the detection limit of some EDXRF instruments and might lead to wrong conclusions. This samples might be identified by EDXRF if barium is taken into consideration or with LA-ICP-MS using the plots presented from Figures 2–4.

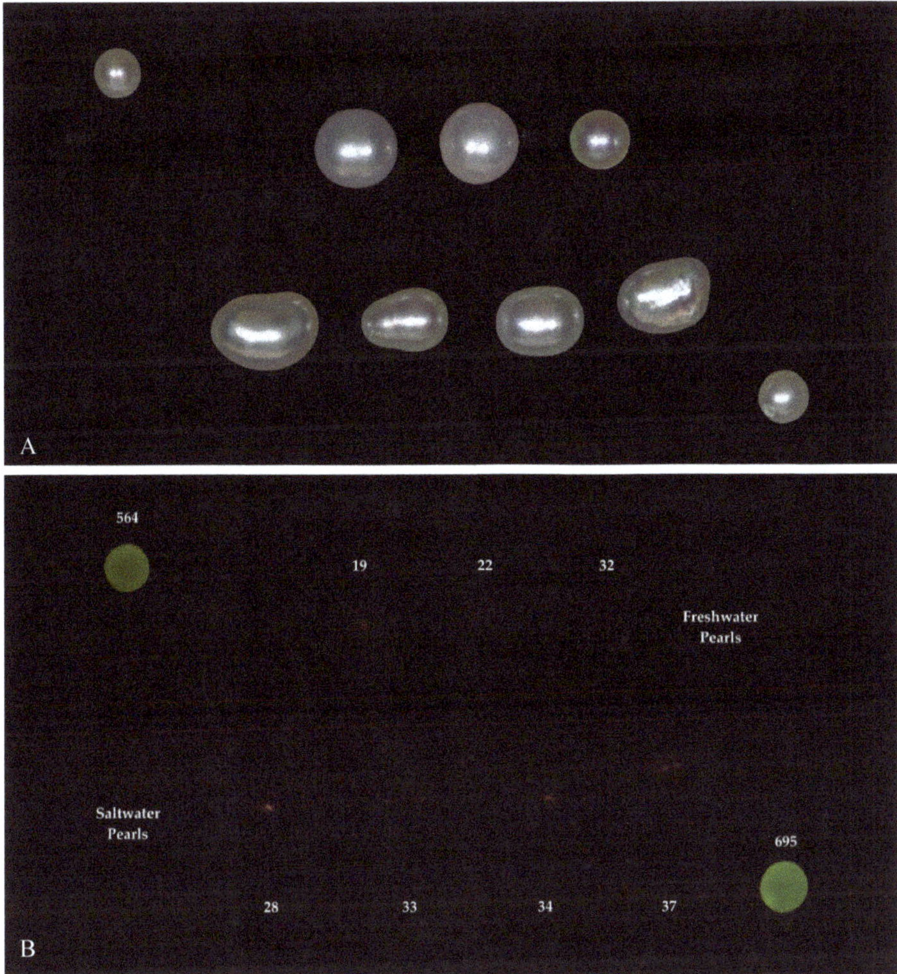

Figure 9. Three freshwater (upper row) and four saltwater (bottom row) samples under daylight (**A**) and under X-rays (**B**). The two freshwater samples upper left (diameter: 2.85 mm) and bottom right (diameter: 2.95 mm) used as reference. All samples present similar luminescence intensity and separation of freshwater samples from saltwater is challenging. The concentrations of ^{55}Mn in ppmw are mentioned. For the samples 3 analyses with LA-ICP-MS that were acquired, the average values are presented (rounded to the nearest one).

One of the studied natural freshwater pearls presented a section with intense orange luminescence under X-rays; and phosphorescence which lasted for about 3 seconds (Figure 10). Orange luminescence activated by CL was previously observed on some cultured freshwater pearls, without bead, from China and this was attributed to calcite emitting Mn^{2+} [6]. Raman spectra acquired on the section where orange luminescence presents (and which, macroscopically and under optical microscope, appears with different lustre than the rest of the sample; refer to Figure 10A) have shown Raman bands at around 1091, 1081 and 1075 cm^{-1} due to ν_1 symmetric stretching of carbonate ions in vaterite (Figure 11) and not at 1086 cm^{-1} due to ν_1 symmetric stretching of carbonate ions in aragonite as on the section where medium yellow-green luminescence was observed [45]. CL and X-ray induced luminescence spectra of

Mn^{2+} in calcite and vaterite are very close, which suggest a 6-fold coordination of oxygen around both calcite and vaterite [42]. Sections of vaterite were already previously identified in freshwater cultured pearls without bead from Japan and China [45–47]; but not in natural freshwater pearls.

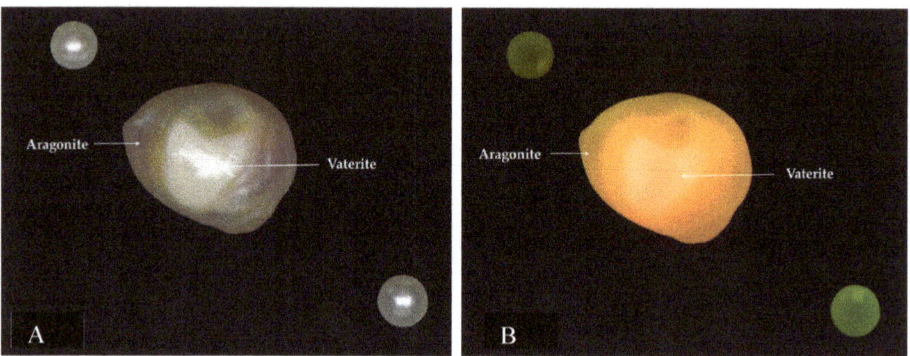

Figure 10. Natural freshwater pearl under daylight (**A**) and under X-rays (**B**); orange coloured part is made of vaterite (verified by Raman; see Figure 11). The two freshwater samples upper left (diameter: 2.85 mm) and bottom right (diameter: 2.95 mm) were used as references.

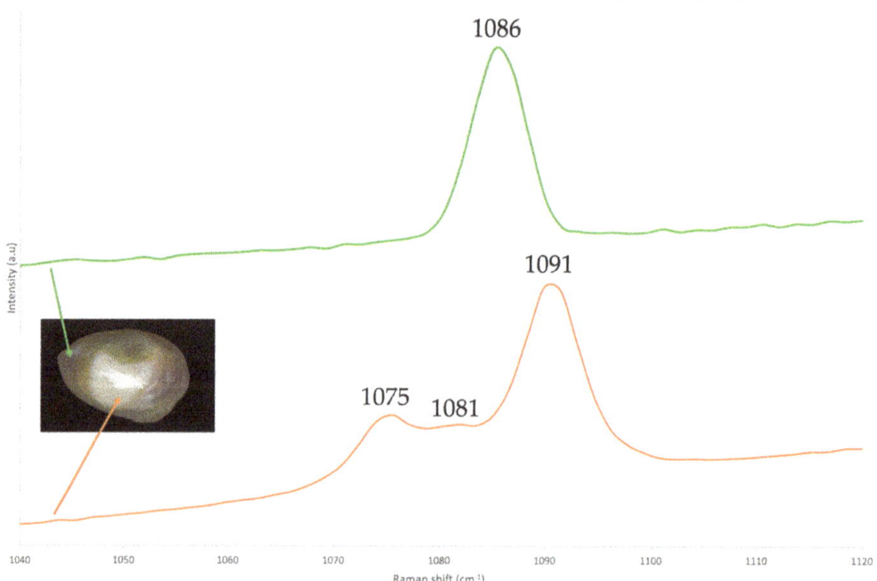

Figure 11. Raman spectra on the white part with low lustre (bottom spectrum) and on the cream coloured part with medium lustre (top spectrum) showing respectively bands due to vaterite and calcite. Note that the spectra are stacked and shifted for clarity.

4. Conclusions

LA-ICP-MS measurements on freshwater samples reveal higher [55]Mn and [137]Ba and lower [23]Na, [24]Mg and [88]Sr concentrations than the studied saltwater samples. Only [137]Ba concentrations of the studied samples do not present any overlap; the other concentrations slightly overlap. Plots combining [137]Ba with [55]Mn, [23]Na and [88]Sr clearly separate freshwater samples from saltwater samples. Very little correlation between the studied chemical elements of natural and cultured pearls and host animal's species was found. Saltwater samples from *P. maxima* are the only studied saltwater samples which contain ^{55}Mn >20 ppmw. On the other hand, only some saltwater samples from *P. radiata* could present ^{24}Mg <65 ppmw and saltwater samples from *P. imbricata* could present ^{137}Ba >2.5 ppmw. It seems that all the chemical elements studied here are linked to a combination of environmental and physiological factors.

Most studied freshwater samples can be separated from the saltwater samples as X-ray luminescence intensity is correlated with Mn^{2+} content in aragonite. Some of the studied freshwater samples present similar manganese content with some saltwater samples as well as similar low intensity luminescence under X-rays. The growth environment of these samples could only inadequately be identified by LA-ICP-MS or a well calibrated EDXRF with a good limit of detection for manganese and barium.

Studies on a bigger number of samples from known animals and various regions (with known geographic coordinates) that take into account the local environmental factors, along with statistical analysis (e.g., principal components), should be performed in order to confirm the abovementioned differences and better understand the link between the chemical elements and the natural and cultured pearls. Studying natural pearls is not an easy task because they are very rarely found and in parallel rarely used for scientific purposes. Additional measurements of isotopes might shed more light on a possible link between environmental factors and natural (or cultured) pearls. In order to inspect the possible role of organic matter, TGA analysis and chemical analysis would also be useful. It is also important to further study the possible role of elements such us Fe^{2+}, Cu^{2+}, Co^{2+} and Ni^{2+} which could quench Mn^{2+} luminescence and their effect on the separation of saltwater from freshwater (natural and cultured) pearls. Orange coloured X-ray luminescence of freshwater samples should also be further investigated and its link with vaterite should be confirmed with spectroscopic means.

Supplementary Materials: The following are available online at http://www.mdpi.com/2075-163X/9/6/357/s1, Table S1: LA-ICP-MS analysis in ppmw of the studied natural pearls samples from P. radiata (Heirats, Bahrain), Table S2: LA-ICP-MS analysis in ppmw of the studied natural pearls samples from P. margaritifera (Red Sea, Egypt), Table S3: LA-ICP-MS analysis in ppmw of the studied natural pearls samples from P. imbricata (Venezuela), Table S4: LA-ICP-MS analysis in ppmw of the studied cultured pearls samples from P. maxima (Indonesia), Table S5: LA-ICP-MS analysis in ppmw of the studied cultured pearls samples from P. maxima (Burma), Table S6: LA-ICP-MS analysis in ppmw of the studied cultured pearls samples from P. fucata (Halong Bay, Vietnam), Table S7: LA-ICP-MS analysis in ppmw of the studied natural pearls samples from *Margaritifera margaritifera* (Spey river, Scotland), Table S8: LA-ICP-MS analysis in ppmw of the studied natural pearls samples from Unionidae indet. (North American rivers and lakes), Table S9: LA-ICP-MS analysis in ppmw of the studied cultured pearls samples from *Hyriopsis* sp. (Chinese rivers and lakes).

Author Contributions: S.K. formulated the paper designed the experiments, participated to the data interpretation and wrote the manuscript. F.M. prepared the experiments for X-ray luminescence analysis, did part of LA-ICP-MS analysis, data reduction, edited the manuscript and draw its figures and tables. H.A. prepared all the images of the manuscript. F.A. and L.F. performed most of LA-ICP-MS analysis on the samples and S.S. has designed, supervised and evaluated LA-ICP-MS analysis. A.A. selected the samples, designed the experiments and edited the manuscript.

Funding: This research received no external funding.

Acknowledgments: The authors wish to thank the National Museum of Bahrain which loaned the natural saltwater pearls from *P. radiata* fished off Bahrain, Peter Balogh for the natural saltwater pearls from *Pinctada imbricata* fished off Venezuela and Peter Piva for the natural saltwater pearls from *Pinctada margaritifera* fished off Egypt. The rest of the samples are from DANAT reference collection.

Conflicts of Interest: The authors declare no conflict of interest.

References

1. Gauthier, J.P.; Karampelas, S. Pearls and corals: "Trendy biomineralizations". *Elements* **2009**, *5*, 179–180. [CrossRef]
2. Cartier, L.; Krzemnicki, M.S. New developments in cultured pearls production: Use of organic and baroque shell nuclei. *Aust. Gemmol.* **2013**, *25*, 6–13.

3.   Schiffmann, C.A. Pearl identification: Some laboratory experiments. *J. Gemmol.* **1971**, *12*, 284–296. [CrossRef]

4.   Lorenz, I.; Schmetzer, K. Possibilities and limitations in radiographic determination of pearls. *J. Gemmol.* **1986**, *20*, 114–123. [CrossRef]

5.   Hänni, H.A.; Kiefert, L.; Giese, P. X-ray luminescence, a valuable test in pearl identification. *J. Gemmol.* **2005**, *29*, 325–329. [CrossRef]

6.   Habermann, D.; Banerjee, A.; Meijer, J.; Stephan, A. Investigation of manganese in salt- and freshwater pearls. *Nucl. Instrum. Methods* **2001**, *181*, 739–743. [CrossRef]

7.   Gütmannsbauer, W.; Hänni, H.A. Structural and chemical investigations on shell and pearls of nacre forming salt- and freshwater bivalve mollusks. *J. Gemmol.* **1994**, *24*, 241–252. [CrossRef]

8.   Karampelas, S.; Kiefert, L. Gemstones and minerals. In *Analytical Archaeometry: Selected Topics*, 1st ed.; Edwards, H.G.M., Vandenabeele, P., Eds.; Royal Society of Chemistry: London, UK, 2012; pp. 291–317. ISBN 978-1-84973-162-1.

9.   Zhou, C.; Hodgins, G.; Lange, T.; Saruwatari, K.; Sturman, N.; Kiefert, L.; Schollenbruch, K. Saltwater pearls from the pre- to early Columbian era: A gemological and radiocarbon dating study. *Gems Gemol.* **2017**, *53*, 286–295. [CrossRef]

10.   Goebel, M.; Dirlam, D.M. Polynesian black pearls. *Gems Gemol.* **1989**, *25*, 130–148. [CrossRef]

11.   Elen, S. Spectral reflectance and fluorescence characteristics of natural-color and heat-treated "Golden" south sea cultured pearls. *Gems Gemol.* **2001**, *37*, 114–123. [CrossRef]

12.   Jacob, D.E.; Wehrmeister, U.; Hager, T.; Hofmeister, W. Identifying Japanese freshwater cultured pearls from Lake Kasumigaura. *J. Gemmol.* **2006**, *22*, 539–541.

13.   Scarratt, K.; Bracher, P.; Bracher, M.; Atawi, A.; Safar, A.; Saeseaw, S.; Homkrajae, A.; Sturman, N. Natural pearls from Australian *Pinctada maxima*. *Gems Gemol.* **2012**, *48*, 236–261. [CrossRef]

14.   Hänni, H.A.; Cartier, L.E. Tracing cultured pearls from farm to consumer: A review of potential methods and solutions. *J. Gemmol.* **2013**, *33*, 185–191. [CrossRef]

15.   Sturman, N.; Homkrajae, A.; Manustrong, A.; Somsaard, N. Observations on pearls reportedly from the Pinnidae family (pen pearls). *Gems Gemol.* **2014**, *50*, 202–215. [CrossRef]

16.   Homkrajae, A.; Sun, Z.; Blodgett, T.; Zhou, C. Provenance discrimination of freshwater pearls by laser ablation–inductively coupled plasma–mass spectrometry (LA-ICP-MS) and linear discriminant analysis (LDA). *Gems Gemol.* **2019**, *55*, 47–60.

17.   Perkins, W.T.; Fuge, R.; Pearce, N.J.G. Quantitative analysis of trace elements in carbonates using laser ablation inductively coupled plasma mass spectrometry. *J. Anal. Atom. Spectr.* **1991**, *6*, 445–449. [CrossRef]

18.   Sinclair, D.J.; Kinsley, L.P.J.; McCulloch, M.T. High resolution analysis of trace elements in corals by laser-ablation ICP-MS. *Geochim. Cosmochim. Acta* **1998**, *62*, 1889–1901. [CrossRef]

19.   Gillikin, D.P. Geochemistry of Marine Bivalve Shells: The Potential for Paleoenvironmental Reconstruction. Ph.D. Thesis, Vrije University, Brussel, Belgium, 2005; 265p.

20.   Jacob, D.E.; Soldati, A.L.; Wirth, R.; Huth, J.; Wehrmeister, U.; Hofmeister, W. Nanostructure, composition and mechanisms of bivalve shell growth. *Geochim. Cosmochim. Acta* **2009**, *72*, 5401–5415. [CrossRef]

21.   Mertz-Kraus, R.; Brachert, T.C.; Jochum, K.P.; Reuter, M.; Stoll, B. LA-ICP-MS analyses on coral growth increments reveal heavy winter rain in the eastern Mediterranean at 9 Ma. *Palaeogeogr. Palaeoclimatol. Palaeoecol.* **2009**, *273*, 25–40. [CrossRef]

22.   Phung, A.T.; Baeyens, W.; Leermakers, M.; Goderis, S.; Vanhaecke, F.; Gao, Y. Reproducibility of laser ablation–inductively coupled plasma–mass spectrometry (LA–ICP–MS) measurements in mussel shells and comparison with micro-drill sampling and solution ICP–MS. *Talanta* **2013**, *396*, 42–50. [CrossRef]

23.   Strasser, C.A.; Mullineaux, L.S.; Walther, B.D. Growth rate and age effects on Mya Arenaria shell chemistry: Implications for biogeochemical studies. *J. Exp. Mar. Biol. Ecol.* **2008**, *355*, 153–163. [CrossRef]

24.   Jochum, K.P.; Scholz, D.; Stoll, B.; Weis, U.; Wilson, S.A.; Yang, Q.; Schwalb, A.; Borner, N.; Jacob, D.E.; Andreae, M.O. Accurate trace element analysis of speleothems and biogenic calcium carbonates by LA-ICP-MS. *Chem. Geol.* **2012**, *318–319*, 31–44. [CrossRef]

25.   Gordon, C.M.; Carr, R.A.; Larson, R.E. The influence of environmental factors on the sodium and manganese content of barnacle shells. *Limnol. Oceanogr.* **1970**, *15*, 461–466. [CrossRef]

26.   Soldati, A.L.; Jacob, D.E.; Glatzel, P.; Swarbrick, J.C.; Geck, J. Element substitution by living organisms: The case of manganese in mollusc shell aragonite. *Sci. Rep.* **2016**, *6*, 22514. [CrossRef] [PubMed]

27. Geeza, T.J.; Gillikin, D.P.; Goodwin, D.H.; Evans, S.D.; Watters, T.; Warner, N.R. Controls on magnesium, manganese, strontium, and barium concentrations recorded in freshwater mussel shells from Ohio. *Chem. Geol.* **2019**. [CrossRef]

28. Carroll, M.; Romanek, C.S. Shell layer variation in trace element concentration for the freshwater bivalve *Elliptio complanta*. *Geo-Mar. Lett.* **2008**, *28*, 369–381. [CrossRef]

29. Kelemen, Z.; Gillikin, D.P.; Bouillon, S. Relationship between river water chemistry and shell chemistry of two tropical African freshwater bivalve species. *Chem. Geol.* **2019**. [CrossRef]

30. Poulain, C.; Gillikin, D.P.; Thebault, J.; Munaron, J.M.; Bohn, M.; Robert, R.; Paulet, Y.M.; Lorrain, A. An evaluation of Mg/Ca, Sr/Ca, and Ba/Ga ratios as environmental proxies bivalve shells. *Chem. Geol.* **2015**, *396*, 42–50. [CrossRef]

31. Wanamaker, D.; Gillikin, D.P. Strontium, magnesium and barium incorporation in aragonitic shells of junevile *Arctica islandica*: Insights from temperature controlled experiments. *Chem. Geol.* **2019**. [CrossRef]

32. Stecher, H.A.; Krantz, D.E.; Lord, C.J.; Luther, G.W.; Bock, K.W. Profiles of strontium and barium in *Mercenaria mercenaria* and *Spisula solidisima* shells. *Geochim. Cosmochim. Acta* **1996**, *60*, 3445–3456. [CrossRef]

33. Schroeder, H.A.; Tipton, I.H.; Nason, A.P. Trace metals in man: Strontium and barium. *J. Chronic Dis.* **1972**, *25*, 491–517. [CrossRef]

34. Gillikin, D.P.; Dehairs, F.; Lorrain, A.; Steenmans, D.; Baeyens, W.; Andre, L. Barium uptake into the shells of the common mussel (*Mytilus edulis*) and the potentials for estuarine paleo-chemistry reconstruction. *Geochim. Cosmochim. Acta* **2006**, *70*, 395–407. [CrossRef]

35. Gillikin, D.P.; Lorrain, A.; Paulet, Y.M. Synchronous barium peaks in high-resolution profiles of calcite and aragonite marine bivalve shells. *Geo-Mar. Lett.* **2008**, *28*, 351–358. [CrossRef]

36. Zhao, L.; Schone, B.R.; Mertz-Kraus, R.; Yang, F. Controls on strontium and barium incorporation into freshwater bivalve shells (*Corbicula fluminea*). *Palaeogeogr. Palaeoclimatol. Palaeoecol.* **2017**, *465*, 386–394. [CrossRef]

37. Zhao, L.; Schone, B.R.; Mertz-Kraus, R. Sodium provides unique insights into transgenerational effects of ocean acidification on bivalve shell formation. *Sci. Total Environ.* **2017**, *577*, 360–366. [CrossRef] [PubMed]

38. Herath, D.; Jacob, D.E.; Jones, H.; Fallon, S.J. Potential of shells of three species of Eastern Australian freshwater mussels (Bivalvia: Hyriidae) as environmental proxy archives. *Mar. Freshw. Res.* **2018**, *70*, 255–269. [CrossRef]

39. Foster, L.C.; Finch, A.A.; Allison, N.; Andersson, C.; Clarke, L.J. Mg in aragonitic bivalve shells: Seasonal variations and mode of incorporation in *Arctica islandica*. *Chem. Geol.* **2008**, *254*, 113–119. [CrossRef]

40. Gillikin, D.P.; Dehairs, F.; Baeyens, W.; Navez, J.; Lorrain, A.; André, L. Inter- and intra-annual variations of Pb/Ca ratios in clam shells (*Mercenaria mercenaria*): A record of anthropogenic lead pollution. *Mar. Pollut. Bull.* **2005**, *50*, 1530–1540. [CrossRef]

41. Collins, D. Thousands of Pearls Found in Shipwreck. Available online: https://www.cbsnews.com/news/thousands-of-pearls-found-in-shipwreck/ (accessed on 20 March 2019).

42. Sommer, S.E. Cathodoluminescence of carbonates 1. Characterization of cathodoluminescence from carbonate solid solutions. *Chem. Geol.* **1972**, *9*, 257–273. [CrossRef]

43. Götte, T.; Richter, D.K. Quantitative aspects of Mn activated cathodoluminescence of natural and synthetic aragonite. *Sedimentology* **2009**, *56*, 483–492. [CrossRef]

44. Hainschwang, T.; Karampelas, S.; Fritsch, E.; Notari, F. Luminescence spectroscopy and microscopy applied to study gem materials: A case study of C centre containing diamonds. *Mineral. Petrol.* **2013**, *107*, 393–413. [CrossRef]

45. Wehrmeister, U.; Jacob, D.E.; Soldati, A.L.; Häger, T.; Hofmeister, W. Vaterite in freshwater cultured pearls from China and Japan. *J. Gemmol.* **2007**, *31*, 269–276. [CrossRef]

46. Qiao, L.; Feng, Q.L.; Li, Z. Special vaterite in freshwater lackluster pearls. *Cryst. Growth Des.* **2006**, *7*, 275–279. [CrossRef]

47. Ma, H.; Su, A.; Zhang, B.; Li, R.K.; Zhou, L.; Wang, B. Vaterite or aragonite observed in the prismatic layer of freshwater-cultured pearls from South China. *Prog. Nat. Sci.* **2009**, *19*, 817–820. [CrossRef]

Article

Zircon Xenocrysts from Cenozoic Alkaline Basalts of the Ratanakiri Volcanic Province (Cambodia), Southeast Asia—Trace Element Geochemistry, O-Hf Isotopic Composition, U-Pb and (U-Th)/He Geochronology—Revelations into the Underlying Lithospheric Mantle

Paula C. Piilonen [1,*], F. Lin Sutherland [2], Martin Danišík [3], Glenn Poirier [1], John W. Valley [4] and Ralph Rowe [1]

[1]   Mineralogy Section, Beaty Centre for Species Discovery, Research & Collections, Canadian Museum of Nature, P.O. Box 3443, Station D, Ottawa, Ontario K1P 6P4, Canada; gpoirier@nature.ca (G.P.); rrowe@nature.ca (R.R.)

[2]   Geoscience, Australian Museum, 1 William Street, Sydney NSW 2010, Australia; linsutherland1@gmail.com

[3]   John de Laeter Centre, The Institute for Geoscience Research (TIGeR), Curtin University, Perth WA 6845, Australia; m.danisik@curtin.edu.au

[4]   Wisconsin Secondary Ion Mass Spectrometer Laboratory (WiscSIMS), Department of Geoscience, University of Wisconsin, Madison, Wisconsin 53706, USA; valley@geology.wisc.edu

*   Correspondence: ppiilonen@nature.ca

Received: 26 October 2018; Accepted: 19 November 2018; Published: 30 November 2018

Abstract: Zircon xenocrysts from alkali basalts in Ratanakiri Province, Cambodia represent a unique low-Hf zircon within a 12,000 km long Indo-Pacific megacryst zone. Colorless, yellow, brown, and red crystals ({100}, {101}, subordinate {211}, {1103}), with hopper growth and corrosion features range up to 20 cm in size. Zircon chemistry indicates juvenile, Zr-saturated, mantle-derived alkaline melt (Hf 0.6–0.7 wt %, Y <0.2 wt %, U + Th + REE (Rare-Earth Elements) < 600 ppm, Zr/Hf 66–92, Eu/Eu*$_N$ ~1, positive Ce/Ce*$_N$, HREE (Heavy REE) enrichment). Incompatible element depletion with increasing Yb/Sm$_N$ from core to rim at ~ constant Hf suggests single stage growth. Ti-in-zircon temperatures (~570–740 °C) are lower than predicted by crystal morphology (800–900 °C) and decrease from core to rim (ΔT = 10–50 °C). The $\delta^{18}O$ values (4.88 to 5.01‰ VSMOW (Vienna Standard Mean Ocean Water)) are relatively low for xenocrysts from the zircon Indo-Pacific zone (ZIP). The $^{176}Hf/^{177}Hf$ values (+ εHf 4.5–10.2) give T$_{Depleted\ Mantle}$ model source ages of 260–462 Ma and T$_{Crustal}$ ages of 391–754 ma. The source magmas reflect variably depleted lithospheric mantle with little supracrustal input. Zircon U-Pb (0.88–1.56 Ma) and (U-Th)/He (0.86–1.02 Ma) ages are older than host basalt ages (~0.7 Ma), which suggests limited residence before transport. Zircon genesis suggests Zr-saturated, Al-undersaturated, carbonatitic-influenced, low-degree partial melting (<1%) of peridotitic mantle at ~60 km beneath the Indochina terrane.

Keywords: zircon; xenocryst; alkali basalt; Ratanakiri Volcanic Province; trace elements; O-isotopes and Hf-isotopes; U-Pb; (U-Th)/He

---

1. Introduction

Southeast (SE) Asia is an important source of gem corundum and zircon [1–3]. Recent and paleo-alluvial gem deposits occur in Thailand, Cambodia, Vietnam, and Laos, as deposits in a belt that extends >12,000 km along the west Pacific continental margins south to Australia and north to eastern China and Russia. This belt is known as the zircon megacryst (xenocryst) Indo-Pacific zone (ZIP, Figure 1) [4–14]. The western Pacific margin is marked by dynamic tectonic settings in which plate motions carry both continental and former sea-floor lithosphere over zones of hot upwelling asthenosphere. All SE Asian deposits are associated with late Cenozoic (<30 Ma) intraplate alkaline basaltic volcanism, which is the result of onset of decompression melting and an extensional tectonic regime in SE Asia following the Himalayan orogeny [15,16]. In SE Asia, intraplate alkali basalt magmas erupted through the late Precambrian to early Cenozoic along roughly northward-trending rifted blocks and fold belt terranes flanking the Indochina cratonic block.

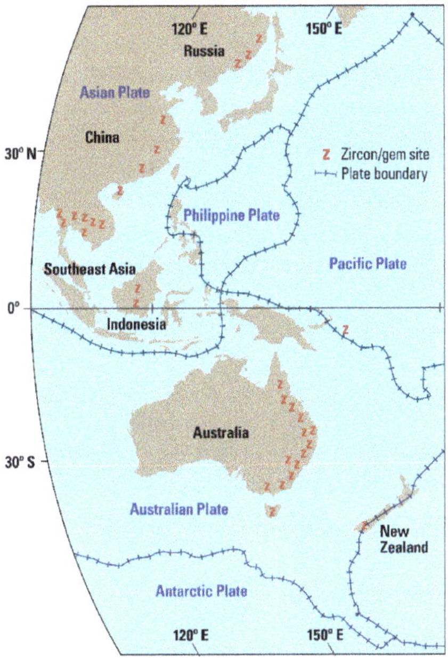

Figure 1. Map showing the 12,000 km long zircon megacryst Indo-Pacific zone (ZIP) along the western Pacific continental margins in Eastern Australasia, Asia, and Russia. The intraplate basalts hosting the megacryst suites are dominantly alkaline in composition and involve deeper mantle generation of magmas that may intersect pre-existing gem-bearing felsic or metamorphic bodies [11]. Consult this reference for zircon locality details.

Gem zircon and corundum xenocrysts, which have been eroded out of the host alkali basalts during the production of lateritic soils are concentrated into economic-grade deposits by secondary processes [17]. Mining for xenocrystic gems has taken place since the early 1400s in Thailand (Chanthaburi-Trat) and the late 1880s in Cambodia (Pailin and Ratanakiri provinces). In some deposits, zircon dominates over corundum, as is the case within the Ratanakiri Volcanic Province (RVP). Mechanized mining continues today in parts of Thailand but in Cambodia, small-scale artisanal mining by Vietnamese, Burmese and local indigenous peoples is common, often utilizing the same primitive techniques used hundreds of years ago.

Zircon is a ubiquitous accessory mineral in a wide range of rock types including mafic and felsic rocks derived from both crustal and mantle sources, lunar rocks, tektites, and metamorphic rocks [18]. It is chemically resistant and can survive weathering and transport processes, which allow it to be concentrated in secondary placer heavy mineral deposits. Zircon is highly refractory and has a low solubility in most melt and fluid compositions, which allows it to survive almost any crustal process including high temperature metamorphism and anatexis. Although the abundance of zircon is low, it has a strong effect on the behavior of many trace elements during crystallization and, thus, is an important accessory mineral in understanding petrogenesis. Furthermore, the low diffusion rates of rare-earth elements (REE), Th, U, Pb, Hf, and O in zircon under most geological conditions conserves both trace element composition zoning and the isotopic signature, which offers a window into its growth, evolution, and recycling events.

Xenocrystic zircon are known from many Cenozoic alkaline basalt provinces world-wide. However, their origins, petrogenesis, and relationship with their host basaltic rocks are still widely debated. Most authors are in agreement with a mantle source for some xenocrystic suites. However, the exact geochemical, isotopic, and mineralogical characteristics of this source are not completely known. Boehnke et al. [19] performed experimental studies on Zr saturation in mafic (basaltic) melts and concluded that such melt compositions require an unrealistically high concentration of Zr (>5000 ppm) to directly crystallize zircon. As such, zircon within ZIP basalts must have crystallized from late-stage, evolved magmas and become entrained in the rising alkali basalt host. Additional experimental studies on Zr saturation in alkaline melts, silicate, or carbonatitic are limited and focused on intermediate to felsic crustal compositions (e.g., [20–22]). More information is needed concerning Zr saturation under upper mantle conditions before inferences can be made about the petrogenesis of zircon xenocrysts. By utilizing trace element geochemistry, isotopic systems that reflect original magmatic conditions (i.e., O and Hf) and comparisons of the crystallization and eruption ages of the xenocrysts within the host basalts. Insights can be gained on the origin of these zircons.

This study presents field and laboratory studies started in 2012 investigating the zircon megacrysts which are being mined from the RVP basaltic gem fields in Northeastern Cambodia.

A recent study on similar zircons [23] has provided valuable new data on one set of RVP zircon megacrysts. However, the present study differs in reporting on a wider range of zircon samples and their color and form variants from several well-defined localities as well as providing further analytical and literature results. Some detailed comparisons of the two data sets are incorporated, which allows for wider discussion and interpretations on the origin of these Cambodian zircons. This study is the first of its kind in Cambodia and provides insight into understanding the composition and the evolutionary history of the subcontinental lithospheric mantle (SCLM) in Northeast Cambodia.

2. General Geology

Southeast Asia including Thailand and Cambodia is an assemblage of distinct terranes amalgamated together to form two dominant crustal blocks known as the Indochina and the Shan-Thai (or Sibumasu) terranes (Figure 2). These two terranes are separated and overprinted by Mesozoic and younger arc systems of the Sukhothai Zone (Fold Belt) and the Sa Kaeo Suture Zone (SKS). Although these two terrains have separate geological histories, they both have their origins at the margin of Gondwana prior to collision during the Mesozoic [24]. The Indochina terrain, which extends from Eastern Thailand, Laos, across Cambodia, and into Southern Vietnam, rifted away from the northern edge of Gondwanaland during the Devonian. This resulted in the opening of Paleotethys [25]. Little is known about the Precambrian tectonic history of Thailand and Cambodia even though gneisses and schists of the Proterozoic Kontum massif in Vietnam extend into Northeast Cambodia near the border with Thailand [26]. Similar high-grade metamorphic rocks have been found in Pailin along the border with Thailand [26,27].

Collision and convergence of the Indochina block with the Asian continent (Shan-Thai terrane in Thailand) started at ~250 Ma and culminated around the Triassic-Jurassic boundary (~210–200 Ma) with the closure of Paleotethys and the end of the Indosinian orogeny [25,26]. The Indochina terrain is composed of continental crust, which has remained intact and stable since the end of the Indosinian orogeny, and is surrounded by younger, post-Jurassic fold belts. Cambodia lies entirely in the Indochina terrain between the Truongson fold belt to the north and the Loei fold belt to the west. Both of these regions contain Carboniferous to Triassic collisional arc-type volcanosedimentary sequences.

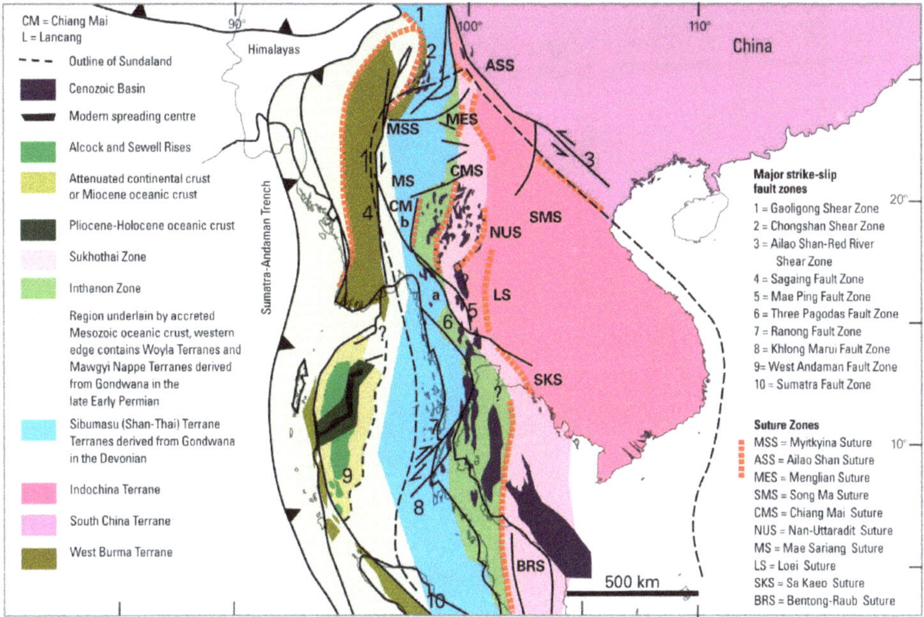

Figure 2. Tectonic map of SE Asia [25]. Cambodia is entirely contained within the Indochina terrain bounded on the northeast by the Song Ma Suture (SMS).

The Cenozoic tectonic evolution of Cambodia and Thailand is related to the collision between the Indian and Eurasian plates during the Himalayan Orogeny (55–45 Ma, [28]). Compressional stresses dominating in the northern part of the collisional zone resulted in the Southeast China block and adjacent parts of Southeast Asia (Indosinia) to be rotated and forced to the southeast [29], which results in uplift and transtension and graben structures with predominantly a NW-SE strike along with the opening of the Gulf of Thailand, the South China Sea, and the Andaman Sea [30]. Uplift, extensional tectonics and rifting throughout SE Asia resulted in lithospheric thinning, which triggered decompressional melting and upwelling of the underlying asthenosphere along deep-seated faults to form late Cenozoic flood basalts [16,29]. ZIP fields across SE Asia are flanked to the east by extinct marginal spreading basins and offset rifts that formed behind the Pacific island arc-subduction system [31]. These basins involved thermal rifting and could provide a continuous source of sub-lithospheric melting to promote zircon (and corundum) crystallization during prolonged basaltic events [11,32,33].

Cenozoic basalts occur throughout SE Asia (red regions, Figure 3) ranging in age from 24 Ma to as recently as a 1923 eruption off the Vietnamese coast near the island of Poulo Cécir [34,35]. Known collectively as the SE Asian Volcanic Province [36–38] or the Central-East Asian basaltic province [8], the basalts ascended along rift structures bounded by strike-slip faults within Archean to Paleozoic terrains and were more extensive in Vietnam, Eastern Cambodia, and Southern Laos than elsewhere in the region even though no temporal and spatial correlation has been established [39]. In most areas, two separate periods of volcanism are noted: (1) an early phase consisting of SiO_2-rich, Fe-depleted, and Ti-depleted quartz and olivine tholeiites and rare trachyandesites, representative of lithospheric sources, and (2) a later, highly alkaline phase consisting of lower SiO_2, high Fe, and Ti olivine tholeiites, alkali basalts, basanites, and rare trachybasalts and trachytes, representative of asthenospheric mantle sources [16,40]. Zircon and corundum xenocrysts are associated with late-stage, alkaline phases of volcanism [16,29,41].

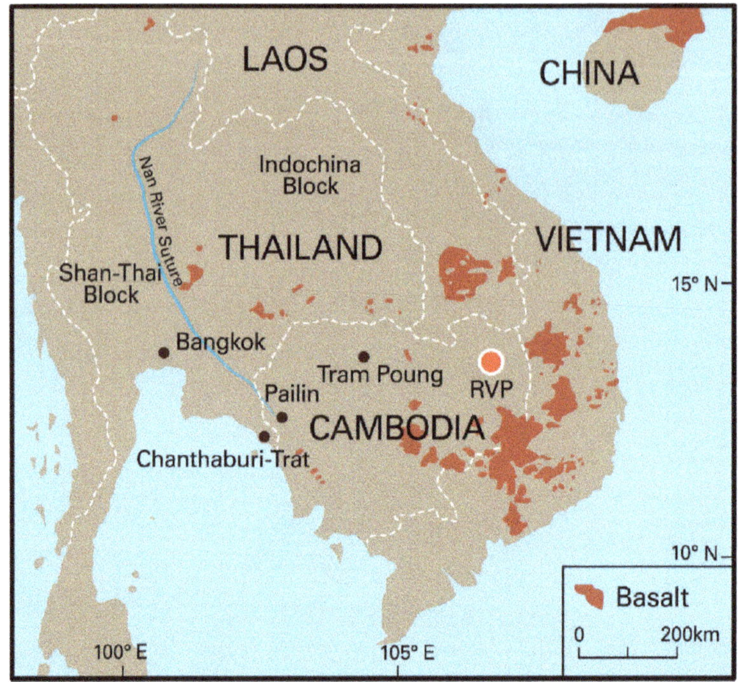

Figure 3. Distribution of Cenozoic alkali basalt provinces (red) in SE Asia (modified after [16]). The location of the Ratanakiri Volcanic Province (RVP) is shown as a red circle.

2.1. Ratanakiri Volcanic Province (RVP)

Detailed geological information about Cambodia is difficult to obtain as a result of the destruction of all academic material by the Khmer Rouge. Early work by Lacombe [42,43] concerning the geology, geochemistry, and mineralogy of the RVP basalts is still the most definitive work in existence. Previous 1:500,000 maps by both France (1970s) and Russia (1990) are out of date. Little work has been completed in the country since that time. Detailed mapping, age dating, and geochemical analyses of the rocks in the RVP is currently underway by the lead author.

Quaternary basaltic rocks cover approximately 10,000 km^2 of the country and overlie Quaternary alluvium and Mesozoic sediments. Small flows and vents occur in the west near Pailin and Poipet and in the central northern province of Preah Vihear. Two distinctly larger basaltic areas are found in Ratanakiri province (Figure 4) and comprise the fifth largest basaltic plateau in Indochina: (1) the Bokeo plateau is located between the Sre Pok and the Sesan rivers, herein referred to as the Ratanakiri Volcanic Province (RVP), and it is host to xenocrystic gem zircon deposits and (2) the smaller, zircon-free Ban Chay plateau, northeast of Bokeo and north of the Sesan river near the Vietnamese border [42]. Lacombe [42] estimated the thickness of the basalt plateaus to 80 m near the center and 40 m closer to the extremities. Initially the flows would have covered an area of around 3200 km^2, but now weathered down to under 2000 km^2 (RVP: 1500 km^2, Ban Chay Plateau: 200 km^2). The initial topographic relief was quickly filled by the early flows, which allows later flows to spread in all directions and cover a wide area with thin, individual flows (often <2 m thick). Lacombe [42] observed only alkaline basalts in the Ban Lung area (olivine basalts, alkali basalts, basanites, nephelinites, and trachyandesites). He also noted that early flows preceded the last paleomagnetic inversion (0.7 Ma) and the later, more explosive basaltic volcanism occurred during and after the geomagnetic reversal. This falls within the range (17.6–0 Ma) established by Hoang & Flower [40] for related basaltic volcanism in Vietnam and other Cenozoic volcanism throughout the Northwestern Pacific continental margin.

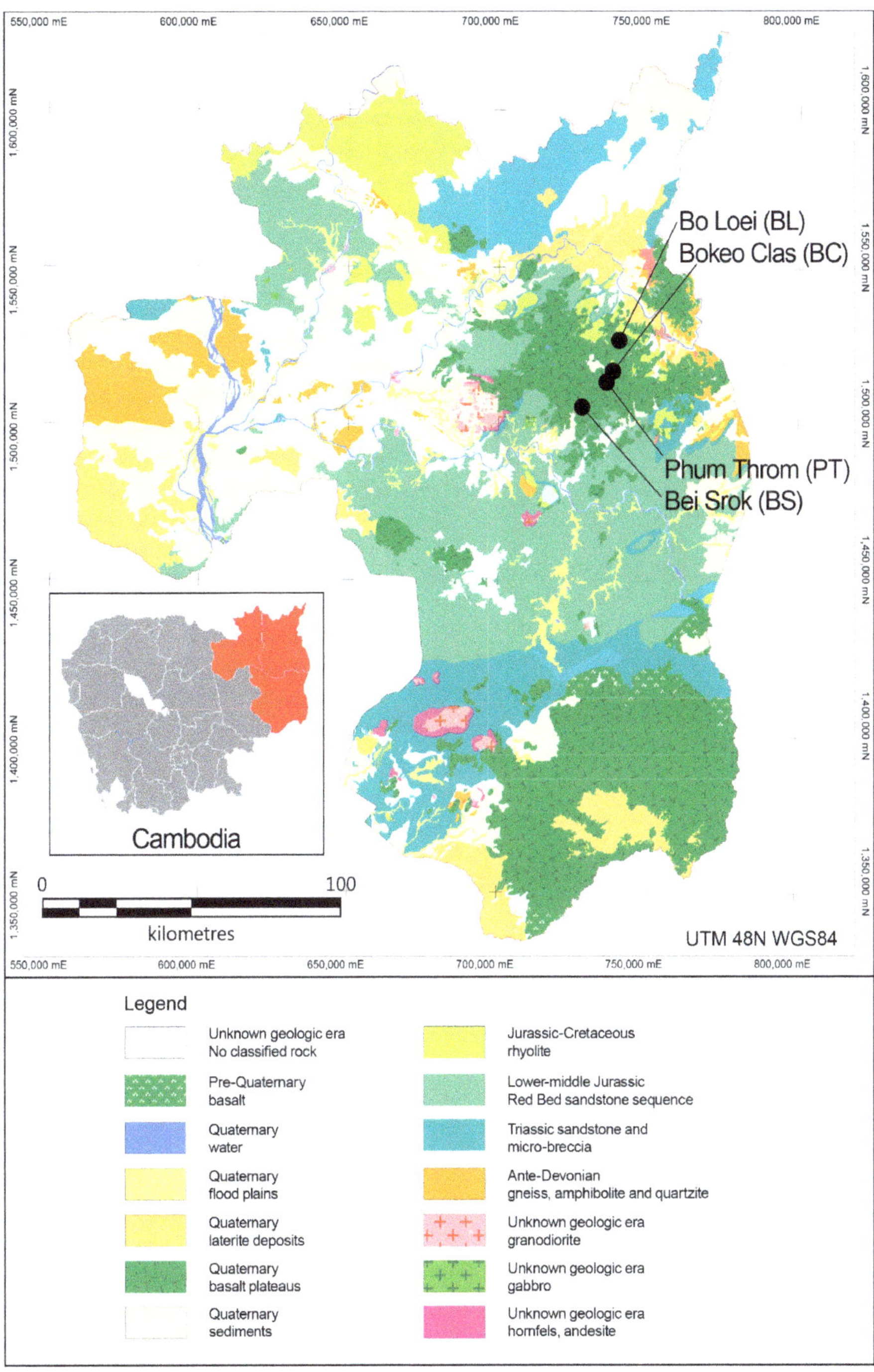

Figure 4. Geological map of Stung Treng, Ratanakiri, and Mondulkiri provinces, Northeast Cambodia. The Ratanakiri Volcanic Province (RVP) and Ban Chay Plateau Quaternary basalts are indicated in green with hatching. Zircon xenocryst localities (black dots) include Bokeo Clas (BC), Phum Throm (PT), Bei Srok (BS), and Bo Loei (BL).

Basaltic volcanism began with fissure flows. A brief, erosional hiatus was followed by increasingly more explosive volcanism from both old and new fissures [42]. The final stage was even more explosive and resulted in craters and deposits of mafic scoria, ignimbrite, welded tuff, tuff, ash flows, and pumice. Volcanic features include: (1) scoria cones (1–3 km wide × 150 m tall), (2) horseshoe craters, which have been opened as a result of extensive erosion, (3) crater cones, (4) explosive craters, and (5) rift valleys and volcanotectonic depressions [42,44]. Many of these structures are only recognizable from air photos and satellite imagery since complete volcanic profiles in the RVP have extensive lateritization and vegetative cover.

2.2. Zircon Deposits

In situ zircon is extremely rare and has only been collected by the senior author in alkali basalt at Phnom Dang, Bokeo (Figure 5), which is an alkali basalt scoria outcrop on the flank of the Phnom Dang scoria cone, and from a tephriphonolite flow 3 km SE of Phnom Dang, Bokeo. Lacombe [43] also noted the presence of xenocrysts in trachybasalt/alkali basalts and associated pyroclastics at Phnom Dang, Bo Tum and Bo Loei. At all localities, zircon occurs as single crystals in the alkali basalt without other xenocrystic minerals. Gem zircon, which is derived from the alkali basalts, is mined from secondary deposits within a roughly NE–SW trending region ~30 km in strike, 10 km wide, east of the provincial capital of Ban Lung (Figure 4). Unlike deposits in other west Pacific continental margin regions including Western Cambodia and Southeastern Thailand and gem deposits in the RVP contain dominantly zircon with only trace corundum. Zircon mines in Ratanakiri are small-scale affairs-gem-quality stones are found in the residual soils and gravels derived from weathered basalts. The gemstones are mined by digging vertical pits (10 m deep, 1 m wide) to bedrock and hauling to surface the lateritic soils and basalt gravels in which loose xenocrysts are concentrated. On the surface, other miners sift through the soil with their hands to extract the zircons. In rare cases, during the wet season, pressured water is used to loosen the lateritic soil and wash out the zircons. Only 20% of zircon are actually valuable as gem stones. Many of the stones are fractured, contain inclusions, or are an unsuitable color. Zircon from Bei Srok (BS) has historically been the most sought after by the gem industry due to its dark red color, which, when heat-treated in a reduced environment, changes to a dark, more brilliant blue color (Figure 5). An electron-related or hole-related color center is thought to be the cause of dark red-brown color in the RVP zircon while authors have not yet been able to unequivocally determine the mechanism for the blue coloration upon heat-treating [45]. The highest quality gem zircon is found at the Bei Srok deposit, which is 16 km south of the provincial capital of Ban Lung at the southern end of the gem-bearing region. The Bokeo Clas (BC) and Phum Throm (PT) deposits are located 23 km east of Ban Lung near the village of Bokeo and represent the most prolific mining areas. Other less significant occurrences are known throughout the Bokeo plateau where locals pan for zircon in the creeks and riverbeds. Rare blue-green corundum has been found in zircon concentrates from Bei Srok.

The Bo Loei (BL) deposit is located 24 km ENE of Ban Lung, 28 km W from the Vietnamese border, and 28 km NNE of Bei Srok. BoLoei marks the northern-most gem occurrence and comprises an alluvial wash containing xenocrysts of zircon, corundum (blue, blue-green, and yellow-orange), and spinel as well as fragments of alkali basalt, felsic volcanics (rhyolite), and other country rock. Bo Loei is principally a gem zircon deposit with corundum as a by-product (<3% of the total yield, [12]). The water table at Bo Loei is higher than that at other Ratanakiri zircon deposits, which allows for panning and concentration of the zircon ore in nearby streams.

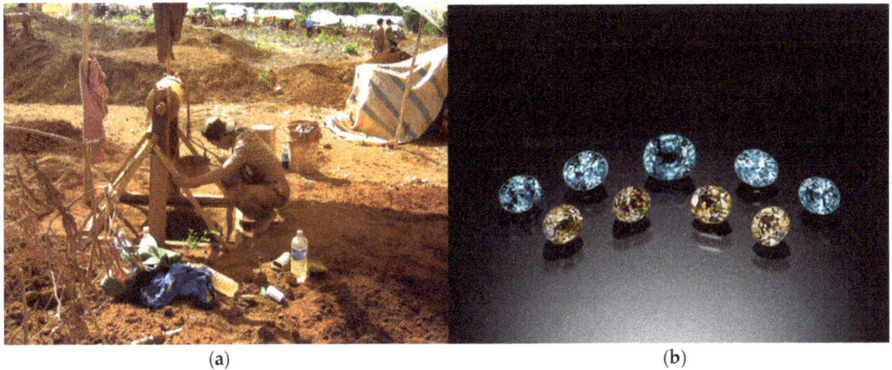

(a)   (b)

Figure 5. (**a**) Zircon miner at Phum Throm, Ratanakiri, Cambodia. (**b**) Natural and heat-treated faceted RVP zircon (each stone is ~0.75 carats). Photo copyright M. Bainbridge, Canadian Museum of Nature collection.

3. Materials and Methods

3.1. Sample Descriptions

Zircon crystals from the RVP range in size from 2 × 2 × 4 mm to 15 × 20 × 20 cm long (average: 7 × 8 × 8 mm). Crystals are subhedral to euhedral, dominantly vitreous to adamantine, transparent grey-yellow, honey yellow, pinkish-orange, light orange-brown, and dark red with minor translucent to opaque milky-grey and yellow crystals. Squat prismatic crystals dominated by {100} and {101} are most typically euhedral. Less common crystal forms include the {211} bipyramid and {110} prism and more often occur in subhedral grains. Prismatic zircon ranges in color from transparent light yellow to dark red to highly-included, translucent milky-yellow and red-brown. Subhedral to anhedral multiform zircon are most often transparent and range in color from light orange-yellow to orange-red to dark red. Sharp, concentric color zoning is evident optically in many crystals, which corresponds to zoning observed in back-scattered electron (BSE) and cathodoluminescence (CL) images. In situ zircon occurs solely as isolated grains within the alkali basalt (Figure 6a,b). The interface between the zircon and the host alkali basalt is characterized by a very thin (1–2 μm) glassy reaction rim with numerous fine fractures, which is indicative of contraction of the glass upon cooling. The presence of such a discrete, quenched glassy interface without evidence for chemical interaction suggests a wide temperature difference between the entrained zircon and the host basaltic magma.

The majority of Ratanakiri zircon display striking resorption and corrosion features along with hopper-type growth patterns (Figure 6c) indicative of disequilibrium with their host magmas. Subhedral to euhedral zircon are often dominated by a short prism capped by a bipyramid on one termination and a hopper-type growth feature on the bottom termination. This hopper-type or corroded habit appears to represent an attachment point with an associated xenocrystic phase on the growth medium. However, multimineralic xenoliths containing zircon have not been found to confirm these possible intergrowths. Late-stage dendritic etching in-filled by baddeleyite is also present on many crystals (Figure 6d), which indicates late-stage interaction with a highly-alkaline corrosive melt or fluid, and remobilization of Zr. Fluid inclusions dominated by H_2O with rare CO_2 bubbles were found by Zeug et al. [45] in a number of RVP zircon xenocrysts. Partially healed fissures surrounded many of the fluid inclusions.

Zircon from Bo Loei is commonly subhedral to anhedral and does not display the etching, dissolution, and corrosion features that are common to other Ratanakiri zircon, which is possibly a result of increased transport in alluvial or fluvial systems.

Figure 6. Zircon xenocrysts from the RVP: (**a**) euhedral zircon (7 × 4.5 × 4 mm) in situ from Phnom Dang, Bokeo. (**b**) Glassy reaction rim (1–2 μm wide lining the 5 mm hole) between zircon xenocryst and the host alkali basalt. (**c**) Hopper-type growth feature on the subhedral zircon. (**d**) Late-stage dendritic etching in-filled by baddeleyite on the surface of a zircon xenocryst.

All zircon samples were purchased directly from miners at each locality. Representative zircon grains ranging in size from 4 × 4 × 4 mm to 19 × 11 × 8 mm were hand-picked under a binocular microscope and mounted in epoxy blocks for analysis. The blocks were cut in half, laterally, on a thin section saw to expose the interior center of the grains prior to being polished.

3.2. Cathodoluminescence Microscopy and Electron Microprobe Analyses

Cathodoluminescence (CL) and back-scattered electron imaging (BSEI) was done on a JEOL 6610Lv scanning electron microscope (JEOL USA Inc., Peabody, MA, USA) operating at 20 kV with a spot size of 55 μm and a working distance of 20 mm. The SEM was equipped with a monochromatic Gatan miniCL detector (Gatan Inc., Pleasanton, CA, USA). Chemical analyses of the zircon xenocrysts were done with a JEOL Superprobe 8230 (JEOL USA Inc., Peabody, MA, USA) at the University of Ottawa. Operating conditions were as follows: beam diameter of 5 μm, operating voltage 20 kV, and a beam current of 40 nA. A total of six elements were sought and the following standards and X-ray lines were employed: synthetic YIG garnet (Y $L\alpha$), zircon (Zr $L\alpha$, Si $K\alpha$), synthetic UO_2 (U $M\alpha$), synthetic ThO_2 (Th $M\alpha$), and hafnon (Hf $M\alpha$). Count times for Zr, Th, U, and Si were 20 s on peak and 10 s on the background and count times on Hf and Y were 50 s on peak and 25 s on the background. Raw intensities were corrected by using the PAP routine [46]. The Hf concentrations obtained by EMPA (Table 1) were used as an internal standard for trace element determination by laser ablation inductively-coupled plasma mass spectrometry (LA-ICP-MS).

3.3. Trace-Element Analysis

Trace element contents were analyzed by using an Agilent 7700 quadrupole ICP-MS instrument (Agilent Technologies, Santa Clara, CA, USA) attached to a Photon Machines Excimer 193 nm laser system (Excite, Photon machines Inc., Redmond, WA, USA) at the Macquarie University. The analyses were carried out by using the same laser conditions as for U-Pb dating. Detailed descriptions of analytical and calibration procedures have been given by Belousova et al. [5]. Quantitative results for 25 elements reported in this paper were obtained through calibration of relative element sensitivities using the NIST-610 standard glass and the GEMOC GJ-1 and Mud Tank zircon standards [47] as the external calibration standard as well as normalization of each analysis to the EMPA data for Hf as an internal standard. The precision and accuracy of the NIST-610 analyses are 1–2% for REE, Y, Sr, Nb, Hf, Ta, Th, and U at the ppm concentration level, 4% for Ti, 5% for Pb, and 20% for P at ppm concentrations.

3.4. Hf-Isotope Analysis

Methodology and an analytical condition for the Lu-Hf isotope analysis are provided by Griffin et al. [48]. Hf-isotope analyses were carried out in situ using a New Wave/Merchantek UP-213 laser-ablation microprobe (New Wave Research, Inc., Petach Tikva, Israel), attached to a Nu Plasma multi-collector ICP-MS (Nu Instruments Ltd., Wrexham, UK) at Macquarie University. The analyses were carried out with a beam diameter 55 μm and a 5 Hz repetition rate. Typical ablation times were 100–120 s and resulted in pits 30–40 μm deep. The ablated sample was transported by He carrier gas from the laser-ablation cell via a mixing chamber to the ICP-MS torch.

Interference of ^{176}Lu on ^{176}Hf is corrected by measuring the intensity of the interference-free ^{175}Lu isotope and using ^{176}Lu/^{175}Lu = 0.02669 [49] to calculate ^{176}Lu/^{177}Hf. Similarly, the interference of ^{176}Yb on ^{176}Hf has been corrected by measuring the interference-free ^{172}Yb isotope and using ^{176}Yb/^{172}Yb to calculate ^{176}Yb/^{177}Hf. The appropriate value of ^{176}Yb/^{172}Yb was determined by spiking the JMC475 Hf standard with Yb and finding the value of ^{176}Yb/^{172}Yb (0.58669) required to yield the value of ^{176}Hf/^{177}Hf obtained on the pure Hf solution. Detailed discussions regarding the overlap corrections for ^{176}Lu and ^{176}Yb are provided in Pearson et al. [50]. Precision and accuracy obtained on the ^{176}Hf/^{177}Hf ratio are illustrated by analyses of standard zircons in Griffin et al. [48] and Pearson et al. [50]. The typical 2SE precision on the ^{176}Hf/^{177}Hf ratios presented here is about 0.00002, which is equivalent to +0.7 eHf unit. The Mud Tank and Temora zircon were used as independent control on reproducibility and instrument stability. Most of the data and the mean value are within 2 s.d. of the recommended values reported for Mud Tank [^{176}Hf/^{177}Hf = 0.282522 ± 42 (2 s.d.)], [51]) and Temora reference material (0.282680 ± 15; [52]).

In order to calculate εHf values, the chondritic values of Bouvier et al. [53] were adopted: ^{176}Lu/^{177}Hf (Chondrite Uniform Reservoir (CHUR), today) = 0.0336, ^{176}Hf/^{177}Hf (CHUR, today) = 0.282785, and the decay constant for ^{176}Lu of 1.865 x 10^{-11} yr^{-1} ([54]). To calculate model ages (T$_{DM}$) based on a depleted-mantle source, we assume that the depleted mantle (DM) reservoir developed from an initially chondritic mantle is complementary to the crust extracted over time. T$_{DM}$ ages, which are calculated using the measured ^{176}Lu/^{177}Hf of the zircon, can only give a minimum age for the source material of the magma from which the zircon crystallized. Therefore, we have also calculated a "crustal" model age (T$_{DM}$^{C} in data tables) for each zircon that assumes its parental magma was produced from an average continental crust (^{176}Lu/^{177}Hf = 0.015, Geochemical Earth Reference Model database, http://www.earthref.org/), which was derived from a depleted mantle.

3.5. Geochronometry

3.5.1. U-Pb Dating by LA-ICP-MS

Zircon U-Pb ages were measured by using an Agilent 7700 quadrupole ICP-MS attached to a Photon Machines Excimer 193 nm laser system. The analyses were carried out with a beam diameter of 50 μm with 5 Hz repetition rate and energy of 8 J/cm^2. The analytical procedures for the U-Pb dating have been detailed previously [55]. A very fast scanning data acquisition protocol was employed to minimize signal noise. Data acquisition for each analysis was 3 min (1 min background, 2 min signal). Ablation was carried out in He to improve sample transport efficiency, provide more stable signals, and give more reproducible U/Pb fractionation. Provided that constant ablation conditions are maintained, accurate correction for U/Pb fractionation can then be achieved.

Sample analyses were bracketed by pairs of analyses of the GEMOC GJ-1 zircon standard [47]. The other well-characterized zircon standard 91500 was analyzed within the run as an independent control on reproducibility and instrument stability (see data tables). U-Pb ages were calculated from the raw signal data by using the online software package GLITTER (version 4.4.4, ARC National Key Centre, Sydney, Australia, www.glitter-gemoc.com; [56]). GLITTER calculates the relevant isotopic ratios for each mass sweep and displays them as time-resolved data. This allows isotopically homogeneous segments of the signal to be selected for integration. GLITTER then corrects the integrated ratios for ablation related fractionation and instrumental mass bias by calibration of each selected time segment against the identical time segments for the standard zircon analyses.

The common-Pb correction procedure of Andersen [57] was employed and the analyses presented here have been corrected assuming recent Pb-loss with a common-Pb composition corresponding to present-day average orogenic Pb as given by the second-stage growth curve of Stacey and Kramers [58] for $^{238}U/^{204}Pb=9.74$. No correction has been applied to analyses that are concordant within 2σ analytical error in $^{206}Pb/^{238}U$ and $^{207}Pb/^{235}U$ or which have less than 0.2% common lead.

3.5.2. (U-Th)/He Thermochronology

The age of eruption of the zircon host basalt was determined by (U-Th)/He thermochronology. This method, when applied to zircon, has a closure temperature of ~180 °C (for cooling rate of 10 °C/Ma, ~60 μm diameter diffusion domain equivalent sphere radius, [59]). In the simplest scenario (i.e., simple cooling without subsequent reheating), the ages measured by this method can be interpreted as true 'eruption ages' recording the passage of zircon-bearing magma to the surface and associated cooling. It is important to note that the traditionally used zircon U-Pb geochronology has a closure temperature in excess of ~900 °C and records the time of zircon crystallization in the magma chamber as well as provides a maximum limit for the eruption age, but it cannot directly date the eruption age.

The (U-Th)/He dating of zircon was conducted at the University of Waikato (New Zealand) by following the protocols described in Danišík et al. [60,61]. The large zircon megacrysts (mm–cm sized) could not be fit into the Nb microtubes (i.e., cylinders ~0.9 mm long, with internal diameter of ~0.6 mm) and were, thus, first abraded (thereby removing the uppermost ~30 microns of the surface) using an air-abrasion cell with pyrite as abrasion medium. This step circumvented the need of alpha ejection correction ([62]). Abraded megacrysts were then crushed in a steel mortar. The 60–150 μm fraction was separated using sieves and cleaned in an ultrasonic bath with ethanol. Clean shards (3–5 per sample) were individually loaded into Nb microtubes, degassed at ~1250 °C under ultra-high vacuum by using a diode laser, and analyzed for ^{4}He by isotope dilution on a Pfeiffer Prisma QMS-200 mass spectrometer. Following He measurements, the zircon shards in Nb microtubes were spiked with ^{235}U and ^{230}Th, and dissolved in hydrofluoric, nitric, and hydrochloric acids. The solutions were analyzed by isotope dilution for U and Th and by external calibration for Sm on a Perkin-Elmer SCIEX ELAN DRC II ICP-MS (PerkinElmer, Inc., Waltham, MA, USA). The total analytical uncertainty (TAU) was calculated as the quadratic addition on He and weighted uncertainties on U, Th, Sm, and He measurements and is typically ~2% (1σ). The raw zircon (U-Th)/He ages were not corrected for alpha ejection (Ft correction) given the abrasion step described above. Replicates (3 to 5 per sample) with associated uncertainties were used to calculate the geometric mean ([63]) and error-weighted standard deviation as representative eruption age for each sample. Replicate analyses of the Fish Canyon Tuff zircon (n = 18) measured over the period of this study as an internal standard yielded a mean (U-Th-Sm)/He age of 28.2 ± 0.6 Ma, which is in excellent agreement with the (U-Th-Sm)/He age of the Fish Canyon Tuff zircon mineral standard dated at 28.3 $\pm$ 1.3 Ma [64]. All replicates in each sample overlap within 1σ error bars and total analytical errors for individual replicates are significantly below 5%, which is typical for zircon dating procedures.

3.6. $\delta^{18}O$ Isotopic Analyses

Zircons were analyzed by laser fluorination at the University of Wisconsin, Madison. Zircon concentrates were prepared by standard crushing as well as gravimetric and magnetic techniques and soaked in cold HF overnight to remove radiation-damaged domains and contamination Bulk samples of ~2 mg were pre-treated in BrF_5 overnight, fluorinated by laser heating with a 32 W CO_2 laser operating at a wavelength of 10.6 µm, and analyzed by a gas-source mass spectrometer. Data are standardized against multiple analyses of the UWG-2 garnet standard performed on the same day ($\delta^{18}O$ = 5.80‰ VSMOW, [65,66]).

4. Results

4.1. Zircon Crystal Morphology and CL Zonation

Morphological studies were done on intact zircon grains as well as by CL after the crystals had been cut and polished to expose their centers. The recognition of the dominant forms, {100} prism and {101} bipyramid, as well as less pronounced {110} prisms and {211} pyramids in each crystal allows for discrimination of zircon morphological types by using the Pupin [67] classification. On the basis of empirical observations, Pupin [67] argued for a relationship between zircon morphology and the composition of the medium in which the zircon crystallized. Those crystallizing from a peraluminous melt are dominated by {211} pyramids and those crystallizing from peralkline melts are dominated by {101} pyramids. The alkalinity index [Al/(Na+K)] is, thus, designated as index A. The temperature of the growth medium (index T) is the dominant factor in controlling the relative development of prism faces with higher temperatures favoring {100} prisms and lower temperatures favoring {110} prisms.

RVP zircon xenocrysts occupy a narrow range of forms on the Pupin [67] typological correlation diagram and fall into two categories: (1) those dominated by the {100} prism and {101} bipyramid belonging to subgroup J5 and (2) those with additional moderately developed {110} prisms and {211} dipyramids belonging to subgroups S19–20 and S24–25. There is no evidence in CL images to suggest that rapid changes in the magma composition or temperature took place during zircon crystallization. Both morphological populations suggest crystallization in an alkaline environment at temperatures of 800–900 $\pm$ 50 °C.

All zircon display strong, primary, magmatic oscillatory growth zoning when imaged by CL (Figure 7), which is a reflection of the non-equilibrium conditions during crystallization [68]. Oscillatory growth areas are very fine with bands on the order of <1–25 µm (Figure 7a,b). A number of grains also display distinctive sector zoning upon which the oscillatory zoning is imposed (Figure 7c–e). Zonation patterns are generally very sharp and crystallographically-controlled. With the exception of one grain, all zircon are non-metamict, unaltered, and do not contain relict cores, which confirms their primary igneous origin. Studies by Witter et al. [69] also found RVP zircon to have negligible radiation damage with time-integrated self-irradiation doses on the order of 10^{16} alpha-decay events/gram (α/g) (note that detectable radiation damage requires a minimum of 0.1–0.2 $\times$ 10^{18} α/g). One grain (BL53-1, Figure 7f) contains a core that is black in CL and does not display any discernible zoning pattern in either CL, BSEI, or optically. Limited yet distinct magmatic zoning is observed in BSE images of RVP zircon. However, when the CL and BSE images can be compared, dark regions in CL correspond to bright areas (highest mean atomic weight) in BSE images. Chemical analyses of the individual zones reveal Th and U to be the most important influence, measurable by EMPA, on the observed zonation patterns—regions rich in Th + U are darkest in CL (both sector and oscillatory zoning) and brightest in BSE images. Optically, zones enriched in Th + U are darker in color and range from dark orange-brown to dark red.

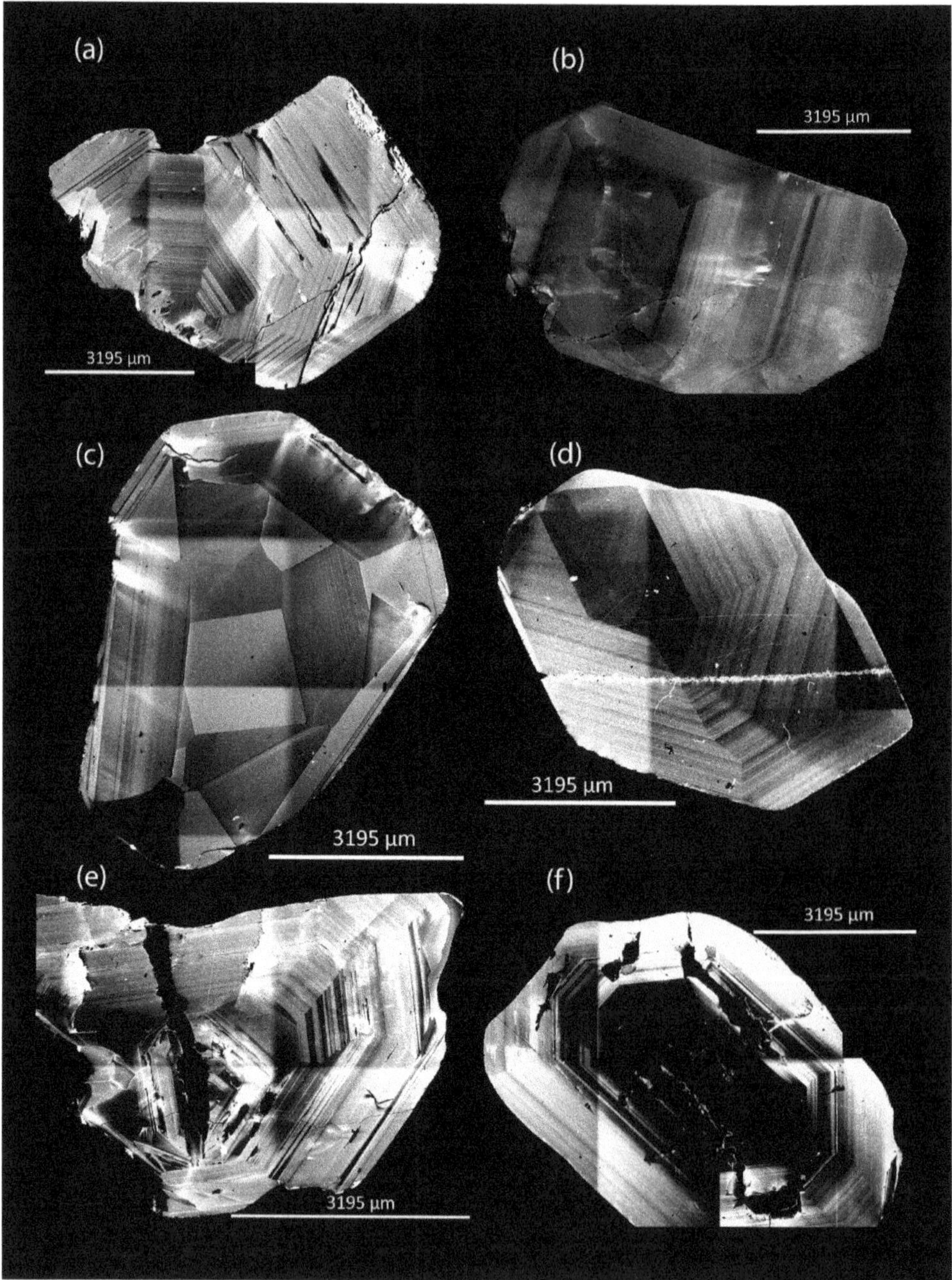

Figure 7. Cathodoluminescence images of RVP zircon xenocrysts showing both oscillatory and sector zonation. Darker zones are enriched in Th, U, and REE: (**a**) concentric zoning in BC43-2, (**b**) concentric zoning in PT49-2, (**c**) sector zoning in PT48-3, (**d**) sector and concentric zoning in BL54-3, (**e**) sector and concentric zoning in BL53-2, and (**f**) dark, CL-inactive core in BL53-1.

These results are in direct contrast to those presented by Cong et al. [23] for six alluvial zircon from an unknown locality in the RVP. Cong et al. [23] indicate that none of the zircon display oscillatory magmatic zoning in CL images and, that in BSE images, the samples are homogeneous. Cathodoluminescence images presented in the Cong et al. [23] study show very weak growth zoning, which is contrary to statements made within the text. In total, more than 50 zircons from the RVP deposits were examined by CL imaging in this study and all showed strong growth zoning. This discrepancy between the two sample sets may simply be the result of incorrect settings during CL imaging by Cong et al. [23]. However, the authors later use these results incorrectly as evidence for thermal homogenization of the zircons in the mantle.

4.2. Zircon Chemistry

Average EMPA data are shown in Table 1 and representative trace element data (ppm) are shown in Table 2. Rim and core analyses are noted when applicable. Figure 8 depicts box plot diagrams for relevant trace elements (a) and ratios (b) in zircon from each locality. Rare-earth element (REE) data were normalized to C-1 chondrite values [70] and plotted on logarithmic multi-element diagrams (La-Lu, Figure 9). Figure 10 depicts the chondrite-normalized trace element spidergram for the RVP zircon (in shaded orange). Data from Cong et al. [23] as well as for zircon from a variety of rock types [5].

Table 1. Average EMPA analyses for RVP zircon xenocrysts.

Sample	BC45	BC46-1	BC46-2	BS47	PT48-1	PT48-2	PT48-3	PT49-1	PT49-2	PT50-1	PT50-2	PT50-3
n *	6	7	7	8	7	4	10	8	7	8	7	7
HfO_2	0.74	0.72	0.80	0.69	0.74	0.77	0.73	0.77	0.80	0.71	0.76	0.79
ZrO_2	66.87	66.04	66.57	66.58	67.04	66.67	67.07	67.50	67.52	67.46	67.39	67.34
Y_2O_3	0.04	0.01	0.03	0.10	0.02	0.03	0.06	0.02	0.03	0.03	0.02	0.03
ThO_2	0.01	0.00	0.01	0.01	0.00	0.01	0.01	0.01	0.01	0.00	0.00	0.01
UO_2	0.01	0.00	0.00	0.02	0.01	0.02	0.01	0.01	0.00	0.01	0.01	0.01
SiO_2	32.34	31.69	32.05	32.62	32.47	32.38	32.49	32.35	32.26	32.32	32.30	32.22
Total	100.01	98.47	99.46	100.01	100.28	99.86	100.37	100.66	100.62	100.53	100.47	100.40

Sample	BL51-1	BL51-2	BL52	BL53-1	BL53-2	BL54-1	BL54-2	BL54-3	BL55-1	BL55-2	BL55-3	BL55-4
n *	6	6	8	16	10	10	5	11	5	6	7	8
HfO_2	0.79	0.72	0.69	0.75	0.73	0.71	0.72	0.75	0.75	0.71	0.71	0.70
ZrO_2	67.66	67.84	68.17	65.79	66.31	66.04	65.80	65.39	66.93	67.20	66.98	67.11
Y_2O_3	0.03	0.04	0.07	0.11	0.03	0.03	0.04	0.17	0.08	0.03	0.06	0.05
ThO_2	0.01	0.00	0.01	0.05	0.01	0.01	0.01	0.07	0.01	0.01	0.01	0.01
UO_2	0.01	0.01	0.01	0.04	0.01	0.01	0.01	0.05	0.03	0.02	0.01	0.01
SiO_2	32.08	32.06	31.99	32.67	32.68	32.38	32.30	32.26	32.28	32.27	32.27	32.35
Total	100.58	100.67	100.94	99.41	99.78	99.19	98.88	98.69	100.07	100.24	100.04	100.23

n * = number of analyses.

Table 2. Average trace element composition, ratios, and sums (ppm) of RVP zircon xenocrysts.

Locality (Code)	Bokeo Clas (BC)			Bei Srok (BS)	Phum Throm (PT)					BoLoei (BL)	
Sample # (n) *	45 (5)	46-1 (4)	46-2 (4)	47 (5)	49-1 (4)	49-2 (4)	50-1 (4)	50-2 (4)	50-3 (3)	53-1 (3)	53-2 (4)
HfO_2 wt %	0.73	0.71	0.80	0.68	0.78	0.79	0.71	0.76	0.80	0.75	0.72
Hf wt %	0.62	0.60	0.67	0.58	0.66	0.67	0.60	0.64	0.68	0.64	0.61
P	58	47	51	91	34	43	51	51	54	53	47
Ti	5.1	4.5	3.7	7.7	3.1	3.1	4.6	3.7	3.8	3.8	4.0
Nb	4.0	1.7	2.3	11.7	1.3	2.0	2.7	2.4	3.2	3.9	2.5
Ta	1.7	0.9	1.3	3.3	0.6	1.1	1.2	1.3	1.6	1.8	1.0
Pb^{204}	1.05	1.16	1.02	1.48	0.97	1.27	1.19	1.18	1.61	1.58	1.28
Pb^{206}	0.05	0.05	0.02	0.13	0.06	0.07	0.03	0.03	0.06	0.18	0.04
Pb^{207}	0.03	0.02	0.02	0.04	0.02	0.02	0.00	0.02	0.03	0.01	0.04
Pb^{208}	0.02	0.06	0.01	0.02	0.02	0.03	0.02	0.01	0.07	0.02	0.01
Th	42	19	13	168	9	17	25	25	30	51	22
U	83	41	39	187	23	43	55	58	61	90	42
Th/U	0.46	0.46	0.33	0.88	0.35	0.35	0.43	0.44	0.38	0.54	0.41
Nb/Ta	2.2	2.0	1.8	3.5	2.3	2.0	2.2	1.8	2.1	2.1	2.4
Yb/Sm_N	71	77	86	25	83	79	59	67	97	52	79
Lu/Gd_N	20	21	24	7	23	22	15	19	28	13	20
Y/Hf	0.00	0.00	0.00	0.00	0.00	0.00	0.00	0.00	0.00	0.00	0.00
Eu/Eu^*_N	1.00	1.01	0.99	0.99	1.04	1.00	1.00	1.03	0.95	1.02	0.99

Table 2. *Cont.*

Locality (Code)	Bokeo Clas (BC)			Bei Srok (BS)	Phum Throm (PT)					BoLoei (BL)	
Sample # (n) *	45 (5)	46-1 (4)	46-2 (4)	47 (5)	49-1 (4)	49-2 (4)	50-1 (4)	50-2 (4)	50-3 (3)	53-1 (3)	53-2 (4)
Y	365.9	150.4	169.9	921.8	119.7	196.1	259.3	245.1	222.7	299.8	206.2
La	bdl	bdl	bdl	bdl	bdl	bdl	bdl	bdl	bdl	bdl	bdl
Ce	1.7	0.7	0.9	5.3	0.6	1.0	1.0	1.1	1.2	2.3	1.2
Pr	0.1	0.0	0.0	0.2	0.0	0.0	0.0	0.0	0.0	0.0	0.0
Nd	0.9	0.3	0.3	3.8	0.3	0.6	0.5	0.7	0.3	0.6	0.3
Sm	1.7	0.6	0.6	6.7	0.6	1.0	1.2	1.3	0.7	1.2	0.8
Eu	1.3	0.5	0.5	4.6	0.4	0.7	0.9	0.9	0.6	1.0	0.6
Gd	8.8	3.5	3.4	29.7	2.8	4.8	6.4	6.1	4.2	7.4	4.8
Tb	2.9	1.1	1.2	8.8	0.9	1.5	2.1	1.9	1.5	2.4	1.6
Dy	37	15	16	103	12	19	27	24	20	31	21
Ho	12	5	6	30	4	7	9	8	7	10	7
Er	55	23	26	120	18	30	39	37	35	43	30
Tm	11	5	5	20	4	6	7	7	7	8	6
Yb	100	44	54	166	36	57	63	70	69	68	52
Lu	18	8	10	25	7	11	11	13	13	12	9
ΣREE	250	107	124	524	86	139	168	172	160	186	135
ΣREE+Y	616	257	294	1446	205	335	427	417	383	486	341
Ce/Ce*$_N$	18	11	23	24	14	21	19	13	38	31	37

bdl = below detection limit.

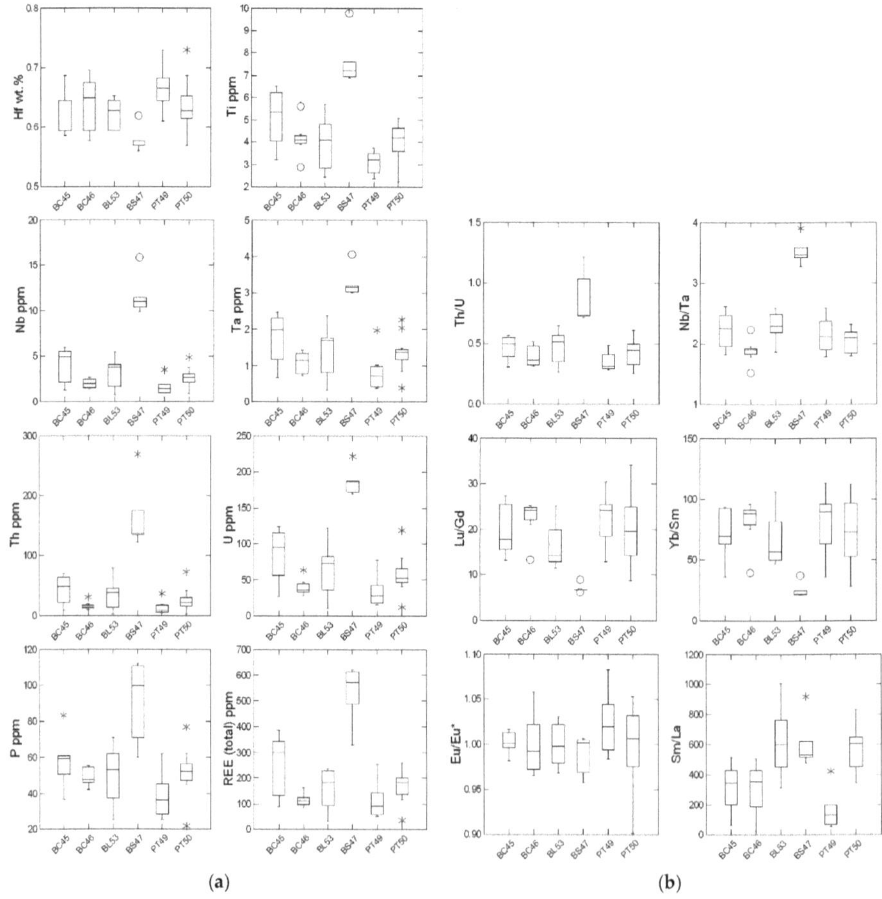

(a)   (b)

Figure 8. Boxplot diagrams for relevant trace element (**a**) and trace element ratios (**b**) in RVP zircon xenocrysts. The vertical line and whiskers represent the range in the data values with the horizontal line representing the median of the data set. The box is defined by the first and third quartiles.

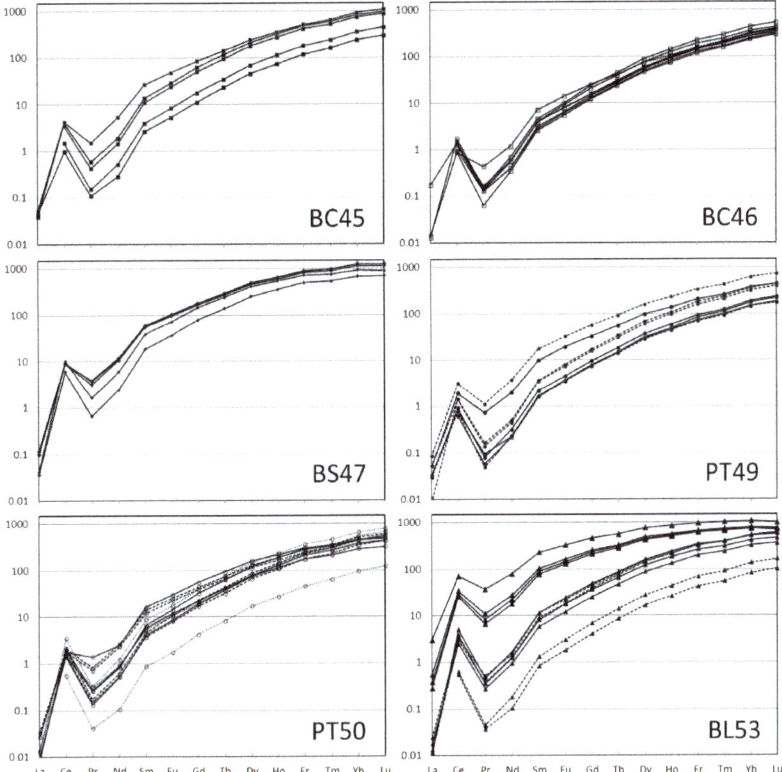

Figure 9. Average chondrite-normalized REE patterns for RVP zircon xenocrysts.

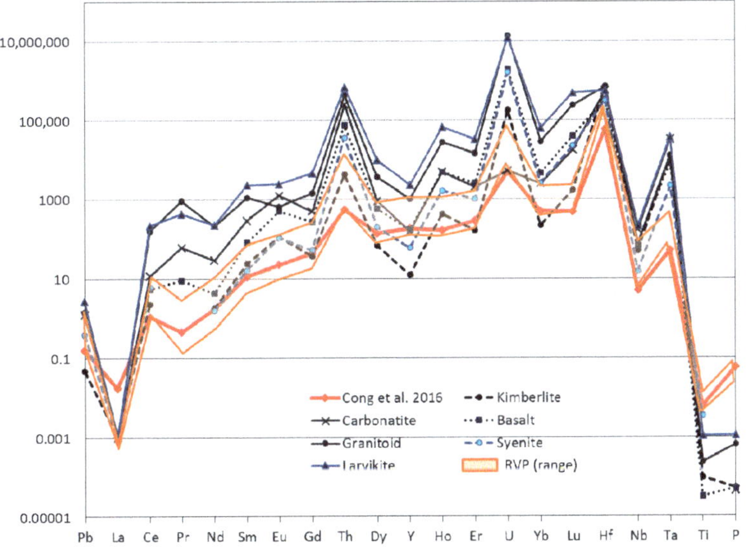

Figure 10. Chondrite-normalized trace element spidergram for zircon from the RVP (orange shaded region) including average data from Cong et al. (red [23]) and zircon from other rock types [5].

Electron microprobe analysis of the zircon xenocrysts gave major element compositional ranges of SiO_2 from 30.93 to 32.87 wt % and ZrO_2 65.06 to 68.36 wt %. Hafnium contents range from 0.66 to 0.86 wt % with an average of 0.74 wt % and do not show significant variation from core to rim. Y_2O_3 contents range up to 0.23 wt % (average: 0.05 wt %). Zircon from Bei Srok has lower HfO_2 than other RVP zircon (range: 0.66 to 0.73 wt %, average: 0.68 wt %). The Zr/Hf ratios (elemental) range from 66 to 88 (average: 79), which is higher than expected for mantle zircon (Zr/Hf = 60–68, Pupin [71]). Zircon from the RVP show wide variations in Ti, Nb, Ta, U, Th, P, Y, and REE contents between both grains and localities. In general, all RVP zircon show heavy REE (HREE) enrichment, have a positive Ce anomaly, and lack a negative Eu anomaly with ΣREE+Y, U, Th, and Ti contents exhibiting a decrease from core to rim in most samples. No compositional difference is observed between morphological types of the xenocrysts. Two distinct compositional populations exist: (1) the main population of RVP zircon from Bokeo Clas, Phum Throm, and BoLoei and (2) zircon from Bei Srok, which have higher overall trace element concentrations than the main RVP population.

Zircon from the main population have trace element concentrations as follows [ppm range (average)]: P 22–88 (average 51), Ti 1.6–6.5 (average 3.7), Nb 0.8–6 (average 2.6), Ta 0.3–3 (average 1.3), Th 2.6–135 (average 28), U 10–157 (average 55), and Y 43–560 (average 227). Average Nb/Ta and Th/U ratios are 2.11 and 0.44, respectively (Figure 5b). The bulk of the zircon are characterized by low ΣREE contents (31–386 ppm, average 154 ppm), which is typical of xenocrysts from alkaline mantle sources [5]. Total REE(+Y) contents decrease from the core to the rim in all samples. The chondrite-normalized REE patterns (Figure 10) show strong light REE (LREE) depletion, which is a pronounced positive Ce anomaly (Ce/Ce*$_N$ = 11–38, average 27, Eu/Eu*$_N$ = 1, and heavy REE (HREE) enrichment characterized by a steep, straight slope with Yb/Sm$_N$ ranging from 28–113 (average 75) and Lu/Gd$_N$ = 9–34 (average 20). Normalized Ce and Eu anomalies were calculated as Ce/Ce*$_N = \frac{Ce}{\sqrt{La \times Pr}}$ and Eu/Eu*$_N = \frac{Eu}{\sqrt{Sm \times Gd}}$, respectively.

Zircon from Bei Srok (BS47) has higher overall trace element contents as compared to those from other RVP localities [Figure 8, range (average) ppm]: P 60–112 (average 91), Ti 7–10 (average 8), Nb 10–16 (average 12), Ta 3–4 (average 3), Th 123–269 (average 168), U 170–221 (average 187), and Y 574–1076 (average 922). Total REE contents range from 328 to 621 ppm (average 524 ppm) and decrease from core to rim. Bei Srok zircon has significantly higher Nb/Ta (average 3.53) and Th/U (average 0.88) as compared to zircon from BC, PT, and BL (Figure 9b). The chondrite-normalized REE pattern for BS zircon (Figure 9) shows strong LREE depletion with a pronounced positive Ce anomaly (Ce/Ce*$_N$ = 13–38, average 24), Eu/Eu*$_N$ = 0.99, and middle REE (MREE)/HREE enrichment characterized by a steep, slightly concave-down curvature with Yb/Sm$_N$ = 21–37 (average 25) and Lu/Gd$_N$ = 6–9 (average 7).

Ti-in-Zircon Thermometer

Watson and Harrison [72] and Watson et al. [73] used high-pressure/high-temperature experimental methods coupled with analyses of natural zircons to determine the exact relationship between Ti content and crystallization temperature in zircon TiO_2. The resultant Ti-in-zircon thermometer (Equation (1)) has the capability to provide temperature with a precision of ± 5 °C depending on the Ti concentration and the analytical method employed [73].

$$T\,(^{\circ}C)_{zircon} = [5080/(6.01 - \log(Ti_{ppm})] - 273 \tag{1}$$

A number of limiting factors must be considered when utilizing the Ti-in-zircon thermometer including the variation of TiO_2 or SiO_2 activity in the melt, the pressure fluctuation, changes in the Ti content of the zircon as a result of solid state diffusion or post-crystallization alteration, and the assumption that the temperature dependence of Ti in zircon is an equilibrium process and obeys Henry's Law [74]. Cherniak and Watson [75] demonstrated that Ti is strongly retained by the zircon structure in both anhydrous (1 atm) and H_2O-CO_2-bearing (1.1–1.2 GPa) systems at temperatures between 1350 and 1550 °C.

Titanium contents in the analyzed zircon are all above the detection limit (0.1 ppm) and range from 2 to 10 ppm (Table 3). All zircon are primary magmatic phases with oscillatory zonation, which is the result of intrinsic non-linear feedback between the parental melt and the growing crystal [68]. Although this zonation suggests local non-equilibrium during growth, which is a result of the continuous process of depletion and replenishment of solute at the crystal-melt boundary, there is no evidence for large-scale disequilibrium due to extrinsic mechanisms in the system. This indicates that Ti contents in the zircon are a primary magmatic feature. In addition, there is no evidence for metamictization or alteration which could be responsible for Ti diffusion. As such, the temperatures obtained by the Ti-in-zircon thermometer are considered representative of the difference in relative crystallization temperatures between zircon from the RVP localities.

Table 3. Ti-in-zircon temperatures for RVP zircon xenocrysts.

Locality	Sample # (n) *	Ti ppm $\pm$ 2 SD	T °C $\pm$ 2 SD
Bokeo Clas	45 (5)	5.07 $\pm$ 2.83	682 $\pm$ 46
-	46-1 (4)	4.53 $\pm$ 1.45	675 $\pm$ 23
-	46-2 (4)	3.75 $\pm$ 1.20	661 $\pm$ 26
Bei Srok	47 (5)	7.68 $\pm$ 2.38	718 $\pm$ 24
Phum Throm	49-1 (4)	3.05 $\pm$ 1.04	646 $\pm$ 24
-	49-2 (4)	3.14 $\pm$ 1.07	648 $\pm$ 26
-	50-1 (4)	4.64 $\pm$ 0.81	677 $\pm$ 14
-	50-2 (4)	3.67 $\pm$ 0.36	660 $\pm$ 7
-	50-3 (3)	3.81 $\pm$ 2.87	659 $\pm$ 63
BoLoei	53-1 (3)	3.77 $\pm$ 1.23	661 $\pm$ 26
-	53-2 (4)	4.03 $\pm$ 3.52	662 $\pm$ 69

* n = number of analyses, SD = standard deviation.

Average Ti-in-zircon temperatures are 663 $\pm$ 28 °C (range: 601–739 °C, Table 3). In general, there is a decrease in the calculated temperature from core to rim in all samples with ΔT ranging from 10–55 °C. Wider variation in temperature between core and rim may simply reflect the larger size of the zircon grains themselves, which is a possible reflection of residence time within the parental melt. No difference in calculated temperatures is observed between the various morphological types, which is contrary to what is predicted based on Pupin's T versus A diagram [67]. Temperatures based on morphological studies suggest crystallization at 800 to 900 °C, which is significantly higher than that predicted by Ti-in-zircon temperatures. The lower T for Ti-in-zircon results than predicted for natural zircon may result from pressure and Ti^{4+} substitution effects [76]. The relatively low Ti-in-zircon for RVP samples may reflect the likely high-pressure mantle origin of these xenocrysts, which has also been suggested for kimberlite zircons [77].

4.3. Geochronology Results

U-Pb age dates were obtained for 11 zircon grains from the four mining regions with three to nine analyses per grain (Table S1). All samples lack any significant U-Pb age differences between core and rim, which confirms the lack of inherited cores and a single growth event for all the zircon. The young zircon xenocrysts are depleted in ^{207}Pb and subsequently ^{207}Pb/^{235}U ages are very poor with large associated errors. As a result, only ^{206}Pb/^{238}U is reported in this case (Table 4). Weighted mean ^{206}Pb/^{238}U ages range from 0.88 $\pm$ 0.22 Ma (PT50) to 1.56 $\pm$ 0.21 Ma (PT49) at the 95% confidence level (1σ error, Table 4). Seven U-Pb ages determined by ID-TIMS on RVP zircon during a gemological study [45] gave an average age of 0.92 $\pm$ 0.07 Ma (range: 0.83–1.03 Ma), which is in agreement with analyses in this study (Table 4). Analyses of a zircon inclusion in corundum from the BL deposit gave a U-Pb age of 0.855 $\pm$ 0.098 Ma, which suggests a temporal, if not a genetic, relationship between the two xenocrystic phases [12]. Similar age ranges are noted for zircon xenocrysts from Eastern

Gondwana margins including Western Cambodia (Table 4). Fission track dating of zircon from Bokeo by Carbonnel et al. [78] yielded ages ranging from 1.10–1.41 Ma, which is also in agreement with the ages from this study.

Table 4. U-Pb, fission track and (U-Th)/He age dates for ZIP xenocrysts.

Locality	U-Pb	Fission Track/(U-Th)/He	Reference
Ratanakiri Volcanic Province (This Study)	-	-	-
Bokeo Clas (BC45)	0.98 ± 0.12	-	This study
Bokeo Clas (BC46)	1.07 ± 0.19	0.91 ± 0.02	-
Bei Srok (BS47)	1.143 ± 0.073	1.02 ± 0.02	-
Phum Throm (PT49)	1.56 ± 0.21	0.93 ± 0.02	-
Phum Throm (PT50)	0.88 ± 0.22	1.02 ± 0.02	-
-	-	0.86 ± 0.02	-
BoLoei (BL53)	0.978 ± 0.054	0.96 ± 0.06	-
-	-	0.96 ± 0.09	-
RVP (unknown locality)	0.92 ± 0.07	-	[45]
-	-	-	-
Pailin, Cambodia	2.74 ± 0.47	2.14 ± 0.02	This study
	2.27 ± 0.15–2.60 ± 0.20		[78]
-	1.77 ± 0.15–2.0 ± 0.20	-	[79]
Chanthaburi-Trat, Thailand	1.19 ± 0.29–2.22 ± 0.22	0.90 ± 0.04–2.13 ± 0.04	This study
-	2.57 ± 0.20	-	[78]
Dak Nong, Vietnam	1.05 ± 0.05	-	[7]
-	7.13 ± 0.88	-	-
Xuan-Loc, Vietnam	0.57–0.70	-	[78]
-	0.2–0.9	-	-
Ban Huai Sai, Laos	1.14 ± 0.07–1.46 ± 0.06	-	[38]
-	2.4 ± 0.7–4.3 ± 1.0	-	[11]
Penglai, Hainan Island, China	4.4 ± 0.1	-	[80]
-	-	4.06 ± 0.35	[81]
Mingxi, China	1.2 ± 0.1	-	[13]
NE Australia	2.61 ± 0.16–2.92 ± 0.16	-	[10]

A total of 37 crystal shards of zircon crystals from the four mining regions were dated by the (U-Th)/He method (Table S2). Reproducibility of (U-Th)/He ages (ZHe) from each locality (typically 5-6 ages) was excellent with all replicates overlapping within analytical uncertainty. Final ZHe ages, calculated for each locality as error-weighted means and standard deviation as uncertainty, range between 0.86 ± 0.02 and 1.02 ± 0.02 Ma (Table 4). These are slightly younger or overlap with uncertainty via the corresponding U-Pb ages measured by LA-ICP-MS, which represent a minimum time for zircon crystallization in accordance with the closure temperature concept [82] and also confirms the accuracy of both dating methods. The age and similarity of ZHe and U-Pb ages points to an extremely fast cooling of the crystals from magmatic temperatures through the zircon helium closure isotherm and suggests that the dated crystals erupted to the surface approximately in the same time between 1.02 and 0.86 Ma. As with the U-Pb ages, there seems to be no correlation between the ZHe eruption ages and the spatial distribution of dated samples.

4.4. Hf Isotope Signatures

Zircon from the RVP deposits show a moderate range in the Hf isotope composition (Table S3)—the $^{176}Hf/^{177}Hf$ ratio ranges from 0.282919–0.283065 (average 0.282982 ± 15, n = 20), which corresponds to positive εHf values ranging from 4.77 to 9.92 (average 7.00 ± 0.25). Zircon from Bokeo Clas, Bei Srok, and BoLoei are the most radiogenic with the highest $^{176}Hf/^{177}Hf$ ratios (BC45 = 0.283065, εHf 9.92, BS47 = 0.283039, εHf 9.01, and BL53 = 0.283021, εHf 8.37), which all exhibit within-grain variation greater than their 2SE uncertainties. While variations in $^{176}Hf/^{177}Hf$ within single grains are observed, there is no consistent relationship between core and rim compositions.

Estimated minimum model ages (T_{DM}) for the source material range from 260 to 401 Ma with crustal model ages ($T_{DM}{}^{C}$) ranging from 391 to 633 Ma.

Zircon from Phum Throm (PT) are the least radiogenic and exhibit less within-grain and between-grain variation (Figure 8). The $^{176}Hf/^{177}Hf$ ratio ranges from 0.282919 to 0.282985 (average 0.282964 $\pm$ 6) with εHf 4.77–7.11 (average 6.34 $\pm$ 0.59). Estimated minimum model ages (T_{DM}) for the source material are higher for PT zircon and range from 372 to 462 Ma (average 403 Ma) with crustal model ages ($T_{DM}{}^{C}$) ranging from 605 to 754 Ma (average 654 Ma).

4.5. Oxygen Isotope Ratios

Oxygen isotope ratios provide valuable information about the melt from which the zircon crystallized, which often allows for the discrimination between mantle and crustal sources. Values of $\delta^{18}O$ in RVP zircon are homogeneous with $\delta^{18}O$: 4.93 $\pm$ 0.05‰ VSMOW (n = 5, Table 5). These values are consistent with those obtained by SIMS for zircon megacrysts from an unknown RVP locality: $\delta^{18}O$ = 4.5–5.5‰ VSMOW, weighted mean = 5.0 $\pm$ 0.18‰ VSMOW, n = 210 [23]. The slow rate of diffusion of ^{18}O in non-metamict zircon, coupled with the large crystal size and primary magmatic compositional zoning indicate the $\delta^{18}O_{zrn}$ to represent initial magmatic values [83,84]. These values are low, but, within the $\delta^{18}O_{zrn}$ range for igneous zircon in high temperature equilibrium with a mantle source ($\delta^{18}O_{zrn}$ = 4.7–5.7‰ VSMOW) and olivine from mantle xenoliths ($\delta^{18}O_{ol}$ = 5.15 $\pm$ 0.13‰ [85]). Similarly, RVP $\delta^{18}O_{zrn}$ are lower than those observed for other alkali basalt-hosted zircon xenocrysts in SE Asia including in Western Cambodia at Pailin (5.95 $\pm$ 0.02‰), Chanthaburi-Trat and Bo Phloi (Thailand, 5.82–6.16‰, Table 5) and Penglai, Hainan Island (South China, $\delta^{18}O_{zrn}$ = 5.31 $\pm$ 0.18‰ [80]). The lower $\delta^{18}O$ values observed in RVP zircon are akin to values observed in zircon from monominerallic anorthoclase xenoliths, which are called anorthoclasites, in alkali basalt at Elie Ness, Scotland (4.49–5.03‰, average 4.78 $\pm$ 0.16‰; [86]) and in zircon from kimberlite [77,87]. The $\delta^{18}O_{zrn}$ values for RVP zircon suggest crystallization from a homogeneous igneous reservoir in the SCLM underlying the RVP with little to no contamination by supracrustal lithologies.

Table 5. $\delta^{18}O$ data for xenocrystic zircon from the RVP and Thailand.

Sample	mg	μml	μml/mg	$\delta^{18}O$ raw	$\delta^{18}O$ VSMOW
Cambodia-RVP	-	-	-	-	-
Bokeo Clas, BC46	2.85	32.7	11.5	4.93	4.95
Bei Srok, BS47	3.14	34.8	11.1	4.87	4.89
Phum Throm, PT49	2.82	30.7	10.9	4.99	5.01
Phum Throm, PT50	3.40	36.7	10.8	4.86	4.88
Bo Loei, BL55	3.26	35.1	10.8	4.90	4.92
Cambodia-Pailin	2.97	32.1	10.8	5.93	5.95
Thailand	-	-	-	-	-
Khao Ploi Waen, Chanthaburi-Trat	3.80	41.0	10.8	6.14	6.16
Tok Phrom, Chanthaburi-Trat	3.05	32.3	10.6	5.80	5.82
Bo Phloi, Kanchanaburi	3.46	37.1	10.7	5.86	5.88
Standards	-	-	-	-	-
UWG-2, garnet	3.39	45.6	13.5	5.78	-
UWG-2, garnet	2.29	31.4	13.7	5.79	-
UWG-2, garnet	2.16	28.3	13.1	5.80	-
UWG-2, garnet	2.70	36.0	13.3	5.77	-

* n = 4, x = 5.78, $\pm$ 0.02‰ 2 std. dev. Note: Thailand analyses performed under the same analytical conditions as RVP zircon.

5. Discussion

5.1. Classification and Comparisons of RVP Zircon Xenocrysts

The extensive substitution of trace elements (up to 50) into the zircon structure is useful in studies of the compositional variation in igneous zircons as a petrogenetic indicator. Such data can reveal fractionation processes, the nature of source rocks, and provenance of zircons from secondary deposits [5,88–92]. Shnukov et al. [92] and Pupin [93] utilized Hf and Y concentrations to classify magmatic zircons by focusing on early-stage and late-stage magmatic phases from both mantle and crustal rocks of orogenic and anorogenic origins. Belousova et al. [5] analyzed trace element compositions of zircons from nine different igneous rock types ranging from ultramafic to granitic and syenitic in nature as a way to statistically discriminate between zircon populations and assist in the identification of their source regions. The study suggested that compositional variation in HfO_2, Y, Nb, Ta, Lu, Ce/Ce^*_N, ΣREE, Yb/Sm_N, and Th/U are the most useful petrogenetic indicators across the represented range of rock types and that the provenance or parent rock type of a zircon can be determined at confidence levels of 75% or more.

Zircon xenocrysts from the RVP typify zircons that crystallize from a juvenile, Zr-saturated, REE+Y-enriched alkaline magma or melt with low HfO_2 (<0.9 wt %), Y_2O_3 (<0.2 wt %), U, Th, ΣREE (<600 ppm), $Eu/Eu^*_N \approx 1$, and steep, and HREE-enriched REE profiles [4,5,94]. RVP zircon xenocrysts exhibit the depletion of $\Sigma REE+Y$, U, and Th with increasing Yb/Sm_N from core to rim at almost a constant Hf content. Such zoning favors zircon growth as a single growth stage in an alkaline environment rather than growth involving fractional crystallization [22,94,95]. In depolymerized alkaline melts, the partition coefficients for Hf and Zr between zircon and melt are approximately equal over a wide T range and melt compositions [22]. As such, crystallization of zircon will not fractionate Zr and Hf significantly, which results in near constant Zr/Hf in the zircon/melt. An accompanying trend of decreasing REE, Y, U, and Th and increasing Yb/Sm_N suggests that RVP zircon crystallized during a single magmatic event with no influx of additional melt components. The lack of a negative Eu anomaly in these zircons also suggests crystallization in a system where plagioclase did not crystallize or at pressures above the plagioclase stability zone (~0.80 GPa in fertile mantle lherzolite [96]), which lends further support to a sodium-potassium-rich melt. Increasing Yb/Sm_N from core to rim may indicate the depletion of the melt in LREE as a result of a co-crystallizing phase (i.e., apatite) or selective absorption of HREE during the crystallization period [94]. RVP zircon xenocrysts found as isolated crystals within the host alkali basalt, contain corroded REE-rich apatite, and REE carbonates as rare inclusions in larger crystals. Lacombe [43] also noted a carbonate mineral within the RVP zircon as a further sink for LREE and a likely growth environment enriched in CO_2.

Utilizing the classification by Pupin [93] based on HfO_2 and Y_2O_3 contents, all RVP zircons plot in field 1c, mantle-derived hawaiite and alkali basalt, which lies within a narrow range of Hf contents and a spread of Y contents (18–2011 ppm). In Shnukov's Y versus Hf diagram, RVP zircons almost entirely fall within the carbonatite field (VII), with rare samples in the alkaline rocks and alkaline metasomatite field (VI) [92]. Such affinities also apply to zircon xenocrysts from nearby Ban Huai Xai, Laos [11], and Pailin, and Cambodia (Piilonen, unpublished data, Figure 11). The statistical classification scheme by Belousova et al. [5] predicts a cabonatitic provenance for almost all RVP zircon xenocrysts in agreement with the Y versus Hf discrimination diagram except for BS47 in which a higher Lu content (>20.7 ppm) suggests an origin from a basaltic parental melt.

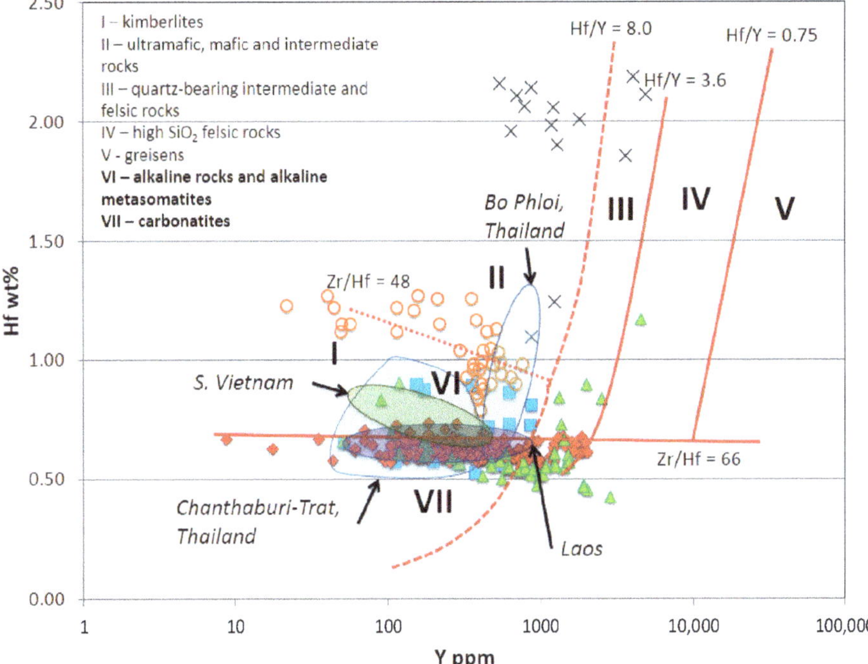

Figure 11. Y versus Hf discrimination diagram for RVP zircon, zircon xenocrysts from ZIP localities, and zircon inclusions in corundum (modified after [92]). Red diamonds: RVP zircon, blue squares: Pailin, Cambodia (Piilonen, unpublished data), blue field: Thailand (Piilonen, unpublished data), green triangles: Australia [11], orange circles: zircon xenocrysts from carbonatites [5], black crosses: zircon inclusions in corundum xenocrysts [10,12,38,97].

Isolated RVP zircon xenocrysts differ from many other zircon xenocryst suites that are commonly spatially associated with corundum xenocrysts either in multimineralic xenoliths or as discrete crystals in heavy mineral separates, e.g., Bo Phloi and Chanthaburi-Trat regions in Thailand (Piilonen, unpublished data, [98,99]), South Vietnam [100], Loch Roag, Scotland [97], and Lava Plains, NE Australia [10,12]. Such zircons associated with corundum, either as xenocrysts or as corundum-hosted inclusions, have consistently higher Hf (>1 wt % and up to 4.5 wt % HfO₂) and often negative Eu anomalies, which suggests evolved, likely peraluminous corundum-bearing sources (Figure 11). Zircon inclusions in blue-green-yellow magmatic corundum xenocrysts (BGY suite) not only show enriched Hf and noticeable negative Eu anomaly but also have enrichments in Y (up to 1 wt % Y₂O₃), U (up to 1 wt % UO₂), Th (up to 2.1 wt % ThO₂), and REE up to 12,060 ppm [11,101–103]. These corundum-hosted zircons are dominated by the {110} prism over the {100} prism unlike the RVP zircon xenocrysts and are a morphology thought to be related to high U, Th, Y, REE, and P contents in the parental melt [104]. Plotted on the Y versus Hf discrimination diagram (Figure 12), the zircon compositions hosted in RVP corundum fall across Type II and III sources (II—ultramafic, mafic, and intermediate rocks, III—quartz-bearing intermediate and felsic rocks).

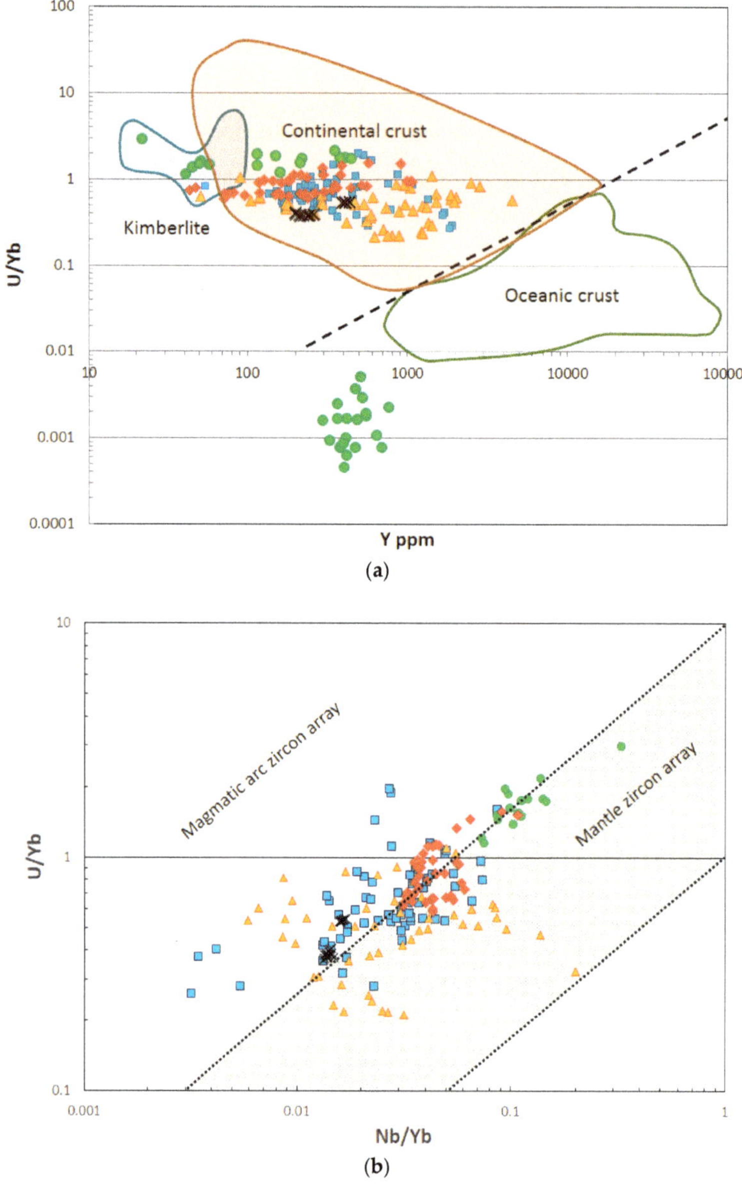

Figure 12. Y versus U/Yb (**a**) and Nb/Yb versus U/Yb. (**b**) Discrimination diagrams for zircon from the RVP (red diamonds, black X's [23]), Thailand (blue squares, Piilonen, unpublished data), other ZIP localities (orange triangles, Australia, Laos, Vietnam [11]), and carbonatites (green circles [5]). Fields for continental and oceanic crust and kimberlite from Grimes et al. [105]. Fields for mantle and magmatic arc arrays from Grimes et al. [106].

5.2. Source Affinities

All RVP zircon xenocrysts have trace element and O-Hf isotopic compositions indicative of a primary, alkaline magma derived from a variably-depleted lithospheric mantle source with limited supracrustal contamination. Their Hf isotopic compositions lie between those for depleted, juvenile mantle and a chondritic reservoir (CHUR, Figure 13). This supports either (1) a juvenile mantle source mixed with older crustal material or (2) recycled material with a longer crustal residence time. Their Y versus Yb/U are more chemically aligned to continental crust rather than recycled oceanic crust values (Figure 11), which is observed for zircon xenocrysts from other mantle-derived rocks including some kimberlites and carbonatites [105]. The variance in the Hf isotope compositions between RVP samples and within grains (εHf = 4.77–9.92) may, therefore, reflect localized source variation or unmixed crustal or metasomatic components. Griffin et al. [48] suggested that low and variable εHf in some kimberlitic zircon may be due to crystallization from DM-derived or the ocean island basalt (OIB)-type magmas undergoing assimilation/fractionation reactions with a non-radiogenic lithosphere. The SCLM underneath the RVP is a complex region of multiple rift-related and back-arc-related collisional remnants, the Song Ma suture zone, and back margins of the Himalayan collision zone. As the zircon xenocrysts did not crystallize from the host alkali basalt magma, an additional Mantle-derived melt is needed. A DM-derived melt interacting with SCLM underneath the RVP would encounter various reservoirs with non-radiogenic Hf, widely varying ages, and Lu/Hf ratios will result in variable ^{176}Hf/^{177}Hf in the crystallizing zircon.

Similarities of U-Pb (0.88–1.56 Ma) and (U-Th)/He ages (0.86–1.02 Ma) yielded by RVP zircon crystals suggest their minimal mantle residence time (0.18–0.86 Ma) and a rapid ascent to the near-surface temperatures facilitated by the alkali basalts. The U-Pb and (U-Th)/He ages, however, are clearly older than the geomagnetic reversal at 0.7 Ma that predates the later, explosive phase of basaltic volcanism in the area [42]. The U-Pb and (U-Th)/He, thus, suggest an earlier onset of basaltic volcanism in the area starting at least at ~1.02 $\pm$ 0.02 Ma. Crystal morphology, compositional zoning, and REE patterns indicate zircon growth from a single magma without fractional crystallization or injection of additional melts.

Model Hf ages for RVP zircon give dates significantly older than given by U-Pb and (U-Th)/He methods and correspond to major tectonic events in the SE Asia region. The estimated minimum T_{DM} model ages (333–403 Ma) correspond to the Alleghenian orogeny and suturing of the Indochina and South China blocks (~340 Ma). The older estimated minimum $T_{DM}{}^{C}$ model ages suggest that the source material for RVP zircon crystallization separated from the depleted mantle around 500–650 Ma during a major crust-building collision between East and West Gondwana to form late Neoproterozoic Pannotia (~650 Ma), which then broke-up in the early Paleozoic (~550 Ma).

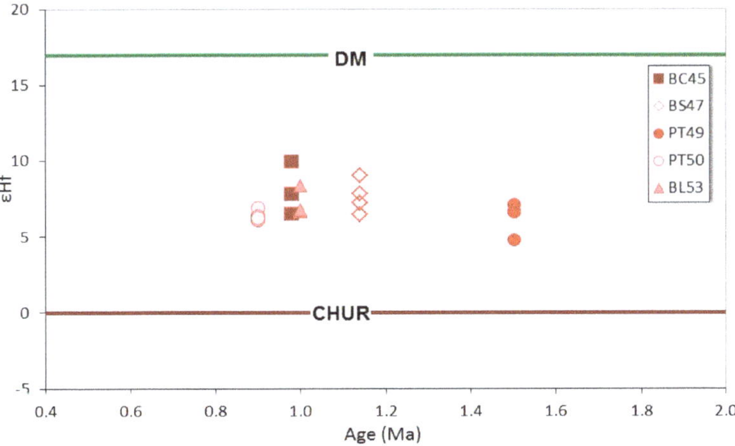

Figure 13. U-Pb Age (Ma) versus εHf for RVP zircon. CHUR: Chondritic uniform reservoir, DM: Depleted mantle.

5.3. Carbonatitic Signatures in Ratanakiri Zircons

Geochemical signatures in RVP zircon suggest that the crystallizing melts had carbonatitic links. Chemical ranges for carbonatite-derived zircons, however, are loosely defined and are much wider than the main carbonatite field (VII) in their Hf-Y discrimination diagram. They also fall within fields (II) ultramafic, mafic, and intermediate rocks as well as (VI) alkaline rocks and alkaline metasomites (Figure 9) [5]. Some zircons analyzed from carbonatitic dikes proved to represent crystallization from a silicate melt instead of a carbonatitic melt and are probably xenocrysts from wall rock contamination [107]. Strict chemical characterization of 'carbonatitic zircons' is clearly complex and partly unreliable [108,109]. Saava et al. [109] surveyed more than 100 carbonatitic zircons and found only 45% could be assigned to a carbonatitic source and that large compositional variation existed between both individual grains and localities. This variation may represent a constantly-evolving carbonatitic melt and its interaction with co-magmatic silicate sources or reflect variable generation in continental carbonatites from the SCLM and, in cases of oceanic island carbonatites, generation in the asthenosphere [110,111].

Carbonatites seem absent in Cambodian surface geology even though they appear in adjacent Northwestern Vietnam where rift-related carbonatites are dated at 28–44 Ma (biotite K-Ar ages) and 30–32 Ma (U-Th-Pb isochron ages) in South Nam Xe [112]. Geochemically, these carbonatites suggest involvement of an enriched mantle component in their genesis. These HSFE-deficient rocks, however, are low in Zr and lack zircon. Cenozoic carbonatite intrusions (12–40 Ma) lie 1600 km to the north within the SW China craton (Mianning-Denchang intrusive belt [113]). For the RVP magmatic sources, with a predisposition to crystallize carbonatitic-type zircon, their exact nature remains uncertain.

The absence of surficial carbonatites does not preclude a carbonatized mantle source [114]. Several types of continental carbonatitic-influenced mantle sources are known to have the potential to generate zircon megacrysts of an RVP type. Asian examples include: (1) stratified mantle deduced from ultramafic xenoliths that show initial silicate and then carbonatitic metasomatism in kamafugites from Western Qinlan, Central China [115], and (2) Pacific slab-induced carbonatitic metasomatism in garnet-bearing lherzolite xenoliths from Cenozoic basalts in NE China [116–118]. The latter is related to alkali basalts at Changle where zircon megacrysts have Hf-Y ranges similar to those in RVP zircons and the carbonatitic-modified mantle is recognized [13]. Elsewhere, upper mantle xenoliths with carbonatitic reactions are recorded in young basalt fields in SE Australia [119]. Vent deposits in this field include zircon megacrysts that show transitional silicate-carbonatitic affinities [120].

The extent to which carbonatitic altered mantle sources have contributed to zircon megacryst suites along Western Pacific continental margins remain uncertain. This uncertainty stems from the patchy spread of carbonatitic fields within the classification and regression trees CART discrimination model and the availability of inclusion and geochemical data furnished from the zircon-related host basalts [11]. A study of mantle xenoliths from young basalts on the Leizhou Peninsula, which is adjacent to the Pengali, Hainan Island mantle zircon xenocryst suites detailed a range of mantle melting and metasomatic effects [13,121]. The study identified trace element diffusion effects within the mantle and showed that HFSE depletions in metasomatized lherzolite do not necessarily require a carbonatitic metasomatizing agent.

5.4. Variation in Zircon-Bearing Xenocryst Assemblages in SE Asia

The enigmatic associations of zircon and corundum xenocrysts in intraplate alkaline basalt provinces continue to raise questions about their petrogenesis. The highly zircon-dominated RVP xenocryst suite, with minor corundum, is rare among the many basaltic gem deposits along West Pacific continental margins; corundum is common elsewhere in Cambodia (Pailin), through much of SE Asia and along Eastern Australia [8]. The lack of corundum in RVP alkali basalts suggests that particular mantle conditions favored RVP zircon genesis.

Some clues to these conditions may be offered by a study of zircon and corundum (sapphire)-bearing albititic dykes exposed within mantle assemblages in the French Pyrenees [122,123].

Some dykes contain only zircon megacrysts, some contain zircon + corundum, and some contain only corundum. Zircon in corundum-free dykes have systematically lower Hf (0.46–0.68 wt % Hf) and lower Y (740–1850 ppm), which is similar to that observed in the RVP zircon while zircon accompanied by corundum has higher values (Hf up to 1.8 wt %, Y up to 13,000 ppm). Geochemically, these dykes indicate a mantle origin, which is the result of very low degrees of partial melting (<1%) of a harzburgite source that has undergone metasomatism by a carbonatitic fluid phase prior to crystallization of the mega-crysts. To answer this dichotomy in megacryst generation, Pin et al. [123] attributed zircon formation to more CO_2-rich fluid conditions relative to more hydrous conditions for corundum formation within the feldspathic melts in the dykes during high pressure crystallization of the assemblage on the liquidus.

5.5. Models for Generation of RVP Zircon-Rich, Corundum-Poor Gem Suite

Explanations for the distinct zircon generations (low-Hf isolated zircon versus high-Hf zircon associated with corundum) along with associated megacrysts (anorthoclase) include low degrees of partial melting of the upper mantle coupled with metasomatism by a carbonatitic melt/fluid, the immiscibility of a silico-carbonatitic melt, or the interaction of carbonatitic and syenitic melts within the SCLM [86,100,124]. Both alkali basalt magmas and their mantle source regions are known sources of CO_2 and, in intraplate continental basalts, CO_2 is the main volatile [125].

At pressures ~20 kbar, CO_2 solubility increases dramatically, which promotes low-T melting of peridotite (<1000 °C) as well as significantly lowering of SiO_2 contents in resultant partial melts [114]. Very small degrees of partial melting (<1%) of a peridotite or a harzburgite at upper mantle conditions can produce early liquids strongly enriched in Si, Na, and Al and depleted in Fe, Mg, Ca, and Ti, which is similar in composition to syenite melts [122,123]. If such alkali-rich partial melt further undergoes enrichment by an H_2O-rich or CO_2-rich metasomatizing fluid, the behavior and activity of Si and Al are affected. The addition of H_2O to the metasomatizing fluid increases corundum saturation but does not affect Zr solubility [19]. The degree of partial melting, Zr content, alkalinity, and the degree of Al saturation of this silicate melt will be the determining factors in producing low-Hf or high-Hf zircon and corundum. Any carbonatitic melt fraction will reject HFSE (Zr, Hf) and preferentially incorporate LREE elements, which will result in an HFSE-HREE-enriched silicate melt. Foley et al. [114] performed partial melting experiments on peridotite with H_2O and CO_2 in which early melts are carbonate-rich and progress to carbonated silicateelts with further melting. Hafnium along with Nb and Ta show limited compatibility with early carbonate-dominant melts but are enriched in carbonated silicate melts at higher degrees of partial melting. This suggests that low-Hf zircon may crystallize in a carbonate-dominant environment while high-Hf zircon and associated Nb-Ta oxide phases are products of a carbonated silicate melt that has undergone Al-saturation. Saturation of a peralkaline melt in zircon requires orders of magnitude higher Zr content than in peraluminous melts [22,126]. Higher Zr contents in early peralkaline, carbonate-dominated melts may be responsible for the dominance of zircon over corundum in certain deposits such as in the RVP gem field.

The minor high-Hf zircon included in rare RVP corundum xenocrysts distinctly differs in morphology and geochemistry to the low-Hf mantle-derived zircon and suggests a separate, limited genesis. The high Hf-zircon resemble examples in Dak Nong, Vietnam sapphires that have crystal O isotope values and contain fluid inclusions suggestive of lower pressure crystallization from a hydrous CO_2/CO_3^{2-}-saturated melt [7,100]. As seen in the RVP zircon relationship, the Dak Nong corundum accompanies lower-Hf zircon xenocrysts, which were assigned to higher-pressure origin from melt enriched in but not saturated in CO_2.

6. Conclusions

Xenocrystic zircon from Cenozoic alkali basalts of the Ratanakiri Volcanic Province represent a unique suite within the larger zircon megacryst Indo-Pacific zone (ZIP). Crystal morphology, compositional zoning, and trace element geochemistry indicate single-stage growth within a primary, mantle-derived alkaline melt at temperatures between 601 and 739 °C. The RVP zircon have lower $\delta^{18}O$ isotope compositions than observed in xenocrysts from other ZIP suites and a moderate range of Hf isotopic compositions, which suggest crystallization from a juvenile, alkaline lithospheric mantle magma source with limited supracrustal contamination. The xenocrysts have been dated by U-Pb (0.88–1.56 Ma) and (U-Th)/He (0.86–1.02 Ma) methods and give ages that are similar to those obtained for zircon xenocrysts from other alkali basalt provinces in SE Asia. The RVP zircon predate some of the alkali basalts (~0.7 Ma), which indicates very short mantle residence times before entrainment in the erupting magma and an earlier beginning of basaltic volcanism. The SCLM underneath the RVP is a complex region consisting of remnants of multiple rift-arc-related and back-arc-related collisional margins including the Song Ma suture zone and back margins of the Himalayan collision zone. The overriding conclusion is of crystallization from a metasomatized "carbonatitic"-influenced melt from very low partial melting of a peridotite SCLM source at about 60 km beneath the Indochina terrain.

Supplementary Materials: The following are available online at http://www.mdpi.com/2075-163X/8/12/556/s1, Table S1: U-Pb isotopic analyses of RVP zircon xenocrysts, Table S2: (U-Th)/He age dates for RVP zircon xenocrysts, Table S3: Hf-Lu isotope data and model ages for xenocrystic zircon.

Author Contributions: Conceptualization, P.C.P. Methodology, P.C.P. and M.D. Formal Investigation, P.C.P., M.D., G.P., J.W.V., and R.R. Resources, P.C.P. Data Curation, P.C.P. Writing-Original Draft Preparation, P.C.P. Writing-Review & Editing, P.C.P., F.L.S., and M.D. Project Administration, P.C.P. Funding Acquisition, P.C.P. and M.D.

Funding: This research was funded by the Canadian Museum of Nature. JWV is funded by the US National Science Foundation (EAR-1524336). MD was supported by Australian Research Council Discovery funding scheme (DP160102427) and a Curtin Research Fellowship. The Angkor Gold Corporation provided in-kind support for logistics in the field.

Acknowledgments: The authors would like to thank Mike & Delayne Weeks, JP Dau, Dennis Oullette, Craig Tammas Richardson, Adrian Mann, Samath Ma and Viseth Keo of Angkor Gold Corporation and all their dedicated staff for help with logistics, translation, accommodations, transportation, coffee, access to drill core and geologic maps, and for making Ban Lung a second home while working in Cambodia. Thanks to all the zircon miners and their families in Ratanakiri for their help in obtaining zircons for study and for taking the time to discuss their operations. Graphical assistance provided by Reni Barlow. This paper is dedicated to Votha Un (1976–2014) who was a fantastic gem guide and friend to all of us who choose to work and travel in Cambodia. Thanks are extended to two anonymous reviewers whose input helped to improve this manuscript.

Conflicts of Interest: The authors declare no conflict of interest. The funders had no role in the design of the study, in the collection, analyses, or interpretation of data, in the writing of the mnuscript, or in the decision to publish the results.

References

1. Shor, R.; Weldon, R. Ruby and sapphire production and distribution: A quarter century of change. *Gems Gemol.* **2009**, *45*, 236–259. [CrossRef]

2. Smith, M.H.; Balmer, W.A. Zircon Mining in Cambodia. *Gems Gemol.* **2009**, *45*, 152–154.

3. Schwarz, D. Coloured gemstones—Mines and Markets. In Proceedings of the 34th International Gemmological Conference, Vilnius, Lithuania, 25–30 August 2015; pp. 110–112.

4. Abduriyim, A.; Sutherland, F.L.; Belousova, E.A. U-Pb age and origin of gem zircon from the New England sapphire fields, New South Wales, Australia. *Aust. J. Earth Sci.* **2012**, *59*, 1067–1081. [CrossRef]

5. Belousova, E.A.; Griffin, W.L.; O'Reilly, S.Y.; Fisher, N. Igneous zircon: Trace element composition as an indicator of source rock type. *Contrib. Mineralog. Petrol.* **2002**, *143*, 602–622. [CrossRef]

6. Chen, T.; Ai, H.; Yang, M.; Zheng, S.; Liu, Y. Brownish Red Zircon from Muling, China. *Gems Gemol.* **2011**, *47*, 36–41. [CrossRef]

7. Garnier, V.; Ohnenstetter, D.; Guliani, G.; Fallick, A.E.; Trong, T.P.; Quang, V.H.; Van, L.P.; Schwarz, D. Basalt petrology, zircon ages and sapphire genesis from Dak Nong, southern Vietnam. *Mineral. Mag.* **2005**, *69*, 21–38. [CrossRef]

8. Graham, I.; Sutherland, L.; Zaw, K.; Nechaev, V.; Khanchuk, A. Advances in our understanding of the gem corundum deposits of the West Pacific continental margins intraplate basaltic fields. *Ore Geol. Rev.* **2008**, *34*, 200–215. [CrossRef]

9. Qiu, Z.; Wu, F.; Yu, Q.; Xie, L.; Yang, S. Hf isotopes of zircon megacrysts from the Cenozoic basalts in eastern China. *Chin. Sci. Bull.* **2005**, *50*, 2602–2611. [CrossRef]

10. Sutherland, F.L.; Coenraads, R.R.; Abduriyim, A.; Meffre, S.; Hoskin, P.W.O.; Giuliani, G.; Beattie, R.; Wuhrer, R.; Sutherland, G.B. Corundum (sapphire) and zircon relationships, Lava Plains gem fields, NE Australia: Integrated mineralogy, geochemistry, age determination, genesis and geographical typing. *Mineral. Mag.* **2015**, *79*, 545–581. [CrossRef]

11. Sutherland, F.L.; Graham, I.; Yaxley, G.; Armstrong, R.; Giuliani, G.; Hoskin, P.; Nechaev, V.; Woodhead, J. Major zircon megacryst suites of the Indo-Pacific lithospheric mrgin (ZIP) and their petrogenetic and regional implications. *Mineral. Petrol.* **2016**, *110*, 399–420. [CrossRef]

12. Sutherland, F.L.; Piilonen, P.C.; Zaw, K.; Meffre, S.; Thompson, J. Sapphire within zircon-rich gem deposits, Bo Loei, Ratanakiri Province, Cambodia: Trace elements, inclusions, U-Pb dating and genesis. *Aust. J. Earth Sci.* **2015**, *62*, 761–773. [CrossRef]

13. Yu, Y.; Xu, X.; Chen, X. Genesis of zircon megacrysts in Cenozoic alkali basalts and the heterogeneity of subcontinental lithospheric mntle, eastern China. *Mineral. Petrol.* **2010**, *100*, 75–94. [CrossRef]

14. Zaw, K.; Sutherland, F.L.; Dellapasqua, F.; Ryan, C.G.; Yui, T.-F.; Mernagh, T.P.; Duncan, D. Contrasts in gem corundum characteristics, eastern Australian basaltic fields: Trace elements, fluid/melt inclusions and oxygen isotopes. *Mineral. Mag.* **2006**, *70*, 669–687. [CrossRef]

15. Fan, P.-F. Tectonic patterns and Cenozoic basalts in the western mrgin of the South China Sea. *Geol. Soc. Malays. Bull.* **1995**, *37*, 91–97.

16. Fedorov, P.I.; Koloskov, A.V. Cenozoic volcanism of southeast Asia. *Petrology* **2005**, *13*, 352–380.

17. Taylor, J.G.C.; Buravas, S. Gem deposits at Khao Ploi Waen and Bang Ka Cha, Chanthaburi Province. In *Geologic Reconnaissance of the mineral Deposits of Thailand*; United States Government Printing Office: Washington, WA, USA, 1951; Volume 984, pp. 144–148.

18. Harley, S.L.; Kelly, N.M. Zircon Tiny but Timely. *Elements* **2007**, *3*, 13–18. [CrossRef]

19. Boehnke, P.; Watson, E.B.; Trail, D.; Harrison, T.M.; Schmitt, A.K. Zircon saturation re-revisited. *Chem. Geol.* **2013**, *351*, 324–334. [CrossRef]

20. Watson, E.B. Zircon saturation in felsic liquids: Experimental results and applications to trace element geochemistry. *Contrib. Mineral. Petrol.* **1979**, *70*, 407–419. [CrossRef]

21. Watson, E.B.; Harrison, T.M. Zircon saturation revisited: Temperature and composition effects in a variety of crustal mgma types. *Earth Planet. Sci. Lett.* **1983**, *64*, 295–304. [CrossRef]

22. Linnen, R.L.; Keppler, H. Melt composition control of Zr/Hf fractionation in mgmatic processes. *Geochim. Cosmochimi. Acta* **2002**, *66*, 3293–3301. [CrossRef]

23. Cong, F.; Li, S.-Q.; Lin, F.-C.; Shi, M.-F.; Zhu, H.-P.; Siebel, W.; Chen, F. Origin of Zircon Megacrysts from Cenozoic Basalts in Northeastern Cambodia: Evidence from U-Pb Age, Hf-O Isotopes, and Inclusions. *J. Geol.* **2016**, *124*, 221–234. [CrossRef]

24. Ridd, M.F.; Barber, A.J.; Crow, M.J. Introduction to the geology of Thailand. In *Geology of Thailand*; Ridd, M.F., Barber, A.J., Crow, M.J., Eds.; The Geological Society: London, UK, 2011; pp. 1–17.

25. Searle, M.P.; Morley, C.K. Tectonic and thermal evolution of Thailand in the regional context of SE Asia. In *Geology of Thailand*; Ridd, M.F., Barber, A.J., Crow, M.J., Eds.; The Geological Society: London, UK, 2011; pp. 539–571.

26. Workman, D.R. Geology of Laos, Cambodia, South Vietnam and the eastern part of Thailand. *Overseas Geol. Miner. Resour.* **1977**, *50*, 1–34.

27. Sotham, S. Geology of Cambodia. *CCOP Tech. Bull.* **1997**, *26*, 13–23.

28. Khain, V.E. *Tectonics of the Continents and Oceans*; Nauchny Mir: Moscow, Russia, 2001.

29. Mukasa, S.B.; matthew Fischer, G.; Barr, S.M. The Character of the Subcontinental Mantle in Southeast Asia: Evidence From Isotopic and Elemental Compositions of Extension-Related Cenozoic Basalts in Thailand. In *Earth Processes: Reading the Isotopic Code*; Basu, A., Hart, S., Eds.; Geophysical Monograph-American Geophysical Union: Washington DC, WA, USA, 1996; Volume 95, pp. 233–252.

30. Le Pichon, X.; Fournier, M.; Jolivet, L. Kinematics, topography, shortening, and extrusion in the India-Eurasia collision. *Tectonics* **1992**, *11*, 1085–1098. [CrossRef]

31. Miller, M.S.; Kennett, B.L.N.; Toy, V.G. Spatial and temporal evolution of the subducting Pacific plate structure along the western Pacific margin. *J. Geophys. Res.* **2006**, *111*, B02401. [CrossRef]

32. Sutherland, F.L. Alkaline rocks and gemstones, Australia: A review and synthesis. *Aust. J. Earth Sci.* **1996**, *43*, 323–343. [CrossRef]

33. Sutherland, F.L.; Fanning, C.M. Gem-bearing basaltic volcanism, Barrington, New South Wales: Cenozoic evolution, based on basalt K-Ar ages and zircon fission track and U-Pb isotope dating. *Aust. J. Earth Sci.* **2001**, *48*, 221–237. [CrossRef]

34. Patte, E. Etude de l'ile des Cendres, volcan apparu au large de la côte d'Annam. *Bull. Serv. Géol. l'Indoch.* **1925**, *13*, 1–19.

35. Saurin, E. La néotectonique de l'Indochine. *Rev. Géogr. Phys. Géol. Dyn.* **1967**, *9*, 143–151.

36. Barr, S.M.; MacDonald, A.S. Geochemistry and petrogenesis of Late Cenozoic alkaline basalts of Thailand. *Geol. Soc. Malays. Bull.* **1978**, *10*, 25–52.

37. Barr, S.M.M.; Alan, S. Geochemistry and Geochronology of Late Cenozoic Basalts of Southeast Asia. *GSA Bull.* **1981**, *92*, 1069–1142. [CrossRef]

38. Sutherland, F.L.; Bosshart, G.; Fanning, C.M.; Hoskin, P.W.O.; Coenraads, R.R. Sapphire crystallization, age and origin, Ban Huai Sai, Laos: Age based on zircon inclusions. *J. Asian Earth Sci.* **2002**, *20*, 841–849. [CrossRef]

39. Barr, S.M.J.; Dodie, E. Trace element characteristics of Upper Cenozoic basaltic rocks of Thailand, Kampuchea and Vietnam. *J. Southeast Asian Earth Sci.* **1990**, *4*, 233–242. [CrossRef]

40. Hoang, N.; Flower, M. Petrogenesis of Cenozoic Basalts from Vietnam: Implication for Origins of a 'Diffuse Igneous Province'. *J. Petrol.* **1998**, *39*, 369–395. [CrossRef]

41. Barr, S.M.; Charusiri, P. Volcanic rocks. In *The Geology of Thailand*; Ridd, M.F., Barber, A.J., Crow, M.J., Eds.; The Geological Society: London, UK, 2011; pp. 415–438.

42. Lacombe, P. La mssif basaltique quaternaire à zircons-gemmes de Ratanakiri (Cambodge nord-oriental)—Premiére partie. *Bull. Rech. Géol. min.* **1969**, *3*, 31–91.

43. Lacombe, P. La mssif basaltique quaternaire à zircons-gemmes de Ratanakiri (Cambodge nord-oriental)-Troisiéme partie: Minéralogie et géologie des gisements de zircons. *Bull. Rech. Géol. Min.* **1970**, *4*, 33–79.

44. Whitford-Stark, J.L. A survey of Cenozoic volcanism on mainland Asia. In *Geological Society of America Special Paper*; Geological Society of America: Boulder, CO, USA, 1987; Volume 213.

45. Zeug, M.; Nasdala, L.; Wanthanachaisaeng, B.; Balmer, W.A.; Corfu, F.; Wildner, M. Blue Zircon from Ratanakiri, Cambodia. *J. Gemmol.* **2018**, *36*, 112–132. [CrossRef]

46. Pouchou, J.L.; Pichoir, F. *'PAP'j(rZ) Procedure for Improved Quantitative Microanalysis*; San Francisco Press: San Francisco, CA, USA, 1984.

47. Elhlou, S.; Belousova, E.A.; Griffin, W.L.; Pearson, N.J.; O'Reilly, S.Y. Trace element and isotopic composition of GJ-red zircon standard by laser ablation. *Geochim. Cosmochim. Acta* **2006**, *70*, A158. [CrossRef]

48. Griffin, W.L.; Pearson, N.J.; Belousova, E.; Jackson, S.E.; van Achterbergh, E.; O'Reilly, S.Y.; Shee, S.R. The Hf isotope composition of cratonic mntle: LAM-MC-ICPMS analysis of zircon megacrysts in kimberlites. *Geochim. Cosmochi. Acta* **2000**, *64*, 133–147. [CrossRef]

49. DeBievre, P.; Taylor, P.D.P. Table of the isotopic composition of the elements. *Int. J. Mass Spectrom. Ion Process.* **1993**, *123*, 149–166. [CrossRef]

50. Pearson, N.J.; Griffin, W.L.; O'Reilly, S.Y. Mass fractionation correction in laser ablation-multiple collector ICP-MS: Implications for overlap corrections and precise and accurate in situ isotope ratio measurement. In *Laser-Ablation ICP-MS in the Earth Sciences*; Sylvester, P.J., Ed.; mineralogical Association of Canada: Quebec City, QC, Canada, 2008; Volume 40, pp. 93–116.

51. Griffin, W.L.; Pearson, N.J.; Belousova, E.A.; Saeed, A. Reply to "Comment to short-communication 'Comment: Hf-isotope heterogeneity in zircon 91500' by W.L. Griffin, N.J. Pearson, E.A. Belousova and A. Saeed [Chemical Geology 233 (2006) 358-363]" by F. Corfu. *Chem. Geol.* **2007**, *244*, 354–356. [CrossRef]

52. Kemp, A.I.S.; Wormald, R.J.; Whitehouse, M.J.; Price, R.C. Hf isotopes in zircon reveal contrasting sources and crystallization histories for alkaline to peralkaline granites of Temora, southeastern Australia. *Geology* **2005**, *33*, 797–800. [CrossRef]

53. Bouvier, A.; Vervoort, J.D.; Patchett, P.J. The Lu-Hf and Sm-Nd isotopic composition of CHUR: Constraints from unequilibrated chondrites and implications for the bulk composition of terrestrial planets. *Earth Planet. Sci. Lett.* **2008**, *273*, 48–57. [CrossRef]

54. Scherer, E.; Münker, C.; Mezger, K. Calibration of the Lutetium-Hafnium Clock. *Science* **2001**, *293*, 683–687. [CrossRef] [PubMed]

55. Jackson, S.E.; Pearson, N.J.; Griffin, W.L.; Belousova, E.A. The application of laser ablation microprobe-inductively coupled plasma-mass spectrometry (LAM-ICPMS) to in situ U-Pb zircon geochronology. *Chem. Geol.* **2004**, *211*, 47–69. [CrossRef]

56. Griffin, W.L.; Powell, W.J.; Pearson, N.J.; O'Reilly, S.Y. GLITTER: Data reduction software for laser ablation ICP-MS. In *Laser Ablation ICP-MS in the Earth Sciences*; Sylvester, P.J., Ed.; Mineralogical Association of Canada: Quebec City, QC, Canada, 2008; pp. 204–207.

57. Andersen, T. Correction of common lead in U-Pb analyses that do not report 204Pb. *Chem. Geol.* **2002**, *192*, 59–79. [CrossRef]

58. Stacey, J.S.T.; Kramers, J.D. Approximation of terrestrial lead isotope evolution by a two-stage model. *Earth Planet. Sci. Lett.* **1975**, *26*, 207–221. [CrossRef]

59. Reiners, P.W.; Spell, T.L.; Nicolescu, S.; Zanetti, K.A. Zircon (U-Th)/He thermochronometry: He diffusion and comparisons with ^{40}Ar/^{39}Ar dating. *Geochim Cosmochim. Acta* **2004**, *68*, 1857–1887. [CrossRef]

60. Danišík, M.; Kuhlemann, J.; Dunkl, I.; Evans, N.J.; Székely, B.; Frisch, W. Survival of Ancient Landforms in a Collisional Setting as Revealed by Combined Fission Track and (U-Th)/He Thermochronometry: A Case Study from Corsica (France). *J. Geol.* **2012**, *120*, 155–173. [CrossRef]

61. Danišík, M.; Štěpančíková, P.; Evans, N.J. Constraining long-term denudation and faulting history in intraplate regions by multisystem thermochronology: An example of the Sudetic Marginal Fault (Bohemian Massif, central Europe). *Tectonics* **2012**, *31*. [CrossRef]

62. Farley, K.A.; Wolf, R.A.; Silver, L.T. The effects of long alpha-stopping distances on (U-Th)/He ages. *Geochim. Cosmochim. Acta* **1996**, *60*, 4223–4229. [CrossRef]

63. Vermeesch, P. HelioPlot, and the treatment of overdispersed (U-Th-Sm)/He data. *Chem. Geol.* **2010**, *271*, 108–111. [CrossRef]

64. Reiners, P.W. Zircon (U-Th)/He Thermochronometry. *Rev. Mineral. Geochem.* **2005**, *58*, 151–179. [CrossRef]

65. Valley, J.W.; Chiarenzelli, J.R.; McLelland, J.M. Oxygen isotope geochemistry of zircon. *Earth Planet. Sci. Lett.* **1994**, *126*, 187–206. [CrossRef]

66. Valley, J.W.; Kitchen, N.; Kohn, M.J.; Niendorf, C.R.; Spicuzza, M.J. UWG-2, a garnet standard for oxygen isotope ratios: Strategies for high precision and accuracy with laser heating. *Geochim. Cosmochim. Acta* **1995**, *59*, 5223–5231. [CrossRef]

67. Pupin, J.-P. Zircon and granite petrology. *Contrib. Mineral. Petrol.* **1980**, *73*, 207–220. [CrossRef]

68. Shore, M.; Fowler, A.D. Oscillatory zoning in minerals: A common phenomenon. *Can. Mineral.* **1996**, *34*, 1111–1126.

69. Witter, A.; Nasdala, L.; Wanthanachaisaeng, B.; Bunnag, N.; Škoda, R.; Balmer, W.A.; Geister, G.; Zeug, M. Mineralogical characterization of gem zircon from Ratanakiri, Cambodia. In Proceedings of the CORALS 2013, Vienna, Austria, 3–6 July 2013.

70. McDonough, W.F.; Sun, S.S. The composition of the Earth. *Chem. Geol.* **1995**, *120*, 223–253. [CrossRef]

71. Pupin, J.P. *Transactions of the Royal Society of Edinburgh, Earth Sciences*; RSE Scotland Foundation: Edinburgh, UK, 2000.

72. Watson, E.B.; Harrison, T.M. Zircon Thermometer Reveals minimum Melting Conditions on Earliest Earth. *Science* **2005**, *308*, 841–844. [CrossRef] [PubMed]

73. Watson, E.B.; Wark, D.A.; Thomas, J.B. Crystallization thermometers for zircon and rutile. *Contrib. Mineral. Petrol.* **2006**, *151*, 413–433. [CrossRef]

74. Fu, B.; Page, F.Z.; Cavosie, A.J.; Fournelle, J.; Kita, N.T.; Lackey, J.S.; Wilde, S.A.; Valley, J.W. Ti-in-zircon thermometry: Applications and limitations. *Contrib. Mineral. Petrol.* **2008**, *156*, 197–215. [CrossRef]

75. Cherniak, D.J.; Watson, E.B. Ti diffusion in zircon. *Chem. Geol.* **2007**, *242*, 470–483. [CrossRef]

76. Crisp, L. The effect of pressure on Ti-in-zircon and Zr-in-rutile. In Proceedings of the Petrology and geochemistry seminars, Research School of Earth Sciences, Canberra, Australia, 29 May 2015.

77. Page, F.Z.; Fu, B.; Kita, N.T.; Fournelle, J.; Spicuzza, M.J.; Schulze, D.J.; Viljoen, F.; Basei, M.A.S.; Valley, J.W. Zircons from kimberlite: New insights from oxygen isotopes, trace elements, and Ti in zircon thermometry. *Geochim. Cosmochim. Acta* **2007**, *71*, 3887–3903. [CrossRef]

78. Carbonnel, J.-P.; Duplaix, S.; Selo, M. La méthode des traces de fission de l'U apliquée á la géochronologic. Datation du magmatisme récente de l'Asie du Sud-Est. *Rev. Géogr. Phys. Geolo. Dyn.* **1972**, *14*, 29–46.

79. Carbonnel, J.-P.; Selo, M.; Poupeau, G. Fission track age of the gem deposit of Pailin (Cambodia) and recent tectonics in the Indochinan province. *Mod. Geol.* **1973**, *4*, 61–64.

80. Li, X.-H.; Long, W.-G.; Li, Q.-L.; Liu, Y.; Zheng, Y.-F.; Yang, Y.-H.; Chamberlain, K.R.; Wan, D.-F.; Guo, C.-H.; Wang, X.-C.; et al. Penglai Zircon Megacrysts: A Potential New Working Reference Material for Microbeam Determination of Hf-O Isotopes and U-Pb Age. *Geostand. Geoanal. Res.* **2010**, *34*, 117–134. [CrossRef]

81. Li, Y.; Zheng, D.; Wu, Y.; Wang, Y.; He, H.; Pang, J.; Wang, Y.; Yu, J. A Potential (U-Th)/He Zircon Reference Material from Penglai Zircon Megacrysts. *Geostand. Geoanal. Res.* **2017**, *41*, 359–365. [CrossRef]

82. Dodson, M.H. Closure temperature in cooling geochronological and petrological systems. *Contrib. Mineral. Petrol.* **1973**, *40*, 259–274. [CrossRef]

83. Page, F.Z.; Ushikubo, T.; Kita, N.T.; Riciputi, L.R.; Valley, J.W. High-precision oxygen isotope analysis of picogram samples reveals 2 μm gradients and slow diffusion in zircon. *Am. Mineral.* **2007**, *92*, 1772–1775. [CrossRef]

84. Bowman, J.R.; Moser, D.E.; Valley, J.W.; Wooden, J.L.; Kita, N.T.; Mazdab, F.K. Zircon U-Pb isotope, $\delta^{18}O$ and trace element response to 80 m.y. of high temperature metamorphism in the lower crust: Sluggish diffusion and new records of Archean craton formation. *Am. J. Sci.* **2011**, *311*, 719–772. [CrossRef]

85. Mattey, D.; Lowry, D.; Macpherson, C. Oxygen isotope composition of mntle peridotite. *Earth Planet. Sci. Lett.* **1994**, *128*, 231–241. [CrossRef]

86. Upton, B.G.J.; Hinton, R.W.; Aspen, P.; Finch, A.; Valley, J.W. Megacrysts and Associated Xenoliths: Evidence for Migration of Geochemically Enriched Melts in the Upper Mantle beneath Scotland. *J. Petrol.* **1999**, *40*, 935–956. [CrossRef]

87. Valley, J.W.; Kitchen, N.; Kohn, M.J.; Niendorf, C.R.; Spicuzza, M.J. Oxygen Isotopes in Zircon. In *Reviews in Mineralogy and Geochemistry*; Hanchar, J.M., Hoskin, P.W.O., Eds.; Mineralogical Society of America: Chantilly, VA, USA, 2003; Volume 53, pp. 343–385.

88. Belousova, E.A.; Griffin, W.L.; O'Reilly, S.Y. Zircon Crystal Morphology, Trace Element Signatures and Hf Isotope Composition as a Tool for Petrogenetic Modelling: Examples From Eastern Australian Granitoids. *J. Petrol.* **2006**, *47*, 329–353. [CrossRef]

89. Heaman, L.M.; Bowins, R.; Crocket, J. The chemical composition of igneous zircon suites: Implications for geochemical tracer studies. *Geochim. Cosmochim. Acta* **1990**, *54*, 1597–1607. [CrossRef]

90. Hoskin, P.W.O.; Ireland, T.R. Rare earth element chemistry of zircon and its use as a provenance indicator. *Geology* **2000**, *28*, 627–630. [CrossRef]

91. Seifert, W.; Rhede, D.; Tietz, O. Typology, chemistry and origin of zircon from alkali basalts of SE Saxony (Germany). *N. Jahrb. Mineralo. Abh.* **2008**, *184*, 299–313. [CrossRef]

92. Shnukov, S.E.; Andreev, A.V.; Savenok, S.P. Admixture elements in zircons and apatites: A tool for provenance studies of terrigenous sedimentary rocks. In Proceedings of the European Union of Geosciences (EUG 9), Strasbourg, France, 23–27 March 1997.

93. Pupin, J.-P. Granite genesis related to geodynamics from Hf-Y in zircon. *Earth Environ. Sci. Trans. R. Soc. Edinb.* **2000**, *91*, 245–256. [CrossRef]

94. Visonà, D.; Caironi, V.; Carraro, A.; Dallai, L.; Fioretti, A.M.; Fanning, M. Zircon megacrysts from basalts of the Venetian Volcanic Province (NE Italy): U-Pb ages, oxygen isotopes and REE data. *Lithos* **2007**, *94*, 168–180. [CrossRef]

95. Hoskin, P.W.O.; Schaltegger, U. The Composition of Zircon and Igneous and Metamorphic Petrogenesis. *Rev. Mineral. Geochem.* **2003**, *53*, 27–62. [CrossRef]

96. Borghini, G.; Fumagalli, P.; Rampone, E. The Stability of Plagioclase in the Upper Mantle: Subsolidus Experiments on Fertile and Depleted Lherzolite. *J. Petrol.* **2010**, *51*, 229–254. [CrossRef]

97. Hinton, R.W.; Upton, B.G.J. The chemistry of zircon: Variations within and between large crystals from syenite and alkali basalt xenoliths. *Geochim. Cosmochim. Acta* **1991**, *55*, 3287–3302. [CrossRef]

98. Coenraads, R.R.; Vichit, P.; Sutherland, F.L. An unusual sapphire-zircon-magnetite xenolith from the Chanthaburi Gem Province, Thailand. *Mineral. Mag.* **1995**, *59*, 465–479. [CrossRef]

99. Khamloet, P.; Pisutha-Arnond, V.; Sutthirat, C. Mineral inclusions in sapphire from the basalt-related deposit in Bo Phloi, Kanchanaburi, western Thailand: Indication of their genesis. *Russ. Geol. Geophys.* **2014**, *55*, 1087–1102. [CrossRef]

100. Izokh, A.E.; Smirnov, S.Z.; Egorova, V.V.; Tuan Anh, T.; Kovyazin, S.V.; Phuong, N.T.; Kalinina, V.V. The conditions of formation of sapphire and zircon in the areas of alkali-basaltoid volcanism in Central Vietnam. *Russ. Geol. Geophys.* **2010**, *51*, 719–733. [CrossRef]

101. Guo, J.; O'Reilly, S.Y.; Griffin, W.L. Corundum from basaltic terrains: A mineral inclusion approach to the enigma. *Contrib. Mineral. Petrol.* **1996**, *122*, 368–386. [CrossRef]

102. Guo, J.; O'Reilly, S.Y.; Griffin, W.L. Zircon inclusions in corundum megacrysts: I. Trace element geochemistry and clues to the origin of corundum megacrysts in alkali basalts. *Geochim. Cosmochim. Acta* **1996**, *60*, 2347–2363. [CrossRef]

103. Sutherland, F.L.; Hoskin, P.W.O.; Fanning, C.M.; Coenraads, R.R. Models of corundum origin from alkali basaltic terrains: A reappraisal. *Contrib. Mineral. Petrol.* **1998**, *133*, 356–372. [CrossRef]

104. Benisek, A.; Finger, F. Factors controlling the development of prism faces in granite zircons: A microprobe study. *Contrib. Mineral. Petrol.* **1993**, *114*, 441–451. [CrossRef]

105. Grimes, C.B.; John, B.E.; Kelemen, P.B.; Mazdab, F.K.; Wooden, J.L.; Cheadle, M.J.; Hanghøj, K.; Schwartz, J.J. Trace element chemistry of zircons from oceanic crust: A method for distinguishing detrital zircon provenance. *Geology* **2007**, *35*, 643–646. [CrossRef]

106. Grimes, C.B.; Wooden, J.L.; Cheadle, M.J.; John, B.E. "Fingerprinting" tectono-magmatic provenance using trace elements in igneous zircon. *Contrib. Mineral. Petrol.* **2015**, *170*, 92–117. [CrossRef]

107. Liu, Y.; Williams, I.S.; Chen, J.; Wan, Y.; Sun, W. The significance of Paleoproterozoic zircon in carbonatite dikes associated with the Bayan Obo REE-Nb-Fe deposit. *Am. J. Sci.* **2008**, *308*, 379–397. [CrossRef]

108. Campbell, L.S.; Compston, W.; Sircombe, K.N.; Wilkinson, C.C. Zircon from the East Orebody of the Bayan Obo Fe-Nb-REE deposit, China, and SHRIMP ages for carbonatite-related mgmatism and REE mineralization events. *Contrib. Mineral. Petrol.* **2014**, *168*, 1041. [CrossRef]

109. Saava, E.V.; Belyatsky, B.V.; Antonov, A.B. Carbonatitic Zircon-Myth or Reality: Mineralogical-Geochemical Analyses. Available online: http://alkaline09.narod.ru/abstracts/Savva_Belyatsky.htm (accessed on 26 October 2018).

110. Bell, K.; Simonetti, A. Source of parental melts to carbonatites—Critical isotopic constraints. *Mineral. Petrol.* **2010**, *98*, 77–89. [CrossRef]

111. Chakmouradian, A. The geochemistry of continental carbonatites revisited: Two mjor types of continental carbonatites and their trace-element signatures. In Proceedings of the European Geosciences Union General Assembly, Vienna, Austria, 19–24 April 2009; p. 10806.

112. Thi, T.N.; Wada, H.; Ishikawa, T.; Shimano, T. Geochemistry and petrogenesis of carbonatites from South Nam Xe, Lai Chau area, northwest Vietnam. *Mineral. Petrol.* **2014**, *108*, 371–390.

113. Hou, Z.; Liu, Y.; Tian, S.; Yang, Z.; Xie, Y. Formation of carbonatite-related giant rare-earth-element deposits by the recycling of mrine sediments. *Sci. Rep.* **2015**, *5*, 1–10. [CrossRef] [PubMed]

114. Foley, S.F.; Yaxley, G.M.; Rosenthal, A.; Buhre, S.; Kiseeva, E.S.; Rapp, R.P.; Jacob, D.E. The composition of near-solidus melts of peridotite in the presence of CO_2 and H_2O between 40 and 60 kbar. *Lithos* **2009**, *112*, 274–283. [CrossRef]

115. Su, B.-X.; Zhang, H.-F.; Sakyi, P.A.; Ying, J.-F.; Tang, Y.-J.; Yang, Y.-H.; Qin, K.-Z.; Xiao, Y.; Zhao, X.-M. Compositionally stratified lithosphere and carbonatite metasomatism recorded in mntle xenoliths from the Western Qinling (Central China). *Lithos* **2010**, *116*, 111–128. [CrossRef]

116. Deng, L.L.; Liu, Y.; Gao, S. Pacific slab-induced carbonatite mntle metasomatism in the eastern North China Craton. In Proceedings of the American Geophysical Union Fall Meeting, San Francisco, CA, USA, 15–19 December 2014.

117. Zeng, G.; Chen, L.-H.; Xu, X.-S.; Jiang, S.-Y.; Hofmann, A.W. Carbonated mntle sources for Cenozoic intra plate alkaline basalts in Shandong, North China. *Chem. Geol.* **2010**, *273*, 35–45. [CrossRef]

118. Deng, L.; Liu, Y.; Zong, K.; Zhu, L.; Xu, R.; Hu, Z.; Gao, S. Trace element and Sr isotope records of multi-episode carbonatite metasomatism on the eastern mrgin of the North China Craton. *Geochem. Geophys. Geosyst.* **2017**, *18*, 220–237. [CrossRef]

119. Yaxley, G.M.; Green, D.H.; Kamenetsky, V. Carbonatite Metasomatism in the Southeastern Australian Lithosphere. *J. Petrol.* **1998**, *39*, 1917–1930. [CrossRef]

120. Sutherland, F.L.; Graham, I.T.; Hollis, J.D.; Meffre, S.; Zwingmann, H.; Jourdan, F.; Pogson, R.E. Multiple felsic events within post-10 Ma volcanism, Southeast Australia: Inputs in appraising proposed magmatic models. *Aust. J. Earth Sci.* **2014**, *61*, 241–267. [CrossRef]
121. Yu, J.-H.; O'Reilly, S.Y.; Zhang, M.; Griffin, W.L.; Xu, X. Roles of Melting and Metasomatism in the Formation of the Lithospheric Mantle beneath the Leizhou Peninsula, South China. *J. Petrol.* **2006**, *47*, 355–383. [CrossRef]
122. Monchoux, P.; Fontan, F.; De Parseval, P.; Martin, R.F.; Wang, C.R. Igneous albitite dikes in orogenic lherzolites, western Pyréneés, France: A possible source for corundum and alkali feldspar xenocrysts in basaltic terraines. I. mineralogical associations. *Can. Mineral.* **2006**, *44*, 817–842. [CrossRef]
123. Pin, C.; Monchoux, P.; Paquette, J.L.; Azambre, B.; Wang, C.R.; Martin, R.F. Igneous albititic dikes in orogenic lherzolites, western Pyréneés, France: A possible source for corundum and alkali feldspar xenocrysts in basaltic terraines. II. Geochemical and petrogenetic considerations. *Can. Mineral.* **2006**, *44*, 843–856. [CrossRef]
124. Long, A.M.; Menzies, M.A.; Thirlwall, M.F.; Upton, B.G.J.; Aspen, P. Carbonatite-mantle interaction: A possible origin for megacryst/xenolith suites in Scotland. In Proceedings of the 5th International Kimberlite Conference, Araxá, Brazil, 18 June–4 July 1991; pp. 467–477.
125. Bakumenko, I.T.; Tomilenko, A.A.; Bazarova, T.Y.; Yarmolyuk, V.V. The conditions of formation of volcanics in the Late Mesozoic–Cenozoic West Trans-Baikalian volcanic area (from results of study of melt and fluid inclusions in minerals). *Geokhimiya* **1999**, *12*, 1352–1356.
126. Gervasoni, F.; Klemme, S.; Rocha-Júnior, E.R.V.; Berndt, J. Zircon saturation in silicate melts: A new and improved model for aluminous and alkaline melts. *Contrib. Mineral. Petrol.* **2016**, *171*, 21. [CrossRef]

Article

Gem-Quality Zircon Megacrysts from Placer Deposits in the Central Highlands, Vietnam—Potential Source and Links to Cenozoic Alkali Basalts

Vuong Bui Thi Sinh [1,*], Yasuhito Osanai [2], Christoph Lenz [3,4], Nobuhiko Nakano [2], Tatsuro Adachi [2], Elena Belousova [4] and Ippei Kitano [2]

[1] Graduate School of Integrated Sciences for Global Society, Kyushu University, 744 Motooka, Nishi-ku, Fukuoka 819-0395, Japan

[2] Division of Earth Sciences, Faculty of Social and Cultural Studies, Kyushu University, 744 Motooka, Nishi-ku, Fukuoka 819-0395, Japan; osanai@scs.kyushu-u.ac.jp (Y.O.); n-nakano@scs.kyushu-u.ac.jp (N.N.); t-adachi@scs.kyushu-u.ac.jp (T.A.); i.kitano@scs.kyushu-u.ac.jp (I.K.)

[3] Institut für Mineralogie und Kristallographie, Universität Wien, 1090 Wien, Austria; christoph.lenz@univie.ac.at

[4] Australian Research Council (ARC) Centre of Excellence for Core to Crust Fluid Systems (CCFS) and National Key Centre for Geochemical Evolution and Metallogeny of Continents (GEMOC), Department of Earth and Planetary Sciences, Macquarie University, Sydney, NSW 2109, Australia; elena.belousova@mq.edu.au

[*] Correspondence: 3GS17013Y@s.kyushu-u.ac.jp

Received: 30 December 2018; Accepted: 29 January 2019; Published: 1 February 2019

Abstract: Gem-quality zircon megacrysts occur in placer deposits in the Central Highlands, Vietnam, and have euhedral to anhedral crystal shapes with dimensions of ~3 cm in length. These zircons have primary inclusions of calcite, olivine, and corundum. Secondary quartz, baddeleyite, hematite, and CO_2 fluid inclusions were found in close proximity to cracks and tubular channels. LA-ICP-MS U-Pb ages of analyzed zircon samples yielded two age populations of ca. 1.0 Ma and ca. 6.5 Ma, that were consistent with the ages of alkali basalt eruptions in the Central Highlands at Buon Ma Thuot (5.80–1.67 Ma), Pleiku (4.30–0.80 Ma), and Xuan Loc (0.83–0.44 Ma). The zircon geochemical signatures and primary inclusions suggested a genesis from carbonatite-dominant melts as a result of partial melting of a metasomatized lithospheric mantle source, but not from the host alkali basalt. Chondrite-normalized rare earth element patterns showed a pronounced positive Ce, but negligible Eu anomalies. Detailed hyperspectral Dy^{3+} photoluminescence images of zircon megacrysts revealed resorption and re-growth processes.

Keywords: zircon megacrysts; placer deposits; rare earth elements (REE); carbonatite-dominant melts; Central Highlands; Vietnam; hyperspectral photoluminescence imaging; LA-ICP-MS

---

1. Introduction

Zircon ($ZrSiO_4$, tetragonal, $I4_1/amd$) is an accessory mineral in most types of igneous and metamorphic rocks [1]. Zircon megacrysts are often found in placer deposits derived from intraplate basaltic fields as xenocrysts or xenolith debris in alkali basaltic rocks [2]. Alkali basalt fields in South-East Asia commonly contain mantle xenoliths that include garnet lherzolites, spinel lherzolites, and harzburgite, as well as mantle- and/or crust-derived megacrysts of pyroxene, olivine, plagioclase, garnet, zircon, and corundum [3]. Therefore, zircon placers that are related to basaltic magmatism are often associated with megacrysts of other important gem materials (e.g., sapphire, garnet, and spinel), as reported from various localities, including Australia, Vietnam, Cambodia, Thailand, China, Myanmar, Sri Lanka, and Tanzania [4–11]. Zircon material from the Central Highlands, Vietnam, shows brownish-red colors and blue, upon heat-treatment under reducing conditions [12]. This color effect is similar to zircon found in placers in the Ratanakiri district, Cambodia [13]. The Vietnamese zircon deposits are known since the late 1980s, and sporadic mining activities are conducted by local people at numerous small sites. These deposits are exploited by digging to a depth of approximately 1–2 m, then picking the gems by hand after washing the alluvial material. The exploration of placer deposits of gem-quality zircons and other gem megacrysts is of high economic importance in many South-East Asian countries that supply the global gem market [14]. The source characteristics and forming processes of placer zircon play an important role for further exploration of this kind of gem deposits in geologically related regions.

The genesis of zircon megacrysts associated with alkali basalt has been addressed previously, but it is still vigorously debated [2,15–18]. Some authors agree that large zircon and corundum crystals represent xenocrysts that are transported by alkaline-basalt magmas and do not crystallize primarily from them (e.g., Hinton and Upton [2]), while others consider crystallization directly from basaltic magma [19,20]. Various hypotheses have been reported for the origin of zircon megacrysts from various sites in different geological contexts: (1) crystallization from melts derived from a metasomatized upper mantle, as supported by O-isotope studies [15,17,18,21]; (2) formation within a late-stage fractional crystallization process of oceanic island basalt magma [16,19]; and (3) crystallization from a primitive alkaline mafic magma, which later evolved to a less alkaline host magma [20]. According to Cong et al. [10], zircon megacrysts from Cenozoic basalts in northeastern Cambodia crystallized in the mantle during metasomatic events caused by phosphate-rich fluids and/or silicate melts enriched in zirconium. Piilonen et al. [22] recently reported findings of megacrystic zircons as single crystals enclosed in the alkali basalt from Ratanakiri, Cambodia, and suggested a single stage growth in a carbonate-influenced environment.

In the Central Highlands of Vietnam, gem quality zircon is commonly accompanied by corundum (sapphire) in alluvial deposits that are considered eroded from Cenozoic alkali basalt fields in close proximity (e.g., Garnier et al. [17]). In this study, zircon megacrysts from various alluvial deposits in the Central Highlands were systematically studied to investigate their origin and potential links to Cenozoic basalt eruptions. In addition to trace element data and U-Pb geochronological data, photoluminescence (PL) imaging methods were applied to reveal internal growth and potential secondary alteration textures. In this paper, we used mineral abbreviations as given by Whitney and Evans [23].

2. Geological Setting

South-East Asia was formed by an amalgamation of several crustal blocks, including South China, Indochina, Siumasu, Inthanon, the West Burma block, and the Trans Vietnam Orogenic Belt (TVOB) (Figure 1A) [24]. The TVOB was first proposed by Osanai et al. [24] as a zone of Permo-Triassic metamorphic rocks in Vietnam, which were formed by continent-continent collision between South China and Indochina blocks. This orogenic belt is characterized by numerous shear zones with strong deformation, such as the Red River shear zone, Song Ma suture zone, Tam Ky-Phuoc Son shear zone, and the Dak To Kan shear zone (Kontum Massif). The northern extension of these shear zones reaches the Yuan Nan Province through Ailaoshan in China. The Central Highlands lies entirely within the Indochina block and are located in the southern part of the TVOB, partly extending into eastern Vietnam and Laos (e.g., Hutchison [25]) (Figure 1A).

Miocene-Pliocene alkaline and subalkaline basalts are exposed over a vast region of Thailand, Laos, and Vietnam [26]. In southern and central Vietnam, this province stretches across an area of approximately 23,000 km^2 with a thickness up to several hundred meters [3] (Figure 1A). The basalt plateau is accompanied by pull-apart structures composed of short extensional rifts bounded by strike-slip faults [27]. At least two eruptive episodes ("early" and "late") have been reported by Hoang et al. [28]. Tholeiitic, and rarely, alkali basaltic flows represent the "early" episode at Dalat (17.60–7.90 Ma), whereas in the "late" episodes, olivine tholeiite, alkali basalt, basanite, and (rarely) nephelinite erupted at Phuoc Long (<8.00–3.40 Ma), Buon Ma Thuot (5.80–1.67 Ma), Pleiku (4.30–0.80 Ma), Xuan Loc (0.83–0.44 Ma), and the Re Island centers (0.80–0 Ma) [27] (Figure 1B). Hoang et al. [28] proposed that tholeiitic basalts are the most common basalt type in the region and build up much more volumetric mass in comparison to the alkali basalts and the rarely occurring nephelinites which erupted from small volcanoes.

Erosion of these Cenozoic (Neogene-Quaternary) basalts formed placer deposits that represent a major source of gem-quality corundum, zircon, olivine, garnet, pyroxene, and plagioclase [4,14,26]. In South-Central Vietnam, gem-quality zircons have been found in alluvial deposits from six provinces, including Kontum, Gia Lai, Dak Lak, Dak Nong, Lam Dong, and Binh Thuan [12] (Figure 1A,B).

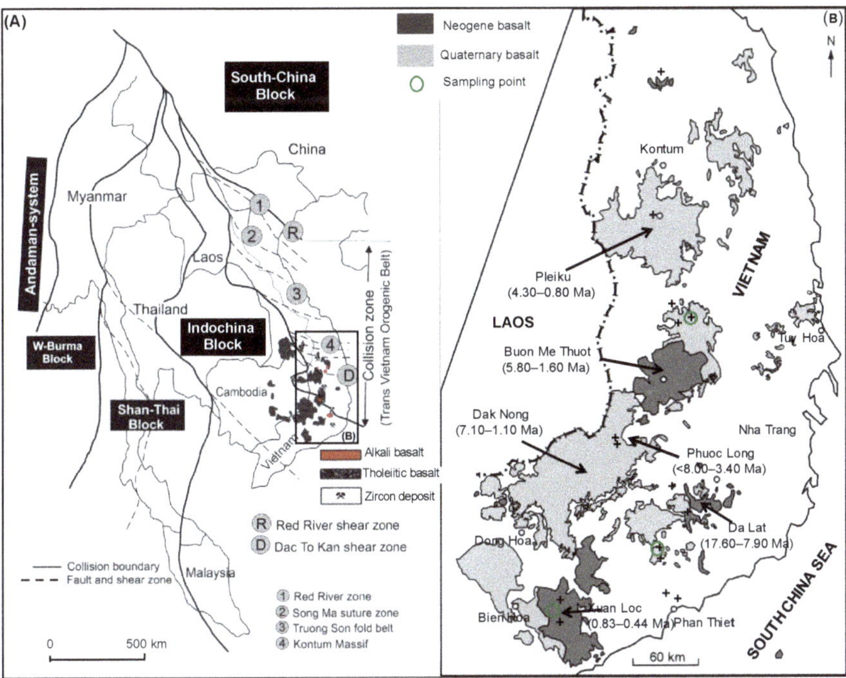

Figure 1. (**A**) Location of various alluvial zircon deposits in Cambodia and Vietnam within South-East Asia [12,24]. (**B**) Distribution of Neogene and Quaternary basalts with their K-Ar and Ar-Ar ages (after [4,6]). Reproduced with permission from all authors. Green circles indicate the sampling localities.

3. Sample Description and Methods

In this study, representative samples of zircon megacrysts collected from several alluvial deposits in the Central Highlands, Vietnam, were investigated in more detail. A representative zircon sample (Rata) from the Ratanakiri district, Cambodia was included for comparison. Large, gem-quality zircon specimens have dimensions up to several centimeters (Figure 2). Their color varies from colorless, orange, brownish-orange to dark brown and dark red (Figure 2). Some crystals have anhedral to subhedral shapes with rounded termination, while most grains are euhedral crystals with a typical combination of bipyramid and tetragonal prism (Figure 2B). Occasionally, internal colored zones that comprise an oscillatory change of orange and brown bands are visible to the naked eye (Figure 2B, sample C16).

Selected samples were cut along their long axis for petrographic analyses. Inclusions in zircon samples were observed and identified using an optical microscope attached to a confocal laser Raman spectrometer system (JASCO NSR-3100, JASCO (USA)) at Kyushu University. A 100× objective (with a numerical aperture NA = 0.90) and a green, continuous 532 nm frequency-doubled YAG:Nd laser (with an energy output of approx. 10 mW at the sample surface) was used to perform the Raman spot analyses. Note that potential bias of Raman spectra from the photoluminescence (PL) of Er^{3+} in zircon has been reported when using this laser wavelength [29]. A very minor PL contribution of Er^{3+} was identified in most recorded Raman spectra, but did not hamper the identification of typical Raman bands as obtained from various inclusion phases. Inclusions that were accessible via the polished sample surface were further identified qualitatively using a JEOL JSM-5310S-JED2140 scanning electron microprobe (SEM, JEOL, Ltd., Tokyo, Japan) equipped with an energy-dispersive X-Ray (EDX) detector (Li- doped Si semiconductor) at Kyushu University.

The internal texture of the prepared zircon crystals was investigated using a back-scattered electron (BSE) detector system and a GATAN MiniCL panchromatic cathodoluminescence (CL) detector attached to the above-mentioned SEM system at Kyushu University. In addition, we applied laser-induced photoluminescence spectroscopy using a HORIBA LabRam HR800 Evolution spectrometer equipped with a Peltier-cooled Si detector and a grating of 600 lines per millimeter (Horiba, Ltd., Kyoto, Japan). Confocal PL spot measurements were performed using an Olympus BX80 microscope (manufactured by Olympus Corporation, Tokyo, Japan) with a solid-state continuous laser, operating at 473 nm, to excite the most prominent REE^{3+} photoluminescence emissions in zircon (e.g., Dy^{3+}; Lenz et al. [29]). Hyperspectral images of polished, large zircon single-crystals were produced by measuring a multitude of spots in a point-by-point raster with a step-width of 10–20 μm using a software-controlled Maerzhauser mechanical x-y table. A software-based, automated data treatment procedure for each single spectrum was applied, and the plotted spectral parameters of interest were color-coded. With respect to the large size of investigated zircon crystals, and the comparatively large mapping step-widths, we used a 50× objective (NA = 0.5) and a confocal hole of 200 μm to adjust the lateral spatial resolution to be approximately 8–10 μm. However, we note that a diffraction-limited maximum spatial resolution of ~1 μm (planar) and 2–3 μm in depth may be achieved using the combination of a small confocal hole and a 100× objective.

Zircon U-Pb dating of samples C05, C07, and C16 was carried out by applying laser-ablation inductively coupled plasma mass-spectrometry (LA-ICP-MS) using an Agilent 7500cx quadrupole ICP-MS with a New Wave Research UP-213 YAG:Nd laser at the Kyushu University, that produced a laser-ablation spot of 100 μm. Details of the analytical procedure are presented by Adachi et al. [30]. The zircon standards Temora (417 Ma; [31]) and FC-1 (1099 Ma; [32]) were used for calibration and accuracy checks, respectively. The NIST SRM-611 glass standard was applied to determine the Th/U ratios. Raw data were treated using GLITTER software (Version 4.4.2, Glitter sold through Access Macquarie Ltd., Sydney, Australia) [33] and plotted in concordia diagrams using the Isoplot/Ex 3.7 software (Berkeley Geochronology Center, Berkeley, CA, USA) [34]. Trace element concentrations were obtained from similar zones in close proximity to spots used for U-Pb dating. Details of the analytical procedure are given in Nakano et al. [35] using the theoretical Si concentration in zircon as an internal standard.

Dating of zircon samples DL1, DL2, and DL3 was carried out using an Agilent 7700 quadrupole ICP-MS instrument attached to a Photon Machines Excimer 193 nm laser system at GEMOC, Macquarie University, using a beam diameter of ca. 50 μm with a 5 Hz repetition rate, and energy of around $0.06\,J/cm^2$ to $8\,J/cm^2$. Ablation was carried out in He gas to improve the sample transport efficiency, and to provide stable signals with reproducible Pb/U fractionation. Sample analyses were run along with analyses of the GEMOC GJ-1 zircon standard [36]. This standard is slightly discordant, and has a TIMS $^{207}Pb/^{206}Pb$ age of 608.5 Ma [37]. The other well-characterized zircon standard 91500 and Mud Tank were analysed within the run as an independent control on reproducibility and instrument stability. Individual time-resolved data analysis provides isotopically homogeneous segments of the signal to be selected for integration. We corrected the integrated ratios for ablation related fractionation and instrumental mass bias by the calibration of each selected time segment against the identical time segments for the standard zircon analyses. Furthermore, we employed the common-Pb correction procedure described by Andersen [38]. The analyses presented here were corrected assuming recent lead-loss with a common-lead composition corresponding to the present-day average orogenic lead, as given by the second-stage growth curve of Stacey and Kramers [39] for $^{238}U/^{204}Pb = 9.74$. No correction was applied to the analyses that were concordant within a 2σ analytical error in $^{206}Pb/^{238}U$ and $^{207}Pb/^{235}U$, or which had less than 0.2% common lead. Trace element concentrations of samples DL1, DL2, and DL3 were obtained using the same LA-ICP-MS system. Calibration of the relative element sensitivities were done using the NIST-610 standard glass as external calibration. Zircon BCR-2g and GJ-1 reference material were analyzed along the measurement runs as an independent control on reproducibility and instrument stability. Theoretical Zr content of the zircon was used for internal calibration of unknown zircon samples. The precision and accuracy of the NIST-610 analyses were 1–2% for REE, Y, Nb, Hf, Ta, Th, and U at the ppm concentration level, and at 5–10% for Ti (further details see in Belousova et al. [16]).

Figure 2. (**A**) Typical large, gem-quality, cut zircon samples from the Central Highlands are available on the gem market. (**B**) Selected zircon megacrysts used in this study from the same region. Samples C05, C07, and C16 are from the Buon Ma Thuot and Xuan Loc alkali-basaltic field. Samples DL1, DL2, and DL3 (not shown) are from deposits in the Dak Lak alkali-basaltic field (Figure 1B).

4. Results

4.1. Inclusion Features

Zircon crystals investigated in this study contained various fluid and mineral inclusions throughout the grains, especially along the cracks (Figure 3A–G). Tubular channels, bifurcating cavities, and vesicles in association with fluid inclusions were observed along the 2-dimensional planes (Figure 3A–G). Typical fluid inclusions along the fractures or fissures consisted of two phases, liquid (H_2O) with CO_2 or O_2 bubbles. Mineral inclusions were identified as calcite, hematite, corundum, olivine, baddeleyite, quartz, and feldspar. We identified two different paragenetic types of mineral inclusions: (1) the inclusions that were distributed throughout the grains without fluid inclusions in close proximity; and (2) others that were distributed along the cracks or were associated with tubular channels, bifurcating cavities, and vesicles with fluid inclusions nearby. Calcite inclusions were found in both areas: ovoid and droplet shaped ones sat inside the tubular channels or were associated with fluid inclusions (Figure 3B); whilst square or rectangular shaped calcite inclusions were accumulated along the chains or in groups that did not occur next to the fluid inclusions or within the tubular channels (Figure 3I). We rarely found corundum (Figure 3H) and olivine inclusions. They were of an euhedral and subhedral shape, and appeared in areas where no cracks, fluid inclusions, or tubular channels were present. Quartz, baddeleyite, hematite, and feldspar were found with irregular or ovoid shapes with rounded termination. They were located along the fissures, within tubular channels, cavities or associated with fluid inclusions (Figure 3C–G).

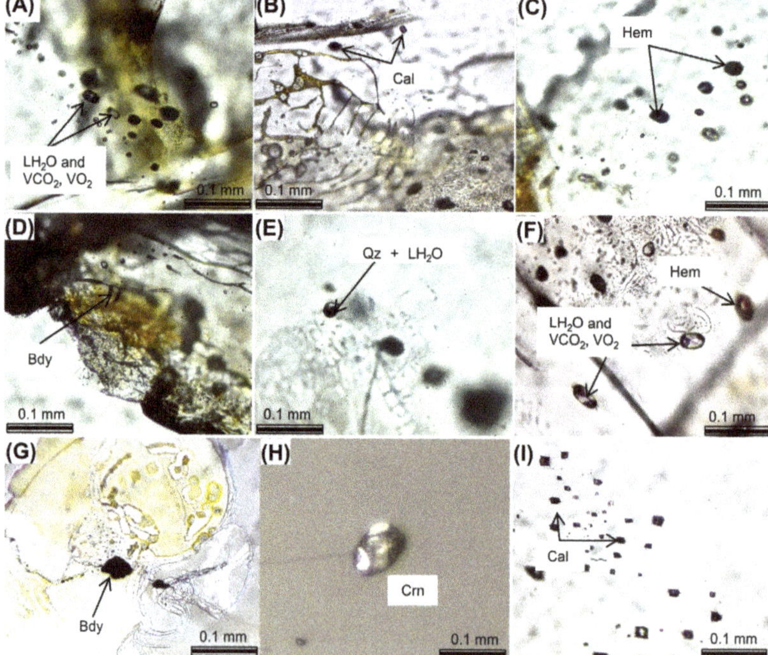

Figure 3. Representative inclusions found in placer zircon megacrysts: (**A**) fluid inclusions (LH$_2$O: aqueous H_2O VCO$_2$ and VO$_2$: vapor CO_2 and O_2, respectively); (**B**) Calcite (Cal); (**C**) hematite (Hem); (**D**) baddeleyite (Bdy) within tubular channels; (**E**) quartz (Qz) within tubular channels; (**F**) hematite and fluid (Hem + Fld) within bifurcating cavities and vesicles; (**G**) baddeleyite (Bdy) in tubular channels, bifurcating cavities, and vesicles; (**H**) corundum (Crn); and (**I**) calcite (cal) inclusion groups and along chains.

4.2. Internal Texture of Zircon Megacrysts

Backscattered-electron images obtained from the analysis of polished sections revealed no internal textural features although high contrast levels were applied (not shown). The latter indicated a homogeneous distribution and/or low concentration levels of trace elements. In contrast, cathodoluminescence images shown in Figure 4, revealed complex internal textures comprised of oscillatory zoning and sector zoning, although overall CL intensities were comparably low, except for the colorless zircon (Zrn-C05, see Figure 4). Anhedral grains were found to be split or broken into sections of former even larger grains (e.g., F03, F04), whereas the euhedral grains were characterized by multiple CL growth bands that retraced multiple distinct growth stages within a single crystal (e.g., C16 in Figure 4). To obtain minor differences in the CL zonation, high intensity contrast-levels were applied. This caused white stripes to be present in the CL images of some samples that did not correlate with the crystal's internal zonation (Sample C01, C07, and F01). They were most prominent along cracks and break-outs of fine material that produced mechanically-induced defects on the sample surface during polishing. Those structural defect centers mostly caused broad and intense CL signals in the UV spectral region, which were detected efficiently using panchromatic CL detectors.

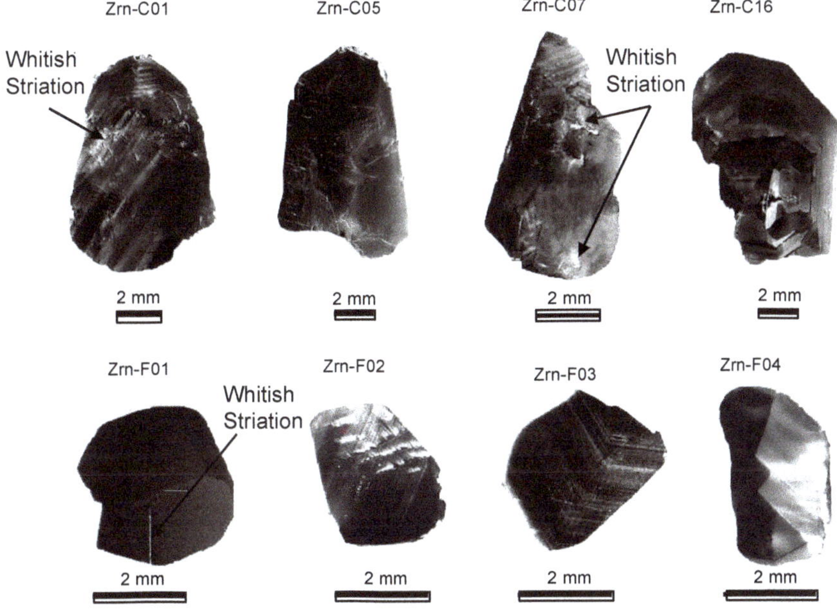

Figure 4. Cathodoluminescence images of the selected zircon samples from the Central Highlands, resemble zircon internal textural features that are typical for zircon of magmatic growth, e.g., oscillatory, sector, and growth zoning. Note, however, that whitish striations (indicated by black arrows) in CL contrasts obtained in the images of samples C01, C07, and F01 are due to mechanically-induced structural defect centers emanating from cracks and break-outs during polishing.

We applied laser-induced PL hyperspectral mapping to the large euhedral crystal grains with the aim of visualizing the internal luminescence distribution patterns as specifically caused by the emissions of REEs (Figure 5). The advantage of applying the latter technique is that the PL of specific REE species may be excited effectively using the appropriate laser sources in the visible spectral range without being obscured by intensive broad-band luminescence features as observed by CL spectroscopy and imaging [29]. Figure 5 shows hyperspectral images of the zircon samples from the Central Highlands (C16, DL2) and from Ratanakiri, Cambodia (Rata). The integrated intensity of the most prominent PL emission of Dy^{3+} ($^4F_{9/2} \rightarrow {}^6H_{13/2}$) was used as a plotted spectral parameter (see Figure 6A; and Figure 1 in Lenz et al. [29]). Note that the Dy sub-level bands of Vietnamese and Cambodian (Rata) zircon samples were characterized by exceptionally narrow band-widths (sublevel at 581 nm had a band-width of 11.0 cm^{-1}; see Figure 6A), which indicated a high degree of crystallinity and the absence of structural radiation-damage that might be caused by decay of radioactive U and Th. This was in accordance to findings by Lenz and Nasdala [40] and Zeug et al. [13] who reported very low PL and Raman band widths for zircon from Ratanakiri, Cambodia (Raman ν_3 [SiO_4] ~1.8 cm^{-1}). This has been interpreted to be the result of very low α-doses due to low U and Th concentrations in combination with a very young age (ca. 1 Ma).

In this study, we found a systematic linear correlation of the integrated PL intensity of Dy with its concentration as obtained from multiple LA-ICP-MS measurement spots (Figure 6B and Table S1). Furthermore, the latter laser-ablation reference spots of Dy concentrations and their correlation with PL intensity were used as an external calibration for the PL hyperspectral images to infer Dy concentrations from the PL signal in regions of unknown trace-element chemistry (see color-coded scales in Figure 5). Especially in the euhedral crystals (e.g., C16 and DL3), we identified several growth stages that were characterized by abrupt changes in concentrations of trace Dy (REE). Typically, the crystal's cores were found to be enriched in Dy, followed by a decreasing oscillatory REE substitutional budget (see inset of sample C16 in Figure 5). PL patterns further revealed episodical resorption of pre-existent crystals and fast re-growth. Growth zones that were characterized by low REE concentrations cross-cut former zonation patterns and occurred along with large holes and cavities (see arrows in Figure 5, samples C16 and DL3).

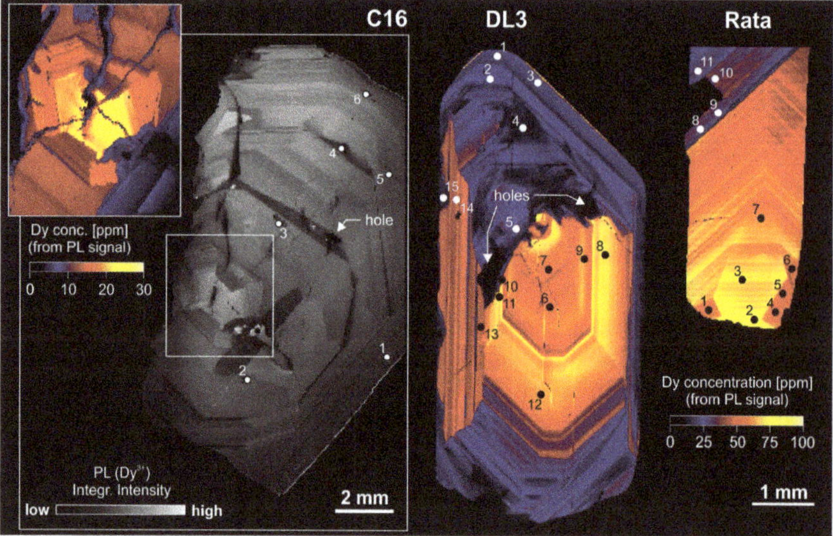

Figure 5. Laser-induced photoluminescence hyperspectral images of zircon samples from the Central Highlands (C16 and DL3) and from Ratanakiri, Cambodia (Rata). The integrated intensity of the Dy^{3+} ($^4F_{9/2} \rightarrow {}^6H_{13/2}$) emission (compare Figure 6) is the plotted spectral parameter (grey-scale). Trace-element concentrations of Dy from multiple LA-ICP-MS spots on the grains, were used for external calibration to correlate the Dy concentrations with its PL response. Concentrations given color-coded were back-calculated using the latter correlation (see Figure 6) to extrapolate the Dy distribution in regions of unknown trace-element chemistry. Besides oscillatory and sector zoning, several growth zones may be distinguished that were interpreted to result from resorption and re-growth events, as indicated by the cross-cut zonation patterns, abruptly changing (decreasing) Dy concentrations, and the appearance of large holes and cavities (see arrows).

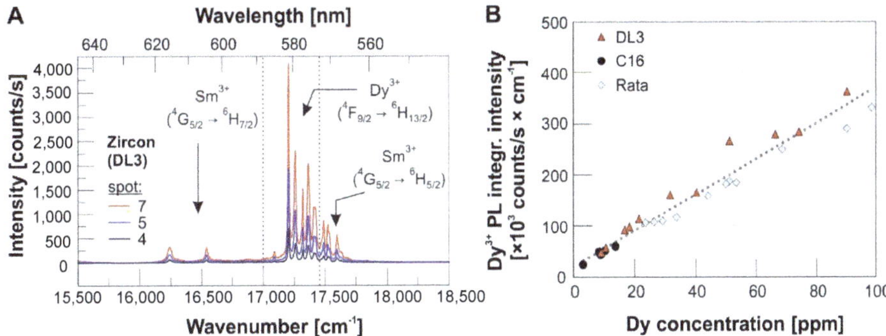

Figure 6. (**A**) Laser-induced PL spectra (laser-excitation wavelength λ_{exc} = 473 nm) obtained from three different measurement spots on the zircon sample DL3 (spots 4, 5, and 7; Figure 5). (**B**) The integrated intensity of the most intensive emission of Dy^{3+} in the spectral range 17,000–17,450 cm^{-1} (see dotted lines in sub-image (**A**)) from multiple spots on samples DL3, C16, and Rata plotted against the Dy trace-element concentration as obtained from LA-ICP-MS (Table S1). Linear regression from the data pairs PL vs. Dy conc. (R^2 = 0.95) was used to calibrate hyperspectral maps presented in Figure 5.

4.3. Geochemical Characteristics

Results from trace-element analyses of Vietnamese zircon samples, in addition to a Cambodian sample from Ratanakiri (Rata) are summarized in Table S1. Representative samples that comprised a colorless (C05), a dark-red (C07), an orange (C16), and three heavily zoned samples (DL1, DL2, DL3), were selected for detailed analyses. All zircon megacrysts had low total Y + REE concentrations typically within a 50–550 ppm range. Somewhat higher concentrations were detected at measurement spots placed in the core region of the heavily zoned samples DL2 and DL3 that were up to 2300 ppm Y + REE (see Table S1). The chondrite-normalized REE concentrations revealed a steep slope from La to Lu (Figure 7). Note that the REE geochemical characteristics of samples from the Central Highlands closely coincided with those of the samples from Ratanakiri (see samples Rata and data from Cong et al. [10]). All samples were characterized by a pronounced positive Ce-anomaly and a lack of Eu-anomaly, which is typical of magmatic zircon crystallized from more evolved magma sources with co-crystallization of feldspars that substitute Eu^{2+} on Ca-sites. The average Eu/Eu* ratio, that reflects the deviation of the measured Eu concentration from a theoretical Eu* concentration—extrapolated from chondrite-normalized Sm and Gd abundance (Eu* = Eu_N / [Sm_N + Gd_N]$^{0.5}$)—was found to be close to 1 (Figure 8A). The obtained REE compositional characteristics excluded crustal derived rocks or granitoids as a potential source of the zircon samples studied, and were more typical of the zircon of kimberlitic and/or carbonatitic origin (Figure 8A). Similar to REEs, the overall U and Th concentrations were found to be comparably low, with typical variability at 10–60 ppm and 3–30 ppm, respectively. Core regions of samples DL2 and 3 were found to be more enriched in U and Th with concentrations up to 500 ppm. Despite the pronounced variation in U and Th concentrations among the different growth zones within the samples, their U/Th ratios showed a clear correlation (Figure 8B). Likewise, the Y and U data pairs in a Y vs. U discrimination diagram scattered appreciably amongst the different growth zones within and among the samples, but coincided well with the compositional fields that were reported for basic, kimberlitic, and carbonatitic rocks (Figure 8C; Belousova et al. [16]). A source rock discrimination based on the Hf composition as proposed by Shnukov et al. [41], is presented in Figure 8D. The samples' Hf composition was found to be very low and shared compositional similarities with zircon from carbonatitic rocks, which had the lowest reported Hf concentrations (<0.7 wt %).

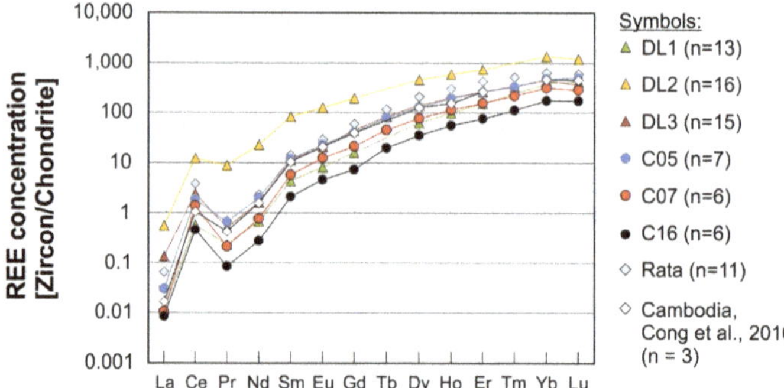

Figure 7. Chondrite-normalized plot of REE concentrations of zircon samples from the Central Highlands compared to samples from Ratanakiri, Cambodia. REE composition of C1 chondrite is described by Sun and McDonough [42]. All samples showed a steep slope rise from La to Lu, towards heavy REEs. A positive pronounced Ce-anomaly, but no Eu-anomaly was detected.

4.4. Geochronology

Results of the measured U-Th-Pb isotopic ratios and calculated ages of samples DL1, DL2, C05, C07, C16 are summarized in Tables S2 and S3 (see Supplementary Materials) and Figure 9. Vietnamese zircon samples from the Central Highlands were found to have Cenozoic, fairly concordant (~50% of discordance in average), and U-Pb ages (Figure 9A). Detailed comparison of the more robust ^{206}Pb/^{238}U ages indicated two distinct events of zircon megacrysts formation. Samples DL1, DL2, and C16 had a U-Pb age of around ca. 6.5 Ma, whereas samples C05 and C07 were even younger with an age of ca. 1 Ma. Note that the LA-ICP-MS measurement spots of individual single crystals were from different growth zones among the entire grains, which were clearly discernable in CL imaging, as well as PL hyperspectral mapping, and they had significant differences in trace-element concentrations (results above). We, however, found no systematic difference in U-Pb ages between the growth zones of individual samples within the standard errors.

5. Discussion

Zircon megacrysts from the Central Highlands that were investigated in this study had various colors, ranging from colorless, orange, brownish-orange, and dark brown to dark red. Some were characterized by colored zonation visible to the naked eye. Most of the selected grains showed a euhedral crystal shape with typical combinations of bipyramid and tetragonal prisms. Other grains had a subhedral to anhedral shape with rounded termination (Figure 2B). The latter might result from magmatic resorption on crystal faces and/or erosion during weathering and transportation into the placer. The internal texture of single crystals revealed by CL imaging and PL hyperspectral mapping was characterized by wide to narrow oscillatory and/or sector zoning (Figures 4 and 5). Typical crystal shape and textural features were conclusive hints that the megacrysts had a magmatic origin [1]. In heavily zoned samples (DL1–3, C16), dark-brownish colored zones correlated with the elevated concentrations of REE, U, and Th. On the other hand, a colorless zircon sample (C05, Figure 2B) was found to be more enriched in REEs, Th, and U, than the non-transparent dark-red sample C07 (Figure 2B, Figure 7 and Table S2). Hence, any rigorous interpretation based on color information should be avoided. For example, we found no corroboration for the interpretation that dark-brownish coloration of zircon was related to the presence of structural radiation damage. Multiple zircon samples obtained from placer deposits in Vietnam, including samples from Ratanakiri, Cambodia, had various colors ranging from transparent to non-transparent dark-brown, but all were characterized by very low Raman and PL spectral band-widths that indicated no radiation damage to be present [13].

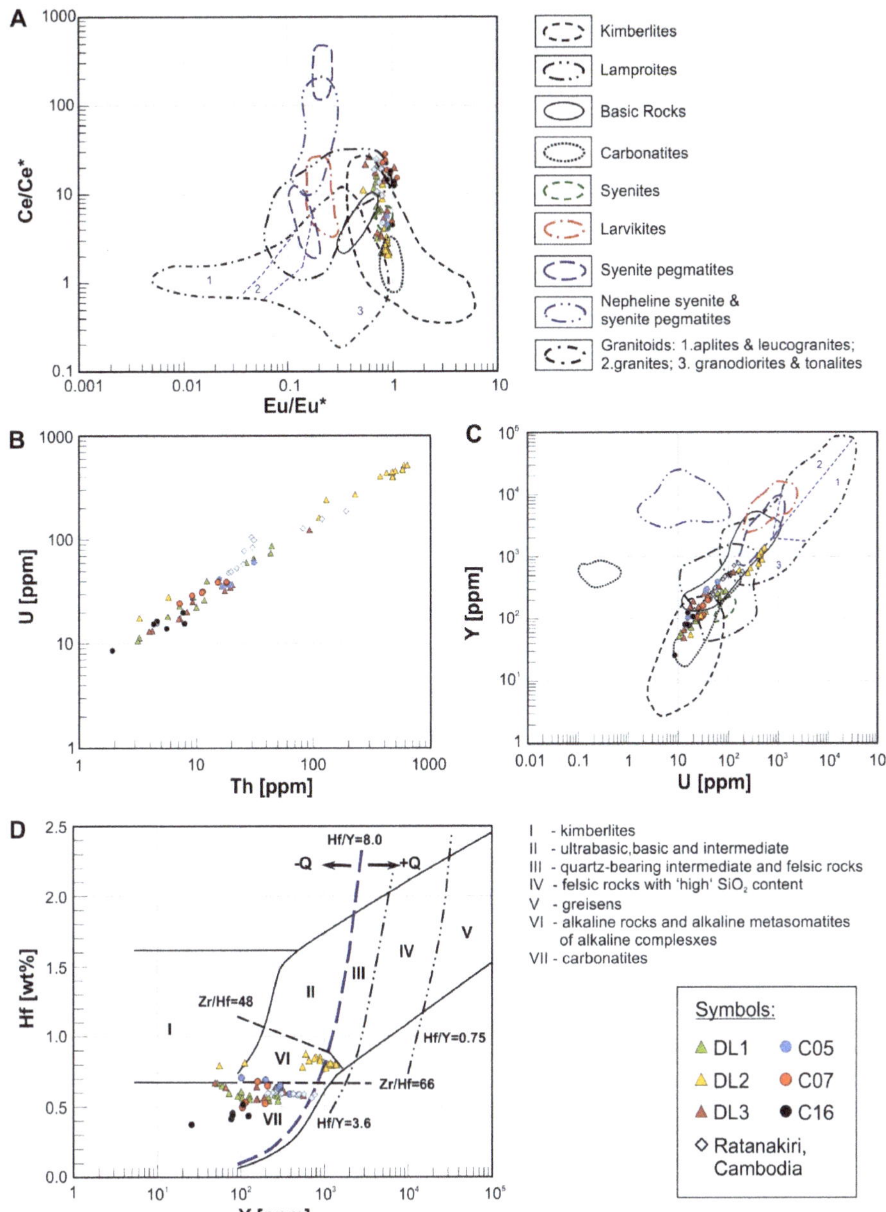

Figure 8. Compositional discrimination diagrams using trace-element concentrations of the studied samples obtained by LA-ICP-MS [16]: Ce/Ce* vs. Eu/Eu* (**A**); U vs. Th (**B**); Y vs. U (**C**); and Hf vs. Y (**D**) [41].

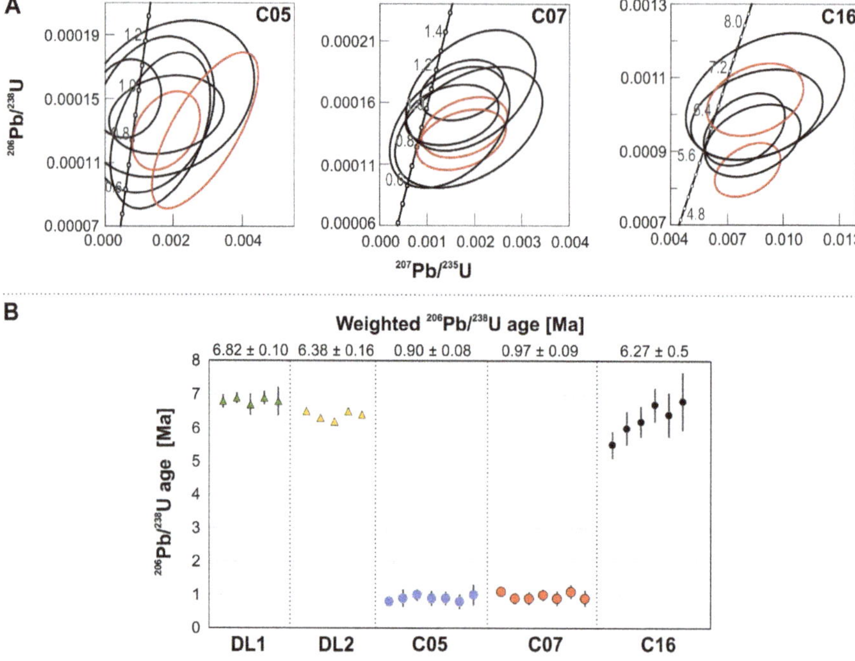

Figure 9. Summary of U-Pb geochronology data obtained by LA-ICP-MS (see details in Tables S2 and S3). (**A**) Representative concordia plots of samples C05, C07, and C16. Ellipses give a 2σ error, and red ellipses correspond to measurement spots placed in the core region of single crystals. (**B**) ^{206}Pb/^{238}U ages of the samples analyzed with the mean weighted age given for each sample. Error bars represent 2σ. Note that the ages obtained do not differ systematically amongst the measurement spots placed in the different growth zones within the individual samples. Two distinct sample populations were clearly differentiated (ca. 6.5 Ma and ca. 1 Ma).

We used PL hyperspectral imaging of the Dy^{3+} emission as an effective tool to visualize the distribution of REEs across the zircon samples (Figure 5). The latter technique was more reliable with respect to specifically exciting and detecting the emissions of REEs, because the panchromatic CL imaging is often strongly affected by other luminescence emissions that are caused by defects that are not coupled to the substitution of REEs (see mechanically induced bright CL striations caused by polishing; Figure 4). We further found that PL integrated intensities of Dy^{3+} used for PL hyperspectral imaging, correlated well with the Dy concentrations determined using spot LA-ICP-MS analyses (Figure 6). We, therefore, used PL hyperspectral images to quantitatively visualize Dy concentrations in regions of unknown chemical compositions as extrapolated from Dy^{3+} integrated emission intensities. Note, however, that the correlation of Dy concentrations with their PL response may be strongly hampered in other zircon samples that show a much higher accumulation of structural radiation damage, e.g., Lenz and Nasdala [40] reported that the presence of radiation damage resulted in substantial quenching of the PL intensities.

In this study, we identified multiple growth stages that included the resorption and re-growth of large volumes of the zircon megacrysts (Figure 5, samples C16 and DL3). Outer growth zones irregularly cut the former zones repeatedly and were characterized by lower REE concentrations, which indicated that the growth condition of megacrysts may have changed spontaneously to a more REE undersaturated growth environment that resulted in re-growth of the zones with depleted REE concentrations. Although absolute REE concentrations changed considerably across the different growth zones, their general chondrite-normalized concentration patterns were qualitatively similar.

Chondrite-normalized REE patterns in the core and overgrowth regions had a steep slope from light REE (LREE) to heavy REE (HREE), with pronounced positive Ce-, but with a lack of an Eu-anomaly (Figure 7). The positive Ce anomaly is typically caused by the significant difference between Ce^{4+} and $LREE^{3+}$ to fit into the zircon's structure, as Ce^{4+} has the same charge and similar ionic radius compared to Zr^{4+}. The Eu anomaly of zircon in REE patterns is generally explained by Eu^{2+} fractionation in the plagioclase and/or alkali feldspar that crystallizes before or during zircon formation from the magma [1,43,44]. Therefore, a pronounced negative Eu anomaly is commonly found in zircons from crustal-derived felsic rocks [16,44], whereas no Eu anomaly is found in zircon from feldspar free rocks, such as little to non-fractionated or mantle-derived rocks [45]. Note also that zircons of mantle origin usually have low REE and Y concentrations (zircon in kimberlite from southern Africa with TREE = 5–39 ppm and Y = 11–74 ppm [46]; zircon from Jwaneng kimberlite TREE up to 12 ppm and Y up to 23 ppm [47]), whereas zircons in crustal rocks are more enriched in REEs ranging from 250 ppm to 5000 ppm, with a 1500–2000 ppm average [1,16,48]. Trace-element concentrations of zircon crystals investigated in this study (TREE = 25–309 ppm, Y = 26–392 ppm) shared chemical characteristics comparable to mantle-derived zircon, and had no Eu anomaly (Figure 7). Moreover, the provenance discrimination plots proposed by Belousova et al. [16] and Shnukov et al. [41] revealed that the trace-element composition of Vietnamese zircon megacrysts were consistent with zircon typically found in kimberlite, syenite, carbonatites (Figure 8A–C), or alkaline rocks and alkaline metasomatites of alkaline complexes (Figure 8D). As the observed trace-element chemical characteristics ambiguously meet with those of several discrimination fields, it is hard to define an exact source composition. In fact, zircon samples from carbonatites often fall into various discrimination fields [49,50]. For example, Saava et al. [49] reported that only 45 out of 100 analyzed zircon grains extracted from a carbonatite body meet the trace-element signatures that indicate a carbonatitic source, while all others suggest an alkaline or ultramafic source. This compositional variation has been interpreted to result from the interaction between carbonatitic melt and co-magmatic silicate sources that broaden the chemical variation of carbonatitic zircon. In this study, total REE contents of zircon typically varied in the range of 25 to 309 ppm. These concentrations were slightly higher than those found in the kimberlitic zircon that had typically less than 50 ppm [46,47], but had a much lower concentration than the one from zircon of corundum bearing syenites (up to 3500 ppm [2]). Low concentrations of U and Th (Figure 8B), and low Nb and Ta concentrations are also typical of the zircon that originates from very little, to non-fractionated, Si-poor melts [16]. Trace-element chemical signatures of the zircon megacrysts of south-central Vietnam were found to be very comparable to those of zircon from Ratanakiri, Northeast Cambodia (Figure 7 in Cong et al. [10]; Figure 7 in this study) and other places such as New Zealand [11], East Australia [18], eastern and northeast China [5,8]. Results of a detailed study by Cong et al. [10] and Piilonen et al. [22] using $\delta^{18}O$ and $^{176}Hf/^{177}Hf$ isotopes clearly demonstrated that zircon from Ratanakiri, Cambodia, had a mantle origin and may be derived from metasomatized, partially melted lithospheric mantle material (such as peridotite, harzburgite), with strong carbonatitic geochemical fingerprint. In fact, experimental studies by Foley et al. [51], have demonstrated that mantle metasomatism, as induced by the presence of CO_2 and H_2O, lowers the solidus temperature at upper mantle conditions and promotes partial melting of the peridotites that results in melts with carbonatitic compositions at low degree, and carbonated silicate melts with higher degrees of melting. Early carbonate-dominant melts were characterized by a low Hf compatibility, that is consistent with low Hf concentrations obtained from zircon in this study (Figure 8D) [5,8,10,11]. With increasing degrees of partial melting, Hf is more compatible in the carbonated silicate melt [51].

Owing to the close geographical, geochemical, and geochronological relation of Cambodian zircon from Ratanakiri with material from the Central Highlands in Vietnam, studied here, we considered a very similar genetic origin for the latter. This was further supported by the presence of primary inclusions. Euhedral inclusions of calcite and corundum, and subhedral olivine are present in the zircon host, that is free of fluid inclusions, tubular channels, which shows no indication of visible cracks or alteration (Figure 3). These findings were in accordance to those of Le Bas [52], who interpreted the presence of abundant primary carbonate and olivine inclusions in the zircon megacrysts to be indicative of the growth in a silica-undersaturated, carbonatite-like melt. The experimental work of Baldwin [53] demonstrates that corundum and olivine likely crystallize from carbonatitic melts during early phases. However, no carbonatite bodies were reported from the studied area in the Central Highlands, Vietnam. In fact, the only known occurrence of carbonatites in Vietnam has been found in South Nam Xe, Northwestern Vietnam. Those rift-related carbonatites have reported ages of 28–44 Ma (biotite K-Ar ages) and 30–32 Ma U-Th-Pb isochron ages [54,55], and hence, may not be considered as a potential source. Instead, we found U-Pb zircon ages that were consistent with periods of basaltic volcanic activities in southern Vietnam. Two distinct zircon populations with U-Pb ages of ca. 6.5 Ma and ca. 1.0 Ma were identified. Although zircon megacrysts may be potentially derived from carbonatitic, metasomatized mantle material and do not primarily relate to alkali basalts, the zircon ages obtained were consistent with bi-episodal eruptive events at Buon Ma Thuot, Dak Lak (5.8–1.67 Ma), Pleiku (4.3–0.8 Ma), and the younger basalt eruptions exposed at Xuan Loc (0.83–0.44 Ma) (Figure 1B) [4]. Zircon megacrysts from North-East Cambodia not only occur geographically close to the megacrysts found in the present study, but also share a similar U-Pb age connected to the younger period of volcanic activity in the region (0.98 ± 0.04 Ma, Cong et al. [10]; 0.88 ± 0.22 Ma to 1.56 ± 0.21 Ma, Piilonen et al. [22]). Obtained zircon U-Pb ages were just slightly older than the eruption events that were dated based on the K-Ar ages for the whole rock [4]. This might be due to the difference in closure temperatures for U-Pb ages of zircon and K-Ar ages from the basalt, that indicated rapid extraction of zircon from its formation source, and the comparably fast uplift by the alkali-basalt eruption events [17,22]. Note, however, that the rim zones of resorption and re-growth (visualized by PL images in Figure 5) gave the very same ages as the zircon core regions within statistical errors. We excluded a potential alteration process to explain this texture. Late-stage, fluid-driven dissolution-reprecipitation typically results in geochronological resetting [56], and geochemical signatures typically indicate a progressive REE fractionation, e.g., a more pronounced negative Eu-anomaly. As both criteria are not verified for the overgrowth zones found in the zircon megacrysts, in this study, we assumed that multiple resorption and re-growth processes took place in the deep sub-crustal "carbonatitic" source magma chambers at temperatures higher than U-Pb closure (>900 °C; Cherniak et al. [57]). Multiple, late-Cenozoic, relatively fast uplifts of basaltic melts induced by tectonic weakening along shear-zones in the tectonically active region, may have either caught up with xenolithic zircon megacrysts from and/or mixed with pre-existent carbonatite-dominated melts generated by CO_2 and/or H_2O metasomatization of the upper mantle material. Inclusions of hematite, baddeleyite, quartz, and feldspar, as well as some ovoid or droplet carbonate inclusions along fractures or along channels, that are associated with fluid inclusions of liquid H_2O and vapor bubbles of CO_2, are considered to be of secondary origin (Figure 3). These phases are well known to occur upon zircon alteration. Particularly, the formation of baddeleyite due to the metasomatic alteration of zircon induced by Ca-bearing fluids has been proven by the experimental works of Lewerentz [58]. Therefore, the occurrences of these phases along fluid inclusions in close proximity to fractures, within tubular channels and/or vesicles are interpreted to result from late stage circulation of oxidative fluids in the ascending alkali-basaltic magma. Finally, weathering processes eroded zircon megacrysts from the alkali-basalt hosts in the Central Highlands and resulted in the deposition and enrichment of the nearby alluvial zircon placers found today.

6. Conclusions

Crystal morphology and internal texture of gem-quality, placer zircon megacrysts of the Central Highlands, Vietnam, are indicative of magmatic origin. The geochemical signatures and primary inclusions of calcite, olivine, and corundum indicate that zircon megacrysts might have crystallized from a carbonatite-dominant melt caused by the low-grade partial melting of metasomatized lithospheric mantle. Zircon resorption and re-growth textural features, observed with PL imaging, indicate an extended residence time in the sub-crustal magma chambers at variable REE-saturation levels, but at temperatures higher than closure of the U-Pb system (>900 °C). Zircon megacrysts were subsequently incorporated into ascending alkali basalts as xenocrysts. Two distinct populations with U-Pb ages at ca. 6.5 Ma and ca. 1.0 Ma that correlate with the eruption of alkali basalt fields in the Central Highlands, were identified. Recent reports of the direct findings of megacrystic zircon in the geographically and genetically related basalts of Ratanakiri, Cambodia [22] strongly supports this close relationship of placer zircon with Cenozoic alkali basalts as a potential host. The basalt magma and/or late stage circulation of carbonate-rich fluids must have resulted in the entrapment of further secondary inclusions like CO_2-H_2O fluids, baddeleyite, quartz, hematite, and feldspar along the cracks and tubular channels.

Supplementary Materials: The following are available online at http://www.mdpi.com/2075-163X/9/2/89/s1, Table S1: Results of LA-ICP-MS trace-element analyses of multiple measurement spots on zircon samples from Central Highlands, Vietnam (DL1, DL2, DL3, C05, C07, and C16) and from Ratanakiri, Cambodia (Rata), Table S2: U-Pb isotopic ratios and ages obtained from LA-ICP-MS analyses of zircon samples from Central Highlands, Vietnam, Table S3: U-Th-Pb isotopic ratios and ages obtained from LA-ICP-MS analyses of zircon samples from Dak Lak, Central Highlands, Vietnam.

Author Contributions: Concept of study, V.B.T.S.; methodology data acquisition and treatment, V.B.T.S., C.L., N.N., T.A. and E.B.; formal investigation, V.B.T.S., Y.O, C.L., N.N., and I.K.; compilation and writing of the manuscript, V.B.T.S.; review and editing of manuscript, C.L., Y.O., N.N., I.K., T.A., and E.B.

Funding: This study was funded by the Monbukagakusho/Ministry of Education, Culture, Sports, Science and a Technology-Japan (MEXT) scholarship at Kyushu University. Y.O. is funded by JSPS KAKENHI grants No. 16H02743, 22244063, and 21253008. C.L. and E.B. gratefully acknowledge the use of instrumentation funded through an honorary-associate agreement with the ARC CFSS/GEMOC at Macquarie University, Sydney. Financial support to C.L. was provided by the Austrian Science Fund (FWF), through project J3662-N19. N.N. is funded by JSPS KAKENHI grants No. 15K05345 and 18H01316.

Acknowledgments: The authors thank Le Thi Thu Huong from Senckenberg museum in Frankfurt, Germany, and Nguyen Thi Minh Thuyet from Hanoi University of Science, Vietnam for their help during fieldwork and sample collection in Vietnam. Special thanks to the local zircon miners in Central Highlands, Vietnam for providing insights into their mining operations. We are indebted to three anonymous reviewers for their critical comments on the manuscript.

Conflicts of Interest: The authors declare no conflict of interest. The funders had no role in the design of the study; in the collection, analyses, or interpretation of data; in the writing the manuscript; or in the decision to publish the results.

References

1. Hoskin, P.W.O.; Schaltegger, J.M. The composition of zircon and igneous and metamorphic petrogenesis. *Rev. Mineral. Geochem.* **2003**, *53*, 27–62. [CrossRef]

2. Hinton, R.W.; Upton, B.G.J. The chemistry of zircon: Variation within and between large crystals from syenite and alkali basalt xenoliths. *Geochim. Cosmochim. Acta* **1991**, *55*, 3287–3302. [CrossRef]

3. Hoang, N.; Han, N.X. Petrochemistry of Quaternary basalts of Xuan Loc area (South Vietnam). In *Geology of Cambodia, Laos, Vietnam*; Geological Survey of Vietnam: Hanoi, Vietnam, 1990; Volume 2, pp. 77–88.

4. Hoang, N.; Flower, M. Petrogenesis of Cenozoic basalts from Vietnam: Implication for origins of a 'Diffuse igneous province'. *J. Petrol.* **1998**, *39*, 369–395. [CrossRef]

5. Qiu, Z.; Yang, J.; Yang, Y.; Yang, S.; Li, C.; Wang, Y.; Lin, W.; Yang, X. Trace element and Hafnium isotopes of Cenozoic basalt-related zircon megacrysts at Muling, Heilongjiang province. Northeast China. *Acta Pet. Sin.* **2007**, *23*, 481–492.

6. Izokh, A.E.; Smirnov, S.Z.; Egorova, V.V.; Anh, T.T.; Kovyazin, S.V.; Phuong, N.T.; Kalinina, V.V. The condition of formation of sapphire and zircon in the areas of alkali-basaltoid volcanism in Central Vietnam. *Russ. Geol. Geophys.* **2010**, *51*, 719–733. [CrossRef]

7. Shigley, J.E.; Laurs, B.M.; Janse, A.J.A.; Elen, S.; Dirlam, D.M. Gem localities of the 2000s. *Gems Gemol.* **2010**, *46*, 188–216. [CrossRef]

8. Yu, Y.; Xu, X.; Chen, X. Genesis of zircon megacrysts in Cenozoic alkali basalts and the heterogeneity of subcontinetal lithospheirc mantle, eastern China. *Mineral. Petrol.* **2010**, *1000*, 75–94. [CrossRef]

9. Chen, T.; Ai, H.; Yang, M.; Zheng, S.; Liu, Y. Brownish red zircon from Muling, China. *Gems Gemol.* **2011**, *47*, 36–41. [CrossRef]

10. Cong, F.; Li, S.Q.; Lin, F.C.; Shi, M.F.; Zhu, H.P.; Siebel, W.; Chen, F. Origin of Zircon Megacrysts from Cenozoic Basalts in Northeastern Cambodia: Evidence from U-Pb Age, Hf-O Isotopes, and Inclusions. *J. Geol.* **2016**, *124*, 221–234. [CrossRef]

11. Sutherland, L.; Graham, I.; Yaxley, G.; Armstrong, R.; Giuliani, G.; Hoskin, P.; Nechaev, V.; Woodhead, J. Major zircon megacryst suites of the Indo-Pacific lithospheric margin (ZIP) and their petrogenetic and regional implications. *Mineral. Petrol.* **2016**, *110*, 399–420. [CrossRef]

12. Zeug, M.; Nasdala, L.; Wanthanachaisaeng, B.; Balmer, W.A.; Corfu, F.; Wildner, M. Blue Zircon from Ratanakiri, Cambodia. *J. Gemmol.* **2018**, *36*, 112–132. [CrossRef]

13. Huong, L.T.T.; Vuong, B.T.S.; Thuyet, N.T.M.; Khoi, N.N.; Somruedee, S.; Bhuwado, W.; Hofmeister, W.; Tobias, H.; Hauzenberger, C. Geology, gemological properties and preliminary heat treatment of gem-quality zircon from the Central Highlands of Vietnam. *J. Gemmol.* **2016**, *35*, 308–318. [CrossRef]

14. Sutherland, F.L.; Bosshart, G.; Fanning, C.M.; Hoskin, P.W.O.; Coenraads, R.R. Sapphire crystallization, age and origin, Ban Huai Sai, Laos: Age based on zircon inclusions. *J. Asian Earth Sci.* **2002**, *20*, 841–849. [CrossRef]

15. Sutherland, F.L. Alkaline rocks and gemstones, Australia: A review and synthesis. *Aust. J. Earth Sci.* **1996**, *43*, 323–343. [CrossRef]

16. Belousova, E.A.; Griffin, W.L.; O'Reilly, S.Y.; Fisher, N.I. Igneous zircon: Trace element composition as an indicator of source rock type. *Contrib. Mineral. Petrol.* **2002**, *143*, 602–622. [CrossRef]

17. Garnier, V.; Ohnenstetter, D.; Giuliani, G.; Fallick, A.E.; Phan Trong, T.; Quang, V.H.; Van, L.P.; Schwarz, D. Basalt petrology, zircon ages and sapphire genesis from Dak Nong, southern Vietnam. *Mineral. Mag.* **2005**, *69*, 21–38. [CrossRef]

18. Sutherland, F.L.; Meffre, S. Zircon megacryst ages and chemistry, from a placer, Dunedin volcanic area, eastern Otago, New Zealand. *N. Z. J. Geol. Geophys.* **2009**, *52*, 185–194. [CrossRef]

19. Griffin, W.L.; Pearson, N.J.; Belousova, E.; Jackson, S.E.; Van Achterbergh, E.; O'Reilly, S.R.; Shee, S.Y. The Hf isotope composition of cratonic mantle: LAM-MC-ICPMS analysis of zircon megacrysts in kimberlites. *Geochim. Cosmochim. Acta* **2000**, *64*, 133–147. [CrossRef]

20. Visonà, D.; Caironi, V.; Carraro, A.; Dallai, L.; Fioretti, A.M.; Fanning, M. Zircon megacrysts from basalts of the Venetian Volcanic Province (NE Italy): U-Pb ages, oxygen isotopes and REE data. *Lithos* **2007**, *94*, 168–180. [CrossRef]

21. Upton, B.G.J.; Hinton, R.W.; Aspen, P.; Finch, A.; Valley, J.W. Megacrysts and associated xenoliths: Evidence for migration of geochemically enriched melts in the upper mantle beneath Scotland. *J. Petrol.* **1999**, *40*, 935–956. [CrossRef]

22. Piilonen, P.C.; Sutherland, F.L.; Danisik, M.; Poirier, G.; Valley, J.W.; Rowe, R. Zircon xenocryst from Cenozoic Alkaline Basalts of the Ratanakiri Volcanic Province (Cambodia), Southeast Asia—Trace Element Geochemistry, O-Hf Isotopic Composition, U-Pb and (U-Th)/He Geochronology—Revelations into the Underlyi Lithospheric Mantle. *Minerals* **2018**, *8*, 556. [CrossRef]

23. Whitney, D.L.; Evans, B.W. Abbreviations for names of rock-forming minerals. *Am. Mineral.* **2010**, *95*, 185–187. [CrossRef]

24. Osanai, Y.; Nakano, N.; Owada, M.; Nam, T.N.; Miyamoto, T.; Minh, N.T.; Nam, N.V.; Tri, T.V. Collision zone metamorphism in Vietnam and adjacent South-eastern Asia: Proposition for Trans Vietnam Orogenic Belt. *J. Mineral. Petrol. Sci.* **2008**, *103*, 226–241. [CrossRef]

25. Hutchison, C.S. Geological Evolution of Southeast Asia. *Oxf. Monogr. Geol. Geophys.* **1989**, *13*, 368.

26. Barr, S.M.; Macdonald, A.S. Geochemistry and geochronology of late Cenozoic basalts of Southeast Asia. *Bull. Geol. Soc. Am.* **1981**, *92*, 1069–1142. [CrossRef]

27. Rangin, C.; Huchon, P.; Le Pichon, X.; Bellon, H.; Lepvrier, C.; Roques, D.; Hoe, N.D.; Quynh, P.V. Cenozoic deformation of central and south. *Tectonophysics* **1995**, *251*, 179–196. [CrossRef]

28. Hoang, N.; Flower, M.F.J.; Carlson, R.W. Major, trace element, and isotopic compositions of Vietnamese basalts: Interaction of enriched mobile asthenosphere with the continental lithosphere. *Geochim. Cosmochim. Acta* **1996**, *60*, 4329–4351. [CrossRef]

29. Lenz, C.; Nasdala, L.; Talla, D.; Hauzenberger, C.; Seitz, R.; Kolitsch, U. Laser-induced REE^{3+} photoluminescence of selected accessory minerals—An "advantageous artefact" in Raman spectroscopy. *Chem. Geol.* **2015**, *415*, 1–16. [CrossRef]

30. Adachi, T.; Osanai, Y.; Nakano, N.; Owada, M. La-ICP-MS U-Pb zircon and FE-EPMA U-Th-Pb monazite dating of pelitic granulites from the Mt. Ukidake area, Sefuri Mountains, northern Kyushu. *J. Geol. Soc. Jpn.* **2012**, *118*, 29–52.

31. Black, L.P.; Kamo, S.L.; Allen, C.M.; Aleinikoff, J.N.; Davis, D.W.; Korsh, R.J.; Foudoulis, C. TEMORA1: A new zircon standard for Phanerozoic U-Pb geochronology. *Chem. Geol.* **2003**, *200*, 155–170. [CrossRef]

32. Paces, J.P.; Miller, J.D.J. U-Pb ages of Duluth Complex and related mafic intrusions, northeastern Minnesota: Geochronological insights to physical, petrogenetic, paleomagnetic, and tectonomagmatic processes associated with the 1.1 Ga midcontinent rift system. *J. Geophys. Res.* **1993**, *98*, 13997–14013. [CrossRef]

33. Griffin, W.L.; Powell, W.J.; Pearson, N.J.; O'Reilly, S.Y. GLITTER: Data reduction software for laser ablation ICP-MS. *Mineral. Assoc. Can. Short Course* **2008**, *40*, 308–311.

34. Ludwig, K.R. User's Mnual for Isoplot 3.70: A geochronological toolkit for Microsoft Excel. *Berkeley Geochronl. Cent. Spec. Publ.* **2008**, *4*, 1–77.

35. Nakano, N.; Osanai, Y.; Adachi, T. Major and trace element zoning of euhedral garnet in high-grade (>900 °C) mafic granulite from the Song Ma Suture zone, northern Vietnam. *J. Mineral. Petrol. Sci.* **2010**, *105*, 268–273. [CrossRef]

36. Elhlou, S.; Belousova, E.; Griffin, W.L.; Pearson, N.J.; O'Reilly, S.Y. Trace element and isotopic composition of GJ-red zircon standard by laser ablation. *Geochim. Cosmochim. Acta Suppl.* **2006**, *70*, A158. [CrossRef]

37. Jackson, S.E.; Longerich, H.P.; Dunning, G.R.; Freyer, B.J. The application of laser-ablation microprobe; inductively coupled plasma-mass spectrometry (LAM-ICP-MS) to in situ trace-element determinations in minerals. *Can. Mineral.* **1992**, *30*, 1049–1064.

38. Andersen, T. Correction of common lead in U-Pb analyses that do not report ^{204}Pb. *Chem. Geol.* **2002**, *192*, 59–76. [CrossRef]

39. Stacey, J.S.; Kramers, J.D. Approximation of terrestrial lead isotope evolution by a two-stage model. *Earth Planet. Sci. Lett.* **1975**, *26*, 207–221. [CrossRef]

40. Lenz, C.; Nasdala, L. A photoluminescence study of REE^{3+} emissions in radiation-damaged zircon. *Am. Mineral.* **2015**, *100*, 1123–1133. [CrossRef]

41. Shnukov, S.E.; Andreev, A.V.; Savenok, S.P. *Admixture Elements in Zircons and Apatites: A Tool for Provenance Studies of Terrigenous Sedimentary Rocks*; European Union of Geosciences (EUG 9): Strasbourg, France, 1997; Volume 65.

42. Sun, S.S.; McDonough, W.F. Chemical and isotopic systematics of oceanic basalts: Implications for mantle composition and processes. *Geol. Soc. Lond.* **1989**, *42*, 313–345. [CrossRef]

43. Snyder, G.A.; Taylor, L.A.; Crozaz, G. Rare earth element selenochemistry of immiscible liquids and zircon at Apollo 14: An iron probe study of evolved rocks on the Moon. *Geochim. Cosmochim. Acta* **1993**, *57*, 1143–1149. [CrossRef]

44. Li, X.H.; Liang, X.R.; Sun, M.; Liu, Y.; Tu, X.L. Geochronology and geochemistry of single-grain zircons: Simultaneous in-situ analysis of U-Pb age and trace elements by LAM-ICP-MS. *Eur. J. Mineral.* **2000**, *12*, 1015–1024. [CrossRef]

45. Borghini, G.; Fumagalli, P.; Rampone, E. The stability of plagioclase in the upper mantle: Subsolidus experiments on fertile and depleted lherzolite. *J. Petrol.* **2010**, *51*, 229–254. [CrossRef]

46. Belousova, E.A.; Griffin, W.L.; Pearson, N.J. Trace element composition and cathodoluminescence properties of southern African kimberlitic zircon. *Mineral. Mag.* **1998**, *62*, 355–366. [CrossRef]

47. Hoskin, P.W.O. Minor and trace element analysis of natural zircon (ZrSiO$_4$) by SIMS and laser ablation ICPMS: A consideration and comparison of two broadly competitive techniques. *J. Trace Microprobe Tech.* **1998**, *16*, 301–326.

48. Pupin, J.P. Granite genesis related to geodynamics from Hf-Y in zircon. *Trans. R. Soc. Edinb. Earth Sci.* **2000**, *91*, 245–256. [CrossRef]

49. Saava, E.V.; Belyatsky, B.V.; Antonov, A.B. Carbonatitic Zircon-Myth or Reality: Mineralogical-Geochemical Analyses. Available online: http://alkaline09.narod.ru/abstracts/Savva_Belyatsky.htm (accessed on 30 December 2018).

50. Campbell, L.S.; Compston, W.; Sircombe, K.N.; Wilkinson, C.C. Zircon from the East Orebody of the Bayan Obo Fe-Nb-REE deposit, China, and SHRIMP ages for carbonatite-related magmatism and REE mineralization events. *Contrib. Mineral. Petrol.* **2014**, *168*, 1041. [CrossRef]

51. Foley, S.F.; Yaxley, G.M.; Rosenthal, A.; Buhre, S.; Kiseeva, E.S.; Rapp, R.P.; Jacob, D.E. The composition of near-solidus melts of peridotite in the presence of CO_2 and H_2O between 40 and 60 kbar. *Lithos* **2009**, *112*, 274–283. [CrossRef]

52. Le Bas, M.J. Carbonatite magmas. *Mineral. Mag.* **1981**, *44*, 133–140. [CrossRef]

53. Baldwin, L.C.; Tomaschek, F.; Ballhaus, C.; Gerdes, A.; Fonseca, R.O.C.; Wirth, R.; Geisler, T.; Nagel, T. Petrogeneis of alkaline basalt-hosted sapphires megacrysts. Petrological and geochemical investigations of in situ sapphire occurrences from the Siebengebirge volcanic field Western Germany. *Contrib. Mineral. Petrol.* **2017**, *172*, 43. [CrossRef]

54. Chi, N.T.; Flower Martin, J.F.; Hung, D.T. Carbonatites in Phong Tho, Lai Chau Province, north–west Vietnam: Their petrogenesis and relationship with Cenozoic potassic alkaline magmatism. In Proceedings of the 33rd International Geological Congress, Oslo, Norway, 6–14 August 2008; pp. 1–45.

55. Thi, T.N.; Wada, H.; Ishikawa, T.; Shimano, T. Geochemistry and petrogenesis of carbonatites from South Nam Xe, Lai Chau area, northwest Vietnam. *Mineral. Petrol.* **2014**, *108*, 371–390.

56. Geisler, T.; Schaltegger, U.; Tomaschek, F. Re-equilibration of zircon in aqueous fluids and melts. *Elements* **2007**, *3*, 43–50. [CrossRef]

57. Cherniak, D.J.; Watson, E.B. Pb diffusion in zircon. *Chem. Geol.* **2000**, *172*, 5–24. [CrossRef]

58. Lewerentz, A. Experimental Zircon Alteration and Baddeleyite Formation in Silica Saturated Systems: Implications for Dating Hydrothermal Events. Master's Thesis, Lund University, Lund, Sweden, 2011.

Article

Gems and Placers—A Genetic Relationship Par Excellence

Harald G. Dill

Mineralogical Department, Gottfried-Wilhelm-Leibniz University, Welfengarten 1, D-30167 Hannover, Germany; haralddill@web.de

Received: 30 August 2018; Accepted: 15 October 2018; Published: 19 October 2018

Abstract: Gemstones form in metamorphic, magmatic, and sedimentary rocks. In sedimentary units, these minerals were emplaced by organic and inorganic chemical processes and also found in clastic deposits as a result of weathering, erosion, transport, and deposition leading to what is called the formation of placer deposits. Of the approximately 150 gemstones, roughly 40 can be recovered from placer deposits for a profit after having passed through the "natural processing plant" encompassing the aforementioned stages in an aquatic and aeolian regime. It is mainly the group of heavy minerals that plays the major part among the placer-type gemstones (almandine, apatite, (chrome) diopside, (chrome) tourmaline, chrysoberyl, demantoid, diamond, enstatite, hessonite, hiddenite, kornerupine, kunzite, kyanite, peridote, pyrope, rhodolite, spessartine, (chrome) titanite, spinel, ruby, sapphire, padparaja, tanzanite, zoisite, topaz, tsavorite, and zircon). Silica and beryl, both light minerals by definition (minerals with a density less than 2.8–2.9 g/cm^3, minerals with a density greater than this are called heavy minerals, also sometimes abbreviated to "heavies". This technical term has no connotation as to the presence or absence of heavy metals), can also appear in some placers and won for a profit (agate, amethyst, citrine, emerald, quartz, rose quartz, smoky quartz, morganite, and aquamarine, beryl). This is also true for the fossilized tree resin, which has a density similar to the light minerals. Going downhill from the source area to the basin means in effect separating the wheat from the chaff, showcase from the jeweler quality, because only the flawless and strongest contenders among the gemstones survive it all. On the other way round, gem minerals can also be used as pathfinder minerals for primary or secondary gemstone deposits of their own together with a series of other non-gemmy material that is genetically linked to these gemstones in magmatic and metamorphic gem deposits. All placer types known to be relevant for the accumulation of non-gemmy material are also found as trap-site of gemstones (residual, eluvial, colluvial, alluvial, deltaic, aeolian, and marine shelf deposits). Running water and wind can separate minerals according to their physical-chemical features, whereas glaciers can only transport minerals and rocks but do not sort and separate placer-type minerals. Nevertheless till (unconsolidated mineral matter transported by the ice without re-deposition of fluvio-glacial processes) exploration is a technique successfully used to delineate ore bodies of, for example, diamonds. The general parameters that matter during accumulation of gemstones in placers are their intrinsic value controlled by the size and hardness and the extrinsic factors controlling the evolution of the landscape through time such as weathering, erosion, and vertical movements and fertility of the hinterland as to the minerals targeted upon. Morphoclimatic processes take particular effect in the humid tropical and mid humid mid-latitude zones (chemical weathering) and in the periglacial/glacial and the high-altitude/mountain zones, where mechanical weathering and the paleogradients are high. Some tectono-geographic elements such as unconformities, hiatuses, and sequence boundaries (often with incised valley fills and karstic landforms) are also known as planar architectural elements in sequence stratigraphy and applied to marine and correlative continental environments where they play a significant role in forward modeling of gemstone accumulation. The present study on gems and gemstone placers is a reference example of fine-tuning the "Chessboard classification scheme of mineral deposits" (Dill 2010) and a sedimentary supplement to the digital maps that form the core of the overview "Gemstones and geosciences in space and time" (Dill and Weber 2013).

Keywords: gemstones; placer; heavy and light minerals; landforms; climate; geodynamic setting

1. Introduction—Coupling Gemstones and Placer Deposits

Inorganic raw materials and mineral deposits are subdivided in the most convenient way using a tripartite classification scheme into (1) ore minerals and metallic resources, (2) industrial minerals and rocks, and (3) gemstones and ornamental stones [1]. Category 1 is self explanatory and serves as a resource for metals like Pb, Zn, and Fe, category 2 comprises mineral raw materials which by virtue of their physical and chemical properties such as insulation capacity or fire-resistant properties are essential for human beings, e.g., perlite, kaolin, bentonite, zeolites, diatomite, or vermiculite, while category 3 stands out by the aesthetic value of its items, like a diamond in a necklace or a statue sculptured out of a block of marble. Gems and gemstones as part of category 3 are known for their elevated hardness expressed by the Mohs hardness numbers. The three top scorers in the hardness scale of Mohs diamond (10), corundum (9), and topaz (8) are renowned gemstones (Table 1). All but three gemstones (varieties of beryl, quartz modifications, amber) are heavy minerals by definition (minerals with a density greater than 2.8–2.9 g/cm^3, which is the density range of the most common rock-forming minerals, quartz, plagioclase, alkaline feldspar, and calcite) (Table 1).

Placer deposits are a subtype of clastic sedimentary rocks composed of rock fragments or mineral particles which were transported from their source by water, wind, ice, or simply by gravity to the depocenter. While common siliciclastic rocks have the light minerals feldspar s.s.s. (=solid solution series), quartz, phyllosilicates, and carbonate minerals as main constituents and contain heavy minerals only as accessory minerals, and in placer deposits, the latter group of minerals prevails over light minerals, in places, to such an extent that they can be mined for the three commodity groups mentioned at the beginning, e.g., magnetite placers (ore), phosphate placers (industrial minerals), and diamond placers (gemstones) (see Section 3.8). Gemstones fulfill all physical requirements to survive the transport from the source to the depositional environment; they have the hardness to resist chemical weathering in the provenance area, to stand the attrition and pass through the comminution processes on transport so as to appear at the site of deposition as a raw material to be mined for a profit. The separation of the gemstones from the diluting thrash or gangue minerals (quartz, feldspar) by their different specific gravities and hardness takes full effect leading to a grain size and quality justifying the minerals being called a gemstone of economic relevance for cutters and jewelers—jeweler quality (Table 1, Figure 1). Lower qualities are recovered, in places, for the showcase attractive to mineral dealers and collectors, and the lowest grade can still be harnessed owing to its hardness as an industrial mineral, used as abrasives in the industry, e.g., corundum.

Table 1. Gem minerals found in placer deposits, their physical properties relevant for their emplacement clastic haloes around their source rocks and the sedimentary traps. The gem placer deposits are subdivided according to the "Chessboard classification scheme of mineral deposits" [1]—See Appendix A, so as to assist their assignment to the overall concentration processes of elements, minerals, and rocks during magmatic, metamorphic, and sedimentary processes.

Element (for General Classification Scheme see [1])	Mineral	Type	Density (Mean) kg/m³	Hardness
Beryllium	Beryl and its varieties emerald, aquamarine, heliodor, morganite, goshenite, pezzottaite (a) Chrysoberyl (b)	Residual, colluvial, alluvial (fluvial). Argillaceous (kaolin) regolith	(a) 2.8 (b) 3.7	(a) 7.5–8.0 (b) 8.5
Boron	Tourmaline (further minerals see Table 2)	Residual to colluvial placers in stream sediments and marine placer deposits as byproduct, useful as proximity indicator/pathfinder	3.2	7.5
Fluorine	Topaz	Residual to colluvial placers in stream sediments useful as proximity indicator/pathfinder	3.6	8.0
Phosphorous	Apatite	Eluvial, colluvial, alluvial placers	3.2	5.0
Zirconium	Zircon (hyacinth)	Gemstone: residual to alluvial placers. Industrial minerals: marine placers	4.7	7.5
Garnet group	Grossular (hessonite, tsavorite), spessartine, pyrope and its variety rhodolite, andradite (demantoid), almandine	Gemstone: residual to alluvial placers. Industrial minerals: also in marine placers	4.2	7.0–8.0
Corundum	Ruby, sapphire, padparaja	Residual, eluvial, colluvial alluvial-fluvial	4.1	9.0
Spinel	Spinel group minerals	See corundum	3.6	8.0
Diamonds	Diamond	Residual, eluvial, colluvial, alluvial-fluvial, marine and aeolian modern and paleoplacers deposits	3.5	10
Silica	Rock crystal, agate, amethyst, citrine, quartz, rose quartz, smoky quartz	Alluvial-fluvial placers (amethyst, agate, rock crystal). Residual placer (double terminated rock crystal)	2.6	7.0
Chromium	Cr titanite (a), Cr diopside (b)	Colluvial-alluvial-fluvial placers (Cr diopside), ((Cr)-titanite) short distance of transport, useful as pathfinder	(a) 3.5 (b) 3.4	(a) 5.0–5.5 (b) 6.0
Olivine s.s.s.	Peridote	Colluvial-alluvial-fluvial-marine placers only in case of a high ratio of uplift/weathering and/or short distance of transport	3.3	6.5–7.0
Epidote s.s.s.	Tanzanite	Colluvial-alluvial	3.3	6.5
Lithium	Kunzite, hiddenite	Residual to alluvial placers, swiftly decomposes to argillaceous material	3.2	6.5–7.0
Organic compounds	Amber	Fluvial, marine placers	1.1	2.0–2.5

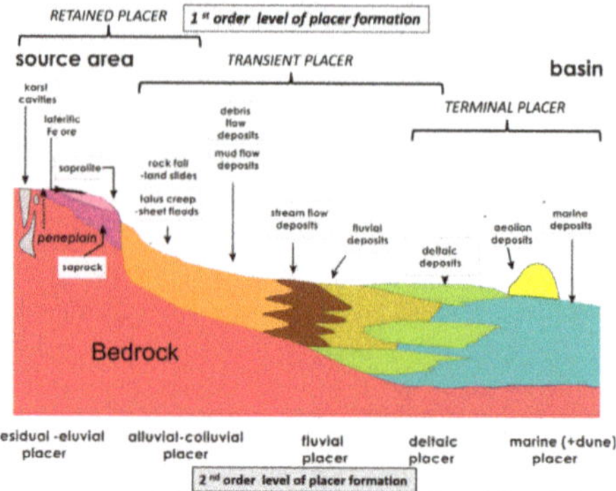

Figure 1. Cartoon to show the different types of placer deposits from the source area to the basin, from the weathering mantle on the top of the exposed bedrock through the continental terrigenous lithofacies types into the marine lithofacies. Glacial and fluvial-glacial lithofacies types because of the minor relevance for placer deposits in terms of separating light and heavy minerals are not shown in the cartoon.

2. Placer—Environments of Formation and Sedimentary Processes

2.1. Subdivision of Placer Deposits

Placer deposits are surficial mineral deposits resulting from mechanical processes operative in various depositional environments from the source to the basin (Figure 1). These accumulations composed, predominantly of heavy minerals have been investigated intensively by different authors paying most attention to the depositional environments [2–17]. Concluding from the aforementioned studies, a first-order subdivision may be achieved into modern placers and paleoplacers. Modern placers are unconsolidated to loosely bound accumulations embedded into mostly Neogene to Quaternary near surface deposits of different clastic lithofacies types and lacking post-sedimentary diagenetic alteration. The conglomerates at Blind River-Elliot Lake, Canada, and Witwatersrand, South Africa occur in stratigraphic series that developed when the Earth's atmosphere was different from today's oxidizing conditions and called paleoplacers. Their quartz pebble conglomerates formed from the Archaean to the Early Proterozoic (3100–2200 Ma), some also extending into the Middle Proterozoic such as the Tarkwa (1900 Ma). World-class deposits are at Witwatersrand, South Africa, Elliot Lake, Canada, Sierra do Jacobina, Brazil, Tarkwa, Ghana, and the Guyana Shield area in Colombia

Milesi et al., Malitc et al., and Frimmel [18–20]. Paleoplacers of Precambrian age may grade into modern placers through a wide range of Mesozoic and Cenozoic heavy mineral repositories which are the subject of exploration geologists mainly targeting upon diamond, as shown in some case histories of gem deposits discussed later in the study. Besides gold they also gave host to diamonds and colored gemstones but predominantly act as a protore and as such agree with the binary classification scheme of Stanaway [21].

The second level of subdivision in the hierarchy does not reflect a function of age, but depicts the depositional environment. A "catena" of placer deposits can be established from source to basin (Figure 1): (1) Residual and eluvial placers, (2) colluvial, (3) alluvial-fluvial placers, (4) deltaic placers, and (5) marine and aeolian placers (Figure 1). In the succeeding sections, the various environments are described in more detail with special reference to the accumulation of gemstones. Due to the bimodal distribution of gemstones, which reveals that one part of them is preferably laid down near the source area, and another at the very end of the "catena" in the marine basins, the environmental analysis in Section 2.2 places its emphasis predominantly on the most proximal and most distal lithofacies types of the "catena", which literally translates into the two groups of gemstones, the colored gems, and the diamonds (Figure 1).

2.2. Depositional Environments of Gemstone Placers

2.2.1. Residual-Eluvial Placers

Residual-eluvial placers mirror the start-up of a placer development. Magmatic, metamorphic, or sedimentary bedrocks (parent material of the gemstones) undergo chemical and/or mechanical weathering leading to a disintegration of the source rocks and a gradual liberation of the rock-forming minerals highly resistant to supergene alteration from those more vulnerable to it. Water removes the "trash minerals" from the weathering mantle and washes the chemical compounds in the meteoric fluids out of the regolith resultant in an in-situ concentration of heavy minerals (Figure 2a–c). The weathering profile in the *statu nascendi* consists of the fresh parent rock that is converted into a saprock, still displaying the structure and texture of the primary rock and that is overlain by a saprolite reflecting together with ferricretes on top the most intensive stage of chemical weathering (Figure 2a). The saprolite on top of an ultrabasic rocks may be enriched in precious corundum, which is accompanied by anatase, brookite, rutile, zircon, and Fe oxide-hydrates after flushing out the phyllosilicates emplaced during supergene alteration (Figure 2b,c). Therefore, to show the intensity of weathering and the transition from residual placers into parautochthonous colluvial ones Ti dioxides and zircon are excellent marker minerals and the chemical compounds of TiO_2 and Zr well reflect this reworking process in the chemolog (Figure 2c). The current chemolog is representative of a composite weathering profile with hiatuses marked by the arrowheads where loose weathering loam was transported away by hill wash downslope. Such breaks in weathering and sedimentation are planar elements guiding precious corundum (Section 3.7) and diamond accumulation (Section 3.8).

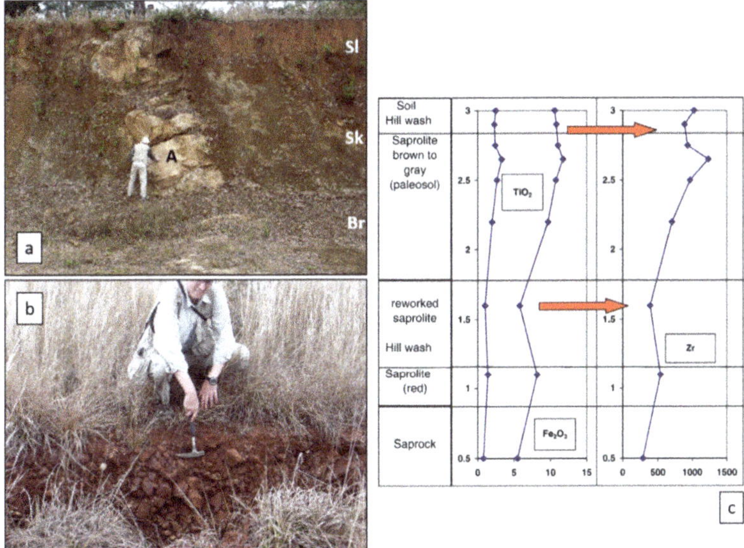

Figure 2. Residual placers from the savannah in Malawi. (**a**) Penetration of weathering as a function of bedrock lithology. The migmatitic K gneiss (Br) is more susceptible to weathering than the aplitic dyke (A) producing a weathering mantle of saprock (Sk) and saprolite (Sl). The meteoric fluids easily penetrated the metamorphic rocks along the planes of foliation which are very narrowly spaced and almost vertical. (**b**) Rusty red saprolite with relic corestones of metapyroxenite. The weathering loam hosts pale green and white sapphire crystals disseminated in a case-hardened duricrust. (**c**) Chemolog along a road cut measuring 3 m in depth illustrating the transition from a residual placer to a colluvial placer. Rutile (see TiO_2 contents) and zircon (see Zr contents) are cast in the role of marker minerals. The dip-slope motion of the weathering material is denoted by the red arrowheads.

2.2.2. Colluvial Placers

Colluvial placers form along hill slopes different with regard to their dip simply as a function of gravity as the most driving force and the availability of water as the decisive modifier of the various processes [22] (Figure 3a). Dip of slope and the extent to which the substrate is soaked with water controlling the speed of deposition. Sediment transport by means of talus or soil creep does not contribute much to the build-up of placers, as is the case with simple heave. Heave is a process causing slope instability of a certain kind of "swelling" soil and overburden that are abundant in clay minerals, particularly smectite. These fine-grained rocks lie on the opposite end of the grain-size scale of the common host rocks of gemstone placers and therefore can be cast aside in a treatment of colluvial placers.

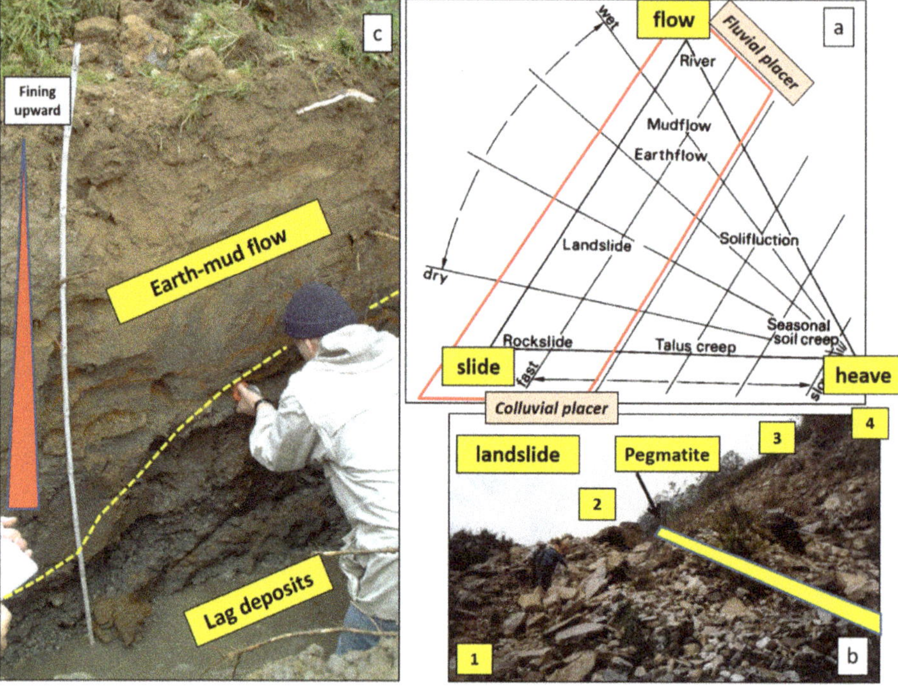

Figure 3. Colluvial placers (**a**) A triplot to show the various forms of colluvial processes and deposits as a function of wetness of the material and the velocity of mass wasting (slightly modified after [22]). The area framed with the red line is the area essential for the accumulation of gemstones attaining a size of economic importance. (**b**) The topmost part of a pegmatite layer which was worked for tourmaline is exposed and partly covered by rubble of a landslide, Nepal. The different zones are marked: (1) Zone of accumulation = colluvial tourmaline placer, (2) transverse ridge, (3) zone of depletion, and (4) main scarp (photograph taken by Tamrakar). The placer deposit is situated on a steep hill slope in the Himalaya, Nepal (**c**) A trench dug into earth-mud flow bearing a colluvial "nigrine" placer (nigrine = intergrowth of rutile and ilmenite). The upper boundary is marked by the yellow dashed line. The image shows sampling for OSL (optically stimulated luminescence) age dating of the colluvial placer, SE Germany. The colluvial processes are operative along the slopes of a small river perpendicular to the thalweg of the fluvial drainage system.

The processes relevant for gemstone placers are fast mass wasting processes, as demonstrated by the rubble in the zone of accumulation of a landslide underneath the tourmaline-bearing pegmatite which is synonymous with the zone of exploitation since it contains the "gemstone placer ore", whereas the primary gemstone deposit is uneconomic because of little or no accessibility (Figure 3a,b). It is a fast mass wasting process with little water involved with the zone of depletion marking the source area and the outcrop of the pegmatite marking the transverse ridge bulldozed over (Figure 3b). In the trench on display in Figure 3c, the geologists are faced with an entirely different situation. The driving force behind these colluvial processes is gravity, but the material of mud and gravel is soaked with water. The earth-mud flow bearing a colluvial "nigrine" and cassiterite accumulation evolved perpendicular to the thalweg of a fluvial drainage system and as a consequence of the steep slope provoked the fast motion of the "slurry" responsible for the coarse lag deposits exposed at the final depth of the trench [23,24]. To clearly demonstrate which process colluvial or fluvial was instrumental in the emplacement of the placer deposits a quantitative recording of the orientation of the longitudinal axes of clasts and plotting them in pole diagrams (situmetry) is recommended. It goes without saying that pure end member types such as pure colluvial or fluvial placer system are the exception rather than the rule in nature particular to the section of retained and transient placers (Figures 1 and 3a). To guarantee the emplacement of a colluvial placer of significance as to the amount of heavy minerals and their particle size the velocity of the mass wasting processes is paramount. Talus creep and heave are not efficacious in the production of significant accumulations of colluvial gemstone placers as stated above.

2.2.3. Alluvial-Fluvial Placers

The alluvial-fluvial placer deposits intertongue with the aforementioned colluvial deposits with stream flow accumulation at their most distal position or overlap with each other during a stepwise uplift of the hinterland when the system becomes wetter (Figure 1, Figure 3a, and Figure 4). The alluvial-fluvial systems play a major part in the formation of sedimentary concentrations of the top-score among the precious metals and gemstones, gold, PGE (Platinum Group Elements), and diamonds, respectively [25–27]. A scarcely to non-vegetated system of alluvial fans debouching into a periodically flooded more or less straight drainage system is displayed in Figure 4a. They often coalesce resulting in an apron of bajada fans in the foreland of rising modern fold belts such as the Andes in Argentina (Figure 4a). It is a sort of continental terrigenous sediments that has been extensively discussed by [28] and Galloway et al. [29]. Moving downslope the drainage systems change from braided, through meandering into anastomosing fluvial drainage system reflecting a decreasing gradient (Figure 4b–d). There are heavy minerals found in all zones of the alluvial-fluvial drainage system, yet the probability of finding economic placer deposits of gemstones wanes downstream, as indicated by arrowhead facing downward (Figure 4b–d). The potential to find gemstone-bearing placer deposits is high in sand and gravel bars of the braided streams as shown by the finds of tourmaline and garnet in the tributaries of Trisuli River in the Ganesh Himal, Nepal (Figure 4e). Meandering rivers, following more downstream, only host heavies in bar sands of the active channels of the drainage system, whereas oxbow lakes indicative of abandoned drainage systems are barren as is also the case with anastomosing fluvial systems, which are rife with suspended load as a consequence of the lower flow regime (Figure 4c). Black sands or fluvial placer deposits are frequently abundant in mafic minerals and magnetite so as to accentuate the textures of the cross-bedding and enabling the exploration geologists in the field to determine the flow regime, flow direction, and accretion, and last but not least, the siting of gemstone concentrations. These black sands, although subeconomic are cast into the role of "pathfinders". Within the channel heavy minerals (such as gold and diamonds) used to be preferentially laid down in the channel lag deposits of braided streams immediately overlying the bedrock (Figure 4g, Section 3.8).

Figure 4. Alluvial-fluvial placers. (**a**) Scarcely to non-vegetated alluvial fans debouching into a periodically flooded more or less straight drainage system. The basement rocks are Permo-Carboniferous in age, the fan system has been developing since the Pleistocene (San Juan Valley, Argentina). Ri: active braided trunk river system, ac: active channels, and ab: abandoned channels. There are at least two stages of fan evolution. (**b**) Braided river system with the sandy bars in yellow. (**c**) Meandering river system with the sandy bars in yellow. Sand bars developed in the active channels at the point bars of the meander bows while the oxbow lakes and abandoned isolated meander belts filled with fine-grained clastic and organic sediment. (**d**) Anastomosing river system with almost no sand bars. The vertical wedge-shaped pattern denotes a decreasing-downstream concentration of heavy minerals by alluvial-fluvial processes. (**e**) Active channel system in the Ganesh Himal, Nepal, with rounded mega-clasts at the boundary between alluvial fans and braided rivers. Imbrication denotes flow direction from left to right. It is the environment most prospective for gemstone placers (in this tourmaline and garnet). (**f**) Black sands (fluvial placers of pyroxene and magnetite) underscore the sedimentary bedding textures in a tributary of the Shire River, Malawi. Sands with distorted bedding in a whirlpool overlain by sands characterized by antidunes. (**g**) Channel lag deposits forming the lowermost concentration of heavy minerals and aggregated immediately on top of the bedrock of the braided streams (Pleystein, SE Germany). In this case the fluvial placer is composed mainly of "nigrine" (rutile-ilmenite intergrowth).

2.2.4. Deltaic Placers

Fluvial drainage systems end up in the sea or in large depressions on land occupied by perennial or ephemeral lakes. Depending upon the three forces operative in these aquatic environments a tripartite subdivision into fluvial-, wave-, and tide-dominated deltas can be achieved for these point-source landforms (Figure 5). As can be deduced from the triplot only mixed-type wave- and fluvial-dominated deltas provide favorable conditions to emplace deltaic placer deposits, because these peculiar placer types benefit from the balanced interaction of rivers, responsible for a continuous supply of detritus from the hinterland, and waves which act as a "natural processing plant". By the back and fro and the longshore drift of the waves the "wheat is separated from the chaff". Irrespective of the final depocenter, there are fluvio-deltaic placers along the shoreline of lakes such as at the Zr-Ti-REE placers of Salimaat Lake Malawi and the placers in the Sulina paleo-delta of the River Danube which evolved from a regression as early as 4000 years ago [12,30]. Apart from the ilmenite, magnetite, chromite, zircon, titanite, staurolite, kyanite, apatite, pyroxene, amphibole, and apatite they also host garnet and tourmaline, yet not of gem quality. The tidal-dominated deltas or estuaries are undernourished with regard to the sediment supply from land and characterized by prevailing fine-grained siliciclastics over course-grained particles and miss natural panning processes that could lead to a concentration of heavy minerals. Excluding the sediment-hosted diamond (Section 3.8) and precious corundum deposits (Section 3.7), deltaic placers play a subordinate role in the accumulation of gemstones.

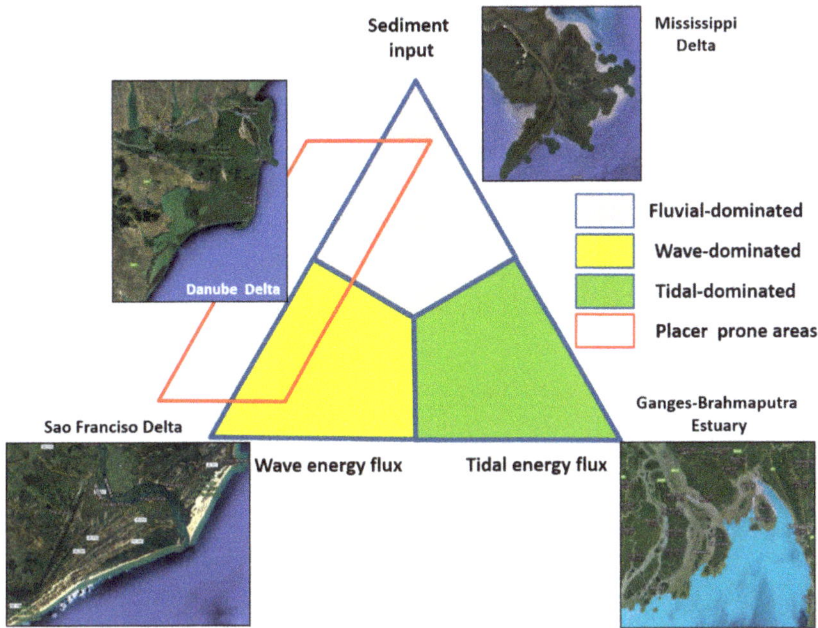

Figure 5. Deltaic placer deposits (area framed by the red line) as a function of wave energy flux, tidal energy flux and sediment supply. Satellite images (source: Google Maps) have been introduced to show the various landform of wave, tidal, and fluvial-dominated deltas. The mixed fluvial to wave-dominated delta of the River Danube has been selected as reference type for mixed-type wave-fluvial-dominated deltaic placer deposits because of its widespread heavy mineral concentration, among others garnet and tourmaline.

2.2.5. Nearshore Marine and Aeolian Placers

The most distal placer deposits relative to the source of minerals are located along the terrigenous shorelines of the sea or large inland lakes where they take advantage of the point-source landforms described in Section 2.2.4 as detrital inlets (Figure 1). The concentration of heavy minerals is accomplished by the wave action in the breaker, surf, and swash zones and by wind in the subaerial zone landward of the three zones called the backshore (Figure 6). The various sedimentological processes and the geomorphological features of this peculiar environment between land and sea are very complex and were dealt with in some comprehensive papers covering the entire tidal range from the micro- to the macrotidal hydrodynamic regimes [31–36]. Concluding from Figure 5, tidal dominated environments, whether it is a point source or linear shoreline environment, are less favorable areas to produce placer deposits. The mechanism shaping the coast in a regressive mixed energy barrier islands has been named as "drumstick model" [36] (Figure 6a). Figure 6a depicts the wave action and how the clastic material is transported from a source area to the depocenter by longshore currents, a process that is very productive, as shown by the zone of accretion in the oblique aerial view of Capitan Sam's inlet, CA, USA (Figure 6b). Near the high-water line of the backshore environment at Kiawah Island, CA, USA, ilmenite placers in coastal antidunes structures evolved (Figure 6c). These bedforms can also occur in fluvial environments given the Froud number exceeds 1 attesting to an upper or supercritical flow regime. The bedforms bearing placer minerals move upstream in the opposite direction of those bedforms of the lower flow regime used to do (Fielding 2006) [37].

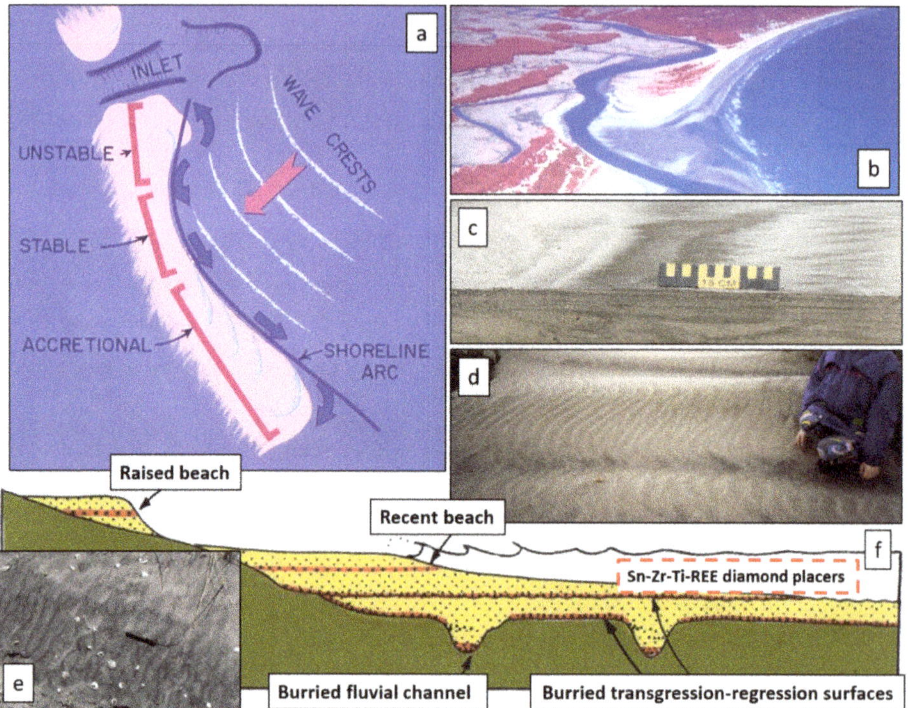

Figure 6. Nearshore marine and aeolian placers. (**a**) The linear barrier island shoreline configuration by M.O. Hayes illustrating the motion of waves and the sediment transport from the unstable zone to the zone of accretion. (**b**) Oblique aerial view of Capitain Sam's inlet. The inlet migrates towards the S (bottom of photograph at a rate of 60 to 70 m per year) (photograph: courtesy of M.O. Hayes). (**c**) Ilmenite placers in coastal antidunes near the high-water line of the backshore environment at Kiawah Island, CA, USA. Arrowhead points towards the sea. (**d**) Sigmoidal-crested sand dunes with bifurcation at Lanzarote Island, Spain, with aeolian placers of magnetite and mafic minerals. (**e**) Two different types of heavy mineral accumulations. First generation heavy mineral accumulation produced during storms resulting in shoreline-parallel layers at the boundary between the dunes and the backshore (see for scale black biro—8 cm-parallel to the shoreline). Second generation heavy mineral redeposition in bifurcating asymmetrical ripples as a consequence of wind and water parallel to the shoreline in NE Patagonia, Argentina. (**f**) Cartoon to show the association of buried fluvial placers and beach placers of different generations [38].

When aquatic placers get exposed, be it in a coastal or fluvial environment (e.g., flood plain deposits) and brought into the reaches of strong winds sigmoidal-crested sand dunes and dune belts come into existence with another accumulation of heavy minerals (Figure 6d). An example is the coast in NE Patagonia, Argentina, which shows a 1st generation heavy mineral accumulation produced during storms resulting in shoreline-parallel layers at the boundary between the dunes and the backshore and a 2nd generation heavy mineral redeposition in bifurcating asymmetrical ripples as a consequence of wind and water parallel to the shoreline (Figure 6e). An idealistic way of placer deposition and redeposition in a composite fluvial and coastal marine environment was described by Evans [38]. Fluvial placers in drowned river valleys are the source for marine placers reworked along transgressive surfaces.

After being raised above the sea level and exposed to wind, deflation may provoke another type of aeolian placers to come into being in the backshore zone. Aeolian placers are the result of one of the three major driving forces (fluvial, marine, and aeolian) in the concentration processes of diamonds in Namibia (Section 3.8). Aeolian processes come into effect as deflation and accumulation. In a dry climate loose and fine particles are lifted up and transported away, leaving behind lag deposits of heavier components which characterize the hamadah or serir desert among the arid landforms, which are equivalent to the eluvial placers. Transportation and deposition of heavy minerals generate aeolian dunes denominated based on their shape and orientation as seif, parabolic, barchan, and transverse dunes or dune belts/beach ridges in the coastal environment (Figure 6c,e)—see Section 3.8 for diamonds.

2.3. The Physical-Chemical Regime of Gemstone Placer Deposits

2.3.1. Physical Parameters

Placer deposits are mechanical deposits which are formed at the interface between the lithosphere (parent rocks) and atmosphere (controlling through the climate weathering and the intensity of the transporting agents water, ice and wind) resulting in the shaping of the landscape and delivery of minerals and lithoclasts. The main transport medium is water, the flow regime of which can be subdivided into a low-energy system encompassing prevalently suspended-load relevant for the formation of clay deposits and a high-energy system characterized by bedload mineral assemblages, among others, those containing gemstones. Depending upon the aridity or humidity of the climate zone hosting the placer deposit, the river discharge and bedload is either higher or lower.

With the exception of the dry continental and the tropical arid climate zones extending parallel to the equator on the Northern and Southern hemisphere, all remaining morphoclimatic zones provide favorable conditions to develop aquatic placer deposits, while the dry continental environments are prone to aeolian placers [39].

It is the ambient influences that control the settling and the accumulation of the heavy minerals. The internal factors controlling the emplacement of placer deposits are the grain morphology and second to none the density of the minerals. Two process hydraulic sorting and entrainment have a strong say on how, where, and when heavy minerals are laid down along the thalweg and by the coastal marine currents [40–42]. Figure 7 illustrates the hydraulic equivalence of some opaque heavy minerals in common placer deposits relative to quartz [41]. It shows the spheres of the selected opaque minerals supplemented with gemstones in the density interval 2.8 kg/m^3 (beryl) to 4.7 kg/m^3 (zircon), which have the hydraulic settling equivalence of quartz or in other words, identical settling velocity unit (Table 1). It was far earlier recognized that hydraulic settling of heavy minerals is not the only way of concentration in the fluvial and coastal marine environments. Considering the entrainment equivalence of placer minerals or interstice trapping in placer formation, the average median size of magnetite-ilmenite comes much closer to the size of a reference quartz grain than considering the hydraulic equivalence of grains only.

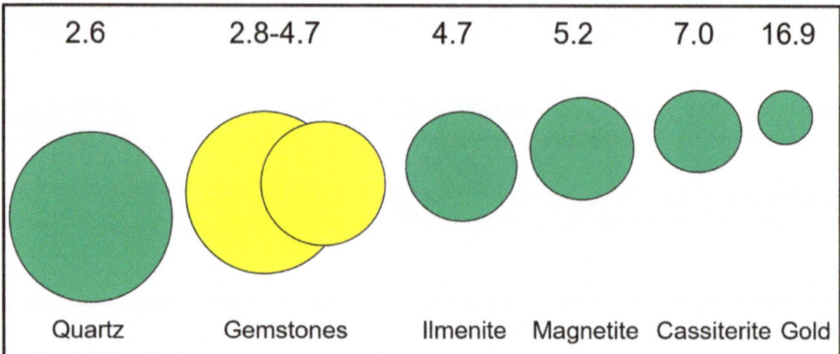

Figure 7. Hydraulic equivalence of heavy minerals relative to quartz (source of data, excluding those of gemstones from Reid and Frostick 1985) [41] with an amendment by the author with regard to gemstones.

2.3.2. Chemical Parameters

The chemical composition and the crystal structure, including cleavage and fractures of the gem minerals, are crucial when it comes to an assessment of the vulnerability to attrition and resistance to chemical weathering (e.g., silicates and phosphates bearing bivalent Fe are very much susceptible to oxidation) of placer minerals in the source area as well as along the thalweg on transport (Figure 8). There are several comprehensive studies by Morton and Lång [43,44] addressing the resistance of the common rock-forming minerals to weathering, which have been amended by Dill [45–47] as to the weathering of clay minerals and some rare minerals from pegmatitic and ultrabasic source rocks. The length of the horizontal bar in Figure 8 reflects the relative stability of the gemstones, but does not necessarily correlate with the type of placer deposits or distance from their source rocks. Olivine is the least resistant transparent rock-forming mineral and its Mg-enriched member is almost exclusively found in very young placers near the source rock. It can be of colluvial, fluvial, and coastal marine type; only the proximity to the source rocks counts as shown by the "Green Beaches" at Papakolea Beach, Hawai-USA where the waves directly cut into the volcaniclastic host rock of olivine. The Fe-enriched end members of this s.s.s. are very rare, but locally, are strongly enriched in stream sediments forming some sort of a fluvial placer. It is the most recent type of placer and is only found next to ancient smelters where fayalite was washed out from dumped Fe slags. This example was selected to raise awareness among exploration geologists of the wide range of artificial chemical compounds in modern stream sediments that have been derived from production sites and may make false claims by simulating gemmy material [48].

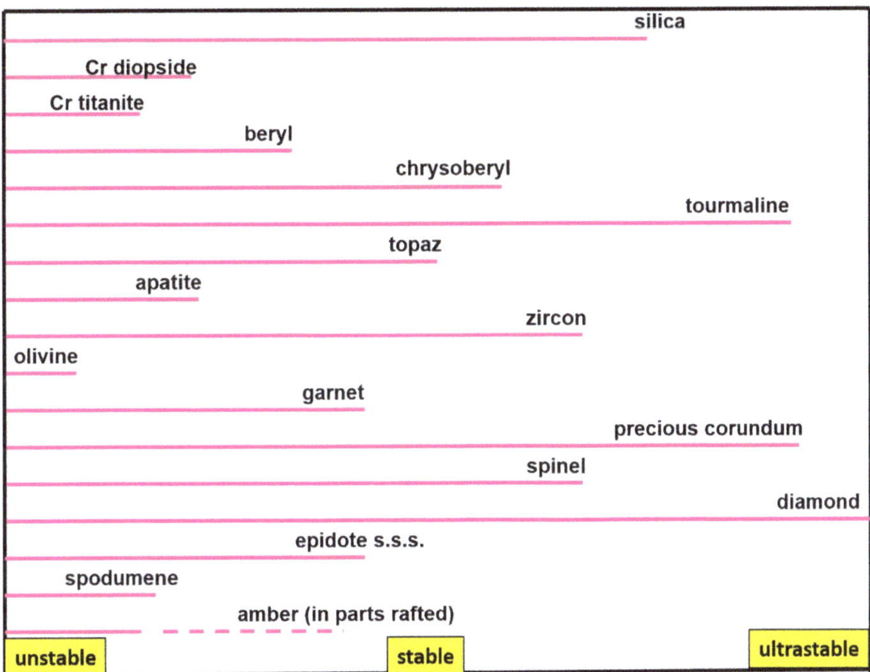

Figure 8. Relative stability of gemstones in placer deposits against attrition and weathering (data from (Morton and Dill [43,45–47])).

3. Gemstone Placers from Beryl to Zircon

The platform for a more detailed discussion of gemstones in placer deposits and their integration into the overall classification scheme of magmatic, structurebound, sedimentary, and metamorphic mineral deposits is the "Chessboard" classification scheme of mineral deposits: Mineralogy and geology from aluminum to zirconium", which was supplemented by a series of digital maps with a description of each gemstone group. All known gemstone groups worldwide were treated by country, geology, and geomorphology on 99 digital maps [1,49]—see Appendix A. For the current placer-type deposits the distribution of gemstone deposits the reader is referred to the geomorphological maps as the most relevant tool to get an insight into the enrichment of gemstones worldwide. Notwithstanding the common practice of dealing gemstones corresponding to their economic assessment, in the current review, the prevailing elements or chemical compounds are used for classification.

3.1. Beryllium-Bearing Gemstone Placers—Beryl and Chrysoberyl

There are several Be minerals that can be ranked among the gemstones such as euclase, phenakite and taaffeite, but only the true heavy mineral chrysoberyl and the modifications of transparent precious beryl (aquamarine/blue, morganite/pink, heliodor/yellow, emerald/green, goshenite/colorless, and pezzottaite/raspberry-red) are attractive for jewelers and mineral dealers [50–59]. The various varieties of precious beryl, although not a heavy mineral in the strict sense occur in alluvial (fluvial) placers and residual placers close to sites where the primary deposits are exposed to weathering and denudation (Figure 9a–c). Beryl can under intensive tropical chemical weathering also decompose into kaolinite-group phyllosilicates. A typical primary-secondary association of beryl gemstones is the side-by-side of pegmatite hosted aquamarine and alluvial placer-hosted aquamarine, e.g., Teofilo Otoni, Brazil.

Figure 9. Varieties of beryl from different placer deposits. (**a**) Aquamarine and multicolored beryl from alluvial (fluvial) deposits in Tanzania. (**b**) Emerald fragments at the most proximal site to the current open pit. (**c**) Schist-hosted pegmatite related emerald deposit in Zambia. The chlorite-talc schists (metabasic rocks are overlain by muscovite-biotite schists. The emerald deposit gradually passes into the regolith forming the topstratum (residual placer) (Photograph: T. Häger).

Chrysoberyl is also exploited from alluvial-fluvial placer deposits, e.g., in Sri Lanka and in Brazil [60]. In some mining districts, the gemstones form a by-product of alluvial gold exploration and exploitation. From this secondary placer accumulation in gravels the diggers moved along the thalweg upstream and found their way to the primary deposits, mostly pegmatites [61]. The pits were opened up in alluvial-fluvial placers and the most favorable layers encountered approximately three meters under the valley floor in clay and sand [62]. It is the channel lag deposits that developed immediately on top of the discontinuity between the drainage system and the bedrock leading to the zone to be targeted upon with the most success. Proctor [63] described colluvial-alluvial gem gravels, termed by the locals "cascalhos" which alternate with different layers of red, gray and black clay, and sand. Chrysoberyl, which is the gemstone of interest is associated with beryl (heliodor, aquamarine), topaz, rhodolite garnet, andalusite, and also mixed up with considerable amounts of quartz and olivine. In another alluvial-colluvial placer alexandrite was recovered which originated from a pervasively kaolinized weathering mantle. Chrysoberyl is also known as a minor constituent from placer deposits operated for precious corundum such as the New England Gem Fields in New South Wales, Australia [64].

3.2. Boron-Bearing Gemstone Placers-Tourmaline

Tourmaline s.s.s. are among the most widespread heavy minerals owing to the strong resistance to all kinds of alteration and they are besides zircon and rutile ubiquitous in clastic sedimentary rocks, but tourmaline of jewelry quality is rarely mined from placer deposits, excluding some sites in Sri Lanka, Brazil, and Nepal [65–68]. The rugged terrain favors the concentration of these gemstones in dislodged blocks scattered in colluvial deposits and grade downstream in alluvial (fluvial) drainage systems (Figure 3b). Basset (1987) [69] provided a good overview of these occurrences. Multihued elbaite is known from Hyakule and Phakuwa in Eastern Nepal. Further information on Nepalese gemstones has been published by UN-ESCAP, Aryal, and Kaphle et al. [70–72]. High mountainous areas are promising target areas to harness gravity driven mass wasting processes shaping the landscape for placer exploitation. Tourmaline s.s.s is present in the entire chain of placer deposits from the residual through to the marine placer deposits (Figure 10a,b). The rare colorful varieties of elbaite called indicolite (blue), rubellite (red), verdellite (green), and the most precious blue one called Paraiba Tourmaline are frequently exploited from primary pegmatitic deposits [61] excluding the peculiar highly reliefed terrains (Figure 10c–e). Owing to its outstanding stability, panned concentrates of tourmaline can be used as pathfinder minerals. Multicolored concentrates like those from Tanzania (Figure 10a) are positive indicators, whereas monotonous tourmaline heavy mineral suites composed of elbaite-schorl series or dravite-schorl series sensu Henry et al. [73] only like those met in SE Germany do not signal a high potential of finding precious tourmaline in the catchment area and in the source rocks around (Figure 10b). Apart from tourmaline there are numerous B-bearing minerals some of which may also achieve gemstone quality (Table 2). Studies of these B-bearing gem minerals are only focused on their primary deposits so that any integration into a review like that devoted to their secondary deposits is fraught with difficulties.

Table 2. Boron-bearing minerals, achieving locally, gem quality [1].

Axinite	$Ca_2MgAl_2(BO_3)Si_4O_{12}(OH)$
Danburite	$CaB_2Si_2O_8$
Datolite	$CaB(SiO_4)(OH)$
Dravite	$NaMg_3Al_6(BO_3)_3Si_6O_{18}(OH)_4$
Dumortierite	$Al_{6.9}(BO_3)(SiO_4)_3O_{2.5}(OH)_{0.5}$
Elbaite Elbaite (Paraiba)	$NaLi_{2.5}Al_{6.5}(BO_3)_3Si_6O_{18}(OH)_4$
Elbaite (Indicolite)	$Na(Li,Al)_3Al_6(BO_3)_3Si_6O_{18}(OH)_4$
Jeremejevite	$Al_6B_5O_{15}F_{2.5}(OH)_{0.5}$
Kornerupine	$Mg_{3.5}Fe^{2+}_{0.2}Al_{5.7}(SiO_4)_{3.7}(BO_4)_{0.3}O_{1.2}(OH)$
Painite	$Ca_{0.77}Na_{0.19}Al_{8.8}Ti_{0.19}Cr_{0.03}Zr_{0.04}B_{1.06}O_{18}$
Elbaite (Rubellite)	$Na(Li,Al)_3Al_6(BO_3)_3Si_6O_{18}(OH)_4$
Serendibite	$Ca_2Mg_{4.5}Al_{1.5}Si_{3.6}Al_{1.8}B_{0.}$
Sinhalite	$MgAl(BO_4)$

Figure 10. Tourmaline s.s.s. from placer and pegmatite deposits. (**a**) Panned concentrate for subrounded tourmaline from a fluvial placer in Tanzania. (**b**) Tourmaline concentrate from a colluvial placer SE Germany. (**c**) Pink and blue elbaite tourmaline laths in the Batalha Pegmatite, Brazil. (**d**) Paraiba tourmaline from the Capoeira Pegmatite, Brazil. (**e**) Pink and blue and elbaite and green verdellite from the Mendoza Pegmatite Province, Argentina.

3.3. Fluorine-Bearing Gemstone Placers—Topaz

Topaz accommodates as much as 11 wt % F in its lattice and falls short of fluorite as a source of F, the most common ore mineral mined for fluorine (48.67 wt % F). It is dealt with here as the only relevant F-bearing gemstones found in placer deposits. Topaz and tourmaline are common associates in highly-differentiated granites and are also found in some rare-element pegmatites [1,61,74]. They accompany cassiterite and wolframite in granite-related Sn–W deposits. Seldom has such well-crystallized topaz appeared elsewhere as in the mineral aggregates of Thomas Range, Utah, USA (Figure 11a). Topaz takes an intermediate position as to the mineral stability and its grains swiftly see their edge beveled and grinded to a grain size neither of relevance for jewelers nor for exploration geologists who might make use of them as pathfinder to primary gemstone or ore deposits (Figures 8 and 11b). Topaz is a heavy mineral common in the clastic dispersion halo around highly differentiated granites, but at around four–to–five km, it disappears from the heavy mineral assemblage of stream sediments and looses its role as a pathfinder mineral to Sn–W deposits. Eluvial placer deposits in Brazil and alluvial-(fluvial) placer deposits in Brazil, USA, Sri Lanka and Australia are the mayor sedimentary sites to recover topaz from. Topazes from Ouru Preto, Brazil, are washed out of a thick lateritic saprolite, a gentle method to save the most precious specimens of their kind. Fluorite, although a member of the group of heavy minerals, rarely shows up in stream sediments, and owing to its perfect cleavage, it is met only as tiny splinters without any value.

3.4. Phosphate-Bearing Gemstone Placers—Apatite-Group Minerals

Apatite of gem quality is rare, even if this phosphate shows up in a great variety of colors (Figure 12a). It is less attractive for gem dealers than for mineral collectors who put facetted specimens on display in their showcase. It is a common phosphate mineral in sedimentary, metamorphic and magmatic rocks, and therefore, ranks among the most widespread heavy minerals, although its stability is moderately high, particularly under the attack of acidic meteoric solutions during weathering (Figure 8). In residual deposits originating from long-term pervasive chemical weathering it has a low conservation potential and gets decomposed to almost full completeness in the topmost saprolite [33,34]. The phosphates newly formed in the course of supergene alteration are part of the APS minerals (aluminum-phosphate-sulfate minerals) when the pH significantly drops below seven. Dependent on the soluble compounds associated with Al and P in the meteoric solutions different APS minerals come into existence, such as turquoise which may develop in meteoric fluids abundant in Cu [75]. They pertain to a wide range of supergene gemstones categorized as duricrusts, or in the current example, termed as apicretes or phoscretes. If at all present in placers, these chemical sediments only occur in stream sediments proximal to the site of formation.

Figure 11. Topaz crystals—from the primary deposit to the secondary deposit in placers. (**a**) Euhedral well-crystallized transparent topaz from Thomas Range, Utah, USA. (**b**) Subhedral crystal with beveled edges, Brazil. (**c**) Anhedral detrital topaz from eluvial placer deposits in the Rondonia tin mining district, Brazil.

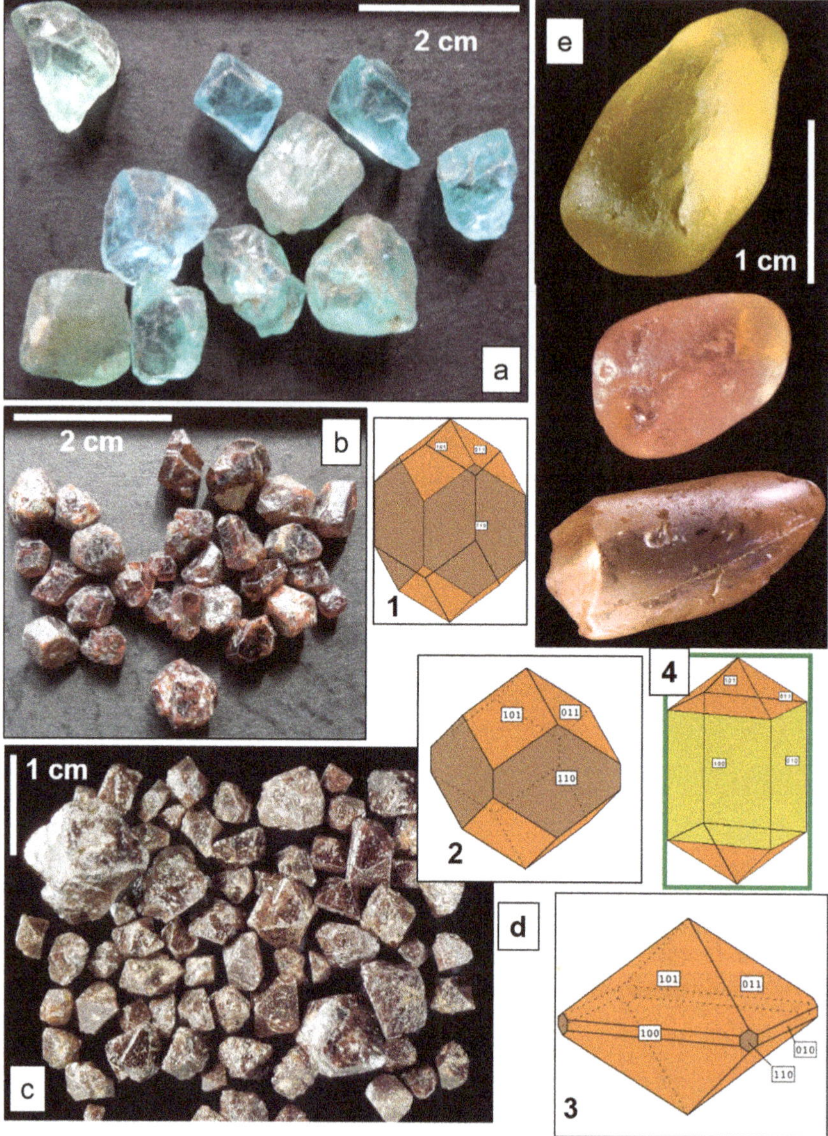

Figure 12. Apatite and zircon from placer deposits. (**a**) Greenish and blue apatite from alluvial placer deposits in Madagascar. (**b**) Brownish red translucent zircon from alluvial placer deposits in Madagascar. (**c**) Brown opaque zircon from alluvial placer deposits in SE Germany. (**d**) Crystal morphologies typical of zircon from pegmatitic deposits (1–3) and from their country rocks (4). (**e**) Pinkish and yellow well rounded zircon grains from an alluvial placer in SE Germany (photograph: B. Weber).

Two processes, tectonic uplift and unroofing, controlled the variation of heavy minerals (Dill 1995) [76]. When the relief is low, the chemical weathering intensive and the erosion is slow meteoric fluids can attack these minerals for a long period of time and thus eradicate heavy minerals of low stability as is the case with apatite. To provoke apatite to show up in the alluvial fans and its placers the uplift must be accelerated so that the evolution of a full-blown weathering profile is impeded [76]. Given there are source rocks such as pegmatites fostering the growth of mega-crystal exposed in the hinterland apatite are likely to be part of alluvial placer deposits at a considerable size (Figure 12a). The majority of apatite, however, is observed in the sand fraction of various placers and only of stratigraphic value.

3.5. Zirconium-Bearing Gemstone Placers—Zircon

In contrast to the aforementioned phosphate mineral apatite, zircon takes quite a different position in the stability diagram of Figure 8, being placed near the ultra-stable contenders. In combination with its ubiquity in the metamorphic and magmatic rocks, it is second to none among the heavy minerals. Together with rutile, ilmenite, and "leucoxene", zircon forms the major constituents of the marine placer deposits mined along the coast of Australia, Canada, South Africa, India, and USA, which constitute the sole economic source of zircon of industrial grade [3,11,13,17,77,78]. Its grain size in these depositional environments is far too small to attract the attention of jewelers. Zircon, for example, hyacinth of jeweler's grade as to size, transparency and color can only be expected next to magmatic rocks that contain zircon at sufficient size. Zirconium-bearing pegmatites are closely related to mantle derived magmatic intrusions such as peralkaline miaskitic and agpaitic rocks, e.g., Zomba, Malawi [10,61] (Figure 12b–d). Another source of multicolored gem-quality zircon is found in alkali basalts in Myanmar and Cambodia from which they get released during weathering [79,80]. Only recently in SE Germany such multicolored well-rounded zircon grains were identified as xenocrysts in alkaline volcanic rocks (basaltic lavas, nepheline basanites, and olivine nephelinites) genetically related to the Eger/Ohře Rift [81]. The gem-quality zircon from the alkaline basaltic source rocks is not only different with regard to its color, but also with regard to grain morphology. The zircon grains originating from pegmatitic source rocks in alluvial placer deposits appear with their crystal morphology well-preserved, whereas the zircon from the basaltic source rocks are well rounded or sub-rounded. No faces or edges are still preserved. It is not a suite of water-worn grains smoothed by the running water. Their shape has been preserved on release from the basalt and on transport to the proximal placer as can be deduced from a simple comparison of xenocrysts in the basaltic source rock and in the placer deposits, where zircon is associated with ilmenite, spinel, and corundum [81]. The high preservation potential of grain morphology sparked an intensive study whether the crystal morphology of zircon is a function of its physical-chemical conditions of [82–86]. The investigations of crystal morphology of zircon are very encouraging and the results can also be applied to heavy mineral accumulations in order to identify the source rock [87]. The three crystal morphologies 1–3 feature stubby crystals with a predominance of the pyramid over the prism, which becomes almost nil in type 3 (Figure 12d). Types I and II are indicative of a temperature around 600 °C, while type 3 points to a temperature interval 500 to 550 °C. The morphological types 1 to 3 of the opaque minerals correspond to zircon which has been derived from pegmatites. The multicolored zircon grains are rounded and do not allow us to label the faces with the Miller's indices in the common way for an attribution of the zircon grains to a peculiar type of zircon morphology. If an attempt is made to approximate the prismatic crystals of the yellow and red zircon grains a shape like that on display in Figure 12d provides the best fit (Figure 12d). Using the data from Pupin [82] the temperature of formation of this elongated zircon crystal stands at a T value of more than 900 °C, a fact in full agreement with the origin of these placer-type minerals sourced by basic magmatic rocks which are also of attraction to gemologists. What does the color of the zircon grains in the placer deposits tell us about the origin of the heavy minerals in question? Rozendaal and Philander [77] listed some chemical elements which act as chromophores in zircon. Colorless zircon has low contents of Fe, U, and Th. Pink to red zircon is enriched in REE (Rare Earth Elements), U, and Th. Yellow zircon is abundant in REE, U, Th, and abundant in Fe, Al, P, and Ca. Metamictic zircon grains contain U, locally, in the range of as much as some wt % and they are very much abundant in REE, Fe, and Y. These zircons suffering from radiation damage of their lattice have no high preservation potential and even if they might be used as a pathfinder to primary radioactive deposits, their application is limited in practice to the proximal clastic dispersion halo of U and Th mineralization.

3.6. Garnet-Bearing Gemstone Placers

Garnet-group minerals belong to the most common heavy minerals of intermediate stability. They have been derived from a wide range of source rocks (Figures 8 and 13) [88]. Pyrope is common to basic and ultrabasic magmatic rocks and their metamorphic equivalents, as exemplified by the well-rounded grains and subangular aggregates of pyrope, which has been released during weathering and denudation from alkaline basanites and olivine basalts, Mongolia [7] (Figure 13a,b). A pink pyrope-almandine garnet-group mineral is called rhodolite, which is among others mined from placer deposits in British Columbia, Canada (Figure 13c) [89]. The spessartine (a member of garnet-group minerals, with Mn as major element) frequently shows up in well-shaped crystal morphologies in placer deposits (Figure 13e). Primarily, it forms in pegmatites, granite-related mineralization, as well as low to medium grade mica schists, provided they are abundant in Mn [9,10]. A well known example of this manganese variety named, due to its color, as "Mandarin Garnet" and after the most prominent river in the region as "Kunene Garnet" was mined in the Marienfluss-Hartmannberge claims, an area where metamorphosed Mn-bearing rocks of probable sedimentary-exhalative origin occur [90]. It was found in alluvial-fluvial sediments of the Cuvelai-Etosha Basin, Namibia, close by [87]. Grossular originating from metacarbonate rocks is also a common garnet in placer deposits (Figure 13g). Grossular was originally named "cinnamon stone" ("Kanelstein" in German) and renamed grossularite. The current name which originates from the color of gooseberries (Ribes grossularium)) normally has a greenish tint, while a rare member of this grossular subgroup called hessonite stands out by it brown mineral color, which renders it of great interest for jewelers (Figure 13d). The most widespread red to brown garnet species originating from metapelitic rocks is almandine. It is the most resistant member of this garnet heavy mineral suite which was arranged in increasing order of abundance and resistance to weathering. Garnet is almost present in all placer deposits from the residual placers down to the coastal placer deposits (Figure 13h–j). The grain morphology of garnet is an excellent mirror image of these different types of placers, yet the outward appearance cannot be used when considering the provenance of the various garnet members. Gemstones can only be expected near their source rocks—see above—in residual to alluvial placers, while grinding of garnet grains on fluvial transport and reworking them by the waves along lake or sea shores significantly reduce the grain size so that they can no longer be categorized as jewelers' grade but only recovered as an abrasive grade by-product (Figure 13j). Many of these modern marine garnet placers were formed as interstadial placer deposits in the shore-face to inner-shelf, e.g., Cape Ortegal, NW Spain [91]. The major garnet production derives from placer deposits along the Southern coast of Tamil Nadu, India [92]. Western Australia is also among the major supplier of industrial garnet with most garnet being produced from the mining operation at Port Gregory. Almandine is in this site associated with ilmenite and zircon [93]. All of them have to be denominated as industrial minerals in the strict sense. All members of the garnet group minerals may be observed in the various placer deposits, however, dominated by the red and brown varieties. In the aeolian placer deposits at Biluut uul and Elsteingol, Mongolia, almandine prevails. It has to be noted that in the majority of cases garnet is won as a by-product in placer mining, while the entrepreneurs keep an eye on precious corundum, spinel or diamonds, a group of gemstones that are used to fetch a higher price than garnet, excluding some very rare varieties of the grossular such as tsavorite or hessonite.

Figure 13. Garnet from placer deposits. (**a**) Well rounded pyrope-enriched garnet grains from fluvial placer, about 1 km downstream of the pyrope aggregate of Figure 13b. (**b**) Pyrope-enriched garnet cluster from a colluvial placer, Mongolia. (**c**) Rhodolite fragments from an alluvial placer, Madagascar. (**d**) Hessonite grains with beveled edges from an alluvial-fluvial placer, Tanzania. (**e**) Spessartine-enriched almandine garnet-group minerals, derived from pegmatitic and aplitic rocks from a fluvial placer (small creek), SE Germany. (**f**) Almandine garnet derived from paragneisses reworked into an alluvial-fluvial placer, SE Germany. (**g**) Grossular aggregates and grains from an alluvial placer, Malawi. (**h**) Micrograph of a well-rounded garnet from a beach placer (thin section, plane-polarized light, size approx. 300 μm), Bight of Riga, Latvia. (**i**) Micrograph of an angular garnet with etch pits from a residual to alluvial placer, (thin section, plane-polarized light, size approximately 300 μm), Malawi. (**j**) Small arenaceous strand plain with a cliff carved out of Devonian sandstones (Veczemji Cliff, Latvia). The garnet accumulation ("red sands") of the beach placer in the backshore is marked by a red dashed line from the seaward clean white beach sand spread across the foreshore.

3.7. Corundum and Spinel-Bearing Gemstone Placers

In nature, the "worst enemy" for Al_2O_3 to form corundum or spinel in primary deposits is silicon, which in almost all environments is present in excess and consequently fosters the formation of alumo-silicates such as sillimanite or staurolite and hence contributes to the rarity of corundum deposits. However, a desilicified environment undergoing high P and T conditions would not yet yield precious but at the best opaque corundum. It needs chromophores like trivalent chromium to make a ruby or sapphire, the involvement of multiple ions such as in sapphire as well as color centers. This accounts for why precious corundum ruby, sapphire, and padparaja are so rare despite of their extraordinary hardness, which only trails behind diamond by one point in the Mohs hardness scale. The situation is not any better for the common spinel in nature. Considering the formula XY_2O_4 (X = Mg, Zn, Fe, Mn, Y = Al, Cr, Ti, and Fe) may give rise to a sloppy denomination of spinel s.s.s. as an "advanced corundum" and allows for their subdivision into Al, Fe, and Cr spinels. In Figure 14 the various placer types of precious corundum are shown using the morphology of the mineral grains and aggregates to distinguish the various types from each other. In primary deposits, which may be magmatic or metamorphic ruby, tends to crystallize in well-shaped euhedral crystals, as exemplified by the metacarbonates in Vietnam hosting remarkable ruby deposits [94] (Figure 14a). Chemical weathering affecting calcareous sedimentary units triggers karstification resultant in cavities filled with argillaceous material and crystals still well preserving their original shape (Figure 14b). Due to mechanical reworking by mass wasting processes such as landslides or mud flows and the running water the minerals get broken, but their fragments still allow for singling out red spinel grains from the overall red corundum (Figure 14c). In fluvial placers of the same mineral province in Northern Vietnam, the different grains of corundum have to pay tribute to the attrition. The morphology of ruby is much more obliterated than the crystal morphology of spinel whose octahedral faces can still be identified in panned heavy mineral assemblages (Figure 14d). The "brethren" of ruby, the white and blue sapphire follows the same path during weathering and the ensuing placer evolution. A mega-sapphire measuring 12 tons by weight has been smoothed by aquatic processes and constitutes a natural monument today in front of an office building (Figure 14e). A block like that can neither be moved by water worn in a river nor can it be chemically altered in-situ by a simple solution process. It needs a joint action of chemical weathering (residual placer) and a slurry characteristic of colluvial and alluvial processes which smoothed the faces and edges of the gemstone by some sort of natural abrasive paste (Figure 14e). Fragments like those shown in Figure 14f are worth considering as abrasives in the industry. Precious corundum may be expected downstream to the level of fluvial placer deposits in braided streams, where these colored gemstones often are accompanied by other gemstones such as garnet or iolite, a variety of cordierite. The commercially viable placers are found between residual and alluvial placers. Karstification has an upgrading effect on the secondary deposits as shown by the placers in Myanmar and Vietnam. In many mining districts, modern placers develop from paleoplacers, similar to what is also known from secondary diamond deposits [1] (Section 3.9). The mineral assemblage becomes downstream more variegated with heavy minerals of different stability groups, as shown by some examples [95]: Sapphire, spinel, garnet, chrysoberyl, topaz, tourmaline, andalusite, kyanite, and zircon. Sri Lanka recovers a great deal of its revenues from gemstones exploiting precious corundum with almost 90% of gems found in paleoplacers underneath the active river beds in alluvial gravels of Pleistocene to Recent age ("illam") [96–98]. Coastal placers do not play any significant role for corundum, e.g., Hainan. Only the Fe and Cr spinels are, in places, enriched in coastal placer deposits given basic and ultrabasic rocks are exposed in the hinterland [99].

Figure 14. Precious corundum and spinel in placer deposits. (**a**) Euhedral to subhedral crystals of ruby in a pargasite-bearing marble (primary deposit), Vietnam. (**b**) Aggregate of ruby in a karst cavity of the marble, Vietnam, (residual placer). (**c**) Panned heavy minerals from a colluvial to alluvial placer of ruby and spinel (see octahedral crystal fragments), Vietnam. (**d**) Well-rounded grains of ruby with some subrounded octahedral of red spinel in a fluvial placer, Vietnam. (**e**) Sapphire measuring 12 tons by weight from a residual placer deposit, Vietnam. (**f**) Sapphire from a colluvial to alluvial placer, Madagascar.

3.8. Diamond-Bearing Gemstone Placers

Due to the extraordinary position taken by the diamonds in terms of their value and their physical properties, particularly hardness, these gemstones form an ideal heavy mineral that is recovered from the full spectrum of depositional environments to be identified among modern and paleoplacers deposits and has attracted the attention of many researchers from different geoscientific camps [100–107].

Sedimentary diamond deposits are closely linked with a peculiar geotectonic-geological setting, which is marked by three elements. The various placer deposits are in and around an ancient craton, be it exposed or hidden, which is pierced by subcrustal magmatic rocks belonging to the lamproite and/or kimberlite suite that act as host- and source-rock lithologies [108–113].

Rock-forming minerals common to these two source and host-rock lithologies of diamonds are used as pathfinder minerals in stream sediments and different placer types to find either profitable secondary or even primary diamond deposits. Forsterite is recorded in association with pyrope and chromium diopside from heavy mineral accumulations, as a pathfinder to diamondiferous source rocks [114,115]. Titanite is a very widespread Ti-bearing mineral among the heavy minerals, but in the stream sediments it issues, some sort of a "natural disclaimer" for diamonds in the basic source rocks. Ultrabasic host rocks of diamonds used to concentrate Ti in heavy minerals in an apron around like perovskite, titaniferous amphibole, and phlogopite, as well as ilmenite, all of which can show up in the placer deposits [7].

Paleoplacers are common repositories of diamonds in Africa and South America. Alluvial–fluvial gravel beds in the upper Karoo are operated at Somabula, Zimbabwe, not only for diamonds, but also for topaz, chrysoberyl, sapphire, ruby, and emerald [116]. In neighboring Angola, the upper Cretaceous "Calonda Deposits" are host to alluvial–fluvial diamond concentrations [117]. At a stratigraphically similar level, diamonds were found in the neighboring DR Congo within conglomerates of the upper Cretaceous Kwango Series. Between 100 and 80 Ma a strong uplift of the African Craton triggered the deposition of piedmont conglomerates and reworking of diamonds into different types of modern placer deposits (Figure 15a). Fossil placers in Carboniferous sandstones as well as alluvial deposits were mined in Brazil, Venezuela and Guyana where the primary source of diamonds is still enigmatic [118,119]. They were reworked into braided-stream sediments and deposits, which were derived from ephemeral flash floods [1].

The area most attractive for the recovery of diamonds from sedimentary deposits is without any doubt situated at the SW edge of the African continent where so-called mega-placers evolved around the Kaapvaal Craton [106]. Diamond mega-placers, have been defined as ≥50 million carats at ≥95% gem quality by the aforementioned authors. Production figures were published by Porter GeoConsultancy Pty Ltd. (Linden Park, Australia) [120]. At the end of 2000, Namdeb reported remaining resources and reserves (land + marine) as follows: (1) Probable reserves of 59.4 Mt @ 1.5 carats per hundred metric tons (cpht), (2) indicated total resources of 73.6 Mt @ 2.3 cpht, and (3) inferred total resources of 301.7 Mt @ 1.5 cpht. Porter GeoConsultancy Pty Ltd. (2004) [120] provided a brief overview, based upon which, the principal factors of the main concentration systems may be listed as follows:

(1) Fluvial diamond accumulation took place by the Orange-Vaal River drainage system for some 100 Ma. From the early Paleogene onward, continental uplift caused the drainage system to incise and create several diamond-bearing terraces between 19 and 17 Ma, and subsequently during 5 and 3 Ma.

(2) A wave-dominated delta at the mouth of the Orange River and the variations in sea level during the last 40 million years has produced a number of palaeo-shorelines with onshore "raised beaches" and offshore "submerged beaches"—see also Figure 6f—passing towards more recent periods into the formation of barrier beach and linear beach—see also Figure 6a,b.

(3) Strong winds working their way for 40 million years in the same direction in the arid climate of the Namib Desert caused a Northward-directed longshore current. In context with the marine currents, deflation basins, a combination of ephemeral run-off, salt weathering and aeolian action, in general, the concentration of diamonds was upgraded and aeolian placers were formed within the dune sands of the Namib Desert.

(4) Offshore marine deposits are encountered as drowned placer deposits located in waters ranging from the surf zone near the coast, to a depth as great as 150 m on the shelf. Evidence to date suggests that the marine environments host the same complex range of beach-barrier, beach, deflation, and aeolian systems seen onshore [120].

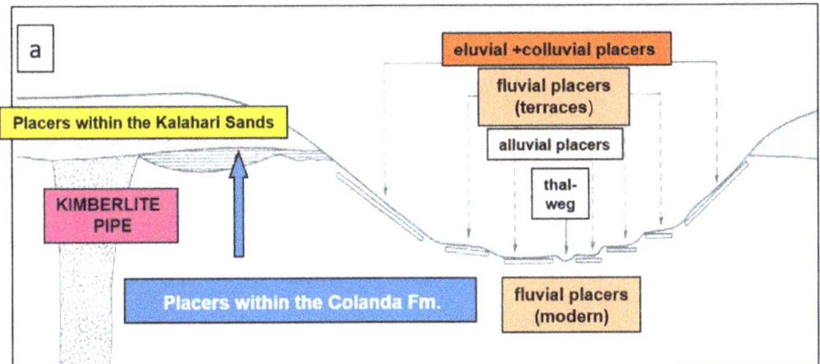

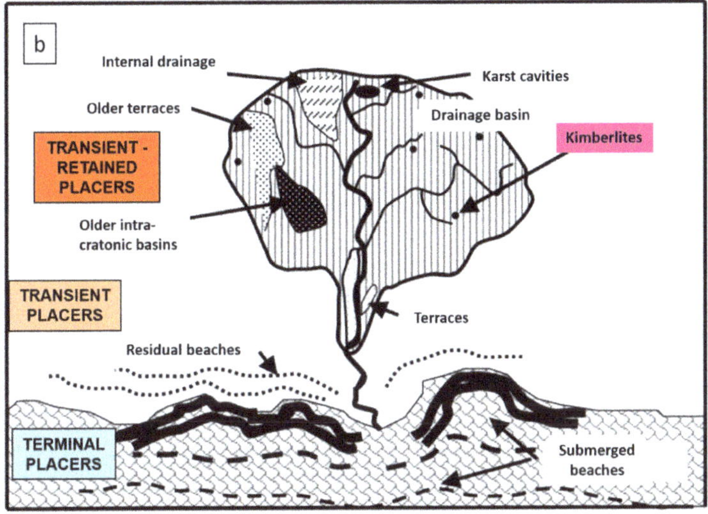

Figure 15. Diamond placer deposits in cross section [121]. (**a**) From the diamond-bearing pipe to the placer deposits. An idealized cross section illustrating the primary (kimberlite pipe), intermediate paleoplacers (Calonda and Kalahari placer sands) and final modern placer deposits of diamonds at midslope, footslope positions, and along the thalweg in the drainage systems in Angola (modified from [122]. (**b**) Cartoon to illustrate the tripartite subdivision of diamond placers into retained, transient, and terminal placer deposits leading to the modern-day mega-placers in Southern Africa (modified from [106]).

A different picture can be drawn based upon the studies of Bluck et al. [106] who placed more emphasis on a higher level in the hierarchy of placer deposition and provided a tripartite categorization of these diamond placers into (1) retained placers, (2) transient placers, and (3) terminal placers which reflect a combination of modern and paleoplacers systems and take into consideration the peculiar physical-chemical characteristics of diamonds (Figure 15b). The different levels can be combined and are also used to illustrate how the placer aprons evolve around a basement high of metacarbonate or metasilicate rocks (Figure 1). The latter properties established by Bluck et al. [106] take particular effect during the variation of the geodynamic setting which the diamond grains went through on their transit from the primary magmatic kimberlitic rocks more than 1 Ga ago to the final modern-day placer deposits. The tripartite classification scheme of Bluck et al. [106] does not replace the existing classification scheme applied in the current study, but combines some sedimentological processes with the geodynamic setting with a clear focus on diamonds—see also Figure 1.

Retained placers: The two major sedimentary traps to retain diamonds released from the kimberlites, in places, >2.5 Ga ago, are intracratonic basins and karst cavities resembling those traps described for precious corundum deposits in Myanmar and Vietnam (Figure 15b). It is the potholes and paleoriver systems which contain the diamonds scattered between gravels, mainly entrained in lag deposits similar to what has been shown in Figure 4g. Some of the gravel deposits stand out from the surrounding landscape like esker in a glacial terrain reflecting an inversion of the relief. In a glacial landscape, ice has melted away causing ridges of coarse sediments to emerge and stand out like a railway embankment. In the diamondiferous landscape it is the less soluble material, the calcareous rock which when being dissolved in the course of karstification provoke siliceous sedimentary material to stand out in a similar way [123]. Large intracratonic basins are the second important depocenters covering a wide variety of terrigenous sediments mainly of coarse grain size and chemical sediments such as calcretes marking hiatuses. Since the break-up of Gondwana thick siliciclastic series with a Middle Mesozoic through a Cenozoic stratigraphic record evolved there. Extensional periods affecting the Archean-Proterozic cratons, on one hand, were conducive to large basins acting as trap-sites for placers, and on the other hand, concealed diamond-bearing kimberlites. An example for this category of diamond placers is given in Figure 15a grading from the retained into the transient placers.

Transient placers: According to Bluck et al. [106] transient placers are diamond-bearing sediment piles stored along the dispersal route or within the active drainage basin. This category of placers occurs in terraces and they are strongly controlled by the aforementioned ratio of uplift and weathering which affects the dispersal and concentration of diamonds in an indirect way due to its outstanding physical-chemical features compared to the surrounding country rocks. The transient placers, exemplified by the Orange River drainage system in SW Africa can be considered as an intermediate repository or protoplacer for the final step or mega-placer to come. However, it can also upgrade to a placer system in its own right, depending on the efficiency of its fluvial drainage system in collecting the diamonds. Jacob et al. [124] investigated the Lower Orange River drainage system and its terraces demonstrated the change of size and grade along with the stratigraphic evolution. The highest grade was reached during the Oligocene (?) with high background grades extending into the terminal placer.

Terminal placers: The terminal placer marks the end of the "flow path of the natural processing plant of diamonds". The quality of the concentration process in the marine environments depends upon different parameters enhancing, on one side, the accumulation of diamonds while reducing the amount of the dilutors both of which are provided by the nearby craton in the hinterland [106]. The essential parameters for the built-up of the diamond mega-placers in Southern African are listed as some bullet points according to the aforementioned authors:

The potential dilutor of the diamond placers, the Karoo overburden was removed during the Cretaceous and dumped in subsiding basins

Incision of a pre-existing (mainly Cretaceous) drainage systems to access newly exposed primary and pre-concentrated diamond deposits during the Tertiary

A high-energy wave-dominated coastal regime with a unidirectional wind over a shelf and a point-source delivery system active at the same time (Figures 5 and 6)

The final dispersal should be young and unlithified to facilitate mining.

Large basins hosting known or potential trap-sites of diamonds along and around ancient cratons are not only located in South Africa, but are also found in Angola, Niger, DR Congo, Chad, Senegal, Mauretania, Western Australia, Siberia, North, and South America, mainly in Brazil, where glaciation is held to be a principal exogenous process for the deposition of diamonds [108]. The authors singled out three periods relevant for the accumulation of diamonds, the first one during the Early Proterozoic, designated the Jequital glaciation, the second one was dated as Cambrian, which is represented by the Santa Fé Tillites, and the last one operative during the Carboniferous being called the Itarar.

3.9. Quartz-, Diopside-, Titanite-, Olivine-, Epidote- and Spodumene-Bearing Gemstone Placers

There are several minerals that fall in a grey area, whether these minerals are true gemstones by definition or not and accordingly their accumulation in sedimentary rocks forms part of what might be called a gemstone placer. The most common contenders of them are siliceous minerals and treated in this overview for the sake of completeness as only silicates are capable of surviving the chemical and physical processes involved the formation of placers. Even if some sulfides and oxides such as pyrite, sphalerite or hematite are exploited for their aesthetic value in some deposits, none of them ends up in placers in a quality coming close to what might be called "gemmy" [1].

Quartz is not a heavy mineral in the strict sense, quite the contrary, it is the "standard mineral" among the light minerals, but nevertheless, some of its varieties warrant mentioning because they are found in placer deposits and can be won from the stream sediments for a profit. One of these varieties is famous for its well-shaped clear crystals, called "rock crystals", another one is known for its blue to purple color called amethyst, and the third in the row is the multicolored agate, a variety of chalcedony (fine-grained quartz). (Figure 16a). Especially rock crystals may form placer deposits of economic interest for the purity of their silica grains [125]. Double-terminating ones, however, may suffer from transport when their edges get beveled and the pointed pyramids broken. Therefore, they are better preserved in the loamy substrate of residual placers where those minerals are recovered to satisfy the demands of gem collectors. In the Donghai County of Eastern China, rock crystal-bearing quartz veins occur in the Su–Lu ultra-high pressure metamorphic belt. Uplift and denudation of this metamorphic host complex created a large apron of alluvial–fluvial placer deposits during the Quaternary hosting these quartz crystals [126,127]. Near-source alluvial placers of rock crystals are also known from Madagascar and from Sri Lanka. The most well known alluvial-fluvial deposits of amethyst in Brazil occur on the banks and in the fluvial terraces of the Araguaia River [112]. Agate has been found among the gravel of the Mormora River in the Sidoma Province, Ethiopia, and mined at an economic grade.

Two other silicate minerals also deserve treatment among the group of gemstones concentrated in alluvial to fluvial placer deposits. It is diopside and titanite with their vivid green varieties accommodating chromium in their lattice which are the most popular ones with gemstone collectors and exploration geologists who use them as pathfinder minerals during diamond exploration (Figure 16b,c). Apart from its marker mineral character for kimberlite pipes, the chrome diopside is also found at metal deposits devoid of diamonds, e.g., Outukumpu, Finland. In general, titanite is not very common in placer deposits, mainly due to its limited occurrence in primary deposits and moderate mineral strength (Figure 8). To demonstrate the variation of the amount of titanite, together with its associated heavy minerals along transport, in Table 3 the different heavy mineral suites and lithoclasts communities are listed as a function of the different placer types from the source rock through the tributary river to the trunk river in Central Mongolia (see also the section on diamonds—Section 3.8). The distance of transport from the basic source rocks to the trunk river measures approximately 1500 m and the difference in height between the two sites is about 250 m. The reason why titanite together with a series of rather unstable heavy minerals was able to survive in these placer types, different to their sedimentological processes of formation is explained by the high ratio of uplift vs. chemical weathering. The latter process is controlled by a rather cold climate in Central Mongolia that did not favor a pervasive chemical, but rather physical weathering. All heavy minerals observed in the placers are rather well preserved, excluding the least resistant mineral olivine the grains of which in the placer show strongly serrated rims as being compared to its source-rock-hosted subrounded phenocrysts (Figure 16d,e). The distance of transport also plays a significant role in the marine placer deposits resulting in the "green beaches" of Hawai, USA. Chemical weathering cannot keep pace with the newly vented amounts of pyroclastic and volcanic parent material and the coastal erosion in the swash zone both of which are working hand in hand in exposing permanently new material for the reworking into beach placer deposits. The two examples of olivine placers referred to in the current study furnish clear evidence that the formation of the landscape by exogenic and endogenic processes cannot be cast aside and give the sole answer why heavy minerals from the opposite ends of the scale of stability appear in the same placer type (Figure 8).

Figure 16. Silicates of rare occurrence in placer deposits. (**a**) "Rock crystals" of double-terminated quartz washed out from placer deposits, Madagascar. (**b**) Panned diopside crystals from placer deposits, Madagascar. (**c**) Chromium-bearing titanite from the Ural Mts., Russia (rotated for display by 90°). (**d**) Micrograph of subrounded phenocrysts of olivine besides angular microphenocrysts of plagioclase floating in a basanitic host rock, Mongolia (crossed polars). (**e**) Olivine strongly corroded during transport and weathering in a fluvial placer deposit of a tributary stream proximal to the basanitic host rock—see Figure 16c, Mongolia (panned heavy mineral concentrate, plane polarized light). (**f**) Green spodumene gradually alters to cookeite at the edge of a pegmatite in Minas Gerais, Brazil. See biro for scale.

Table 3. Lithoclast- and heavy mineral variation along the thalweg from the primary source through the different colluvial, alluvial and fluvial placer deposits in a diamondiferous olivine-garnet placer deposit, Mongolia (modified from Dill et al.) [7].

Environment	Lithoclasts	Heavy Minerals
Primary deposit	70% volcanites (olivine basalt, bassanite, trachybasalts), 30% tuffaceous breccia composed mainly of fragments of shales, sandstones, crystalline rocks and volcaniclastic material	**olivine**, titanite, garnet (pyrope-enriched), garnet (almandine-enriched), *zircon, clinozoisite, epidote (pistazite), amphibole*
Secondary colluvial to alluvial placer deposit	50% volcanites, 20% tuffaceous breccia, 20% crystalline rocks, 10% olivine-garnet aggregates	**titanite, garnet (pyrope-enriched), garnet (almandine-enriched), olivine**, zircon green amphibole, brown amphibole, tremolite-actinolite, *apatite, epidote (pistazite), clinozoisite, orthopyroxene (bronzite, hypersthene), clinopyroxene, biotite, spinel*
Fluvial placer deposits (proximal tributary stream)	20% volcanites, 10% tuffaceous breccia, 30% crystalline rocks, 20% granitic fragments, 20% shales/slates	**titanite, green amphibole, brown amphibole**, olivine, epidote (pistazite), garnet (almandine-enriched), zircon, *apatite, orthopyroxene (bronzite, hypersthene), spinel, andalusite*
Fluvial placer deposits (intermediate tributary stream)	20% volcanites, 10% tuffaceous breccia, 30% crystalline rocks, 20% granitic fragments, 20% shales/slates	**titanite, green amphibole, brown amphibole**, zircon, epidote (pistazite) *apatite, garnet (almandine-enriched), monazite, tourmaline, fayalite-enriched olivine, hypersthene,*
Fluvial placer deposits (distal tributary stream)	10% volcanites, 5% tuffaceous breccia, 50% crystalline rocks, 10% granitic fragments, 25% shales/slates	**titanite, green amphibole, brown amphibole** epidote (pistazite), *zircon, apatite, garnet (almandine-enriched), (pyrope-enriched), orthite*
Fluvial placer deposits (trunk river)	10% volcanites, 60% crystalline rocks, 10% granitic fragments, 20% shales/slates	**titanite, green amphibole brown amphibole**, epidote (pistazite), garnet (almandine-enriched), (pyrope-enriched), fayalite-enriched olivine, zircon, *apatite, forsterite-enriched olivine, hypersthene*

Key: Secondary colluvial to alluvial placer deposits are most relevant for gemstones including diamonds (diamonds are not listed here as they are only sporadic). Amount of heavy minerals: Abundant (bold-faced), common (normal font), rare (italics).

Epidote-clinozoisite s.s.s. (s.s.s. = solid solution series) are common heavy minerals of intermediate stability and found in many siliciclastic sediments, but only the blue tanzanite attracted the interest of gem dealers. As to its presence in placer deposits, it closely resembles garnet-group minerals, as to its abundance of mineral grains attaining gem quality, however, it trails far behind garnet. Profitable finds are only to be expected in proximity to the primary gemstone deposits (Figure 8).

Another gemstone is referred to only for completeness in this review of gemstone placer deposits. Spodumene a Li silicate typical of pegmatitic source rock has two varieties, kunzite, and hiddenite that qualify for being ranked among gemstones. Spodumene is only present in the clastic apron close to the source pegmatite [34]. Similar to its companion petalite it decomposes easily into phyllosilicates such as cookeite, smectite and kaolinite depending upon the physical-chemical conditions (Figure 16f). Not surprisingly, precious spodumene is a very rare species in placer deposits and rose-colored kunzite was only reported from Minas Gerais to be mined from some alluvial deposits. The majority of this gemological variety of spodumene with the largest crystal ever found near Conselheiro Pena weighing 7.410 kg (=37,050 carats) has been recovered from pegmatites [112].

3.10. Amber Placer Deposits

Baltic amber originates from resin of a pinaceous conifer similar to Pseudolarix [128]. The original deposits formed from the Eocene to the Early Oligocene in a nearshore marine environment along the eastern Baltic Sea coast (Jantarnyj and Svetlogorsk in East Prussia on the Semba Peninsula, Russia), while another site is located near Darłovo, Poland [1,129,130].

The amber units were reworked and re-deposited during the Quaternary and placer-like accumulations are now also found along the Polish coast [131]. Holocene amber deposits are widespread along the coast of the Baltic Sea up to the Kuršių Marios lagoon. Nirgi et al. (2017) [132] reported amber-bearing layers from the Holocene coastal plain on the SW Saaremaa Island where amber is not known in sedimentary successions, but is common in Stone Age and Bronze Age archaeological sites. Amber cannot be grouped among the heavy minerals and the rules of settling and entrainment equivalence can hardly be applied to the majority of these organic compounds (Figure 7). Due to its specific gravity of between 1.05 and 1.08 amber can float on saltwater and frequently occurs in marine environments after having been transported over a long distance from its original site of formation [133]. Its transport does not need strong longshore currents, although strong surge eases long distance-transport and its release from pre-existing amber beds. Amber is also found in fluvial deposits, such as oxbow lakes and slack-water facies [133]. Both subenvironments of meandering stream deposits are not the preferred loci to concentrate heavy minerals and create fluvial placer deposits owing to its low speed of the water currents. After being released from its primary beds such as coal, it can survive reworking provided it is protected from being exposed to oxygen for a long period of time. In places, amber is rafted bound to waterlogged wood and concentrated together with this drift wood. Due to the moderate hardness placer-like accumulations of amber are found mainly in young sedimentary units from the Paleogene through to the Holocene.

4. Synopsis and Conclusions

Of the approximately 150 gemstones, roughly 40 can be recovered from placer deposits for a profit after having passed through the natural processing plant of weathering, transport, attrition, and deposition (agate, almandine, amber, amethyst, apatite, aquamarine, beryl, (chrome) diopside, (chrome) tourmaline, chrysoberyl, citrine, demantoid, diamond, emerald, enstatite, hessonite, hiddenite, kornerupine, kunzite, kyanite, morganite, peridote, pyrope, quartz, rhodolite, rose quartz, smoky quartz, spessartine, (chrome) titanite, spinel, ruby, sapphire, tanzanite, zoisite, topaz, tsavorite, and zircon). Going downhill from the source area to the basin means in effect separating the wheat from the chaff, the showcase from the jeweler's quality, because only the flawless and strongest contenders among the gemstones survive it all. Some of these gemstones can also be used as pathfinder minerals for primary or secondary gemstone deposits of their own together with a series of other non-gemmy materials which are genetically linked to these gemstones.

All placer types known to be relevant for the accumulation of non-gemmy material such as gold, cassiterite, ilmenite, "coltan"—group s.s.s. etc. are also found as trap-sites of gemstones (residual, eluvial, colluvial, alluvial, deltaic, aeolian, and marine shelf deposits) (Figure 1). Running water and wind can separate minerals according to their physical-chemical features; ice can only transport minerals and rocks. Only in the glacial-fluvial sands and gravels of the outwash plain in front of the terminal moraine and between this moraine and the retreating glacier the sorting and separating influence of running water can take effect. This does not mean that glacial or periglacial landforms have to be confined to play a subordinate role in this matter [134]. The techniques used in till exploration to delineate an ore body of whatever commodity it may be, can also be used for gemstones shown in Table 1 and in the previous paragraph, but in practice it is only worth the trouble for diamonds and in some places for modifications of precious corundum. In addition to the marker minerals listed for diamond exploration in Section 3.8, it is particularly the Cr-enriched varieties of pyrope and almandine, pyroxene, spinel, Mg ilmenite, and Mg olivine that count [135]

The general parameters that matter during accumulation of gemstones in placers are their intrinsic value controlled by the size and hardness and the extrinsic factors controlling the evolution of the landscape through time such as weathering, erosion and vertical movements, and fertility of the hinterland as to the minerals targeted upon. The impact of both parameters can be demonstrated by the most unstable gemstone peridote (olivine) and the ultrastable gemstone diamond (Figures 1 and 8). Olivine does not survive long-term pervasive chemical weathering widespread in tropical morpho-climatic zones where this Mg–Fe silicate ends up in a great variety of Ni-, Co, and Cr-bearing serpentine, chlorite, smectite, talc, and sepiolite-group phyllosilicates called "garnierites" [33,34]. Chemical weathering has two maxima, a first-order one in the humid tropical zone and a second-order one in the humid mid-latitude zones. It has two minima in the periglacial/glacial zone and the high-altitude/mountain zone, where mechanical weathering prevails over chemical weathering, erosion, and the gradients are high. In those environments typical of a high rate of erosion vs. chemical weathering even the unstable peridote can appear in marine and fluvial heavy mineral assemblages in a proximal placer. Due to its stability, diamonds are left unaffected by those issues mentioned above and can be recycled in a multi-stage process. A critical aspect during deposition of ultrastable gemstones such as diamonds is the covering and diluting effect exerted by the sediments laid down concomitantly with the gemstones. High energetic aquatic and aeolian winnowing processes can remove the dilutors from the various trap-sites, a fact which is also underscored in different placer deposits by lag deposits residing immediately on top of the bedrock. Some tectono-geographic elements such as unconformities, hiatuses, and sequence boundaries (often with incised valley fills and karstic landforms) and known as common planar architectural elements in sequence stratigraphy applied to marine and correlative continental environments can play a significant role in forward modeling of placer deposition.

Funding: This research received no external funding.

Acknowledgments: I acknowledge with thanks the fruitful comments made by the three reviewers. My sincere thanks are extended to the guest editors of the special volume "Mineralogy and Geochemistry of Gems" Panagiotis Voudouris, Stefanos Karampelas, Vasilios Melfos, and Ian Graham for their invitation to make a scientific contribution to the special volume and their editorial assistance. I extend my gratitude also to the team of "MINERALS" for the technical support and for the final layout. I would like to thank also my former colleagues W. Hofmeister and T. Häger (Johannes-Gutenberg University Mainz, Section: Geo-Materials and Gemstone Research).

Conflicts of Interest: The author declares no conflict of interest.

Appendix A. See Text for Reference

Studies of economic geology *sensu lato* are centered on the emplacement of the ore body and the development of its minerals and rocks. As a consequence the "Chessboard Classification Scheme of Mineral Deposits" uses mineralogy and geology as x- and y-coordinates of a classification chart of mineral resources resembling a "chessboard" (or "spreadsheet"). Magmatic and sedimentary lithologies together with tectonic structures (1-D/pipes, 2-D/veins) are plotted along the x-axis in the header of the diagram representing the columns in this chart diagram. 63 commodity groups, encompassing minerals and elements are plotted along the y-axis, forming the lines of the spreadsheet. These commodities are subjected to a tripartite subdivision into ore minerals, industrial minerals and gemstones/ornamental stones. Further information on the various types of mineral deposits, as to the major ore and gangue minerals, current models and mode of formation or when and in which geodynamic setting these deposits mainly formed throughout the geological past may be obtained from the text of "Chessboard Classification Scheme of Mineral Deposits" by simply using the code of each deposit in the chart. This code can be created by combining the commodity (lines) shown by numbers plus lower caps with the host rocks or structure (columns) given by capital letters. Each commodity has a small preface on the mineralogy and chemistry and ends up with an outlook into its final use and the supply situation of the raw material on a global basis, which may be updated by the user through a direct link to databases available on the internet, e.g., the database of the US Geological Survey. The internal subdivision of each commodity section corresponds to the common host rock lithologies (magmatic, sedimentary, metamorphic) and structures. Cross sections and images illustrate the common ore types of each commodity.

Table of contents: Chromium, nickel, cobalt, platinum group elements (PGE/platinum palladium-osmium-iridium-rhodium-ruthenium), titanium, vanadium, iron, manganese, copper, selenium-tellurium, molybdenum-rhenium, tin-tungsten, niobium-tantalum-scandium beryllium, lithium-cesium-rubidium, lead-zinc-germanium-indium-cadmium, silver, bismuth, gold, antimony, arsenic, thallium, mercury, rare earth elements (REE/lanthanum, cerium, praseodymium, neodymium, promethium, samarium, europium, gadolinium, terbium, dysprosium, holmium, erbium, thulium, ytterbium, lutetium)-yttrium, uranium-radium, thorium, aluminum-gallium, magnesium, calcium, boron, sulfur-calcium sulfate, fluorine, barium, strontium, potassium-sodium-chlorine-bromine, nitrogen-iodine, sodium carbonate- sodium sulfate, phosphorus, zirconium-hafnium, silica, feldspars, feldspathoids, zeolites, amphibole-asbestos (asbestiform minerals), olivine-dunite, pyroxene-inosilicates, garnet-group minerals, epidote-group minerals, sillimanite-group minerals, corundum-spinel, diamond, graphite, allophane-imogolite, halloysite, kaolinite-group minerals(kandites), talc-pyrophyllite group, smectite-group minerals, vermiculite, mica-group minerals, chlorite-group minerals, sepiolite-palygorskite (hormites), jet (lapidary coal), amber.

To allow for a direct correlation of the "Chessboard classification scheme of mineral deposits" [1] and its special edition of digital maps of gemstone deposits [49]; with the current study of placer deposits which is designed as a more detailed treatment of those commodities also of significance for gems and gemstones, the pertinent tables from the "Chessboard classification scheme of mineral deposits" are listed below. The gemstone placer deposits under study in the current review can be linked to the primary magmatic, metamorphic and structure-bound deposits which are the potential source of the gems and gemstones in the placer deposits around the globe.

Classification Scheme of Beryllium Deposits

1. Magmatic beryllium deposits
(1) Beryl–emerald–euclase–hambergite-bearing granite pegmatites (14a D)
(2) Taaffeite- and emerald-bearing skarns (14c D)
(3) Replacement deposits in volcaniclastic deposits and granites (14b D)
(4) Be-bearing alkaline intrusive rocks (nepheline syenite) (14a E)
(5) Tugtupite-bearing alkaline intrusions (14b E)
(6) Chrysoberyl within rare element pegmatites (14c E)
(7) Beryl-bearing pegmatoids (14d D)

2. Structure-related beryllium deposits

Aquamarine veins (14a G)

3. Sedimentary beryllium deposits
(1) Regolith-hosted emerald deposits (gemstone) (14a H)
(2) Alluvial-fluvial chrysoberyl placer (14a l)
(3) Black shale-hosted emerald deposits (gemstone) (14a J)

4. Metamorphic beryllium deposits
(1) Schist-related emerald deposits with or without pegmatitic mobilizates (gemstone) (14a AB)
(2) Chrysoberyl in pegmatitic mobilizates in the contact zone
    (1) Metaultrabasic rocks (14b A)
    (2) Metapelites (14b J)

Classification Scheme of Boron Deposits

1. Magmatic boron deposits
(1) Pegmatites bearing tourmaline–danburite–dumortierite–kornerupine mineralization (30a D)
(2) Skarn bearing axinite–serendibite–sinhalite mineralization (30b D)
(3) Contact metasomatic/skarn Fe–B deposits and datolite–danburite mineralization (30b CD)
(4) Metaultrabasic to basic rock bearing datolite–kornerupine mineralization (30d A)
(5) Poudretteite-bearing alkaline magmatic complexes (30a E)

2. Sedimentary boron deposits
(1) Playa lakes and lagoons with B–(As–Sb–Li) brines (hot brine-related) (30b J)
(2) Playa lake with B–(As–Sr–Li) (lacustrine-syn(dia)genetic) (30c J)
(3) Salars with Li–K–Cs–(B) (30a L)
(4) Residual deposits (30b L)
(5) Evaporites/byproduct (30c L)
(6) Volcano-sedimentary B deposits/metamorphosed SEDEX deposit (30e J-30e L)
(7) Boron in plant remains (30a N)

3. Metamorphic boron deposits
(1) Al-enriched gneisses bearing dumortierite (30e HI)
(2) Mg-enriched schists (whiteschists) bearing kornerupine (30d KL)
(3) Cr-bearing tourmalinites in serpentinite–limestone contact zones (30b A)
(4) Datolite–danburite marble (30e K)

Classification Scheme of Fluorine Deposits

1. Magmatic deposits
(1) Fluorite deposits related to granitic intrusions and fluorite skarn W–Sn and Pb–Zn deposits (32e D)
(2) Fluorite deposits related to U-REE carbonatites and alkaline intrusive rocks (32a E)
(3) Granite-related Be–Nb–Ta–fluorite deposits (32d D)
(4) Pegmatite-hosted F–(Sc) deposits (32a D)
(5) Volcanic-hosted F–U–Mo deposits (32g CD)
(6) Volcanic-hosted topaz (32f D)
(7) Granite-related topaz deposits (gemstone) (32b D)
(8) Intragranitic fluorite deposits (32c D)
(9) Cryolite deposit related to metasomatic A-type granites (32b E)

2. Structure-related deposits
(1) F-bearing mega breccia (32c F)
(2) Epithermal ("hot brine") fluorite breccia veins (32a G)
(3) Epithermal F–(Ba) deposits (maars) (32a F)

(4) Unconformity-related (shallow veins) Pb–Zn–Ba fluorite deposits (32b G)

(5) Thrustbound and replacement topaz deposits (gemstone) (32c G)

3. Sedimentary deposits

(1) Stratabound Pb–Zn–Ba fluorite deposits in carbonate rocks (MVT) including replacement deposits ("mantos") (32a K)

(2) Stratabound fluorite-celestite deposits in carbonate rocks (32b K)

(3) Residual and karst-related fluorite deposits

(4) Fluorite cavity-fillings and calcretes (32a H)

(5) Fluorite replacement/calamine deposits (32b H)

(6) Phosphorites with F apatite (32c UK)

(7) Topaz placers (32a I)

Classification Scheme of Phosphate Deposits

1. Magmatic phosphate deposits

(1) Pegmatite-hosted Li–Fe–Mn–Mg phosphates (in places of gem-quality) (38a D)

(2) Apatite-bearing Fe deposits (Kiruna-type) (38b CD)

(3) Apatite-bearing titanomagnetite-magnetite deposits (Kola-type) (38a E)

(4) Apatite-bearing REE-Fe deposit in alkaline and carbonatite complexes (Sokli-type) (38b E)

2. Structure-related phosphate deposits

(1) Apatite veins (38a G)

(2) Vein-type remobilisation of apatite of phosphorite deposits (38b G)

3. Sedimentary phosphate deposits

(1) Continental phosphates

    (1) Alluvial placers of apatite possessing gem-quality (38a l)

    (2) Nb–P–Ti laterites and bauxites (38a H)

    (3) Phoscretes

        (1) Apicretes (38c H)

        (2) APS-mineral bearing duricusts (38b H)

    (4) Lacustrine phoscretes

        (1) (Fluvial-) lacustrine and bog iron ores (38g H–38g K)

        (2) Perennial or organic lakes (38a JK)

        (3) Ephemeral lakes (38b I)

(2) Marine phosphates

    (1) Guano deposits (38a M)

    (2) Carbonate-hosted phosphorites (38e K)

    (3) Siliciclastic-hosted phosphorites (38e J)

    (4) Phosphate sandstones (38e l)

    (5) Marine phosphate-bearing ironstones

        (1) Phosphatic bonebeds in oolithic ironstones (Minette-/Wabana types) (38f IJ)

        (2) Phosphatic bonebeds in detrital iron ore deposits (38g l)

(3) P-bearing biological gem material (ivory) (38a N)

4. Metamorphic phosphate deposits

(1) Apatite-bearing skarn-like mineralization (regionally metamorphosed limestone) (38c K)

(2) Regionally and contact-metamorphosed phosphorites (38c J)

(3) Lazulite-bearing metasedimentary rocks (38c l)

Classification Scheme of Zirconium and Hafnium Deposits

1. Magmatic zirconium deposits
(1) Zircon–sapphire-bearing alkali basalt (39 B)
(2) Zr-REE–P–Nb–Ta–F–(Be–Th) carbonatite-alkaline igneous complex (39 E)

2. Sedimentary zirconium deposits
(1) Zircon (hyacinth) placer deposits (39 1)

Classification Scheme of Garnet Deposits

1. Magmatic garnet deposits
(1) Mg-enriched garnet s.s.s. in kimberlites (47a A)
(2) Garnets in volcanic rocks
    (1) Mg-enriched garnet s.s.s. in basic volcanic rocks (47a B)
    (2) Fe-enriched garnet s.s.s. in dacite and andesite (47a C)
    (3) Mn–Fe-enriched garnet s.s.s. in rhyolite (47a D)
(3) Fe–Mn enriched garnet s.s.s. in (granite) pegmatites (47b D)
(4) Ti-enriched garnet s.s.s. in alkaline igneous rocks and carbonatites (47a D)
(5) Fe–Ti–Ca-enriched (hydro) garnet s.s.s. in skarn (47c CD)

2. Structure-related garnet deposits
(1) Mg-enriched garnet s.s.s. in kimberlites (47b EF)
(2) Ca–Fe-enriched garnet s.s.s. in Cu deposits (47b G)
(3) Sedimentary garnet deposits
    Mg–Fe–Mn–Ca-enriched garnet s.s.s. in alluvial through coastal placer
    deposits (47b I)
(4) Metamorphic garnet deposits
    (1) Mg–Cr-enriched garnet s.s.s. in meta(ultra)basic igneous rocks (47c A)
    (2) Fe–Al-enriched garnet s.s.s. in metapelites (47a IJ)
    (3) Ca–Mn-enriched garnet s.s.s. in calcsilicate rocks (47cd K)
    (4) Mn-enriched garnet s.s.s. in manganiferous BIF ore deposits (47b J)
    (5) V-enriched garnet s.s.s. in carbonaceous slates/schists (47d J)

Classification Scheme of Corundum and Spinel (including Gemstones) Deposits

1. Magmatic corundum and spinel deposits
(1) High-sulfidation-type deposits adjacent to acidic igneous rocks (50d CD)
(2) Corundum xenolites and sapphire in alkali basalt and anorthoclasites (50a B)
(3) Chrome spinel with ruby and sapphire in ultramafic volcanics (50b A)
(4) Corundum (nepheline) syenite and pegmatites (50b E)
(5) Ruby-spinel Skarn (50c D)
(6) Sapphire in lamprophyres (50d E)
(7) Corundum pegmatites (plumasite + marundite) (49c CD–50a CD)

2. Sedimentary corundum and spinel deposits
(1) Precious corundum and spinel in modern alluvial to alluvial-fluvial placers (50b 1)
(2) Paleoplacer (50d 1)

3. Metamorphic corundum and spinel deposits
(1) Sapphirine-corundum meta(ultra)basic rocks (49c A–50c A)
(2) Ruby in Al-enriched meta(ultra) basic igneous rocks
    (1) Ruby metaultrabasites-serpentinites (50a A)
    (2) Ruby in zoisite amphibolite (50c B)
(3) Ruby in marble (50c K)
(4) Corundum/emery-bearing metaduricrusts

(1) Corundum-diaspore-spinel metabauxite in marble (50b K)

(2) Sillimanite-corundum metapalaeosol (49c H)

(3) Corundum-spinel metabauxite (50a H)

(5) Ruby and sapphire in gneisses (50b J)

Classification Scheme of Diamond Deposits

1. Magmatic deposits

(1) Diamond in komatiites, lamprophyres and other ultrabasic rocks (51b A)

(2) Diamond in peridotites of ophiolite sequences (51c A)

(3) Diamond deposits bound to kimberlites (51a A–51a F)

(4) Diamond deposits bound to lamproites (51a E)

2. Sedimentary deposits

(1) Placer deposits

    (1) Alluvial-fluvial and near-shore-marine modern diamond placer deposits (51a I)

    (2) Palaeoplacer diamond deposits (51b I)

    (3) Alluvial-fluvial carbonado placer (51c I)

3. Metamorphic deposits

(1) Microdiamonds in impact structures (51a N)

(2) Microdiamonds in high-pressure zones (51b N)

(3) Microdiamonds in garnet gneiss (51b J)

(4) Diamond in graphite schist (51c J)

Classification Scheme of Silica Deposits

1. Magmatic deposits

(1) deposits in basic to acidic volcanic rocks

    (1) Hypogene agate-amethyst-rock crystal-opal deposits in basic magmatic rocks (40a B)

    (2) Thundereggs in felsic magmatic rocks (40a D)

    (3) Supergene opal deposits on top of basic and felsic magmatic rocks (40b B–40b H)

    (4) Hypogene zeolite-celadonite-bearing opal deposits in basic to acidic magma tic rocks (40f BCD)

(2) Opal and quartz in volcanic sinter, geyserites and high-sulfidation replacement deposits (40c CD)

(3) Quartz- and semi-precious gemstone pegmatites (40d D)

(4) Glass-bearing pyroclastic and volcanic rocks

    (1) Pumice (40g CDE)

    (2) Scoria (40g B)

    (3) Perlite (40h CD)

2. Structure-related deposits

(1) Unconformity-related shallow quartz veins (40a G)

(2) Deep-seated lineamentary quartz veins (40c G)

3. Sedimentary deposits

(1) Silcretes

    (1) Quartz (40a H)

    (2) Chalcedony (40c H)

    (3) Opal (40d H)

(2) Rock crystal placers (40a l)

(3) Quartz sand and arenites (40c l)

(4) Limnoquartzite (transitional into volcano-sedimentary deposits) (40e J)

(5) Biogenic silica deposits

    (1) Diatomite (40c J)

    (2) Radiolarite (40d J)

  (3) Chert (40g J)
  (4) Rottenstone/ tripolite (40c K)
4. Metamorphic deposits
(1) (Meta) quartzites (40d l)
(2) Opal concretions in graphite schist (40h J)
(3) Onyx within marble (40d K)
(4) Tektites (40a N)
(5) Zebra rock (40a J)

Classification Scheme of Chromium Deposits

1. Magmatic chromium deposits
(1) Stratiform chromite deposits (la A)
(2) Podiform chromite deposits (1b A)
(3) Concentric chromite deposits (lc A)
(4) Chrome diopside mineralization in kimberlites (Id A)

2. Sedimentary chromium deposits
(1) Cr concentration in laterites and saprolites (la H)
(2) Mtorolite mineralization (1b H)
(3) Chromite placer deposits (la l)

3. Metamorphic chromium deposits
(1) Uvarovite and sphene gem deposits (le A)

Classification Scheme of Olivine Deposits

1. Magmatic olivine deposits
(1) Dunite (olivine) deposits (45a A)
(2) Peridote (chrysolite) gemstone deposits in
        (1) Ultrabasic volcanic rocks (451r A)
        (2) Basic volcanic rocks (45b B)
        (3) Forsterite Skarn (45d CD)
2. Sedimentary olivine deposits
(1) Peridote placer (45a I)

3. Metamorphic olivine deposits
(1) Olivine in serpentinite (45d A)
(2) Olivine (forsterite) in meta-limestones (45d K)
(3) Ophicalcite (45a K)

Classification Scheme of Epidote Deposits

1. Magmatic epidote deposits
(1) "Unakite"-bearing granite (48a D)
(2) "Saussurite"-bearing basalt (48a B)
(3) Skarn-like tanzanite deposits (48c CD)
(4) Thulite-piedmontite skarn (48d CD)
(5) Thulite-bearing granite (48b D)

2. Metamorphic epidote-group deposits
(1) Epidote in regional and contact metamorphic epidosite (48c B)
(2) Thulite-piedmontite in Mn-bearing metapelites(48b J)
(3) Thulite-nephrite amphibolite (48b B)

(4) Zoisite-epidosite in metacarbonates (48b K)

(5) Tanzanite in calcsilicates (48c K)

Classification Scheme of Lithium Deposits

1. Magmatic lithium and cesium deposits

(1) Pegmatitic Li (including gem spodumene) and Cs deposits (15a D)

(2) Li–Cs–Rb in rhyolitic tuffs with Be and F (15b D)

2. Sedimentary lithium and cesium deposits

(1) Brine deposits and salars (15a L)

    (1) Geothermal waters and oil-field formation waters

    (2) Li brines within playas in Chile

(2) Clay deposits (15a J)

    Hectorite in altered volcaniclastic rocks related to hot-spring activity

Classification Scheme of Amber Deposits

1. Sedimentary jet and amber deposits

(1) Jet deposits (62 J)

(2) Primary amber deposits (63 J)

(3) Reworked amber deposits (63 I)

The codes (e.g. 14aD, 15aJ) refer to the pertinent sections in the "Chessboard classification scheme of mineral deposits" [1].

References

1. Dill, H.G. The "chessboard" classification scheme of mineral deposits: Mineralogy and geology from aluminum to zirconium. *Earth Sci. Rev.* **2010**, *100*, 1–420. [CrossRef]
2. Buck, S.G.; Minter, W.E.L. Placer formation by fluvial degradation of an alluvial fan sequence: The Proterozoic Carbon Leader placer, Witwatersrand Supergroup, South Africa. *J. Geol. Soc.* **1985**, *142*, 757–764. [CrossRef]
3. Roy, P.S. Heavy mineral beach placers in southeastern Australia: Their nature and genesis. *Econ. Geol.* **1999**, *94*, 567–588. [CrossRef]
4. Lalomov, A.V.; Tabolitch, S.E. Age determination of coastal submarine placer, Val'cumey, northern Siberia. *Creat. Ex Nihilo Tech. J.* **2000**, *14*, 83–90.
5. Burton, J.P.; Fralick, P. Depositional placer accumulations in coarse-grained alluvial braided river systems. *Econ. Geol.* **2003**, *98*, 985–1001. [CrossRef]
6. Corbett, I.; Burrell, B. The earliest Pleistocene (?) Orange River fan-delta: An example of successful exploration delivery aided by applied Quaternary research in diamond placer sedimentology and paleontology. *Quat. Int.* **2003**, *82*, 63–73. [CrossRef]
7. Dill, H.G.; Khishigsuren, S.; Majigsuren, Yo.; Bulgamaa, J.; Hongor, O.; Hofmeister, W. The diamondiferous peridot olivine-garnet deposit Shavryn Tsaram, Central Mongolia, with special reference to its placer deposits. *Z. Gemmol.* **2004**, *53*, 87–104.
8. Dill, H.G.; Melcher, F.; Fuessl, M.; Weber, B. The origin of rutile-ilmenite aggregates ("nigrine") in alluvial-fluvial placers of the Hagendorf pegmatite province, NE Bavaria, Germany. *Mineral. Petrol.* **2007**, *89*, 133–158. [CrossRef]
9. Dill, H.G. A review of mineral resources in Malawi: With special reference to aluminum variation in mineral deposits. *J. Afr. Earth Sci.* **2007**, *47*, 153–173. [CrossRef]
10. Dill, H.G. Grain morphology of heavy minerals from marine and continental placer deposits, with special reference to Fe–Ti oxides. *Sediment. Geol.* **2007**, *198*, 1–27. [CrossRef]
11. Gujar, A.R.; Ambre, N.V.; Mislankar, P.G. *Onshore Heavy Mineral Placers of South Maharashtra, Central West Coast of India National Seminar on Exploration, Exploitation, Enrichment and Environment of Coastal Placer Minerals (PLACER-2007)*; Macmillan India: New Delhi, India, 2007.

12. Dill, H.G.; Ludwig, R.-R. Geomorphological–sedimentological studies of landform types and modern placer deposits in the savanna (Southern Malawi). *Ore Geol. Rev.* **2008**, *33*, 411–434. [CrossRef]

13. Jones, G.; O'Brien, V. Aspects of resources estimation fro mineral sands deposits Transactions of the Institutions of Mining and Metallurgy, Section B. *Appl. Earth Sci.* **2014**, *123*, 86–94. [CrossRef]

14. Abzalov, M. Mineral Sands. In *Modern Approaches in Solid Earth Sciences*; Springer: Heidelberg, Germany, 2016; Volume 12, pp. 427–433.

15. Bern, C.R.; Shah, A.K.; Benzel, W.M.; Lowers, H.A. The distribution and composition of REE-bearing minerals in placers of the Atlantic and Gulf coastal plains, USA. *J. Geochem. Explor.* **2016**, *162*, 50–61. [CrossRef]

16. Sitdikova, L.M.; Ibragimov, E.A.; Badrutdinov, O.R.; Khasanova, N.M.; Mukhamatdinov, I.I. Material composition of coastal marine placer deposits of the Arabian Sea Coast (Kollam, Kerala, India). *Int. Multidiscip. Sci. GeoConf. Surv. Geol. Min. Ecol. Manag. SGEM* **2016**, *1*, 361–368.

17. Hou, B.; Keeling, J.; Van Gosen, B.S. Geological and exploration models of beach placer deposits, integrated from case-studies of Southern Australia. *Ore Geol. Rev.* **2017**, *80*, 437–459. [CrossRef]

18. Milesi, J.P.; Ledru, P.; Marcoux, E.; Mougeot, R.; Johan, V.; Lerouge, C.; Sabate, P.; Bailly, L.; Respaut, J.P.; Skipwith, P. The Jacobina Paleoproterozoic gold-bearing conglomerates, Bahia, Brazil: A hydrothermal shear-reservoir model. *Ore Geol. Rev.* **2002**, *19*, 95–136. [CrossRef]

19. Malitc, K.N.; Merkle, R.K.W. Ru-Os-Ir-Pt and Pt-Fe alloys from the Evander Goldfield, Witwatersrand Basin, South Africa: Detrital origin inferred from compositional and osmium-isotope data. *Can. Mineral.* **2004**, *42*, 631–650. [CrossRef]

20. Frimmel, H.E. Archean atmospheric evolution: Evidence from the Witwatersrand gold fields, South Africa. *Earth Sci. Rev.* **2005**, *70*, 1–46. [CrossRef]

21. Stanaway, K.J. Heavy mineral placers. *Min. Eng.* **1992**, *44*, 352–358.

22. Summerfield, M.A. *Global Geomorphology*; John Wiley & Sons Inc.: New York, NY, USA, 1991; p. 537.

23. Dill, H.G.; Melcher, F.; Fuessl, M.; Weber, B. Accessory minerals in cassiterite: A tool for provenance and environmental analyses of colluvial-fluvial placer deposits (NE Bavaria, Germany). *Sediment. Geol.* **2006**, *191*, 171–189. [CrossRef]

24. Dill, H.G.; Techmer, A.; Weber, B.; Fuessl, M. Mineralogical and chemical distribution patterns of placers and ferricretes in Quaternary sediments in SE Germany: The impact of nature and man on the unroofing of pegmatites. *J. Geochem. Explor.* **2008**, *96*, 1–24. [CrossRef]

25. Dill, H.G. Geogene and anthropogenic controls on the mineralogy and geochemistry of modern alluvial-(fluvial) gold placer deposits in man-made landscapes in France, Switzerland and Germany. *J. Geochem. Explor.* **2008**, *99*, 29–60. [CrossRef]

26. Dill, H.G.; Klosa, D.; Steyer, G. The "Donauplatin": Source rock analysis and origin of a distal fluvial Au-PGE (gold-platinum-group-element) placer in Central Europe. *Mineral. Petrol.* **2009**, *96*, 141–161. [CrossRef]

27. Dill, H.G.; Steyer, G.; Weber, B. Morphological studies of PGM grains in alluvial-fluvial placer deposits from the Bayerischer Wald, SE Germany: Hollingworthite and ferroan platinum. *Neues Jahrb. Mineral. Abh.* **2010**, *187*, 101–110. [CrossRef]

28. Miall, A.D. *Principles of Sedimentary Basin Analysis*, 3rd ed.; Springer: Berlin, Germany, 1999; p. 616.

29. Galloway, W.E.; Hobday, D.K. *Terrigenous Clastic Depositional Systems: Applications to Fossil Fuel and Groundwater Resources*; Springer: Berlin, Germany, 1996; p. 489.

30. Panin, N.; Panin, S. Sur la genèse des accumulations des mineraux lourds dans le delta du Danube. *Rev. Géogr. Phys. Géol. Dyn.* **1969**, *11*, 511–522.

31. Boyd, R.; Dalrymple, R.; Zaitlin, B.A. Classification of clastic coastal depositional environments. *Sediment. Geol.* **1992**, *80*, 139–150. [CrossRef]

32. Dalrymple, R.W. Tidal depositional systems. In *Facies Models 4*; James, N.P., Dalrymple, R.W., Eds.; Geological Association of Canada: St. John's, NL, Canada, 2010; pp. 201–232.

33. Dalrymple, R.W.; Choi, K. Morphologic and facies trends through the fluvial-marine transition in tide-dominated depositional systems: A schematic framework for environmental and sequence-stratigraphic interpretation. *Earth Sci. Rev.* **2007**, *81*, 135–174. [CrossRef]

34. Davis, R.A., Jr.; Dalrymple, R.W. *Principles of Tidal Sedimentology*; Springer: New York, NY, USA, 2011; p. 621.

35. Longhitano, S.G.; Mellere, D.; Steel, R.J.; Ainsworth, R.B. Tidal depositional systems in the rock record: A review and new insights. *Sediment. Geol.* **2012**, *279*, 2–22. [CrossRef]

36. Hayes, M.O.; FitzGerald, D.M. Origin, Evolution, and Classification of Tidal Inlets. *J. Coast. Res.* **2013**, *69*, 14–33. [CrossRef]

37. Fielding, C.R. Upper flow regime sheets, lenses and scour fills: Extending the range of architectural elements for fluvial sediment bodies. *Sediment. Geol.* **2006**, *190*, 227–240. [CrossRef]

38. Evans, A.M. *Ore Geology and Industrial Minerals—An Introduction*; Blackwell: Oxford, UK, 1993; p. 358.

39. Peel, M.C.; Finlayson, B.L.; McMahon, T.A. Updated world map of the Köppen–Geiger climate classification. *Hydrol. Earth Syst. Sci.* **2007**, *11*, 1633–1644. [CrossRef]

40. Slingerland, R. The effects of entrainment on the hydraulic equivalence relationships of light and heavy minerals in sands. *J. Sediment. Petrol.* **1977**, *47*, 753–770. [CrossRef]

41. Reid, I.; Frostick, L.E. Role of settling entrainment and dispersive equivalence and of interstice trapping in placer formation. *J. Geol. Soc. Lond.* **1985**, *142*, 739–746. [CrossRef]

42. Hughes, M.G.; Keene, J.B.; Joseph, R.G. Hydraulic sorting of heavy-mineral grains by swash on a medium-sand beach. *J. Sediment. Res.* **2000**, *70*, 994–1004. [CrossRef]

43. Morton, A.C. Stability of detrital heavy-minerals in Tertiary Sandstones from the North Sea Basin. *Clay Miner.* **1984**, *19*, 287–308. [CrossRef]

44. Lång, L.-O. Heavy mineral weathering under acidic soil conditions. *Appl. Geochem.* **2000**, *15*, 415–423. [CrossRef]

45. Dill, H.G. Kaolin: Soil, rock and ore: From the mineral to the magmatic, sedimentary, and metamorphic environments. *Earth Sci. Rev.* **2016**, *161*, 16–129. [CrossRef]

46. Dill, H.G. An overview of the pegmatitic landscape from the pole to the equator—Applied geomorphology and ore guides. *Ore Geol. Rev.* **2017**, *91*, 795–823. [CrossRef]

47. Dill, H.G. Residual clay deposits on basement rocks: The impact of climate and the geological setting on supergene argillitization in the Bohemian Massif (Central Europe) and across the globe. *Earth Sci. Rev.* **2017**, *165*, 1–58. [CrossRef]

48. Günther, B. *Bestimmungstabellen für Edelsteine, Synthesen, Imitationen*; Lenzen: Kirchweiler, Germany, 1988; p. 162.

49. Dill, H.G.; Weber, B. Gemstones and geosciences in space and time. Digital maps to the "Chessboard classification scheme of mineral deposits". *Earth Sci. Rev.* **2013**, *127*, 262–299. [CrossRef]

50. Besaire, H. *Gites Mineraux de Madagascar*; Fascicule; Annales Geologique de Madagascar: Tananarive, Madagascar, 1966; Volume XXXIV, p. 135.

51. Cairncross, B.; Campbell, I.C.; Huizenga, J.M. Topaz, aquamarine, and other beryls from Klein Spitzkoppe, Namibia. *Gems Gemol.* **1998**, *34*, 114–125. [CrossRef]

52. Pezzotta, F. *Madagaskar Ein Paradies voll mit Mineralien und Edelsteinen*; Weise: München, Germany, 1999; Volume 17, pp. 1–96.

53. Grundmann, G. Die Smaragde der Welt. *ExtraLapis* **2001**, *21*, 26–37.

54. Kievlenko, E.Y. *Geology of Gems*; Ocean Publications Ltd.: Littleton, CO, USA, 2003; p. 432.

55. Laurs, B.M.; Simmons, W.B.; Rossman, G.R.; Quinn, E.P.; McClure, S.F.; Peretti, A.; Armbruster, T.; Hawthorne, F.C.; Falster, A.U.; Günther, D.; et al. Pezzottaite from Ambatovita, Madagascar: A new gem mineral. *Gems Gemol.* **2003**, *39*, 284–301. [CrossRef]

56. Cairncross, B. *Field Guide to Rocks and Minerals of Southern Africa*; Struik New Holland Publishing: Cape Town, South Africa, 2004; p. 297.

57. Seifert, A.V.; Žáček, V.; Vrána, S.; Pecina, V.; Zachariáš, J.; Zwaan, J.C. Emerald mineralization in the Kafubu area, Zambia. *Czech Geol. Surv. Bull. Geosci.* **2004**, *79*, 1–40.

58. Kuo, C.S. The Mineral Industry of Sri Lanka. In *U.S. Geological Survey Minerals Yearbook 2005*; U.S. Geological Survey: Reston, FL, USA, 2005; pp. 25.1–25.2.

59. Shigley, J.E.; Laurs, B.M.; Janse, A.J.A.; Elen, S.; Dirlam, D. Gem localities of the 2000s. *Gems Gemol.* **2010**, *46*, 188–216. [CrossRef]

60. Walton, L. *Exploration Criteria for Colored Gemstone Deposits in the Yukon*; Open File 2004-10; Yukon Geological Survey: Whitehorse, YT, Canada, 2004; p. 184.

61. Schumann, W. *Gemstones of the World*; Sterling Publishing Co.: New York, NY, USA, 1997; p. 280.

62. Gübelin, E.; Erni, F. *Gemstones, Symbols of Beauty and Power*; Geoscience Press: Tucson, AZ, USA, 2000; p. 240.

63. Siebel, W.; Schmitt, A.K.; Danišík, M.; Chen, F.; Meier, S.; Weiß, S.; Ero, S. Prolonged mantle residence of zircon xenocrysts from the western Eger rift. *Geosci. Nat.* **2009**, *2*, 886–890. [CrossRef]

64. Pupin, J.P. Zircon and granite petrology. *Contrib. Mineral. Petrol.* **1980**, *73*, 207–220. [CrossRef]

65. Dill, H.G. Pegmatites and aplites: Their genetic and applied ore geology. In *Ore Geol. Rev.*; 2015; Volume 69, pp. 417–561.

66. Proctor, K. Gem pegmatites of Minas Gerais, Brazil: Exploration, occurrence, and aquamarine deposits. *Gems Gemol.* **1984**, *1984*, 78–100. [CrossRef]

67. Proctor, K. Chrysoberyl and alexandrite from the pegmatite districts of Minas Gerais, Brazil. *Gems Gemol.* **1988**, *1988*, 16–32. [CrossRef]

68. Schmetzer, K.; Caucia, F.; Gilg, H.A.; Coldham, T.S. Chrysoberyl from the New England Placer Deposits, New South Wales, Australia. *Gems Gemol.* **2016**, *52*. [CrossRef]

69. Henry, D.J.; Guidotti, C.V. Tourmaline as a petrogenetic indicator mineral: An example from the staurolite—Grade metapelites of NW Maine. *Am. Mineral.* **1985**, *70*, 1–15.

70. Henry, D.J.; Dutrow, B.L. Metamorphic tourmaline and its petrologic applications. *Rev. Mineral.* **1996**, *33*, 503–557.

71. Hawthorne, F.J.; Henry, D.J. Classification of the minerals of the tourmaline group. *Eur. J. Mineral.* **1999**, *11*, 201–216. [CrossRef]

72. Castañeda, C.; César-Mendes, J.; Pedrosa-Soares, A.C. Turmalinas. *Soc. Bras. Geol.* **2001**, 152–179.

73. Basset, A.M. Nepal gem tourmaline. *J. Nepal Geol. Soc.* **1987**, *4*, 31–41.

74. UN-ESCAP Geology and Mineral Resources of Nepal. *Atlas of Mineral resources of the ESCAP region. UN/ESCAP in coordination with DMG*; Commission for the Asia and the Pacific: New York, NY, USA, 1993; Volume 9, p. 107.

75. Aryal, R.K. *Current Status of Precious and Semi-Precious Stones of Nepal*; Unpublished Report; Ministry of Industry, Department of Mines and Geology: Kathmandu, Nepal, 2001.

76. Kaphle, K.P.; Einfalt, H.C. Prospects of Precious and Semiprecious stones in Nepal Himalaya and their Mining Opportunities. In Proceedings of the 29th Himalaya-Karakoram-Tibet Workshop, Lucca, Italy, 2–4 September 2014; pp. 77–78.

77. Henry, D.J.; Novak, M.; Hawthorne, F.C.; Ertl, A.; Dutrow, B.L.; Uher, P.; Pezzotta, F. Nomenclature of the tourmaline-supergroup minerals. *Am. Mineral.* **2011**, *96*, 895–913. [CrossRef]

78. Menzies, M.A. The mineralogy, geology and occurrence of topaz. *Mineral. Rec.* **1995**, *25*, 5–53.

79. Dill, H.G.; Busch, K.; Blum, N. Chemistry and origin of veinlike phosphate mineralization, Nuba Mts. (Sudan). *Ore Geol. Rev.* **1991**, *6*, 9–24. [CrossRef]

80. Dill, H.G. Heavy mineral response to the progradation of an alluvial fan: Implications concerning unroofing of source area, chemical weathering, and paleo-relief (Upper Cretaceous Parkstein fan complex/SE Germany). *Sediment. Geol.* **1995**, *95*, 39–56. [CrossRef]

81. Rozendaal, C.; Philander, A. Mineralogy of heavy mineral placers along the west coast of South Africa. In Proceedings of the 6th International Congress on Applied Mineralogy ICAM 2000, Göttingen, Germany, 17–19 July 2000; pp. 417–420.

82. Philander, C.; Rozendaal, A. Geology of the Cenozoic Namakwa Sands Heavy Mineral Deposit, West Coast of South Africa: A World-Class Resource of Titanium and Zircon. *Econ. Geol.* **2015**, *110*, 1577–1623. [CrossRef]

83. Pupin, J.P.; Turco, G. Le zircon, minéral commun significatif des roches endogènes et exogènes. *Bull. Minéral.* **1981**, *104*, 724–731.

84. Bossart, P.J.; Meier, M.; Oberli, F.; Steiger, R.H. Morphology versus U-Pb systematics in zircon: A high-resolution isotopic study of a zircon population from a Variscan dyke in the Central Alps. *Earth Planet. Sci. Lett.* **1986**, *78*, 339–354. [CrossRef]

85. Benisek, A.; Finger, F. Factors controlling the development of prism faces in granite zircons: A microprobe study. *Contrib. Mineral. Petrol.* **1993**, *114*, 441–451. [CrossRef]

86. Bingen, B.; Davis, W.J.; Austrheim, H. Zircon U-Pb geochronology in the Bergen arc eclogites and their Proterozoic protoliths, and implications for the pre-Scandian evolution of the Caledonides in western Norway. *Geol. Soc. Am. Bull.* **2001**, *113*, 640–649. [CrossRef]

87. Dill, H.G.; Kaufhold, S.; Lindenmaier, F.; Dohrmann, R.; Ludwig, R.; Botz, R. Joint clay-heavy-light mineral analysis: A tool to investigate the hydrographic-hydraulic regime of the Late Cenozoic deltaic inland fans under changing climatic conditions (Cuvelai-Etosha Basin, Namibia). *Int. J. Earth Sci.* **2012**, *102*, 265–304. [CrossRef]

88. Grew, E.S.; Locock, A.J.; Mills, S.J.; Galuskina, I.O.; Galuskin, E.V.; Hålenius, U. Nomenclature of the garnet supergroup. *Am. Mineral.* **2013**, *98*, 785–811. [CrossRef]

89. Wilson, B.S. Colored Gemstones from Canada. *Rocks Miner.* **2010**, *85*, 24–43. [CrossRef]

90. Lind, T.; Henn, U.; Bank, H. Spessartine aus Namibia. *Neues Jahrb. Mineral. Monatsh.* **1993**, *1993*, 569–576.

91. Gent, M.R.; Alvarez, M.M.; Iglesias, J.J.M.G.; Álvarez, J.T. Offshore occurrences of heavy-mineral placers, Northwest Galicia, Spain. *Mar. Georesour. Geotechnol.* **2005**, *23*, 39–59. [CrossRef]

92. Angusamy, N.; Rajamanickam, G.V. Depositional environment of sediments along the southern coast of Tamil Nadu, India. *Oceanologia* **2006**, *48*, 87–102.

93. Harben, P.W.; Kužvart, M. *Industrial Minerals. A Global Geology*; London Industrial Minerals Information Ltd.: London, UK, 1996; p. 462.

94. Dill, H.G. Placer deposits—Sedimentary, mineralogical and economic aspects. In *Material Characterization by Solid State Spectroscopy*; Hofmeister, W., Dao, N.Q., Quang, V.X., Eds.; The Minerals of Vietnam: Hanoi, Vietnam, 2001; pp. 105–123.

95. Laurs, B.M. Update on some Madagascar gem localities. *Gems Gemol.* **2000**, *36*, 165–167.

96. Schwarz, D.; Petsch, E.J.; Kanis, J. Sapphires from Andranondambo region, Madagascar. *Gems Gemol.* **1996**, *32*, 80–99. [CrossRef]

97. Yager, T.R. The mineral industry of Madagascar. In *U.S. Geological Survey Minerals Yearbook 2003*; U.S. Geological Survey: Reston, FL, USA, 2003; pp. 21.1–21.5.

98. Rakotondrazafy, A.F.M.; Giuliani, G.; Ohnenstetter, D.; Fallick, A.E.; Rakotosamizanany, S.; Andriamamonjy, A.; Ralantoarison, T.; Razanatseheno, M.; Offant, Y.; Garnier, V.; et al. Gem corundum deposits of Madagascar: A review. *Ore Geol. Rev.* **2008**, *34*, 134–154. [CrossRef]

99. Dill, H.G.; Goldmann, S.; Cravero, F. Zr-Ti-Fe placers along the coast of NE Argentina: Provenance analysis and ore guide for the metallogenesis in the South Atlantic Ocean. *Ore Geol. Rev.* **2018**, *95*, 131–160. [CrossRef]

100. Hall, A.M.; Thomas, M.F.; Thorp, M.B. Late Quaternary alluvial placer development in the humid tropics: The case of the Birim Diamond Placer, Ghana. *J. Geol. Soc.* **1985**, *142*, 777–787. [CrossRef]

101. Thomas, M.F.; Thorp, M.B.; Teeuw, R.M. Palaeogeomorphology and the occurrence of diamondiferous placer deposits in Koidu, Sierra Leone. *J. Geol. Soc.* **1985**, *142*, 789–802. [CrossRef]

102. Boxer, G.L.; Deakin, A.S. Argyle alluvial diamond deposits. *AusIMM* **1990**, *14*, 1655–1658.

103. Fazakerley, V.W. Bow River alluvial Diamond deposit. *Geol. Miner. Depos. Aust. Papua New Guinea* **1990**, *14*, 1659–1664.

104. Thomas, M.F. Landscape sensitivity in time and space—An introduction. *Catena* **2001**, *42*, 83–98. [CrossRef]

105. Teeuw, R.M. Regolith and diamond deposits around Tortiya, Ivory Coast, West Africa. *Catena* **2002**, *46*, 111–127. [CrossRef]

106. Bluck, B.J.; Ward, J.D.; De Wit, M.C.J. Diamond mega-placers: Southern Africa and the Kaapvaal craton in a global context. *Geol. Soc. Lond. Spec. Publ.* **2005**, *248*, 213–245. [CrossRef]

107. Rau, T.K. Incidence of diamonds in the beach sands of the Kanyakumari Coast, Tamil Nadu. *J. Geol. Soc. India* **2006**, *67*, 11–16.

108. Tompkins, L.A.G.; Gonzaga, M. Diamonds in Brazil and a proposed model for the origin and distribution of diamonds in the Coromandel Region, Minas Gerais, Brazil. *Econ. Geol.* **1989**, *84*, 591–602. [CrossRef]

109. Mitchell, R.H. Kimberlites and Lamproites: Primary sources of diamond. *Geosci. Can.* **1991**, *18*, 1–16.

110. Nixon, H. The morphology and nature of primary diamondiferous occurrences. *J. Geochem. Explor.* **1995**, *53*, 41–71. [CrossRef]

111. Svisero, D. Distribution and origin of diamonds in Brazil: An overview. *J. Geodyn.* **1995**, *20*, 493–514. [CrossRef]

112. Delaney, P.J.V. *Gemstones of Brazil: Geology and Occurrences*; Revista Escola de Minas, Praca Tiradentes; REM: Ouro Preto, Brazil, 1996; Volume 20, p. 125.

113. Tappert, R.; Stachel, T.; Harris, J.W.; Muehlenbachs, K.; Brey, G.P. Placer diamonds from Brazil: Indicators of the Composition of the Earth's Mantle and the distance to their kimberlitic sources. *Econ. Geol.* **2006**, *101*, 453–470. [CrossRef]

114. Kopylova, M.G.; Russell, J.K.; Stanley, C.; Cookenboo, H. Garnet from Cr- and Ca-saturated mantle: Implications for diamond exploration. *J. Geochem. Explor.* **2000**, *68*, 183–199. [CrossRef]

115. Mafound, R.F. Presence of diamond in the pyrope peridotite, Drekeesh area, Tartous province, NW Syria: A new theory on the origin of diamond. *Microchem. J.* **2002**, *73*, 265–271.

116. Schulz, K.-H.; Adelhardt, W.; Wagner, H. Simbabwe. *Rohst. Länderber.* **1993**, *37*, 1–224.

117. Steiner, L. Angola. *Rohst. Länderber.* **1992**, *35*, 1–135.

118. Green, T. *The World of Diamonds*; Willian Morrow and Co.: New York, NY, USA, 1981; p. 300.

119. Erlich, E.I.; Hausel, W.D. *Diamond Deposits: Origin, Exploration, and History of Discovery*; Society for Mining Metallurgy: Littleton, CO, USA, 2008; p. 374.

120. Porter GeoConsultancy. Namdeb, Sperrgebiet, SW Namibian On-shore and Marine Placer Diamonds. 2004. Available online: http://www.portergeo.com.au/database/mineinfo.asp?mineid=mn945 (accessed on 15 October 2018).

121. Mcdonald, I.; Boyce, A.J.; Butler, I.B.; Herrington, R.T.; Polya, D.A. (Eds.) *Mineral Deposits and Earth Evolution*; Geological Society: London, UK, 2005; Volume 248, pp. 213–245.

122. Helmore, R. Diamond mining in Angola. *Mineral. Mag.* **1984**, *7*, 530–537.

123. Bamford, M.K. Cenozoic macroplants. *Oxf. Monogr. Geol. Geophys.* **2000**, *40*, 351–356.

124. Jacob, J.; Ward, J.D.; Bluck, B.J.; Scholz, R.A.; Frimmel, H.E. Some observations on diamondiferous bedrock gully trap sites on Late Cenozoic, marine-cut platforms of the Sperrgebiet, Namibia. *Ore Geol. Rev.* **2006**, *28*, 493–506. [CrossRef]

125. Patyk-Kara, N.G. Placers in the system of sedimentogenesis. *Lithol. Miner. Resour.* **2002**, *37*, 429–441. [CrossRef]

126. Li, X.F.; Chen, Z.Y.; Wang, Y.A.; Zhu, H.P. The primarily study on the genesis of rocked quartz in Donghai, Jiangsu Province: Evidence from fluids inclusions and the Si, O isotope data. *Acta Petrol. Sin.* **2006**, *22*, 2018–2028, (In Chinese with English abstract).

127. Li, X.F.; Watanabe, Y.; Wang, C.; Hirano, H.; Zhang, Y. The age of the Donghai rock crystals (clear quartz), eastern China: Constraint from biotite Ar–Ar geochronology. *Bull. Geol. Survey Jpn.* **2007**, *58*, 1–6. [CrossRef]

128. Anderson, K.B.; LePage, B.A. Analysis of fossil resins from Axel Heiberg Island, Canadian Arctic. In *Amber, Resinite and Fossil Resins*; Anderson, K.G., Crelling, J.C., Eds.; American Chemical Society Symposium Series; American Chemical Society: Washington, DC, USA, 1995; Volume 617, pp. 170–192.

129. Paškevicius, J. *The Geology of the Baltic Republics*; Geological Survey and University of Lithuania: Vilnius, Lithuania, 1997; p. 387.

130. Kharin, G.; Emelyanov, E.M.; Zagorodnich, A.V. Paleogene mineral resources of the SE Baltic Sea and Sambian Peninsula. *Z. Angew. Geol.* **2004**, *2*, 64–71.

131. Kosmoswska-Cevanowicz, B. Quaternary amber bearing deposits on the Polish Coast. *Z. Angew. Geol.* **2004**, *2*, 73–84.

132. Nirgi, T.; Rosentau, A.; Ots, M.; Vahur, S.; Kriiska, A. A Buried amber finds in the coastal deposits of Saaremaa Island, eastern Baltic Sea—Their sedimentary environment and possible use by Bronze Age islanders. *Boreas* **2017**, *46*, 725–736. [CrossRef]

133. Tyson, R. *Sedimentary Organic Matter: Organic Facies and Palynofacies*; Springer: Dordrecht, The Netherlands, 2012; p. 615.

134. McClenaghan, M.B.; Kjarsgaard, B.A. Indicator mineral and surficial geochemical exploration methods for kimberlite in glaciated terrain: Examples from Canada. *Geol. Assoc. Can.* **2007**, *4*, 983–1006.

135. Hutchison, M.T.; Frei, D. Diamondiferous kimberlite from Garnet Lake, West Greenland II: Diamonds and the mantle sample. In Proceedings of the 9th International Kimberlite Conference, Frankfurt, Germany, 10–15 August 2008. Extended Abstract No. 9IKCA-00182.

Article

Gem-Quality Green Cr-Bearing Andradite (var. Demantoid) from Dobšiná, Slovakia

Ján Štubňa [1,*], Peter Bačík [2], Jana Fridrichová [2], Radek Hanus [3], Ľudmila Illášová [1], Stanislava Milovská [4], Radek Škoda [5], Tomáš Vaculovič [6] and Slavomír Čerňanský [7]

1   Gemmological Institute, Faculty of Natural Sciences, Constantine the Philosopher University in Nitra, Nábrežie mládeže 91, 949 74 Nitra, Slovakia; lillasova@ukf.sk

2   Department of Mineralogy and Petrology, Faculty of Natural Sciences, Comenius University in Bratislava, Ilkovičova 6, Mlynská dolina, 842 15 Bratislava, Slovakia; peter.bacik@uniba.sk (P.B.); fridrichova.jana2@gmail.com (J.F.)

3   Gemological Laboratory of e-gems.cz, 110 00 Prague, Czech Republic; kakt@centrum.cz

4   Earth Science Institute of the Slovak Academy of Sciences, Ďumbierska 1, 974 01 Banská Bystrica, Slovakia; milovska@savbb.sk

5   Department of Geological Sciences, Masaryk University, Kotlářská 2, 611 37 Brno, Czech Republic; rskoda@sci.muni.cz

6   Central European Institute of Technology, Masaryk University, Kamenice 5, 625 00 Brno, Czech Republic; tomas.vaculovic@ceitec.muni.cz

7   Department of Environmental Ecology, Faculty of Natural Sciences, Comenius University in Bratislava, Ilkovičova 6, Mlynská dolina, 842 15 Bratislava, Slovakia; slavomir.cernansky@uniba.sk

*   Correspondence: janstubna@gmail.com

Received: 11 February 2019; Accepted: 4 March 2019; Published: 8 March 2019

Abstract: Andradite, variety demantoid, is a rare gem mineral. We describe gem-quality garnet crystals from serpentinized harzburgites from Dobšiná, Slovakia which were faceted. Both the andradite samples were transparent, with a vitreous luster and a vivid green color. They were isotropic with refractive indices >1.81. The measured density ranged from 3.82 to 3.84 $g \cdot cm^{-3}$. Andradite var. demantoid appeared red under Chelsea filter observation. Both samples contained fibrous crystalline inclusions with the typical "horsetail" arrangement. The studied garnet had a strong Fe^{3+} dominant andradite composition with 1.72–1.85 apfu Fe^{3+}, Cr^{3+} up to 0.15 apfu, Al^{3+} 0.03 to 0.04 apfu, V^{3+} up to 0.006 apfu substituted for Fe^{3+}, Mn^{2+} up to 0.002 apfu, and Mg up to 0.04 apfu substituted for Ca. Raman spectrum of garnet showed three spectral regions containing relatively strong bands: I—352–371 cm^{-1}, II—816–874 cm^{-1}, and III—493–516 cm^{-1}. The optical absorption spectrum as characterized by an intense band at 438 nm and two broad bands at 587 and 623 nm and last one at 861 nm, which were assigned to Fe^{3+} and Cr^{3+}. Transmission was observed in the ultraviolet spectral region (<390 nm), near the infrared region (700–800 nm), and around 530 nm in the green region of visible light, resulting in the garnet's green color.

Keywords: andradite; demantoid; gemstone; Raman spectroscopy; UV-Vis-NIR spectroscopy; X-ray fluorescence spectroscopy; gem-quality; garnet

---

1. Introduction

Slovakia has only a few occurrences which produce gem-quality minerals. For a long time in history, precious opal from Červenica-Dubník was the only mineral considered as a gemstone [1–7]. Only a few Slovak minerals, including quartz var. smoky quartz [8,9], sphalerite [8,10], pyrite [8], garnet var. almandine [8], hematite [8], and corundum var. sapphire [11–13] are suitable for faceting.

Andradite garnet, $Ca_3Fe_2Si_3O_{12}$, is a rock-forming mineral typically found in metamorphic rocks such as skarns and rodingites. The andradite crystal structure consists of alternating SiO_4 tetrahedra and FeO_6 octahedra, and Ca with eight-fold coordination. Several studies have been carried out on the crystal chemistry and the thermodynamic properties of andradite garnets [14–22]. Green to yellow-green andradite garnets, demantoid variety, represent one of the most appreciated and precious gemstones among the garnet-group minerals, due to their color, brilliance, and rarity [23–25]. Gem-quality andradite-garnet sources occur in Russia [26–29], Italy [30], Iran [31–33], Pakistan [25,34–37], Namibia [38–41], Madagascar [42–47], Afganistan [24], Switzerland [24], Canada [48], Mexico [49], and China [50].

Our goal is a mineralogical and crystal-chemical investigation of andradite var. demantoid from a serpentinite body in Dobšiná, Slovakia using various analytical methods.

Geological Setting

The serpentinized harzburgites from Dobšiná are part of the Jurassic mélange complex (Meliata Unit). This inner Western-Carpathian unit forms small nappes over the Central Western Carpathians [51]. The Meliata Ocean formed after the continental break up in the Anisian [52], probably as a back-arc basin, due to northward subduction of the Paleotethys oceanic crust beneath the Eurasian continent [53]. The back-arc basin (BAB) basalts and mid-ocean ridge basalts (MORB) were dominant in the Meliatic oceanic rift [54,55]. The association of eupelagic and deepwater pelagic sediments, basalts, dolerite dikes, gabbros, and serpentinized mantle peridotites found in the Meliata Unit clearly indicates an oceanic suite [56,57]. The geochemistry of the basalts reveals a normal mid-ocean ridge basalts (N-MORB) [(La/Sm) n = 0.6–0.9] composition [54,55]. Traces of obducted oceanic crust (blocks of basalts alternating with radiolarite layers) were also identified from deepwater sedimentary breccias with a radiolaritic matrix [58]. The Neotethyan Meliata Oceanic back-arc basin closed in the Late Jurassic, according to ^{40}Ar–^{39}Ar ages of white mica from the blueschists and U/Pb LA-ICP-MS age of metamorphic-metasomatic perovskite [59,60]. The Meliatic subduction–accretionary wedge [61] was incorporated into the Cretaceous orogenic wedge of the Central Western Carpathians [62] between 130 and 50 Ma dated by ^{40}Ar–^{39}Ar method on white micas [63] in the form of mostly small nappe-sheeted bodies, thrust over the Gemeric and Veporic units of the Central Western Carpathians [64,65]. The zircon (U-Th)/He data recorded three evolutionary stages: (i) cooling through the ~180 °C isotherm at 130–120 Ma related to the startling collapse of the accretionary wedge, following exhumation of the high-pressure slices in the Meliatic accretionary wedge; (ii) postponed exhumation and cooling of some fragments through the ~180 °C isotherm from 115 to 95 Ma due to ongoing collapse of this wedge; and (iii) cooling from 80 to 65 Ma, postdating the thrusting (~100–80 Ma) during the Late Cretaceous compression related to formation of the Central Western Carpathians orogenic wedge [66].

The Dobšiná quarry (48°49.622′ N, 20°21.977′ E) (Figure 1) is in a tectonic fragment of the Meliata tectonic unit overlying the Gemeric tectonic unit. The locality is famous for the occurrence of yellow-green Cr-rich andradite (demantoid) in serpentinite veins [67–69]. Andradite-bearing serpentinites are part of the Late Jurassic (to Early Cretaceous) serpentinite–blueschist mélange complex. The mélange complex contains talc–phengite–glaucophane schists, marbles, metaconglomerates, blueschists of magmatic and sedimentary origin, and serpentinites. All rocks form decimeter- to 100 m-sized fragments within the Late Jurassic very low-temperature greenschist-facies metasediments [70].

Partly-preserved magmatic structures of harzburgite consist of olivine (~70 vol. %), orthopyroxene (~25 vol. %), accessory spinel (~4 vol. %), and rare clinopyroxene (~1 vol. %). Magmatically corroded and partly dissolved porphyric orthopyroxene (±clinopyroxene) is enclosed in olivine matrix with spinel. Both pyroxenes show exsolution lamellae, of clinopyroxene in orthopyroxene, or orthopyroxene in clinopyroxene, respectively. Orthopyroxene often forms sigmoidally rotated porphyroclasts, including clinopyroxene exsolution lamellae. The serpentinized harzburgite contains serpentinite group minerals (antigorite replaced by chrysotile and lizardite), Cr-Fe spinel, perovskite, pyrophanite, Ti andradite, magnetite, and hematite, and rarely carbonates with talc, quartz, and characteristic relics of olivine, orthopyroxene, clinopyroxene, and spinel [70,71].

Figure 1. Locations of andradite var. demantoid source (small point—capital town of region) (CZ—Czech Republic, AT—Austria, HU—Hungary, UA—Ukraine, PL—Poland).

2. Materials and Methods

2.1. Sample Description

Two pieces of transparent andradite var. demantoid were separated from the host rock and faceted into round brilliant cuts with dimensions of 2.80 × 2.80 × 1.64 mm (0.11 carat) and 1.71 × 1.74 × 1.14 mm (0.025 carat) dimensions (Figure 2).

Figure 2. Andradite var. demantoid from Dobšiná, Slovakia (0.11 and 0.025 carat). Photo by J. Štubňa.

2.2. Analytical Methods

The faceted samples were examined by standard gemmological methods at the Gemmological Institute, Faculty of Natural Sciences, Constantine the Philosopher University in Nitra, to determine their optical properties (optical character and refractive indices), hydrostatic specific gravity, ultraviolet fluorescence, and microscopic features. The refractive indices were measured with a refractometer Kruss and Presidium Refractive Index Meter II. A Kern ABT 120-5DM with extension KERN ABT A01 hydrostatic balance was used to determine the density of the gemstones. The ultraviolet fluorescence was investigated with a short (254 nm) and long (366 nm) wavelength ultraviolet lamp. We used a Chelsea filter.

The chemical composition of andradite garnet was measured using a Cameca SX100 electron microprobe operated in wavelength-dispersion mode at the Masaryk University, Brno, Czech Republic, under the following conditions: accelerating voltage 15 kV, beam current 20 nA, and beam diameter 5 μm. The samples were analyzed with the following standards: fluorapatite (PKα), sanidine (SiKα, AlKα, KKα), titanite (TiKα), chromite (CrKα), vanadinite (VKα), YAG (YLα), almandine (FeKα), spessartine (MnKα), forsterite (Mg Kα), gahnite (ZnKα), wollastonite (CaKα), albite (NaKα), and topaz (FKα). Lower detection limits of the measured elements varied between 200–400 ppm except for Na (530–570 ppm), Fe (670–730 ppm), Zn (880–900 ppm), and Y (470–500 ppm). The measured content of Y, Mn, Zn, Na, and K was always below the detection limit. Analytical times were 15 to 40 s during measurement depending on the expected concentration of the element in the mineral phase. Major elements were measured using shorter times, whereas longer times were applied for elements with a low concentration. The andradite chemical formula was calculated based on 12 anions assuming all Fe as ferric.

Instrumentation for laser ablation inductively coupled plasma mass spectrometry (LA-ICP-MS) at the Department of Chemistry, Masaryk University, Brno, consists of a UP 213 laser ablation system (New Wave Research, Inc., Fremont, CA, USA) and Agilent 7500 CE ICP-MS spectrometer (Agilent Technologies, Santa Clara, CA, USA). A commercial Q-switched Nd-YAG laser ablation device worked at the 5th harmonic frequency corresponding to 213 nm wavelength. The ablation device was equipped with a programmable XYZ-stage to move the sample along a programmed trajectory during ablation. Target visual inspection and photographic documentation were achieved by the built-in microscope/CCD-camera system. A sample was enclosed in the SuperCell and was ablated by the laser beam, which was focused onto the sample surface through a quartz window. The ablation cell was flushed with a helium carrier gas which transported the laser-induced aerosol to the inductively coupled plasma (1 L·min^{-1}). Sample argon gas flow was admixed with the helium carrier gas flow after the laser ablation cell to 1.6 L·min^{-1} total gas flow. NIST SRM 610 silicate glass reference material was used to optimize the gas flow rates, the sampling depth, and MS electrostatic lens voltage in LA-ICP-MS conditions. This provided maximum signal-to-noise ratio and minimum oxide formation (ThO$^+$/Th$^+$ count ratio 0.2%, U$^+$/Th$^+$ counts ratio 1.1%). Laser ablation required a 100 μm laser spot diameter, 8 J·cm^{-2} laser fluence, and a 20 Hz repetition rate. The fixed sample position during laser ablation enabled 60 s hole-drilling duration for each spot. All element measurements were normalized on ^{28}Si in the investigated andradite. Analyses were gained from 10 spots in a line across the crystal.

Raman analyses were performed on the LabRAM-HR Evolution (Horiba Jobin-Yvon, Kyoto, Japan) spectrometer system with a Peltier cooled CCD detector and Olympus BX-41 microscope (Department of Geological Sciences, Masaryk University). Raman spectra excited by blue diode laser (473 nm) in the range of 50–4000 cm^{-1} were collected from each stone using 50× objectives. The acquisition time of 10 s per frame and 2 accumulations were used to improve the signal-to-noise ratio.

Optical absorption spectra of cut stones in the region (400–750 nm) were measured with the GL Gem SpectrometerTM at room temperature in the Gemmological Institute of Constantine the Philosopher University, Nitra. Both Raman and UV/Vis/NIR spectra were processed in Seasolve PeakFit 4.1.12 software. Raman and absorption bands were fitted by Lorentz function with the automatic background correction and Savitsky-Golay smoothing.

3. Results

Both the andradite samples were transparent, with a vitreous luster and a yellow-green color, which defines the andradite variety. They were all isotropic with refractive indices >1.81. The measured density ranged from 3.82 to 3.84 g·cm^{-3}, and the gemstones were inert in shortwave and longwave ultraviolet radiation. Andradite var. demantoid appeared red under Chelsea filter observation, which is typical for Cr-bearing gemstones. Both samples contained fibrous crystalline inclusions with the typical "horsetail" arrangement (Figure 3).

Figure 3. Fibrous crystalline inclusions of serpentine group minerals in andradite var. demantoid from Dobšiná, Slovakia. Photomicrograph by J. Štubňa; magnified 50×.

The studied garnet had a strongly Fe^{3+} dominant andradite composition with 1.72–1.85 apfu Fe^{3+} (Table 1 and Figure 4). Only Cr^{3+} (up to 0.15 apfu), Al^{3+} (0.03 to 0.04 apfu), and Mg (up to 0.04 apfu) substituted for Fe^{3+}. From the trace elements, Cr was the most abundant, along with moderate to minor concentrations (in relative decreasing order of abundance) of V, Ti, and Mn. The other trace elements measured (Sc, Ge, As, Y, and Gd) were near or below their respective detection limit (Table 2). There was no specific chemical zoning of crystal observed in major or trace elements.

Table 1. Electron microprobe analyses of demantoids, first provided in wt % of elements and then recalculated on the basis of 12 anions.

	1	2	3	4	5	6	7	8
P_2O_5	0.02	0.02	0.02	0.02	0.04	0.00	0.02	0.02
SiO_2	35.21	35.41	35.38	35.33	35.47	35.43	35.74	35.72
TiO_2	0.05	0.05	0.04	0.03	0.00	0.03	0.08	0.04
Al_2O_3	0.33	0.34	0.31	0.32	0.40	0.34	0.37	0.32
V_2O_3	0.09	0.07	0.07	0.08	0.07	0.09	0.06	0.05
Cr_2O_3	1.24	1.50	2.31	2.14	1.54	1.34	0.97	0.63
Fe_2O_3	29.15	29.15	27.69	28.34	28.64	28.82	29.16	29.65
MgO	0.20	0.23	0.20	0.17	0.34	0.23	0.20	0.21
CaO	33.88	33.80	33.74	33.72	33.85	33.85	33.74	33.65
F	0.08	0.11	0.09	0.11	0.09	0.08	0.09	0.10
O=F	−0.04	−0.05	−0.04	−0.06	−0.04	−0.04	−0.05	−0.05
Total	100.19	100.63	99.80	100.20	100.40	100.16	100.38	100.34
P^{5+}	0.002	0.001	0.001	0.001	0.003	0.000	0.001	0.001
Si^{4+}	2.975	2.977	2.993	2.981	2.984	2.990	3.003	3.004
F^-	0.021	0.028	0.023	0.030	0.023	0.022	0.024	0.026
Σ	2.997	3.006	3.018	3.012	3.010	3.012	3.029	3.031
Ti^{4+}	0.003	0.003	0.003	0.002	0.000	0.002	0.005	0.003
Al^{3+}	0.033	0.034	0.031	0.032	0.040	0.034	0.037	0.032
V^{3+}	0.006	0.005	0.005	0.006	0.005	0.006	0.004	0.003
Cr^{3+}	0.083	0.100	0.155	0.143	0.102	0.089	0.064	0.042
Fe^{3+}	1.845	1.837	1.757	1.793	1.806	1.823	1.840	1.873
Σ	1.969	1.979	1.950	1.975	1.953	1.954	1.950	1.952
Mg^{2+}	0.025	0.029	0.026	0.022	0.043	0.028	0.025	0.026
Ca^{2+}	3.067	3.044	3.059	3.049	3.052	3.060	3.037	3.032
Σ	3.092	3.073	3.084	3.071	3.095	3.089	3.063	3.058

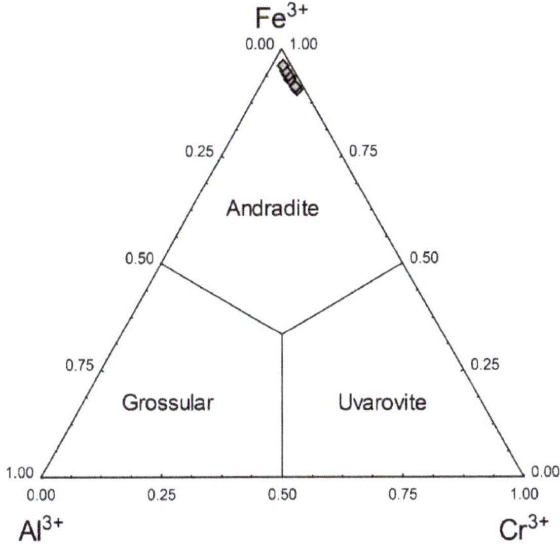

Figure 4. Ternary classification diagram of garnets.

Table 2. The trace-elements contents (in ppm) in demantoid from Dobšiná measured by laser ablation inductively coupled plasma mass spectrometry (LA-ICP-MS). (LOD—limit of detection).

	K	Sc	Ti	V^{51}	Cr^{53}	Mn	Ge^{72}	As^{75}	Y^{89}	Mo^{95}	Gd^{157}
Min	392	16	103	305	3814	103	4	11	2	0	0
Max	787	40	389	572	13331	224	13	22	5	1	2
Average	523	25	244	444	9199	143	9	16	3	1	1
LOD	48.7	2.1	11.4	0.7	9.2	7.8	3.8	9.1	0.1	0.1	0.1

From the calculated crystal-chemical formulae, it is obvious that Z site occupancy by Si is slightly lower than three. It could be filled to three by Al or Fe^{3+}, but their content is sufficient only to accommodate the Y site. Even after adding all possible octahedral cations, the Y-site sum is always slightly below two which, however, could result from the analytical error. However, the small deficiency at the Z site can be justified by the presence of F. The X site is almost uniformly occupied by Ca with a very small admixture of Mg.

Raman spectrum of garnet (Figure 5 and Table 3) showed three spectral regions containing relatively strong bands: I—352–371 cm^{-1}, II—816–874 cm^{-1}, III—493–516 cm^{-1}. The other bands exhibited relatively lower intensities. Measured bands were assigned to various molecular motions (Table 3): translation of SiO$_4$—T(SiO$_4$), rotation of SiO$_4$—R(SiO$_4$), translation of YO$_6$ (Y = Fe^{3+}, Cr^{3+})—T(M), symmetric band ν_2, asymmetric stretching frequency ν_3, asymmetric bending frequency ν_4 of SiO$_4$ group. Our spectrum is consistent with published andradite spectra [72], except the 343 cm^{-1} band, which is not present in published data. Only bands with similar Raman shift are in almandine and pyrope spectra associated with rotation of SiO$_4$ [72].

The optical absorption spectrum (Figure 6 and Table 4) is characterized by small absorption features below 370 nm, an intense band at 438 nm with shoulders at 470 and 488 nm, and two broad bands at 587 and 623 nm, and 861 nm with a shoulder at 964 nm. These spectral features were all assigned to Fe^{3+} and Cr^{3+}. Transmission was observed in the near the infrared region (700–800 nm), and around 530 nm in the green region of visible light resulting in the garnet's green color.

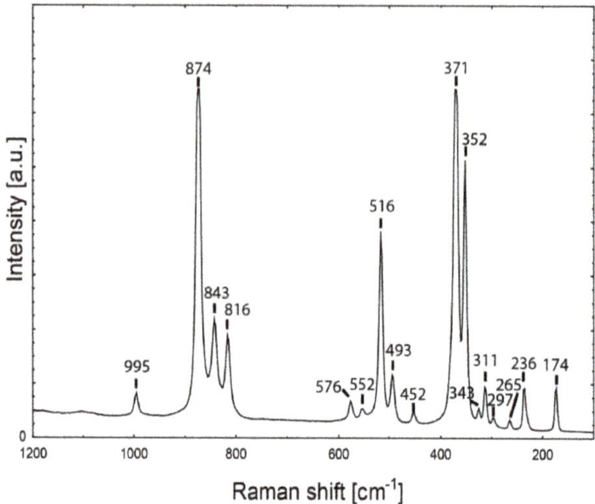

Figure 5. Raman spectrum of the andradite var. demantoid from Dobšiná, Slovakia.

Table 3. Raman shifts (cm^{-1}) of Raman-active bands in the studied sample. Assignment of band modes was made according to Reference [72].

Symmetry	Assignment	Demantoid Dobšiná	Andradite [72]
$T_{2g} + T_{1u}$	T(M)	174	173
$T_{2g} + T_{1u}$	T(M)	236	235
T_{2g}	$T(SiO_4)_{mix}$	265	264
E_g	$T(SiO_4)$	297	296
$T_{2g} + T_{1u}$	$T(SiO_4)$	311	311
?	?	343	-
E_g	ν_2	352	352
A_{1g}	$R(SiO_4)$	371	370
$T_{2g} + T_{1u}$	ν_2	452	452
E_g	ν_2	493	494
A_{1g}	ν_2	516	516
$T_{2g} + T_{1u}$	ν_4	552	553
E_g	ν_4	576	576
$T_{2g} + T_{1u}$	ν_3	816	816
$T_{2g} + T_{1u}$	ν_3	843	842
E_g	ν_3	874	874
$T_{2g} + T_{1u}$	ν_3	995	995

Table 4. The optical spectral bands (nm) in the studied sample. Assignment of bands was made according to Reference [22] and references therein.

Chromophores	Demantoid Dobšiná
Fe^{3+}	438
Fe^{3+}	470
Cr^{3+}	488
Fe^{3+}	587
Cr^{3+}	623
Fe^{3+}	861
Fe^{3+}	964

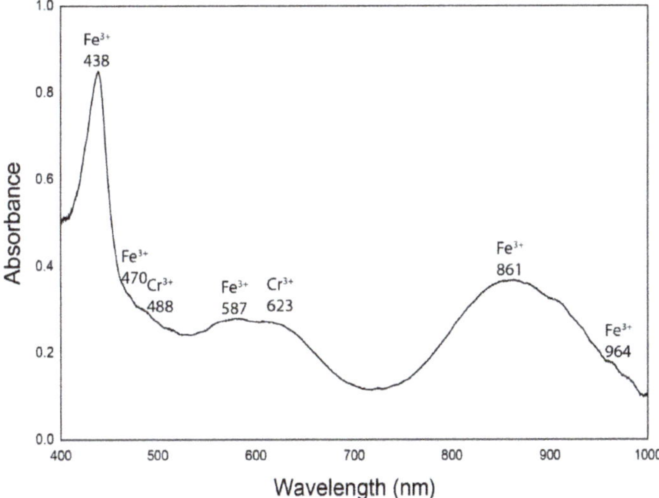

Figure 6. Optical absorption spectrum of the andradite var. demantoid from Dobšiná, Slovakia.

4. Discussion and Conclusions

Chromium-bearing garnets are relatively rare and are associated with specific geological environments and geochemical conditions. In the Western Carpathians, only a few occurrences have been described, including Pezinok–Rybníček [73], Čierna Lehota [74], and Dobšiná [67].

The garnet from Dobšiná occurs as the three following types [67]. The first type is present as relatively evenly dispersed minute grains in the rock. It originated preferentially at the expense of pseudomorphs after rhombic pyroxene. The garnet also forms lenticular or elongated layers 1 to 2 mm thick and several centimeters long. The second-type of garnet occurs as infilling of chrysotile veinlets. The veinlets' thickness is up to 3 mm. The third type of garnet forms monomineral fills in 0.3 to 3 mm-thick fissures. The garnet crystals display usually euhedral habit. However, only the third type of garnets are emerald-green color. All three garnet types occur in pale-colored, pale-green varieties of serpentinite.

Andradite color differences are caused by the variable Cr content. The studied garnet belongs to the third type which is generally enriched in Cr compared to the other two types. In some crystals of andradite, the determined value exceeded 6 wt % Cr_2O_3 [67], but the studied crystals only contain up to 2.3 wt % Cr_2O_3. However, this content is still significantly higher than in other demantoid occurrences around the world [21,30,47]. Moreover, demantoid from Dobšiná is enriched in V (up to 0.09 wt % V_2O_3 from EMPA, 305–572 ppm from LA-ICP-MS) compared to other localities with V usually below the detection limit of EMPA and below 270 ppm determined by LA-ICP-MS [21,30,47]. The As content between 10–22 ppm (although very close to the detection limit, were below detection limit on other localities [21]) and absence of REE (except very low Y), which are usually between 1–90 ppm [21] could also be a fingerprint feature for Dobšiná demantoid.

Raman mode analysis predicts the existence of 25 active Raman modes in all garnet species $3A_{1g}$, $8E_g$, and $14T_{2g}$. The vibrational spectra of garnet clearly possess two different groups of vibrational modes. The first group extends from 816 to 995 cm^{-1} in the Raman spectrum and possesses three strong bands and one weak one. In the garnet structure, these modes are commonly assigned to the ν_1 Si–O stretching motions within SiO_4 groups [72,75–78]. The second region extends from 174 to 576 cm^{-1} and possesses three strong and ten weak bands. The bands at 352, 452, 493, 516, 552, and 576 cm^{-1} are commonly assigned to the Si–O bending motions within SiO_4 groups [72]. The bands at 343 and

371 cm^{-1} can be assigned to rotation of the $[SiO_4]^{4-}$ tetrahedron $R[SiO_4]^{4-}$ [72]. The bands at 174, 236, 265, 297, and 311 cm^{-1} can be assigned to translation modes of the tetrahedra and octahedra [72].

The optical absorption spectra of studied garnet revealed a few absorption and transmission regions. The absorption feature at 438 nm and its shoulders are attributed to octahedral Fe^{3+}, although a further contribution from Cr^{3+} cannot be excluded [22,50,75,79–82], but this was not very significant due to the low Cr content. The weak absorption band at 488 nm is related to Fe^{3+} [30,47]. Two bands at 587 and 623 nm belong to octahedral Fe^{3+} and Cr^{3+}, respectively [22,50,75,79–82]. Presumed contribution of Cr^{3+} to the band at 438 nm can be assigned, according to the Tanabe–Sugano diagram [83–85], to the dominant $^4A_{2g} \rightarrow {}^4T_{2g}$ transition in Cr^{3+}. The second $^4A_{2g} \rightarrow {}^4T_{1g}$ transition in Cr^{3+} was at 623 nm but was overlapping with the Fe^{3+} absorption band at 587 nm. Moreover, a broad band at about 861 nm was due to octahedral Fe^{3+} [22,50,75,79–82].

Author Contributions: Concept of study, J.Š., R.H., J.F., and P.B.; methodology, J.Š., R.H., J.F., P.B, S.M., R.Š., T.V.; formal analysis, S.M., R.Š., T.V., and P.B.; formal investigation, J.Š., Ľ.I., J.F., and P.B.; compilation and writing of the manuscript, J.Š., J.F., and P.B.; review and editing of manuscript, J.Š., Ľ.I., J.F., S.Č., and P.B.

Funding: This present research was supported by the Ministry of Education of Slovak Republic grant agency under the contracts KEGA 062UKF-4/2016, VEGA-1/0079/15, VEGA-1/0499/16 and VEGA-1/0151/19.

Conflicts of Interest: The authors declare no conflict of interest.

References

1. Kaličiak, M.; Ďuďa, R.; Burda, P.; Kaličiaková, E. Structural geological characteristics of the Dubník opal deposits. *Zborník Vychodoslovenského Múzea v Košiciach* **1976**, *17*, 7–22.

2. Rondeau, B.; Fritsch, E.; Guiraud, M.; Renac, C. Opals from Slovakia ("Hungarian" opals): A re-assessment of the conditions of formation. *Eur. J. Mineral.* **2004**, *16*, 789–799. [CrossRef]

3. Gaillou, E.; Fritsch, E.; Aguilar-Reyes, B.; Rondeau, B.; Post, J.; Barreau, A.; Ostroumov, M. Common gem opal: An investigation of micro- to nano-structure. *Am. Mineral.* **2008**, *93*, 1865–1873. [CrossRef]

4. Gaillou, E.; Delaunay, A.; Rondeau, B.; Bouhnik-le-Coz, M.; Fritsch, E.; Cornen, G.; Monnier, C. The geochemistry of gem opals as evidence of their origin. *Ore Geol. Rev.* **2008**, *34*, 113–126. [CrossRef]

5. Kiefert, L.; Karampelas, S. Use of the Raman spectrometer in gemmological laboratories: Review. *Spectrochim. Acta-Part A Mol. Biomol. Spectrosc.* **2011**, *80*, 119–124. [CrossRef] [PubMed]

6. Caucia, F.; Marinoni, L.; Leone, A.; Adamo, I. Investigation on the gemological, physical and compositional properties of some opals from Slovakia ("Hungarian" opals). *Periodico di Mineralogia* **2013**, *82*, 251–261.

7. Caucia, F.; Ghisoli, C.; Marinoni, L.; Bordoni, V. Opal, a beautiful gem between myth and reality. *Neues Jahrbuch für Mineralogie Abhandlungen* **2013**, *190*, 1–9. [CrossRef]

8. Ďuďa, R. Gemmological and genetic classification of precious and trim stones of Slovakia. *Mineral. Slovaca* **1987**, *19*, 353–362. (In Slovakian)

9. Fridrichová, J.; Bačík, P.; Illášová, Ľ.; Kozáková, P.; Škoda, R.; Pulišová, Z.; Fiala, A. Raman and optical spectroscopic investigation of gem-quality smoky quartz crystals. *Vib. Spectrosc.* **2016**, *85*, 71–78. [CrossRef]

10. Fridrichová, J.; Bačík, P.; Malíčková, I.; Illášová, Ľ.; Pulišová, Z. Optical-spectroscopic study of gem sphalerite from Banská Štiavnica and Hodruša-Hámre. *Esemestník* **2017**, *6*, 13–15. (In Slovakian)

11. Uher, P.; Sabol, M.; Konečný, P.; Gregáňová, M.; Táborský, Z.; Puškelová, Ľ. Sapphire from Hajnáčka Cérová Highland, southern Slovakia. *Slovak Geol. Mag.* **1999**, *5*, 273–280.

12. Uher, P.; Sabol, M.; Konečný, P.; Gregáňová, M.; Táborský, Z.; Puškelová, Ľ. Zafír vo výplni vrchnopliocénneho maaru pri Hajnáčke. *Miner. Slovaca* **2001**, *33*, 307–308.

13. Uher, P.; Giuliani, G.; Szakáll, S.; Fallick, A.; Strunga, V.; Vaculovič, T.; Ozdín, D.; Gregáňová, M. Sapphires related to alkali basalts from the Cerová Highlands, Western Carpathians (southern Slovakia): Composition and origin. *Geol. Carpath.* **2012**, *63*, 71–82. [CrossRef]

14. Quareni, S.; De Pieri, R. La struttura dell'andradite. *Atti. e Memorie dell'. Accademia Patavina di Science Lettere ed Arti* **1966**, *78*, 151–170.

15. Novak, G.A.; Gibbs, G.V. The crystal chemistry of the silicate garnets. *Am. Mineral.* **1971**, *56*, 791–825.

16. Hazen, R.M.; Finger, L.W. High-pressure crystal chemistry of andradite and pyrope: Revised procedure for high-pressure diffraction experiments. *Am. Mineral.* **1989**, *74*, 352–359.

17. Lager, G.A.; Armbruster, T.; Rotella, F.J.; Rossman, G.R. OH substitution in garnets: X-ray and neutron diffraction, infrared, and geometric-modeling studies. *Am. Mineral.* **1989**, *74*, 840–851.

18. Armbruster, T.; Geiger, C.A. Andradite crystal chemistry, dynamic X-site disorder and structural strain in silicate garnets. *Eur. J. Mineral.* **1993**, *5*, 59–71. [CrossRef]

19. Armbruster, T.; Birrer, J.; Libowitzky, E.; Beran, A. Crystal chemistry of Ti-bearing andradites. *Eur. J. Mineral.* **1998**, *10*, 907–921. [CrossRef]

20. Pavese, A.; Diella, V.; Pischedda, V.; Merli, M.; Bocchio, R.; Mezouar, M. Pressure-temperature equation of state of andradite and grossular, by high-pressure and -temperature powder diffraction. *Phys. Chem. Miner.* **2001**, *28*, 242–248. [CrossRef]

21. Bocchio, R.; Adamo, I.; Diella, V. Trace-element profiles in gem-quality green andradite from classic localities. *Can. Mineral.* **2010**, *48*, 1205–1216. [CrossRef]

22. Adamo, I.; Gatta, G.D.; Rotiroti, N.; Diella, V.; Pavese, A. Green andradite stones: Gemmological and mineralogical characterisation. *Eur. J. Mineral.* **2011**, *23*, 91–100. [CrossRef]

23. Balčiūnaitė, I.; Kleišmantas, A.; Norkus, E. Chemical composition of rare garnets, their colours and gemmological characteristics. *Chemija* **2015**, *26*, 18–24.

24. O'Donoghue, M. *Gems*, 6th ed.; Butterworth-Heinemann: Oxford, UK, 2006; p. 936.

25. Adamo, I.; Bocchio, R.; Diella, V.; Caucia, F.; Schmetzer, K. Demantoid from Balochistan, Pakistan: Gemmological and Mineralogical Characterization. *J. Gemmol.* **2015**, *34*, 428–433. [CrossRef]

26. Phillips, W.R.; Talantsev, A.S. Russian demantoid, czar of the garnet family. *Gems Gemol.* **1996**, *32*, 100–111. [CrossRef]

27. Krzemnicki, M.S. Diopside needles as inclusions in demantoid garnet from Russia: A Raman microspectrometric study. *Gems Gemol.* **1999**, *35*, 192–195. [CrossRef]

28. Laurs, B.M. A new find of demantoid at a historic side in Kladovka, Russia. *Gems Gemol.* **2003**, *39*, 54–55.

29. Mayerson, W.M. Lab Notes: Large demantoid of exceptional color. *Gems Gemol.* **2006**, *42*, 261–262.

30. Adamo, I.; Bocchio, R.; Diella, V.; Pavese, A.; Vignola, P.; Prosperi, L.; Palanza, V. Demantoid from Val Malenco, Italy: Review and update. *Gems Gemol.* **2009**, *45*, 280–287. [CrossRef]

31. Douman, M.; Dirlam, D. Update on demantoid and cat's-eye from Iran. *Gems Gemol.* **2004**, *40*, 67–68.

32. Du Toi, G.; Mayerson, W.; van Der Bogert, C.; Douman, M.; Befi, R.; Koivula, J.I.; Kiefert, L. Demantoid from Iran. *Gems Gemol.* **2006**, *42*, 131.

33. Karampelas, S.; Gaillou, E.; Fritsch, E.; Douman, M. Les grenats andradites-démantoïde d'Iran: Zonage de couleur et inclusions. *Rev. Gemmol.* **2007**, *160*, 14–20.

34. Fritz, E.A.; Laurs, B.M. Gem News International: Andradite from Balochistan, Pakistan. *Gems Gemol.* **2007**, *43*, 373.

35. Milisenda, C.C.; Henn, U.; Henn, J. Demantoide aus Pakistan. *Gemmol. Z. Dtsch. Gemmol. Ges.* **2001**, *50*, 51–56.

36. Quinn, E.P.; Laurs, B.M. Demantoid from northern Pakistan. *Gems Gemol.* **2005**, *41*, 176–177.

37. Palke, A.C.; Pardieu, V. Gem News International: Demantoid from Baluchistan Province in Pakistan. *Gems Gemol.* **2014**, *50*, 302–303.

38. Lind, T.; Henn, U.; Bank, H. New occurrence of demantoid in Namibia. *Aust. Gemmol.* **1998**, *20*, 75–79.

39. Cairncross, B.; Bahmann, U. The Erongo Mountains, Namibia. *Mineral. Rec.* **2006**, *37*, 361–470.

40. Koller, F.; Niedermayar, G.; Pintér, Z.; Syabó, C. The demantoid garnets of the Green Dragon Mine (Tubussi, Erongo region, Namibia). *Acta Mineral.-Petrogr.* **2012**, *7*, 72.

41. Kaiser, H.; Enzersdorf, M. Demantoid aus der Green Dragon Mine, Erongo Region, Namibia. *Gemmo News* **2012**, *32*, 5.

42. Danet, F. Gem News International: New discovery of demantoid from Ambanja, Madagascar. *Gems Gemol.* **2009**, *45*, 218–219.

43. Rondeau, B.; Mocquet, B.; Lulzac, Y.; Fritsch, E. Les nouveaux grenats démantoïde d'Ambanja, Province d'Antsiranana, Madagascar. *Le Règne Mineral* **2009**, *90*, 41–45.

44. Rondeau, B.; Fritsch, E.; Mocquet, B.; Lulyac, Y. Ambanja (Madagascar) New source of gem demantoid garnet. *InColor* **2009**, *11*, 16–20.

45. Praszkier, T.; Gajowniczek, J. Demantoide aus Antetezambato auf Madagascar. *Miner. Welt* **2010**, *21*, 32–41.

46. Pezzotta, F. Andradite from Antetezambato, North Madagascar. *Mineral. Rec.* **2010**, *41*, 209–229.

47. Pezzotta, F.; Adamo, I.; Diella, V. Demantoid and topazolite from Antetezambato, northern Madagascar: Review and new data. *Gems Gemol.* **2011**, *47*, 2–14. [CrossRef]

48. Wilson, B.S. Coloured gemstones from Canada. *Rocks Miner.* **2009**, *85*, 24–43. [CrossRef]

49. Ostrooumov, M. Mexican demantoid from new deposits. *Gems Gemol.* **2015**, *51*, 450–452.

50. Guobin, L.; Xu, K.; Lin, Z. On the Genesis of Demantoid from Xinjiang, China. *Chin. J. Geochem.* **1986**, *5*, 381–390. [CrossRef]

51. Mock, R.; Sýkora, M.; Aubrecht, R.; Ožvoldová, L.; Kronome, B.; Reichwalder, P.; Jablonský, J. Petrology and stratigraphy of the Meliaticum near the Meliata and Jaklovce Villages, Slovakia. *Slovak Geol. Mag.* **1998**, *4*, 223–260.

52. Kozur, H. The Evolution of the Meliata-Hallstatt ocean and its significance for the early evolution of the Eastern Alps and Western Carpathians. *Palaeogeogr. Palaeoclimatol. Palaeoecol.* **1991**, *87*, 109–135. [CrossRef]

53. Stampfli, G.M. The Intra-Alpine terrain: A Paleotethyan remnant in the Alpine Variscides. *Eclogae Geol. Helv.* **1996**, *89*, 13–42.

54. Ivan, P. Relics of the Meliata Ocean crust: Geodynamic implications of mineralogical, petrological and geochemical proxies. *Geol. Carpath.* **2002**, *53*, 245–256.

55. Faryad, S.W.; Spišiak, J.; Horváth, P.; Hovorka, D.; Dianiška, I.; Józsa, S. Petrological and geochemical features of the Meliata mafic rocks from the sutured Triassic Oceanic Basin, Western Carpathians. *Ofioliti* **2005**, *30*, 27–35.

56. Kozur, H.; Mock, R. New paleogeographic and tectonic interpretations in the Slovakian Carpathians and their implications for correlations with the Eastern Alps and other parts of the Western Tethys, Part II: Inner Western Carpathians. *Miner. Slovaca* **1997**, *29*, 164–209.

57. Kozur, H.; Mock, R. Erster Nachweis von Jura der Meliata-Einheit der südlichen Westkarpaten. *Geologisch-Paläontologische Mitteilungen Innsbruck* **1985**, *13*, 223–238.

58. Putiš, M.; Radvanec, M.; Hain, M.; Koller, F.; Koppa, M.; Snárska, B. 3-D analysis of perovskite in serpentinite (Dobšiná quarry) by X-ray micro-tomography. In *Proceedings of the Petros Symposium*; Ondrejka, M., Šarinová, K., Eds.; Univerzita Komenského: Bratislava, Slovakia, 2011; pp. 33–37.

59. Dallmeyer, R.D.; Neubauer, F.; Handler, R.; Fritz, H.; Müller, W.; Pana, D.; Putiš, M. Tectonothermal evolution of the internal Alps and Carpathians: ^{40}Ar/^{39}Ar mineral and whole rock data. *Eclogae Geol. Helv.* **1996**, *89*, 203–277.

60. Putiš, M.; Yang, Y.-H.; Koppa, M.; Dyda, M.; Šmál, P. U/Pb LA-ICP-MS age of metamorphic-metasomatic perovskite from serpentinized harzburgite in the Meliata Unit at Dobšiná, Slovakia: Time constraint of fluid-rock interaction in an accretionary wedge. *Acta Geol. Slovaca* **2015**, *7*, 63–71.

61. Faryad, S.W.; Henjes-Kunst, F. Petrological and K-Ar and 40Ar/39Ar age constraints for the tectonothermal evolution of the high-pressure Meliata unit, Western Carpathians (Slovakia). *Tectonophysics* **1997**, *280*, 141–156. [CrossRef]

62. Plašienka, D. Cretaceous tectonochronology of the central western Carpathians, Slovakia. *Geol. Carpath.* **1997**, *48*, 99–111.

63. Putiš, M.; Frank, W.; Plašienka, D.; Siman, P.; Sulák, M.; Biroň, A. Progradation of the Alpidic Central Western Carpathians orogenic wedge related to two subductions: Constrained by ^{40}Ar/^{39}Ar ages of white micas. *Geodin. Acta* **2009**, *22*, 31–56. [CrossRef]

64. Dal Piaz, G.; Martin, S.; Villa, I.M.; Gosso, G.; Marschalko, R. Late Jurassic blueschist facies pebbles from the Western Carpathians orogenic wedge and paleostructural implications for Western Tethys evolution. *Tectonics* **1995**, *14*, 874–885. [CrossRef]

65. Putiš, M.; Gawlick, H.J.; Frisch, W.; Sulák, M. Cretaceous transformation from passive to active continental margin in the Western Carpathians as indicated by the sedimentary record in the Infratatric Unit. *Int. J. Earth Sci.* **2008**, *97*, 799–819. [CrossRef]

66. Putiš, M.; Danišík, M.; Ružička, P.; Schmiedt, I. Constraining exhumation pathway in an accretionary wedge by (U-Th)/He thermochronology—Case study on Meliatic nappes in the Western Carpathians. *J. Geodyn.* **2014**, *81*, 80–90. [CrossRef]

67. Fediuková, E.; Hovorka, D.; Greguš, J. Compositional zoning of andradite from serpentinite at Dobšiná (West Carpathias). *Věst. Ústr. Úst. Geol.* **1976**, *51*, 339–345.

68. Hovorka, D.; Jaroš, J.; Kratochvíl, M.; Mock, R. The Mesozoic ophiolites of the Western Carpathians. *Krystalinikum* **1984**, *17*, 143–157.

69. Hovorka, D.; Ivan, P.; Jaroš, J.; Kratochvíl, M.; Reichwalder, P.; Rojkovič, I.; Spišiak, J.; Turanová, L. *Ultramafic rocks of the Western Carpathians, Czechoslovakia*; Geological Institute of Dionýz Štúr Publishers: Bratislava, Czechoslovakia, 1985; p. 258.

70. Putiš, M.; Koppa, M.; Snárska, B.; Koller, F.; Uher, P. The blueschist-associated perovskite-andradite-bearing serpentinized harzburgite from Dobšiná (the Meliata Unit), Slovakia. *J. Geosci.* **2012**, *57*, 221–240. [CrossRef]

71. Putiš, M.; Yang, Y.-H.; Vaculovič, T.; Koppa, M.; Li, X.-H.; Uher, P. Perovskite, reaction product of a harzburgite with Jurassic-Cretaceous accretionary wedge fluids (Western Carpathians, Slovakia): Evidence from the whole-rock and mineral trace element data. *Geol. Carpath.* **2016**, *67*, 133–146. [CrossRef]

72. Hofmeister, A.M.; Chopelas, A. Vibrational spectroscopy of end-member silicate garnets. *Phys. Chem. Miner.* **1991**, *17*, 503–526. [CrossRef]

73. Uher, P.; Kováčik, M.; Kubiš, M.; Shtukenberg, A.; Ozdín, D. Metamorphic vanadian-chromian silicate mineralization in carbon-rich amphibole schists from the Malé Karpaty Mountains, Western Carpathians, Slovakia. *Am. Mineral.* **2008**, *93*, 63–73. [CrossRef]

74. Bačík, P.; Uher, P.; Kozakova, P.; Števko, M.; Ozdín, D.; Vaculovič, T. Vanadian and chromian garnet- and epidote-supergroup minerals in metamorphosed Paleozoic black shales from Cierna Lehota, Strazovske vrchy Mountains, Slovakia: Crystal chemistry and evolution. *Mineral. Mag.* **2018**, *82*, 889–911. [CrossRef]

75. Peters, T. A water-bearing andradite from the Totalp serpentine (Davos, Switzerland). *Am. Mineral.* **1965**, *50*, 1482–1486.

76. Passaglia, E.; Rinaldi, R. Katoite, a new member of the $Ca_3Al_2(SiO_4)_3$-$Ca_3Al_2(OH)_{12}$ series and a new nomenclature for the hydrogrossular group of minerals. *Bulletin de la Société Française de Minéralogie et de Cristallographie* **1984**, *107*, 605–618. [CrossRef]

77. Galuskina, I.O.; Galuskin, E.V. Garnets of the hydrogrossular-"hydroandradite"-"hydroschorlomite" series. *Spec. Pap. Mineral. Soc. Pol.* **2003**, *22*, 54–57.

78. Galuskin, E. *Minerals of the Vesuvianite Group from the Achtarandite Rocks (Wiluy River, Yakutia)*; University of Silesia Publishing House: Katowice, Poland, 2005; p. 192.

79. Moore, R.K.; White, W.B. Electronic spectra of transition metal ions in silicate garnets. *Can. Mineral.* **1972**, *11*, 791–811.

80. Manning, P.G. Optical absorption spectra of Fe^{3+} in octahedral and tetrahedral sites in natural garnets. *Can. Mineral.* **1972**, *11*, 826–839.

81. Amthauer, G. Crystal chemistry and colour of chromium bearing garnets. *Neues Jahrbuch für Mineralogie Abhandlungen* **1976**, *126*, 158–186.

82. Stockton, C.M.; Manson, D.V. Gem andradite garnets. *Gems Gemol.* **1983**, *19*, 202–208. [CrossRef]

83. Tanabe, Y.; Sugano, S. On the absorption spectra of complex ions II. *J. Phys. Soc. Jpn.* **1954**, *9*, 766–779. [CrossRef]

84. Tanabe, Y.; Sugano, S. On the absorption spectra of complex ions I. *J. Phys. Soc. Jpn.* **1954**, *5*, 753–766. [CrossRef]

85. Tanabe, Y.; Sugano, S. On the absorption spectra of complex ions III. *J. Phys. Soc. Jpn.* **1956**, *11*, 864–877. [CrossRef]

Article

A Review of the Classification of Opal with Reference to Recent New Localities

Neville J. Curtis [1,2], Jason R. Gascooke [2], Martin R. Johnston [2] and Allan Pring [1,2,*]

[1] South Australian Museum, North Terrace, Adelaide, SA 5000, Australia; neville.curtis@flinders.edu.au
[2] College of Science and Engineering, Flinders University, Sturt Rd, Bedford Park, SA 5042, Australia;
jason.gascooke@flinders.edu.au (J.R.G.); martin.johnston@flinders.edu.au (M.R.J.)
* Correspondence: Allan.Pring@flinders.edu.au; Tel.: +61-8-8201-5570

Received: 9 April 2019; Accepted: 13 May 2019; Published: 15 May 2019

Abstract: Our examination of over 230 worldwide opal samples shows that X-ray diffraction (XRD) remains the best primary method for delineation and classification of opal-A, opal-CT and opal-C, though we found that mid-range infra-red spectroscopy provides an acceptable alternative. Raman, infra-red and nuclear magnetic resonance spectroscopy may also provide additional information to assist in classification and provenance. The corpus of results indicated that the opal-CT group covers a range of structural states and will benefit from further multi-technique analysis. At the one end are the opal-CTs that provide a simple XRD pattern ("simple" opal-CT) that includes Ethiopian play-of-colour samples, which are not opal-A. At the other end of the range are those opal-CTs that give a complex XRD pattern ("complex" opal-CT). The majority of opal-CT samples fall at this end of the range, though some show play-of-colour. Raman spectra provide some correlation. Specimens from new opal finds were examined. Those from Ethiopia, Kazakhstan, Madagascar, Peru, Tanzania and Turkey all proved to be opal-CT. Of the three specimens examined from Indonesian localities, one proved to be opal-A, while a second sample and the play-of-colour opal from West Java was a "simple" Opal-CT. Evidence for two transitional types having characteristics of opal-A and opal-CT, and "simple" opal-CT and opal-C are presented.

Keywords: opal; hyalite; silica; X-ray diffraction; Raman; Infrared; ^{29}Si nuclear magnetic resonance; SEM; provenance

---

1. Introduction

Opal is a generic term for a group of amorphous and paracrystalline silica species, containing up to 20% "water" as molecular H_2O or silanol (R_3SiOH) or both [1,2]. Opals, both common opal and precious opal, have been the subject of considerable study over the last five decades or so [3]. Australia has long been the dominant sources of precious opal, exhibiting play-of-colour (POC) [4], but recently, opals from new fields, such as Ethiopia [5], Madagascar [6], Indonesia [7], Tanzania [8] and Turkey [9], have appeared on the market. This influx of new material on the market makes a re-examination of the classification opal timely.

Conventionally, opals are classified into three types [10]: opal-A (further divided into opal-AG (opal) and opal-AN (hyalite)), opal-CT and opal-C [11,12]. For nearly 50 years, the classification of opals proposed by Jones and Segnit [1], based on X-ray powder diffraction (XRD), has been widely adopted, e.g., [13–15]. The pertinent features of the XRD classification are:

- Opal-A (both AG and AN): broad absorption only, centred on 4.0 Å.
- Opal-CT: two prominent peaks at ~4.1 Å and 2.5 Å with a further peak showing variable degrees of separation at ~4.27 Å.
- Opal-C: prominent peaks at 4.04 Å and 2.5 Å.

The notation refers to "amorphous" (A) or is based on the similarity of the XRD reflection positions for α-cristobalite (C) and α-tridymite (T) [16,17]. The exact nature of atomic structures of opal-CT and opal-C remain largely unresolved. Opal-CT appears to be not just a fine-grained intergrowth of layers of cristobalite and tridymite, but a paracrystalline form of silica that has some structural characteristics of these minerals. Separate from the structural features revealed by XRD, some opals exhibit a play-of-colour. This is a textural feature related to the regular packing of silica spheres which are of a size to diffract visible light [4,10,12,15,18]. Both opal-A and opal-CT can show play-of-colour, but the atomic structure of the silica in the spheres is clearly different.

Other techniques such as Raman spectroscopy [13,19–21], [29]Si nuclear magnetic resonance (NMR) [22–27], near infra-red spectroscopy [28–30] and neutron scattering [31] have also been used to try to unravel the complex structural relationships between the types of opal. Trace chemical analysis both of opal and associated minerals [7,15,18,32–36] may also provide evidence of provenance [37].

In this paper, we explore whether the XRD classification system is still valid for the newer-sourced material, particularly the precious or play-of-colour opals and that the terms opal-AG, opal-AN, opal-CT and opal-C represent homogenous structural groups. Transitions from opal-A to opal-CT to opals-C to quartz have been reported [38–43] and we take the opportunity to try to identify specimens showing evidence of intermediate forms. In addition to XRD, we will present results of techniques that focus on Si–O bonding to see if they provide evidence for homogeneity of the opal types. Techniques comprise Raman spectroscopy, far and medium infra-red (IR) spectroscopy, and single-pulse ^{29}Si NMR. The key to this approach is the large suite of samples with all opals measured under similar conditions so that trends or differences readily come to light. We hope that this suite of opal samples will also be used by other researchers in their investigations.

The form of this paper is to focus on the results of the characterization of some 48 samples from new or unusual localities out of a total sample suite of some 230 samples that we have examined. A primary classification of all samples into groups according to the XRD methodology of Jones and Segnit [1] (opal-A, opal-CT and opal-C) was undertaken. Then selected samples were subjected to further study using Raman spectroscopy, infra-red spectroscopy and [29]Si nuclear magnetic resonance spectroscopy.

2. Materials and Methods

Over 230 opal samples (opal-A, 67 samples; opal-CT, 161 samples; opal-C, 4 samples and 4 samples which appear to be intermediates between forms) were sourced from the South Australian Museum (G prefix), Flinders University (E), the Tate Collection (T) of the University of Adelaide, Museum Victoria (M), the Smithsonian National Museum of Natural History (NMNH) and through recent acquisitions (G NEW, OOC and SO). Where ambiguity exists, such as multiple samples in a single catalogued specimen lot, obvious differences between subsamples are indicated. About 250 individual specimens were analysed. We used well-documented specimens where possible and assumed that locality details were correct. A full list of the specimens examined is given in Supplementary Materials.

X-ray powder diffraction patterns (Bruker D8 Advance machine, Co source K_α = 1.78897 Å) were recorded with a scan speed of 0.0195° per second over the 2θ range 10 to 65°. Samples were ground under acetone before use. At least half of the samples were free from any obvious impurity. Literature [25,44,45] d-spacings of the major lines in the XRD patterns of the crystalline silica polymorphs were as follows: quartz: 4.25 Å and 3.33 Å, moganite: 4.43 Å, 4.38 Å, 3.39 Å (obscuring 3.33 Å), 3.10 Å and 2.86 Å, α-cristobalite: 4.04 Å, 3.12 Å, 2.83 Å and 2.48 Å and α-tridymite: 4.38 Å, 4.14 Å, 3.75 Å, 2.98 Å and 2.51 Å.

The major "impurity" was quartz, and this varied from minute traces to overwhelming amounts. A range of clays and related layer silicates was also noted. Only specimens with no or only trace amounts of impurities in the XRD pattern were included in the study.

Curve fitting for XRD data (d-spacing) gave peak positions, full-width half-maximum (FWHM) and relative proportions. A baseline spline was calculated using data read from the relatively flat portions of the pattern and the Microsoft Excel Solver software was used to calculate the minimised least squares fit (about 1000 data points up to 45° 2θ). Since the major peaks spanned a wide range of 2θ, baselines showed variability (particularly for opal-A and some of the opal-CT samples) and only a generalised curve fitting regime was followed. The literature suggests a mix of Lorentzian and Gaussian types (pseudo-Voigt formalization) [46]. Fits could be obtained using either form or a mixture of forms, though this introduced extra variables but showed no obvious gain, and we adopted a pure Lorentzian peak shape for this initial study. Patterns were not corrected for $K_{\alpha 2}$. Consistent results from the large number of samples analysed for opal-A and opal-CT gave confidence in this approach.

Raman spectra were collected in the 100 to 1500 wavenumber (cm^{-1}) region, using a XplorRA Horiba Scientific Confocal Raman microscope. Spectra were acquired using a 50× objective (numerical aperture 0.6) at an excitation wavelength of 786 nm (27 mW measured at the sample) and spectrometer resolution of 4.5 cm^{-1} FWHM. Typical integrations times for the spectra were 30 s and averaged from 6 (for chip samples) to 60 (for powdered samples). The instrument was calibrated using the 520.7 cm^{-1} line of silicon and spectra were corrected to account for absorptions by the edge filter used to suppress the Rayleigh scattering peak. The spectrum of each sample was confirmed at several locations on the sample. Additional runs were made with laser wavelengths of 640 nm (3.8 mW) and 532 nm (7.3 mW) to confirm the Raman spectra recorded at 786 nm. Raman spectra of reference compounds are found in References [21,47–51]. There is, however, inconsistency in reported values for tridymite [52] with several types of Raman spectra despite having similar XRD patterns. Silanole [53] may also present a complicating Raman band at around 500 cm^{-1}. Several samples showed overwhelming fluorescence or gave a weak and largely featureless spectrum indicating that Raman is not as universal as XRD for characterisation. Because of the relatively low Raman scattering cross-section of opal, small amounts of "impurities" could produce misleadingly large peaks. All opals showed a significant baseline component that was most intense below 500 cm^{-1}. Baseline correction was not undertaken since the observed peaks were broad and the exact form of the baseline is unknown.

Attenuated total reflectance (ATR) infra-red (IR) spectra using a diamond ATR crystal were collected in two wavenumber ranges (100–600 cm^{-1} and 400–4000 cm^{-1}). Spectra in the low wavenumber region were recorded at the Australian Synchrotron's THz-Far Infrared beamline using a Bruker IFS 125/HR spectrometer at a spectral resolution of 4 cm^{-1}. Four-hundred to five-hundred scans of the powdered samples were averaged to generate the final spectrum. For mid-IR, ATR spectra were collected using a Perkin–Elmer Frontier spectrometer and recorded at a 2 cm^{-1} resolution. Overlapping wavenumber regions were examined and found to be comparable. The extended range provides full cover of the Si–O bonding derived peaks. Peaks consistent with water stretching (3000–3500 cm^{-1}) and bending (1600–1650 cm^{-1}) were noted but not further investigated for the current study. A simple intensity correction was applied by assuming the ATR penetration depth was directly proportional to the wavelength. We note that the spectra will give slightly different peak positions and band shapes to those obtained via transmission IR spectra or ATR spectra recorded with different crystals due to the anomalous dispersion (change of refractive index) across an absorption band.

The ^{29}Si solid-state Magic Angle Spinning (MAS) NMR spectra were obtained using a Bruker Avance III 400 MHz spectrometer equipped with either a Bruker 4- or 7-mm probe with rotors spinning at 5 or 2.5 kHz, respectively. All spectra were collected at ambient temperature. Single-pulse (SP) experiments were typically carried out using a 90° pulse, high-power decoupling during acquisition (TPPM or SPINAL-64), followed by a recycle delay of 60 s. Spectra were referenced to 4,4-dimethyl-4-silapentane-1-sulfonic acid (DSS) at 0 ppm. It was noted that the opal-A samples required less sample or time to achieve good ^{29}Si NMR spectra compared to other opal forms. The combination of the insensitivity of ^{29}Si and the requirement for relatively large amounts of powdered sample (100 mg) limited our ability to measure numerous samples, thus only selected samples were used. The Q_4 (RO_4Si), Q_3 (RO_3SiOH) and Q_2 ($RO_2Si(OH)_2$) peaks were centred at around −112, −102 and −93 ppm [22] and overlapped owing to the broadness of the resonances (up to 10 ppm). The presence of a comparatively large Q_3 peak will affect the Q_4 peak position and, for this reason, the spectra were deconvoluted. We have also measured ^1H and ^{29}Si [22] cross-polarisation

(CP) MAS NMR spectra at various contact times allowing analysis of CP dynamics, and we will report these in a further paper (Curtis, Gascooke, Johnston and Pring, to be published). Reported chemical shifts (Q_4) [27,28] were: quartz −107.2 ppm to −107.1 ppm, cristobalite: −108.1 ppm and tridymite: −111.4 ppm and −109.3/−110.7/−114.0 ppm.

3. Results

The results are presented in two parts. The first is a summary of the results in terms of the overall classification of the opal family from the various techniques. The second presents the results for each opal group, focusing on the variation within the opal-CT group.

3.1. Overview

Figure 1 shows an overall appreciation of the XRD patterns, Raman, mid-IR spectra and NMR of the four groups of opal-AG, opal-AN (hyalite), opal-CT and opal-C using a set of exemplar specimens. As can be seen there are clear differences between opal-A, opal-CT and opal-C, though as will be discussed below the situation is more complex than this figure might suggest. Table 1 presents specimen data for the samples we have designated as exemplars (or typical examples) with a fuller list in the Supplementary Materials.

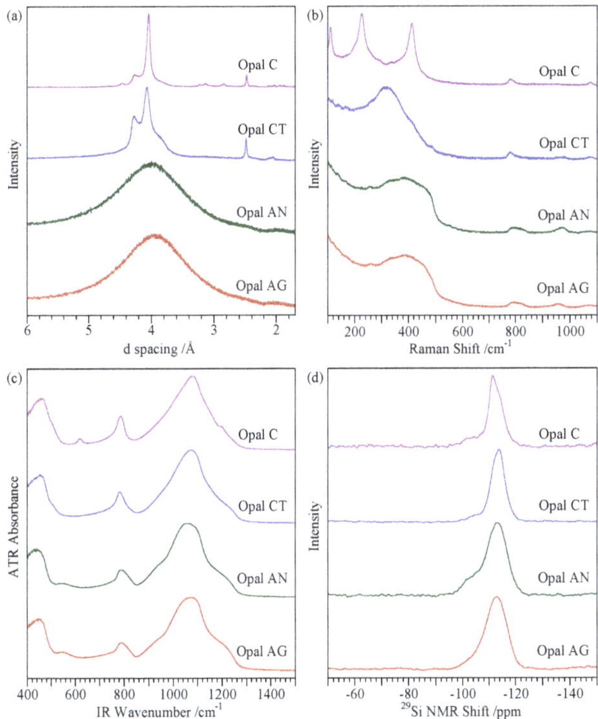

Figure 1. Patterns and spectra of typical samples showing (**a**) XRD patterns, (**b**) Raman spectra, (**c**) attenuated total reflectance mid-IR spectra and (**d**) single pulse ^{29}Si MAS NMR spectra. In ascending order, the samples are: opal-AG (White Cliffs, Australia G13771) (red); opal-AN/hyalite (Valec, Czech Republic G32740) (green); opal-CT (Angaston, Australia, G9942) (blue) and opal-C (Iceland M5081) (purple). Spectra were scaled and offset for comparison.

Table 1. List of exemplar opal samples examined in detail in the study [a].

Country	Location	Sample ID	Appearance	Type
Australia	White Cliffs, NSW	G1401	Translucent no POC	A
Australia	Eurolowie, NSW	G1425	Translucent pale brown glossy	CT
Australia	Iron Monarch, SA	G9620	White glossy opaque	CT
Australia	White Cliffs, NSW	G8608	White opaque POC	A
Australia	Unknown	G9260	White to grey opaque	A
Australia	Two mile Coober Pedy, SA	G9594	Translucent milky glossy minor POC	A
Australia	Four miles S of Angaston, SA	G9942	Translucent white glossy	CT
Australia	Near Murwillumbah, NSW	G9964	Slightly cloudy, clear POC	CT
Australia	Lightning Ridge, NSW	G13769	Black, glassy band in matrix	A
Australia	White Cliffs, NSW	G13771	Bag of samples	A
Australia	Angaston, SA	G24346	Brown opaque	CT
Australia	Springsure, Qld	M8736	Glassy (hyalite like)	A
Australia	Yinnar, Vic	T19006	Glassy grey-brown	CT
Czech Rep	Valec, Bohemia	OOC11	Glassy clear	A
Czech Rep	Valec, Bohemia	G32740	Hyalite outgrowth, colourless	A
Ethiopia	Mezezo	G25374	Deep-brown translucent	CT
Ethiopia	Afar	G32752	Brown glass, some with POC	CT
Ethiopia	Mezezo	NMNH Eth1	Pinkish POC on white	CT
Ethiopia	Yita Ridge, Menz-Gishe	G31892	Nodules with clear orange centres	CT
Ethiopia	Mezezo	NMNH Eth 2	Transparent brown POC	CT
Ethiopia	Wello	NMNH Eth 3	Milky transparent POC	CT
Honduras	Unknown	G1441	Milky transparent some POC	CT
Iceland	Unknown	M5081	Opaque white	C
Indonesia	Cilayang Village, West Java	G34240	Colourless with POC	CT
Indonesia	Mangarrai Prov, Flores	OOC6	Translucent white	CT
Indonesia	Mamuju, West Sulawesi	OOC13	Blue-green matrix of "grape agate"	A
Kazakhstan	Voznesenovka, Martuk	M53407	Orange glass	CT
Kazakhstan	Zelinograd	G32925	Translucent vermilion glassy	CT
Madagascar	Bemi, Befotaka District	G NEW05	Clear yellow	CT
Madagascar	Bemi, Befotaka District	G NEW07	Translucent pale brown	CT
Mexico	La Trinidad Queretaro	G31851	Single piece with opal inclusions	CT
Namibia	Khorixas district	G NEW29	Blue to white opaque	CT
Peru	Acari	G33912	Massive blue	CT
Spain	Mazarron, Murcia	OOC4	Composite with green zones	CT
Tanzania	Kigoma, Region	G NEW19	Pale orange shades glassy	CT
Tanzania	Haneti	G NEW03	Opaque green	CT
Tanzania	Haneti	G NEW04	Opaque green, some glassy zones	CT
Tanzania	Arusha	G34238	Transparent green layer	CT
Turkey	Kutahya	G NEW24	Translucent green and brown	CT
Turkey	Eskisehir	G NEW25	Opaque white with indigo speckles	CT
Turkey	Anatolia	G NEW26	Opaque white transparent green inside	CT
Turkey	Yozgat, Anatolia	G NEW27	Blue-green transparent glass	CT
Turkey	Yozgat, Anatolia	G NEW28	Olive-green transparent glass	CT
USA	Opal Butte Mine, Oregon	G NEW18	Glassy white	CT-C
USA	Manzano Mtns. New Mexico	G NEW30	White opaque mass	CT
USA	Virgin Valley, Nevada	G31852	Milky and translucent zones	CT
USA	Virgin Valley, Nevada	G32263	Translucent brown	CT
USA	Virgin Valley, Nevada	M19717	Opaque glassy POC	CT
USA	Virgin Valley, Nevada	OOC5	White and POC zones	CT

[a] Some samples yielded more than one experimental sample.

The form of the XRD patterns shown in Figure 1a matches those given in the original opal classification by Jones and Segnit [1] published in 1971. We found that for the most part samples taken from the same specimen, though differing in colour or texture, showed similar, but not necessarily identical, XRD patterns. We note that the difference between opal-CT and opal-C may be subtle when faced with a single sample, as assignment to one group or the other can be difficult without detailed comparison.

The major absorption for the Raman spectra for exemplar specimens (Figure 1b) was in the 200–500 cm^{-1} region with isolated smaller peaks up to 1100 cm^{-1}, and the spectra for opal-A, opal-CT and opal-C were distinct. Spectra were consistent between the unground and the powdered samples (which was also used in XRD). Generally, samples were mostly homogenous but, in some cases, extra peaks were observed in the spectra. These were probably non-opal impurities such as inclusions of silicate minerals. The most likely "impurity", quartz, has a characteristic spectrum of a relatively strong, isolated, sharp peak at 459 cm^{-1} and lesser ones at 108 cm^{-1} (sharp) and 227 cm^{-1} (broad).

All ATR-IR spectra (Figure 1c) for opals show a common set of peaks at around 470, 790 and 1080 cm^{-1}. Below 400 cm^{-1}, peaks were either non-existent or very weak, and thus measurement of spectra in the 400-1600 cm^{-1} range is needed to provide unambiguous differentiation of opal-A, opal-CT and opal-C. The IR spectra of reference compounds are found in References [48–50,54]. Quartz may be identified by peaks at 697 and 780 cm^{-1} which are at lower energy to the common peak at 795 cm^{-1}, seen for both quartz and all the opal samples. Both opal-A and opal-C showed distinct peaks that were absent for opal-CT (see later). A complication occurred with the far-IR as a range of spectra were seen (Figure 2a). The peak at around 470 cm^{-1} showed a progressive trend of broadening, shifting to lower wavenumber and the appearance of a second peak at around 440 cm^{-1}. This was found to be common for all types of opal. There was no obvious correlation with the other spectral methods or with the behaviour in the mid-IR. Figure 2b also shows variation in the mid-IR range (in this case for opal-A for which it is most pronounced).

The ^{29}Si MAS NMR spectra are potentially discriminating for the opal types, although the average Q$_4$ peak positions were close for all the forms. Some difference was seen for FWHM and the positions of the Q$_3$ peaks. Peaks may be readily curve-fitted to differentiate the Q$_4$ and Q$_3$ peaks (Figure 3).

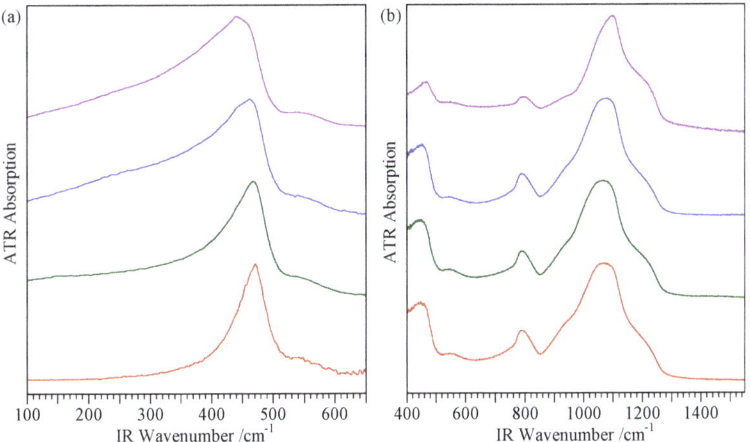

Figure 2. Far IR spectra of opal-AG samples showing the range of types. In ascending order: (**a**) White Cliffs, Australia (G8608), Lightening Ridge, Australia (G13769), Coober Pedy Australia (G9594) and Iron Monarch, Australia (G9260). (**b**) Mid-IR spectra of opal-A samples showing the range of types. In ascending order: Valec, Czech Republic (OOC11) (opal-AN), Coober Pedy Australia (G9594) (opal-AG), Springsure, Australia (M8736) (opal-AN) and White Cliffs, Australia (G1401) (opal-AG). Spectra were scaled and offset (*y*-axis) for comparison.

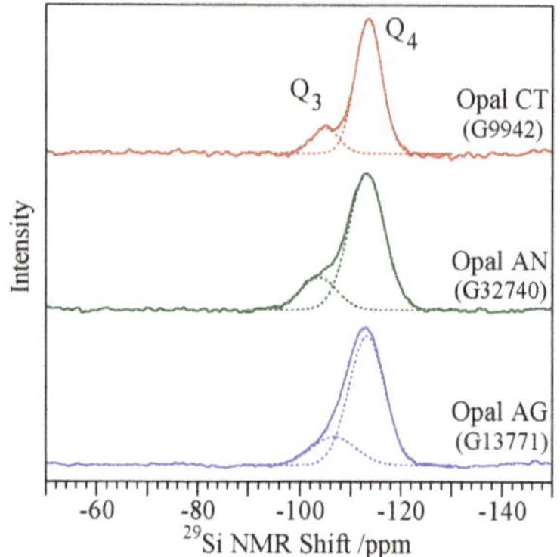

Figure 3. Experimental and fitted curves for ^{29}Si MAS NMR spectra. The exemplar opal-AG, opal-AN and opal-CT samples: opal-AG (White Cliffs, Australia G13771) (red), opal-AN/hyalite (Valec, Czech Republic G32740) (green) and opal-CT (Angaston, Australia, G9942) (blue).

3.2. Opal-A

All samples showed broad and weak XRD patterns with absorption centred on ~4.0 Å (Figure 1a) and had a poorly defined baseline compared to the other opal types. There was also a weak broad band at around 2.0 Å. Consistency was demonstrated through curve fitting with peak maxima of 3.98 ± 0.04 Å and FWHM 0.53 ± 0.02 Å. There was little, if any, difference between the opal-AG and opal-AN samples.

Opal-A Raman spectra were distinct from the spectra for opal-C and opal-CT in the 100–600 cm^{-1} range, so this provided a secondary delineation of the opal type. There were no obvious differences between Raman spectra for opal-AG and opal-AN in the 100–600 cm^{-1} range. Raman spectra between 700 and 1200 cm^{-1} (Figure 4) showed several weak peaks and a broad peak (probably two peaks) in the range 760–860 cm^{-1} that was characteristic of both opal-AG and opal-AN and which also separated these from most examples of opal-CT and opal-C. In general, the middle peak in Figure 4 was at 960–965 cm^{-1} for opal-AG and 970–975 cm^{-1} for opal-AN, though the peaks were weak. The IR spectra (Figure 1c) for opal-A were characteristic and distinct to those for opal-CT and opal-C with the peak at around 550 cm^{-1} being specific for opal-A, but there were no obvious differences between opal-AG and opal-AN.

The ^{29}Si MAS NMR spectra (Figure 1d) were potentially discriminating for opal-A, as the FWHMs tended to be larger for it than for other types. The Q_4 peak positions were -113.3 ± 0.2 ppm with FWHM of 8.5 ± 0.2 ppm for opal-AG and -113.4 ± 0.3 ppm and 8.7 ± 0.6 ppm for opal-AN. These FWHMs were higher than the 6.5 ± 0.6 ppm seen for opal-CT. All samples showed significant amounts of Q_3 peaks within the range 10–40%. The visual difference of opal-AG and opal-AN in Figure 1d can be traced to the placement of the Q_3 peaks with the former at -106.5 ± 0.9 ppm and the latter at -103.8 ± 0.8 ppm (see Figure 3).

Of the more recent finds, the massive blue-green opal that occurred as the base of some of the purple "grape agate" from near Mamuju, West Sulawesi, Indonesia was an opal-AG. A previous report [37] suggested that this material might have been a clay.

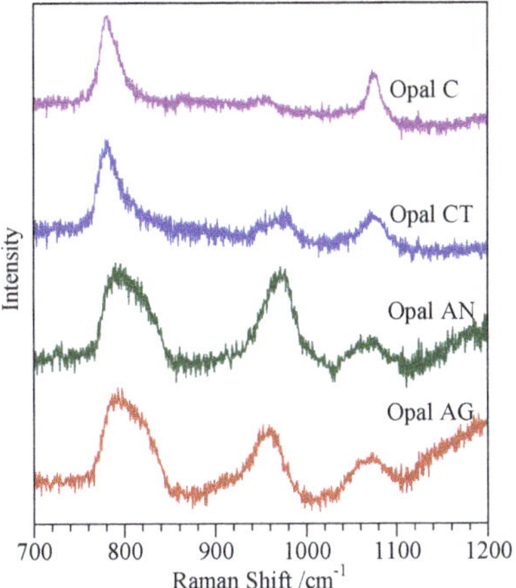

Figure 4. Raman spectra (700 to 1200 cm^{-1}) of opal samples. In ascending order: White Cliffs, Australia (G13771) (opal-AG), Valec, Czech Republic (G32740) (opal-AN), Angaston, Australia (G9942) (opal-CT) and Iceland (M5081) (opal-C). Spectra were scaled and offset for comparison.

3.3. Opal-CT

All samples showing substantial XRD peaks at ~4.1 Å and 2.5 Å may be considered as opal-CT according to the Jones and Segnit [1] classification. However, we believe that this is an oversimplification. The maximum was distinct from that for the opal-C grouping (~4.1 Å to 4.04 Å, respectively), though this was only apparent when two samples were directly compared. The increasing complexity of the main peak ~4.1 Å in the XRD pattern (Figure 5), presented in terms of increasing sharpness of the peak at 2.5 Å) suggests that the term opal-CT does not represent a homogenous group. We noted a trend whereby a peak at 4.28 Å was sometimes absent, sometimes present as a shoulder and sometimes as a separate peak. This was noted in a number of previous studies [55–57]. Complexity of the peak at ~4.1 Å appears associated with the peak width at 2.5Å. It is also apparent from the large number of samples measured here (161 specimens) that the patterns varied according to the relative peak heights and widths in the composite at around ~4.1 Å (see later). Sample G32752 (Afar, Ethiopia) (Figure 5) showed a simple, near symmetric XRD peak at 4.1 Å, as well a broad peak at 2.5 Å. At the other extreme, sample G NEW19 (Kigoma, Tanzania) showed two sharp peaks and a prominent shoulder in the composite at around ~4.1 Å and a sharp peak at 2.5 Å.

We note that the more "simple" opal-CT specimens include POC specimens from Ethiopia (G25374, G31892, G32752 (three distinct samples from the same specimen), NMNH Ethiopia samples 1, 2 and 3), Honduras (G1441), the USA (M19717, OOC5), Mexico (NMNH 117414) and Australia (G9964). The group also includes mostly transparent samples from worldwide localities such as Turkey (G NEW24, G NEW26, G NEW27, G NEW28), Madagascar (G NEW05), Mexico (G34738, NMNH115816, NMNHR1694), Australia (T19006, T23363, M12495, M20970), Peru (G33912), Brazil (T1152), Indonesia (OOC6), Spain (G NEW12) and the USA (G9116, G31852, G34243). The converse is not, however, true, with for instance G NEW07 from the same site in Madagascar as G NEW05 showing a more complex pattern. Some samples showing POC also have more complex XRD patterns.

We also noted a distinct change in the shape of the background of the XRD pattern. In the simpler opal-CT patterns, the background was distinctly lower on the low angle side and higher on the high-angle side of the peak at 4.1 Å. In contrast, the background around this peak became more uniform as the peak at 4.1 Å became more complex.

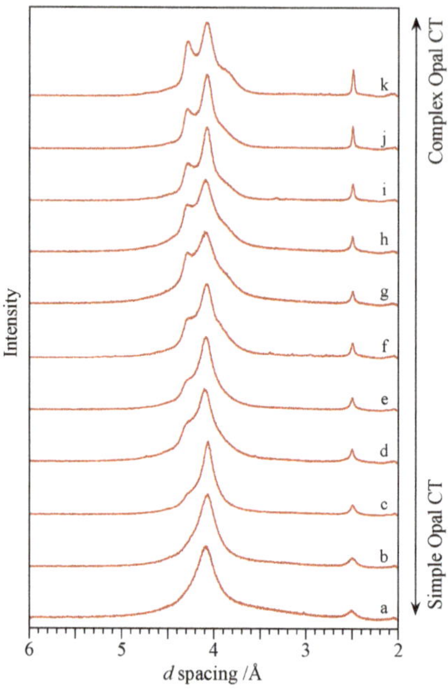

Figure 5. A series of XRD patterns illustrating the structural changes across the opal-CT group. The patterns were arranged in order of increasing complexity of the main peak at ~4.1 Å. Any subdivision between "simple" opal-CT and "complex" opal-CT is arbitrary, but patterns *a* and *b* are clearly distinct from patterns *i*, *j* and *k*. Note the progressive change in the sharpness and shape of the reflections at 4.1 and 2.5 Å. The specimens are (in ascending order from simplest to most complex): (**a**) Afar, Ethiopia (G32752), (**b**) Mezezo, Ethiopia (NMNH Eth 1), (**c**) Murwillumbah, Australia (G9964), (**d**) Acari, Peru (G33912), (**e**) Honduras (G1441), (**f**) Indonesia (OOC6), (**g**) Kazakhstan (M53407), (**h**) Nevada, USA (G32263), (**i**) Tanzania (G34238), (**j**) Nevada, USA (G31851) and (**k**) Tanzania (G NEW19).

We explored peak fitting to investigate the heterogeneity seen in opal-CT to see if quantifiable support may be gained for the continuum of "simple" and "complex" opal-CT. As Figure 6 shows, a reasonable proposition is that there were three peaks present (P1 to P3) in the 4.1 Å complex with a further one at 2.5 Å (P4). Well-fitting deconvolutions were obtained for all relatively pure (i.e., no visible or only a very minor amounts of quartz or other impurities) samples and confirms three peaks with positions at 4.27 ± 0.01 Å (P1), 4.08 ± 0.01 Å (P2) and 3.89 ± 0.03 Å (P3). The peak at 2.50 ± 0.01 Å (P4) was also constant.

The peak areas were normalized to P2 and the FWHM examined for each peak used to explore trends. A strong correlation in FWHM was for P1 and P4 (Figure 7a) where the linear trend between "simple" and "complex" extremes is manifest. The "simple" opal-CTs plot in the top right-hand corner are Ethiopian opals with POC. Other POC opals are distributed though the plot suggesting that this textural effect is independent of crystal–chemical features. It is also worth noting that other textural features such as the alignment and size of spheres are also implicated in POC.

There is also a linear correlation for the relative areas of P1 and P3 peaks as shown in Figure 7b, implying that, in simplistic terms, that this could be considered to represent the increase in the "tridymite" component comparted to the "cristobalite" (P2 coinciding with the overlap of the C and T contribution). The presence of trace amounts of quartz does not seem to affect this trend, but the problem associated with fitting caused by the asymmetric baseline may affect the ratio and over-represent P1 or P3.

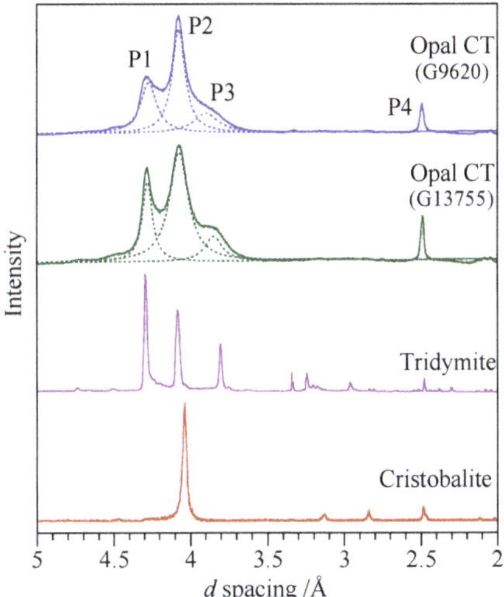

Figure 6. Comparison of mixed XRD patterns for mineral samples of tridymite (G1395) and cristobalite (RRUFF Database ID R060648) with Ross, Tasmania, Australia (G13755) and Iron Monarch, Australia (G9620) showing curve-fitting elements and actual pattern. Spectra were scaled and offset for comparison.

The implication of the separate relationships shown in Figure 7 is that the XRD pattern for any particular opal-CT will be hard to predict. Peaks may be large or small and sharp or narrow. We think it unlikely that a single parameter may be derived to quantify an opal-CT. Principal component analysis may provide a solution, but this would need to have a chemical or structural basis rather than mathematical manipulation. We noted that the two points at the top right of Figure 7b (G9960 from Iron Monarch, Australia and G NEW17 from Turkey) lie in the middle of the plot in Figure 7a. Similarly, G13755 (Ross, Tasmania, Australia, bottom left of Figure 7a) which shows very prominent and sharp XRD peaks (see Figure 6) lies in the centre of Figure 7b. It is also unlikely that the XRD pattern will allow prediction for the potential of POC from that locality.

The Raman spectra Figure 8 also demonstrated that opal-CT is not a homogenous group as shown by the spectra in Figure 5. An arbitrary classification may be made according to the rationales: (i) featureless spectrum with a maximum above 300 cm^{-1}, (ii) signs of structure with maximum at or below 300 cm^{-1} and (iii) more developed spectrum with partially delineated peaks. It is not clear if the trend was due to the change in structure with different spectra or merely a sharpening of existing peaks. We noted that those samples with few features in the Raman were mostly coincident with the "simple" opal-CT points at the upper right of Figure 7b (see Figure 8b). The most Raman-structured samples tended to lie at the lower left of Figure 8b and thus correlated with the more complex XRD patterns.

The 700 cm^{-1} to 1000 cm^{-1} Raman region showed potential differences between the "simple" and "complex" opal-CT examples in the 850–1000 cm^{-1} region. Unfortunately, the spectra were generally weak and were indicative rather than definitive. This is further complicated as not all samples were amenable to Raman due to the fluorescence, and this limits the applicability of the technique.

The ^{29}Si MAS NMR data showed that SP opal-CT Q_4 peak positions were marginally downfield of those for opal-A at -113.7 ± 0.3 ppm with FWHM 6.5 ± 0.6 ppm. In general, the Q_3 peaks were visible and were at -105.3 ± 0.8 ppm. The FWHMs of 6.5 ± 0.6 ppm were smaller than for opal-A at 8.4 ± 0.4 ppm. We also found that some of the "simple" opal-CT types gave visually different spectra that could not be deconvoluted satisfactorily.

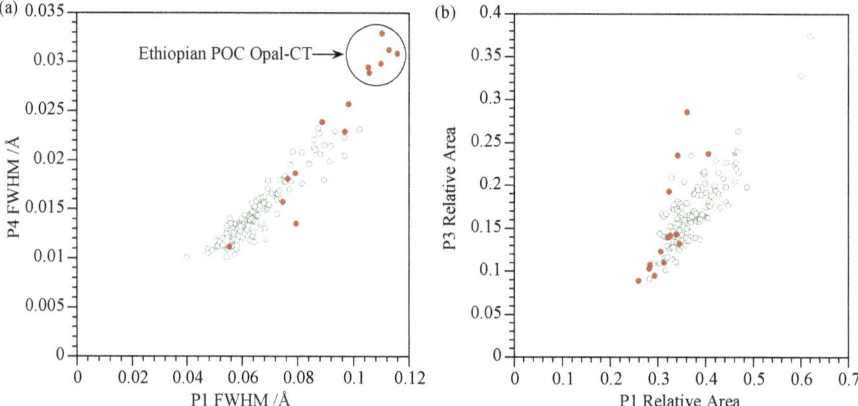

Figure 7. Correlations of XRD curve fitting data: (**a**) FWHM for P1 (*x*-axis) and P4 (*y*-axis), and (**b**) relative amounts of the P1 (*x*-axis) and P3 (*y*-axis) peaks (P2 is set at unity). Samples showing play-of-colour (POC) are shown as red-filled circles, whereas non-POC samples are represented by green-open circles. The subset of samples from Ethiopia displaying POC are circled in panel (**a**).

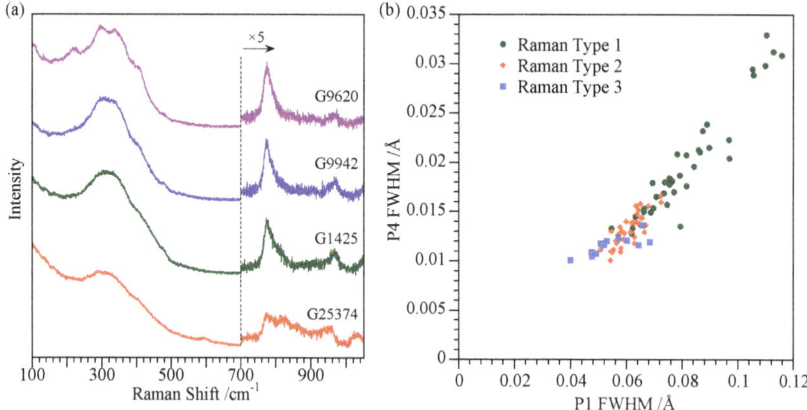

Figure 8. Raman spectra of opal-CT showing progressive structure. In ascending order: (**a**) "simple" opal-CT from Mezezo Ethiopia (G25374), and increasingly complex forms from Eurolowie, Australia (G1425), Angaston, Australia (G9942) and Iron Monarch Australia (G9620). (**b**) Plot of the XRD pattern FWHM of P4 versus P1 separated into different Raman types. See text for details regarding the definitions of the Raman types observed in this study. Not all samples yielded a Raman spectrum due to the problems with fluorescence.

3.4. Opal-C

Only four discrete samples were identified as opal-C in this study. Two of these had zones of both transparency and opaqueness though having identical XRD patterns. All showed very small peaks for quartz. The common feature was a large peak at 4.04 Å with a smaller one at 4.28 Å. The small pair of equally sized peaks at 3.11 Å and 2.84 Å was diagnostic of cristobalite [44].

Raman spectra (Figure 1b) showed characteristic spectra similar to that of cristobalite [51] at low wavenumber (peaks at 107, 222 and 409 cm^{-1}), while the medium wavenumber spectrum shown in Figure 4 is possibly diagnostic but again weak. The combined far- and mid-IR for the opal-C samples showed these features: 300 cm^{-1} (sh), 385 cm^{-1} (sh), 480 cm^{-1} (m), 625 cm^{-1} (w), 795 cm^{-1}

(m), 1090 cm^{-1} (s) and a shoulder at 1200 cm^{-1}. The most delineating feature was the band at around 625 cm^{-1} which was visible both via far- and mid-IR. We did not, however, find any evidence of an IR band at 145 cm^{-1} as was reported in a previous study [50].

The ^{29}Si MAS NMR spectra of the opal-C samples showed variability and the Iceland sample (M5081) shown in Figure 1d should be treated as an example rather than a typical spectrum. This had characteristics of a peak position of −114.4 ppm and an FWHM of 5.7 ppm. Spectra suggest more complexity than those seen for opal-A and opal-CT with the likelihood of additional peaks.

3.5. Transitional Samples

3.5.1. Samples Showing Opal-A and Opal-CT Characteristics

Three samples had XRD patterns which had characteristics of both opal-A and the simpler form of opal-CT, with a very broad peak at 4.1 Å and a relatively small and broad peak at 2.5 Å. Sample OOC4 from Mazarron, Spain had a major peak FWHM estimate of about 6.3°, while that for T22824, from Megyasro, Hungary, was 4.3°. Neither gave a satisfactory XRD curve fitting. Not shown is E1950 (Canungra Mts. Australia) which had an FWHM of around 3° and which could be XRD curve fitted (and plottted as a "simple" opal-CT). This contrasts with a value of about 8° for a typical opal-A and less than 2° for the POC opal-CT from Mezezo, Ethiopia (G25374) and the remainder of the simple opal-CT samples. Only T22824 yielded a Raman spectrum and this was consistent at both low and medium wavenumbers with opal-A. The orange opal-CT from Voznesenovka, Kazakhstan (T22824) and OOC4 from Mazarron, Spain showed a peak at 550 cm^{-1} in the IR though this was not readily apparent in E1950 (Figure 9).

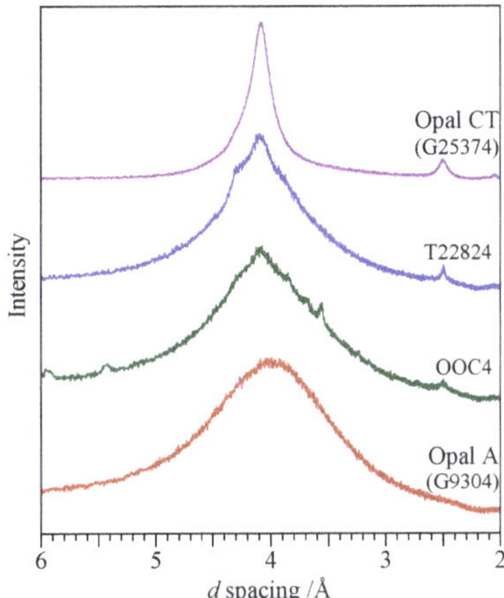

Figure 9. Transitional opal XRD pattern (lower-middle OOC4 from Mazarron, Murcia, Spain, upper-middle T22824 from Megyasro, Hungary). Shown with G9304 (opal-A, lower) and G25374 (simple opal-CT, upper). Scaled and offset (*y*-axis) for comparison.

Raman spectra showed broad peaks at 760–860 cm^{-1} (as in opal-A) and 975 cm^{-1} (as in hyalite). While there were coincidences in position, it is difficult to assert that the XRD peak at 2.5 Å represents a narrowing of the broad and weak second peak in opal-A. Critically, the XRD patterns and Raman spectra showed no evidence of cristobalite, so a reasonable assumption was that these represent transitional opal-A/opal-CT. The SEM images (Figure 10) were not consistent with opal-AG.

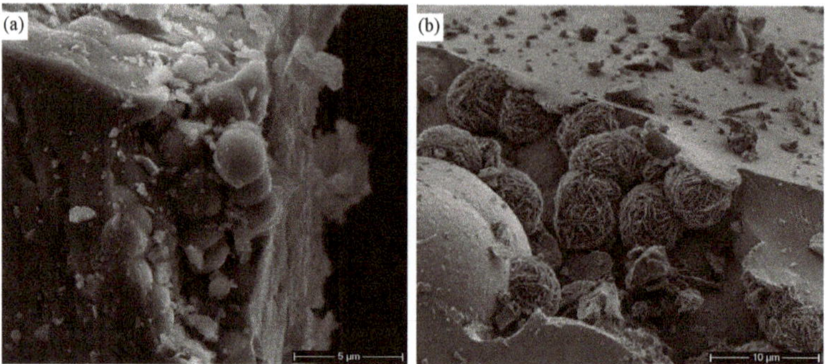

Figure 10. SEM of: (**a**) OOC4 AND (**b**) T22824 (RHS) showing large spheres and bundles of plates.

3.5.2. Samples Showing Opal-CT and Opal-C Characteristics

The sample from the Opal Butte Mine, Oregon USA (G NEW18) appears to be a transitional form of opal-CT to opal-C as the characteristic cristobalite peaks are present in both the XRD pattern and Raman spectrum (Figure 11). The XRD maximum is at 4.02 Å which is consistent with the cristobalite-like patterns of opal-C as are the two small peaks between 3.14 Å and 2.85 Å. The form and position of the Raman absorption at low wavenumbers are not characteristic of opal-A and are more like that seen for the "simple" opal-CT samples with an overlay of cristobalite peaks. At medium wavenumbers only one peak is seen at around 800 cm^{-1} while the peak at just below 1000 cm^{-1} is perhaps more significant than is seen for the opal-C samples. A weak response at 625 cm^{-1} in the IR is also consistent with opal-C as are the patterns at lower wavenumber. Overall this suggests a transitional "simple" opal-CT to opal-C species.

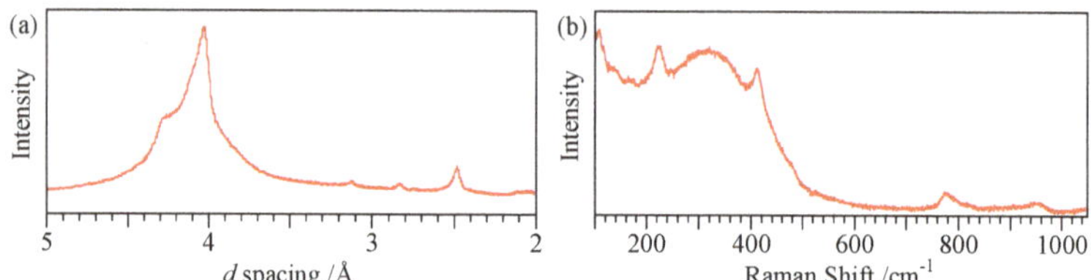

Figure 11. Transitional opal CT to C from Opal Butte Mine, Oregon USA (G NEW18). (**a**) XRD pattern and (**b**) Raman spectrum.

4. Discussion

4.1. Applicability of XRD for Primary Classification

We found that all opal samples, including those from "newer" sources, including Ethiopia, Indonesia, Kazakhstan, Madagascar, Peru and Tanzania could be readily classified using XRD into one of the Jones and Segnit [1] groups. All of the opals for newer localities were opal-CT, except for two of the Indonesia samples which were Opal-A (Table 1). XRD provides a ready and informative delineation of opal-A, opal-CT and opal-C but the identification of transitional forms requires additional data from spectroscopic techniques. Differentiation of opal-AG and opal-AN is, however, not possible from XRD, but can usually be readily seen visually. The two types of precious opal (opal-A and "simple" opal-CT) can be readily separated by XRD.

Further detail may be gleaned, particularly of opal-CT, through semi-quantitative curve fitting.

The work described here provides an alternative analysis to previous studies [14,17,55,58] where the maximum and width of the composite peak at 4.1 Å were interpreted in terms of contributions from cristobalite-like and tridymite-like components.

4.2. Homogeneity and Characterisation of Opal Groups

4.2.1. Opal-A

This group was clearly identified by the distinctive broad XRD pattern, the featureless but broad Raman pattern centred at 370 cm^{-1} (uncorrected) and a specific IR peak at around 530 cm^{-1}. The ^{29}Si NMR will show a relatively broad (8.5 ppm) and symmetric Q_4 peak at around −113.3 ppm hiding the Q_3 component. Hyalites (opal-AN) were visually different from the opal-AG group being botryoidal and gel-like though with similar XRD patterns and Raman and IR. The major ^{29}Si NMR spectra will show resolved peaks for Q_4 and Q_3 unlike for opal-AG.

4.2.2. Opal-CT

The XRD patterns for this group are characterized by peaks at 4.1 Å and 2.5 Å. With the 4.1 Å reflection range from a "simple" asymmetric peak accompanied by broad weak reflections at 2.5 Å, to a "complex" broad group of three peaks centred on 4.1 Å and a sharp reflection at 2.5 Å. The opal-CT that gave the simple X-ray patterns were more or less transparent, though they may have been coloured and showed POC to some extent, while those with complex patterns tended to be opaque and included material that was generally considered common opal, though there were examples with POC. The Raman spectra for those that gave "simple" XRD patterns were generally weak and featureless showing only a peak around 300 cm^{-1}, while those for the opal-CTs with more complex XRD patterns (if obtainable) showed broad absorption in the 200 to 500 cm^{-1} range and varying degrees of structure, possibly with discrete peaks at 220, 295, 340 and 410 cm^{-1}. The IR spectra and ^{29}Si MAS NMR across the range of opal-CTs did not provide differentiation.

These "simple" opal-CTs included the Ethiopian play-of-colour opals as well as some Australian specimens, but were not from the current active opal fields. For instance, G9964 is labelled as "jelly opal" from Murwillumbah in New South Wales and has a cloudy but transparent appearance. It has been proposed that Australian POC opal is sedimentary if opal-A and volcanic if opal-CT [59]. Other sources for "simple" opal-CT without play-of-colour include: Brazil, France, Honduras, Indonesia, Madagascar, Mexico, Peru, USA (Nevada). Published spectra show further examples of these simple XRD patterns, e.g., [7,13,19,60].

Recent measurements suggest that POC in opal-CT is also caused by the presence of diffracting patterns of spheres [5,61] similar in concept to those for opal-A [4]. Our work on the chemical characteristics of opal-CT, however, can provide no light for prediction of POC, though we note that an Ethiopian opal with a wide peak at 2.5Å is likely to show the effect. Other POC examples do not show any correlation with XRD patterns or spectroscopic measurements.

4.2.3. Opal-C

There is no visible difference between these and opal-CT samples. Some show the feature of transparent specimens merged with white translucent areas [49] which have the same XRD patterns, Raman and IR spectra. The distinct features are XRD pattern peaks at 4.04 Å and 2.5 Å coupled with a pair of small peaks between the major ones. Raman shows a characteristic pattern with peaks 107, 222 and 409 cm^{-1}. The IR spectra were distinct with a small but clearly defined peak at 625 cm^{-1}. The ^{29}Si MAS NMR showed 6 ppm FWHM peak at −113 ppm.

4.3. Spectroscopic Characterisation Techniques

Raman spectroscopy [6,7,13,20,32] can be used in non-destructive mode with data gained in the far- and mid-range providing information related to Si–O bonding, although direct assignment of bands to the opal structure is difficult. The spectra for opal-A, opal-CT and opal-C were distinct and provided ready differentiation. With some care, hyalite may also be identified if it is not already apparent from its distinctive appearance. Raman spectroscopy is not as viable a technique as XRD for two major reasons [62]. Firstly, opal has a relatively low cross-section for Raman scattering resulting in weak signals, thus requiring long collection times. Second, many samples exhibited a large degree of fluorescence which swamps the signal. The "complex" opal-CT samples were more likely to be fluorescent than the "simple" types, indicating possibly higher content of metal impurities in "complex" opal-CT. Opal-C samples, and some proposed transitional forms, had similar peaks to that of α-cristobalite and this is probably the preferred means for identification. The current work is consistent with previous studies.

ATR IR has been used previously to examine silica species [1,54] but has been little exploited recently for opal. It does, however, represent an alternative to XRD as opal-A, opal-CT and opal-C can be readily discriminated, though without the additional information relating to "simple" or "complex" forms. While we used ground samples, it could be a non-destructive method in ATR mode. Although not performed here, we suggest that reflectance IR spectroscopy may also be a valuable tool for non-destructive examination of opal samples. The ^{29}Si NMR is of more interest for investigation of the chemical structure of opal rather than as a differentiation technique.

4.4. Comments on Nature of Opal-CT

This has been the subject of controversy for some years. The term opal-CT can be interpreted as zones of cristobalite and tridymite or as an intimate intergrowth, whether regular or disordered. In our opinion, we feel that the term "opal-CT" is a misnomer. The notation was based on a similarity of XRD peak positions of α-cristobalite and α-tridymite with those in opal samples [1,25,44]. This has been complemented by modelling studies of XRD patterns [63] and Raman spectra [52,64]. The XRD patterns have been analysed in terms of α-cristobalite to α-tridymite ratio [14]. The TEM images have been interpreted in terms of the intergrowth of domains of cristobalite and tridymite and as tridymite-like stacking faults in cristobalite [55,56]. As Figure 6 implies, a reasonable proposal based on peak positions of reference compounds is that the opal-CT peaks derive from cristobalite and tridymite, with P1 due to tridymite, P2 due to cristobalite and tridymite and P3 due to tridymite (and possibly cristobalite). While the XRD curve fitting suggests complexity in the structure, it is not clear how many discrete species may be involved, as correlation was noted between the intensities of the P1 and P3 peaks in one instance and the sharpness of the P1 and P4 peaks in another. The trending evidence (Figure 7) did not imply this. If tridymite was present, then we might expect the P1, P2 and P3 peaks to be linked in some way. P3 did not correlate in sharpness with P1 and P4 and was shifted compared to α-tridymite. We also believe that the Raman evidence was equivocal, as while we find (baseline uncorrected) peaks at 220, 295, 340 and 410 cm^{-1} in the most structured opal-CT samples, we do not believe that the spectra were of sufficient quality to add anything significant to the issue.

The notion that opal-CT is a disordered, intimate mix of cristobalite and tridymite has also been questioned on the basis on XRD and Raman data [52,57,64,65]. The lack of evidence for the presence of cristobalite led to the proposal for "opal-T" based on interpretation of the Raman data [52,57,65]. Whether this is a "not cristobalite" rather than a "positive tridymite" assignment is a moot point. Supporting evidence for tridymite, however, presents a major problem with the potential multitude of stacking variations of this structure [44]. We found no evidence for the triplet reported for tridymite in ^{29}Si studies [24,27]. The topology of the silica structural frameworks in opal-CT remain a matter for debate, and the changes we observed between "simple" and "complex" opal-CTs may represent different structural states rather than different structural intergrowths.

4.5. Comments on Opal Formation and Transitions Between Opal-A, Opal-CT, Opal-C and Quartz

A recent paper [39] has proposed that temperature of formation ($\leq$45° for opal-AG and >160° for opal-CT) is the prime determinant of opal type rather than type of deposit, i.e., volcanic versus sedimentary [40,49] sources for opal-CT and opal-A, respectively. Our results are consistent with this proposition, our simple opal-CT showing a play-of-colour all come from deposits associated with volcanism.

It is possible that the transition between the different forms of opal derives from a similar process as the initial formation: for example, opal-A dissolution (partially hydrated silica to silicic acid) followed by deposition of opal-CT [41]. Analysis of dated sinter samples from hot springs sites [42] show the presence of opals with XRD patterns consistent with opal-A, a transitional form between opal-A and opal-CT, opal-CT, opal-C and quartz. The SEM images also show a change in form. While the opal-CT sample was probably consistent with "simple" opal-CT, verification is difficult owing to a significant amount of quartz. The SEM evidence for transition is also noted for geysers [38]. Changes in XRD pattern and near-IR have also been noted in accelerated aging at 300 °C of deep sea deposits [43].

We see two types of transitional forms that could be interpreted as opal-A to opal-CT and "simple" opal-CT to opal-C. However, these transitional samples are very uncommon and possibly far less than would be expected if this was a routine occurrence, but this could depend on the kinetics of the process and the geological age of the samples. The transformation of one form of opal to another is most probably a dissolution/reprecipitation reaction, given that opal is associated with flow of aqueous crustal fluids and these processes are known to be relatively rapid in terms of geological time [66,67].

Table 2 gives a brief summary of the defining characteristics found for the different opal types described in this work. The presence of "impurities" may cause misidentification for single samples.

Table 2. Summary of differentiating opal properties (this work).

Opal Type	XRD	Raman [a]	IR [b] (ATR)	^{29}Si NMR [b] (Single Pulse)
Opal-AG	Very broad peak between ~2.2 Å and ~6.5 Å with maximum at 3.9–4.0 Å	Broad peak between ~230 and ~530 cm^{-1} with maximum at ~370 cm^{-1}; 760–860 cm^{-1} 970–975 cm^{-1}	Peak at 530 cm^{-1}	Q_4 FWHM 8.5 ppm Q_3 peak(s) not prominent
Opal-AN (hyalite)	Very broad peak between ~2.2 Å and ~6.5 Å with maximum at 3.9–4.0 Å	Broad peak between ~230 and ~530 cm^{-1} with maximum at ~370 cm^{-1} 760–860 cm^{-1} 960–965 cm^{-1}	Peak at 530 cm^{-1}	Q_4 FWHM 8.3 ppm Q_3 peak visible as a shoulder
Opal-CT [c]	All have peak at 2.50 Å. Simpler types have a single peak at 4.08 Å. More complex types also show a peak or shoulder at 4.28 Å and a shoulder at 3.89 Å	Broad peak between ~180 and ~500 cm^{-1} with maximum at ~300 cm^{-1} to more defined maxima at 220, 295, 340 and 410 cm^{-1}	*Absence of peaks at 530 and 625 cm^{-1}*	Q_4 FWHM 6.5 ppm Q_3 peak visible
Opal-C	4.04 Å and 2.50 Å	Sharp peaks at 107, 222 and 409 cm^{-1}	Peaks at 300, 385, 470 and 625 cm^{-1}	*No common feature*

[a] Without baseline correction. [b] Only unique features are noted. [c] Trend discussed in text.

4.6. Summary

This study provides a classification of examples from many sites, both gem quality and other samples, and incorporates a number of techniques. The large number of spectra and XRD patterns presented here illustrate the range of opals that may be found under each heading.

XRD remains the primary analytic method of choice as all samples, including those from newer sources, can be readily classified as opal-A, opal-CT, opal-C or a reasonable case may be made for a transitional form. We note that mid-IR ATR spectroscopy also fulfils this role. The XRD patterns for Opal-CT exhibited a range of forms from quite simple patterns to more complex ones, but this range was a continuum. At the "simple" end, the Ethiopian POC occurred while the "complex" forms included the common opals. Play-of-colour opals may belong to either the opal-A or opal-CT groups. Thus, the terms "precious opal", "play-of-colour", "potch", "common opal" and "fire opal" are best treated with caution, possibly only to be used within the trade, rather than in scientific studies.

The large body of samples examined in this work has allowed us to identify exemplars or typical specimens for the various opal groups. These provide authentic references that could be used in provenance authentication, in comparing purity and for providing characterised samples for further studies, such as chemical or thermal modification. New samples can be compared against these established and well-characterised examples. They may also be used for other studies such as geological (e.g., volcanic versus sedimentary) settings or trace element content.

Supplementary Materials: The following are available online at http://www.mdpi.com/2075-163X/9/5/299/s1.

Author Contributions: Conceptualization, A.P. and N.J.C.; Data curation, N.J.C.; Formal analysis: N.J.C., J.R.G. and M.R.J.; Investigation, N.J.C., J.R.G. and M.R.J.; Methodology N.J.C., J.R.G. and M.R.J.; Validation, N.J.C., J.R.G., M.R.J. and A.P.; Visualization, N.J.C. and A.P.; Writing—original draft preparation, N.J.C.; Writing—review and preparation, N.J.C., J.R.G. and A.P.

Funding: This research received no external funding.

Acknowledgments: The authors thank Ben McHenry of the South Australian Museum and Tony Milne of the Tate collection at the University of Adelaide for their unstinting assistance in locating samples for this study. The collection mangers of the Flinders University of South Australia, Museum Victoria and the Smithsonian National Museum of Natural History are thanked for the provision of samples and useful conversations. The authors acknowledge the expertise, equipment and support provided by Microscopy Australia and the Australian National Fabrication Facility (ANFF) at the South Australian nodes under the National Collaborative Research Infrastructure Strategy. We acknowledge access to the facilities at the Australian Synchrotron and the technical support provided by Dominique Appadoo for the Far-IR data collection. The constructive comments of the two anonymous referees and the associate editor are gratefully acknowledged.

Conflicts of Interest: The authors declare no conflict of interest.

References

1. Jones, J.B.; Segnit, E.R. The Nature of Opal I. Nomenclature and Constituent Phases. *J. Geol. Soc. Aust.* **1971**, *18*, 57–68. [CrossRef]

2. Murray, M.J.; Sanders, J.V. Close packed structures of spheres of two different sizes II. The packing densities of likely arrangements. *Philos. Mag. A* **1980**, *42*, 721–740. [CrossRef]

3. Caucia, F.; Ghisoli, C.; Marinoni, L.; Bordoni, V. Opal, a beautiful gem between myth and reality. *Neues Jahrbuch für Mineralogie Abhandlungen J. Miner. Geochem.* **2013**, *190*, 1–9. [CrossRef]

4. Sanders, J.V. Colour of precious opal. *Nature* **1964**, *204*, 1151–1153. [CrossRef]

5. Chauvire, B.; Rondeau, B.; Mazzero, F.; Ayalew, D. The precious opal deposit at Wegel Tena, Ethiopia: Formation via successive pedogenesis events. *Can. Miner.* **2017**, *55*, 701–723. [CrossRef]

6. Simoni, M.; Caucia, F.; Adamo, I.; Galinetto, P. New occurence of fire opal from Bemia, Madagascar. *Gems Gemol.* **2010**, *46*, 114–121. [CrossRef]

7. Ansori, C. Model mineralisasi pembentukan opal banten. *Jurnal Geol. Indones.* **2010**, *5*, 151–170. [CrossRef]

8. Shigley, J.E.; Laurs, B.M.; Renfro, N.D. Chrysoprase and prase opal from Haneti, Central Tanzania. *Gems Gemol.* **2009**, *45*, 271–279. [CrossRef]

9. Hatipoglu, M.; Kibici, Y.; Yanik, G.; Ozkul, C.; Demirbilek, M.; Yardmici, Y. Nano-structure of the Cristobalite and Tridymite Staking Sequences in the Common Purple Opal from the Gevrekseydi Deposit, Seyitomer-Kutahka, Turkey. *Orient. J. Chem.* **2015**, *31*, 35–49. [CrossRef]

10. Pewkliang, B.; Pring, A.; Brugger, J. The formation of precious opal: Clues from the opalization of bone. *Can. Miner.* **2008**, *46*, 139–149. [CrossRef]

11. Gauthier, J.-P.; Fritsch, E.; Aguilar-Reyes, B.; Barreau, A.; Lasnier, B. Phase de Laves dans la première opale CR disperse. *Mineralogie* **2004**, *336*, 187–196.

12. Sanders, J.V. Close-packed structures of spheres of two different sizes I. Observations of natural opal. *Philos. Mag. A* **1980**, *42*, 705–720. [CrossRef]

13. Sodo, A.; Municchia, A.C.; Barucca, S.; Bellatreccia, F.; Ventura, G.D.; Butini, F.; Ricci, M.A. Raman, FT-IR and XRD investigation of natural opals. *J. Raman Spectrosc.* **2016**, *47*, 1444–1451. [CrossRef]

14. Ghisoli, C.; Caucia, F.; Marinoni, L. XRPD patterns of opals: A brief review and new results from recent studies. *Powder Diffr.* **2010**, *25*, 274–282. [CrossRef]

15. Liesegang, M.; Milke, R. Australian sedimentary opal-A and its associated minerals: Implications for natural silica sphere formation. *Am. Miner.* **2014**, *99*, 1488–1499. [CrossRef]

16. Jones, J.B.; Segnit, E.R. Genesis of Cristobalite and Tridymite at Low Temperature. *J. Geol. Soc. Aust.* **1972**, *18*, 419–422. [CrossRef]

17. Elzea, E.L.; Odom, I.E.; Miles, W.J. Distinguishing well ordered opal-CT and opal-C from high temperature cristoblite by x-ray diffraction. *Anal. Chim. Acta* **1994**, *286*, 106–116. [CrossRef]

18. Gaillou, E.; Delaunay, A.; Rondeau, B.; Bouhnik-le-Coz, M.; Fritsch, E.; Corren, G.; Monnier, C. The geochemistry of gem opals as evidence of their origin. *Ore Geol. Rev.* **2008**, *34*, 113–126. [CrossRef]

19. Ostrooumov, N.; Fritsch, E.; Lasnier, B.; Lefrant, S. Spectres Raman des opales: Aspect diagnostique et aide a la classification. *Eur. J. Miner.* **1999**, *11*, 899–908. [CrossRef]

20. Smallwood, A.G.; Thomas, P.S.; Ray, A.S. Characterisation of sedimentary opals by Fourier transfrom Raman spectroscopy. *Spectrochim. Acta Part A* **1997**, *53*, 2341–2345. [CrossRef]

21. Kingma, K.J.; Hemley, R.J. Raman spectroscopic study of microcrystalline silica. *Am. Miner.* **1994**, *79*, 269–273.

22. Paris, M.; Fritsch, E.; Aguilar-Reyes, B. ^{1}H, ^{29}Si and ^{27}Al NMR study of the destabilization process of a paracrystalline opal from Mexico. *J. Non-Cryst. Solids* **2007**, *353*, 1650–1656. [CrossRef]

23. Brown, L.D.; Ray, A.S.; Thomas, P.S. ^{29}Si and ^{27}Al NMR study of amorphous and paracrystalline opals from Australia. *J. Non-Cryst. Solids* **2003**, *332*, 242–248. [CrossRef]

24. de Jong, B.W.H.S.; van Hoek, J.; Veeeman, W.S.; Manson, D.V. X-ray diffraction and ^{29}Si magic-angle-spinning NMR of opals: Incoherent long- and short-range order in opal-CT. *Am. Miner.* **1987**, *72*, 1195–1203.

25. Graetsch, H.; Gies, H.; Topalovic, I. NMR, XRD and IR study on microcrstalline opal. *Phys. Chem. Miner.* **1994**, *21*, 166–175. [CrossRef]

26. Graetsch, H.; Mosset, A.; Gies, H. XRD and ^{29}Si MAS-NMR study of some non-crystalline silica minerals. *J. Non-Cryst. Solids* **1990**, *119*, 173–190. [CrossRef]

27. Smith, J.V.; Blackwell, C.S. Nuclear magnetic resonance of silica polymorphs. *Nature* **1983**, *303*, 223–225. [CrossRef]

28. Day, R.; Jones, B. Variations in water content in opal-A and opal-CT from geyser discharge aprons. *J. Sediment. Res.* **2008**, *78*, 301–315. [CrossRef]

29. Bobon, M.; Christy, A.A.; Kluvanec, D.; Illasova, L. State of water molecules and silanol groups in opal minerals: A near infrared spectrscopic study of opals from Slovakia. *Phys. Chem. Miner.* **2011**, *38*, 809–818. [CrossRef]

30. Chauviré, B.; Rondeau, B.; Mangold, N. Near infrared signature of opal and chalcedony as a proxy for their structure and formation conditions. *Eur. J. Miner.* **2017**, *29*, 409–421. [CrossRef]

31. Eckert, J.; Gourdon, O.; Jacob, D.E.; Meral, C.; Monteiro, P.J.M.; Vogel, S.C.; Wirth, R.; Wenk, H.-R. Ordering of water in opals with different microstructures. *Eur. J. Miner.* **2015**, *27*, 203–213. [CrossRef]

32. Rondeau, B.; Fritz, E.; Mazzero, F.; Gauthier, J.-P.; Cencki-Tok, B.; Bekele, E.; Gaillou, E. Play-of-color opal from Wegel Tena, Wollo Province, Ethiopia. *Gems Gemol.* **2010**, *46*, 90–105. [CrossRef]

33. McOrist, G.D.; Smallwood, A. Trace elements in precious and common opals using neutron activation analysis. *J. Radioanal. Nucl. Chem.* **1997**, *223*, 9–15. [CrossRef]

34. Brown, L.D.; Ray, A.S.; Thomas, P.S. Elemental Analysis of Australian amorphous banded opals by laser-ablation ICP-MS. *Neues Jahrbuch für Minerologie Monatshefte* **2004**, *2004*, 411–424. [CrossRef]

35. Dutkiewicz, A.; Landgrebe, T.C.W.; Rey, P.F. Origin of silica and fingerprinting of Australian sedimentary opals. *Gondwana Res.* **2015**, *27*, 786–795. [CrossRef]

36. Rondeau, B.; Cenki-Tok, B.; Fritsch, E.; Mazzero, F.; Gauthier, J.-P.; Bodeur, Y.; Bekele, E.; Gaillou, E.; Ayalew, D. Geochemical and petrological characterizarion of gem opals from Wegel Tena, Wolo, Ethiopia: Opal formation in an Oligocene soil. *Geochem. Explan. Environ. Anal.* **2012**, *12*, 93–104. [CrossRef]

37. Ivey, J. Grape agate from West Sulawesi, Indonesia. *Miner. Rec.* **2018**, *49*, 827–836.

38. Jones, B.; Renaut, R.W. Microstructural changes accompanying the opal-A to opal-CT transition: New evidence from the siliceous sinters of Geysir, Haukadalur, Iceland. *Sedimentology* **2007**, *54*, 921–948. [CrossRef]

39. Martin, E.; Gaillou, E. Insight on gem opal formation in volcanic ash deposits from a supereruption: A case study through oxygen and hydrogen isotopic composition of opals from Lake Tecopa, California, U.S.A. *Am. Miner.* **2018**, *103*, 803–811. [CrossRef]

40. Fritsch, E.; Gaillou, E.; Rondeau, B.; Barreau, A.; Albertini, D.; Ostroumov, M. The nanostructure of fire opal. *J. Non-Cryst. Solids* **2006**, *352*, 3957–3960. [CrossRef]
41. Williams, L.A.; Crerar, D.A. Silica diagenesis, II. General mechanisms. *J. Sediment. Petrol.* **1985**, *55*, 312–321.
42. Lynne, B.Y.; Campbell, K.A.; James, B.; Browne, P.R.L.; Moore, J. Siliceous sinter diagenesis: Order among the randomness. In Proceedings of the 28th NZ Geothermal Workshop, Auckland, New Zealand, 15–17 November 2006.
43. Rice, S.B.; Freund, H.; Huang, W.-L.; Clouse, J.A.; Isaacs, C.M. Application of Fourier transform infrared spectrscopy to silica diagenis: The opal-A to opal-CT transformation. *J. Sediment. Res.* **1995**, *A65*, 639–647.
44. Smith, D.K. Opal, cristobalite and tridymite: Noncrystallinity versus crystallinity, nomenclature of the silica minerals and bibliography. *Powder Diffr.* **1998**, *13*, 2–19. [CrossRef]
45. Anthony, J.W.; Bideaux, R.A.; Bladh, K.W.; Nichols, M.C. *Handbook of Mineralogy, vol 2 Silica, Silicates*; Mineral Data Publishing: Tucson, AZ, USA, 1995.
46. Pecharsky, V.L.; Zavalij, P.Y. *Fundamentals of Powder Diffraction and Structural Characterization of Materials*; Springer Verlag: Secaucus, NJ, USA, 2005.
47. Kihara, K.; Hirose, T.; Shinoda, K. Raman spectra, normal modes and disorder in monoclinic tridymite and its higher temperature orthorhombic modification. *J. Miner. Petrol. Sci.* **2005**, *100*, 91–103. [CrossRef]
48. Schmidt, P.; Bellot-Gurlet, L.; Sciau, P. Moganite detection in silica rocks using Raman and inrared spectroscopy. *Eur. J. Miner.* **2013**, *25*, 797–805. [CrossRef]
49. Ostrooumov, M. A Raman, infrared and XRD analysis of the instability in volcanic opals from Mexico. *Spectrochim. Acta Part A* **2007**, *68*, 1070–1076. [CrossRef]
50. Etchepare, J.; Merian, M.; Kaplan, P. Vibrational normal modes of SiO_2, II Cristobalite and tridymite. *J. Chem. Phys.* **1978**, *68*, 1531–1537. [CrossRef]
51. Bates, J.B. Raman Spectra of alpha and beta cristobalite. *J. Chem. Phys.* **1972**, *57*, 4042–4047. [CrossRef]
52. Ilieva, A.; Mihailova, B.; Tsintov, Z.; Petrov, O. Structural state of microcrystalline opals: A Raman spectroscopic study. *Am. Miner.* **2007**, *92*, 1325–1333. [CrossRef]
53. Schmidt, P.; Bellot-Gurlet, L.; Slodczyk, A.; Froehlich, F. A hitherto unrecognised band ub the Raman spectra of silica rocks: Influence of hydroxylated Si-O bonds (silanole) on the Raman moganite band in chalcedony and flint (SiO_2). *Phys. Chem. Miner.* **2012**, *39*, 455–464. [CrossRef]
54. Lippincott, E.R.; Valkenbur, A.v.; Weir, C.E.; Bunting, E.N. Infrared studies of polymorphs of silicon dioxide and germaniun dioxide. *J. Res. Natl. Bur. Stand.* **1958**, *61*, 61–70. [CrossRef]
55. Elzea, J.M.; Rice, S.B. TEM and X-Ray diffraction evidence for cristobalite and tridymite stacking sequences in opal. *Clays Clay Miner.* **1996**, *44*, 492–500. [CrossRef]
56. Nagase, T.; Akizura, M. Texture and structure of opal-CT and opal-C in volcanic rocks. *Can. J. Miner.* **2007**, *35*, 947–958.
57. Wilson, M.J. The structure of opal-CT revisited. *J. Non-Cryst. Solids* **2014**, *405*, 68–75. [CrossRef]
58. Esenli, F.; Sans, B.E. XRD studies of opals (4A peak) in bentonites from Turkey: Implications for the origin of bentonites. *Neues Jahrbuch für Minerologie Abhandlungen* **2013**, *191*, 45–63. [CrossRef]
59. Smallwood, A.G.; Thomas, P.S.; Ray, A.S. Comparative Analysis of Sedimentary and Volcanic Opals from Australia. *J. Aust. Ceram. Soc.* **2008**, *44*, 17–22.
60. Thomas, P.S.; Ray, A. The thermophysical properties of australian opal. In Proceedings of the 9th International Congress for Appplied Mineralogy, Brisbane, Australia, 8–10 September 2008; pp. 557–565.
61. Gaillou, E. An overview of gem opals: From the geology to color and microstructure. In Proceedings of the Thirteenth Annual Sinkankas Symposium—Opal, Carlsbad, CA, USA, 18 April 2015.
62. Kiefert, L.; Karampelas, S. The use of the Raman spectrometer in gemmological laboratories: Review. *Spectrochim. Acta Part A* **2011**, *80*, 119–124. [CrossRef]
63. Guthrie, G.D.; Bish, D.L.; Reynolds, R.C. Modeling the X-ray diffraction pattern of opal-CT. *Am. Miner.* **1995**, *80*, 869–872. [CrossRef]
64. Ivanov, V.G.; Reyes, B.A.; Fritsch, E.; Faulques, E. Vibrational States in Opal Revisited. *J. Phys. Chem. C* **2011**, *115*, 11968–11975. [CrossRef]
65. Eversull, L.G.; Ferrell, R.E. Disordered silica with tridymite-like structure in the Twiggs clay. *Am. Miner.* **2008**, *93*, 565–572. [CrossRef]

66. Altree-Williams, A.; Pring, A.; Ngothai, Y.; Brugger, J. Textural and compositional complexities resulting from coupled dissolution–reprecipitation reactions in geomaterials. *Earth-Sci. Rev.* **2015**, *150*, 628–651. [CrossRef]
67. Xia, F.; Brugger, J.; Ngothai, Y.; O'Neill, B.; Chen, G.; Pring, A. Three-dimensional ordered arrays of zeolite nanocrystals with uniform size and orientation by a pseudomorphic coupled dissolution–reprecipitation replacement route. *Cryst. Growth Des.* **2009**, *9*, 4902–4906. [CrossRef]

Article

On the Color and Genesis of Prase (Green Quartz) and Amethyst from the Island of Serifos, Cyclades, Greece

Stephan Klemme [1,*], **Jasper Berndt** [1], **Constantinos Mavrogonatos** [2], **Stamatis Flemetakis** [1], **Ioannis Baziotis** [3], **Panagiotis Voudouris** [2] and **Stamatios Xydous** [3]

[1] Institut für Mineralogie, Westfälische Wilhelms Universität Münster, Corrensstrasse 24, 48149 Münster, Germany; jberndt@uni-muenster.de (J.B.); stam.flemetakis@uni-muenster.de (S.F.)
[2] Faculty of Geology and Geoenvironment, National and Kapodistrian University of Athens, 15784 Athens, Greece; kmavrogon@geol.uoa.gr (C.M.); voudouris@geol.uoa.gr (P.V.)
[3] Laboratory of Mineralogy and Geology, Agricultural University of Athens, 11855 Athens, Greece; ibaziotis@aua.gr (I.B.); stxydous@aua.gr (S.X.)
* Correspondence: stephan.klemme@uni-muenster.de; Tel.: +49-(0)251-8333047

Received: 10 September 2018; Accepted: 24 October 2018; Published: 26 October 2018

Abstract: The color of quartz and other minerals can be either caused by defects in the crystal structure or by finely dispersed inclusions of other minerals within the crystals. In order to investigate the mineral chemistry and genesis of the famous prase (green quartz) and amethyst association from Serifos Island, Greece, we used electron microprobe analyses and oxygen isotope measurements of quartz. We show that the color of these green quartz crystals is caused by small and acicular amphibole inclusions. Our data also shows that there are two generations of amphibole inclusions within the green quartz crystals, which indicate that the fluid, from which both amphiboles and quartz have crystallized, must have had a change in its chemical composition during the crystallization process. The electron microprobe data also suggests that traces of iron may be responsible for the amethyst coloration. Both quartz varieties are characterized by isotopic compositions that suggest mixing of magmatic and meteoric/marine fluids. The contribution of meteoric fluid is more significant in the final stages and reflects amethyst precipitation under more oxidizing conditions.

Keywords: green quartz; prase; amethyst; color; amphibole; actinolite; skarn; Serifos; Greece

---

1. Introduction

Colored varieties of quartz (SiO_2) are ubiquitous in many igneous, metamorphic and sedimentary rocks [1–3]. They are well-loved collectors' items and comprise a number of semi-precious gems such as violet amethyst, yellow citrine, pinky rose quartz, and dark smoky quartz. Color in quartz is caused either by electronic defects of the crystal structure, transition metals or other elements incorporated in the crystal structure. Furthermore, color in quartz, and in other minerals, can also be caused by small, often nano-sized, well-dispersed mineral inclusions [3]. Blue quartz, pink rose quartz or jasper are known to be colored because of small inclusions of other minerals. Whilst it was thought for a long time that the color of rose quartz was caused by small rutile (TiO_2) inclusions, an electron microscopy study [4] showed that the rose quartz crystals from Montana, USA, contained small inclusions of dumortierite [$(Al,Fe^{3+})_7(BO_3)(SiO_4)_3O_3$]. Furthermore, there is evidence that other rose quartz types are colored by the incorporation of Mn, Ti or P in the quartz structure [3].

Compared to yellow, violet or pink quartz, green varieties of quartz are rare. There are some descriptions of natural green quartz which are commonly referred to as prasiolite [5,6]. However, most green prasiolite is prepared from natural amethyst by heating to 300–600 °C [7,8]. Some early studies found that the color of green prasiolite quartz was caused by Fe^{2+} on an interstitial site in octahedral coordination within the quartz structure [2,9] but more recent studies using Mössbauer Spectroscopy questioned these results and concluded that the green color of prasiolite was caused by Fe^{3+} [6]. Detailed studies illustrate how color changes from amethyst to prasiolite during heating [3,7,8]. Rare natural examples of prasiolite have been reported from Reno, Nevada, USA [10] and Thunder Bay, Ontario, Canada [11,12].

Another green variety of quartz is commonly referred to as chrysoprase. The color of chrysoprase varies from olive-green to pale sea-green [3]. As far as we know, such chrysoprase has been described from Australian serpentinites [13] and jadeite-rich veins in lower Silesia [14]. Rossman concludes in his review [3] that the color of chrysoprase is caused by "fine-grained nickel compounds in the silicate matrix rather from substitutional Ni in the silica itself". Possible Ni-minerals include bunsenite (NiO) or other complex hydrous Ni-silicates [3]. Recent studies of chrysoprase samples from Szklary (Poland) and Sarykul Boldy (Kazakhstan) showed that the "apple-green" color is caused by nano-scaled inclusions of Ni-kerolite and minor pimelite but they found no evidence for bunsenite inclusions [15]. Finally, Yang et al., [16] reported green quartzites (also known as "Guizhou Jade") from the Quinglong antimony deposit, in the Guizhou province, China, and claimed that green coloration is due to the presence of Cr^{3+}- and V^{3+}-bearing dickite inclusions.

Amethyst, on the other hand, is a relatively common variety of quartz that forms in a wide variety of environments [17]. It is well known as a gemstone since antiquity, as it is mentioned by Theofrastus, the disciples of Aristotle (4th century B.C.). Amethyst comes in an attractive violet color with reddish or bluish tint, that was originally interpreted as a result of the presence of Fe^{3+}, which turns into Fe^{4+} after irradiation and substitutes for Si^{4+} in a deformed tetrahedral position [3]. However, recent studies, questioned this result and by using modern analytical techniques, like EPR, Mössbauer and synchrotron X-ray absorption spectroscopy attributed the coloration of amethyst in the presence of Fe^{3+} [6,18–20].

Amethyst is not very common in skarn deposits. When present, it is usually only of mineralogical interest, as it rarely comes in gem quality in this geological environment [21]. Amethyst has been described from the Fe-rich skarn deposits in Angara-Ilim in Russia [21,22]. Collector specimens of amethyst have been reported from the Dashkesan skarn deposit in Azerbaijan [21]. Sceptre-shaped amethyst crystals are also found in the skarns of Denny mountain, Washington, USA [21] and finally, amethyst crystals have also been found in Hässellkulla, Sweden, in the Örebro Fe-rich skarn deposit [21,23,24].

Here we will present some results of an electron microprobe and oxygen isotope study on green quartz crystals and associated amethyst from the skarn of Serifos Island in Greece, which comprises, to our knowledge, the only locality that produces specimens with both quartz varieties. These green quartz and amethyst crystals are well known to collectors and fetch rather high prices during auctions [25,26]. However, not much is known about the origin of these quartz crystals or the origin of the green color in Serifos quartz. Similar-looking green quartz crystals are available for sale from internet mineral dealers and are sourced from Dalnegorsk, Russia [27]. As to the color, there are, to our knowledge, no peer-reviewed articles that investigated the nature of the green Serifos or similar green quartz crystals in any detail. Hyrsl and Niedermayr [28] reported actinolite from XRD analyses of prase from Serifos, but presented no further analytical data. Similarly, Maneta and Voudouris [29] reported actinolite inclusions to be responsible for the green color of quartz from Serifos, and Voudouris and Katerinopoulos [30] described similar green quartz crystals from the skarn occurrences at the Xanthi and Kresti/Drama areas in northern Greece, but without any detailed analytical data. Some recent published studies suggested inclusions of hedenbergite as potential candidates that may be responsible for the color of green quartz crystals in Serifos quartz [27,31].

2. Materials and Methods

We analyzed representative cm-sized prismatic crystals of prase (green quartz), and amethyst. To characterize both the quartz varieties and to search for possible mineral inclusions, we used optical microscopy and electron microprobe analyses (EMPA). Electron microprobe analyses on prase and its inclusions were performed with a JEOL 8530F field emission microprobe at the University of Münster (Germany). This microprobe enables us to take high-resolution back-scattered electron images and also to analyze small crystals with high accuracy. Analytical conditions for amphiboles were 15 kV accelerating voltage, 15 nA beam current, 5 µm spot size, and counting times of 10 s for peak and 5 s for the background signal except for Na and K (5 s for peak and 2 s for background). Prior to quantitative analyses all elements were standardized on matrix matched natural (Na, Mg, K, Si, Al, Ca, Fe, Mn) and synthetic (Ti, Cr) reference materials. Although quartz is known to be a robust mineral in regards to structure, hardness and weathering, it is nevertheless very sensitive to electron irradiation. Thus, special care must be taken analyzing quartz and its trace elements like Al, Fe, Ti, and K. We followed a recently published analytical protocol [32] using a beam current of 80 nA at 15 kV, 60 s peak and 30 s background counting time with a probe diameter of 15 µm. A blank quartz sample was measured along with the unknowns in order to check for "true-zero" concentrations. The average detection limits (3σ) for Al, K, Ti, and Fe is in the range of 0.011 wt % oxides. The phi-rho-z correction was applied to all data. To monitor accuracy and precision over the course of this study micro-analytical reference materials were analyzed and obtained results match published values within error.

Amethyst in polished sections embedded in resin, was analyzed using a JEOL JXA 8900 Superprobe equipped with four wavelength-dispersive spectrometers (WDS) and one energy-dispersive spectrometer (EDS) at the Agricultural University of Athens (Greece). All analyses were performed with an accelerating voltage of 15 kV, 15 nA beam current, slightly defocused beam (2 µm), 20 s counting time on peak position and 10 s for each background. Natural mineral standards used were jadeite (Na), quartz (Si), corundum (Al), diopside (Ca), olivine (Mg), fayalite (Fe), spessartine (Mn), microcline (K), ilmenite (Ti), chromite (Cr), apatite (P) and Ni-oxide (Ni).

Stable isotope analyses were performed at the Stable Isotope and Atmospheric Laboratories, Department of Geology, Royal Holloway, University of London (UK). The oxygen isotope composition of quartz was obtained using a CO_2 laser fluorination system similar to that described by Mattey [33]. Each mineral separate is weighed at 1.7 mg $\pm$ 10%. These were loaded into the 16-holes of a nickel sample tray, which was inserted into the reaction chamber and then evacuated. The oxygen was released by a 30 W Synrad CO_2 laser in the presence of BrF_5 reagent. The yield of oxygen was measured as a calibrated pressure based on the estimated or known oxygen content of the mineral being analyzed. Low yields result in low $\delta^{18}O$ values for all mineral phases, so accurate yield calculations are essential. Yields of >90% are required for most minerals to give satisfactory $\delta^{18}O$ values. The oxygen gas was measured using a VG Isotech (now GV Instruments, Wythenshawe, UK) Optima dual inlet isotope ratio mass spectrometer (IRMS). All values are reported relative to the Vienna Standard Mean Ocean Water (V-SMOW). The data are calibrated to a quartz standard (Q BLC) with a known $\delta^{18}O$ value of +8.8‰ V-SMOW from previous measurements at the University of Paris-6 (France). This has been further calibrated for the RHUL laser line by comparison with NBS-28 quartz. Each 16-hole tray contained up to 12 sample unknowns and 4 of the Q BLC standard. For each quartz sample a small constant daily correction, normally less than 0.3‰, was applied to the data based on the average value for the standard. Overall, the precision of the RHUL system based on standard and sample replicates is better than $\pm$0.1‰.

3. Geological Setting

Serifos Island is located in the NW edge of the Cyclades Island complex. In the Cyclades area, three major tectono-metamorphic units have been distinguished, namely the Cycladic Continental Basement Unit (CCBU), the Cycladic Blueschists Unit (CBU) and the Upper Cycladic Unit (UCU) [34–37].

Serifos is occupied by a metamorphic sequence which comprises [38–40]: (a) a basal gneiss, mylonitic schists and marbles unit, with a maximum thickness of about 200m, that has been interpreted as member of the CCBU; (b) an intermediate rock unit made of amphibolites and overlying greenschists with rare marble intercalations, occupying the northern part of the Island and containing relics of glaucophane, that are typical of the CBU [41]; (c) and finally, an uppermost unit with marbles, ankeritised protocataclastic shales and minor serpentinite [31]. Contacts between the rock units are marked by tectonic structures (Figure 1), namely the Megalo Livadi and Kavos Kyklopas detachment faults, which belong to the West Cycladic Detachment System, (WCDS [42]) and are characterized by top-to-the SSW kinematics on the footwalls.

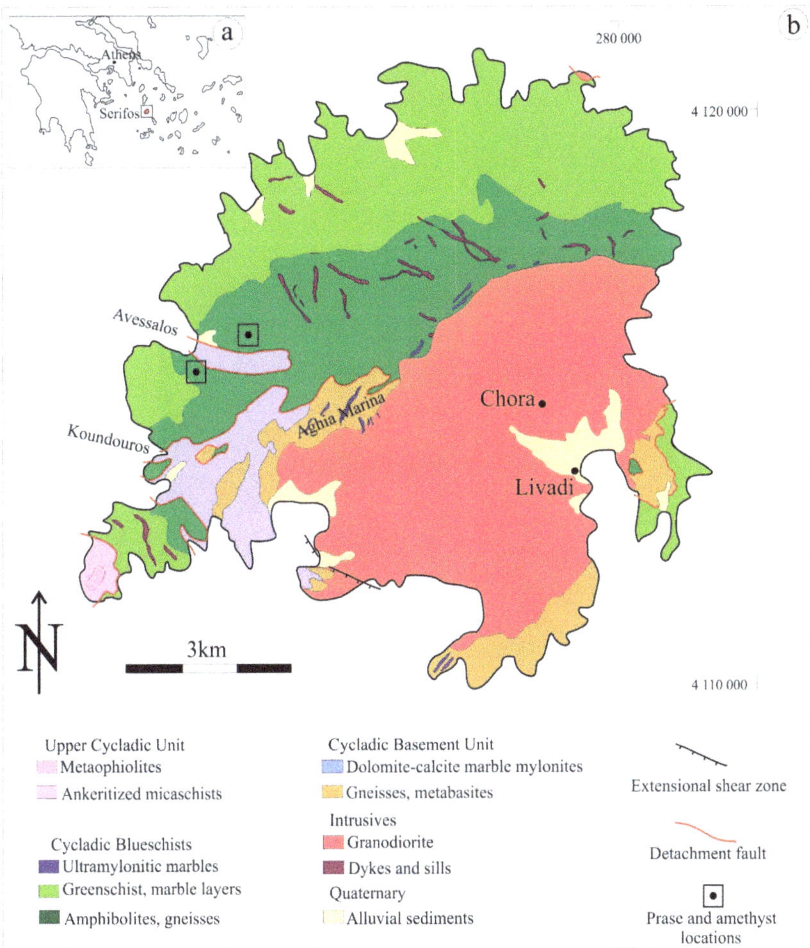

Figure 1. (**a**) Serifos Island is located in the NW edge of the Cyclades Island complex in the Aegean Sea, Greece. (**b**) Simplified geological map of Serifos Island (modified after Ducoux et al. [31]). Black squares indicate the study areas of Avessalos (South) and Neroutsika (North).

The above-mentioned rock units are intruded by an I-type, hornblende-biotite granodiorite which displays two variants [43]: the inner part of the intrusion comprises a fine-grained equigranular, unfoliated rock, while the outer part is composed of a coarse-grained equivalent, characterized by large biotite flakes and mafic enclaves. Many dykes and sills of granodiorite, microgranite and rhyodacite/dacite are scattered around the intrusion [44].

A number of skarns have formed close to the granodiorite intrusion [39–41,43] and have been long known for their Fe-rich mineralisation, that was exploited until the 1960's. Mines are widespread over the Island and were targeted at magnetite-rich skarn zones or in hematite/limonite±barite orebodies in marbles [31]. These skarns are also the general area where the unique Serifos green quartz, mostly known as prase (or prasem), is found (Figure 2). Except for the well-known green quartz, a great variety of other aesthetic minerals can also be found in the skarn zones of Serifos [25].

Figure 2. Field and prase/amethyst specimen photographs: (**a**); The Avessalos location (view towards SW), famous for its prase; (**b**) The Neroutsika location (view towards NE) from where unique combinations of prase with amethyst were recovered (Figure 2g); (**c**) Geode within hedenbergitic skarn, filled with euhedral prase crystals, Avessalos locality; (**d**,**e**) aesthetic prase crystals combinations from Avessalos locality; (**f**) negative scepter, formed by an upper part of amethystine quartz that grows atop a basal part of prase, Avessalos locality; (**g**) bi-colored crystal consisting of a basal prase part and an upper amethyst part of gem quality, Neroutsika locality. Figures and specimen 3d and 3e–f are courtesy of C. Mavrogonatos and P. Voudouris respectively.

Among them, ilvaite, a rare sorosilicate mineral is mainly found in Koundouros area, which is located a few kilometers north from Avessalos. Samples from this locality, are considered to be among the best of its kind.

Andraditic garnets, often zoned are found in euhedral translucent crystals on top of hedenbergite in geodes with varying dimensions in the areas of Avessalos and Agia Marina. Finally, tabular translucent barite and rhombohedral calcite have been found in exceptional crystals in places close to the Fe-rich mineralization [25,30].

The green quartz from Serifos is usually translucent, with a characteristic green color in different intensities. Typical are cigar-shaped crystals which extend from a fine-grained matrix consisting of hedenbergite. Often smaller crystals sprout from the prism of larger crystals, resulting in very aesthetic aggregates. Green quartz is extracted from a few localities in the SW part of Serifos Island. The most important of these are the localities of Avessalos (Figures 1 and 2 and Neroutsika (Figures 1 and 2b). Our samples are from the western part of Serifos (Figure 1), within amphibolite facies rocks and the skarn mineralization [25,26,29,30].

Minor green quartz occurrences can also be found in Agia Marina and Koundouros. The studied samples come from the well-known locality of Avessalos (Figure 2a,c–f), from where the best green quartz samples have been found. Crystals from this site are usually lustrous with vivid green color, in contrast to other sites, from where green quartz is usually not translucent, but with varying intensity of its green color. Green quartz is often found in geodes between brecciated, up to tens of meters-sized blocks of skarnified amphibolites [25,26,29,30]. Brecciation of the amphibolites has been caused by the activation of the WCDS and created pathways through which hydrothermal fluids arising from the granodiorite created the widespread skarns on the Island [31].

Empty spaces between the more-or-less skarnified amphibolite blocks, commonly occur in triangular shapes that may vary significantly in size (Figure 2c), sometimes reaching a width of a few meters across. These spaces are filled by hedenbergite, followed by ilvaite, prase ± amethyst, hematite and carbonate (calcite) deposition towards the centre of the open spaces. Inside these geodes, the prase (green quartz) crystals develop in a variety of crystals shapes. Figure 2d–g displays some aesthetic aggregates with sharply terminated crystals, sceptres and reverse sceptres of prase combined with amethyst from both Avessalos and Neroutsika areas. In many cases, prase is accompanied by calcite and iron-roses composed of hematite [25,26]. Its crystal shape is usually spindle-like in the basal part and evolves into hexagonal on the top of the crystal. Prismatic green quartz crystals are also found, as well as bi-colored, rocket-like shapes, with amethyst in the base and green quartz on top. The sceptres found in the area are normal or inverse with green quartz in the base and an often trigonal amethystine termination (Figure 2f–g). An unusual type of crystals encountered in the area is the "interrupted" prase or amethyst crystals which form when platy calcite crystals are deposited together with quartz in successive layers [21,26].

4. Results

4.1. Mineral Chemistry

Detailed microscopy and back-scattered electron (BSE) images reveal that the Serifos green quartz contains numerous small inclusions. We find two general types of mineral inclusions. The first type comprises myriads of acicular amphibole crystals which are often 10–100 µm long, but their thickness is limited to below a few µm. These thin actinolite needles are practically found everywhere and seem to be randomly oriented. We did not observe any obvious relationship between amphibole orientation and the crystallographic orientation of the quartz crystals. Some of the needles (Figure 3a,b) are clearly curved, a fact that supports random orientation of the amphibole needles in relation to the host quartz crystals. With our electron microprobe we could only identify the needles as actinolitic amphibole but accurate analysis was not possible, due to the small size of the inclusions. We, therefore, assume that the chemical compositions of these thin amphibole needles are identical to the rims of the larger amphibole inclusions, a fact that also seems to be supported by electron microprobe maps (Figure 3e–f).

The second type consists of larger inclusions (Figure 3a–c) with sizes that reach a few tens of µm in length (commonly between 10 and 60 µm), and that always display zonation. EPMA analyses revealed that these inclusions are Ca-Fe-rich actinolitic amphiboles with compositions very close to the end-member ferro-actinolite. These minerals contain about 5 wt % MgO, almost no Na or K, and about 2 wt % Al_2O_3. All the studied larger actinolite inclusions contain a core, which is significantly lighter in BSE images. Microprobe analyses (Table 1) reveal that these cores are also actinolite, with some differences compared to the rims. MnO content is almost doubled compared to the rim (0.56 and 0.25 wt % respectively) and FeO is slightly higher. Subsequently, the cores contain lower MgO and (3.8 wt %) and much less Al_2O_3 (0.3 wt %).

Table 1. Electron microprobe measurements of amphibole inclusions, prase and amethyst from Serifos Island.

	Amphibole					Prase		Amethyst	
n.o.a.	3 (core)		6 (rim)			11		14	
	wt %	2σ	wt %	2σ		wt %	2σ	wt %	2σ
SiO_2	51.1	0.04	50.6	0.57		99.3	0.56	99.9	0.44
TiO_2	0.05	0.00	0.01	0.02		-	-	0.005	0.01
Al_2O_3	0.38	0.01	1.54	0.57		0.26	0.12	0.123	0.11
Cr_2O_3	n.a.	-	n.a.	-		n.a.	-	0.006	0.01
FeO	29.4	0.40	28.3	1.15	Fe_2O_3	0.027	0.019	0.021	0.03
MnO	0.56	0.02	0.25	0.12		n.a.	-	0.006	0.01
MgO	3.81	0.07	4.98	0.72		n.a.	-	0.01	0.01
CaO	11.6	0.04	11.3	0.09		n.a.	-	0.008	0.01
Na2O	0.03	0.02	0.26	0.11		n.a.	-	0.01	0.01
K2O	0.09	0.02	0.19	0.05		0.008	0.007	0.006	0.01
P_2O_5	n.a.	-	n.a.	-		n.a.	-	0.012	0.02
Total	96.9		97.5			99.6		100.2	

	Structural formulae				
Si	8.08	7.85	Si	0.998	0.998
AlVI	0.00	0.15	Ti	0.000	0.000
M1-3 site			Al	0.003	0.000
AlVI	0.07	0.13	Fe^{3+}	0.000	0.0002
Ti	0.01	0.00	Mn	-	-
Fe^{3+}	0.00	0.13	Mg	-	0.000
Mg	0.89	1.15	Ca	-	-
Fe^{2+}	3.87	3.55	Na	-	-
Mn	0.08	0.03	K	0.000	-
Σ M1-3	4.92	5.00	P	-	0.000
M4 site			Σ All	1.001	1.009
Mg	0.00	0.00			
Fe^{2+}	0.00	0.00			
Mn	0.00	0.00			
Ca	1.95	1.88			
Na	0.01	0.08			
Σ M4	1.96	1.96			
A site					
Na	0.00	0.00			
K	0.02	0.04			
Σ A	0.02	0.04			
Σ All	15.0	15.0			

n.a. = not analysed; (-) = below detection, n.o.a. = number of analyses.

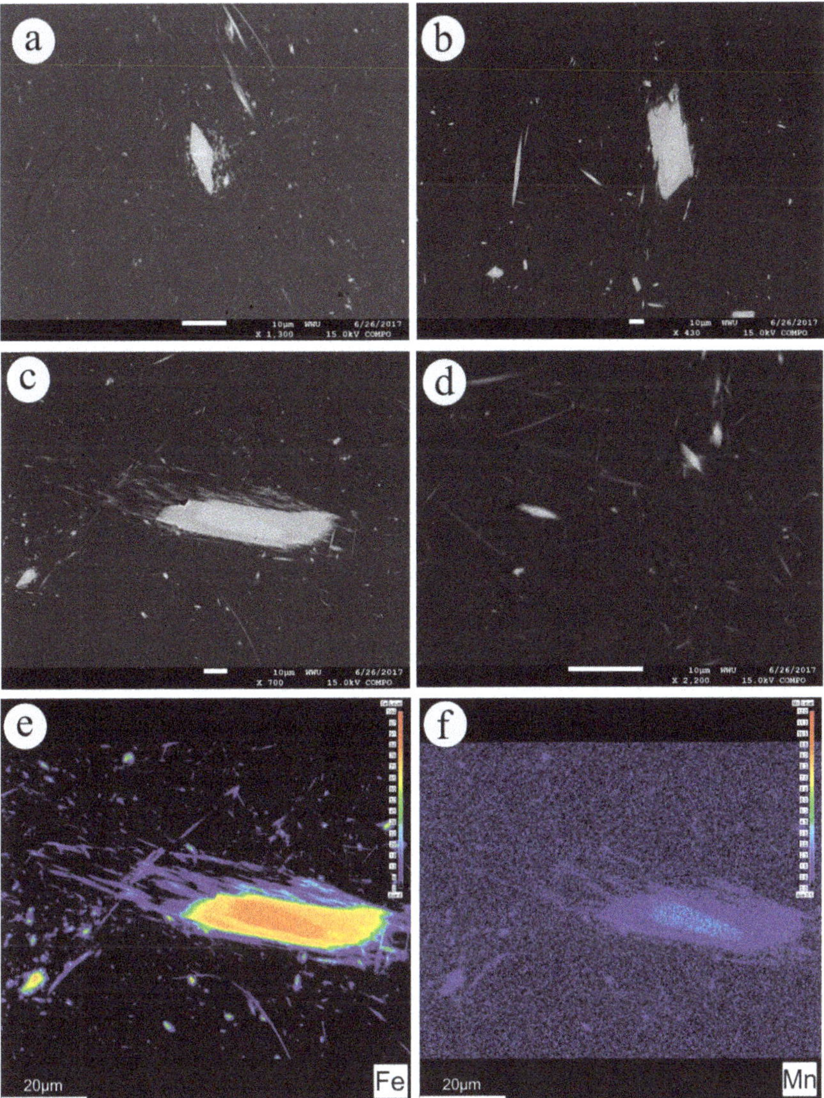

Figure 3. Back-scattered electron (BSE) images and quantitative electron microprobe element maps. (**a**): thin, nm-sized needles of amphibole together with a larger amphibole. The latter larger actinolite contains a core of more calcic actinolite; (**b**) Large amphibole with no clearly defined edges, as the rapid growth of the crystal has stopped. Furthermore, it seems that this amphibole has already determined the outer edge of the crystal which it would have grown into if growth had not been interrupted. The origin of the observed textures is unknown; (**c**) Similar larger amphibole with a fringed edge also indicating rapid growth; (**d**) high-resolution image of the fine-grained matrix. The dark blebs are empty and are interpreted as fluid inclusions that were opened during sample preparation; (**e**) Quantitative electron microprobe map of Fe. The cores of the larger amphiboles contain significantly more FeO (Table 1) than the rims. The color bar on the right-hand side indicates chemical composition (wt % FeO). Note that the very thin needle-shaped (blueish in this Figure) actinolite crystals are too small to be analyzed so that the FeO-content is underestimated; (**f**) Quantitative electron microprobe map of Mn. The cores contain significantly more Mn than the rim actinolite, indicating two different generations of amphibole.

The CaO content is quite similar in both rim and core with values around 11.5 wt %. Finally, traces of Na_2O and K_2O were detected in both rim and core actinolites. Values for both elements increase towards the rims of the crystals, reaching values of up to 0.26 and 0.19 wt % respectively. Average chemical formulae for the compositions from the cores and the rims of the crystals correspond to $[Ca_{1.95}Na_{0.01}K_{0.02}]_{1.98}$ $[Mg_{0.89}Fe^{2+}_{3.87}Mn_{0.08}Al_{0.07}Ti_{0.01}]_{4.92}$ $[Si]_{8.08}$ $O_{22}(OH)_2$ and $[Ca_{1.88}Na_{0.08}K_{0.04}]_2$ $[Mg_{1.15}Fe^{2+}_{3.55}Fe^{3+}_{0.13}Mn_{0.03}Al_{0.13}]_{4.99}$ $[Si_{7.85}Al_{0.15}]_8$ $O_{22}(OH)_2$, respectively. Analytical data from prase and amethyst yielded traces of Al_2O_3, FeO, MgO, MnO, CaO, Na_2O, K_2O and P_2O_5. The highest FeO and Al_2O_3 contents were found in prase (up to 0.26 and 0.024 wt % respectively). Amethyst Fe_2O_3 content reaches up to 0.020 wt % (calculated from the total FeO of the analyses).

4.2. Quartz Oxygen Isotopes

The quartz oxygen isotopic compositions were analyzed in hand-picked prase and amethyst crystals. The compositions of the fluid in equilibrium with the prase and amethyst crystals were calculated following Sharp et al. [45], using temperatures of 350 °C for prase and 250 °C for amethyst, based on the respective mineral assemblages of Salemink [43].

The studied prase samples yielded isotopic $\delta^{18}O_{Qz}$ values of 12.72 and 12.74‰, which correspond to calculated $\delta^{18}O_{Fl}$ values of 7.32 and 7.14‰ (Table 2). Amethyst samples yielded $\delta^{18}O_{Qz}$ values of 9.94 and 9.85‰, which correspond to calculated $\delta^{18}O_{Fl}$ values of 1.04 and 0.95‰ respectively. The isotopic signature of prase is comparable to those of quartz samples from different assemblages of the skarn zone, but amethyst displays significantly lighter isotopic signature (Figure 4).

Table 2. Oxygen isotope compositions of quartz from the skarn of Serifos Island [δ values in‰ relative to Standard Mean Ocean Water (SMOW)]. The oxygen isotope values in the fluid ($\delta^{18}O_{Fl}$) in equilibrium with quartz have been calculated according to Sharp et al., [45].

Sample	Variety	Associated Minerals	$\delta^{18}O_{Qz}$	T (°C)	$\delta^{18}O_{Fl}$
SR2a	prase	ac	12.72	350	7.32
SR2b	prase	ac	12.54	350	7.14
SR1a	amethyst	hem	9.94	250	1.04
SR1b	amethyst	hem	9.85	250	0.95
20–11 *	quartz	ep	11.4	400	6.8
27–30 *	quartz	ep	12.6	400	8.0
26–49 *	quartz	Mt + hem	11.6	410	7.2
26–48 *	quartz	Mt + hem	12.3	400	7.7
26–87 *	quartz	Ep + ac	13.3	405	13.3
137 *	quartz	Ac + cc	14.7	390	14.7
55 *	quartz	iv	12.6	350	6.8
55-(1) *	quartz	iv	12.5	325	6.0
56 *	quartz	iv	13.2	325	6.7
26–35 *	quartz	mt	14.4	325	7.8
ML-2.2 *	quartz	Ac + mt	16.6	265	7.8
ML-1 *	quartz	Ac + mt	16.5	260	7.5
27-7B *	quartz	Py + hem + cc	14.8	250	5.2

Samples marked with (*) are from Salemink [43]. Abbreviations: ac = actinolite; hem = hematite; ep = epidote; mt = magnetite; iv = ilvaite; cc = calcite; py = pyrite.

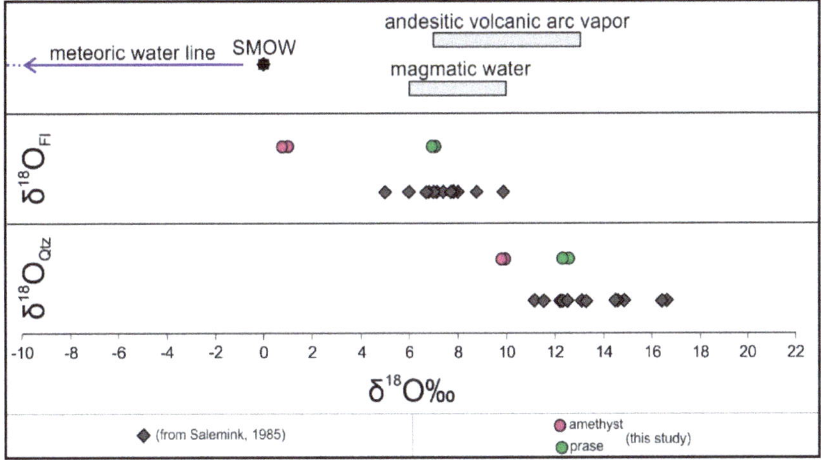

Figure 4. $\delta^{18}O$ values of quartz from the skarn of Serifos Island, Greece. Values for magmatic water and andesitic volcanic arc vapor are from Taylor [46] and Giggenbach [47] respectively.

5. Discussion

Prase or prasem, a green-colored variety of quartz has been long known to collectors because of its rarity, compared to other colorful varieties. Worldwide, the occurrences of natural green quartz are only a few [6].

Among them, prase crystals form the Island of Serifos are believed to be the best of its kind. In rare cases, amethyst, that grows on top of a basal part of prase, creates a unique combination of these two quartz varieties that has not been described elsewhere than Serifos. Despite its high collective and commercial value, little research has been carried about what causes the green color and how this rare variety of quartz is formed together with amethyst in the skarn zones of Serifos. In addition, there is some information in the older literature on color caused by defects in quartz, summarized in an excellent review paper by Rossman [3], but there is very little modern data on the chemical and mineralogical composition of inclusions in quartz in general, and not only in the case of green quartz from Serifos Island.

The majority of previous studies suggested that the color of the Serifos green quartz was caused by hedenbergite-rich pyroxene inclusions (e.g., [27,31]), and only Hyrsl and Niedermayr [28] reported actinolite, based on XRD analyses, but did not present any analytical data. Our new data suggests that it is indeed amphiboles which are responsible for the typical green color. To our knowledge, these are the first quantitative measurements of amphibole inclusions in green quartz and first quantitative data to explain the cause of the color of green quartz from Serifos Island.

Acicular and needle-shaped actinolite inclusions are scattered throughout the quartz crystals. Two types of inclusions were found: a plethora of very small (nanometer-sized) actinolites and a smaller number of larger, chemically zoned actinolite crystals, both without any obvious orientation in relation to the crystallographic orientation of the host quartz crystals.

Chemical analyses suggest that the cores of the larger actinolites is chemical different from the rims, the latter of which are of a similar chemical compositions compared to the very small acicular inclusions. All the larger actinolite inclusions exhibit typical amphibole crystal shapes, but often crystal faces are not well developed. In some cases, amphibole crystals are characterized by fuzzy crystal faces or fringed edges, likely indicating very rapid growth, which has been interrupted probably during rapid depressurization of the fluid-filled vein. This is a process which is typical for pegmatitic veins in which pressure drops rapidly during crack formation and, as a consequence, precipitation of the solutes. In the case of the Serifos green quartz, we surmise that the precipitation of the core actinolites must have happened before the main event, in which the rim actinolite and the quartz itself crystallized. Whilst the core amphibole could have grown relatively slowly from the fluid, the crystallization of the actinolite needles and the overgrowth rim amphibole must have happened very fast indeed. Rapid depressurization and non-equilibrium conditions are in accordance with boiling conditions in the geodes, as suggested by the presence of platy calcite, which is intergrown with prase and amethyst [30,48].

In addition, we identified traces of iron using electron microprobe measurements of the Serifos green quartz which could also contribute to the green color, as suggested by Lehmann and Bambauer [2], Henn, Schultz-Guttler [9] and Czasa et al., [6]. On the other hand, traces of iron, found in amethyst from the same localities are the only impurity that could justify the light violet color of Serifos' amethysts. Our EPMA data revealed a mean content of 0.02 wt % Fe_2O_3, a quantity that is thought to be sufficient to justify the pale violet color of the studied crystals [49].

Previous stable isotope studies revealed a significant contribution of magmatic water for most of the samples of quartz that were studied by Salemink [43]. Oxygen isotopic values in prase are comparable to those of (colorless and milky) quartz, but display a trend towards lighter isotopic signature, which probably corresponds to the evolution of the skarn in its final stages. We find evidence for a major contribution of meteoric (and/or marine) water in case of the amethysts, thus recording a transition to more oxidizing conditions, as the formation of amethyst goes together with incorporation of Fe^{3+} [3] into its structure. The required oxidizing conditions were probably a result of mixing oxidized meteoric or marine water with upwelling hydrothermal fluids [50,51]. Amethyst was deposited in the final stages of the skarn, at temperatures around 250 °C. This is in good agreement with Kievlenko [21], who stated that amethyst in garnet-pyroxene-magnetite skarn deposits crystalizes in late or re-opened fractures and cavities.

Formation of prase and especially amethyst, is therefore, likely to be temporally related to the final stages of the Serifos' skarn, which according to Ducoux et al. [31], is contemporaneous with the intrusion of the granodiorite and the extensional deformation along the West Cycladic detachment system.

6. Conclusions

We conclude that the green color of the Serifos green quartz is caused by nano- to micrometer-sized actinolite inclusions. We find two generations of actinolite crystals, with the earlier and larger actinolite inclusions containing less MgO (and consequently more FeO) and Al_2O_3, but more MnO compared to the later acicular actinolite, which is rimming the earlier larger crystals. Average chemical formulae correspond to $[Ca_{1.95}Na_{0.01}K_{0.02}]_{1.98}$ $[Mg_{0.89}Fe^{2+}_{3.87}Mn_{0.08}Al_{0.07}Ti_{0.01}]_{4.92}$ $[Si]_{8.08}$ O_{22} $(OH)_2$ for the early cores and $[Ca_{1.88}Na_{0.08}K_{0.04}]_2$ $[Mg_{1.15}Fe^{2+}_{3.55}Fe^{3+}_{0.13}Mn_{0.03}Al_{0.13}]_{4.99}$ $[Si_{7.85}Al_{0.15}]_8$ O_{22} $(OH)_2$ for the rims, respectively. The color of Serifos' amethysts is caused of the presence of trace Fe^{3+}. Our conclusions highlight the need for further microanalytical data of quartz crystals, regardless of the color, in order to better understand the procedures under which this unique mineralogical assemblage of Serifos Island was generated. The existence two different generations of actinolite could be used to monitor the chemical composition of the fluids from which the amphibolies and the quartz crystals in the geodes were formed. Future work should include other analytical techniques (e.g., Mössbauer spectroscopy, LA-ICP-MS, EPR etc.) that could also help to define the cause of the green color in prase and amethyst. Moreover, fluid inclusions studies in all quartz varieties from the Serifos Island skarn deposits, should be conducted, as to monitor temperature and salinity of the skarn-forming fluids.

Author Contributions: C.M. collected the samples. J.B. assisted by S.F., I.B. and S.X., evaluated the mineralogical data. P.V. and C.M. evaluated the isotopic analyses. S.K., C.M. and P.V. wrote the manuscript.

Funding: This project was partially funded by a DAAD/IKY Program between University of Münster and the Agricultural University of Athens.

Acknowledgments: The authors wish to thank M. Trogisch at Münster University for preparation of the thin sections and B. Schmitte for her help with the EMPA measurements Two anonymous reviewers are thanked for their constructive comments on the manuscript.

Conflicts of Interest: The authors declare no conflict of interest.

References

1. Hutton, D. Defects in the coloured varieties of quartz. *J. Gemmol.* **1974**, *14*, 156–166. [CrossRef]
2. Lehmann, G.; Bambauer, H.U. Quarzkristalle und ihre Farben. *Angew. Chem.* **1973**, *85*, 281–289. [CrossRef]
3. Rossman, G.R. Coloured Varieties of the Silica Minerals. *Rev. Mineral.* **1994**, *29*, 433–467.
4. Applin, K.R.; Hicks, B.D. Fibers of Dumortierite in quartz. *Am. Mineral.* **1987**, *72*, 170–172.
5. Chauhan, M. Prasiolite with inclusion influenced by Brazil-law twinning. *Gems Gemol.* **2014**, *50*, 159–160.
6. Czaja, M.; Kądziołka-Gaweł, M.; Konefał, A.; Sitko, R.; Teper, E.; Mazurak, Z.; Sachanbińskiet, M. The Mossbauer spectra of prasiolite and amethyst crystals from Poland. *Phys. Chem. Miner.* **2017**, *44*, 365–375. [CrossRef]
7. Neumann, E.; Schmetzer, K. Farbe, Farbursache und Mechanismen der Farbumwandlung von Amethyst. *Zeitschrift der Deutschen Gemmologischen Gesellschaft* **1984**, *33*, 35–42.
8. Neumann, E.; Schmetzer, K. Mechanism of thermal-conversion of colour and colour-centers by heat treatment of amethyst. *N. Jahr. Miner. Monatshefte* **1984**, *6*, 272–282.
9. Henn, U.; Schultz-Güttler, R. Review of some current coloured quartz varieties. *J. Gemmol.* **2012**, *33*, 29–43. [CrossRef]
10. Paradise, J.R. The natural formation and occurrence of green quartz. *Gems Gemol.* **1982**, *18*, 39–42. [CrossRef]
11. Hebert, L.B.; Rossman, G. Greenish quartz from the Thunder Bay Amethyst Mine Panorama, Thunder Bay, Ontario, Canada. *Can. Min.* **2008**, *46*, 61–74. [CrossRef]
12. McArthur, J.R.; Jennings, E.A.; Kissin, S.A.; Sherlock, R.L. Stable-isotope, fluid-inclusion, and mineralogical studies relating to the genesis of amethyst, Thunder Bay Amethyst Mine, Ontario. *Can. J. Earth Sci.* **1993**, *30*, 1955–1969. [CrossRef]
13. Brooks, J. Marlborough Creek chrysoprase deposits, Rockhampton district, central Queensland. *Gems Gemol.* **1965**, *11*, 323–330.
14. Visser, J. Nephrite and chrysoprase of Silesia. *Gemmologist* **1947**, *16*, 229–230.
15. Sachanbinski, M.; Janeczek, J.; Platonov, A.; Rietmeijer, F.J.M. The origin of colour of chrysoprase from Szklary (Poland) and Sarykul Boldy (Kazakhstan). *N. Jahrb. Miner.* **2001**, *177*, 61–76. [CrossRef]
16. Yang, L.; Mashkovtsev, R.I.; Botis, S.M.; Pan, Y. Multi-spectroscopic study of green quartzite (Guizhou Jade) from the Qinglong antimony deposit, Guizhou Province, China. *J. China Univ. Geosci.* **2007**, *18*, 327–329.
17. Lieber, W. *Amethyst: Geschichte, Eigenschaften, Fundorte*; Christian Weise Verlag: München, Germany, 1994; 188p.
18. Dedushenko, S.K.; Makhina, I.B.; Mar'in, A.A.; Mukhanov, V.A.; Perfiliev, Y.U.D. What oxidation state of iron determines the amethyst colour? *Hyperfine Interact.* **2004**, *156*, 417–422. [CrossRef]
19. Di Benedetto, F.; Innocenti, M.; Tesi, S.; Romanelli, M.; D'Acapito, F.; Fornaciai, G.; Montegrossi, G.; Pardi, L.A. A Fe K-edge XAS study of amethyst. *Phys. Chem. Miner.* **2010**, *37*, 283–289. [CrossRef]
20. SivaRamaiah, G.; Lin, J.; Pan, Y. Electron paramagnetic resonance spectroscopy of Fe^{3+} ions in amethyst: thermodynamic potentials and magnetic susceptibility. *Phys. Chem. Miner.* **2011**, *38*, 159–167. [CrossRef]
21. Kievlenko, E.Y. *Geology of Gems*; Ocean Pictures Ltd.: Romsey, UK, 2003; 468p.
22. Tatatrinov, A.V. Silic Minerals and Amethyst formation Conditions in Skarn-Magnetite Fields of the South of Siberian Platform. In *Mineralogy and Genesis of East Siberia Gems*; Nauka: Novosibirsk, Russia, 1983; pp. 34–41. (In Russian)
23. Flink, G. Bidrag till Sveriges mineralogy. *Arkiv Kemi. Mineral. Geol.* **1910**, *3*, 163–166.
24. Geijer, P.; Magnusson, N.H. De mellansvenska järnmalmernas geologi. *Sver. Geol. Und.* **1944**, *35*, 558.
25. Vogt, M. Serifos-die Mineralien-Insel in der griechischen Ägäis. *Lapis* **1991**, *16*, 26–37. (In German)
26. Gauthier, G.; Albandakis, N. Minerals from the Serifos skarn, Greece. *Mineral. Rec.* **1991**, *22*, 303–308.
27. Dibble, H.L. *Quartz: An Introduction to Crystalline Quartz*; Dibble Trust Fund Publ.: New York, NY, USA, 2002; 100p.
28. Hyrsl, J.; Niedermayr, G. *Magic World: Inclusions in Quartz/Geheimnisvolle Welt: Einschlüsse im Quarz*; Bode Verlag GmbH: Haltern, Germany, 2003.
29. Maneta, V.; Voudouris, P. Quartz megacrysts in Greece: Mineralogy and environment of formation. *Bull. Geol. Soc. Greece* **2010**, *43*, 685–696. [CrossRef]
30. Voudouris, P.; Katerinopoulos, A. New occurences of mineral megacrysts in tertiary magmatic-hydrothermal and epithermal environments in Greece. *Doc. Nat.* **2004**, *151*, 1–21.
31. Ducoux, M.; Branquet, Y.; Jolivet, L.; Arbaret, L.; Grasemann, A.; Rabillard, C.; Gumiaux, C.; Drufin, S. Synkinematic skarns and fluid drainage along detachments: The West Cycladic Detachment System on Serifos Island (Cyclades, Greece) and its related mineralization. *Tectonophysics* **2017**, *695*, 1–26. [CrossRef]

32. Kronz, A.; van den Kerkhof, A.M.; Müller, A. Analysis of low element concentrations in Quartz by Electron Microprobe. In *Quartz: Deposits, Mineralogy, and Analytics*; Götze, J., Möckel, R., Eds.; Springer: New York, NY, USA, 2012; pp. 191–217.

33. Mattey, D.P. LaserPrep: An Automatic Laser-Fluorination System for Micromass "Optima" or "Prism" Mass Spectrometers. *Micromass Appl. Note* **1997**, *207*, 8.

34. Jolivet, L.; Brun, J.P. Cenozoic geodynamic evolution of the Aegean region. *Int. J. Earth Sci.* **2010**, *99*, 109–138. [CrossRef]

35. Ring, U.; Glodny, J.; Will, T.; Thomson, S. The Hellenic subduction system: high-pressure metamorphism, exhumation, normal faulting, and large-scale extension. *Annu. Rev. Earth Planet. Sci.* **2010**, *38*, 45–76. [CrossRef]

36. Jolivet, L.; Faccenna, C.; Huet, B.; Labrousse, L.; Le Pourhiet, L.; Lacombe, O.; Lecomte, E.; Burov, E.; Denèle, Y.; Brun, J.-P.; et al. Aegean tectonics: Strain localisation, slab tearing and trench retreat. *Tectonophysics* **2013**, *597–598*, 1–33. [CrossRef]

37. Maluski, H.; Bonneau, M.; Kienast, J.R. Dating the metamorphic events in the Cycladic area; 39Ar/40Ar data from metamorphic rocks of the Island of Syros (Greece). *Bull. Soc. Géol. Fr.* **1987**, *3*, 833–842.

38. Grasemann, B.; Petrakakis, K. Evolution of the Serifos metamorphic core complex. *J. Virtual Explor.* **2007**, *27*, 1–18. [CrossRef]

39. Rabillard, A.; Arbaret, L.; Jolivet, L.; Le Breton, N.; Gumiaux, C.; Augier, R.; Grasemann, B. Interactions between plutonism and detachments during metamorphic core complex formation, Serifos Island (Cyclades, Greece). *Tectonics* **2015**, *34*, 1080–1106. [CrossRef]

40. Schneider, D.A.; Senkowski, C.; Vogel, H.; Grasemann, B.; Iglseder, Ch.; Schmitt, A.K. Eocene tectonometamorphism on Serifos (western Cyclades) deduced from zircon depth-profiling geochronology and mica thermochronology. *Lithos* **2011**, *125*, 151–172. [CrossRef]

41. Salemink, J. On the Geology and Petrology of Serifos Island (Cyclades, Greece). *Annu. Geol. Pays Hell.* **1980**, *30*, 342–365.

42. Grasemann, B.; Schneider, D.A.; Stockli, D.F.; Iglseder, C. Miocene bivergent crustal extension in the Aegean: Evidence from the western Cyclades (Greece). *Lithosphere* **2012**, *4*, 23–39. [CrossRef]

43. Salemink, J. Skarn and Ore Formation at Serifos, Greece. Ph.D. Thesis, University of Utrecht, Utrecht, The Netherlands, 1985.

44. Stouraiti, C.; Baziotis, I.; Asimow, P.D.; Downes, H. Geochemistry of the Serifos calc-alkaline granodiorite pluton, Greece: Constraining the crust and mantle contributions to I-type granitoids. *Int. J. Earth Sci. (Geol. Rundsch)* **2018**, *107*, 1657. [CrossRef]

45. Sharp, Z.D.; Gibbons, J.A.; Maltsev, O.; Atudorei, V.; Pack, A.; Sengupta, S.; Shock, E.L.; Knauth, L.P. A calibration of the triple oxygen isotope fractionation in the SiO_2–H_2O system and applications to natural samples. *Geochim. Cosmochim. Acta* **2016**, *186*, 105–119. [CrossRef]

46. Taylor, H.P., Jr. Oxygen and hydrogen isotope relationships in hydrothermal mineral deposits. In *Geochemistry of Hydrothermal Ore Deposits*, 2nd ed.; Barnes, H.L., Ed.; Wiley: New York, NY, USA, 1979; pp. 236–318.

47. Giggenbach, W.F. The origin and evolution of fluids in magmatic-hydrothermal systems. In *Geochemistry of Hydrothermal Ore Deposits*, 3rd ed.; Barnes, H.L., Ed.; Wiley: New York, NY, USA, 1997; pp. 737–796.

48. Simmons, S.F.; Browne, P.R.L. Hydrothermal minerals and precious metals in the Broadlands-Ohaaki geothermal system: Implications for understanding low-sulfidation epithermal environments. *Econ. Geol.* **2001**, *95*, 971–999. [CrossRef]

49. Dana, J.D.; Dana, E.S.; Frondel, C. *The System of Mineralogy, Vol. 3: Silica Minerals*; John Wiley & Sons: New York, NY, USA, 1962.

50. Fournier, R.O. The behaviour of silica in hydrothermal solutions. Geology and geochemistry of epithermal systems. *Rev. Econ. Geol.* **1985**, *2*, 45–61.

51. Voudouris, P.; Melfos, V.; Mavrogonatos, C.; Tarantola, A.; Götze, J.; Alfieris, D.; Maneta, V.; Psimis, I. Amethyst occurrences in Tertiary volcanic rocks of Greece: Mineralogical, fluid inclusion and oxygen isotope constraints on their genesis. *Minerals* **2018**, *8*, 324. [CrossRef]

Article

Metamorphic and Metasomatic Kyanite-Bearing Mineral Assemblages of Thassos Island (Rhodope, Greece)

Alexandre Tarantola [1,*], Panagiotis Voudouris [2], Aurélien Eglinger [1], Christophe Scheffer [1,3], Kimberly Trebus [1], Marie Bitte [1], Benjamin Rondeau [4], Constantinos Mavrogonatos [2], Ian Graham [5], Marius Etienne [1] and Chantal Peiffert [1]

[1] GeoRessources, Faculté des Sciences et Technologies, Université de Lorraine, CNRS, F-54506 Vandœuvre-lès-Nancy, France; aurelien.eglinger@univ-lorraine.fr (A.E.); christophe.scheffer.1@ulaval.ca (C.S.); kimberly.trebus5@etu.univ-lorraine.fr (K.T.); marie.bitte8@etu.univ-lorraine.fr (M.B.); marius.etienne5@gmail.com (M.E.); chantal.peiffert@univ-lorraine.fr (C.P.)
[2] Department of Geology & Geoenvironment, National and Kapodistrian University of Athens, 15784 Athens, Greece; voudouris@geol.uoa.gr (P.V.); kmavrogon@geol.uoa.gr (C.M.)
[3] Département de Géologie et de Génie Géologique, Université Laval, Québec, QC G1V 0A6, Canada
[4] Laboratoire de Planétologie et Géodynamique, Université de Nantes, CNRS UMR 6112, 44322 Nantes, France; benjamin.rondeau@univ-nantes.fr
[5] PANGEA Research Centre, School of Biological, Earth and Environmental Sciences, University of New South Wales, Sydney, NSW 2052 Australia; i.graham@unsw.edu.au
* Correspondence: alexandre.tarantola@univ-lorraine.fr; Tel.: +33-3-72-74-55-67

Received: 27 February 2019; Accepted: 23 April 2019; Published: 25 April 2019

Abstract: The Trikorfo area (Thassos Island, Rhodope massif, Northern Greece) represents a unique mineralogical locality with Mn-rich minerals including kyanite, andalusite, garnet and epidote. Their vivid colors and large crystal size make them good indicators of gem-quality materials, although crystals found up to now are too fractured to be considered as marketable gems. The dominant lithology is represented by a garnet–kyanite–biotite–hematite–plagioclase ± staurolite ± sillimanite paragneiss. Thermodynamic Perple_X modeling indicates conditions of ca. 630–710 °C and 7.8–10.4 kbars. Post-metamorphic metasomatic silicate and calc-silicate (Mn-rich)-minerals are found within (i) green-red horizons with a mineralogical zonation from diopside, hornblende, epidote and grossular, (ii) mica schists containing spessartine, kyanite, andalusite and piemontite, and (iii) weakly deformed quartz-feldspar coarse-grained veins with kyanite at the interface with the metamorphic gneiss. The transition towards brittle conditions is shown by Alpine-type tension gashes, including spessartine–epidote–clinochlore–hornblende-quartz veins, cross-cutting the metamorphic foliation. Kyanite is of particular interest because it is present in the metamorphic paragenesis and locally in metasomatic assemblages with a large variety of colors (zoned blue to green/yellow-transparent and orange). Element analyses and UV-near infrared spectroscopy analyses indicate that the variation in color is due to a combination of Ti^{4+}–Fe^{2+}, Fe^{3+} and Mn^{3+} substitutions with Al^{3+}. Structural and mineralogical observations point to a two-stage evolution of the Trikorfo area, where post-metamorphic hydrothermal fluid circulation lead locally to metasomatic reactions from ductile to brittle conditions during Miocene exhumation of the high-grade host-rocks. The large variety of mineral compositions and assemblages points to a local control of the mineralogy and fO_2 conditions during metasomatic reactions and interactions between hydrothermal active fluids and surrounding rocks.

Keywords: kyanite; Mn-rich silicates; Rhodope; Thassos; amphibolite facies; metasomatism

---

1. Introduction

Dispersed metal ions in substitution for Al and Si may lead to a large variety of colors in calc- and alumino-silicate minerals, e.g., [1,2]. For instance, the most common natural color of kyanite is blue but it can also be green, yellow, orange, white, black, grey or colorless as a function of the nature of elemental substitutions [3,4]. Kyanite is composed of usually >98 wt. % Al_2O_3 and SiO_2. The ~2 wt. % left is generally dominated by Fe, Ti, Mg, Mn, Cr and V. The variation in blue colors is attributed to Ti and/or Fe substitution and intervalence charge transfers within the crystal lattice [5], while the orange color of kyanite is generally attributed to the presence of Mn^{3+} [1,6–9]. The incorporation of trace elements is interpreted as the result of (i) oxygen fugacity during crystal growth (Fe), (ii) metamorphic grade, temperature of formation (Ti), or the nature of the protolith (Cr and V) [4]. In the same way, Mn^{3+} may substitute for Al^{3+} in andalusite [10,11] resulting in the dark green Mn-rich variety of andalusite, formerly known as viridine [2,12]. This may also be the case for epidotes e.g., [13], garnets e.g., [14] and other metamorphic and metasomatic Ca-Al silicates. Among the interesting trace elements to produce gem quality calc- and alumino-silicate minerals, Mn seems to play an important role. High-*PT* manganian silicate and calc-silicate metamorphic assemblages are very rare on Earth and require very specific conditions such as the preservation of high fO_2 during the entire metamorphic cycle, e.g., [15,16]. The presence in the protolith of Mn-oxides capable of buffering fO_2 to high levels during metamorphism as well as low fluid/rock ratio during the whole geodynamic evolution are necessary [17,18]. The best way to produce gem-quality Mn-rich silicates seems thus hydrothermal fluid circulation through an already Mn-rich protolith.

The geology of Greece is marked by a recent alpine belt, which results in outstanding localities for minerals and gemstones owing to high *PT* metamorphic conditions and subsequent magmatism and hydrothermalism, e.g., [19–23]. Among these localities, the island of Thassos, in the Rhodope Belt, represents a unique mineralogical locality for kyanite and uncommon varieties of manganiferous/manganian and magnesian silicate and calc-silicate minerals, including garnet, andalusite and epidote among others [19,24–31]. These crystals are often large (several centimeters long) and vividly colored, translucent but commonly fractured, which makes them usually inappropriate for faceting gemstones, but more proper for cabochon-shaped material. However, considering the extent of the geological units, their occurrence is promising for gemstones exploration in the area. In this article, we use the term "near gem-quality" as a reference to their aesthetic aspect (large size and vivid color) rather than to their potential as facetable material.

Thassos Island is part of the Southern Rhodope Metamorphic Core Complex (SRCC), with *PT* conditions recorded by the garnet–kyanite–biotite–hematite–plagioclase ± staurolite ± sillimanite assemblage of the paragneisses of the Trikorfo area (intermediate unit) in the range 600–650 °C, 4–7 kbars [24]. At the Trikorfo area, four main distinct lithologies, often bearing Ca-, Mg- and Mn-rich silicates, are distinguished with (i) the dominant metamorphic lithology composed of metasedimentary rocks, mostly mica schists and paragneisses, with intercalation of carbonate layers especially towards the base and top of the formation [24], (ii) green to red horizons parallel to the regional foliation enriched in silicate and calc-silicate minerals always showing a typical mineralogical zonation from the contact with the paragneisses/mica schists to the center of the layers, (iii) kyanite-andalusite bearing mica schists, and (iv) kyanite–quartz ± feldspar ± andalusite weakly deformed coarse-grained veins. The transition to brittle deformation conditions is shown by quartz–clinochlore–ilmenite–adularia–albite alpine-type tension gashes cross-cutting the metamorphic foliation [25]. Locally, when cross-cutting Mn-rich layers, these brittle veins can host spessartine and Mn-bearing clinochlore, epidote and hornblende.

On the basis of the description made by Voudouris et al. [20], the aim of the present study is (i) to obtain a complete inventory of the metamorphic and metasomatic minerals and their assemblages found at the Trikorfo site, (ii) to evaluate the cation substitution responsible for the color variations, (iii) to reevaluate the *PT* conditions attained by the gneiss host-rock on the basis of a comprehensive petrological phase equilibria using the Perple_X program, and (iv) to discuss the conditions of formation, metamorphic versus metasomatic, of these exceptional parageneses and assemblages of

(Mn-rich)-silicate and calc-silicate minerals. Kyanite, as a ubiquitous silicate found in both metamorphic and localized metasomatic assemblages, is the referent mineral described in the manuscript.

2. Geological Context

2.1. The Southern Rhodope Core Complex and Thassos Island

The Aegean domain, which extends from the Rhodope Massif in the north to the island of Crete in the south of the Hellenides, is a broad metamorphic domain whose formation started with the subduction of the Adriatic microplate (the northern part of the Africa Plate) below the Eurasia Plate, in late Cretaceous to Eocene times, e.g., [32–35]. The progressive southward retreat of the downgoing slab since the Oligocene is responsible for the exhumation of metamorphic core complexes (MCC) along low-angle detachment faults [36]. The Southern Rhodope Core Complex (SRCC) [37–41] and the Attic-Cycladic Metamorphic Complex (ACMC) [42–44] are the two main metamorphic domains resulting from Aegean extension during the Oligocene to Miocene. The Aegean domain is now in a back-arc position and the subduction front is currently located to the south of the island of Crete [35]. The SRCC, located in the northern part of Greece, displays an almost triangular shape and extends to south to southern Bulgaria [40]. It is mainly constituted by the Vertiskos Gneiss Complex from the Serbo-Macedonian Massif and to the west by the Chalkidiki Peninsula dominated by gneisses from the Sideroneron Massif (Figure 1). The area is marked by Jurassic and Cretaceous age ultra-high-pressure metamorphic events, including kyanite eclogites, followed by Barrovian type metamorphism, e.g., [37–41].

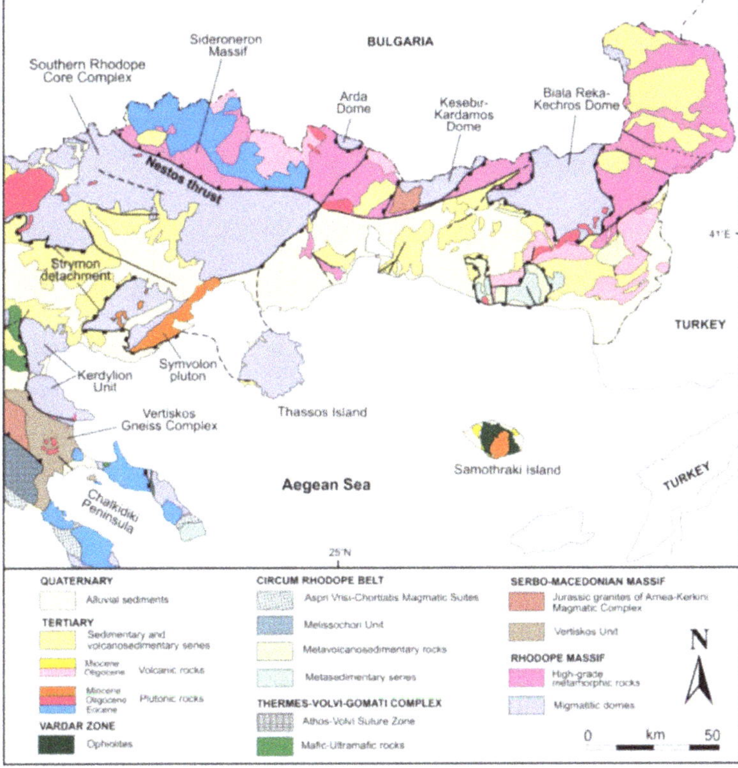

Figure 1. Geological map and regional tectonic framework of northern Greece, modified after Melfos and Voudouris [45].

Thassos Island, with an area of 400 km^2, exposes the south-east end of the SRCC. The geology of the island is dominated by the juxtaposition of shallow-dipping units mainly made of marbles, orthogneisses and paragneisses/mica schists (Figure 2). It is assumed that orthogneisses are the result of metamorphism of Hercynian plutonic units of the Greek Rhodope Massif [36,46] and that the protoliths of marbles and mica schists correspond to the Mesozoic sedimentary cover overlying the Hercynian plutonic basement.

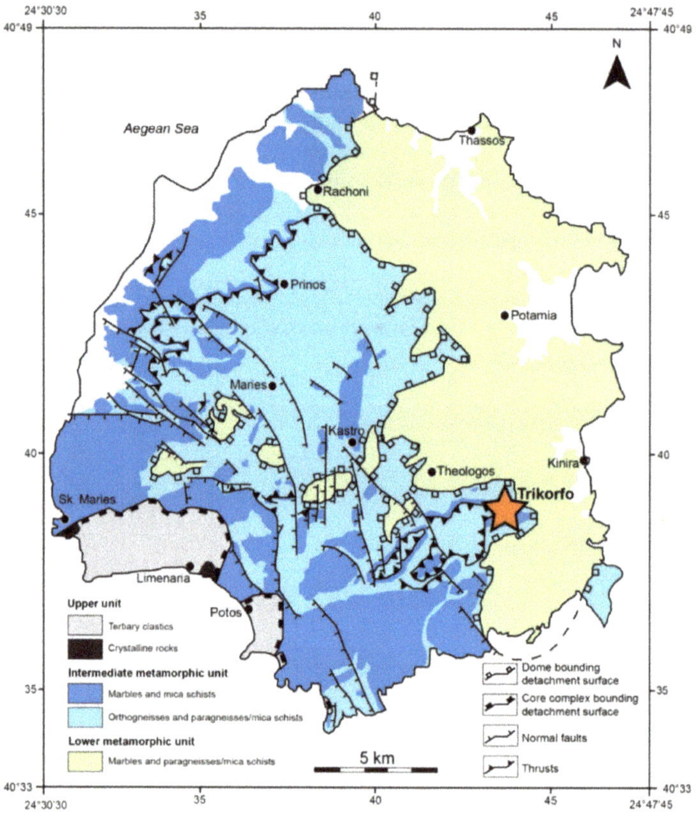

Figure 2. Geological map of Thassos Island with main units, modified from Wawrzenitz and Krohe [36] and Brun and Sokoutis [40]. The location of the Trikorfo area is indicated with an orange star.

Thassos Island is represented by three main metamorphic units with, from base to top, (i) the lower metamorphic unit dominated by mylonitic marbles with intercalations of mica schists and paragneisses forming metamorphic domes to the east of the island, (ii) an intermediate metamorphic unit composed at the base of gneisses overlapped by marbles with paragneiss, mica schist and amphibolite intercalations, (iii) the upper unit is considered as the hanging wall, consisting of non-mylonitic metamorphic rocks dominated by gneisses, migmatites, pegmatites and locally ultramafic rocks and marbles [36,40]. Metamorphic rocks from the lower and intermediate units reached amphibolite facies conditions [24,40,47,48], achieved during crustal thickening [49] at 21–18 Ma and 26–23 Ma (Rb–Sr mica ages from the main metamorphic fabric) for the lower and intermediate units, respectively [36].

Retrograde metamorphism is marked by the development of mylonite marbles, common boudinage along a shallow-dipping foliation with NE–SW stretching lineation with a top to SW sense of shear, and widespread brittle tension gashes [36,40,48,50]. Rb–Sr (on white mica and biotite) and U–Pb (on monazite and xenotime) dating by the above authors revealed a continuous cooling history under ductile to brittle conditions from 700 °C to 300 °C from 26 Ma to 12 Ma. The exhumation *PT* path is interpreted as the result of the progressive uplift of the metamorphic rocks during the formation of the metamorphic core complex [36,51]. The emplacement of the low-angle detachment fault in the Oligocene-Miocene is also associated with the intrusion of syntectonic plutons at the SRCC scale, where the Symvolon pluton crops out 30 km to the NW of Thassos Island [52–54] (Figure 1).

2.2. The Trikorfo Area

The study area at Trikorfo is located along the major SW-dipping low-angle normal detachment separating the lower from the intermediate unit of Thassos Island. Mica schists and paragneisses predominate and tectonically overlay the marbles of Theologos (i.e. lower unit). Muscovite, phlogopite, or two mica quartzo-feldspathic schists (mica schists or paragneisses) with plagioclase (oligoclase) porphyroblasts are the most common lithologies, locally interbedded with carbonate/dolomitic rocks towards the base and top of the formation [24]. In many occurrences, these rocks are rich in minerals from the epidote and garnet groups together with common tourmaline (schorl mainly) and opaque minerals, mostly hematite and ilmenite. Metamorphic Al-silicate minerals (kyanite, fibrolitic sillimanite, andalusite, staurolite) are also documented. These mica schists are thus likely derived from calcareous pelitic sediments [24]. Interestingly, this unit also locally presents uncommon Mn-rich mineral assemblages attributed to post-metamorphic hydrothermal recrystallization [20,24,26]. The garnet–kyanite–biotite–hematite–plagioclase ± staurolite ± sillimanite paragenesis of the mica schists, contemporaneous with the ductile fabric related to extensional deformation, has been specifically interpreted as having attained *PT* conditions of 5.5 ± 1.5 kbars and 600 ± 50 °C, without any evidence of partial melting [24]. Voudouris et al. [20] documented at the top of the formation the presence of an orthogneiss at the contact with the paragneisses/mica schists. Although not exposed, a granitoid is considered to be genetically related to the widespread Miocene gold mineralization in the area [27].

3. Materials and Methods

Representative samples of host-rocks, mineral assemblages and veins were selected for bulk rock analyses and conventional 30 μm polished sections for petrographic observations and electron probe microanalyses (EPMA). Individual kyanite crystals were prepared for bulk rock analyses (powder) and for EPMA and Laser Ablation-Inductively Coupled Plasma Mass Spectroscopy (LA-ICPMS) on 200–300 μm thick sections, and UV-visible-near infrared absorption spectroscopy on 1 mm thick sections. The metamorphic paragenesis of the dominant gneiss of the area was modeled by Perple_X for *PT* conditions estimation. This work represents a compilation of samples collected by different authors during successive field campaigns, resulting in a large variety of sample labelling. All samples come from the Trikorfo area (intermediate unit of Thassos Island) (Figure 2).

3.1. Bulk-Rock Analyses

Host-rock samples CS16_292b and CS16_297 and individual kyanite crystals were powderized so as to obtain a grain-size of about 80 μm. Kyanite crystals were cut as to not contain any apparent alteration minerals, mainly micas, and cleaned with deionized water. Bulk compositions (oxides and trace elements) were obtained by ICP-Optical Emission Spectrometry and ICP-Mass Spectrometry (LiBO$_2$ fusion) at the CRPG-CNRS laboratory (Nancy, France). Sample preparation, analytical conditions and limits of detection are given in Carignan et al. [55].

3.2. Electron Probe MicroAnalyses (EPMA)

The chemical composition of minerals (samples CS16_292b, CS297, KT01-03, KT07 and MB13) was determined using a Cameca SX100 electron microprobe analyzer (EPMA) equipped only with wavelength dispersive spectrometers at the GeoRessources laboratory (Nancy, France) operating with an emission current of 20 nA, an acceleration voltage of 15 kV, and a beam diameter of 1 μm.

The following elements, monochromators, standards, and limits of detection were used: Na (TAP, albite, 515 ppm), Si (TAP, albite, 330 ppm), Mg (TAP, olivine, 265 ppm), Al (TAP, Al_2O_3, 300 ppm), K (LPET, orthoclase, 200 ppm), Ca (PET, andradite, 370 ppm), Ti (LPET, $MnTiO_3$, 250 ppm), Mn (LIF, $MnTiO_3$, 635 ppm), Fe (LIF, Fe_2O_3, 1003 ppm). Three representative samples of blue-green-colorless, yellow and orange kyanite crystals from the same locality as the ones analyzed by bulk geochemistry were selected for EPMA (profiles) analyses and X-ray maps for the elements Al, Si, Fe, Ti and Mn. All analyses were performed with FeO and MnO and converted to Fe_2O_3 and Mn_2O_3 for epidote and andalusite group minerals and Ti-hematite.

Mineral analyses (samples Th01 to Th06) carried-out at Hamburg used a Cameca-SX 100 WDS, accelerating voltage of 20 kV, a beam current of 20 nA and counting time of 20 s. Standards used were andradite (Si, Fe and Ca), Al_2O_3 (Al), MgO (Mg), albite (Na), orthoclase (K), and $MnTiO_3$ (Ti and Mn). Corrections were made using the PAP online program [56].

3.3. Laser Ablation-Inductively Coupled Plasma Mass Spectroscopy (LA-ICPMS)

Kyanite in-situ quantitative LA-ICPMS elemental analyses were performed at GeoRessources laboratory (Nancy, France) using an Agilent 7500c quadrupole ICPMS coupled with a 193 nm GeoLas ArF Excimer laser (MicroLas, Göttingen, Germany). Laser ablation was performed continuously with a speed of 2 µm/s with a constant 5 Hz pulse frequency and a constant fluence of 4.20 J/cm^2 by focussing the beam at the sample surface, from edge to edge of the crystals, along the same profiles analyzed by EPMA. Diameter of the ablation spot was 32 µm. Helium was used as carrier gas to transport the laser-generated particles from the ablation cell to the ICPMS and argon was added as an auxiliary gas via a flow adapter before the ICP torch. Typical flow rates of 0.5 L/min for He and 1 L/min for Ar were used. EPMA analyses of Si and Al served as internal standards. The certified reference materials NIST SRM 610 and 612 (concentrations from Jochum et al. [57]) were used as external standards for calibration of all analyses; they were analyzed twice at the beginning and at the end for each kyanite sample, following a bracketing standardization procedure. ^{27}Al and ^{29}Si were measured with a dwell-time of 10 ms; the following 17 isotopes were measured with a dwell-time of 20 ms for each: ^{7}Li (limit of detection of 20 ppm), ^{23}Na (70 ppm), ^{24}Mg (1 ppm), ^{39}K (50 ppm), ^{43}Ca (6000 ppm), ^{47}Ti (20 ppm), ^{51}V (1 ppm), ^{53}Cr (15 ppm), ^{55}Mn (15 ppm), ^{57}Fe (150 ppm), ^{63}Cu (10 ppm), ^{66}Zn (10 ppm), ^{88}Sr (0.5 ppm), ^{107}Ag (2 ppm), ^{137}Ba (3 ppm), ^{197}Au (1 ppm), and ^{208}Pb (0.5 ppm). All data were reduced off-line with the limits of detection calculated using the commercial version of Iolite (Version 3.71, https://iolite-software.com) data reduction software [58] running with Igor Pro. The uncertainty is calculated for each element for each analysis as a function of the uncertainty of the reference standards, the absolute element content and the time of integration [58]. The resulting value of uncertainty is then varying for a single element between each analysis and is in the order of 10% for high concentration to 80% when the element concentration is measured close to the limit of detection.

3.4. UV-Visible-Near Infrared Absorption Spectra

UV-visible-near infrared absorption spectra were measured at room temperature using a Perkin-Elmer 1050 dual beam spectrophotometer at the Institut des Matériaux Jean Rouxel (University of Nantes, France) on blue-green (sample THA16), yellow (THA08), and orange (THA13) kyanite crystals. Spectra were acquired at a resolution of 1 nm at a rate of 120 nm/min. As our samples were significantly fractured and hence translucent but not transparent, we prepared them as 1 mm thick slices. At this thickness, they became sufficiently transparent so that the light beam could go through, and they also remained colored enough to characterize spectral features related to color.

3.5. Phase Diagram Calculation

PT conditions of the main paragneiss host-rock (sample CS16_292b) were modeled on the basis of the paragenesis described in Section 4.2. The phase diagram was calculated using Perple_X Version 6.8.5 (http://www.perplex.ethz.ch) [59] and the internally consistent end-member data set of Holland

and Powell [60]. Calculations were undertaken in a set of chemical systems ranging from NCKFMASH to MnNCKFMASHT(O) (MnO–Na$_2$O–CaO–K$_2$O–FeO–MgO–Al$_2$O$_3$–SiO$_2$–H$_2$O–TiO$_2$–O$_2$) in order to constrain the effect of TiO$_2$, Fe$_2$O$_3$ and MnO-bearing minerals.

Phases involved in modeling are biotite (Bt), chlorite (Chl), cordierite (Crd), epidote (Ep), garnet (Gt), ilmenite (Ilm), kyanite (Ky), melt (L), plagioclase (Pl), quartz (Qtz), rutile (Rt), sillimanite (Sil), staurolite (St) and white mica (Ms). The solution models utilized for metamorphic minerals and the bulk rock composition are presented in the Table 1 and Table S1, respectively. We used the Ilm(WPH) solution model which enables the calculation of hematite, geikielite and ilmenite end-member abundance [61]. Water was assumed to be saturated.

Table 1. Solution models used for the pseudo-sections (See Figure 15a). See Perple_X documentation (http://www.perplex.ethz.ch) for detailed information.

Phase	Solution Model Label in Perple_X	References
Biotite	Bi(W)	[62]
Chlorite	Chl(W)	[62]
Chloritoid	Ctd(W)	[62]
Cordierite	Crd(W)	[62]
Garnet	Gt(W)	[62]
Ilmenite	Ilm(WPH)	[62]
Melt	melt(W)	[62]
Orthopyroxene	Opx(W)	[62]
Plagioclase	Pl(h)	[63]
Staurolite	St(W)	[62]
White mica	Mica(W)	[62]

4. Results

4.1. Tectonic and Structural Setting of Al-Silicates

The main lithology of the Trikorfo area is dominated by garnet–kyanite–biotite–hematite–plagioclase ± staurolite ± sillimanite bearing paragneisses/mica schists. At the scale of Thassos Island, the orientation of the main foliation varies widely due to post-foliation tectonic events, including large-scale folding. In the Trikorfo area, rare relics of an early foliation S$_{n-1}$ are observed. The paragneisses/mica schists mainly show a sub-horizontal (azimuth N140-300, dip up to 20° W) S$_n$ foliation, bearing kyanite mineral lineation, oriented N255 ± 15°, and gently dipping to the WSW. Late brittle deformation is evidenced by tension gashes that cross-cut S$_n$. These veins are oriented N330 ± 15° and are generally close to vertical. The variation in the orientation of the tension gashes is the same as the regional lineation underlined by synmetamorphic kyanite crystals within 30° that may reflect their continuous formation during counter-clockwise rotation of Thassos Island [40].

4.2. Lithological Units and Rock Sampling

Samples CS16_292b (Figure 3a) and CS16_292a represent the main paragneiss/schist host-rock, apparently unaltered at the outcrop scale. An increasing number of green layers parallel to the gneiss foliation are observed towards the top of the formation (samples CS16_297 (Figure 3b) and Th06). Although not analyzed in our samples, the presence of braunite in the mica schists of the Trikorfo area was shown by previous studies [26,28]. Green to red calc-silicate rocks (up to 80 cm wide) are observed, at the top of the formation, interbedded or at the contacts between quartzo-feldspathic mica schists and metacarbonate rocks (Figure 3c). The rocks typically show a mineral zonation with tremolitic hornblende, diopside, quartz, anorthite/bytownite, epidote, grossular, titanite and Mn-bearing clinozoisite from an outer band adjacent to host mica schists towards the center of the layer (samples Th01, Th02, Th03, KT01 and KT02 (Figure 3c–e)). Sample Th04 is a spessartine–epidote–braunite–hornblende mica schist.

Figure 3. (**a**) Main garnet–kyanite–biotite–hematite–plagioclase ± staurolite ± sillimanite bearing gneiss of the Trikorfo area (sample CS16_292b, N290/65N). (**b**) Green epidote-rich horizon parallel to the main gneiss foliation (sample CS16_297, N270/11N). (**c**) Lense slightly oblique to the subhorizontal gneiss foliation showing a zonation of green epidote and pink clinozoisite together with grossular in a geode (sample KT02). (**d**) Mineralogical zonation with diopside, hornblende, epidote (green), grossular and clinozoisite (pink-red), sample KT02 equivalent to Th04. (**e**) Close up to the core of the geode to near gem-quality pink clinozoisite (clinothulite) and grossular association within locality sample KT02. (**f**) Deep green Mn-rich andalusite (viridine) and kyanite (deep blue and orange) association in sample KT07.

Kyanite is found as a metamorphic mineral, underlining the mineral lineation of the main paragneisses/mica schists rocks as in sample CS16_292b. Locally at the top of the formation, kyanite–piemontite ± andalusite ± muscovite ± spessartine ± braunite mica schists (where kyanite is distributed without any specific orientation) are intercalated within the spessartine–kyanite-bearing mica schists (samples Th05, KT03 (Figure 4a), KT05, and KT07 (Figure 3f)).

Figure 4. Different kyanite color types encountered in the Trikorfo area (Thassos Island). (**a**) Unoriented dark blue kyanite in a mica schist collected towards the top of the area. (**b**) Zoned blue (core) to green-transparent (rim) kyanite in a deformed and folded quartz vein. The pluricentimetric kyanite crystals are found at the interface between the gneiss and the vein. The crystals are weakly deformed and generally oriented parallel to the rock lineation. (**c**) Close-up on zoned-blue kyanite mostly aligned parallel to the rock foliation. (**d**) Field photograph of zoned dark blue-yellow isotropically distributed kyanite crystals in a quartz vein. (**e**) Orange kyanite crystals, generally oriented parallel to gneiss lineation, associated with green Mn-rich andalusite (viridine) and quartz. (**f**) Aggregate of centimetric green Mn-bearing andalusite crystals.

Towards the top of the formation, pluricentimetric kyanite crystals are observed at the contact between metamorphic quartzo-feldspathic kyanite-bearing rocks and coarse-grained quartz-feldspar veins, which are transposed within or secant to the main foliation. In this case, kyanite is generally weakly oriented parallel to the regional lineation. In general the crystals are zoned with dark blue cores and green-transparent (Figure 4b,c) to yellow rims (Figure 4d). Less common are orange kyanite-bearing samples showing association with green andalusite, spessartine and piemontite (Figure 4e,f).

An orthogneiss exposed in the area is composed of orthoclase, oligoclase/andesine along with phlogopite, quartz, hematite, epidote and allanite and is surrounded by a transitional zone with the same mineralogy as the above mentioned calc-silicate layers (i.e. amphibole, plagioclase, and epidote mineral groups from granitoid outwards) [20]. The presence of pluricentimetric crystals of tourmaline (schorl mainly) was also shown in quartz veins and late brittle fissures [20].

Late brittle veins cross-cutting the metamorphic foliation are dominated by euhedral quartz, clinochlore, ilmenite, albite and adularia [20]. Rare veins are enriched in Mn as evidenced by centimetric euhedral crystals of spessartine (Sample MB13; Figure 5).

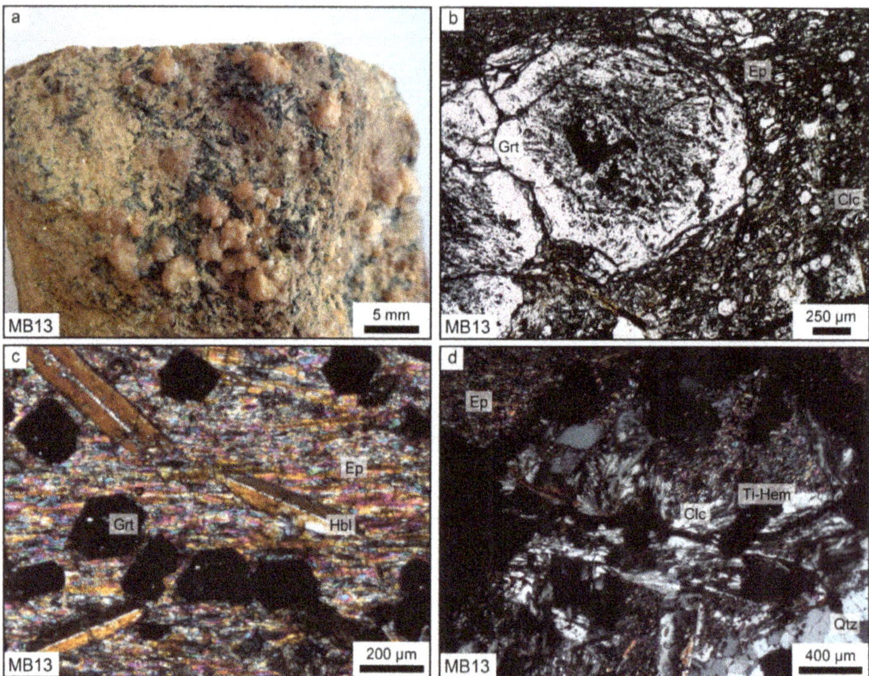

Figure 5. Late brittle fracture sample MB13. (**a**) Hand specimen showing mainly spessartine (orange), Mn-bearing clinochlore and hornblende (green) and quartz. (**b**) Plane polarized light (PPL) photomicrograph with euhedral spessartine (Grt), Mn-rich epidote (Ep) and Mn-bearing clinochlore (Clc) assemblage. (**c**) Crossed-polarized light (XPL) photomicrograph with secondary spessartine (Grt) and hornblende (Hbl) overprinting the metamorphic foliation underlined by Mn-rich epidote (Ep). (**d**) XPL photomicrograph showing Mn-rich epidote (Ep), Mn-bearing clinochlore (Clc), Ti-hematite (Ti-Hem) and quartz (Qtz) association.

4.3. Mineralogy and Mineral Chemistry of Paragneisses/Mica Schists of the Metamorphic Unit at Trikorfo

The composition of two host-rock samples (CS16_292a and _292b) is shown in Table S1a. The structure of the rock is granolepidoblastic with quartz-plagioclase-dominated layers and a schistosity underlined by biotite locally retrogressed to chlorite. C'/S structures indicate top N228 sense of shear. The mineralogy is made of biotite, garnet, quartz, kyanite, staurolite, plagioclase, apatite and Ti-hematite (Figure 6a). Rare late muscovite and epidote are observed without any preferential orientation (Figure 6a).

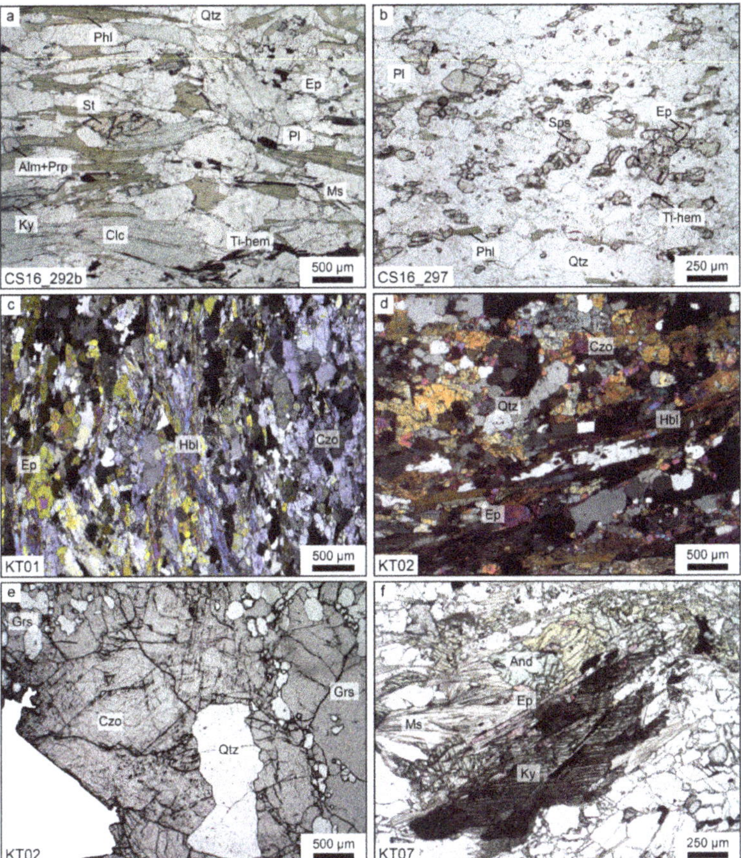

Figure 6. Main mineralogical assemblages from the Trikorfo area. (**a**) Biotite–garnet–kyanite–staurolite–plagioclase-quartz–Ti–hematite paragenesis of the main paragneiss unit modeled with Perple_X. Chlorite (clinochlore/amesite) is a retrogressed phase of phlogopite. Scarce late muscovite and epidote are observed. (**b**) Phlogopite–spessartine–epidote association of the quartz-dominated green horizon parallel to the gneiss foliation. (**c**) Clinozoisite and epidote layers and secondary amphibole in sample KT01 under XPL. (**d**) Clinozoisite–epidote–hornblende in sample KT02 under XPL. (**e**) Grossular-quartz–clinozoisite (clinothulite) microscopic view (PPL) of sample KT02. (**f**) Secondary Mn-rich epidote, orange kyanite and green andalusite assemblage of sample KT07 under PPL. Mineral abbreviation: Almandine + pyrope (Alm + Prp), Andalusite (And), Clinochlore/amesite (Clc), Clinozoisite (Czo), Epidote (Ep), Grossular (Grs), Hornblende (Hbl), Kyanite (Ky), Muscovite (Ms), Phlogopite (Phl), Plagioclase (Pl), Quartz (Qtz), Staurolite (St) and Ti-hematite (Ti-Hem).

Biotite has the composition of phlogopite with X_{Mg} in the range 0.73–0.75 (Table S2a). Plagioclase is oligoclase/andesine with a composition between An$_{23}$ and An$_{34}$ (Table S3a). Garnet and staurolite crystals are isolated in the mica-rich layers. Garnet grains are always corroded remnants of preexisting porphyroblasts (Figure 6a). Garnet chemistry (Table S4a) is dominated by X_{Fe} (up to 39.25) and X_{Mn} (up to 44.06); X_{Mg} is significant, up to 18.82, and X_{Ca} is never higher than 7.72. Garnet composition thus lies close to the border between almandine-(pyrope) and spessartine fields (Figure 7a). No significant and systematic chemical variation was noticed from border to border among all analyzed crystals (Table S4a). Staurolite, also corroded, contains significant amounts of MnO, up to 1.47 wt. % (Table S5). Although Dimitriadis [24] indicated that kyanite seems to postdate garnet growth (no kyanite inclusions in garnet and garnet early porphyroblast remnants might be enveloped by kyanite crystals), the relations are not so clear in sample CS16_292b because most of the crystals are isolated from each other. Our observations did not provide convincing evidence that kyanite postdates garnet growth and both minerals are thus considered as cogenetic. Fe-rich (up to 1.23 wt. % FeO) syn-metamorphic kyanite crystals are oriented parallel to the main foliation S_n. Chlorite analyses revealed a composition between clinochlore and amesite (Figure A1; Table S11). All opaque minerals analyzed by EPMA were Ti-rich hematite (Table S6a). Rare ilmenite was identified by optical microscopy.

4.4. Mineralogy and Mineral Chemistry of Green-Colored Horizons

Sample CS16_297 is representative of the many green-colored horizons parallel to the foliation of the main schist/paragneiss unit of the Trikorfo area (Figure 3b). Element analyses are reported in Table S1a. The structure is mainly granoblastic with the Sn foliation locally underlined by phlogopite alignment. Identified minerals include biotite, quartz, plagioclase, garnet, epidote, Ti-hematite, titanite, rutile and apatite (Figure 6b). Magnetite with hematite exsolution was noticed. The mineralogy is dominated by quartz with layers rich in secondary spessartine and/or epidote/clinozoisite. Biotite has the composition of phlogopite with X_{Mg} in the range 0.72–0.74 (Table S2b). Plagioclase is oligoclase with a composition of An_{15} to An_{21}, slightly depleted in CaO compared to CS16_292b (Table S3b). The content of TiO_2 in Ti-hematite does not exceed 8.15 wt. % (Table S6b). The composition of titanite and rutile is given in Table S9a. Chlorite, staurolite, kyanite and sillimanite are not found in this rock.

The shape of garnet crystals is usually irregular. X_{Fe} and X_{Mg} are much lower (up to 21.69 and 3.24, respectively), and X_{Mn} and X_{Ca} (up to 58.00 and 37.29, respectively) significantly higher than garnet from sample CS16_292b. No significant and systematic chemical variation was noticed from border to border among all analyzed crystals (Table S4b) and garnet composition falls in the spessartine field (Figure 7a). Epidote is abundant and shows a composition depleted in Mn^{3+} ranging from epidote end-member to Fe^{3+}-rich clinozoisite (Figure 7b; Table S7a,j) and is responsible for the green coloration within these rocks.

4.5. Mineralogy and Mineral Chemistry of Calc-Silicate-Dominated Horizons

Moving towards the top of the formation, the thickness of the green layers increases and green-red horizons up to 80 cm thick are visible (Figure 3c,d). These beds are intercalated within the main foliation of the kyanite-bearing paragneisses/mica schists at proximity of carbonate/dolomitic rocks. They typically show calc-silicate assemblages more or less enriched in Mg- and Mn-bearing minerals with typical mineralogical zonation showing tremolitic hornblende, diopside, quartz, anorthite/bytownite, clinozoisite/epidote, grossular, titanite and pink clinozoisite from an outer band adjacent to host mica schists towards the center of the layer (Figure 3c,d) [20,24].

The dark green external part of the calc-silicate is dominated by Mn-bearing varieties of diopside (up to 1.44 wt. % MnO) and hornblende (up to 2.75 wt. % MnO) (Table S8a,c,d). The progressive change in color from pale green to red (samples KT01-02, Th01-03) is mainly due to variation in epidote composition, whose EPMA analyses are presented in Table S7b,e–g, and reported in the Mn–Al–Fe ternary plot of Figure 7b.

Green epidote is reported in green layers parallel to rock foliation as in sample CS16_297 previously described and at the outer margins of the calc-silicate layers along with hornblende, anorthite/bytownite (Table S3d) and titanite (Table S9b) (Figure 6c,d). The composition evolves from epidote to clinozoisite end-members (Figure 7b). It may contain up to 1.04 wt. % MnO. LREE were not analyzed for epidote mineral group; their presence may mark the low total of some of the analyses.

Mn-poor clinozoisite exhibits subhedral to euhedral crystals. The pale pink crystals are usually found in the intermediate zone of the calc-silicate layers mostly associated with quartz, anorthite, titanite, diopside and hornblende. Available microprobe data revealed Fe_2O_3 content between 1.36 and 6.11 wt. % (sample KT02) and between 1.54 and 11.06 wt. % (samples Th01 and Th03), and minor Mn_2O_3 content below 0.67 wt. % (Table S7b,e,g).

Near gem-quality Mn-poor clinozoisite, with the same optical and chemical properties as clinothulite variety described by Bocchio et al. [64], is locally observed in the inner part of the calc-silicate layers, intergrown with Mn-grossular, titanite and quartz, both in the matrix of the rock and in fissures cross-cutting the foliation. The crystals occur in deep pink to red colors and form translucent subhedral to euhedral crystals up to 10 cm (Figures 3e and 6e; Table S7b,f). EPMA data revealed low contents of Fe_2O_3 (<3.15 wt. % in sample KT02 and <2.90 wt. %, in sample Th02) and Mn_2O_3 (<0.72 wt. % in sample KT02 and <0.21 wt. % in sample Th02), however high enough to be responsible for the pink-red coloration.

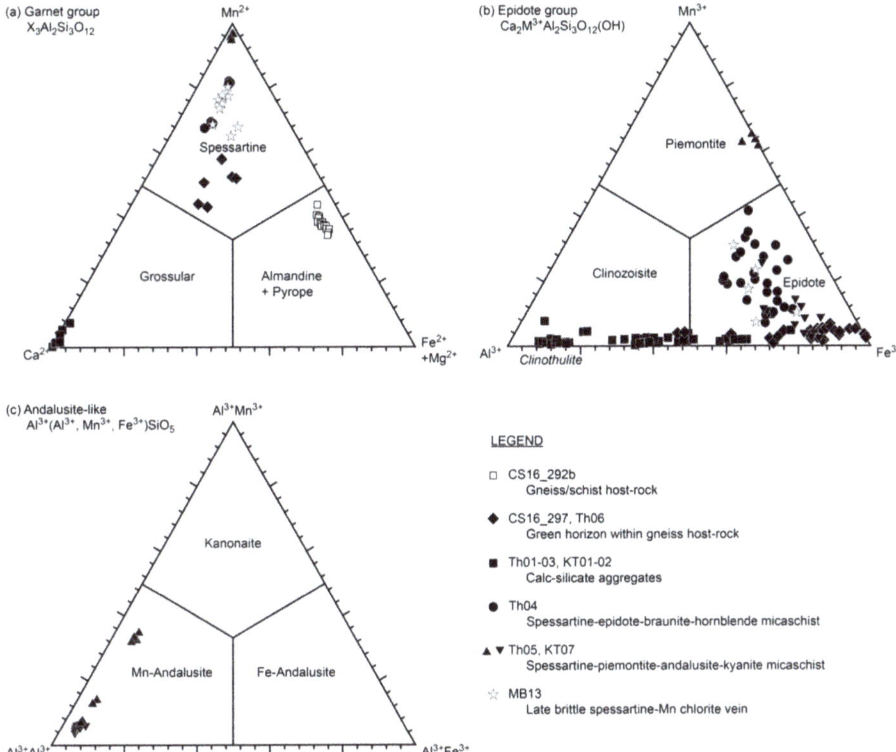

Figure 7. Ternary plots showing compositions of the garnet, epidote and andalusite group minerals. (**a**) Garnet composition considering spessartine, grossular and almandine + pyrope end-members. Garnet composition is $X_3Al_2Si_3O_{12}$ with the X site occupied by divalent cations, Mn^{2+} for spessartine, Ca^{2+} for grossular, Fe^{2+} for almandine and Mg^{2+} for pyrope, e.g., [65]. (**b**) Epidote group minerals, $Ca_2M^{3+}Al_2Si_3O_{12}(OH)$, classified according to one M^{3+} cation substitution with Al in octahedral coordination with $M^{3+} = Al^{3+}$, Mn^{3+} or Fe^{3+}, corresponding to the clinozoisite, piemontite and epidote end-members, respectively [66,67]. Composition analyses for deep pink to red clinozoisite (clinothulite) are shown. (**c**) Andalusite-like minerals, $Al^{3+}(Al^{3+}Mn^{3+}Fe^{3+})SiO_5$, classified according to octahedral Al substitution with Mn^{3+} or Fe^{3+} [2,68].

In the calc-silicate lenses, pure grossular to Mn-bearing grossular form euhedral yellowish crystals of near gem-quality in close association with pink to red colored Mn-bearing clinozoisite and quartz (Figures 3e and 6e). MnO content ranges from 0.08 to 3.60 wt. % (Table S4c,e). All data are plotted in the ternary Mn–Ca–Fe + Mg diagram (Figure 7a).

4.6. Mineralogy and Mineral Chemistry of Mn-rich Schist Layers and Quartz Lenses

Near the top of the Trikorfo area, the formation is dominated by schist layers (samples Th04, Th05 and KT07) and intercalations of quartz lenses (sample KT08) particularly enriched in Mn-bearing minerals as shown by the assemblage of piemontite and Mn-rich epidote together with Mn-rich andalusite and spessartine (Figures 3f, 4d and 5f). Although EPMA analyses were not entirely satisfying (MnO between 27.99 and 45.31 wt. % and maximum total of all analyzed elements of 51.80 wt. %), probably reflecting the heterogeneity and variable Mn oxidation state of the minerals, Mn-oxides could be identified (Table S10). In both lithologies, piemontite and Mn-rich epidote form dark pink/purple-colored prismatic crystals with typical piemontite pleochroism. Microprobe analyses revealed piemontite in the andalusite-bearing mica schists with Mn_2O_3 content up to 12.69 wt. % and Mn-rich epidote in association within the spessartine–braunite–quartz aggregates with Mn_2O_3 in the range of 0.13 to 6.59 wt. % (Figure 7b, Table S7c,h,i). Phlogopite has X_{Mg} of 0.96–0.97 (Table S2d). Plagioclase is oligoclase/andesine with a composition between An_{22} and An_{34} (Table S3c).

Garnet is also marked by enrichment in Mn (Figure 7a). Translucent, orange, euhedral crystals of spessartine up to 1 cm are found (i) in intercalated spessartine–epidote–braunite–hornblende mica schists, and (ii) in spessartine–piemontite–andalusite–kyanite mica schists. Within the spessartine–epidote–braunite–hornblende mica schists, spessartine commonly forms large euhedral crystals with numerous inclusions of epidote and is surrounded by epidote and magnesio–hornblende. Spessartine from the spessartine–piemontite–andalusite–kyanite mica schists is also associated with muscovite, phlogopite, anorthite and hematite. Electron microprobe analyses of spessartine indicate X_{Mn} in the range 64.74–97.19 (Table S4e).

A late brittle vein cross-cutting the Mn-rich schist layers was sampled (MB13; Figure 5). Up to 1 cm near gem-quality euhedral crystals of spessartine (X_{Mn} of 81.23; Figure 7a; Table S4d) are observed together with quartz, Mn-rich epidote (up to 6.67 wt. % Mn_2O_3; Figure 7b; Table S7d), Mn-bearing clinochlore (up to 1.64 wt. % MnO; Figure A1; Table S11d), Mn-bearing hornblende (up to 1.99 wt. % MnO; Table S8b) and almost pure hematite (Table S6c).

Accessory phases include monazite (relatively abundant in discrete layers), titanite and zircon. Tourmaline is very abundant in quartz lenses within mica schists, where it occurs as black crystals (schorl) up to 10 cm and associated with muscovite and chlorite (clinochlore). However not analyzed, olive green-colored tourmaline (crystals up to 3 cm) is observed. Tourmaline is also present in fissures intergrown with clinochlore, titanite and quartz. Finally, up to 10 cm tremolite white crystals were documented in sheared marbles [20].

4.7. Mineralogy and Mineral Chemistry of Kyanite–Quartz Veins

4.7.1. Different Colors of Kyanite Crystals

Among the many spectacular minerals observed at the Trikorfo area, kyanite, found at the contact between quartzo-feldspathic coarse-grained veins and the paragneiss/mica schist rocks, towards the top of the series, is of particular interest. The crystals show a general orientation parallel to the regional foliation but may also be unoriented (Figure 4a). The up to 20 cm crystals are slightly bent as a result of weak plastic deformation. Over a small area of a few km^2, besides the Fe-rich syn-metamorphic kyanite already described in the metamorphic paragenesis (Table S12a), we report at least four dominant types of kyanite based on their color: (i) dark blue-black (Figure 4a, Table S12b), (ii) zoned crystals with dark blue cores and transparent-green rims (Figure 4b,c), (iii) zoned crystals with dark blue cores and yellow rims (Figure 4d) and (iv) orange crystals generally associated with green Mn-rich andalusite (Figure 4e; Table S12c). Most of the crystals are inappropriate for faceting as marketable gemstones due to common fractures, but are at least proper for cabochon-shaped material. In addition, with regard to their overall aesthetic aspect, mainly large size and vivid color, these kyanite crystals are referred to as being of near gem-quality.

4.7.2. Bulk Element Concentration of Individual Kyanite Crystals

When observed down their <100> faces, most kyanite crystals show blue-green-colorless bands parallel to the *c*-axis. Representative kyanite samples have been selected for bulk element concentration analyses (zoned kyanite THA16_1, _2, _3, yellow THA08, pale orange THA12 and deep orange THA13; Figure 8). The elements are classified as oxides and trace elements according to their content in the host rock sample CS16_292b (* symbol in Figures 9a, 10 and 11).

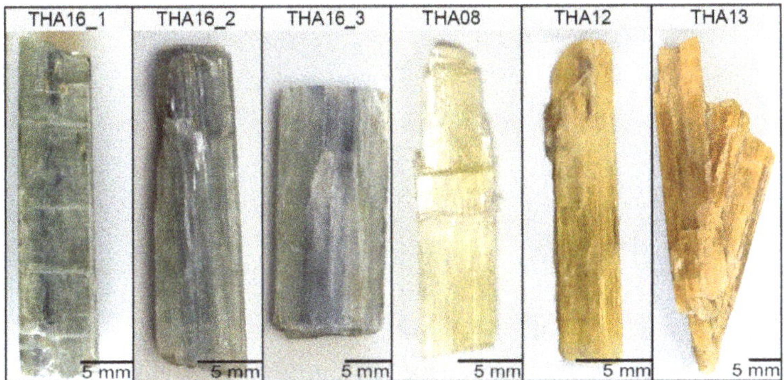

Figure 8. Photographs of individual kyanite crystals prior to bulk rock analyses.

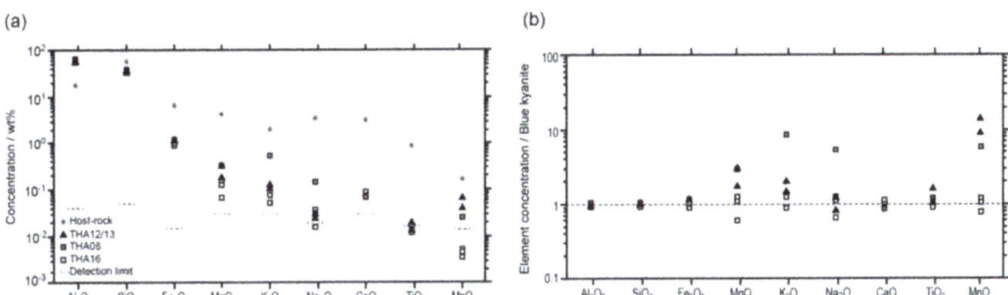

Figure 9. Distribution of measured oxides in CS16_292b host-rock and in blue-green (THA16), yellow (THA08) and orange (THA12 and THA13) kyanite samples. (**a**) Raw concentration. (**b**) Concentration normalized to the average value of the three blue-green THA16 samples.

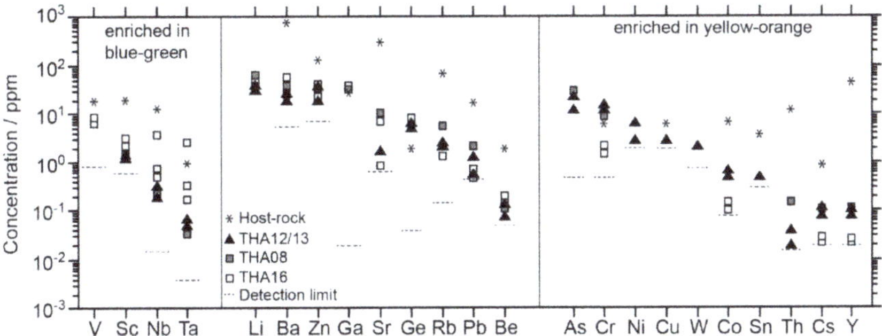

Figure 10. Trace element composition of host-rock CS16_292b and kyanite samples ordered by decreasing concentration. The left part shows elements which are preferentially found in blue-green THA16 kyanite samples, the central part where there is no significant difference among the samples and the right part reports the elements with significant enrichment in yellow THA08 and orange THA12/13 kyanite samples.

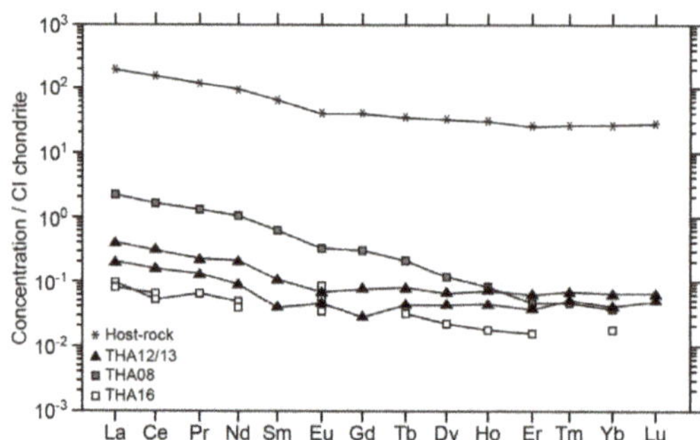

Figure 11. REE spider diagram of blue-green (THA16), yellow (THA08) and orange (THA12 and THA13) kyanite individual crystals and of the host rock sample CS16_292b. The normalizing factors are related to CI chondrite values [69].

The amount of Al_2O_3 and SiO_2 is comparable between all analyzed kyanite samples, higher than 96 wt. % Al_2SiO_5. The measured oxides show a decreasing concentration from Fe_2O_3, MgO, K_2O, Na_2O, CaO, MnO and TiO_2 (Figure 9a). These elements generally show a preferential enrichment in the orange kyanite (Figure 9b), except Na_2O, K_2O, CaO and LOI content (Table S1b), which appear to be correlated but randomly distributed between the samples. This is likely due to alteration phases in microfissures. The total amount of Fe_2O_3, MgO, MnO and TiO_2 increases with coloration from blue-green (1.10–1.21 wt. %), to yellow (1.51 wt. %) and orange (1.42–1.61 wt. %). Yellow and orange samples show markedly higher amounts of MnO (up to one order of magnitude) and MgO (Figure 9b).

Minor elements from the bulk rock analyses are found with amounts lower than 100 ppm in kyanite samples. Ba, Zn and Ga are the most abundant elements. Blue-green kyanite is significantly enriched in V, Nb, Sc and Ta. Yellow and orange samples of kyanite contain predominant amounts of As, Cr, Ni, Cu, W, Co, Sn, Th, Cs and Y. For all other trace element contents, there is no significant difference among the analyzed samples (Figure 10; Table S1c).

The rare earth element concentration measured in each sample was normalized to chondritic abundance (Figure 11). All kyanite samples have comparable chondrite-normalized REE patterns as CS16_292b host-rock with a slight decrease until Gd and then relatively flat towards heavy REE. As for all other elements, REE are less abundant in blue-green samples than in yellow and orange kyanite. The yellow (THA08) sample is characterized by a remarkable enrichment in light REE compared to other samples. This is to be correlated with the higher content of this sample with the elements attributed to alteration phases in microfissures such as K, Na, Li, Sr and Ba.

4.7.3. EPMA X-ray Maps and Element Distribution

For all the samples, Al and Si variations are difficult to interpret. Both elements appear relatively homogeneous throughout the crystals and their maximum/minimum concentration seems to be slightly anti-correlated, likely due to a problem of surface homogeneity. A decrease of Al is noticed on the edges of THA16 marking alteration or re-equilibration with minerals in the surrounding rock. The element Mn also does not show any obvious variation in the different analyzed areas. The only point of interest regarding Mn is that, accordingly with bulk rock analyses, the concentration is lower in blue THA16 (<0.10 wt. % MnO) than in the yellow THA08 (<0.18 wt. % MnO) and orange THA13 (<0.16 wt. % MnO) kyanite samples.

Fe concentration however shows very contrasting behavior. In kyanite samples THA16 (<1.27 wt. % FeO) and THA08 (<1.59 wt. % FeO), the variations are straight and parallel to the *c*-axis. At this scale, the higher concentration seems to be correlated with the clearer parts of the kyanite crystals. In sample THA13 (<1.56 wt. % FeO), Fe distribution seems to be oblique regarding the orientation of the *c*-axis and could be related to plastic deformation of the analyzed crystal (Figure 12).

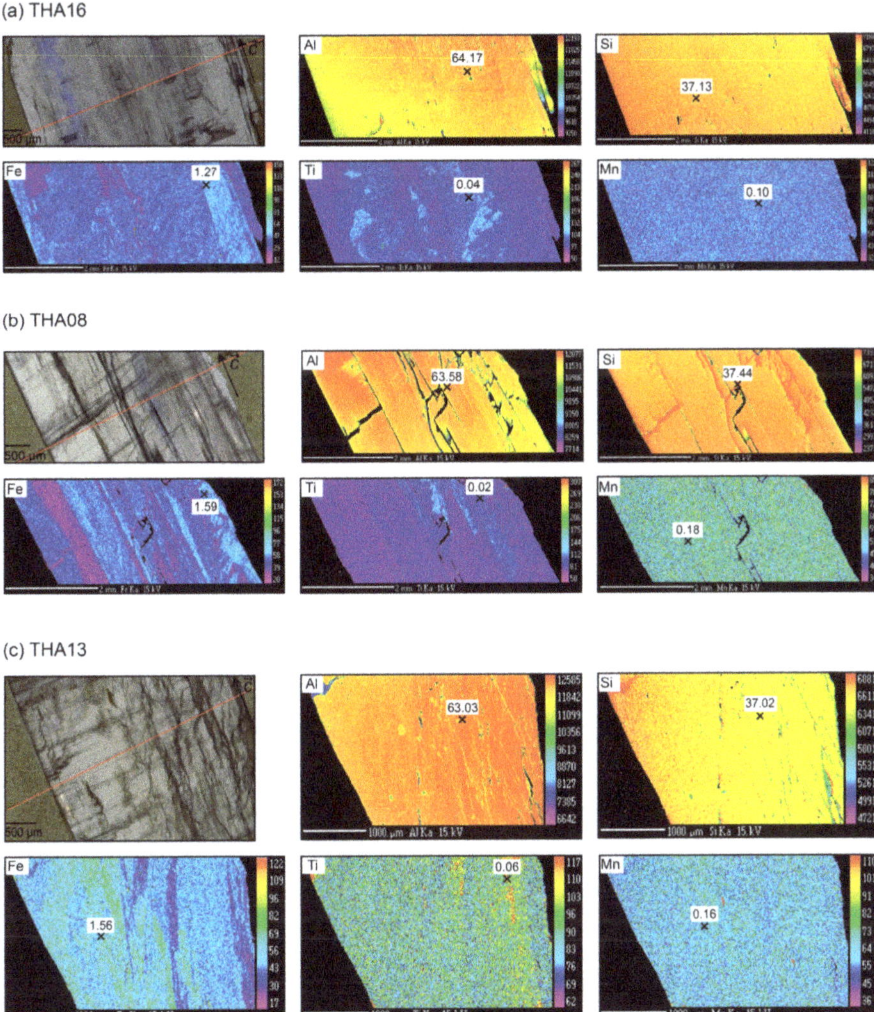

Figure 12. Electron probe microanalyses (EPMA) semi-quantitative X-ray maps of Al, Si, Fe, Ti and Mn on the <100> faces of blue-green THA16 (**a**), yellow THA08 (**b**) and orange THA13 (**c**) kyanite samples, with color scales expressed in counts, not calculated to concentration. The red line corresponds to the EPMA quantitative profile analyses (Table S12d–f). The maximum values (in oxide wt. %) measured along the profile for a given element is reported on each map.

Ti distribution is discontinuous and does not seem to be related with crystallographic orientation. Three marked areas of higher concentration are highlighted in THA16 (<0.05 wt. % TiO_2) and seem to correspond to the blue area of the crystal, at least for the left one. It is also the case to the right of the crystal, but the blue coloration is barely visible. The direct correspondence between blue coloration and Ti concentration is however difficult to establish firmly as most of the blue area is below the sample surface. In sample THA08 (<0.02 wt. % TiO_2) also, the irregular distribution of Ti reveals enrichment at the central right part of the crystal where a thin blue zonation is observed. On THA13 (<0.06 wt. % TiO_2), Ti distribution follows that of Fe, though anti-correlated.

4.7.4. Orange Kyanite and Green Andalusite Assemblages

This atypical polymorphic association is observed towards the top of the formation (i) in kyanite–phlogopite–piemontite–quartz–plagioclase ± muscovite ± spessartine ± braunite mica schists, and (ii) in quartz-feldspar-muscovite lenses. Mn-rich andalusite forms dark green-colored subhedral to euhedral crystals up to 7 cm. In both cases, kyanite and andalusite seem to have grown at equilibrium (Figure 6f). Green andalusite is commonly zoned likely due to twinning, and exhibits deformation lamellae [29]). EPMA data reveal elevated Mn_2O_3 contents, up to 15.98 wt. % (Figure 7c; Table S13a,b).

4.7.5. LA-ICPMS Profiles

The limits of detection of LA-ICMPS are significantly lower than those of EPMA and many elements can be analyzed sequentially during the same analytical run. Edge to edge sections were conducted at the same position, or approximately, as that for the EPMA profiles on the three kyanite samples (Figure 13). Al and Si variations serve as internal standards to provide quantitative elemental evolution through kyanite and are not reported on the figures. All analyses of Ca, Cu, Ag and Au were below the limits of detection. For each sample, the elements are reported according to three different groups with (i) the dominating elements (Fe, Mn, Ti, Mg and Zn) on top, (ii) the elements generally enriched in alteration minerals (K, Na, Li, Sr and Ba) in the middle, and (iii) the other elements, mainly transition metals, only found as traces in the host-rock (Cr, Pb and V) on the lower part.

Kyanite sample THA08 shows large variations in the concentration of the elements K, Na, Li, Sr and Ba. Moreover, it can be noticed that these elements are mainly concentrated either on the sides of the crystals or along fractures inside the crystals, which are common in THA08. Accordingly, these elements are interpreted as the result of alteration or late equilibration with host-rock minerals (feldspars and micas like muscovite and phlogopite).

In all samples, Fe is by far the most common element besides Al and Si. Local concentration decreases are observed where the amount of alteration elements increases, as is clearly visible on sample THA08. Mn and Ti are in the same order of magnitude in THA16 and THA08 but with Ti slightly higher in the blue-green kyanite. Mn concentration does not correlate with any color change, whereas that of Ti shows a relatively good correspondence with the blue color bands, even if they are below the sample surface. This is especially well expressed in THA08 where Ti > Mn at the intersection with the blue band. It is also visible on THA16, where the three Ti highs seen in the EPMA mapping are distinguishable, and correspond to a blue color on the left. However, it must be kept in mind that blue color is mostly below the crystal surfaces. Also, it must be noted that the laser ablation itself down-cuts through the sample (ablation spot of 32 μm) and hence what was actually analyzed may largely be below the surface. In sample THA13, Mn dominates largely over Ti and is one order of magnitude higher as in samples THA16 and THA08. The Mg concentration variation seems to partly correlate with edges and fractures and can be interpreted at least in part due to equilibration with host-rock phlogopite. Along the profiles, Zn is equally present and its concentration evolution is marked by rapid variations up to one order of magnitude. In some parts of the kyanite crystals, Zn is almost as abundant as Mn, Ti and Mg. These variations are not related to any characteristic feature of the crystals. The other elements do not show any obvious correlations. V is found above the limit of detection (1 ppm) only in sample THA16. No systematic evolution of Cr could be observed, because of concentration too close to the limit of detection (15 ppm). Pb shows the same evolution as the elements attributed to alteration and post-crystallization re-equilibration, with higher concentrations at the edges of the crystals and within the fractures.

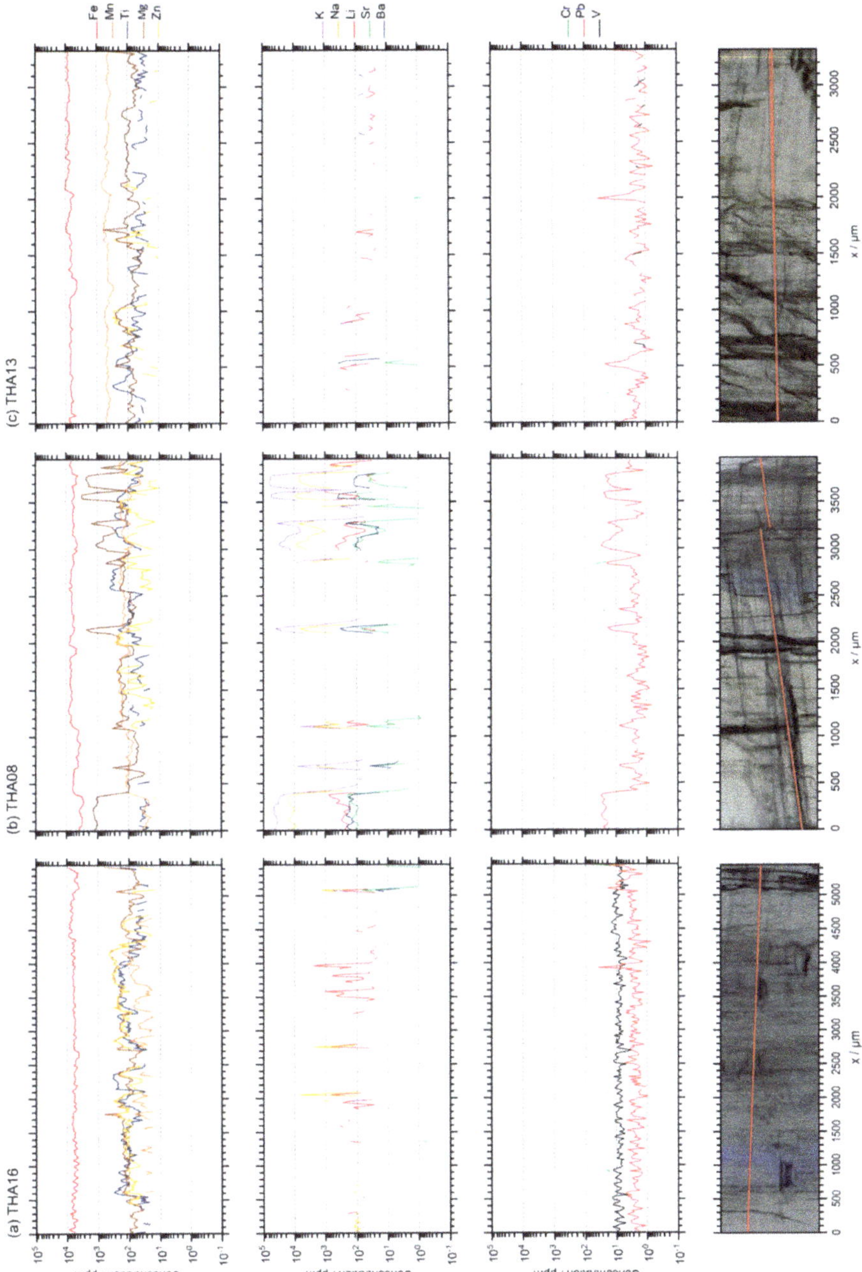

Figure 13. Quantitative Laser Ablation-Inductively Coupled Plasma Mass Spectroscopy (LA-ICPMS) analyses (in ppm) on the <100> faces of blue-green THA16 (**a**), yellow THA08 (**b**) and orange THA13 (**c**) kyanite samples from border to border of the crystals. The location of the profiles (red line) is approximately at the same position as that of EPMA analyses. The element concentration lines might be discontinuous due to values below detection limit.

4.7.6. UV-Near-Infrared Spectroscopy

All spectra (blue, yellow and orange kyanite samples) show an absorption continuum rising from infrared or mid-visible to UV, making absorption range from 315 nm to below (Figure 14). They also all show sharp and weak absorption peaks at 370, 382, 432, 446 nm. Spectra of blue samples are dominated by large absorption bands at 520 and 620 nm that, combined with the previous spectral features, generate a transmission window around 480 nm, in the blue sample. Spectra of orange and yellow kyanite samples are similar. In addition to the common spectral features described initially, both show a band at about 470 nm and a weaker band at 890 nm. These two bands are weaker in the yellow than in the orange kyanite sample. As a consequence, transmission increases from the green (550 nm) toward the infrared, hence the orange and yellow colors of the samples. The allocation of these spectral features to specific ions or defects is proposed in the Discussion section.

4.8. Phase Diagram Calculation of the Paragneiss

The *PT* pseudo-section is presented in Figure 15 for the *PT* range of ca. 550–750 °C and ca. 5.5–10.5 kbars. Calculation results show that the assemblage Grt–St–Bt–Pl–Ky–Qtz–Ti–Hem formed between 630–710 °C and 7.8–10.4 kbars, in the upper amphibolite facies. Calculations in the different systems, ranging from NCKFMASH to MnNCKFMASHT(O), allow for discussing the effects of MnO, TiO$_2$ and Fe$_2$O$_3$. The addition of MnO results in an expansion of the garnet stability field with the garnet-in line shifting to lower temperatures and pressures. Because of the occurrence of Ti-bearing hematite in the mica schists, the role of oxidation on phase relations was investigated through a *T*-XFe$_2$O$_3$ phase diagram section. The addition of TiO$_2$ and Fe$_2$O$_3$ controls the field stability of Fe^{2+}–Fe^{3+}–Ti oxide phases. The minimum XFe$_2$O$_3$ value to stabilize the Ti-bearing hematite instead of rutile is 0.1 (at constant *P* = 8.5 kbars). It appears that the stable assemblage garnet (Grt)–staurolite (St)–biotite (Bt) –plagioclase (Pl)–kyanite (Ky)–quartz (Qtz) with Ti–hematite (Ti–Hem) is stable between ca. 630 and 710 °C for a median XFe$_2$O$_3$ (Fe$_2$O$_3$/(Fe$_2$O$_3$ + FeO)) of 0.5 (Figure A2). In consequence the phase diagram section was computed with this XFe$_2$O$_3$ molar ratio. Considering the *P-T* range used for this pseudosection (550–750 °C; 5.5–10.5 kbars), hematite-rich ilmenite with ca. 50–70% of hematite end-member is stabilized in each stability field of ilmenite and, thus, labelled as "Ti–Hem" in Figure 15. For the specific *PT* field of the described paragenesis (ca. 670 °C and 9.1 kbars), the calculated ilmenite in the ilmenite–hematite solid solution is around 33%, consistent with the 10–35 % measured by EPMA (Table S6a).

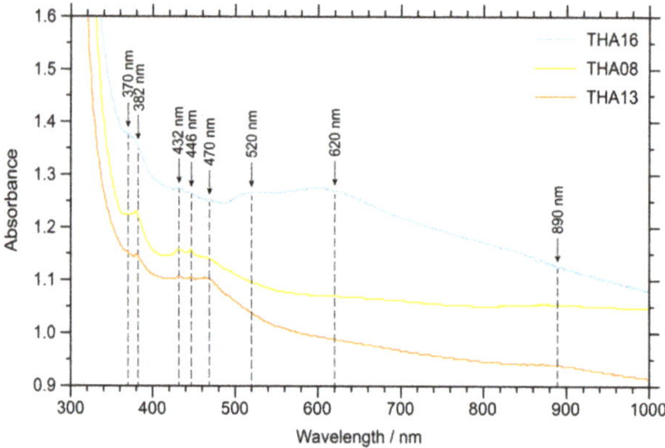

Figure 14. UV near-infrared spectra of kyanite of various colors. Spectra are shifted and re-scaled vertically for clarity. Absorbance values are those of sample THA08.

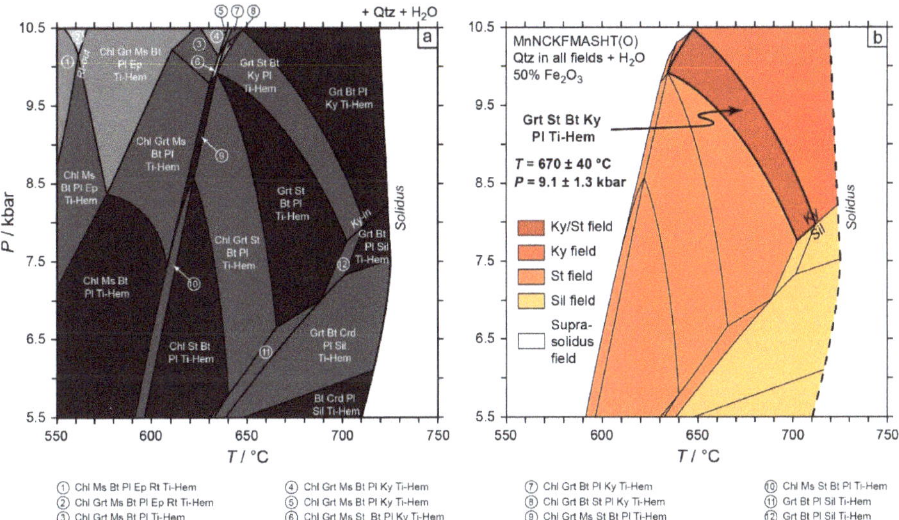

Figure 15. PT pseudo-section calculated for sample CS16_292b. (**a**) Pseudo-section showing labelled metamorphic assemblage fields. Assemblage fields are shaded according to the number of degrees of freedom, with higher-variance assemblages represented by darker shading. Ky-in and Rt-out lines are shown. (**b**) Highlight of the stability fields of kyanite, staurolite and sillimanite index minerals. The interpreted peak metamorphic assemblage field is shaded in red and labelled with bold text. Ky-Sil univariant line is shown. The right white domain corresponds to the supra-solidus field. Dashed line represents the solidus. Mineral abbreviation: biotite (Bt), chlorite (Chl), cordierite (Crd), epidote (Ep), garnet (Grt), kyanite (Ky), muscovite (Ms), plagioclase (Pl), quartz (Qtz), rutile (Rt), sillimanite (Sil), staurolite (St) and Ti-hematite (Ti-Hem).

Chlorite thermometry [70] made on retrogressed phlogopite of the paragneiss CS16_292b and schist sample KT03 indicates temperatures in the range 398–334 °C (Table S2c for phlogopite and Table S11a,b for chlorite composition). The values given by the clinochlore analyzed within sample piemontite–andalusite–muscovite–kyanite schist KT07 are in the range 280–326 °C (Table S11c), while the values from Mn-bearing clinochlore from the late brittle vein MB13 are from 233 to 315 °C (Table S11d).

5. Discussion

The Trikorfo area on Thassos Island represents a unique mineralogical locality where near gem-quality Mn-rich minerals are described. A detailed structural/mineralogical study permits to describe metamorphic parageneses and local metasomatic assemblages. The dominant formation is made of paragneisses/mica schists containing a metamorphic paragenesis, including kyanite. Towards the top of the formation, different units are described parallel to the metamorphic foliation, with some of them bearing large and colored minerals: (i) calc-silicate dominated horizons, (ii) Mn-rich schist layers and quartz lenses, (iii) kyanite-bearing quartz-feldspar coarse-grained veins, and (iv) late brittle veins cross-cutting the metamorphic foliation. The discussion is divided into three parts where (i) the near gem-quality minerals are described and the elemental substitutions leading to their color variation are discussed, (ii) the distinction between metamorphic parageneses and local metasomatic assemblages is made, (iii) a *PT*-deformation exhumation path for the conditions of formation of the different units is proposed. Kyanite is found within the metamorphic paragenesis and as near gem-quality in quartz-feldspar coarse-grained veins; it is therefore the guidance mineral of the study and the discussion.

5.1. Colored Minerals and Cation Substitution in the Trikorfo Area, Thassos Island

Dispersed metal ions in silicate and calc-silicate crystal structures lead to color variation of particular interest for mineral collectors. Silicate and calc-silicate minerals of the Trikorfo area are large and vividly colored, close to gem-quality. Moreover, thanks to trivalent cation substitutions (Mn^{3+} and Fe^{3+}) with Al, the area hosts an outstanding variety of colors within the minerals. This chapter aims to depict the main substitutions observed within kyanite, andalusite, epidote and garnet minerals.

The common natural color of kyanite (Al_2SiO_5) is blue but it can also be green, yellow, orange, white, black, grey or colorless as a function of the nature of elemental substitutions [4]. Moreover, several alternating colors can be found along bands parallel to the *c*-axis in a single kyanite crystal, marking the presence of trace elements in the crystal structure. Kyanite has two Al sites in six-coordination and one Si site in four-coordination. The occupying site and bonding state of transition elements determine the color and physical properties of the crystals [71]. In the Trikorfo area, besides syn-metamorphic crystals, kyanite (zoned blue to green/yellow-transparent, yellow and orange) is also observed in quartz-feldspar coarse-grained veins or in quartz lenses. Bulk chemical composition analyses reveal that each type of color kyanite has its own signature with a characteristic distribution of measured oxides and trace elements in the minerals (Figures 9 and 10). V and Cr distribution is markedly different between zoned blue (up to ~10 ppm and ~2 ppm, respectively) and orange kyanite (below detection limit and up to ~15 ppm, respectively), suggesting a local control from the host-rock composition. However, the low concentrations point to a similar felsic origin [4] that should be attributed to interaction with the main paragneiss unit of the area. Orange kyanite localities are very rare on Earth. The Loliondo deposit (Tanzania) is characterized by elevated MnO content up to 0.23 wt. %. Polarized optical absorption spectra show that the orange coloration is governed by crystal field d-d transitions of Mn^{3+} replacing Al in six-fold coordinated triclinic sites of the kyanite structure [9]. Even if orange kyanite from the Trikorfo area is richer in MnO than the yellow and blue zoned crystals, their MnO content (0.07 wt. % for the bulk sample analysis and locally up to 0.16 wt. % for EPMA analyses) remain far below that of orange kyanite from Tanzania. This is consistent with an orange color less intense in kyanite from Trikorfo than from Loliondo. All kyanite samples show Zn in a significant amount but without any evident correlation with crystal structure or color. The total Cr content is higher in the yellow/orange samples than in the blue kyanite (Figure 10). However, LA-ICPMS sections do not show conclusive information regarding the distribution of Cr due to concentration close to the limit of detection of 15 ppm (Figure 13). Moreover, kyanite colored by Cr^{3+} shows absorption spectra dominated by two broad absorption bands centered at 405–420 nm and 595–625 nm [72,73]. No such absorptions are observed in any of our samples, making Cr^{3+} an unrealistic chromophore in this case. This is consistent with Cr content very low in our samples (typically 1 to 20 ppm; Figure 10; Table S1c) compared to blue kyanite colored by dispersed Cr^{3+} (typically Cr_2O_3 > 1 wt. %; [73]). Hereafter, only Mn^{3+}, Fe^{3+} and Fe^{2+}–Ti^{4+} substitutions with Al^{3+} are thus discussed. The kyanite crystals analyzed by UV-near infrared spectroscopy show weak absorption bands at 370, 381, 432 and 446 nm due to isolated Fe^{3+} [7]. The large absorption bands at 520 and 620 nm, responsible for the blue color of THA16, are due to Fe^{2+}–Ti^{4+} inter-valence charge transfer (IVCT; [74–76]). No spectral features due to the Fe^{2+}-Fe^{3+} IVCT were found. In the spectra of yellow (THA08) and orange (THA13) kyanite samples, the band at 470 nm, that importantly contributes to color, is consistent with the main band attributed to isolated Mn^{3+} [9]. Hence, the origin of the yellow to orange color in our samples is attributed to isolated Mn^{3+}, with some minor contribution of isolated Fe^{3+}. The contribution of the weak 890 nm band remains unattributed.

EPMA X-ray maps and LA-ICPMS profiles show some relation between the blue zones and the Ti content of the zoned kyanite crystals THA16 and THA08 (Figure 12a,b and Figure 13a,b). Interestingly, orange kyanite (THA13) contains more TiO_2 than blue, zoned crystals. This shows that the presence of Ti alone is not sufficient to generate a blue color: Ti^{4+} atoms need to be crystallographically linked to Fe^{2+} atoms for the charge-transfer phenomenon to occur, and hence the blue color to appear. If for any reason, the charge transfer does not occur (titanium atoms may be isolated, or iron ions may be trivalent), the blue color due to IVCT does not appear. A similar behavior is documented in other colored minerals such as corundum: some yellow sapphires sometimes contain more titanium than blue sapphires colored by Fe^{2+}–Ti^{4+} IVCT, e.g., [77]. Again, in this case, titanium concentration alone is not correlated with the blue color.

Orange kyanite is associated with green andalusite in quartz lenses (Figure 3f). Mn^{3+} might substitute for Al^{3+} in andalusite mineral formula, resulting in deep green, Mn-rich andalusite (Al, Mn)$_2$SiO$_5$ [10,78]. This andalusite variety is named viridine in reason of its brilliant green color [12]. Viridine occurrences (Germany, Belgium, Japan, Australia, Brazil and Tanzania) usually provide small or altered crystals [16,78]. Large viridine crystals of the Trikorfo area reach up to ten centimeters large, are vivid-to-deep green and unaltered. Their occurrence is promising for exploration of gem-quality, marketable viridine in the area. Compared to kyanite and sillimanite, andalusite incorporates maximum contents of Mn^{3+} due to its crystal structure [2]. Experimental works quantified that the Mn_2SiO_5 end-member cannot be higher than 6 mol% for kyanite and 1.7 mol % for sillimanite, but can reach 22 mol % for andalusite [2,79]. In the present study, Mn_2O_3 can reach up to 15.98 wt. %, yielding 17.5 mol % $Mn_2Si_2O_5$, close to the maximum possibly accepted by andalusite.

In the Trikorfo area, epidote group minerals are represented by green epidote, pink to red Mn-rich epidote and Mn-rich clinozoisite (variety clinothulite), purple piemontite, and allanite, which is only reported at the thin-section scale by Voudouris et al. [20]. Epidote, clinozoisite and piemontite represent the Fe-, Al- and Mn-rich end-members respectively of the epidote mineral family. Solid solution involves substitution of trivalent cations (Al^{3+}, Fe^{3+} or Mn^{3+}) in three non-equivalent octahedral sites in the basic formula for epidote $Ca_2M^{3+}Al_2Si_3O_{12}(OH)$ [66,67]. True piemontite (up to 12.69 wt. % Mn_2O_3) is restricted to Mn-rich andalusite bearing samples, indicating that these mica schist layers are locally particularly enriched in manganese. Mn-bearing epidote (up to 7.76 wt. % Mn_2O_3) is usually associated with spessartine-Mn oxides-quartz assemblages. The Mn-poor clinozoisite/epidote and Mn-poor clinozoisite (clinothulite) from the Trikorfo area contain up to 0.72 wt. % Mn_2O_3 which is responsible for their pink to red color. In accordance to Bonazzi and Menchetti [80] and Franz and Liebscher [67] the coloration in pink epidote/clinozoisite is due to Mn^{3+} and not Mn^{2+} in the mineral structure.

Four types of garnet are recognizable on the basis of their colors: (i) syn-metamorphic millimetric red almandine (± pyrope)-spessartine, (ii) plurimillimetric Ca–Fe-rich yellow to orange spessartine within the green horizons, (iii) centimetric almost pure pale yellowish brown grossular associated with epidote and clinozoisite in calc-silicate aggregates, and (iv) centimetric deep amber orange almost pure spessartine associated with Mn-rich andalusite, muscovite and orange kyanite. The colors of these garnet crystals directly reflect their main dominant cation (Ca^{2+}, Fe^{2+} or Mn^{2+}). Mn-rich garnet is found within the metamorphic assemblage and the composition to almost pure spessartine is only very localized to certain mica schist layers in association with orange kyanite and green andalusite, and to brittle veins cross-cutting Mn-rich horizons close to the top of the Trikorfo area.

5.2. Metamorphic Parageneses versus Metasomatic Assemblages

The Trikorfo area is dominated by garnet–kyanite–biotite–hematite–plagioclase ± staurolite ± sillimanite paragneisses/mica schists (sample CS16_292b), with intercalations of metacarbonate units near the base and top of the formation and local calc-silicate layers and Mn-rich mica schists, also close to the top of the formation. Field and mineralogical observations suggest that the epidote-grossular-bearing calc-silicate layers represent local inhomogeneities in the bedding of the sedimentary protoliths and/or compositional changes that have been produced by localized metasomatic processes during regional metamorphism, as suggested for similar layers at Therapio, Evros area [81]. In this sense, the mineralogy of the calc-silicate layers could have developed during prograde metamorphism of a Mn-rich, calcareous pelitic protolith, followed by vein formation and local metasomatic reactions during retrograde metamorphism accompanying the exhumation of Thassos Island during the Oligocene-Miocene.

Alternatively, the skarn-similar mineralogy of the calc-silicate layers (e.g., grossular, diopside, hornblende and epidote) could have been formed by fluids released by granitoids during contact metamorphism with the carbonate (calcite and dolomite) rocks found at the top of the Trikorfo area. With the lack of radiometric data, the exact age (i.e. Hercynian?) of the orthogneiss at Trikorfo is not known but is likely older than alpine deformation. Although not exposed on Thassos Island, Oligo-Miocene magmatism is described nearby in the Rhodope area with the intrusion of syntectonic plutons 30 km to the north of Thassos Island, e.g., [52–54]. Melfos and Voudouris [27] considered that widespread gold mineralization in the area is related to this Miocene magmatic activity. Tourmaline at Trikorfo also occurs as a retrograde mineral, resulting in tourmalinite and quartz-tourmaline veins cross-cutting the metamorphic foliation. According to van Hinsberg et al. [82] metasomatic introduction of boron on the retrograde path is most commonly associated with the intrusion of late granites in orogenic belts.

The Trikorfo area shows the superposition of late secondary minerals over the metamorphic paragenesis. These silicates and calc-silcates, including a large variety of garnet and epidote composition, kyanite, andalusite, diopside, hornblende, Ti-hematite, magnetite and tourmaline indicate post-metamorphic metasomatic reactions where hot, chemically active fluids that altered the metamorphic assemblage. Furthermore Mn-rich mineralogical assemblages are found. This type of assemblage is relatively rare on Earth because it requires highly oxidized environments [15–18]. The large diversity of mineral compositions and assemblages observed at the scale of the Trikorfo area points to a local control on the mineralogy and fO_2 conditions during post-metamorphic metasomatic reactions.

For example, five different garnet-epidote assemblages are described with increasing oxidation state as indicated by the amount of Mn_2O_3 in the epidote: (i) Mn-poor grossular (up to 3.60 wt.% MnO) and Mn-poor epidote to clinozoisite (up to 1.01 wt. % Mn_2O_3) within the calc-silicate aggregates, (ii) Mn-poor spessartine (up to 24.91 wt. % MnO) associated with Mn-poor epidote (up to 1.16 wt. % Mn_2O_3) in the green horizons within gneiss host-rock, (iii) Mn-rich spessartine (up to 36.95 wt. % MnO) with Mn-rich epidote (up to 7.76 wt. % Mn_2O_3) within the spessartine–epidote–braunite–hornblende schist, (iv) almost pure spessartine (up to 42.87 wt. % MnO) associated with piemontite (up to 12.69 wt. % Mn_2O_3), Mn-rich andalusite (up to 15.98 wt. % Mn_2O_3) and orange Mn-bearing kyanite in piemontite–andalusite–muscovite–kyanite mica schists, and (v) spessartine-clinochlore-epidote (up to 6.67 wt. % Mn_2O_3). In concordance with the experimental observations of Keskinen and Liou [18], garnet which coexists with piemontite is uniformly more spessartine-rich.

Garnet and epidote are minerals showing large Fe–Al–Mn solid solutions sensitive to *PT* and fO_2 environmental conditions. In the case of epidote composition, oxygen fugacity is more important than temperature [18]. Due to extensive solid solution in terms of Mn/Fe and (Mn + Fe)/Al, Mn-rich garnet and epidote may coexist with element ratio of the different phases sensitive to temperature and degree of oxidation. With decreasing fO_2, piemontite becomes poorer in Mn and garnet and epidote minerals might tend to become more aluminous with increasing temperature [18].

Spessartine can be stable over a wide range of fO_2 and *PT* conditions from greenschist to amphibolite-granulite facies and is evidence of elevated amounts of Mn in the rock [15,83,84]. Natural occurrences of piemontite have shown evidence for metamorphism at high oxygen fugacity [18]. A highly oxidizing environment is shown by the presence of Ti-hematite, Fe^{3+}- and Mn^{3+}-rich epidote, Mn^{3+}-rich andalusite, $Fe^{3+}–Mn^{3+}$-rich kyanite, and so on. The intercalation of decimeter-thick layers of epidote-spessartine-Mn-rich andalusite in less oxidized levels with epidote-spessartine(almandine) assemblages implies (i) significant fO_2 gradients at restricted scale, and (ii) internally controlled fO_2 resulting from oxidized protoliths containing minerals such as Mn-oxides capable of buffering fO_2 to high pressure levels during metamorphism [17,18]. Preservation of the fO_2 gradients supports evidence for a reduced mobility of oxygen throughout high-pressure metamorphism [17] or the localized circulation of highly oxidized fluids during metasomatism. The strong variability of fO_2 dependent mineral assemblages, at all scales, within the Trikorfo area thus indicates a strong local buffering of fO_2 during post-metamorphic metasomatic reactions. As for fO_2, the mineralogical assemblages also suggest a local control in Mn concentration. Indeed, Mn-rich layers, which are mostly found at the top of the series, are scarce and never exceed a few centimeters thick.

Orange kyanite was found coexisting together with Mn-rich andalusite in quartz-feldspar coarse-grained veins and as post-deformation assemblages in mica schists towards the top of the unit in the Trikorfo area. Mn-rich andalusite may also have formed during decompression in veins within host rocks containing kyanite, e.g., [85]. However, petrographic observations suggest growth at equilibrium between kyanite and andalusite. Co-existing Al_2SiO_5 polymorphs occur in various metamorphic rocks and may form either during regional metamorphism or a combination of regional and contact metamorphism [86,87]. When kyanite, sillimanite and andalusite are free of chemical impurities, their equilibrium coexistence is invariant. However, these three polymorphs show extensive solid solution and can incorporate significant amounts of Fe^{3+} and Mn^{3+}, substituting for Al, in highly oxidized rocks [16,88,89]. This results in an increase of the variance of the system and the possible coexistence of kyanite, andalusite and sillimanite in mutual equilibrium over a measurable range of PT conditions [90]. The presence of Mn-bearing andalusite, and especially the association with braunite, quartz and spessartine, indicates high fO_2 [15].

Structural and mineralogical observations thus point to a two-stage evolution of the Trikorfo area, where metamorphic units were affected locally by metasomatic events during exhumation. Fluid circulation leads to fluid-rock interactions and crystallization of large, vividly-colored minerals. The chemistry of the metasomatic mineralogical assemblages is controlled by the chemistry of the protolith more or less enriched in manganese and by local fO_2 buffer.

5.3. PT-Deformation Conditions of Metamorphic Equilibrium and Metasomatic Reactions

The Rhodope domain exposes a large variety of HP-rocks of the Variscan continental crust and Mesozoic sediments that were subjected to Alpine subduction (Early Cretaceous-Eocene) and subsequent exhumation (Early Oligocene-Miocene) [91]. The recorded conditions can be as high as >19 kbars and 750–800 °C with local partially amphibolitized kyanite eclogites, e.g., [37,91].

Thassos Island, belonging to the South Rhodope metamorphic core complex, displays Oligocene to Miocene metamorphic and exhumation history [36,40]. In response to crustal thickening during Alpine tectonics, the geothermobarometric analyses of amphibole gneisses suggest metamorphic conditions of approximately 620 °C/4.7 kbars for the lower unit and 580 °C/2.4 kbars for the intermediate unit [47,48]. On the basis of the observation of a garnet–kyanite–sillimanite assemblage, Dimitriadis [24] estimated the conditions achieved by the mica schists of the Trikorfo area around 5.5 ± 1.5 kbars and 600 ± 50 °C. Brun and Sokoutis [40] reported an unreferenced personal communication of Kostopoulos (2004) stating that the garnet amphibolites yielded temperatures ranging from 545 to 660 °C, as described in Dimitriadis [24], but at significantly higher pressures from 6.5 to 9.5 kbars. On the basis of the garnet–staurolite–kyanite–plagioclase–quartz–phlogopite–Ti–hematite assemblage, our thermodynamic modeling indicates metamorphic conditions at Trikorfo of ca. 9.1 ± 1.3 kbars and 670 ± 40 °C, at significantly higher P and T conditions than previously estimated. The field given in Figure 15 is the only one where kyanite and staurolite are stable together. Instead of staurolite, Dimitriadis [24] reported the presence of sillimanite growing at the expense of kyanite within the metamorphic gneiss. The conditions of the Trikorfo formation should rather stand in the high-temperature/low-pressure part of the domain indicated by our stable metamorphic assemblage, with sillimanite post-dating kyanite. The MnO content of the calculated equilibrated garnet composition is a bit higher than that measured in sample CS16_292b. It is probably due to the fact that the solid solution model used in our thermodynamic calculation is not fully appropriate. The addition of MnO in the system has a small effect on the garnet mode but a great effect on the garnet stability field. Indeed, in agreement with White et al. [92], the incorporation of MnO in our phase diagram calculations (in the MnNCKFMASHT(O) system) leads to a shift to lower PT conditions of the garnet-in line, compared to a system without MnO. The PT conditions obtained for the peak metamorphic assemblage recorded by the kyanite-bearing paragneisses are in good agreement with the geological context and therefore we are confident with the calculated phase diagram, even if the modeled garnet composition is not strictly the same than the measured one. Indeed, this MP-HT metamorphism recorded by paragneisses from the intermediate unit is comparable to those estimated at the Thassos scale. This amphibolite metamorphism, which is associated with the continuous development of the extensive low-angle S_n foliation since ductile to brittle deformation, could mark the beginning of the nappe stack exhumation.

Kyanite and kyanite–andalusite assemblages are found at the interface between metamorphic host rock and quartzo–feldspathic veins. The crystals are generally weakly deformed and aligned parallel to regional lineation. This indicates a local reaction where a SiO_2-rich fluid interacts with the Al_2O_3-rich host-rock [93], as indicated by their Cr/V ratio [4], to form either kyanite or kyanite-andalusite as a function of the *PT* conditions and the variance of the system linked to available MnO [90]. This reaction occurs at ductile conditions while the rock still records stretching lineation. The temperatures derived from the Ti-in kyanite thermometer from Müller et al. [4] are in the range 670–790 °C coherent with the Perple_X modeling. This would then indicate that metasomatic kyanite form at or close to the metamorphic peak conditions. However, EPMA and LA-ICPMS data (Figures 12 and 13) show that Ti distribution is variable within the crystals and these temperatures should be considered with caution. Except for some of the kyanite and andalusite crystals, the other metasomatic silicate and calc-silicate minerals of Thassos are usually not oriented. Dimitriadis [24] suggested retrogression under static conditions as evidenced by undeformed matted fibrolite and unoriented growth of chlorite. However, continuous deformation during the main kinematic event is evidenced by coherent stretching and mineral lineation and brittle vein opening to temperatures as low as 230 °C, as indicated by clinochlore growth in late brittle veins. The presence of non-oriented minerals should be the result of crystal growth from hydrothermal fluids during metasomatic reactions. The presence of calc-silicate minerals (grossular, diopside, hornblende, epidote), especially located to the top of the series where the paragneisses/mica schists are in contact with marble units, implies decarbonation during metasomatic reactions.

6. Conclusions

The garnet–kyanite–biotite–hematite–plagioclase ± staurolite ± sillimanite paragneiss of the Trikorfo area (Thassos Island, Rhodope Massif, northern Greece) recorded metamorphic conditions of ca. 670 ± 40 °C and ca. 9.1 ± 1.3 kbars. Structural, petrographic and mineralogical observations indicate that localized metasomatic reactions occurred during the exhumation of the *HP* unit locally containing layers enriched in Mn with a strong local fO_2 buffering. Metasomatism is likely the result of Miocene intrusion of granitoids, which are related to ore deposits on Thassos Island. Metasomatic reactions first occurred under ductile conditions in an extensive context close to the kyanite-andalusite stability curve and continued until purely brittle conditions as indicated by the presence of late veins cross-cutting the metamorphic foliation. The result is a mineralogical site that could be regarded as a unique geotope with uncommon (Mn-rich) silicate and calc-silicate mineral assemblages similar to what is observed in skarn deposits where hot active hydrothermal fluids interact with carbonate rocks.

Crystals of kyanite, green andalusite, garnet (grossular and spessartine) and red epidote–clinozoisite are large (up to several centimeters), show vivid colors, and are suitable for cabochon-shaped gemstones. As such, the Trikorfo locality can be regarded as a promising area for the exploration of true, facetable gemstones. Their genesis due to metasomatic reactions also underlines the important role of metasomatism for gemstone formation in general, as previously noted in the literature [94–96].

Our combined petrographic, geochemical and spectroscopic observations permit us to demonstrate that the orange color of kyanite in the Trikorfo area is mainly due to Mn^{3+} substitution. Although the Mn content is significantly lower than that in the Loliondo (Tanzania) deposit, Thassos Island (Rhodope, Greece) can now be classified as the second locality worldwide where Mn-rich orange kyanite is reported.

Supplementary Materials: The following are available online at http://www.mdpi.com/2075-163X/9/4/252/s1, Table S1: Bulk rock analyses, Table S2: EPMA analyses PHLOGOPITE, Table S3: EPMA analyses PLAGIOCLASE, Table S4: EPMA analyses GARNET, Table S5: EPMA analyses STAUROLITE, Table S6: EPMA analyses Ti-HEMATITE, Table S7: EPMA analyses EPIDOTE, Table S8: EPMA analyses DIOPSIDE-HORNBLENDE,

Table S9: EPMA analyses TITANITE-RUTILE, Table S10: EPMA analyses MnOXIDES, Table S11: EPMA analyses CHLORITE, Table S12: EPMA analyses KYANITE, Table S13: EPMA analyses ANDALUSITE.

Author Contributions: Conceptualization, A.T.; Data curation, A.T.; Funding acquisition, A.T. and P.V.; Investigation, A.T., P.V., A.E., C.S., K.T., M.B., B.R., C.M., I.G., M.E. and C.P.; Software, A.E.; Writing—original draft, A.T. and P.V.

Funding: This research was partly funded by LABEX ANR-10-LABX-21-Ressources21, Nancy, France, TelluS Program of CNRS/INSU, and OTELo (Observatoire Terre et Environnement de Lorraine, Université de Lorraine, France).

Acknowledgments: The authors warmly acknowledge A. Flammang, J. Moine and O. Rouer (GeoRessources lab., Univ. Lorraine, France) and P. Stutz and S. Heidrich (Institute of Mineralogy and Petrology, University of Hamburg, Germany) for the quality of sample preparation and EPMA analyses. We are grateful to R. Mosser-Ruck (GeoRessources lab., Univ. Lorraine, France) for fruitful discussions. Three anonymous journal reviewers are thanked for their helpful and constructive comments.

Conflicts of Interest: The authors declare no conflict of interest.

Appendix A

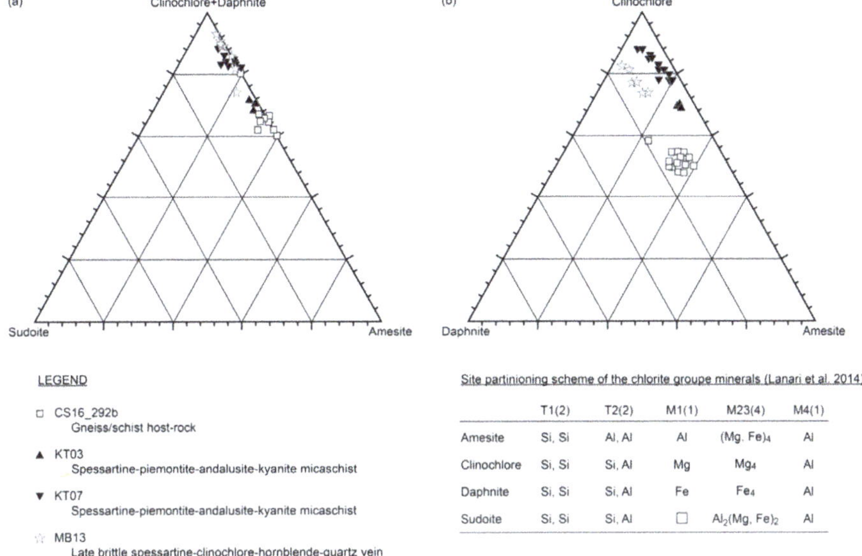

	T1(2)	T2(2)	M1(1)	M23(4)	M4(1)
Amesite	Si, Si	Al, Al	Al	(Mg. Fe)₄	Al
Clinochlore	Si, Si	Si, Al	Mg	Mg₄	Al
Daphnite	Si, Si	Si, Al	Fe	Fe₄	Al
Sudoite	Si, Si	Si, Al	☐	Al₂(Mg, Fe)₂	Al

Figure A1. Composition of chlorite-group minerals analyzed by EPMA in (**a**) Clinochlore + Daphnite–Sudoite–Amesite and (**b**) Clinochlore–Daphnite–Amesite ternary plots. The mineral names are according the site partitioning scheme used in Lanari et al. [97].

Appendix B

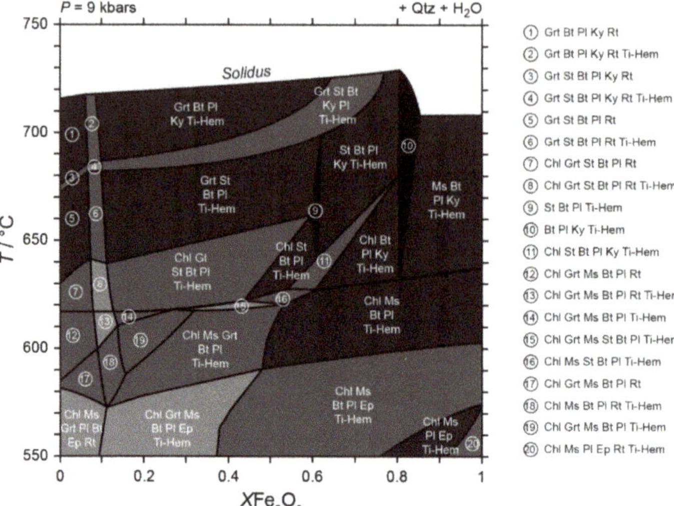

Figure A2. T-XFe$_2$O$_3$ pseudosection calculated for 9 kbars and 550–750 °C for sample CS16_292b. Assemblage fields are shaded according to the number of degrees of freedom, with higher-variance assemblages represented by darker shading. Mineral abbreviation: biotite (Bt), chlorite (Chl), epidote (Ep), garnet (Grt), kyanite (Ky), muscovite (Ms), plagioclase (Pl), quartz (Qtz), rutile (Rt), staurolite (St) and Ti-hematite (Ti-Hem).

References

1. Abs-Wurmbach, I.; Langer, K. Synthetic Mn^{3+}-kyanite and viridine, (Al$_{2-x}$Mn$_x$$^{3+}$)SiO$_5$, in the system Al$_2O_3$-MnO-MnO$_2$-SiO$_2$. *Contrib. Mineral. Petrol.* **1975**, *49*, 21–38. [CrossRef]
2. Abs-Wurmbach, I.; Langer, K.; Seifert, F.; Tillmanns, E. The crystal chemistry of (Mn^{3+}, Fe^{3+})-substituted andalusites (viridines and kanonaite), (Al$_{1-x-y}$Mn$_x$$^{3+}Fe_y$$^{3+}$)$_2$(O|SiO$_4$): Crystal structure refinements, Mössbauer, and polarized optical absorption spectra. *Z. Für Krist.* **1981**, *155*, 81–113.
3. Pearson, G.R.; Shaw, D.M. Trace elements in kyanite, sillimanite and andalusite. *Am. Mineral.* **1960**, *45*, 808–817.
4. Müller, A.; van den Kerkhof, A.M.; Selbekk, R.S.; Broekmans, M.A.T.M. Trace element composition and cathodoluminescence of kyanite and its petrogenetic implications. *Contrib. Mineral. Petrol.* **2016**, *171*, 70. [CrossRef]
5. Faye, G.H.; Nickel, E.H. On the origin of colour and pleochroism of kyanite. *Can. Mineral.* **1969**, *10*, 35–46.
6. Meinhold, K.D.; Frisch, T. Manganese-silicate-bearing metamorphic rocks from central Tanzania. *Schweiz. Mineral. Petrogr. Mitt.* **1970**, *50*, 493–507.
7. Chadwick, K.M.; Rossman, G.R. Orange kyanite from Tanzania. *Gems Gemol.* **2009**, *45*, 146–147.
8. Gaft, M.; Nagli, L.; Panczer, G.; Rossman, G.R.; Reisfeld, R. Laser-induced time-resolved luminescence of orange kyanite Al$_2$SiO$_5$. *Opt. Mater.* **2011**, *33*, 1476–1480. [CrossRef]
9. Wildner, M.; Beran, A.; Koller, F. Spectroscopic characterisation and crystal field calculations of varicoloured kyanites from Loliondo, Tanzania. *Mineral. Petrol.* **2013**, *107*, 289–310. [CrossRef]
10. Hålenius, U. A spectroscopic investigation of manganian andalusite. *Can. Mineral.* **1978**, *16*, 567–575.
11. Smith, G.; Hålenius, U.; Langer, K. Low temperature spectral studies of Mn^{3+}-bearing andalusite and epidote type minerals in the range 30,000–5000 cm^{-1}. *Phys. Chem. Miner.* **1982**, *8*, 136–142. [CrossRef]
12. Klemm, G. Uber Viridin, eine Abart des Andalusites. *Not. Ver Erdk Darmstadt* **1911**, *32*, 4–13.
13. Burns Roger, G. *Mineralogical Applications of Crystal Field Theory*, 2nd ed.; Cambridge University Press: Cambridge, UK, 1970; ISBN 0-521-43077-1.

14. Manning, P.G. The optical absorption spectra of the garnets almandine-pyrope, pyrope and spessartine and some structural interpretations of mineralogical significance. *Can. Mineral.* **1967**, *9*, 237–251.

15. Abs-Wurmbach, I.; Peters, T. The Mn-Al-Si-O system: An experimental study of phase relations applied to parageneses in manganese-rich ores and rocks. *Eur. J. Mineral.* **1999**, *11*, 45–68. [CrossRef]

16. Jöns, N.; Schenk, V. Petrology of whiteschists and associated rocks at Mautia Hill (Tanzania): Fluid infiltration during high-grade metamorphism. *J. Petrol.* **2004**, *45*, 1959–1981. [CrossRef]

17. Reinecke, T. Phase relationships of sursassite and other Mn-silicates in highly oxidized low-grade, high-pressure metamorphic rocks from Evvia and Andros islands, Greece. *Contrib. Mineral. Petrol.* **1986**, *94*, 110–126. [CrossRef]

18. Keskinen, M.; Liou, J.G. Stability relations of Mn–Fe–Al piemontite. *J. Metamorph. Geol.* **1987**, *5*, 495–507. [CrossRef]

19. Voudouris, P.; Graham, I.; Melfos, V.; Sutherland, L.; Zaw, K. Gemstones in Greece: Mineralogy and Crystallizing Environment. In Proceedings of the 34th IGC Conference, Brisbane, Australia, 5–10 August 2012.

20. Voudouris, P.; Graham, I.; Mavrogonatos, K.; Su, S.; Papavasiliou, K.; Farmaki, M.-V.; Panagiotidis, P. Mn-andalusite, spessartine, Mn-grossular, piemontite and Mn-zoisite/clinozoisite from Trikorfo, Thassos island, Greece. *Bull. Geol. Soc. Greece* **2016**, *50*, 2068–2078. [CrossRef]

21. Klemme, S.; Berndt, J.; Mavrogonatos, C.; Flemetakis, S.; Baziotis, I.; Voudouris, P.; Xydous, S. On the color and genesis of prase (green quartz) and Amethyst from the island of Serifos, Cyclades, Greece. *Minerals* **2018**, *8*, 487. [CrossRef]

22. Ottens, B.; Voudouris, P. *Griechenland: Mineralien, Fundorte, Lagerstätten*; Christian Weise Verlag: München, Germany, 2018.

23. Voudouris, P.; Melfos, V.; Mavrogonatos, C.; Tarantola, A.; Götze, J.; Alfieris, D.; Maneta, V.; Psimis, I. Amethyst Occurrences in Tertiary Volcanic Rocks of Greece: Mineralogical, Fluid Inclusion and Oxygen Isotope Constraints on Their Genesis. *Minerals* **2018**, *8*, 324. [CrossRef]

24. Dimitriadis, E. Sillimanite grade metamorphism in Thasos island, Rhodope massif, Greece and its regional significance. *Geol. Rhodopica* **1989**, *1*, 190–201.

25. Voudouris, P.; Constantinidou, S.; Kati, M.; Mavrogonatos, C.; Kanellopoulos, C.; Volioti, E. Genesis of alpinotype fissure minerals from Thasos island, Northern Greece—Mineralogy, mineral chemistry and crystallizing environment. *Bull. Geol. Soc. Greece* **2013**, *47*, 468–476. [CrossRef]

26. Vason, M.; Martin, S. Metamorphosed iron-manganese deposits from the island of Thassos (Western Rhodope region, northern Greece). *Ofioliti* **1993**, *18*, 181–186.

27. Melfos, V.; Voudouris, P. Fluid evolution in Tertiary magmatic-hydrothermal ore systems at the Rhodope metallogenic province, NE Greece. A review. *Geol. Croat.* **2016**, *69*, 157–167. [CrossRef]

28. Zachos, S. Geological Map of Greece, Thassos Sheet. 1982. Available online: https://www.worldcat.org/title/geological-map-of-greece-thassos-sheet-phyllo-thasos-institogto-geologikon-kai-metalletikoi-ereunon-i-geologiki-chartografioi-egiye-apo-to-s-zachos/oclc/492645828 (accessed on 24 April 2019).

29. Su, S.; Graham, I.; Voudouris, P.; Mavrogonatos, K.; Farmaki, M.V.; Panagiotidis, P. *Viridine, Piemontite and Epidote Group Minerals from Thassos Island, Northern Greece*; AGI: Alexandria, VA, USA, 2016; T37.P2; p. 1715.

30. Voudouris, P. Minerals of Eastern Macedonia and Western Thrace: Geological framework and environment of formation. *Bull. Geol. Soc. Greece* **2005**, *37*, 62–77.

31. Voudouris, P.; Melfos, V.; Katerinopoulos, A. Precious stones in Greece: Mineralogy and geological environment of formation. In Proceedings of the Understanding the Genesis of Ore Deposits to Meet the Demand of the 21st Century, Moscow, Russia, 21–24 August 2006; p. 6.

32. Le Pichon, X.; Bergerat, F.; Roulet, M.J. Plate kinematics and tectonics leading to the Alpine belt formation: A new analysis. In *Processes in Continental Lithospheric Deformation*; Clark, S.P., Burchfiel, B.C., Suppe, J., Eds.; Geological Society of America: Boulder, CO, USA, 1988; Volume 218, pp. 111–131.

33. Dewey, J.F.; Helman, M.L.; Knott, S.D.; Turco, E.; Hutton, D.H.W. Kinematics of the western Mediterranean. *Geol. Soc. Lond. Spec. Publ.* **1989**, *45*, 265–283. [CrossRef]

34. Bonneau, M. La Tectonique de L'arc égéen Externe et du Domaine Cycladique. Ph.D. Thesis, Université Pierre et Marie Curie, Paris, France, 1991; p. 443.

35. van Hinsbergen, D.J.J.; Hafkenscheid, E.; Spakman, W.; Meulenkamp, J.E.; Wortel, R. Nappe stacking resulting from subduction of oceanic and continental lithosphere below Greece. *Geology* **2005**, *33*, 325–328. [CrossRef]

36. Wawrzenitz, N.; Krohe, A. Exhumation and doming of the Thasos metamorphic core complex (S Rhodope, Greece): Structural and geochronological constraints. *Tectonophysics* **1998**, *285*, 301–332. [CrossRef]

37. Liati, A.; Seidel, E. Metamorphic evolution and geochemistry of kyanite eclogites in central Rhodope, northern Greece. *Contrib. Mineral. Petrol.* **1996**, *123*, 293–307. [CrossRef]

38. Liati, A.; Gebauer, D. Constraining the prograde and retrograde P-T-t path of Eocene HP rocks by SHRIMP dating of different zircon domains: Inferred rates of heating, burial, cooling and exhumation for central Rhodope, northern Greece. *Contrib. Mineral. Petrol.* **1999**, *135*, 340–354. [CrossRef]

39. Liati, A. Identification of repeated Alpine (ultra) high-pressure metamorphic events by U–Pb SHRIMP geochronology and REE geochemistry of zircon: The Rhodope zone of Northern Greece. *Contrib. Mineral. Petrol.* **2005**, *150*, 608–630. [CrossRef]

40. Brun, J.-P.; Sokoutis, D. Kinematics of the Southern Rhodope Core Complex (North Greece). *Int. J. Earth Sci.* **2007**, *96*, 1079–1099. [CrossRef]

41. Moulas, E.; Kostopoulos, D.; Connolly, J.A.D.; Burg, J.-P. P-T estimates and timing of the sapphirine-bearing metamorphic overprint in kyanite eclogites from Central Rhodope, northern Greece. *Petrology* **2013**, *21*, 507–521. [CrossRef]

42. Dürr, S.; Altherr, R.; Keller, J.; Okrusch, M.; Seidel, E. The Median Aegean Crystal-line Belt: Stratigraphy, Structure, Metamorphism, Magmatism. *IUCG Sci. Rep.* **1978**, *38*, 455–477.

43. Jacobshagen, V.; Dürr, S.; Kockel, F.; Kopp, K.O.; Kowalczyk, G.; Berckhemer, H. Structure and Geodynamic Evolution of the Aegean Region. In *Alps, Apennines, Hellenides*; Cloos, H., Roeder, D., Schmidt, K., Eds.; IUCG: Stuttgart, Germany, 1978; pp. 537–564.

44. Lister, G.S.; Banga, G.; Feenstra, A. Metamorphic core complexes of Cordilleran type in the Cyclades, Aegean Sea, Greece. *Geology* **1984**, *12*, 221–225. [CrossRef]

45. Melfos, V.; Voudouris, P. Cenozoic metallogeny of Greece and potential for precious, critical and rare metals exploration. *Ore Geol. Rev.* **2017**, *89*, 1030–1057. [CrossRef]

46. Liati, A.; Fanning, C.M. Eclogites and Country rock orthogneisses representing upper Permian Gabbros in Hercynian Granitoids, Rhodope, Greece: Geochronological Constraints. *Mitt. Österr. Mineral. Ges.* **2005**, *150*, 88.

47. Schulz, B. Syntectonic heating and loading-deduced from microstructures and mineral chemistry in micaschists and amphibolites of the Pangeon complex (Thassos island, Northern Greece). *Neues Jahrb. Für Geol. Paläontol. Abh.* **1992**, *184*, 181–201.

48. Bestmann, M.; Kunze, K.; Matthews, A. Evolution of a calcite marble shear zone complex on Thassos Island, Greece: Microstructural and textural fabrics and their kinematic significance. *J. Struct. Geol.* **2000**, *22*, 1789–1807. [CrossRef]

49. Dinter, D.A. Late Cenozoic extension of the Alpine collisional orogen, northeastern Greece: Origin of the north Aegean basin. *Gsa Bull.* **1998**, *110*, 1208–1230. [CrossRef]

50. Kounov, A.; Wüthrich, E.; Seward, D.; Burg, J.-P.; Stockli, D. Low-temperature constraints on the Cenozoic thermal evolution of the Southern Rhodope Core Complex (Northern Greece). *Int. J. Earth Sci.* **2015**, *104*, 1337–1352. [CrossRef]

51. Peterek, A.; Polte, M.; Wölfl, C.; Bestmann, M.; Lemtis, O. Zur jungtertiären geologischen Entwicklung im SW der Insel Thassos (S-Rhodope, Nordgriechenland). *Erlanger Geol. Abh.* **1994**, *124*, 29–59.

52. Dinter, D.A.; Royden, L. Late Cenozoic extension in northeastern Greece: Strymon Valley detachment system and Rhodope metamorphic core complex. *Geology* **1993**, *21*, 45–48. [CrossRef]

53. Sokoutis, D.; Brun, J.P.; Driessche, J.V.D.; Pavlides, S. A major Oligo-Miocene detachment in southern Rhodope controlling north Aegean extension. *J. Geol. Soc.* **1993**, *150*, 243–246. [CrossRef]

54. Kilias, A.A.; Mountrakis, D.M. Tertiary extension of the Rhodope massif associated with granite emplacement (Northern Greece). *Acta Vulcanol.* **1998**, *10*, 331–337.

55. Carignan, J.; Hild, P.; Mevelle, G.; Morel, J.; Yeghicheyan, D. Routine Analyses of Trace Elements in Geological Samples using Flow Injection and Low Pressure On-Line Liquid Chromatography Coupled to ICP-MS: A Study of Geochemical Reference Materials BR, DR-N, UB-N, AN-G and GH. *Geostand. Newsl.* **2001**, *25*, 187–198. [CrossRef]

56. Pouchou, J.-L.; Pichoir, F. *Quantitative Analysis of Homogeneous or Stratified Microvolumes Applying the Model "PAP."* In *Electron Probe Quantitation*; Heinrich, K.F.J., Newbury, D.E., Eds.; Springer US: Boston, MA, USA, 1991; pp. 31–75. ISBN 978-1-4899-2617-3.

57. Jochum, K.P.; Weis, U.; Stoll, B.; Kuzmin, D.; Yang, Q.; Raczek, I.; Jacob, D.E.; Stracke, A.; Birbaum, K.; Frick, D.A.; et al. Determination of Reference Values for NIST SRM 610–617 Glasses Following ISO Guidelines. *Geostand. Geoanalytical Res.* **2011**, *35*, 397–429. [CrossRef]

58. Paton, C.; Hellstrom, J.; Paul, B.; Woodhead, J.; Hergt, J. Iolite: Freeware for the visualisation and processing of mass spectrometric data. *J. Anal. At. Spectrom.* **2011**, *26*, 2508–2518. [CrossRef]

59. Connolly, J.A.D. Computation of phase equilibria by linear programming: A tool for geodynamic modeling and its application to subduction zone decarbonation. *Earth Planet. Sci. Lett.* **2005**, *236*, 524–541. [CrossRef]

60. Holland, T.J.B.; Powell, R. An improved and extended internally consistent thermodynamic dataset for phases of petrological interest, involving a new equation of state for solids. *J. Metamorph. Geol.* **2011**, *29*, 333–383. [CrossRef]

61. White, R.W.; Powell, R.; Holland, T.J.B.; Worley, B.A. The effect of TiO_2 and Fe_2O_3 on metapelitic assemblages at greenschist and amphibolite facies conditions: Mineral equilibria calculations in the system K_2O–FeO–MgO–Al_2O_3–SiO_2–H_2O–TiO_2–Fe_2O_3. *J. Metamorph. Geol.* **2000**, *18*, 497–511. [CrossRef]

62. White, R.W.; Powell, R.; Holland, T.J.B.; Johnson, T.E.; Green, E.C.R. New mineral activity–composition relations for thermodynamic calculations in metapelitic systems. *J. Metamorph. Geol.* **2014**, *32*, 261–286. [CrossRef]

63. Newton, R.C.; Charlu, T.V.; Kleppa, O.J. Thermochemistry of the high structural state plagioclases. *Geochim. Cosmochim. Acta* **1980**, *44*, 933–941. [CrossRef]

64. Bocchio, R.; Diella, V.; Adamo, I.; Marinoni, N. Mineralogical characterization of the gem-variety pink clinozoisite from Val Malenco, Central Alps, Italy. *Rend. Lincei* **2017**, *28*, 549–557. [CrossRef]

65. Geller, S. Crystal chemistry of garnets. *Z. Krist.* **1967**, *125*, 1–45. [CrossRef]

66. Armbruster, T.; Bonazzi, P.; Akasaka, M.; Bermanec, V.; Chopin, C.; Gieré, R.; Heuss-Assbichler, S.; Liebscher, A.; Menchetti, S.; Pan, Y.; et al. Recommended nomenclature of epidote-group minerals. *Eur. J. Mineral.* **2006**, *18*, 551–567. [CrossRef]

67. Franz, G.; Liebscher, A. Physical and Chemical Properties of the Epidote Minerals: An introduction. *Rev. Mineral. Geochem.* **2004**, *56*, 1–81. [CrossRef]

68. Schreyer, W.; Bernhardt, H.-J.; Fransolet, A.-M.; Armbruster, T. End-member ferrian kanonaite: An andalusite phase with one Al fully replaced by $(Mn, Fe)^{3+}$ in a quartz vein from the Ardennes mountains, Belgium, and its origin. *Contrib. Mineral. Petrol.* **2004**, *147*, 276–287. [CrossRef]

69. Sun, S.S.; McDonough, W.F. Chemical and isotopic systematics of oceanic basalts: Implications for mantle composition and processes. *Geol. Soc. Lond. Spec. Publ.* **1989**, *42*, 313–345. [CrossRef]

70. Cathelineau, M. Cation site occupancy in chlorites and illites as a function of temperature. *Clay Miner.* **1988**, *23*, 471–485. [CrossRef]

71. Furukawa, Y.; Yoshiasa, A.; Arima, H.; Okube, M.; Murai, K.; Nishiyama, T. Local Structure of Transition Elements (V, Cr, Mn, Fe and Zn) in Al_2SiO_5 Polymorphs. *Aip Conf. Proc.* **2007**, *882*, 235–237.

72. Rossman, G.R. Optical spectroscopy. In *Spectroscopic Methods in Mineralogy and Geology*; Hawthorne, F.C., Ed.; De Gruyter: Vienna, Austria, 1988; Volume 18, pp. 205–254.

73. Platonov, A.N.; Tarashchan, A.N.; Langer, K.; Andrut, M.; Partzsch, G.; Matsyuk, S.S. Electronic absorption and luminescence spectroscopic studies of kyanite single crystals: Differentiation between excitation of FeTi charge transfer and Cr^{3+} dd transitions. *Phys. Chem. Miner.* **1998**, *25*, 203–212. [CrossRef]

74. White, E.W.; White, W.B. Electron Microprobe and Optical Absorption Study of Colored Kyanites. *Science* **1967**, *158*, 915–917. [CrossRef] [PubMed]

75. Rost, F.; Simon, E. Zur Geochemie und Färbung des Cyanits. *Neues Jahrb. Mineral.-Mon.* **1972**, *9*, 383–395.

76. Parkin, K.M.; Loeffler, B.M.; Burns, R.G. Mössbauer spectra of kyanite, aquamarine, and cordierite showing intervalence charge transfer. *Phys. Chem. Miner.* **1977**, *1*, 301–311. [CrossRef]

77. Bonizzoni, L.; Galli, A.; Spinolo, G.; Palanza, V. EDXRF quantitative analysis of chromophore chemical elements in corundum samples. *Anal. Bioanal. Chem.* **2009**, *395*, 2021–2027. [CrossRef]

78. Novák, M.; Škoda, R. Mn^{3+}-rich andalusite to kanonaite and their breakdown products from metamanganolite at Kojetice near Třebíč, the Moldanubian Zone, Czech Republic. *J. Geosci.* **2007**, *52*, 161–167. [CrossRef]

79. Abs-Wurmbach, I.; Langer, K.; Schreyer, W. The Influence of Mn^{3+} on the Stability Relations of the Al_2SiO_5 Polymorphs with Special Emphasis on Manganian Andalusites (Viridines), $(Al_{1-x}Mn_x{}^{3+})_2(O/SiO_4)$: An Experimental Investigation. *J. Petrol.* **1983**, *24*, 48–75. [CrossRef]

80. Bonazzi, P.; Menchetti, S. Manganese in Monoclinic Members of the Epidote Group: Piemontite and Related Minerals | Reviews in Mineralogy and Geochemistry. *Rev. Min. Geochem.* **2004**, *56*, 495–551. [CrossRef]

81. Kassoli-Fournaraki, A.; Michailidis, K. Chemical composition of tourmaline in quartz veins from Nea Roda and Thasos areas in Macedonia, northern Greece. *Can. Mineral.* **1994**, *32*, 607–615.

82. van Hinsberg, V.J.; Henry, D.J.; Dutrow, B.L. Tourmaline as a Petrologic Forensic Mineral: A Unique Recorder of Its Geologic Past. *Elements* **2011**, *7*, 327–332. [CrossRef]

83. Symmes, G.H.; Ferry, J.M. The effect of whole-rock MnO content on the stability of garnet in pelitic schists during metamorphism. *J. Metamorph. Geol.* **1992**, *10*, 221–237. [CrossRef]

84. Geiger, C.A.; Armbruster, T. $Mn_3Al_2Si_3O_{12}$ spessartine and $Ca_3Al_2Si_3O_{12}$ grossular garnet: Structural dynamic and thermodynamic properties. *Am. Mineral.* **1997**, *82*, 740–747. [CrossRef]

85. Kerrick, D.M. *The Al_2SiO_5 Polymorphs*; Reviews in Mineralogy; Mineralogical Society of America: Washington, DC, USA, 1990; ISBN 978-0-939950-27-0.

86. Larson, T.E.; Sharp, Z.D. Stable isotope constraints on the Al_2SiO_5 'triple-point' rocks from the Proterozoic Priest pluton contact aureole, New Mexico, USA. *J. Metamorph. Geol.* **2003**, *21*, 785–798. [CrossRef]

87. Sepahi, A.A.; Whitney, D.L.; Baharifar, A.A. Petrogenesis of andalusite–kyanite–sillimanite veins and host rocks, Sanandaj-Sirjan metamorphic belt, Hamadan, Iran. *J. Metamorph. Geol.* **2004**, *22*, 119–134. [CrossRef]

88. Abraham, K.; Schreyer, W. Minerals of the viridine hornfels from Darmstadt, Germany. *Contrib. Mineral. Petrol.* **1975**, *49*, 1–20. [CrossRef]

89. Kramm, U. Kanonaite-rich viridines from the Venn-Stavelot Massif, Belgian Ardennes. *Contrib. Mineral. Petrol.* **1979**, *69*, 387–395. [CrossRef]

90. Grambling, J.A.; Williams, M.L. The Effects of Fe^{3+} and Mn^{3+} on Aluminum Silicate Phase Relations in North-Central New Mexico, U.S.A. *J. Petrol.* **1985**, *26*, 324–354. [CrossRef]

91. Mposkos, E.; Krohe, A. Petrological and structural evolution of continental high pressure (HP) metamorphic rocks in the Alpine Rhodope Domain (N. Greece). In *Proceedings, Third International Conference on the Geology of the Eastern Mediterranean*; Panayides, I., Xenophontos, C., Malpas, J., Eds.; Geological Survey Department: Nicosia, Cyprus, 2000; pp. 221–232.

92. White, R.W.; Powell, R.; Johnson, T.E. The effect of Mn on mineral stability in metapelites revisited: New a–x relations for manganese-bearing minerals. *J. Metamorph. Geol.* **2014**, *32*, 809–828. [CrossRef]

93. El Mahi, B.; Zahraoui, M.; Hoepffner, C.; Boushaba, A.; Meunier, A.; Beaufort, D. Kyanite-quartz synmetamorphic veins: Indicators of post-orogenic thinning and metamorphism (Western Meseta, Morocco). *Pangea* **2000**, *33/34*, 27–47.

94. Simonet, C.; Fritsch, E.; Lasnier, B. A classification of gem corundum deposits aimed towards gem exploration. *Ore Geol. Rev.* **2008**, *34*, 127–133. [CrossRef]

95. Groat, L.A.; Laurs, B.M. Gem Formation, Production, and Exploration: Why Gem Deposits Are Rare and What is Being Done to Find Them. *Elements* **2009**, *5*, 153–158. [CrossRef]

96. Yakymchuk, C.; Szilas, K. Corundum formation by metasomatic reactions in Archean metapelite, SW Greenland: Exploration vectors for ruby deposits within high-grade greenstone belts. *Geosci. Front.* **2018**, *9*, 727–749. [CrossRef]

97. Lanari, P.; Wagner, T.; Vidal, O. A thermodynamic model for di-trioctahedral chlorite from experimental and natural data in the system $MgO–FeO–Al_2O_3–SiO_2–H_2O$: Applications to P–T sections and geothermometry. *Contrib. Mineral. Petrol.* **2014**, *167*, 968. [CrossRef]

Article

Gemstones of Greece: Geology and Crystallizing Environments

Panagiotis Voudouris [1,*], Constantinos Mavrogonatos [1], Ian Graham [2], Gaston Giuliani [3], Alexandre Tarantola [4], Vasilios Melfos [5], Stefanos Karampelas [6], Athanasios Katerinopoulos [1] and Andreas Magganas [1]

[1] Faculty of Geology and Geoenvironment, National and Kapodistrian University of Athens, GR-15784 Athens, Greece
[2] PANGEA Research Centre, School of Biological, Earth and Environmental Sciences, University of NSW, 2052 Sydney, Australia
[3] CRPG/CNRS, Université Paul Sabatier, GET/IRD et Université de Lorraine, 54501 Vandœuvre cedex, France
[4] GeoRessources, Faculté des Sciences et Technologies, Université de Lorraine, CNRS, F-54506 Vandœuvre-lès-Nancy, France
[5] Faculty of Geology, Aristotle University of Thessaloniki, 54124 Thessaloniki, Greece
[6] Bahrain Institute for Pearls & Gemstones (DANAT), WTC East Tower, P.O. Box 17236 Manama, Bahrain
* Correspondence: voudouris@geol.uoa.gr; Tel.: +30-210-727-4129

Received: 28 June 2019; Accepted: 26 July 2019; Published: 29 July 2019

Abstract: In the Hellenides Orogen, minerals of various gem quality occur in various rock types from mainly four tectono-metamorphic units, the Rhodope, Pelagonian, and the Attico-Cycladic massifs, and the Phyllites-Quartzites unit of Crete Island. In crystalline rocks, gemstones are related to both regional metamorphic-metasomatic processes (e.g., gem corundums, Mn-andalusite, thulite/clinothulite, spessartine, titanite, jadeite), and to the formation of late alpine-type fissures, such as, for example, quartz, albite, adularia and titanite. The Tertiary (and Mesozoic) magmatic-hydrothermal environments provide gem-quality sapphire, beryl, garnet, vesuvianite, epidote, fluorite, and SiO_2 varieties. The supergene oxidation zone of the Lavrion deposit hosts gem-quality smithsonite and azurite. Coloration in the studied material is either due to various chromophore trace elements present in the crystal structure, or due to inclusions of other mineral phases. Future modern exploration methods combined with gemological investigations (such as treatment and faceting of selected stones), are necessary in order to obtain a better knowledge of the gemstone potential of Greece for its possible exploitation.

Keywords: gemstones; corundum; beryl; jadeitite; garnet; quartz varieties; Greece

---

1. Introduction

By definition, gems are materials used for adornment or decoration that must be relatively rare, hard, and tough enough (shock resistant) to resist "normal" wear and withstand corrosion by skin contact and cosmetics [1]. Gems have been prized for thousands of years for their color, luster, transparency, durability, and high value-to-volume ratio [1,2]. A natural gem is one that has been fashioned (or faceted) after having been found in nature, even if it later undergoes treatment processes. Gem materials cover a large variety of products found in the jewelry market today. Most natural gems are single crystals of natural minerals, although others are amorphous (some varieties of opal, natural glass), some are solid solutions (garnets, peridot, etc.), others are rocks (jade, lapis), and some are composed partly or wholly of organic materials (amber, pearl, coral, etc.) [1].

According to Groat [3,4] and Stern et al. [5], any mineral or stone beautiful enough to be sought, mined, and sold for its beauty alone is a gemstone. Among the most important gemstones are diamond, ruby (red-colored corundum) and sapphire (blue-colored corundum), emerald and other gem forms of the mineral beryl, chrysoberyl, tanzanite (blue zoisite), tsavorite (green grossular garnet), "paraiba-type" tourmaline (copper-colored "neon"-blue tourmaline), topaz and jadeite-"jade" (jadeitite) [4]. More common gem materials include, among others, amber, silica gems, feldspar, tourmaline, spinel, garnets, zircon etc. It is not the mineral itself that makes a gemstone, it is the characteristics of a specific sample [2].

Understanding the geological conditions that give rise to gem deposits is of great importance because it can provide guidelines for exploration [2,3,6]. Specific geological conditions include the availability of uncommon major constituents, the presence of adequate chromophores and open space in order to form crystals of sufficient size and transparency [2], and the absence of dramatic post-growth events (e.g., fracturing).

Traditionally, Greece has not been regarded as a source country for gemstones [7]. Known gem material included rubies from Xanthi [8], sapphire and beryl from Naxos Island [9], red-colored spessartine from Paros Island [10], smithsonite from the oxidation zone of the Lavrion mines [11], and the green quartz variety (prase) from the Serifos skarn [12]. Exploration work in Greece over the last three decades has resulted in the discovery of new occurrences of mineral megacrysts many of them being gem-quality [13–19]. The gemstones of Greece are set within diverse geological settings, the study of which will increase our knowledge on the conditions necessary for their crystallization and concentration into economically-viable deposits. The aim of this work is to review all available information on the geology, mineralogy, geochemistry, and fluid characteristics involved in the formation of various gemstones in Greece. The occurrences of gemstones in Greece are presented in Figures 1 and 2 and described with respect to their geological framework and the conditions of their formation.

2. Materials and Methods

Analytical methods included optical and electron microscopy, X-ray powder diffraction studies, and electron microprobe analyses, in the Section of Mineralogy-Petrology at University of Athens and at the Institute of Mineralogy and Petrology, University of Hamburg. X-ray powder diffraction measurements were obtained using a SIEMENS D-500 diffractometer with Cu tube and Co filter. For the mineral analyses, a scanning electron microscope JEOL JSM-5600 combined with an energy dispersive X-ray microanalysis system Oxford Link Isis 300 system were used. Thin and thin-and-polished sections of mineral samples and host rocks were studied by a JEOL JSM 5600 scanning electron microscope equipped with back-scattered imaging capabilities, respectively, at the Department of Mineralogy and Petrology at the University of Athens (Greece). Quantitative analyses were carried-out at the Institute of Mineralogy and Petrology, University of Hamburg using a Cameca-SX 100 WDS. Analytical conditions were as follows: accelerating voltage of 20 kV, a beam current of 20 nA and counting time of 20 s for Al, Si, Ca, Fe and Mg, 60 s for Mn and 120 s for Cr. The X-ray lines used were: AlKα, SiKα, TiKα, FeKα, MnKα, MgKα, CrKα, and CaLα. The standards used were: andradite (for Si, Ca and Fe), and synthetic Al_2O_3 (for Al), MnTiO3 (for Mn and Ti), Fe_2O_3 (for Fe), Cr_2O_3 (for Cr), and MgO (for Mg). Corrections were applied using the PAP online program [20].

LA-ICP-MS analyses presented in this study are from Voudouris et al. [18]. The analyses were conducted at the CODES ARC Centre of Excellence in Ore Deposits of the University of Tasmania, Australia, and the Institute of Mineralogy, University of Münster, Germany. For analytical conditions see Voudouris et al. [18].

Stable isotope analyses were performed at the Stable Isotope and Atmospheric Laboratories, Department of Geology, Royal Holloway, University of London (London, UK). The oxygen isotope composition of quartz was obtained using a CO_2 laser fluorination system similar to that described by Mattey [21]. Each mineral, separate or standard, is weighed at 1.7 mg $\pm$ 10%. These were loaded into the 16-holes of a nickel sample tray, which was inserted into the reaction chamber and then evacuated. The oxygen was released by a 30W Synrad CO_2 laser in the presence of BrF_5 reagent. The yield of oxygen was measured as a calibrated pressure based on the estimated or known oxygen content of the mineral being analyzed. Low yields result in low $\delta^{18}O$ values for all mineral phases, so accurate yield calculations are essential. Yields of >90% are required for most minerals to give satisfactory $\delta^{18}O$ values. The oxygen gas was measured using a VG Isotech (now GV Instruments) Optima dual inlet isotope ratio mass spectrometer (IRMS). All values are reported relative to Vienna Standard Mean Ocean Water (V-SMOW). The data are calibrated to a quartz standard (Q BLC) with a known $\delta^{18}O$ value of +8.8‰ V-SMOW from previous measurements at the University of Paris-6 (France). This has been further calibrated for the RHUL laser line by comparison with NBS-28 quartz. Each 16-hole tray contained up to 12 sample unknowns and 4 of the Q BLC standard. For each quartz run a small constant daily correction, normally less than 0.3‰, was applied to the data based on the average value for the standard. Overall, the precision of the RHUL system based on standard and sample replicates is better than ±0.1‰.

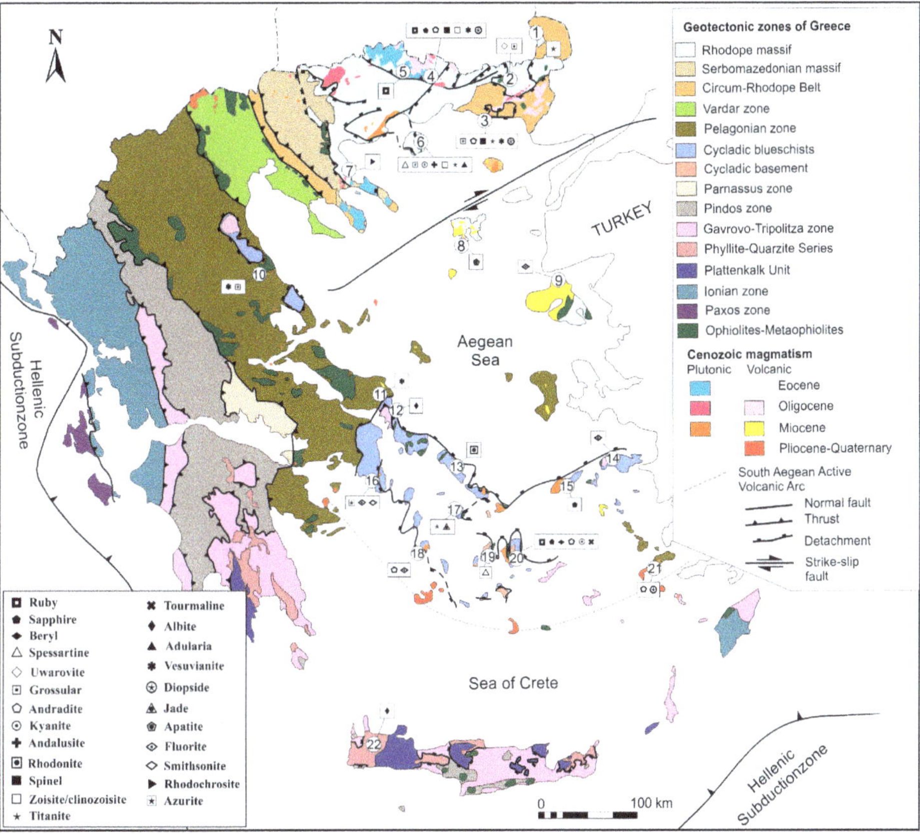

Figure 1. Occurrences of various gemstones in metamorphic and igneous rocks of Greece (Geological map of Greece modified after Ottens and Voudouris [17]): 1. Therapio/Evros, 2. Vyrsini/Evros, 3. Maronia/Rhodopi 4. Gorgona-Kimmeria/Xanthi, 5. Paranesti/Drama, 6. Trikorfo/Thassos, 7. Stratoni-Olympiada/Chalkidiki, 8. Fakos/Limnos Island, 9. Megala Therma/Lesvos Island, 10. Larissa, 11. Kymi/Evia Island, 12. Krieza-Koskina/Evia Island, 13. Petalo/Andros Island, 14. Ampelos/Samos Island, 15. Pounta/Ikaria Island, 16. Lavrion/Attica, 17. Kampos/Syros Island, 18. Agia Marina-Avessalos/Serifos Island, 19, Thapsana/Paros Island, 20. Kinidaros-Kavalaris/Naxos Island, 21. Kardamena/Kos Island, 22. Lakkoi/Crete Island.

3. Geological Setting

The Hellenides constitute part of the Alpine-Himalayan Orogen and formed when Apulia collided with Europe during the Late Cretaceous to Tertiary. They are subdivided into several units: the Rhodope Massif, Servo-Macedonian Massif, Vardar Zone, Pelagonian Zone (Internal Hellenides) and the External Hellenides built-up by Mesozoic and Cenozoic rocks [22–25] (Figures 1 and 2). The Hellenides can be considered an accretionary orogen, resulted from thrusting and SW-verging nappe-stacking of the Rhodopes, Pelagonia and Adria continental blocks, and closure of the Vardar and Pindos oceanic domains of the Neotethys [22]. A Permo-Carboniferous igneous event (known from the Pelagonian Zone, the Rhodope Massif, the Attico-Cycladic Zone, Peloponnese and Crete) documents an active continental margin evolution in the Precambrian-Silurian basement of the Hellenides. Final collision between Europe and Pelagonia at the end of the Cretaceous closed the Neotethys Ocean along the Vardar Suture Zone, as evidenced by obducted Jurassic ophiolites on the Pelagonia continental block [26].

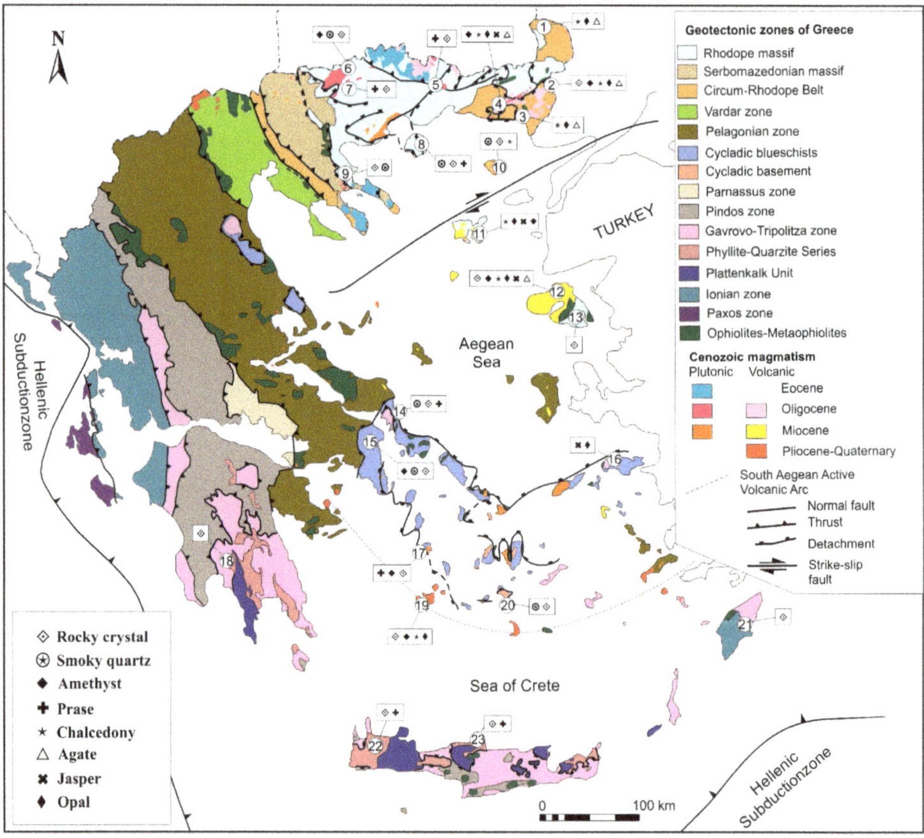

Figure 2. Occurrences of gem-quality silica varieties in metamorphic and igneous rocks of Greece (Geological map of Greece modified after Ottens and Voudouris [17]): 1. Dikea/Evros, 2. Kornofolia/Evros, 3. Aetochori/Evros 4. Sapes/Rhodopi, 5. Kimmeria/Xanthi, 6. Dassoto/Drama, 7. Cresti/Drama, 8. Trikorfo/Thassos, 9. Stratoni-Olympiada/Chalkidiki, 10. Fengari/Samothraki, 11. Moudros-Roussopouli/Limnos Island, 12. Megala Therma/Lesvos Island, 13. Geras/Lesvos, 14. Krieza-Koskina/Evia Island, 15. Penteli Mt/Attika, 16. Ampelos/Samos Island, 17. Avessalos/Serifos Island, 18 Taygetos,. 19. Chondro Vouno-Vani/Milos Island, 20. Mylopotas/Ios Island, 21. Rhodes Island, 22. Prases/Crete Island, 23. Agia Pelagia/Crete Island.

Shortening and syn-orogenic exhumation of HP-LT rocks occurred during the late Cretaceous-Eocene, before an acceleration of slab retreat changed the subduction regime and caused the collapse of the Hellenic mountain belt and thinning of the Aegean Sea from the middle Eocene/late Oligocene to the present [22]. During this post-orogenic episode, large-scale detachments formed, which exhumed metamorphic core complexes in a back-arc setting. Tertiary magmatism in the Aegean region occurred mostly in a post-collisional setting behind the active Hellenic subduction zone [26]. The Pliocene to recent volcanic rocks in the active Aegean volcanic arc formed as a consequence of active subduction beneath the Hellenic trench. In the Hellenides Orogen, gemstones occur in various rock types of mainly three tectono-metamorphic units, the Rhodope- and the Attico-Cycladic massifs and the Phyllites-Quartzites unit of Crete Island.

4. Formation Environments of Gems: An Overview

4.1. Metamorphic-Metasomatic

In crystalline rocks, two groups of gemstones are distinguished: those formed during regional metamorphism-metasomatism and those associated with late alpine-type fissures [14–16,18,19,27–29] (Figures 1 and 2). The first group includes ruby, sapphire, jadeite "jade" (jadeitite), Mn-andalusite, spessartine, Mn-grossular, Mn-zoisite/clinozoisite, Fe-Mn-bearing kyanite and rhodonite. Corundums in the Xanthi-Drama areas (Rhodope massif) and Naxos-Ikaria Islands (Attico-Cycladic massif), are hosted in pargasitic schists, marbles and metabauxites. Host rocks are Carboniferous orthogneisses, metapelites and skarns for Mn-andalusite, spessartine and thulite at Thassos Island, Permian - Middle Triassic schists for rhodonite in Andros Island (Attico-Cycladic massif, [30]) and orthogneisses and marbles in Paros Island [10]. Jadeitites, eclogites and blueschists are host rocks for jadeite "jade" and titanite at Syros Island.

Alpine-type fissure minerals [31,32] in Greece are related to tension gashes formed under brittle-ductile to ductile conditions during the retrograde, late-stage exhumation of metamorphic core complexes [14,33,34]. The alpine-type fissures in Greece contain gem-quality quartz and albite together with adularia, chlorite, epidote, actinolite, hematite, muscovite, rutile, tourmaline and pyrite. Host lithologies are ortho- and paragneisses in the Rhodope Massif (Drama, Thassos Island) and Attico-Cycladic Massif (Penteli Mt, Evia, Ios Islands), amphibolites (Evros, Thassos, Evia and Andros Islands) and finally phyllites (Lesvos Island) and metaquartzites (Crete Island). Quartz in both transparent smoky and colorless (rock crystal) varieties were found in all the above localities and/or host rocks—deep violet amethyst occurs in orthogneisses at Drama area [34]. Albite crystals of gem-quality are abundant in metabasites from Crete Island and orthogneisses from Evia Island.

4.2. Magmatic-Hydrothermal

The Tertiary (and Mesozoic) magmatic-hydrothermal environments in Greece (granitoids, pegmatites, skarns, non-skarn carbonate-replacement deposits, volcanic and ophiolitic rocks) also provide gem-quality material of beryl, corundum, garnet, vesuvianite, diopside, epidote, titanite, spinel, fluorite, rhodochrosite, quartz varieties, and silica microcrystalline species (Figures 1 and 2). Miarolitic cavities and quartz veins cross-cutting granitoids host gem-quality quartz varieties and apatite. Samothraki and Limnos Islands in the northern Aegean, and Kimmeria/Xanthi, Maronia/Rhodopi in the mainland of northern Greece and Tinos Island in the Cyclades are the most important localities. The Greek pegmatites are mostly deficient in miarolitic cavities, and thus contain only matrix embedded mineral crystals. On Naxos Island, blue sapphires are associated with granite pegmatites intruding ultramafic lithologies (plumasites). Blue beryl crystals (var. aquamarine) are typical constituents of Naxos pegmatites both within the migmatitic domes as well as in pegmatites cross-cutting the surrounding metamorphic rocks. Black tourmaline crystals occur in pegmatites near Nevrokopi-Drama and at Naxos Island.

The skarns of Serifos, Kimmeria/Xanthi and Drama, are characterized by abundant quartz. Epidote crystals are associated with quartz at Lefkopetra, near Kimmeria within metasomatized granodiorite and gneiss-hosted exoskarn bodies. Vesuvianite occurs at Kimmeria and Maronia exoscarns, and garnet is a major constituent of the Greek skarns and represented by several gemmy varieties at Maronia, Kimmeria, Kresti, Kos and Serifos. Titanite and spinel occur in the Maronia endo- and exoskarns respectively. The carbonate-replacement Pb-Zn-Au-Ag deposits of Stratoni, Olympias, Lavrion and Serifos host a large variety of non-metallic minerals of gem-quality of both hypogene and supergene origin. Primary, gem-quality minerals from the Chalkidiki mines include: quartz and rhodochrosite. The Lavrion and Serifos deposits are well-known localities for fluorite. Rodingitized gabbros and dolerites at Evros, Evia, Larissa and Othrys Mesozoic ophiolites (Rhodope massif and Pelagonian zone) host gem-quality hessonite and vesuvianite [17].

Hydrothermal-altered volcanic rocks throughout Greece host several silica varieties most of them in gem-quality. Amethyst and chalcedony occur at Kornofolia/Evros area, in Sapes-Rhodopi region and in Lesvos and Milos Islands in epithermal veins accompanying calcite, and/or adularia and barite [35]. Opal occurs in several varieties and colors at Lesvos and Limnos Islands being a constituent of fossilized wood, as well as at various localities at Evros and Milos, Limnos and Lesvos Island and is considered to be part of either silica sinters or steam-heated alteration zones. Fluorite megacrysts occur at Samos and Lesvos Islands, in both cases in the form of monominerallic fluorite veins, cross-cutting epithermally-altered silicified zones and propylitically altered lavas, respectively.

5. Mineralogy of Gems

5.1. Corundum

Greece contains gem corundums mostly within the Rhodope (Xanthi and Drama areas) and Attico-Cycladic (Naxos and Ikaria Islands) tectono-metamorphic units (Figures 1 and 3). In the Xanthi area (Gorgona-Stirigma localities) sapphire deposits are stratiform, occurring within marble layers alternating with eclogitic amphibolites [18,36,37] (Figure 3a). Rubies in the Paranesti-Drama area are restricted to boudinaged lenses of pargasitic schists alternating with amphibolites and gneisses [28,29]. Both occurrences are oriented parallel to the UHP-HP Nestos suture zone. Sapphire from Xanthi marbles is of pink, orange, purple to blue color, usually of tabular or barrel-shaped euhedral form and reaches sizes of up to 4 cm (Figure 3b,c). In some cases, blue corundum alters to spinel. The corundums from Xanthi are transparent with very clear parting and fine cracks. Blue sapphires are zoned, with alternating deep blue and colorless domains. This zoning, or irregular color distribution in the sapphires, is attributed to different Fe and Ti contents in the crystals [18]. The Xanthi sapphires are associated with calcite, dolomite, brown or blue spinel, margarite, and nickeloan tourmaline.

In the Paranesti/Drama area, rubies are associated with kyanite and pargasitic hornblende, and are rimmed by margarite, muscovite, chlorite and chromian spinel. Ruby crystals, ranging in size up to 5 cm and of pale pink to deep red color [28], are mainly flat tabular and less commonly prismatic and barrel-shaped (Figure 3d). The Paranesti rubies are opaque to transparent exhibiting clear parting and lamellar twinning.

On Naxos Island, about 2 km East and Southeast of Kinidaros, blue sapphires up to 3 cm are associated with granite pegmatites intruding ultramafic lithologies (plumasites), occurring either within the pegmatites themselves or the surrounding metasomatic reaction zones [18,37,38] (Figure 3e). Within the plumasites, colorless to blue, purple, and pink corundum may occur either as isolated crystals within the plagioclase matrix, and/or associated with tourmaline and phlogopite (Figure 3f). In the blackwalls that developed at the contacts between the pegmatites and the meta-peridotite country rock purple and pink sapphires are enveloped by phlogopite. The Naxos plumasite sapphire crystals are barrel-shaped, display macroscopic color zoning from a blue core surrounded by a white rim or as a blue-zoned outer rim surrounding a colorless core, or pink cores to purple rims [18]. They are transparent with inner fractures. In the southern part of the Island, close to Kavalaris Hill, a rock termed "corundite" by Feenstra and Wunder [39], which was formed by the dissociation of former diasporites, in meta-bauxites during prograde regional metamorphism, is composed almost entirely of blue corundum. The corundum from this locality does not occur in well-shaped crystals, it is mostly massive, and translucent to opaque, appropriate only to be cut as cabochon.

Figure 3. Field and hand specimens photographs demonstrating gem corundum and beryl occurrences/crystals from Greece: (**a**) Corundum-bearing marbles alternating with amphibolites along the Nestos suture zone, Gorgona/Xanthi area; (**b,c**) Pink to purple sapphires within Xanthi marbles; (**d**) Ruby within pargasite schist from Paranesti/Drama; (**e**) Ultrabasic bodies rimming the migmatite core of Naxos Island. Plumasite formation took place after metasomatic reaction between ultrabasites and pegmatites, (**f**) Blue and colorless sapphires within desilicated pegmatite (plumasite) from Kinidaros, Naxos Island; (**g**) Beryl-bearing pegmatite penetrating migmatite at Kinidaros, Naxos Island, (**h,i**) Blue beryl (var. aquamarine) associated with orthoclase and muscovite within pegmatite from Kinidaros, Naxos Island. Photographs 3c,d,i are courtesy of Berthold Ottens.

On Ikaria Island, blue sapphires occur in the metabauxites of Atheras Mt, hosted within marbles, which lie on top of gneisses. The corundum megacrysts fill together with margarite, extensional fissures and networks of veins discordant to the metabauxite foliation [18,37]. The corundums are deep-blue in color, tabular to well-shaped, reach sizes up to 4 cm and are accompanied by Fe-chlorite, hematite, rutile and diaspore. The corundum contains inclusions of ilmenite, hematite, ulvospinel, rutile and zircon.

The LA-ICP-MS results for averages and ranges for chromophores and genetic indicator elements (Fe, Cr, Ti, V, Ga and Mg) are listed in Table 1. Colorless to blue sapphires from Gorgona/Xanthi display significant variations in their Ti content which reflects their zoned coloration. Bluish areas display high Ti values up to 6462 ppm, and pink varieties are characterized by much less Ti (up to 810 ppm). Iron content reaches values up to 1339 ppm in both colorless and blue areas and up to 1782 ppm in the pink varieties. Pink corundums are characterized by significantly higher Cr (up to 1082 ppm) concentrations, compared to the colorless/blue areas (Cr < 298 ppm). Maximum values for V and Ga content of up to 227 and 121 ppm, respectively are fixed regardless of the color. Mg values (up to 536 ppm) are generally higher in the colorless to blue grains, compared to the pink grains (<65 ppm). Rubies from Paranesti/Drama display very high Cr contents (up to 15347 ppm), and Fe (up to 4348 ppm). The Ti, Mg, Ga and V values are low (up to 148, 31, 24 and 6 ppm, respectively).

Table 1. Chromophores and key trace elements (Mg and Fe) LA-ICP-MS analyses (ppm) of the Greek corundum crystals (data from Voudouris et al. [18]).

Sample (Number of Analyses)	Locality	Color	ppm	Mg	Ti	V	Cr	Fe	Ga
Dr1a-b (n = 21)	Paranesti/Drama	red	aver	17	41	2.9	9142	26,558	16
			min	8	8	2	2431	1799	12
			max	31	148	5.7	15,347	4384	24
Go1a-b (n = 26)	Gorgona/Xanthi	blue-colorless	aver	322	4255	78	84	1003	89
			min	78	1007	23	35	363	29
			max	601	6462	207	251	1339	121
Go5a (n = 8)	Gorgona/Xanthi	pink-purple	aver	28	102	20	104	1291	47
			min	13	30	5	4	962	9
			max	49	205	60	298	1782	100
Go5b (n = 14)	Gorgona/Xanthi	pink	aver	35	509	65	297	424	78
			min	17	39	47	105	294	72
			max	53	810	79	1082	494	87
Ik1a (n = 24)	Ikaria isl.	blue	aver	15	1263	116	223	4326	90
			min	5	267	31	103	2710	68
			max	54	4508	164	313	7324	114
Nx1-4 (n = 77)	Naxos isl./Kinidaros	blue-colorless	aver	58	594	22	214	3400	60
			min	16	10	9	1	1377	42
			max	208	3222	42	851	6361	184
Nx1b (n = 14)	Naxos isl./Kinidaros	purple	aver	64	520	22	43	4677	63
			min	39	124	15	36	3324	55
			max	199	848	31	48	6670	76
Nx5b (n = 12)	Naxos isl./Kinidaros	pink	aver	52	181	20	428	3096	45
			min	17	60	14	274	2138	37
			max	133	348	43	548	4716	52
Nx5a (n = 8)	Naxos isl./Kavalaris	blue	aver	5	462	52	262	3706	87
			min	2	238	43	227	3301	84
			max	14	784	65	339	4268	90

Sapphires from Naxos Island are characterized by high Fe concentrations (up to 6678 ppm), related to blue-colored domains. Ti is considerably lower (up to 966 ppm), with the exception of one colorless to blue corundum, where values up to 3222 ppm were detected. Chromium in blue to colorless varieties is low and reaches values of up to 851 ppm in grains with pink and purple hues. Vanadium content in the majority of the samples measured varies in the range of 20–40 ppm. Mg and Ga values range between 2–208 ppm and 42–184 ppm, respectively. The higher Ga concentrations occur in the metabauxite-hosted sapphires. Sapphires from Ikaria display high Fe and Ti values (up to 7324 ppm and 4508 ppm, respectively). Cr, Mg and V values reach maximum values of up to 313, 54 and 164 ppm, respectively. Gallium shows elevated concentrations, up to 114 ppm. Figure 4 distinguishes different primary sources for the studied corundums based on the chromophore and genetic indicator elements. In the (Cr + V)/Ga versus Fe/Ti diagram (Figure 4a, [40,41]), the majority of the samples plot in the field of metamorphic corundum, exhibiting a large variation in Fe/Ti ratios.

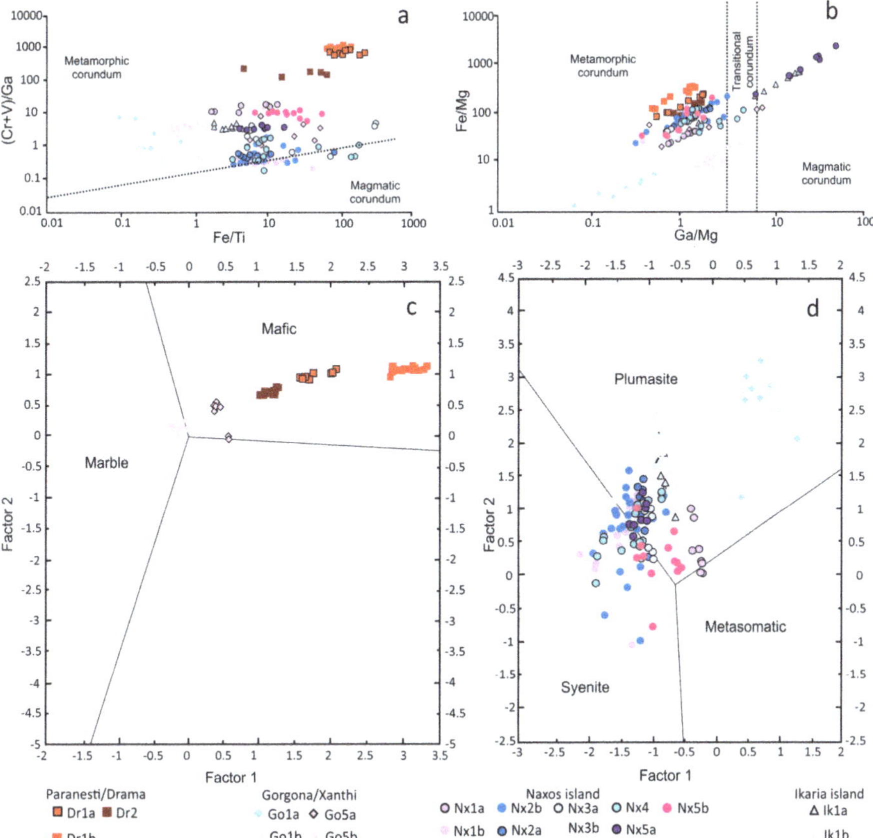

Figure 4. Greek corundum LA-ICP-MS analyses plotted on (**a**) (Cr+V)/Ga versus Fe/Ti discrimination diagram separating the fields for magmatic and metamorphic corundums (adapted from Sutherland et al. [40] and Harris et al. [41]; (**b**) Fe/Mg versus Ga/Mg discrimination separating the fields for magmatic, transitional and metamorphic corundums (adapted from Peucat et al. [42], Sutherland et al. [40] (both **a** and **b** are from Voudouris et al. [18]); (**c**) plot of Greek rubies (and Xanthi pink sapphires) within mafic rocks and marble field in discriminant factors diagram adapted after Giuliani et al. [43]; (**d**) plot of Greek sapphires within plumasite and syenite field in discriminant factors diagram adapted after Giuliani et al. [43].

Rubies from Paranesti/Drama show high Cr/Ga and Fe/Ti ratios, followed by the pink and purple sapphires from Naxos Island. In the Fe/Mg versus Ga/Mg plot (Figure 4b, [40,42]), most samples plot in the metamorphic corundum field, except for sapphires from the metabauxites of Naxos and Ikaria Islands. A few blue sapphires from Naxos Island and pink sapphires from Gorgona/Xanthi plot in the area of transitional corundum. In the discriminant factor diagrams of Giuliani et al. [43], the Paranesti rubies within the pargasitic schists, plot in the "mafic" field and pink corundums (e.g. sapphires because Fe > Cr + V), from Xanthi in the "marble" field. In our study we considered pink corundum for which Cr + V < Fe as sapphires. From Figure 4c we can observe that the marble-hosted pink corundum from Xanthi (sample GO5b) plots in the marble area, while those of Go5a (pink purple) plot in the metasomatite field, suggesting that for sapphires the "marble" field (as defined for rubies) may extend to the right into the "metasomatic" field. Only part of the plumasitic sapphires from Naxos plot in the "plumasitic" field, the rest plotting in the "syenite" field (Figure 4d).

5.2. Beryl

The Greek pegmatites are mostly deficient in miarolitic cavities, and thus only contain matrix-embedded mineral crystals. Blue beryl crystals (var. aquamarine) up to 5cm long are typical constituents of Naxos pegmatites both within the migmatitic dome as well as in pegmatites cross-cutting the surrounding metamorphic rocks (Figure 3g–i).

5.3. Jadeitite

Jadeitite, a rock that is also termed jadeite "jade", consists almost entirely of the pyroxene mineral jadeite. Its green color is attributed to iron substituting for aluminum in the jadeite [Na(Al,Fe^{3+})Si$_2$O$_6$] crystal structure [4]. Jadeitite together with omphacitite occurs within the Kampos mélange, Syros Island in contact relationship or enveloping eclogite [27,44,45] as shown in Figure 5a,b. A production of polished jade axe heads on Syros took place at least since the Neolithic period [46].

Figure 5. Field photographs (**a**) and hand specimens (**b**) of jadeitite from Kampos, Syros Island.

5.4. Al$_2$SiO$_5$ Polymorphs

Kyanite in blue, green, yellowish to orange colored crystals up to 20 cm long, is found in quartz ± feldspar boudins intercalated within metapelites at Trikorfo, Thassos Island (Figure 6a–d) [17,19,47]. Orange kyanite also occurs in association with spessartine, and muscovite. Distinct dark blue color zoning is a common feature in the centre of some light blue kyanite; this color zoning can also be observed in some yellow kyanite. Microprobe analyses for orange kyanite indicate MnO up to 0.1 wt. % contributing to the coloration of the crystals [19] and up to 1.5 wt. % FeO. In a similar occurrence on Naxos Island in the Cyclades, kyanite forms up to 10 cm blue-colored crystals in quartz lenses and veins.

Mn-Andalusite (formerly viridine) in dark green-colored euhedral to subhedral crystals up to 7 cm is found in quartz-feldspar boudins (Figure 6e,f) at Trikorfo, Thassos Island. Mn-andalusite also occurs in orange kyanite-mica schists or in association with muscovite. Microprobe data revealed Mn_2O_3 contents, up to 2.90 wt. % (Table 2). Mn^{3+} substituting for Al^{3+} in the mineral formula is responsible for the dark green color [19,47].

5.5. Epidote Group Minerals—Zoisite

Mn-poor zoisite (var. thulite) and Mn-poor clinozoisite in calc-silicate layers at Trikorfo, Thassos Island, are intergrown with Mn-grossular and quartz (Figure 6g–i). They form light pink to red colored translucent subhedral to euhedral crystals up to 10cm long. Available electron microprobe data revealed low Fe_2O_3 content (1.5 wt. %–2 wt. %) and very low Mn_2O_3 values from 0.15 wt. %–0.21 wt. %, but enough to be responsible for the pink-red coloration [19,47]. Deep green colored epidote crystals, up to 10 cm long associated with quartz and garnets at Lefkopetra, near Kimmeria, Xanthi, within metasomatized granodiorite and gneiss-hosted exoskarn bodies (Figure 6j–l) are partly of gem-quality.

Figure 6. Field and hand specimens photographs demonstrating occurrences and crystals of gem-quality Al_2SiO_5-polymorphs and epidote-group minerals of Greece: (**a–d**) Blue, green to orange kyanite within metapelites and quartz veins at Trikorfo, Thassos Island; (**e,f**) Mn-rich andalusite (var. viridine) in quartz from Trikorfo, Thassos Island; (**g–i**) Pink-red, Mn-bearing zoisite/clinozoisite from Trikorfo, Thassos Island. (j) Epidote-bearing skarn of Lefkopetra, Kimmeria/Xanthi. Granodiorite is at the front of the photo; (**k,l**) Green epidote crystals associated with andradite from Lefkopetra; Photographs 6b,d,e,f,i,l are courtesy of Berthold Ottens.

5.6. Garnets

Spessartine and Mn-grossular are crystallized within andalusite-kyanite schists and in the thulite-bearing calc-silicate layers of Trikorfo respectively [17,19,47]. Spessartine from quartz-muscovite veins within mica schists forms translucent, orange-colored euhedral crystals up to 1 cm and of gem-quality (Figure 7a,b). Electron microprobe analyses of spessartine indicate MnO contents up to 42.9 wt. % (Table 2). Mn-bearing grossular from the calc-silicate layers form euhedral yellowish crystals of gem-quality in close association with pink-to red colored Mn-zoisite and quartz. MnO content ranges from 1.9 wt. %–2.4 wt. % (Table 2).

Figure 7. Field and hand specimens photographs demonstrating gem garnet occurrences/crystals of Greece: (**a**) alternating marbles and metapelites at Trikorfo, Thassos Island. Metapelites host spessartine together with other Mn-bearing silicates; (**b**) spessartine crystals associated with muscovite within quartz matrix; (**c**) spessartine within braunite-phlogopite bearing skarn from Thapsana. Paros Island; (**d**) Maronia skarn at the interface between marbles and the Maronia pluton (half right part of the photograph); (**e**) grossular crystals in contact with the Maronia monzogabbro; (**f**) partly transparent pale green grossular crystals from the Maronia skarn; (**g**) andradite-bearing wollastonitic skarn in contact with marbles, Kimmeria, Xanthi; (**h**) Green andradite and calcite from Kimmeria, Xanthi; (**i**) yellow andradite within amphibolite, Kimmeria, Xanthi; (**j**) Panoramic view of skarn at Dikeon Mt, southern Kos Island; (**k**) orange andradite and diopside from the Dikeon Mt skarn, Kos Island; (**l**) red-brown andradite with green quartz from Avessalos, Serifos Island. Photographs 7**f,h,i,l** are courtesy of Berthold Ottens.

Deep red colored spessartine crystals up to 3 cm at Paros Island/Cyclades, are related to Mn-bearing skarn occurrences at Thapsana (Figure 7c). Spessartine at Thapsana contains up to 34.4 wt. % MnO and is associated with rhodonite and Mn-oxides [48]. Garnet is also a major constituent and represented by several varieties at Maronia, Kimmeria, Kresti, Kos and Serifos skarns. The Maronia skarn (Figure 7d) includes dark green-colored Ti-Cr andradite-grossular up to 1 cm, which postdates crystallization of earlier black-colored schorlomite-uvarovite-kimzeyite solid solutions [13]. Late pale green, brown to orange grossular-andradite solid solutions reach sizes up to 5cm (Figure 7e,f). Ti-bearing brown grossular-andradite contains up to 4.6 wt. % TiO_2. Transparent specimens with shiny dodecahedron faces are of gem-quality. A wide spectrum of colors in andradite-grossular garnets also occurs at Kimmeria where dark green-, brown-, yellow- to orange-colored garnets (up to 3 cm) occur (Figure 7g–i). Orange-colored grossular contains up to 1.15 wt. % MnO, whereas dark green-colored andradite-grossular solid solutions are characterized by elevated Ti contents (up to 2.86 wt. % TiO_2; Table 2).

Red-brown Mn-bearing andradite-grossular is present at Kresti in Drama (with up to 3.6 wt. % MnO) and at Kos Island (with up to 1.1 wt. % MnO and 1 wt.% TiO_2, Table 2), where they reach spectacular sizes of up to 20 cm (Figure 7j,k). The Serifos andradites are famous due to their zonal growth with colors ranging from deep brown to orange. The Agia Marina and Avessalos areas at Serifos are characterized by splendid occurrences of red-brown, locally transparent andraditic garnets in massif garnetitic skarns (Figure 7l). The garnets (up to 5 cm in size) accompany quartz and hematite in hedenbergitic skarn. Rodingitized gabbros at Evros, Evia, Larissa and Othrys Mesozoic ophiolites host gem-quality orange to brown grossular (hessonite), vesuvianite and diopside crystals up to 2 cm long [17,49].

5.7. Quartz

Quartz is found in an enormous variety of forms and colors from alpine-type fissures, granite-hosted miarolitic cavities and/or quartz veins, skarn-carbonate replacement deposits and volcanic-hosted epithermal environments [14,17,35,50,51]. Famous localities of gem-quality alpine-type quartz include the Attica and Drama districts, as well as Evia, Ios, Thassos and Crete Islands (Figures 2 and 8a–c). Crystals reach sizes up to 40 cm. Smoky quartz, black quartz (morion), rock crystal, amethyst, chloritized quartz and rutilated quartz are among the varieties found throughout Greece.

Quartz crystals up to 50 cm, in colorless, smoky and black varieties (morion) fill miarolitic cavities in aplitic granites at Samothraki Island (Figure 8d). Colorless quartz crystals (up to 10 cm) also occur in quartz veins cross-cutting granitoids at Samothraki, Kimmeria, Maronia/Rhodopi and Tinos Island.

Quartz is a very common mineral in the skarns of Serifos, Kimmeria/Xanthi and Kresti/Drama. Combinations of amethyst and green quartz (prase) forming sceptre growths at Serifos are worldwide unique specimens. Green-colored (due to actinolite inclusions) quartz crystals, are also developed within wollastonite skarn, at both Kimmeria and Kresti.

The Avessalos area at Serifos Island is one of the best sites in the world in respect to its green quartz (prase) [12,13,52]. The area is characterized by a garnetitic and hedenbergitic skarn and by the development of huge geodes filled by prograde and retrograde skarn minerals. At Neroutsika location of Avessalos, two forms of prase occur (Figure 8e,f): the first variety refers to very deep green colored crystals accompanied by hematite roses. The second variety refers to double-colored crystals of prase-amethyst. The transition between these two crystals is abrupt within the same crystal, where prase occurs at the base and amethyst at the top of the crystal. The amethysts are transparent and of gem-quality [15,16]. In the southern part of the Avessalos area, rare combinations of scepter-shaped crystals contain both prase and amethyst. These crystals commonly comprise a lower prase part that evolves upwards into amethyst and further upwards again into prase. These alternations can be remarked even within a single crystal [13]. Gem-quality quartz, in crystals up to 60 cm also occurs at the carbonate-replacement deposits of Chalkidiki [53].

Table 2. Representative EPMA of various gem-quality minerals from Greece: Andradite-grossular, Kimmeria (1–2); Andradite-grossular, Maronia (3–4); Andradite-Grossular, Kos Island (5); Grossular, Thassos (6–8); Spessartine, Thassos (9–10); Zoisite, Thassos (11–12); Mn-andalusite, Thassos (13–14); Orange kyanite, Thassos (15–16); Blue kyanite, Thassos (17).

Wt%	1	2	3	4	5	6	7	8	9*	10*	11	12	13*	14*	15*	16*	17*
SiO_2	39.77	37.73	38.22	39.03	38.05	38.97	39.05	39.31	35.71	34.31	39.23	39.11	36.08	36.07	36.40	36.68	36.71
Al_2O_3	20.21	6.02	18.79	19.60	8.17	21.74	21.88	22.20	20.78	23.40	32.71	32.11	58.98	58.24	61.90	61.29	61.18
MgO	bd	0.23	0.64	0.39	0.19	0.14	0.17	0.12	0.31	0.41	0.02	0.10	0.11	0.16	bd	bd	0.01
FeO	3.68	19.28	5.62	4.40	19.00	1.03	1.08	0.61	0.08	0.52	-	-	-	-	0.76	1.52	1.49
Fe_2O_3	-	-	-	-	-	-	-	-	-	-	1.55	2.01	1.93	2.14	-	-	-
Cr_2O_3	0.05	0.73	bd	0.09	-	0.03	bd	0.04	-	-	-	-	-	-	-	-	-
MnO	1.15	0.10	0.20	0.07	1.14	1.90	2.18	2.42	42.87	40.40	-	-	-	-	0.08	0.11	0.02
Mn_2O_3	-	-	-	-	-	-	-	-	-	-	0.15	0.21	2.50	2.89	-	-	-
CaO	34.55	33.59	34.92	35.40	32.10	36.00	35.76	35.39	0.55	0.93	24.78	24.73	0.02	bd	bd	bd	0.03
Na_2O	-	0.26	-	-	bd	0.01	0.01	bd	bd	bd	bd	bd	0.00	0.02	0.04	bd	0.04
TiO_2	0.03	2.86	1.10	0.13	0.92	0.42	0.25	0.12	0.09	0.05	0.02	0.04	0.01	0.02	0.02	0.01	bd
Total	99.39	100.54	99.49	99.11	99.57	100.23	100.37	100,21	100.39	100.02	98.44	98.31	99.64	99.52	99.20	99.62	99.48

Formulae					24 (O)							8 cations				5 (O)	
Si	6.076	6.032	5.856	5.974	6.094	5.872	5.876	5.920	5.880	5.619	2.970	2.970	0.993	0.996	0.995	1.001	1.003
Al	3.640	1.134	3.392	3.536	1.542	3.860	3.880	3.940	4.033	4.517	2.918	2.874	1.913	1.895	1.994	1.973	1.971
Mg	0.000	0.054	0.148	0.088	0.046	0.032	0.038	0.026	0.076	0.100	0.003	0.011	0.005	0.007	0.007	0.000	0.000
Fe^{2+}	0.276	0.554	0.078	0.086	0.496	0.000	0.000	0.000	0.011	0.071	0.000	0.000	0.000	0.000	0.017	0.034	0.033
Fe^{3+}	0.196	2.024	0.642	0.478	2.048	0.296	0.312	0.190	0.000	0.000	0.088	0.115	0.040	0.045	0.000	0.000	0.000
Cr^{3+}	0.003	0.092	0.000	0.010	-	0.004	0.000	0.002	-	-	-	-	-	-	-	-	-
Mn^{2+}	0.148	0.014	0.026	0.010	0.154	0.242	0.278	0.308	5.979	5.604	0.000	0.000	0.000	0.000	0.002	0.003	0.000
Mn^{3+}	0.000	0.000	0.000	0.000	-	-	-	-	0.000	0.000	0.010	0.000	0.052	0.061	0.000	0.000	0.000
Ca	5.658	5.752	5.732	5.804	5.508	5.812	5.764	5.710	0.097	0.163	2.010	2.013	0.001	0.000	0.000	0.000	0.000
Na	0.000	0.011	0.000	0.000	-	0.000	0.000	0.000	0.000	0.000	0.000	0.000	0.000	0.002	0.002	0.000	0.011
Ti	0.004	0.344	0.126	0.014	0.110	0.048	0.028	0.014	0.011	0.006	0.001	0.002	0.000	0.000	0.000	0.000	0.000

(-) = not analyzed; bd = below detection. Analyses marked with (*) are from Tarantola et al. [19] (this volume).

Amethyst is the main quartz variety of gem-quality in the volcanic rock-hosted epithermal environments in Greece. It occurs at Kornofolia/Evros area, in Sapes/Rhodopi region and in Lesvos and Milos Islands in epithermal veins accompanying calcite, and/or adularia and barite [13,35]. At Sapes, amethyst occurs in massive form within crustiform banded quartz-chalcedony veins, however cavities in the veins may host deep purple crystals up to 2cm. In Milos and Kornofolia, well-developed amethyst crystals (up to 2 cm) occur as open space filling in the centre of veins cross-cutting propylitic and sericitic altered volcanics (Figure 8g,h). At Megala Therma, Lesvos Island, amethyst is prismatic, up to 10 cm in length and displays sceptre and window growths and shares similarities to the Mexican and Sardinian occurrences.

Figure 8. Hand specimens illustrating the gem quartz crystals of Greece: (**a**,**b**) smoky quartz and amethyst from Dassoto, Drama; (**c**) smoky quartz from Koskina, Evia Island; (**d**) morion from Samothraki Island; (**e**) green quartz from Avessalos, Serifos Island; (**f**) amethyst on green quartz from Avessalos, Serifos Island; (**g**) amethyst on barite from Chondro Vouno, Milos Island; (**h**) amethyst from Kornofolia, Evros district. Photographs 8e,f are courtesy of Berthold Ottens.

The oxygen isotopic composition of quartz from the various geological environments has been analyzed in hand-picked crystals and the results are presented in Table 3. Quartz from the alpine-type fissures yield isotopic $\delta^{18}O$ values between 9.4‰ and 23.7‰ (Figure 9).

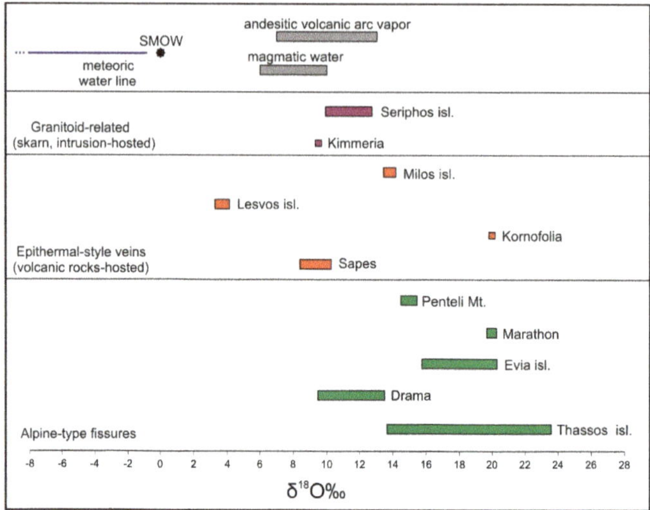

Figure 9. $\delta^{18}O$ values of quartz varieties from Greece. The values for magmatic water and andesitic volcanic arc vapor are from Taylor [54] and Giggenbach [55], respectively.

Table 3. Oxygen isotope composition of quartz from various geological settings in Greece (δ values in ‰ relative to SMOW).

Sample	Locality	Type	$\delta^{18}O_{Qtz}$ (‰)
		Alpine-type fissures	
TH1	Thassos Island	Colorless Qtz, Tessin habit	18.44
TH2	Thassos Island	Smoky Qtz, Tessin habit	13.77
TH3	Thassos Island	Colorless Qtz, skeletal habit	18.82
TH4	Thassos Island	Colorless Qtz, sceptre, prismatic habit	21.41
TH5	Thassos Island	Smoky Qtz, sceptre, prismatic habit	23.66
TH6	Thassos Island	Smoky Qtz, Tessin habit	14.56
TH7	Thassos Island	Choritized Qtz, Tessin habit	18.11
TH8	Thassos Island	Smoky Qtz, Tessin habit	14.66
KN1	Drama	Amethyst, prismatic habit	13.38
KN2	Drama	Amethyst, prismatic habit	13.36
KN3	Drama	Morion, Tessin habit	10.63
KN4	Drama	Morion, Tessin habit	10.83
AL1	Evia Island	Colorless Qtz, prismatic habit	16.80
AL2	Evia Island	Smoky Qtz, prismatic habit	20.77
AL2B	Evia Island	Smoky Qtz, prismatic habit	15.74
AL3	Evia Island	Smoky Qtz, prismatic habit	16.04
AL4	Evia Island	Smoky Qtz, prismatic habit	18.56
AL5	Evia Island	Colorless Qtz, prismatic habit	20.15
MA1	Marathon, Attica	Amethyst, prismatic habit	20.74
MA2	Marathon, Attica	Smoky Qtz, prismatic habit	19.43
PEN1	Penteli Mt, Attica	Colorless Qtz, prismatic habit	15.37
PEN2	Penteli Mt, Attica	Smoky Qtz, prismatic habit	14.14

Table 3. *Cont.*

Sample	Locality	Type	$\delta^{18}O_{Qtz}$ (‰)
	Epithermal veins (volcanic rock-hosted)		
SH1	Sapes, Rhodopi	Smoky Qtz	10.49
SH2	Sapes, Rhodopi	Amethyst	8.24
KIR1	Kirki, Evros	Amethyst	19.10
SF2	Soufli, Evros	Amethyst	20.54
LS2	Lesvos Island	Amethyst	3.33
LS3	Lesvos Island	Colorless Qtz	2.68
LS5	Lesvos Island	Colorless Qtz, Muzo habit	4.13
M1	Milos Island	Amethyst	14.12
M2	Milos Island	Amethyst	13.41
	Granitoid-related (skarn, intrusion-hosted)		
KIM1	Kimmeria, Xanthi	Colorless Qtz	9.49

In the absence of fluid inclusion temperature results, the composition of the fluid in equilibrium with the quartz crystals on the basis of the fractionation equilibrium equation of Sharp et al. [56], cannot be estimated. However, the available data, especially for sceptered quartz crystals from the Rhodope area (Dassoto and Thassos Island), indicate an increase in $\delta^{18}O$ values from the basal Tessin-habit parts of the crystals, towards the amethystine or clear and prismatic-habit sceptres on the upper parts of the crystal. Melfos and Voudouris [57] suggested that smoky quartz and amethyst at Dassoto were formed under different conditions due to mixing of carbonic metamorphic fluids with meteoric waters during the exhumation of the Rhodope core complex. The presence of hematite and pyrite within smoky quartz indicates fS_2/fO_2 conditions at the hematite/pyrite buffer, which were followed by more oxidizing conditions necessary for trivalent Fe to be incorporated into the quartz structure and to the crystallization of amethyst [57].

Similarly, the Thassos colorless and smoky quartz sceptre, display $\delta^{18}O$ values between 21.4 and 23.66‰, heavier than the basal Tessin habit quartz crystals (values from 13.8‰ to 18.8‰, Table 3), thus suggesting involvement of two different fluid types in the crystallization of quartz. This is in accordance to the fluid inclusion results of Bitte et al. [58], who suggested for the quartz crystals from Thassos Island, a continuous event at ductile-brittle to brittle conditions from 360 to 170 °C and at pressures lower than 2 MPa. Carbonic-rich fluids are replaced by surficial fluids, derived from evaporated sea-water during the final period of exhumation.

Smoky, chloritized and colorless quartz crystals in the Attico-Cycladic massif (Evia Island, Penteli Mt including Marathon area) display a similar range of $\delta^{18}O$ values (from 14.1 to 20.8‰) to those from Thassos (Table 3, Figure 9). According to Tarantola et al. [59], metamorphic and basinal fluids were trapped in Evia Island quartz crystals during exhumation of the Attic-Cycladic metamorphic complex. Mineral inclusions of biotite and chlorite in quartz indicate continuous crystallization from 430 to 250 °C.

Volcanic-rock hosted amethysts yield isotopic $\delta^{18}O$ values between 3.3‰ and 20.5‰ (Figure 9; see also Voudouris et al. [35]). Those from Kassiteres-Sapes show $\delta^{18}O$ values of 8.2 to 10.5‰. The amethyst at Kornofolia, Evros district, yielded the highest $\delta^{18}O$ values (20.5‰) and those from Lesvos the lowest $\delta^{18}O$ value of all amethysts analysed (3.3‰). Finally, the Milos amethyst have $\delta^{18}O$ values of 13.4 and 14.1‰. Based on fluid inclusion data, Voudouris et al. [35] suggested that most $\delta^{18}O$ values correspond to a mixing of magmatic and oceanic (and/or meteoric) water, with the highest magmatic component in Kornofolia and the lowest in Lesvos Island.

Quartz from a granite-hosted vein at Kimmeria, Xanthi yielded a $\delta^{18}O$ value of 9.49‰. The isotopic signature of prase at Serifos skarn is comparable to those of quartz samples from different assemblages of the skarn zone at Serifos, but amethyst displays a significantly lighter isotopic signature [52].

5.8. Chalcedony and Other Silica Rich Varieties

Blue agate occurs in veins cross-cutting dacitic lavas and/or as loose fragments at Aetochori, Evros district (Figure 10a) [17]. Agate with various colors is also found in the volcanic environments of Lefkimi/Evros (Figure 10b), and Limnos Island. Chalcedony is present at Kornofolia/Evros and at Petsofas/Lesvos, where it forms typical botryoidal and stalactitic aggregates up to 10 cm, varying in color from pale to deep blue and pink to purple (Figure 10c,d). Red jasper occurs in the Sapes area in the form of up to 40 cm thick horizons hosted in smectite-altered pyroclastics (Figure 10e,f) [60]. Jasper and agate from Lesvos Island have been described by Thewalt et al. [61].

Figure 10. Field and hand specimens photographs demonstrating gem microcrystalline and amorphous silica varieties from Greece: (**a**) blue agate from Aetochori, Evros; (**b**) agate from Lefkimi, Evros; (**c**) chalcedony from Kornofolia, Evros; (**d**) blue chalcedony from Petsofas, Lesvos Island; (**e**) red jasper layer within smectite-altered pyroclastics, Sapes, Rhodopi; (**f**) red jasper from Sapes, Rhodopi; (**g,h**) brown and green opalized wood from Moudros, Limnos Island; (**i**) red opal from Agioi Theodoroi, Milos Island. Photographs 10**a,c** courtesy Berthold Ottens.

5.9. Opal

Opal occurs in several varieties and colors (deep red, yellow, black, orange and green) at Lesvos and Limnos Islands being a constituent of fossilized wood [13,62–64], as well as at various localities at Evros, Milos, Limnos and Lesvos Islands, where it is considered to be part of either silica sinters or steam-heated alteration zones. At the Moudros area, Limnos Island several horizons of opaline silicification within the pyroclastic rocks host opalized wood [64,65]. These horizons represent either thin-bedded, lacustrine-fluviatile intercalations between the pyroclastic formations and were deposited during erosional periods that lasted between the volcanic activity phases, or true silica sinter deposition. In some places the fossiliferous horizons overly an alunitic alteration zone and hydrothermal breccias rich in natroalunite occur. In other cases the opaline horizon is intercalated between fresh to weak argillicaly altered rocks. Silicification within the horizons varies from red, green, white, to black colored, the white color resulting from total depletion in iron oxides (Figure 10g–i). Representative X-ray powder diffraction diagrams from Greek opals are given in Figure 11, demonstrating the transformation of opal-C to quartz in some samples.

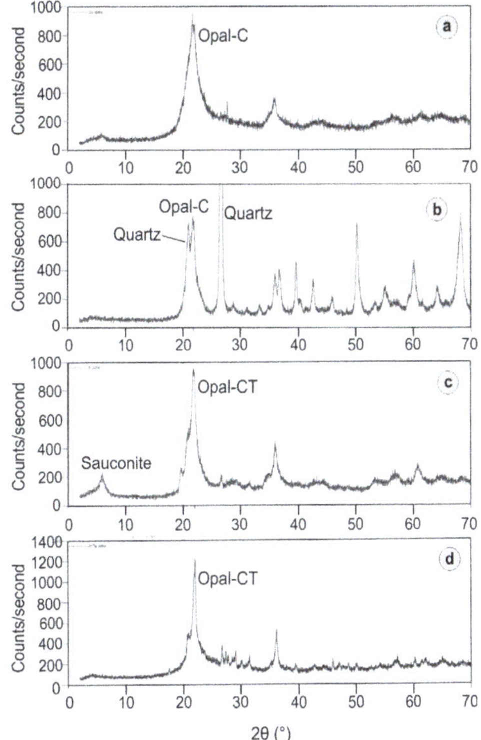

Figure 11. X-ray powder diffraction diagrams of various Greek opals (**a**,**b**) brown-orange opal-C and quartz from Lykofi Evros; (**c**) green opal-CT (color due to sauconite admixtures) from Sapes; (**d**) red-brown opal-CT from Moudros, Limnos Island.

5.10. Feldspars

Gem-quality albite is present as idiomorphic transparent crystals, up to 6 cm in size, grown onto clear and smoky quartz crystals from Evia and Crete Islands (Figure 12a,b). Adularia constitutes well-developed crystals up to 3 cm in size, which occupy alpine-type fissures along with quartz mainly from Evia and Thassos areas.

5.11. Titanite

Titanite in large yellow crystals (up to 3 cm), accompanies zoisite-bearing calc-silicate layers within amphibolites of the Rhodope massif at Therapio, Evros [66]. Green titanite crystals up to 5 cm long at Syros Island, belongs to the retrograde mineral assemblage of eclogites. Titanite in honey colored crystals up to 1 cm, accompany orthoclase and schorlomite in the Maronia endoskarn. Finally, yellow/green colored titanite is associated with adularia in amphibolite-hosted fissures at Thassos Island [51] (Figure 12c). All localities include locally transparent gemmy-material.

5.12. Vesuvianite

The endoscarns from Kimmeria and Maronia host large, up to 10 cm long, partly facetable pale green colored vesuvianite crystals (Figure 12d). Violet-colored vesuvianite in crystals up to 1 cm occur in rodingitized gabbros at Kymi, Evia Island [17].

Figure 12. Field and hand specimens photographs demonstrating occurrences/crystals of various gemstones from Greece: (**a**,**b**) albite from Krieza, Evia Island; (**c**) titanite and quartz from Trikorfo, Thassos Island; (**d**) vesuvianite from the Maronia skarn; (**e**) spinel in marble from Gorgona, Xanthi; (**f**) apatite and K-feldspar from potassically-altered monzonite, Fakos, Limnos Island; (**g**) purple fluorite coated with iron oxides, Ampelos, Samos Island; (**h**) azurite, Kamariza, Lavrion deposit; (**i**) blue smithsonite within marble at the interface with Fe-oxides, Kamariza, Lavrion; (**j**–**l**) colored varieties of smithsonite, Kamariza, Lavrion deposit; Photographs 12b,d,e,f,h are courtesy of Berthold Ottens.

5.13. Spinel

Spinel at Gorgona, Xanthi occurs either as isolated crystals or as rims around corundum. The color of spinel ranges from blue to green and brown (Figure 12e). Transparent blue-colored octahedral spinel crystals up to 1 cm occur at Maronia in association with phlogopite and grossular.

5.14. Tourmaline

Tourmaline is abundant in quartz veins and lenses at Trikorfo and Thymonia/Thassos, in black crystals (schorl) reaching up to 10 cm. However this material is not suitable as a gemstone. Similar crystals occur in pegmatites near Nevrokopi-Drama and at Naxos Island. Nickel-bearing brown tourmaline (dravite) accompanies corundum in the marbles at Gorgona, Xanthi and contains up to 4.4 wt. % NiO, much higher than the Ni content reported in tourmaline from Samos and nickeloan tourmaline from the Berezovskoe gold deposit by Henry and Dutrow [67] and Baksheev and Kudryavtseva [68] respectively. Crystals are generally small (up to 1 cm) but are locally transparent and of gem-quality.

5.15. Apatite, Fluorite, Azurite, Turquoise

Transparent apatite in well-shaped crystals, 1 cm long, accompany phlogopite and orthoclase in the potassic alteration zone at Fakos, Limnos Island (Figure 12f). The carbonate-replacement deposits at Lavrion and Serifos are well-known for large fluorite (up to 20 cm) crystals with colors varying from blue to purple and green. Gem-quality fluorite is also found in volcanic rocks in Samos and Lesvos Islands, where monominerallic fluorite veins cross-cut epithermally altered silicified zones and propylitically altered lavas. The veins are banded and the voids are filled with deep violet and green colored crystals up to 5 cm respectively (Figure 12g). Azurite associated with malachite is widespread in the Lavrion deposit in crystals up to 4 cm as well as in large massive aggregates suitable for cabochon cutting (Figure 12h). Turquoise in gem-quality occurs in the oxidation zone of the Vathi porphyry Cu-Au deposit, Kilkis area [69].

5.16. Smithsonite

The Lavrion smithsonite occurs in many different forms and colorations (Figure 12i–l). Samples have pale blue or green and yellow colors due to both solid solutions and impurities of other minerals [11,70]. The Lavrion smithsonites were intensively exploited during the last one hundred years with a total production of about 1.2 Mt [71]. They occur in cavities in the marbles where they form colorful botryoidal or stalactitic aggregates and/or monominerallic bodies as cavity fillings. The microanalyses suggest that the total content of minor and trace elements in smithsonite is relatively low. The determined elements are Ca, Mg, Fe, Mn, Cu and Cd (Table 4).

Table 4. Fluctuation of trace element content (wt. %) of selected smithsonite samples from Lavrion (modified after Katerinopoulos et al. [11]).

Color	Ca	Mg	Fe	Mn	Cu	Cd	Pb	Visible Inclusions
Yellowish white	0.1–1.3	bd	0.2–0.8	0–0.2	bd	bd	0.6-1.6	-
Yellowish grey	1.1–1.8	bd	0.4–0.9	0–0.2	bd	bd	bd	-
Light yellow	0.3–1.5	bd	0.8–1.9	0–0.2	bd	bd	bd	-
Yellowish brown	0.2–1.5	0.3–1.8	0.3–5.8	0.3–0.5	bd	bd	bd	Chalcophanite
Yellowish green	0.8–1.4	bd	0.4	0.3–0.5	bd	0.2	0–0.7	Greenockite
Greenish yellow	0.3	bd	bd	bd	bd	1.1	2.3	Greenockite
Light blue	0.3–1.2	bd	bd	bd	0.3–2.6	0.3–0.7	bd	-
Bluish white	0.6–0.9	bd	bd	bd	1.1–2.8	bd	bd	-
Light green	0.5–0.7	bd	bd	0.3	0.9–3.0	bd	bd	-
Green	0.2–0.5	bd	bd	0.2–0.3	1.3–2.4	bd	bd	-

As Cd^{2+} is colorless, the yellow ("turkey fat") color of some smithsonites proved to be related only to the presence of greenockite inclusions. Indeed, some samples containing Cd up to 0.7 wt. % but no greenockite inclusions have colors other than yellow. The absence of Cu-rich impurities in blue and green smithsonite and the relatively high amount of Cu indicate that there is a limited solid solution between $ZnCO_3$ and the hypothetical $CuCO_3$ molecule (up to 3.0 wt.% Cu or 6.3 mol % $CuCO_3$) (see also Boni et al. [72]; Frisch et al. [73]). This is in accordance to the findings of Samouchos et al. [74] who suggested that the blue color in Lavrion smithsonite is due to Cu substituting for Zn in the structure. Iron substitution of zinc is limited (up to 1.9 wt. % Fe, except for one sample) and turns the color of smithsonite to a pale yellow. Orange and brown colors are related to impurities of Mn-Fe-Pb oxides and hydroxides. The role of manganese is unknown as its content is very low and the influence on the color is weak.

6. Discussion

Figure 13 represents a hypothetical schematic model, where gemstone occurrences in Greece are related to the various geological environments (regional metamorphic-metasomatic, alpine-type fissures, plutonic-subvolcanic intrusions and pegmatites, zones of contact metamorphism and peripheral volcanic rocks). The hypothetical depths are speculated on the basis of geological criteria. A Late Cretaceous (~ 81.6 ± 3.5 Ma) eclogite-facies metamorphism in the Eastern Rhodopes, confirm previous data that multiple subduction events took place between ~ 200 and ~ 40 Ma along this section of the southern European plate boundary [75]. In the Cyclades, following an initial compressional phase of the Alpine Orogeny related to eclogite-facies metamorphism of the rocks at ~55–49 Ma [45,76], subsequent exhumation of high-P rocks was accompanied by a regional Barrovian-type metamorphism that locally reached partial melting conditions [77,78]. The onset of post-orogenic extensional deformation in the different tectono-metamorphic terranes occurred at ~42 Ma for the Rhodope Massif and ~35 Ma for the Cyclades [23]. These early compressional and later syn-orogenic and post-orogenic extensional events were responsible for the formation of most gemstones found in both regional metamorphic rocks, as well as in late alpine-fissures in Greece. For ruby and jadeite the term "Plate tectonic gemstones" has been proposed by Stern et al. [5], since they generally form as a result of the plate tectonic processes subduction and collision.

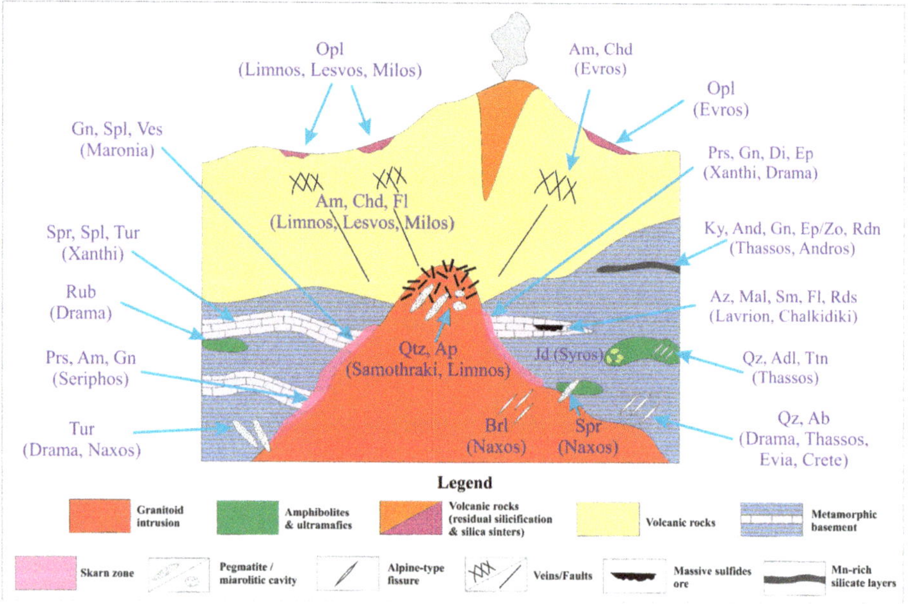

Figure 13. Hypothetical model presenting the various environments related to crystallization of gemstones in Greece. Abbreviations: Ab: Albite, Adl: Adularia; Am: Amethyst, And: Andalusite; Ap: Apatite; Az: Azurite; Brl: Beryl, Chd: Chalcedony, Di: Diopside, Epid: Epidote, Fl: Fluorite, Gn: Garnet, Jd: Jadeite; Ky: Kyanite; Prs: Prase; Rdn: Rhodonite; Rds: Rhodochrosite; Sm: Smithsonite; Spr: Sapphire, Spl: Spinel; Ttn: Titanite; Tur: Tourmaline, Ves: Vesuvianite.

Formation of the jadeitite bodies at Syros Island involved fluid interaction in and around serpentinized peridotite, derived from subduction zone devolatilization [27]. Rodingitization of doleritic and gabbroic dykes in the Rhodope massif and the Pelagonian zone is associated with metasomatic alteration processes, which resulted in the formation of garnet and vesuvianite [79]. According to Koutsovitis et al. [79], rodingitization took place during the exhumation of the mafic–ultramafic mantle wedge rocks within a serpentinitic subduction channel.

Gem corundums in Greece cover a variety of geological environments. The pargasite-schist hosted ruby deposit at Paranesti/Drama area and marble-hosted pink to blue sapphires in Xanthi area occur along the UHP-HP Nestos suture zone and are classified as metamorphic s.s hosted in mafics/ultramafics and marbles respectively [18]. At the Trikorfo area (Thassos Island, Rhodope massif, Northern Greece), which represents a unique mineralogical locality with Mn-bearing minerals such as kyanite, andalusite, garnet and epidote, localized metasomatic reactions occurred during the exhumation of the HP unit locally containing layers enriched in Mn with a strong local fO_2 buffering [19]. Metasomatic reactions first occurred under ductile conditions in an extensive context close to the kyanite-andalusite stability curve and continued until purely brittle conditions as indicated by the presence of late alpine-type veins cross-cutting the metamorphic foliation [19].

Based on the existing data, it can be proposed that the formation of the studied alpine-type fissures in Greece and their mineralogy are closely related to the extensional tectonics accompanying the exhumation of the metamorphic rocks. This was a favored environment for the crystallization of gem-quality quartz and albite. A decrease in pressure and temperature around the fissures during exhumation caused oversaturation in SiO_2 and precipitation of quartz and other mineral constituents in the fissures. Leaching of chemical components from the host rocks after their interaction with the hydrothermal fluids could also explain fluid enrichment and deposition of alpine-type minerals in the fissures studied. The blue sapphires hosted in metabauxites from southern Naxos and Ikaria, fill open-spaces in fissures cross-cutting the metamorphic foliation. They display atypical magmatic signatures indicating a hydrothermal origin and were formed during late extensional conditions.

The development of magmatic-hydrothermal and epithermal systems during the Tertiary, is closely associated with crystallization of a large variety of gemstones. Several gemstones related to the Tertiary magmatic activity are developed within and around the magmatic centres in W. Thrace (Kimmeria, Kassiteres-Sapes, Maronia-Perama, Soufli-Dadia-Levkimi, Drama, Samothraki), Limnos, Lesvos, Serifos and Milos Islands. Initial deep mineralization occurred where plutonic bodies intruded within the basement rocks. Residual fluids, enriched in volatiles, resulted in the emplacement of pegmatitic bodies and to crystallization of beryl (aquamarine) [80,81] and sapphires (in desilicated pegmatites intruding ultramafic rocks) at Naxos Island. Metasomatic processes caused by the release of magmatic-hydrothermal fluids from the intrusive rocks, led to the deposition of various gemstones in the intrusive bodies as well as in the surrounding rocks.

The mineralization in skarns started with the deposition of anhydrous minerals, some of them of gem-quality such as garnets, spinel, titanite and diopside, and ended in the deposition of gem-varieties of quartz, epidote, and vesuvianite in a retrograde stage and under temperature decrease and meteoric water incursion. Contemporaneously, magmatic-hydrothermal fluids circulating within the granitoids were responsible for the deposition of quartz within miarolitic cavities and within quartz veins associated with feldspar, and gemmy apatite.

Later, the systems shifted from a magmatic-hydrothermal dominated stage to a geothermal one, due to the increasing incursion of meteoric waters. Changes in the physico-chemical conditions of the ascending geothermal fluids were the major factors controlling the mineral deposition of the silica varieties (quartz, amethyst, chalcedony and opal) in this geothermal-epithermal environment. Opalized horizons in Thrace and Limnos could represent silica sinter deposition from alkali or neutral geothermal fluids [82]. Amethyst, indicative of oxidizing conditions [83], was probably deposited as a result of mixing of the ascending hydrothermal fluids with meteoric and/or seawater.

Finally, gem-quality smithsonite in the supergene oxidation zone of the Lavrion/Attika carbonate-replacement deposit, was formed as a result of reactivity of carbonate minerals with acidic, zinc-rich solutions, derived from the destruction of sphalerite [11,84,85]. According to the above authors, the neutralization of the acidic solutions deep in the marbles produces enough CO_2 to make smithsonite the stable phase at pH values near neutral.

All Greek corundum, especially the vivid-colored varieties, are translucent to opaque, suitable for cabochons (Figure 14). Greek corundums are characterized by a wide color variation ranging from deep red, pink, purple, and blue to colorless, with crystal sizes of up to 5 cm, homogeneity of the color hues and transparency and should be further examined for their suitability as potential faceted gemstones. Similarly, at Trikorfo/Thassos crystals of kyanite, green andalusite, garnet (grossular and spessartine) and red zoisite–clinozoisite are large (up to several cm), show vivid colors, and are suitable for cabochons.

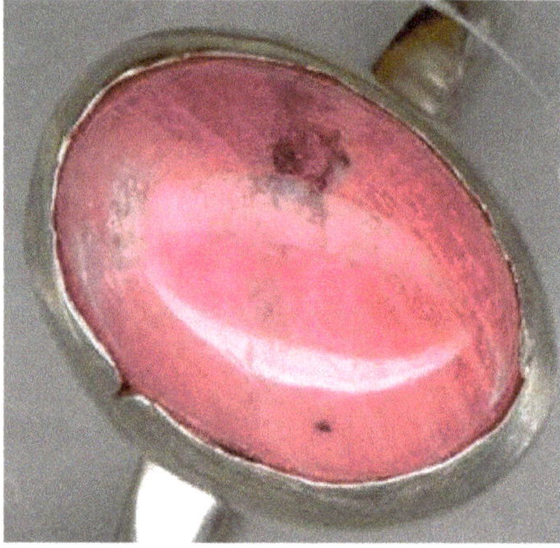

Figure 14. Cabochon quality ruby from Paranesti Drama. Photograph courtesy of V. Melfos.

Figure 15. Various faceted and cabochon gemstones from Greece. Green Mn-andalusite from Thassos (2.2 cm length) in the left, dark green epidote from Kimmeria (1.3 cm) in the center, ruby from Paranesti (4 cm) in the right of the photograph. The rest are silica varieties. Photo courtesy of Anastasios Tsinidis.

As such, the Trikorfo locality can be regarded as a promising area for the exploration of transparent, facetable gemstones. Their genesis due to metasomatic reactions also underlines the important role of metasomatism for gemstone formation in general, as previously noted in the literature [2,19]. According to Tarantola et al. [19], Thassos Island (Rhodope, Greece) can now be classified as the second locality worldwide where Mn-rich orange kyanite is reported, after that of Loliondo (Tanzania) deposit. Finally, silica varieties provide excellent facetable material or cabochons (Figure 15). Future exploration in addition to detailed studies of trace and minor element chemistry using LA-ICP-MS analysis and gemological evaluations are required in order to establish the potential for economic exploitation and to consider them as marketable gems.

7. Conclusions

In the accretionary Hellenides Orogen, gemstones occur in various rock types of mainly four tectonometamorphic units, the Rhodope- and the Attico-Cycladic massifs, the Pelagonian zone, and the Phyllites-Quartzites of Crete Island. In crystalline rocks, two groups of gemstones are distinguished, those formed during regional metamorphism and those associated with late alpine-type fissures. The first group includes Mn-bearing silicates (Mn-andalusite, spessartine, Mn-grossular, Mn-clinozoisite, Mn-zoisite and orange-colored Fe-Mn-kyanite) hosted in Carboniferous ortho- and paragneisses in both Thassos and Paros Islands (Rhodope- and Attico-Cycladic massifs respectively), as well as gem corundums in Xanthi-Drama areas/Rhodope massif and Naxos-Ikaria Islands (Attico-Cycladic massif). In the Xanthi-Drama area, corundum mineralization (sapphires and rubies) is distributed within marbles and eclogitic amphibolites oriented parallel to the (UHP-HP) Nestos suture zone. In the Attico-Cycladic massif, blue sapphires are found within marble-hosted metabauxites (Ikaria and Naxos Island). Alpine-type fissures in Greece contain gem-quality quartz (green quartz, amethyst, smoky and colorless quartz), albite and titanite. Host lithologies are ortho- and paragneisses and metabasites in the Rhodope- (Drama, Thassos Island) and the Attico-Cycladic (Pentelikon Mt, Evia, Ios Islands) Massifs, and metaquartzites (Crete Island). Metamorphic/metasomatic processes within a subduction channel, resulted in the formation of the jadeitite bodies at Syros Island and of garnet and vesuvianite in rodingite bodies at the Rhodope massif and the Pelagonian zone.

The Tertiary magmatic-hydrothermal environments in Greece (granitoids, pegmatites, skarns and carbonate-replacement deposits, and volcanic rocks) may also provide facet-quality material of several species (beryl, sapphire, garnet, vesuvianite, diopside, epidote, fluorite, rhodochrosite, quartz varieties and silica microcrystalline species). The Naxos pegmatites are prospective for aquamarine and blue, purple to pink colored sapphires. Miarolitic cavities and quartz veins cross-cutting granitoids (Samothraki Island, Maronia and, Kimmeria) contain gem-quality colorless and smoky quartz. The endo- and exoskarns of Kimmeria, Kresti and Maronia (Rhodope massif), Serifos Island (Attico-Cycladic massif) and Kos Island, host facetable grossular-andradite garnets, vesuvianite, epidote, as well as prase and amethyst crystals. Rhodochrosite and fluorite occur in the carbonate-replacement deposits of Olympias/Chalkidiki and Lavrion/Attika respectively. Gemstones associated with hydrothermally-altered volcanic rocks include amethyst, chalcedony, opal, fossilized wood and fluorite in Sapes, Soufli areas (Rhodope massif), Lesvos, Limnos and Samos Islands (northeastern and central Aegean volcanic arc) and Milos Island (Attico-Cycladic massif-south Aegean volcanic arc). Finally, the supergene oxidation of the Lavrion/Attika carbonate-replacement deposit contains gem-quality smithsonite in several colorations. Future work aims towards a gemological evaluation of the Greek gemstones and finally estimating the potential for their possible exploitation.

Author Contributions: P.V. and C.M. collected the studied samples. P.V. assisted by C.M., I.C., G.C., A.T., V.M., S.K., A.K., and A.M., evaluated the mineralogical and oxygen isotopic data. P.V. wrote the manuscript.

Funding: This research received no external funding.

Acknowledgments: Authors would like to thank Stefanie Heidrich for her kind help on microanalyses in the Institute of Mineralogy and Petrography, University of Hamburg and Berthold Ottens and Anastasios Tsinidis for kindly providing photographs of Greek mineral specimens and gemstones respectively.

Conflicts of Interest: The authors declare no conflict of interest.

References

1. Fritsch, E.; Rondeau, B. Gemology: The Developing Science of Gems. *Elements* **2009**, *5*, 147–152. [CrossRef]
2. Groat, L.A.; Laurs, B.M. Gem Formation, Production, and Exploration: Why Gem Deposits Are Rare and What is Being Done to Find Them. *Elements* **2009**, *5*, 153–158. [CrossRef]
3. Groat, L.A. (Ed.) *Geology of Gem Deposits*; Mineralogical Association of Canada Short Course 37: Quebec City, QC, Canada, 2007; 276p.
4. Groat, L.A. Gemstones. *Am. Sci.* **2012**, *100*, 128–137.
5. Stern, R.J.; Tsujimori, T.; Harlow, G.; Groat, L.A. Plate tectonic gemstones. *Geology* **2013**, *41*, 723–726. [CrossRef]
6. Kievlenko, E.Y. *Geology of Gems*; Soregaroli, A., Ed.; Ocean Pictures Ltd.: Littleton, CO, USA, 2003; 432p.
7. Tsirambides, A.E. *The Minerals Wealth of Greece*; Giachoudi Pulb.: Thessaloniki, Greece, 2005; 391p.
8. Andronopoulos, V.; Tsoutrelis, C. *Investigation of a Red Corundum Occurrence in Xanthi, Greece*; Unpublished Report; Greek Geol. Survey: Athens, Greece, 1964; 8p.
9. Papastamatiou, I. The Emery of Naxos. *Geol. Geoph. Res.* **1951**, *1*, 37–68.
10. Paraskevopoulos, G. Die Entstehung der Manganlagerstaetten auf der Insel Paros, Griechenland. *N. Jb. Min. Abh.* **1958**, *90*, 367–380.
11. Katerinopoulos, A.; Solomos, Ch.; Voudouris, P. Lavrion smithsonites: A mineralogical and mineral chemical study of their coloration. In *Mineral Deposit Research: Meeting the Global Challenge*; Mao, J., Bierlein, F.P., Eds.; Springer: Berlin, Germany, 2005; pp. 983–986.
12. Gauthier, G.; Albandakis, N. Minerals from the Serifos skarn, Greece. *Mineral. Rec.* **1991**, *22*, 303–308.
13. Voudouris, P.; Katerinopoulos, A. New occurences of mineral megacrysts in tertiary magmatic-hydrothermal and epithermal environments in Greece. *Doc. Nat.* **2004**, *151*, 1–21.
14. Voudouris, P.; Katerinopoulos, A.; Melfos, V. Alpine-type fissure minerals in Greece. *Doc. Nat.* **2004**, *151*, 23–45.
15. Voudouris, P.; Melfos, V.; Katerinopoulos, A. Precious stones in Greece: Mineralogy and geological environment of formation. In Proceedings of the Understanding the Genesis of Ore Deposits to Meet the Demand of the 21st Century, Moscow, Russia, 21–24 August 2006; p. 6.
16. Voudouris, P.; Graham, I.; Melfos, V.; Sutherland, L.; Zaw, K. Gemstones in Greece: Mineralogy and crystallizing environment. In Proceedings of the 34th IGC conference, Brisbane, Australia, 5–10 August 2012.
17. Ottens, B.; Voudouris, P. *Griechenland: Mineralien-Fundorte-Lagerstätten*; ChristianWeise Verlag: Munchen, Germany, 2018; 480p.
18. Voudouris, P.; Mavrogonatos, C.; Graham, I.; Giuliani, G.; Melfos, V.; Karampelas, S.; Karantoni, V.; Wang, K.; Tarantola, A.; Zaw, K.; et al. Gem Corundum Deposits of Greece: Geology, Mineralogy and Genesis. *Minerals* **2019**, *9*, 49. [CrossRef]
19. Tarantola, A.; Voudouris, P.; Eglinger, A.; Scheffer, C.; Trebus, K.; Bitte, M.; Rondeau, B.; Mavrogonatos, C.; Graham, I.; Etienne, M.; et al. Metamorphic and Metasomatic Kyanite-Bearing Mineral Assemblages of Thassos Island (Rhodope, Greece). *Minerals* **2019**, *9*, 252. [CrossRef]
20. Pouchou, J.-L.; Pichoir, F. Quantitative Analysis of Homogeneous or Stratified Microvolumes Applying the Model "PAP". In *Electron Probe Quantitation*; Heinrich, K.F.J., Newbury, D.E., Eds.; Springer: Boston, MA, USA, 1991; pp. 31–75. ISBN 978-1-4899-2617-3.
21. Mattey, D.P. LaserPrep: An Automatic Laser-Fluorination System for Micromass 'Optima' or 'Prism' Mass Spectrometers. *Micromass Appl. Note* **1997**, *207*, 8.
22. Jolivet, L.; Brun, J.P. Cenozoic geodynamic evolution of the Aegean region. *Int. J. Earth. Sci.* **2010**, *99*, 109–138. [CrossRef]
23. Jolivet, L.; Faccenna, C.; Huet, B.; Labrousse, L.; Le Pourhiet, L.; Lacombe, O.; Lecomte, E.; Burov, E.; Denèle, Y.; Brun, J.-P.; et al. Aegean tectonics: Strain localization, slab tearing and trench retreat. *Tectonophysics* **2013**, *597*, 1–33. [CrossRef]
24. Ring, U.; Glodny, J.; Thomson, S. The Hellenic subduction system: High-pressure metamorphism, exhumation, normal faulting and large-scale extension. *Earth Plan. Sci.* **2010**, *38*, 45–76. [CrossRef]

25. Brun, J.-P.; Sokoutis, D. Kinematics of the Southern Rhodope Core Complex (North Greece). *Int. J. Earth Sci.* **2007**, *96*, 1079–1099. [CrossRef]

26. Pe-Piper, G.; Piper, D.J.W. The igneous rocks of Greece. The anatomy of an orogen. In *Beiträge der regionalen Geologie der Erde 30*; Stuttgart: Berlin, Germany, 2002; 573p.

27. Harlow, G.E.; Sorensen, S.S.; Sisson, V.B. Jade. In *The Geology of Gem Deposits*; Groat, L.A., Ed.; Mineralogical Association of Canada Short Course Series: Quebec City, QC, Canada, 2007; Volume 37, pp. 207–254.

28. Wang, K.K.; Graham, I.T.; Lay, A.; Harris, S.J.; Cohen, D.R.; Voudouris, P.; Belousova, E.; Giuliani, G.; Fallick, A.E.; Greig, A. The Origin of A New Pargasite-Schist Hosted Ruby Deposit From Paranesti, Northern Greece. *Can. Mineral.* **2017**, *55*, 535–560. [CrossRef]

29. Wang, K.K.; Graham, I.T.; Martin, L.; Voudouris, P.; Giuliani, G.; Lay, A.; Harris, S.J.; Fallick, A. Fingerprinting Paranesti Rubies through Oxygen Isotopes. *Minerals* **2019**, *9*, 91. [CrossRef]

30. Papanikolaou, D. Contribution to the geology of the Aegean Sea: The Island of Andros. *Ann. Geol Pays Hell.* **1978**, *29*, 477–553.

31. Stalder, H.A. Petrographische und mineralogische Untersuchungen im Grimselgebiet (Mittleres Aarmassiv). *Schweiz Miner. Petr. Mitt.* **1964**, *44*, 187–398.

32. Mullis, J. Growth conditions of quartz crystals from Val d'Illiez (Valais, Switzerland). *Schweiz Miner. Petr. Mitt.* **1975**, *55/3*, 419–430.

33. Niedermayer, G. Alpine Kluefte. *Min. Welt* **1993**, *6*, 57–59.

34. Ottens, B.; Tsinidis, T.; Voudouris, P. Amethyst and smoky quartz: Scepter crystals from Kato Nevrokopi, Greece. *Lapis* **2016**, *41*, 10–21. (In German)

35. Voudouris, P.; Melfos, V.; Mavrogonatos, C.; Tarantola, A.; Götze, J.; Alfieris, D.; Maneta, V.; Psimis, I. Amethyst occurrences in Tertiary volcanic rocks of Greece: Mineralogical, fluid inclusion and oxygen isotope constraints on their genesis. *Minerals* **2018**, *8*, 324. [CrossRef]

36. Liati, A. Corundum-and zoisite-bearing marbles in the Rhodope Zone, Xanthi area (N. Greece): Estimation of the fluid phase composition. *Mineral. Petrol.* **1988**, *38*, 53–60. [CrossRef]

37. Voudouris, P.; Graham, I.; Melfos, V.; Zaw, K.; Lin, S.; Giuliani, G.; Fallick, A.; Ionescu, M. Gem corundum deposits of Greece: Diversity, chemistry and origins. In Proceedings of the 13th Quadrennial IAGOD Symposium, Adelaide, Australia, 6–9 April 2010; Volume 69, pp. 429–430.

38. Graham, I.; Voudouris, P.; Melfos, V.; Zaw, K.; Meffre, S.; Sutherland, F.; Giuliani, G.; Fallick, A. Gem corundum deposits of Greece: A spectrum of compositions and origins. In Proceedings of the 34th IGC Conference, Brisbane, Australia, 5–10 August 2012.

39. Feenstra, A.; Wunder, B. Dehydration of diasporite to corundite in nature and experiment. *Geology* **2002**, *30*, 119–122. [CrossRef]

40. Sutherland, F.L.; Zaw, K.; Meffre, S.; Giuliani, G.; Fallick, A.E.; Graham, I.T.; Webb, G.B. Gem corundum megacrysts from East Australia basalt fields: Trace elements, O isotopes and origins. *Aust. J. Earth Sci.* **2009**, *56*, 1003–1020. [CrossRef]

41. Harris, S.J.; Graham, I.T.; Lay, A.; Powell, W.; Belousova, E.; Zappetini, E. Trace element geochemistry and metasomatic origin of alluvial sapphires from the Orosmayo region, Jujuy Province, Northwest Argentina. *Can. Mineral.* **2017**, *55*, 595–617. [CrossRef]

42. Peucat, J.J.; Ruffault, P.; Fritsch, E.; Bouhnik-Le Coz, M.; Simonet, C.; Lasnier, B. Ga/Mg ratio as a new geochemical tool to differentiate magmatic from metamorphic blue sapphires. *Lithos* **2007**, *98*, 261–274. [CrossRef]

43. Giuliani, G.; Caumon, G.; Rakotosamizanany, S.; Ohnenstetter, D.; Rakototondrazafy, M. Classification chimique des corindons par analyse factorielle discriminante: Application a la typologie des gisements de rubis et saphirs. Chapter mineralogy, physical properties and geochemistry. *Rev. Gemmol.* **2014**, *188*, 14–22.

44. Tsujimori, T.; Harlow, G.E. Petrogenic relationships between jadeitite and associated high-pressure and low-temperature metamorphic rocks in worldwide jadeitite localities: A review. *Eur. J. Min.* **2012**, *24*, 371–390. [CrossRef]

45. Bröcker, M.; Endrers, M. U-Pb zircon geochronology of unusual eclogite-facies rocks from Syros and Tinos (Cyclades, Greece). *Geol. Mag.* **1999**, *136*, 111–118.

46. Higgins, M.D. Neolithic jadeite and eclogite axes from Greece: A possible material source on the Island of Syros, Greece. In *GAC-MAC-CSPG-CSSS*; Hallifax: Nova Scotia, Canada, 2005; Volume 30.

47. Voudouris, P.; Graham, I.; Mavrogonatos, K.; Su, S.; Papavasiliou, K.; Farmaki, M.-V.; Panagiotidis, P. Mn-andalusite, spessartine, Mn-grossular, piemontite and Mn-zoisite/clinozoisite from Trikorfo, Thassos Island, Greece. *Bull. Geol. Soc. Greece* **2016**, *50*, 2068–2078. [CrossRef]

48. Fitros, M.; Tombros, S.F.; Simos, X.C.; Kokkalas, S.; Hatzipanagiotou, K. Formation of Mn-skarn Ores at Thapsana Mines, Paros Island, Attico-Cycladic Metallogenetic Massif, Greece. In *Proceedings of the the 1st Int. Electronic Conference on Mineral Science, 16–31 July 2018*; Sciforum, MDPI AG: Basel, Switzerland.

49. Voudouris, P.; Katerinopoulos, A.; Magganas, A. Mineralogical geotopes in Greece: Preservation and promotion of museum specimens of minerals and gemstones. Sofia Initiative "Mineral diversity preservation". In Proceedings of the IX International Symposium on Mineral Diversity Research and Reservation, Sofia, Bulgaria, 16–18 October 2017; pp. 149–159.

50. Maneta, V.; Voudouris, P. Quartz megacrysts in Greece: Mineralogy and environment of formation. *Bull. Geol. Soc. Greece* **2010**, *43*, 685–696. [CrossRef]

51. Voudouris, P.; Constantinidou, S.; Kati, M.; Mavrogonatos, C.; Kanellopoulos, C.; Volioti, E. Genesis of alpinotype fissure minerals from Thassos Island, Northern Greece—Mineralogy, mineral chemistry and crystallizing environment. *Bull. Geol. Soc. Greece* **2013**, *47*, 468–476. [CrossRef]

52. Klemme, S.; Berndt, J.; Mavrogonatos, C.; Flemetakis, S.; Baziotis, I.; Voudouris, P.; Xydous, S. On the color and genesis of prase (green quartz) and Amethyst from the Island of Serifos, Cyclades, Greece. *Minerals* **2018**, *8*, 487. [CrossRef]

53. Ottens, B.; Voudouris, P. Minerals and gold ores: The mining district of Chalkidiki, Greece. *Lapis* **2007**, *42*, 32–58. (In German)

54. Taylor, H.P., Jr. Oxygen and hydrogen isotope relationships in hydrothermal mineral deposits. In *Geochemistry of Hydrothermal Ore Deposits*, 2nd ed.; Barnes, H.L., Ed.; Wiley: New York, NY, USA, 1979; pp. 236–318.

55. Giggenbach, W.F. The origin and evolution of fluids in magmatic-hydrothermal systems. In *Geochemistry of Hydrothermal Ore Deposits*, 3rd ed.; Barnes, H.L., Ed.; Wiley: New York, NY, USA, 1997; pp. 737–796.

56. Sharp, Z.D.; Gibbons, J.A.; Maltsev, O.; Atudorei, V.; Pack, A.; Sengupta, S.; Shock, E.L.; Knauth, L.P. A calibration of the triple oxygen isotope fractionation in the SiO_2–H_2O system and applications to natural samples. *Geochim. Cosmochim. Acta* **2016**, *186*, 105–119. [CrossRef]

57. Melfos, V.; Voudouris, P. Fluid inclusions and O-isotopes of an amethyst and smoky quartz occurrence in Kato Nevrokopi, Rhodope massif, Northern Greece. In Proceedings of the 11th International Conference on Experimental Mineralogy Petrology Geochemistry, Bristol, UK, 11–13 September 2006.

58. Bitte, M.; Tarantola, A.; Scheffer, C.; Voudouris, P.; Valero, M. *Fluid Inclusion Characterization of Alpine-Type Quartz Tension Gashes of Thassos Island (Rhodope Massif, Greece)*; European Current Research on Fluid and Melt Inclusions (ECROFI): Nancy, France, 2019.

59. Tarantola, A.; Scheffer, C.; Voudouris, P.; Vanderhaeghe, O.; Rakotomanga, S.; Rigaudier, T. *Metamorphic and Basin Fluids Trapped in Evia Island Alpine Quartz Crystals during Exhumation of the Attic-Cycladic Metamorphic Complex (Greece)*; ECROFI 24: Nancy, France, 2017.

60. Thewalt, U.; Voudouris, P. Jaspis von Sykorachi (Thrakien, Griechenland) und seine Einschlüsse. *Miner. Welt* **2019**, *5*, 92–96.

61. Thewalt, U.; Dorfner, G.; Samouchos, M. Chalcedon, Achat und Jaspis vom westlichen Kustenbereich des Golfs von Kalloni, Lesbos, Griechenland. *Miner. Welt* **2018**, *5*, 90–96.

62. Kelepertzis, A.E.; Velitzelos, E. Oligocene swamp sediments of Lesvos Island, Greece (Geochemistry and Mineralogy). *Facies* **1992**, *27*, 113–118. [CrossRef]

63. Velitzelos, E.; Zouros, N. The Petrified Forest of Lesvos. In *Protected Natural Monument, Engineering Geology and the Environment*; Marinos, P., Koukis, G., Tsiambaos, G., Stournaras, G., Eds.; Balkema: Rotterdam, The Netherlands, 1997; pp. 3037–3043.

64. Voudouris, P.; Velitzelos, D.; Velitzelos, E.; Thewald, U. Petrified wood occurences in western Thrace and Limnos Island: Mineralogy, geochemistry and depositional environment. *Bull. Geol. Soc. Greece* **2007**, *40*, 238–250. [CrossRef]

65. Papp, A. Erläuterungen zur Geologie der Insel Lemnos. *Ann. Geol. Pays Hell.* **1953**, *5*, 1–33.

66. Kassoli-Fournaraki, A.; Michailidis, K.; Zanas, I.; Zachos, S. Titanite-rich carbonates from the Therapio area in Thrace, northern Greece: constraints of the mineral assemblage formation. *Schw. Mineral. Petr. Mitt.* **1995**, *75*, 387–398.

67. Henry, D.J.; Dutrow, B.L. Compositional zoning and element partitioning in nickeloan tourmaline from metamorphosed karstbauxite from Samos, Greece. *Am. Mineral.* **2001**, *86*, 1130–1142. [CrossRef]

68. Baksheev, I.A.; Kudryavtseva, O.E. Nickeloan tourmaline from the Berezovskoe gold deposit, Middle Urals, Russia. *Can. Mineral.* **2004**, *42*, 1065–1078. [CrossRef]

69. Sklavounos, S.; Ericsson, T.; Filippidis, A.; Michailidis, K.; Kougoulis, C. Chemical, X-ray and Mossbauer investigation of a turquoise from the Vathi area volcanic rocks, Macedonia, Greece. *Neues Jahrb. Mineral. Mh.* **1992**, 469–480.

70. Voudouris, P.; Tsolakos, A.; Papanikitas, A.; Solomos, C. Neufunde aus Lavrion. *Lapis* **2004**, *29*, 13–15.

71. Marinos, G.; Petrascheck, W.E. Laurium. Geological and geophysical research. *Greece Inst. Geol. Subsurf. Res.* **1956**, *4*, 1–247.

72. Boni, M.; Gilg, H.A.; Aversa, G.; Balassone, G. The "Calamine" of Southwest Sardinia: Geology, Mineralogy, and Stable Isotope Geochemistry of Supergene Zn Mineralization. *Econ. Geol.* **2003**, *98*, 731–748. [CrossRef]

73. Frisch, P.L.; Lueth, V.W.; Hlava, P.F. The colors of smithsonite: A microchemical investigation. New Mexico. *Geology* **2002**, *24*, 132–133.

74. Samouhos, M.; Zavašnik, J.; Rečnik, A.; Godelitsas, A.; Chatzitheodoridis, E.; Sanakis, Y. Spectroscopic and nanoscale characterization of blue-colored smithsonite ($ZnCO_3$) from Lavrion historical mines (Greece). *Per. Mineral.* **2015**, *84*, 373–388.

75. Miladinova, I.; Froitzheim, N.; Nagel, T.J.; Janák, M.; Georgiev, N.; Fonseca, R.O.C.; Sandmann, S.; Münker, C. Late Cretaceous eclogite in the Eastern Rhodopes (Bulgaria): Evidence for subduction under the Sredna Gora magmatic arc. *Int. J. Earth Sci.* **2018**, *107*, 2083–2099. [CrossRef]

76. Laurent, V.; Huet, B.; Labrousse, L.; Jolivet, L.; Monie, P.; Augier, R. Extraneous argon in high-pressure metamorphic rocks: Distribution, origin and transport in the Cycladic Blueschist Unit (Greece). *Lithos* **2017**, *272*, 315–335. [CrossRef]

77. Altherr, R.; Kreuzer, H.; Wendt, I.; Lenz, H.; Wagner, G.A.; Keller, J.; Harre, W.; Höhndorf, A. A late Oligocene/early Miocene high temperature belt in the Attic-Cycladic crystalline complex (SE Pelagonian, Greece). *Geol. Jahrb.* **1982**, *23*, 97–164.

78. Keay, S.; Lister, G.; Buick, I. The timing of partial melting, Barrovian metamorphism and granite intrusion in the Naxos metamorphic core complex, Cyclades, Aegean Sea, Greece. *Tectonophysics* **2001**, *342*, 275–312. [CrossRef]

79. Koutsovitis, P.; Magganas, A.; Pomonis, P.; Ntaflos, T. Subduction-related rodingites from East Othris, Greece: Mineral reactions and physicochemical conditions of formation. *Lithos* **2013**, *172–173*, 139–157. [CrossRef]

80. Jansen, J.B.H.; Schuiling, R.D. Metamorphism on Naxos; petrology and geothermal gradients. *Am. J. Sci.* **1976**, *276*, 1225–1253. [CrossRef]

81. Siebenaller, L. Fluid Circulations during Collapse of an Accretionary Prism: Example of the Naxos Island Metamorphic Core Complex (Cyclades, Greece). Ph.D. Thesis, University of Lorraine, Lorraine, France, 2008.

82. Henley, R.W.; Ellis, A.J. Geothermal systems ancient and modern: A geochemical review. *Earth Sci. Rev.* **1983**, *19*, 1–50. [CrossRef]

83. Fournier, R.O. The behaviour of silica in hydrothermal solutions. Geology and geochemistry of epithermal systems. *Rev. Econ. Geol.* **1985**, *2*, 45–61.

84. Hitzman, M.W.; Reynolds, N.A.; Sangster, D.F.; Allen, C.R.; Carman, C.E. Classification, Genesis, and Exploration Guides for Nonsulfide Zinc Deposits. *Econ. Geol.* **2003**, *98*, 685–714. [CrossRef]

85. Boni, M.; Large, D. Non sulfide Zinc Mineralization in Europe: An Overview. *Econ. Geol.* **2003**, *98*, 715–729. [CrossRef]

MDPI

St. Alban-Anlage 66

4052 Basel

Switzerland

Tel. +41 61 683 77 34

Fax +41 61 302 89 18

www.mdpi.com

Minerals Editorial Office

E-mail: minerals@mdpi.com

www.mdpi.com/journal/minerals